Physics of Structurally
Disordered Solids

NATO ADVANCED STUDY INSTITUTES SERIES

A series of edited volumes comprising multifaceted studies of contemporary scientific issues by some of the best scientific minds in the world, assembled in cooperation with NATO Scientific Affairs Division.

Series B: Physics

RECENT VOLUMES IN THIS SERIES

The series is published by an international board of publishers in conjunction with NATO Scientific Affairs Division

A	Life Sciences	Plenum Publishing Corporation
B	Physics	New York and London
C	Mathematical and Physical Sciences	D. Reidel Publishing Company Dordrecht and Boston
D	Behavioral and Social Sciences	Sijthoff International Publishing Company Leiden
E	Applied Sciences	Noordhoff International Publishing Leiden

Library of Congress Cataloging in Publication Data

NATO Advanced Study Institute on the Physics of Structurally Disordered Solids,
  University of Rhode Island, 1974.
  Physics of structurally disordered solids.

  (NATO advanced study institutes series: Series B, Physics; v. 20)
  Includes indexes.
  1. Solids—Congresses. 2. Amorphous substances—Congresses. 3. Crystals—Defects—
Congresses. I. Mitra, Shashanka Shekhar, 1933-   II. Title. III. Series.
QC176.AlN33 1974   530.4'1   76-18804
ISBN 978-1-4684-0852-2   ISBN 978-1-4684-0850-8 (eBook)
DOI   10.1007/978-1-4684-0850-8

Proceedings of the NATO Advanced Study Institute on
the Physics of Structurally Disordered Solids
held on July 29-August 9, 1974

© 1976 Plenum Press, New York
Softcover reprint of the hardcover 1st edition 1976

A Division of Plenum Publishing Corporation
227 West 17th Street, New York, N.Y. 10011

Physics of Structurally Disordered Solids

Edited by
Shashanka S. Mitra
University of Rhode Island, Kingston

PLENUM PRESS • **NEW YORK AND LONDON**
Published in cooperation with NATO Scientific Affairs Division

PREFACE

     Structurally disordered solids are characterized
by their lack of spatial order that is evidenced by the
great variety of ordered solids. The former class of
materials is commonly termed amorphous or glassy, the
latter crystalline. However, both classes share many
of the other physical properties of solids, e.g., mechanical stability, resistance to shear stress, etc.
The traditional macroscopic distinction between the
crystalline and the glassy states is that while the
former has a fixed melting point, the latter does not.
However, with the availability and production of a large
number of materials in both crystalline and amorphous
states, and their easy inter-convertability, simple definitions are not possible or at best imprecise. For
the present purpose, it is sufficient to say that in
contrast to the crystalline state, in which the positions of atoms are fixed into a definite structure, except for small thermal vibrations, the amorphous state
of the same material displays varying degrees of departure from this fixed structure. The amorphous state
almost always shows no long range order. Short range
order, up to several neighbors, may often be retained,
although averaged considerably around their crystalline
values. It is generally believed that the amorphous
state is a metastable one with respect to the crystalline ordered state, and the conversion to the crystalline state may or may not be easy depending on the nature of the material, e.g., GaAs and Si on the one hand,
and SiO_2 on the other. It also appears that for many
materials the non-crystalline glassy state may be more
normal than the crystalline or even the polycrystalline
state. This latter statement may be appreciated if one
remembers the history of the art of crystal growing.
It is then relized that there are various degrees of
amorphousness, which is also obvious from the fact that
amorphous materials are prepared, depending on their

chemical composition, by a variety of methods, e.g., by
evaporation and condensation on cold substrates, by
cathode sputtering, by irradiating crystalline materials
to very high doses of high energy radiation, by ion implantation, by rapid cooling of the melt, etc.

The basic advances in solid state physics were made
since von Laue's and Bragg's discovery of the periodic
ordered nature of most solids by the X-ray diffraction
technique (ca.1912). The spatial order renders the
prediction and understanding of certain physical properties of crystals relatively easy through the mani-
pulation of their symmetry properties. The periodicity
in real space suggests the concept of the reciprocal
lattice space and the wave vector conservation rule,
giving a special elegance to solid state theory. On
the other hand, the amorphous material with no such
spatial symmetry, and thus lacking computational elegance, has largely been neglected and left out of the
rapid and spectacular advances of solid state physics
in the last forty or so years.

The recent renewed and vigorous interest in the amorphous state of matter curiously stems from the exciting,
but somewhat dubious, prospect of great technological
breakthroughs. Whether the promise of technological
miracles materializes or not, this sudden attention to
the amorphous state has provided the solid state phys-
icists with a challenge of doing solid state physics
without the simplification or the constraint, depending
on ones point of view, of the wave vector selection
rule. This new upsurge has produced the recording of
many fascinating phenomena and explanations.

The present volume thus examines in breadth and
depth the physics of non-crystalline solids. In addition to reviewing recent research results, it has been
our endeavor to make the book as self-sufficient and
complete as possible. The volume is thus not only expected to be of immediate use to current researchers
but also to serve as a basic text in the field and to
be of use as reference material to complement a course
in advanced solid state physics. For the sake of continuity and ease of reading the chapters have been
grouped in the following sequence: structure and modelling, optical properties and electronic band structure, transport properties, phonons, and non-tetrahedrally coordinated solids, although any one of them
may be read independently of the others.

I sincerely thank Dr. B. Kramer, Dr. L. Laude and
Professor W. Paul for their invaluable assistance in
planning the program of this advanced study institute,
and the authors for promptly providing lecture notes
prior to the institute and manuscripts suitable for
publication later. The tireless efforts of Drs. K. V.
Namjoshi, D. K. Paul, Y. F. Tsay and Mrs. J. Lamoureux
and Mr. N. E. Massa made possible the organization and
smooth operation of the institute. Mr. A. Corey, Mr.
J. Goltman and Mr. N. E. Massa were responsible for
proof reading and index preparation. Mrs. Lamoureux
typed the entire manuscript, and without her cooperation, patience and good humor, this book would not have
been possible.

                                        S. S. Mitra
Kingston, Rhode Island

CONTENTS

FORMATION AND TRANSFORMATION BEHAVIOR OF AMORPHOUS SOLIDS

David Turnbull

Division of Engineering and Applied Physics

Harvard University, Cambridge, Mass. 02138

The subject matter of these lectures has been re-viewed extensively in several articles [1,2,3,5,6,7,8,9] written by the author and collaborators and in the Ehrenreich report [4]. There is no point in redigesting these articles here. What follows is a summary of the author's main conclusions on these lecture topics with specific references to the supporting arguments.

A condensed phase, whether crystalline or amorphous, is classified arbitrarily as solid [1] provided its shear viscosity $\eta \gtrsim 10^{15}$ poise. A more fundamental distinction between the fluid and solid states might be based on the nature of the motions of the constituent atoms or molecules [5]. In the fluid these motions are in substantial part translatory, or diffusive, while in the solid they are almost entirely oscillatory. Thus the solid structure is characterized by a definite set of positions about which the atoms or molecules oscillate and it is classified as crystalline or amorphous depending on whether the spatial pattern of these positions exhibits translational symmetry or not.

Oscillatory motions are highly correlated and in amorphous, as well as crystalline, solids these correlations may be associated with relatively low configurational entropy [7,1]. Indeed, such an association is a central feature of the prominent models for atomic transport in viscous melts and amorphous solids [7]. The well correlated motions in amorphous solids may reflect a high

1

degree of development of short range order (SRO) throughout the solid body.

Experience indicates that in their minimum energy
state all materials, excepting He, are crystalline, rather than amorphous, but there is no rigorous theoretical justification for this [2]. In systems of more than
one component, this state is ordered compositionally,
as well as topologically, or it is phase separated. It
is possible that some materials which are highly disordered compositionally and kinetically constrained from
ordering or phase separating will have their lowest energy in an amorphous state, but the experience bearing
on this possibility is still very limited.[5]

In view of these energy relations and provided a-
tomic transport is, indeed, facilitated by increasing
configurational entropy, it is likely, as J. H. Gibbs
pointed out, that the amorphous solid state of any system must generally be metastable relative to one of its
crystallized states. In accord with this expectation
experience shows that the glass temperature, T_g, defined
as the temperature at which $\eta = 10^{13}$ poise, actually is,
in general, well below the thermodynamic crystallization
temperature [3,2].

The procedures most commonly used to put materials
into amorphous solid form are melt quenching, condensation from dilute fluid (e.g. vapor- or electro-deposi-
tion or ion sputtering) or heavy irradiation of crystals.
All of these bring the material through a penultimate
state having higher free energy than the amorphous solid,
as well as the crystalline, state. The conditions on
these methods for bypassing crystallization have been
specified quite comprehensively elsewhere [2,3,7,8].

In melt quenching, a major barrier to crystallization is the very low probability of the topological reconstruction of the short range order necessary for crystal nucleation. Indeed, in many glass forming melts,
this probability is so low that homogeneous nucleation
of crystals is not observed in the finite systems ac-
cessible to investigation [2,3]. Crystallization of these
systems, when it occurs, results from the growth of nuclei formed heterogeneously.

In condensation of amorphous solids, the collecting

surfaces normally contain a considerable density of suitable crystallization centers (heterogeneous nuclei). If the condensate is to remain amorphous under these conditions, it must be kept at temperatures well below those at which the crystal growth rates become substantial. Often the kinetic resistance to crystal growth scales roughly as the shear viscosity [9] and the scaling relation is such that the crystal growth rates are negligible at temperatures below T_g [2].

When the condensate temperature is sufficiently low, the condensing atoms or molecules will be incorporated into the structure at the first binding site they encounter with no subsequent rearrangement [6]. Under these conditions, the predominant SRO will be that most favored by the shortest range atomic interactions. Often this SRO is, by itself, non-crystallographic, so that it might be expected to assemble into a structure which would be amorphous and also quite defective in terms of missing atoms and dangling bonds. These expectations are generally confirmed in computer simulated studies of condensation [6]. In view of the substantial topological reconstruction of the SRO which must attend crystal nucleation, these structures should be able to withstand considerable local atomic rearrangement, as might occur in annealing just below T_g, without crystal nucleation. Such rearrangement should lead generally to a more "ideal" amorphous structure [4].

Amorphous solid condensates crystallize during heating at temperatures, T_c, which are likely to lie well below their thermodynamic crystallization temperatures, T_m. T_c will be lower the greater are the density, ρ_N, of heterogeneous nuclei and the crystal growth rate, u [2]. According to the rough scaling relation between u and $1/\eta$, a film containing e.g. 10^4 nuclei/cm^2 should crystallize within a few hours when held at a temperature a few degrees higher than T_g. u may be decreased sharply by certain impurities or by fine scale porosity in the films. However, in covalently bound films, crystal growth is, sometimes, sharply enhanced by particular impurities [7] or by illumination with photons having energies exceeding some threshold value [4,7].

The following is a detailed outline of the lectures with reference to sections of reviews, already alluded to, in which each particular topic is discussed.

I.   Amorphous Solid Formation

   A.   Solidification [1], pp. 422-426; [2], p. 473-476

      1.   Definition
      2.   Modes - Continuous and Discontinuous

   B.   Thermodynamics [3], pp. 41-44; [4], pp. 7-9; [5], pp. 6-12

      1.   Concept of Metastability
      2.   Fluid - Glass Transition
      3.   Amorphous Solid - Crystal Transition

   C.   Methods [2], pp. 473-488, [4], pp. 17-22

      1.   General Conditions
      2.   Melt Quenching or Compression
      3.   Condensation

         a.   Vapor deposition
         b.   Sputtering
         c.   Electro and chemical deposition

      4.   Irradiation

II. Fluid - Glass Transition [6], pp. 19-23, [5], pp. 6 - 12; [7]

   A.   Characteristics

      1.   Rheological
      2.   Thermal and Volumetric

   B.   Theory

III. Transformation Behavior of Viscous Melts and Amorphous Solids

   A.   Phase Separation of Viscous Melts [4], pp. 13-15; [7]

      1.   Thermodynamics
      2.   Kinetics and Morphology

B. Crystallization

    1. Nucleation [3], pp. 48-54; [2], pp. 479-483; [8], pp. 41-47

        a. Homogeneous
        b. Heterogeneous

    2. Growth and Dissolution [8], pp. 47-50; [2], pp. 482-483; [9], pp. 609-611; [4], pp. 13-15; [7]

        a. Kinetics
        b. Morphology
        c. Impurity and photoeffects
        d. Pressure effects

[1] "The Liquid State and the Liquid Solid Transition", Trans. A.I.M.E. 221, 422 (1961).

[2] "Under what Conditions can a Glass be Formed", Contemp. Phys. 10, 473 (1969).

[3] "Thermodynamics and Kinetics of Formation of the Glass State and Initial Devitrification", pp. 41-56 in "Physics of Non-Crystalline Solids: (edited by J. W. Prins) North Holland, Amsterdam (1964).

[4] Ehrenreich Report - "Fundamentals of Amorphous Semiconductors", Nat. Academy of Sciences, Washington (1972).

[5] "The Liquid State", pp. 1-22 in "Solidification", Am. Soc. Metals, Metals Park, Ohio (1971).

[6] (with D. E. Polk)"Structure of Amorphous Semiconductors", J. Non-Cryst. Solids 8-10, 19 (1972).

[7] (with B. G. Bagley)"Transitions in Viscous Liquids and Glasses", to appear in "Treatise on Solid State Chemistry" (edited by Bruce Hannay) Plenum Press.

[8] (with M. H. Cohen) "Crystallization Kinetics and Glass Formation", pp. 38-62 in Vol. 1, "Modern Aspects of the Vitreous State" (edited by J. D. MacKenzie) Butterworth's, London (1960).

[9] "On the Relation between Crystallization Rate and Liquid Structure", J. Phys. Chem. 66, 609 (1962).

ON THE STRUCTURE ANALYSIS OF AMORPHOUS MATERIALS

J. F. Graczyk

IBM Thomas J. Watson Research Center

Yorktown Heights, New York 10598

I. INTRODUCTION

The determination of atomic arrangements in amorphous materials can be generally accomplished by x-ray and, more frequently, in the past decade by electron and neutron diffraction. Employing x-ray techniques requires the usage of several radiation sources, elimination of Compton scattering either experimentally or during data analysis and accumulation of data for at least several days during which many experimental factors may vary. With conventional electron diffraction, such data can be obtained in a few minutes. This is because electrons are scattered more strongly than x-rays, and the ratio of the intensities can be a factor of 10^3-10^5, depending on the angle of scattering. However, the electron diffraction intensity profiles obtained from conventional photographic-photodensitometer techniques are quite inaccurate. This is due to the high background of incoherently scattered electrons and errors due to exposure and photodensitometer techniques. An accurate method in E.D. which virtually eliminates inelastically scattered electrons is that of scanning electron diffraction with energy filtering. The filter eliminates all inelastically scattered electrons from the scattered intensity and the recorded pattern is an intensity profile in reciprocal space.

Either of the aforementioned diffraction experiments has its advantages as well as disadvantages if one considers all experimental variables. Furthermore as we

shall see later, indeed ideally all three radiations
should be utilized in determining the structure of amorphous materials.

An extensive treatment of the physic of scattering
may be found in several excellent texts on x-ray diffraction, for example B. E. Warren,[1] R. W. James[2] and A.
Guinier.[3]

The atomic arrangement in an amorphous solid is deduced from the analysis of the elastically scattered intensity. The Fourier inversion of it yields the radial
distribution function which gives the average number of
nearest neighbors as well as higher order neighbors and
their average distance from an average atom. The radial
distribution function can then be matched by a model composed of atoms with finite coordinates which represent
the best average fit. The result is a possible model for
the structure of the solid. Depending on the fit of the
theoretical functions to experimental ones, one can eliminate some models. There are cases where an unequi-
vocal fit is not possible.

In these two lectures we consider the fundamental
aspects of the scattering theory for amorphous materials.
We shall consider first the scattering by an atom, define
form factors, and then consider the scattering by a gas
liquid and an amorphous material.

In lecture II we shall discuss some experimental aspects and the radial distribution functions for a solid
composed of single to many element components.

We shall consider mainly scattering of electrons for
both lectures. Experimental details for x-ray will be
referenced.

II. SCATTERING OF ELECTRONS

2.1 Introduction

The collision of an electron beam and an atom is a
many-body problem and solutions to it can only be found
under certain simplifying assumptions. In the collision
process in addition to purely elastic scattering, there
is also an inelastic contribution, which involves a reaction of the scatterer on the scattered beam. This effect complicates to a considerable extent the quantitative
interpretation of scattered intensities and involves the

following processes:

a. Excitation of atoms and molecules by electron impact (usually referred to as a direct inelastic collision).

b. Exchange of electrons between the scattering beam and the scattering atom (here the incident electron is captured and the atomic electron ejected; called a rearrangement inelastic collision).

c. In actual solids particularly metallic we have a collective or long-range excitation of electrons within the solid called plasma excitation.

d. In solids we also have thermal or phonon excitations, but the energy losses involved are much smaller than those of part c.

In what follows we shall only treat the purely elastic scattering of electrons from isolated atoms, gases and condensed matter in a disordered and partially ordered states.

2.2 Elastic Electron Scattering by an Isolated Atom

In experiments on the scattering of a beam of electrons, it is customary to measure the number of scattered electrons per unit time on an element of area dS placed at a distance r from the scattering atom.[4-5] The number of electrons falling on dS will then be proportional to the area dS and inversely to the square of the distance r, or we may say that the number is proportional to the solid angle $d\omega$ subtended by dS at the center of the atom. The differential scattering cross section $D(\omega)$ for the atom is then defined as:

$$\frac{dn}{Io} = i(\theta)d\omega \equiv D(\omega) = \left(\frac{d\sigma}{d\omega}\right) atom \tag{1}$$

for scattering through an angle θ. Io is the intensity of the primary electron beam incident on the atom. σ is the total cross section per atom and as defined here expresses the total fraction of the incident electrons scattered outside a given minimum angle θ_{min} and will be given by:

$$\sigma_{atom} = 2\pi \int_{\theta_{min}}^{\pi} I(\theta) \sin\theta d\theta. \tag{2}$$

The total cross section can be physically interpreted as the effective area of an atom for scattering seen by an approaching electron. This effective cross section de-

pends on the type of collision, and generally may be composed of several contributions. In terms of the differential cross sections, we may write

$$\{D(\omega)\}_{atom} = \sum_i \{D_i(\omega)\}_i \tag{3}$$

where the i stands for elastic, inelastic, plasma, etc.
The elastic differential cross section is related to the
amplitude $f(\Theta)$ of the scattered wave by:[6,7]

$$D_i(\omega) = |f(\Theta)|^2 \tag{4}$$

with wave interference neglected. Θ is the scattering
angle and is related in electron diffraction work to the
wave vector $|k| = \frac{4\pi \sin(1/2\Theta)}{\lambda}$. To find $f(\Theta)$ we must solve

the Schrödinger wave equation for the scattering problem.
To do so we need to know the potential of the atom $V(r)$.
Detailed calculations for several simplified $V(r)$ are
treated in Ref. (5). For a spherically symmetric potential of the type C/r the scattered amplitude is given by:[5]

$$f(\Theta) = \frac{1}{2iK} \sum_{\ell=0}^{\infty} (2\ell+1) \left[e^{2i\eta_\ell}-1\right] P_\ell(\cos\Theta) \tag{5}$$

where $P_\ell(\cos\Theta)$ is the ℓ-th Legendre function. The electron form factor $f(\Theta)$ is therefore a complex quantity,
and the scattered intensity $I(\Theta)$ is:

$$I(\Theta) = f(\Theta)f(\Theta)^* = A^2+B^2$$

$$A = \frac{1}{2K} \sum (2\ell+1)(\cos 2\eta_\ell -1)P\ell \tag{6}$$

$$B = \frac{1}{2K} \sum (2\ell+1)(\sin 2\eta_\ell)P\ell$$

the phase factor η_ℓ, which depends on k, $V(r)$ and the
scattering angle Θ, can only be calculated by numerical
methods. For electron energies $E \gg V(r)$ (fast electrons)
we use the first Born approximation. In the solution we
employ the perturbation or the Green's function method
obtaining an expression for the electron form factor in
which we neglect all phase shifts η_ℓ and is given by:

$$f(\Theta) = -\frac{2m}{h^2} \int_0^{\infty} \frac{\sin kr}{kr} V(r)r^2 \, dr \tag{7}$$

Values of $f(\theta)$ therefore may predict the detail of the form of $V(r)$. It is observed experimentally that the Born approximation gives satisfying results for electron energies greater than Emin, which varies through the periodic table, increasing monotonically with z. The atomic potential $V(r)$ is related to the charge density in the simplest approximation by:

$$V(r) = - \frac{Ze^2}{r} + e^2 \int \frac{\rho(r')dr'}{|r-r'|} \qquad (8)$$

from which we may relate the electron form factor to the x-ray form factor $f_x(k)$ given by:

$$f_x(k) = 4\pi \int_0^\infty \rho(r) \frac{\sin kr}{kr} r^2 \, dr \qquad (9)$$

by use of the Poisson equation $\nabla^2 \phi = -\rho/\varepsilon_0$ with the result:

$$f(\theta) = \frac{8\pi^2 me^2}{h^2} \left(\frac{z-f_x(k)}{k^2}\right) \qquad (10)$$

as $k\to 0$; $f_x(k)\to z$ and the above equation is no longer applicable. To calculate $f(0)$, we must expand $\frac{\sin kr}{kr}$ in a power series and evaluate $f(\theta)$ as $k\to 0$. J. A. Ibers[8] derives for:

$$f(0) = \frac{4\pi^2 me^2}{3h^2} z\langle r^2 \rangle \qquad (11)$$

where $\langle r^2 \rangle$ is the mean square radius of the atom. Because of historical priority, values of $f(\theta)$ were calculated and checked against x-ray form factors by (9).

Values of $f(\theta)$ based on the Born approximation were obtained by Vainshtein[10] using values for $f_x(k)$ of Mc-Weeny[11] for first row elements and those of James and Brindley[12] for second row elements. His value of $f(\theta)$ were obtained by extrapolation of $f(\theta)$ to k=0. For the heavier elements (argon above) Vainshtein used the T.F. potential functions. As values of $f_x(k)$ were recalculated using the Hartree Fock Slater (H.F.S.) improved wave functions for the lighter elements and the Thomas Fermi-Dirac (T.F.D.) potentials, which take into account the effect of electron exchange for the heavier elements; J. A. Ibers et al[13,14] recalculated Vainshtein form factors up to x=80 (Ag).

The latest table of values of $f(\theta)$ consistent with $f_x(k)$ is that of Ibers and Vainshtein and others[10,15-18]. Values of the electron form factor are subject to change as improved methods for finding $V(r)$ are developed. Main difficulties in the theory are due to:

a. Shell structure of the atom.
b. Exchange effects.
c. Errors either of omission or calculation in the contribution to the atomic fields of the outer electrons.
d. Deviations from spherically symmetric potential distributions.
e. Values of $f(o)$ and the variation of $f(\theta)$ near $k=0$ to $(\frac{\sin \theta}{\lambda} < 0.05)$, which is the region of k which shows the largest spread in values.

Errors due to (a) and (b) are as a result of using the T.F.D. potentials, which in fact average out all shell effects.

In summary the form factors for electrons decrease monotonically with scattering angle for isolated atoms. This is a result of the interference of wavelets scattered at varying distance from the center so that the phase difference increases with increasing scattering angle. The electron form factor will strongly depend on the state of aggregation and order of atoms which are packed to form a solid. Since electrons interact very strongly with the atomic potential we expect the experimental values of the form factor to depend on the local atomic potential near the atom. It is therefore possible, and in some cases found experimentally to be so, that the isolated atom scattering factors may not be suitable for either their absolute value or their shape. For large values of the scattering vector k the theoretical values for the atomic $f(\theta)$ should be independent of the local atomic environment, which at least provides a link between experimental measurements and theoretical computations. We must now consider the scattering from a solid which is composed of N atoms with a particular spatial arrangement of these. We may therefore replace each atom by its form factor $f(\theta)$ and consider the contribution of the interference due to the spatial packing of these. In the following section we therefore derive the scattered intensity from an aggregate of atoms.

2.3 The Debye Scattering Equation

Experimentally we measure the scattered intensity

from a sample as a function of the momentum transfer vector $|k| = \frac{4\pi \sin\theta}{\lambda} = |\bar{K} - \bar{K}_o|$. We must therefore calculate the relationship between this intensity and the atomic positions in the sample. To do so we consider the sample composed of N atoms, with scattering power f_a and we denote the atomic position space vector by $R_1 R_2 \cdots R_n$. The scattered intensity in a direction parallel to the vector $\bar{K}$ is calculated by:

$$I_{e.u} = \sum_m f_m e^{i(\bar{K}-\bar{K}_o)\cdot\bar{R}_m} \sum_n f_n e^{-i(\bar{K}-\bar{K}_o)\cdot\bar{R}_n} \qquad (12)$$

The above expression is quite general and can be applied to crystalline and amorphous forms of matter. Equation 12 can be expressed in terms of the difference of $\bar{R}_m - \bar{R}_n = \bar{r}_{mn}$ to obtain the intensity expression:

$$I_{e.u} = \sum_m \sum_n f_m f_n e^{i(\bar{K}-\bar{K}_o)\cdot\bar{r}_{mn}} \qquad (13)$$

Expression 13 is valid for elastic scattering of radiation and for single scattering events. Equation 13 can therefore physically be interpreted as a sum of amplitudes scattered from each atom at a particular point in reciprocal space in terms of their phase difference. To obtain the intensity we simply add first all the amplitudes from all atoms and take the complex conjugate of it. For a large sample the vector $\bar{r}_{mn}$ takes all orientations in space with equal probability so we take the average of the phase term to obtain the Debye equation for scattering:

$$I_{e.u} = \sum_m \sum_n f_m f_n \frac{\sin kr_{mn}}{kr_{mn}} \qquad (14)$$

As examples in which the Debye scattering equation has been very successfully applied we may consider gases and liquids and amorphous solids. For example an ideal monoatomic gas at low density and low pressure will scatter as $f(\theta)$. Experimentally for k=0 we obtain that the intensity is $N^2 f^2$. For a polyatomic gas we obtain molecular scattering, because we have complete incoherency between the scattering by the different molecules, and the diffuse scattering is determined by the distances within the molecule. In the Debye equation we therefore take for gases and liquids the space average as well as time average of the position vectors $\bar{r}_{mn}$ relative to k, whereas for an amorphous solid we only average over space.

One method of determining the particular structure of a noncrystalline solid is to measure the average scattered intensity and compare it to calculated functions for a variety of atomic models. The average local atomic order will then be described by the atomic model which matches the experimental function with highest fit. This method has been applied for example to Si and Ge.[19-21] Because the experiments are performed at room temperature we must take into consideration the thermal motion of the atoms at the measuring temperature. We write the Debye equation in the form:

$$F(k) = k \left(\frac{Ie.u.}{f^2} - 1\right) = \frac{1}{N} \left\{ \sum_m \sum_n \left(\frac{1}{r_{mn}}\right) \sin kr_{mn} \right\} e^{-2M_{mn}}$$

$$m \neq n$$

$$(15)$$

with

$$2M = \frac{12\ h^2 T}{mk\ \Theta_M^2} \left\{ 1 + \left(\frac{\Theta}{T}\right)^2 \right\} \frac{\sin^2\Theta}{\lambda^2} \tag{16}$$

with Θ_M being the Debye temperature of the material.

The multiplication by an average damping factor effectively overdamps the modulations in the reduced intensity function at high values of k. At present reduced values of 2M (where 2M is the crystalline value) have been used in the calculated functions at high values of k to match the experimental fucntions.

We show an example of calculated reduced intensity functions and data for Si and Ge in Figures 1 and 2. In Figure 1 we compare the F(k) for several diamond cubic microcrystallite arrangements and compare to experimental function, from the work of Moss and Graczyk.[19] In Figure 2 we compare the experimental F(s) of Ge to the calculated function for the random network model built by Polk[23] of 519 atoms, from the work of Chaudhari and Graczyk.[24]

The assumption of equal orientational probability of the vector r_{mn} relative to k is generally fulfilled in diffraction experiments. However, for theoretical model calculations it becomes necessary to show that a cluster containing N atoms (where N is not greater than ~500) scatters on the average equal to the experimental volume. Such calculations have been carried out by Betts and Bienenstock[25] and Graczyk and Chaudhari.[22] In experiments on very thin samples the isotropic scattering is no longer present. Chaudhari et al[26] discovered that amorphous materials when very thin show anisotropic scattering. By calculating the structure factor for several clusters it

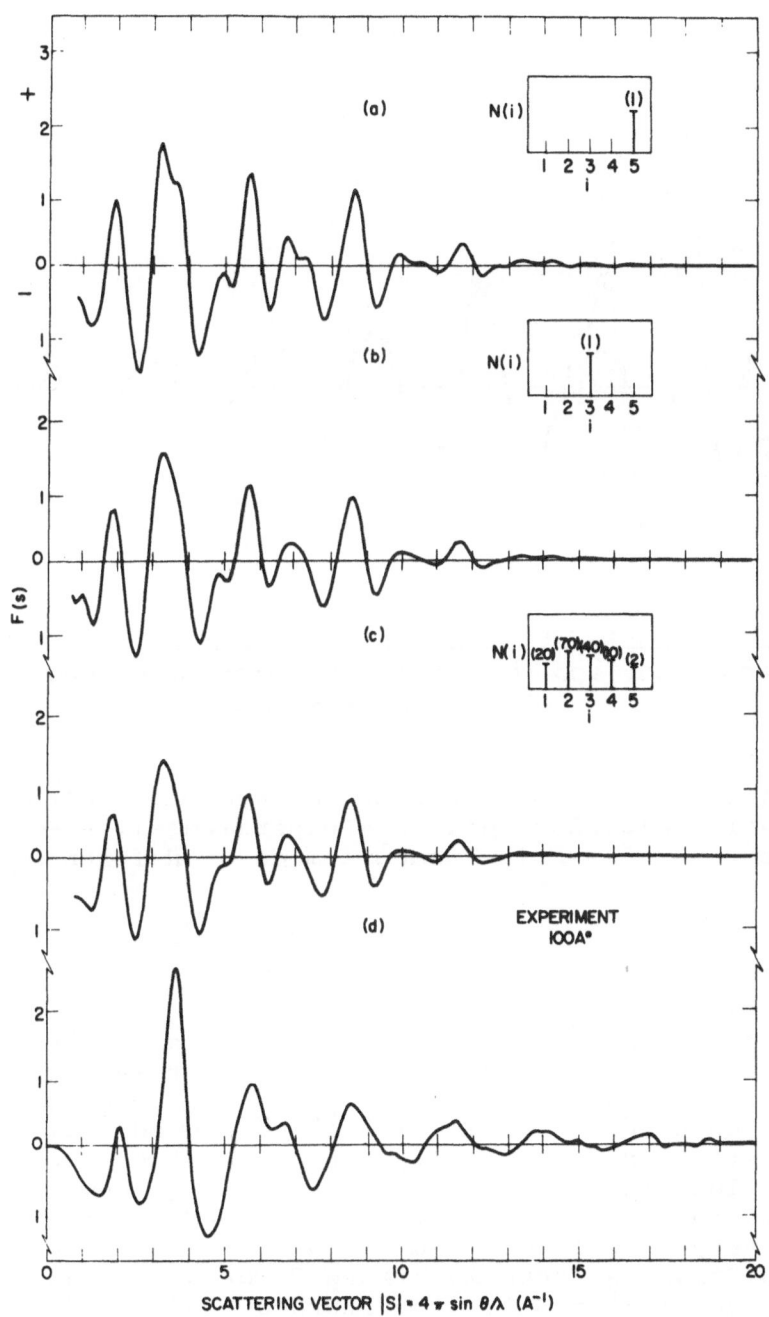

Fig. 1 The reduced intensity function F(k) (a)-(c). Various diamond cubic microcrystallites calculations: (a) for
five neighbor coordination shells about a central origin
(47 atoms); (b) for three neighbor shells (29 atoms); (c)
for the crystallite distribution shown in the insert. (d)
Experimental on a 100Å amorphous Si film.   From ref. 19.

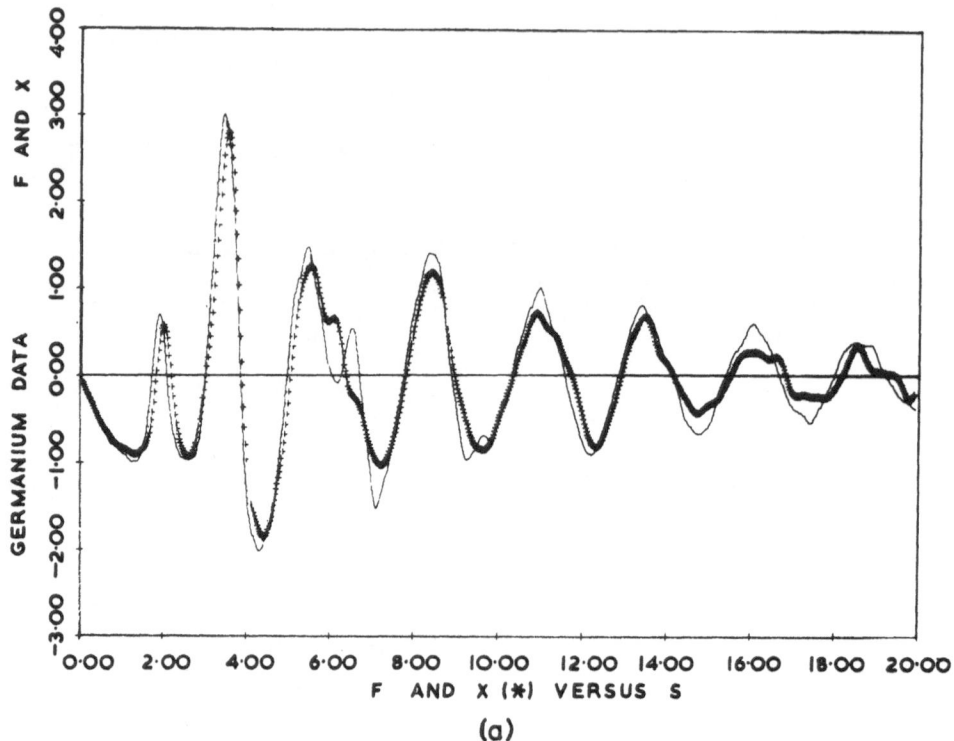

Fig. 2 A comparison of experimental and calculated function F(k) for amorphous Ge. xxx experimental, ——calculated for the 519 Polk-Turnbull model. Reference 24.

has been determined that amorphous models of up to 8,000
atoms[27] scatter anisotropically. To show this we calculate equation 13 as function of Θ, ϕ. The results of calculations on several models[22,24,26,27] showed that the
models scatter anisotropically. For the strong scattering directions the model shows a dramatic alignment of
atoms into rows. This anisotropic property is characteristic of very thin films. An example of a structure factor calculation is shown in Figure 3 for the 61 atom model of Henderson[28] and 68 atom amorphon model of Grigorovici.[29]

Whereas in principle the determination of a possible
structural model can end after obtaining a satisfactory
fit to the reduced intensity function, in practice a

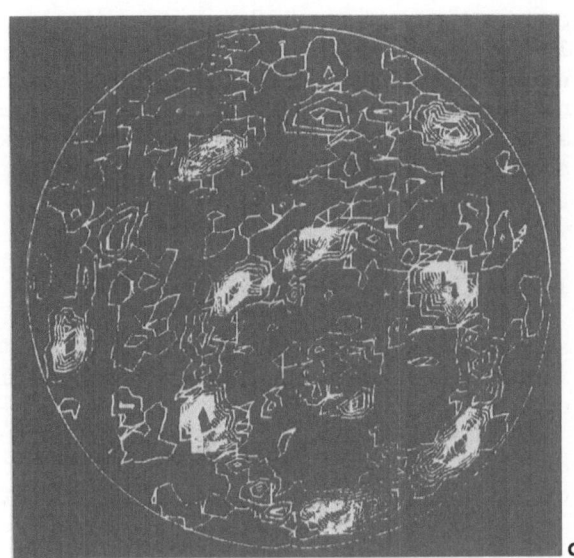

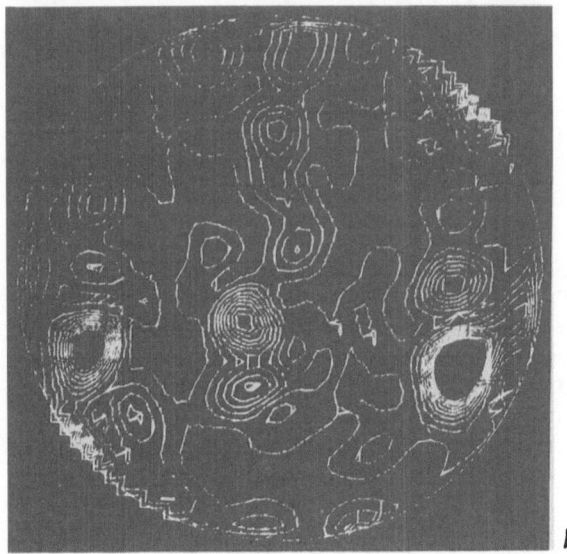

Fig. 3   Equivalent contour maps of the calculated values
of FF* for a fixed value of (k) equal to {111} plotted as
a function of ϕ and Θ. Θ varies radially outward from
0° to 90° and ϕ varies circumferentially from 0° to 360°.
The contours are values of FF* in equal increments of 0.1.
a. is for the 68 atom amorphon model, b. is for the 61
atom cluster of Henderson.   Reference 28.

probable structural model is first constructed from the
radial distribution function, which gives the fundamental structural recipe for the building of the model.
Furthermore we must attempt to match all calculable structural functions before a model for an amorphous solid
can be fully accepted. We consider in the next section
the fundamentals of a radial distribution function determination.

III. RADIAL DISTRIBUTION FUNCTIONS

For gases, liquids and amorphous solids the radial
distribution function $4\pi r^2 \rho(r)$ describes the average number of atom centers between distances r and r+dr from
the center of an average atom. We may also show that it
is a statistical interatomic distance function given by
the convolution of the density distribution of scattering matter in the object with the inversion through the
center of symmetry $\rho^*(r) = \rho(-r)$ of this same density
distribution. The interatomic distance function $P(r)$ is
given by:

$$P(r) = \int \rho(r') \, \rho(r'+r) \, dv_r \tag{17}$$

with the following transform pair:

$$I(k) = \int P(r) e^{i(kr)} dv_r \tag{18}$$

To relate the previous result of $I_{(k)}$ to the radial
distribution function we assume the scattering volume to
consist of N atoms with the atomic fraction of each species given by x_n with scattering atomic factor f_n and we
define the radial distribution function $4\pi r^2 \rho_{ij}(r)$ as
giving the number of j type atoms surrounding an i type
at a distance r. The intensity is then given by: [1-3]

$$I(k) = N \sum_{j=1}^{n} x_j f_j^2 + N \sum_{i=1}^{n} \sum_{j=1}^{n} x_i f_i f_j \int_0^\infty 4\pi r \rho_{ij}(r) \frac{\sin kr}{k} \, dr \tag{19}$$

$\rho_{ij}(r)$ is the density fluctuation which approaches $\bar{\rho}_j$ the
constant mean density of j type atoms. If we substitute
for $\rho_{ij}(r) = \rho_{ij}(r) - \bar{\rho}_j + \bar{\rho}_j$ and rearrange; we obtain:

$$I(k) = N \sum_{j=1}^{n} x_j f_j^2$$

$$+ N \sum_{i=1}^{n} \sum_{j=1}^{n} x_i f_i f_j \int_0^{\infty} 4\pi r (\rho_{ij}(r) - \bar{\rho_j}) \frac{\sin kr}{k} dr$$

$$+ N \sum_{i=1}^{n} \sum_{j=1}^{n} x_i f_i f_j \int_0^{\infty} 4\pi r \bar{\rho_j} \frac{\sin kr}{k} dr \qquad (20)$$

The last term in equation 20 gives rise to very small angle scattering and is neglected. Equation 20 for a monatomic material reduces to:

$$F(k) = k \left(\frac{I_n(k)}{ff^*(k)} - 1 \right) = \int_0^{\infty} 4\pi r (\rho(r) - \rho o) \sin kr \, dr$$

$$= \int_0^{\infty} G(r) \sin kr \, dr \qquad (21)$$

where:

$k = 4\pi \sin\Theta / \lambda$

$I_N(k)$ = elastic electron intensity normalized to ff* corrected for multiple scattering if necessary and for background scattering from sample holder, backing films etc.

$ff^*(k)$ = the product of the theoretical free atom electron scattering factor and its complex conjugate corrected for the appropriate atomic packing effects

r = radial distance

ρ_o = average radial density

$\rho(r)$ = the number of atoms/unit volume at distance r from any arbitrary origin atom

$G(r) = 4\pi r (\rho(r) - \rho_o)$ the reduced radial distribution function.

The radial density function (R.D.F.) $J(r) = 4\pi r^2 \rho(r) = rG(r)$ which we assume a priori to be spherically symmetric describes the number of atoms in a spherical shell i located at r_1 is given by:

$$CN(r) = \int_{r_i - \Delta}^{r_i + \Delta} 4\pi r^2 \rho(r) dr \qquad (22)$$

In principle Eq. (22) is applicable over the full range of $J(r)$ but in practice it is only useful for the

first few nearest neighbors. Beyond these the overlap
as well as asymmetry in the atomic coordination shells
leads to quite arbitrary CN(i) assignments.

The broadening of the peaks in $J(r)$ is due to static displacement and thermal vibrations of the atoms.
We may then represent $J(r)$ as:

$$J(r) = \sum_{i=1}^{\infty} \frac{C_i}{(2\pi\sigma_i^2)^{1/2}} \exp\left[\frac{(r-r_i)^2}{2\sigma_i^2}\right] \qquad (23)$$

where C_i is the number of atoms in the ith shell about
an arbitrary origin atom, r_i is the distance from the
origin to the ith shell and σ_i^2 is the mean square amplitude of vibration plus mean square static displacements.
If we assume these to be independent of each other then
$\sigma_i^2 = \sigma_{ith}^2 + \sigma_{i\,disp}^2$. For a polycrystalline sample containing small and completely random oriented crystallites
(this is a requirement for a true spherical average of
the intensity, and further in the absence of strains and
any atomic distortions) σ_i as obtained from $J(r)$ is solely due to σ_{ith}. The difference in σ_i between amorphous
and polycrystalline material yields $\sigma_{i\,disp}$ for each
shell. For a binary compound the intensity expression
becomes more complex:

$$I(k) = N(x_1 f_1^2 + x_2 f_2^2) + N\,x_1 f_1^2 \int_0^{\infty} 4\pi r (\rho_{11}(r) - \bar{\rho}_1) \frac{\sin kr}{k}\,dr$$

$$+ Nx_1 f_1 f_2 \int_0^{\infty} 4\pi r (\rho_{12}(r) - \bar{\rho}_2) \frac{\sin kr}{k}\,dr$$

$$+ Nx_2 f_2 f_1 \int_0^{\infty} 4\pi r (\rho_{21}(r) - \bar{\rho}_1) \frac{\sin kr}{k}\,dr$$

$$+ Nx_2 f_2^2 \int_0^{\infty} 4\pi r (\rho_{22}(r) - \bar{\rho}_2) \frac{\sin kr}{k}\,dr \qquad (24)$$

Equation 24 for many binary alloys for which $\rho_{ij}/x_j = \rho_{ii}/x_i$ with $\rho_i = \rho_{ij} + \rho_{ii}$ reduces to

$$I'(k) = \langle f^2 \rangle + \frac{\langle f \rangle^2}{k} \int_0^{\infty} 4\pi r (\rho^R(r) - \rho_0) \sin kr\,dr \quad \text{with}$$

$$\rho^R = \frac{\rho_{ii}}{x_i} = \frac{\rho_{jj}}{x_j} = \frac{\rho_{ij}}{x_j} = \frac{\rho_{ji}}{x_i}$$

and

$$\langle f \rangle^2 = \left[\sum_{i=1}^{m} x_i f_i \right]^2$$

$$\langle f^2 \rangle = \sum_{i=1}^{m} x_i f_i^2$$

and

$$F_{(k)} = \left(\frac{I - \langle f^2 \rangle}{\langle f \rangle^2} \right) k = \int_0^\infty 4\pi r (\rho^R - \rho_o) \sin kr \, dr \qquad (25)$$

From equation 24 we note that to determine all the ρ_{ij} values we need at least three experimental R.D.F.'s. These can be determined in only a few cases from experiments, utilizing three different radiations. The discussion of these factors can be found in papers by Pings and Waser[30] and Keating.[31]

IV. EXPERIMENTAL AND DATA ANALYSIS CONSIDERATIONS IN R.D.F. DETERMINATION

Many experimental factors are important in collecting data from an experimental set up.

Very good references for x-rays are the paper by Warren and Mozzi[32] and for electron diffraction by Graczyk and Chaudhari.[20] Once an elastically diffracted intensity trace is obtained which is in arbitrary units, we must normalize the data to ff* or to $\langle f^2 \rangle$ for a binary alloy. For large k the experimental intensity $I_N(k)$ approaches the independent scattering. Normalization of data for large k is likely to introduce an error in the initial value of the normalization constant A. The effect of erroneous A is to introduce large oscillations in $G(r)$ over a range of r from 0. to 0.15Å. To correct A we systematically alter A by $\pm$ ΔA in small increments and minimize the amplitude of the oscillations in $G(r)$ over the aforementioned interval. The correct value of A is such that gives oscillations of minimum amplitude with an r.m.s. average value equal to $-4\pi\rho_o$. The ρ_o has little effect on the final value of the density since $-4\pi\rho_o$ is generally linear over the range from 0 to 1.9Å, whereas in the initial normalization step, one compares the r.m.s. average value of the oscillations to

$-4\pi\rho_o$ over a relatively small interval of r from 0 to
0.15Å . With the data normalized one gets the first
F(s). The transform G(r) still contains oscillations
around $-4\pi\rho_o$ which have no structural interpretation,
plus oscillations in the region around the first and second coordination shells. These oscillations are due to
slowly varying errors and finite data interval. The removal of these errors has been fully discussed by Kaplow,
Strong and Averbach.[33] The removal of termination errors,
by extending the data theoretically has utilized two approaches. One of these, the correlation method was applied by Kakinoki et al[37] to very thin films of amorphous
carbon. In this procedure one calculates only several
components of F(k) for $k>k_{max}$ and compares the experimental F(k) at high k values to these terms. The other method minimizes oscillations in the slope $-4\pi\rho_o$, by extending the F(k) to 2 k_{max} by a self consistent trial
and error procedure.

We show the data of Graczyk and Chaudhari[20] for amorphous Ge thin film, obtained from electron diffraction
with energy filtering experiments in Figure 4. The filter-
ed intensity profile is used as input to a computer program which normalizes the data, and corrects the first
reduced intensity F(k) for the aforementioned errors and
calculates the first reduced density function G(r). To

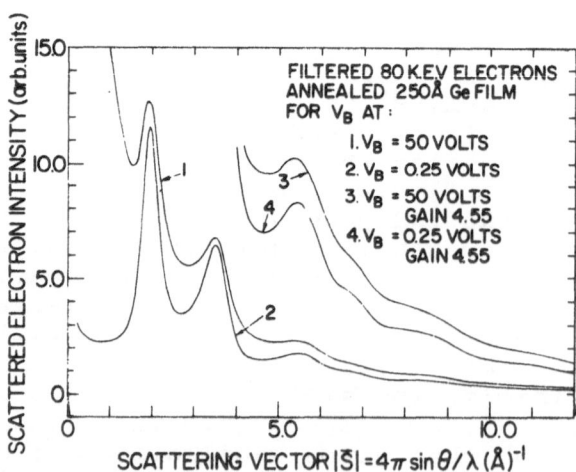

Fig. 4 Scanning electron diffraction pattern from a well
annealed 250Å thick germanium film (Ge-II) with the velocity filter bias set at V_B = 0.25 and 50.0 volts
Reference 20.

illustrate the reliability of the corrections procedures
we have calculated the reduced intensity function $F(k)$
for a diamond cubic crystallite containing 71 atoms;
damped by a single damping factor e^{-2Mmn} shown in Fig.
5a. We then applied the correction procedure to the $F(k)$
and obtained the respective $G(r)$ shown in Fig. 5b. It
is apparent that all peaks in $G(r)$ occur at the proper

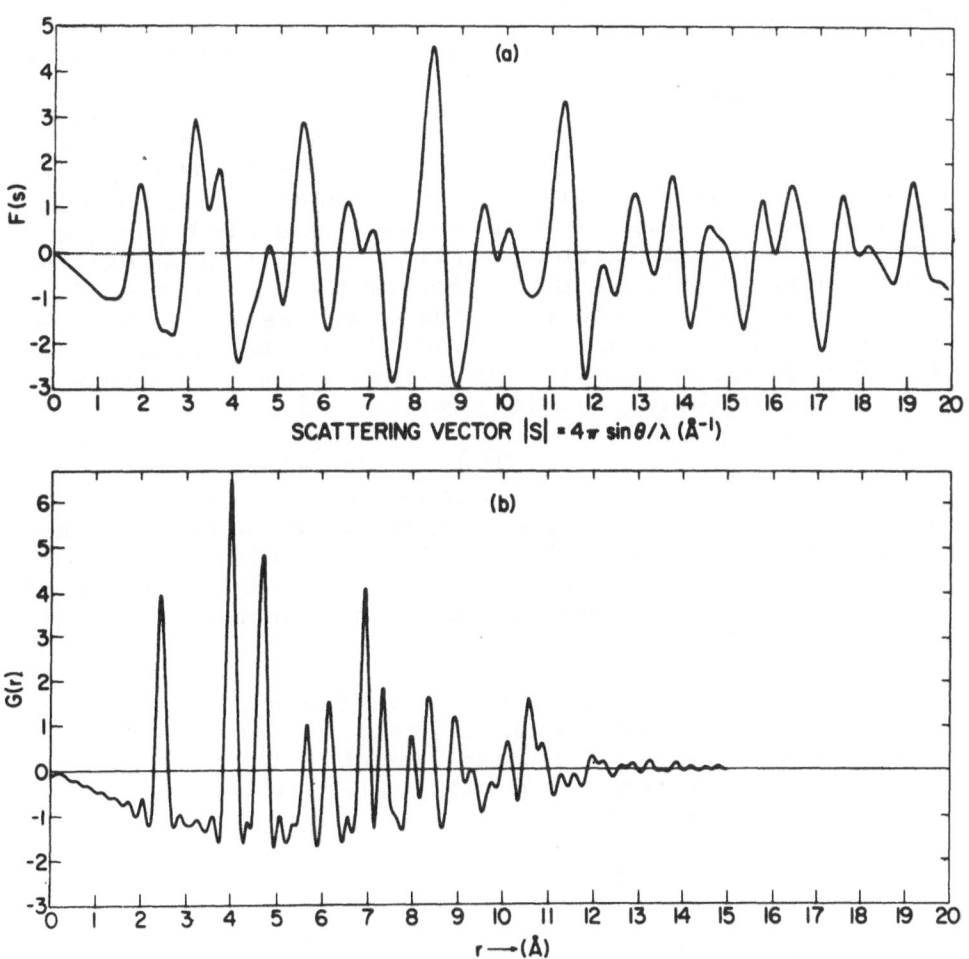

Fig. 5 a. Calculated $F(k)$ for 71 atoms of Ge for a diamond cubic microcrystallite; b. the calculated $G(r)$ for
a., using correction procedure of reference 20.

values of r consistent with a diamond cubic structure
and as expected, have all approximately the same breadth at 1/2 maximum.

In Figure 6a, b and c we show several interesting
results for amorphous Si and polycrystalline Si. We show
the ripple which is introduced in the low r region of
$G(r)$ due to an inadequate atomic scattering factor. We
also show the $G(r)$ for two cases relating to the low angle scattering observed in many amorphous materials Fig.
6a. Removal of the low angle scattering results in a
larger density $-4\pi\rho_o$. In Figure 6b we compare directly
the $G(r)$ for the as deposited amorphous film and the $G(r)$
calculated after the film has been crystallized. Such a
comparison shows directly the differences for both structures. In Figure 6c we compare the $G(r)$ for an as deposited and annealed film at 400°C. In Figure 7 we show
the $J(r)$ for amorphous Si and the three best Gaussian
fits to the first 3 correlation peaks. From this function we calculate the correlation distances and number
of neighbors. In Figure 8 we compare the $J(r)$ for the
amorphous and crystallized Si films. It is this direct
comparison which allows us to determine the static displacements in the amorphous structure. We show in Figure 9 the $J(r)$ for amorphous vapor deposited thin films
of Ge (Ge-I) and a Ge polycrystalline film made amorphous by ion implantation (Ge-IV).

From the experimental $J(r)$ we therefore can determine in principle the coordination numbers C.Ni., the
neighbor distance r_i, static displacements and temperature effects. The following step is to build models
which must generate an R.D.F. consistent with the experimental R.D.F. These have so far been generated by computer methods[38] or hand built models.[39,40]

We must recognize however, that the r.d.f. represents
the local atomic order averaged over all atoms and orientations and a particular model which fits the r.d.f.
must also be consistent with other structural experiments
and physical measurements.

ACKNOWLEDGMENTS

The data presented in these lectures is based on a
series of publications in collaboration with Prof. S. C.
Moss and Dr. P. Chaudhari, and the author is indebted for
their permission to reproduce the figures and many sti-
mulating discussions.

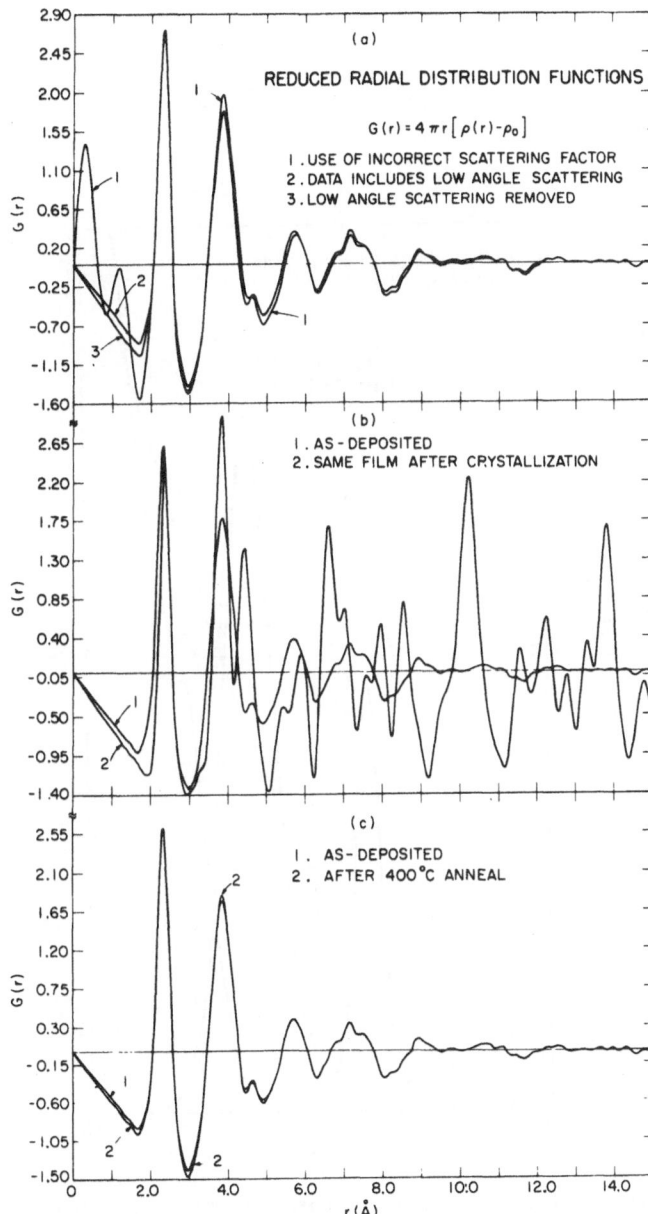

Fig. 6 Reduced radial distribution function G(r) for Si as obtained by Fourier inversion of the reduced intensity function F(k) for: (a) 1) when data is normalized to the experimental fafa* for x-ray transformed for electrons via the Mott equation. 2) the low angle high intensity is left as recorded below S=0.3 3) The intensity is smoothly extrapolated to zero. (b) A comparison of the as deposited and after crystallization. (c) The effects on G(r) after the 400°C anneal.

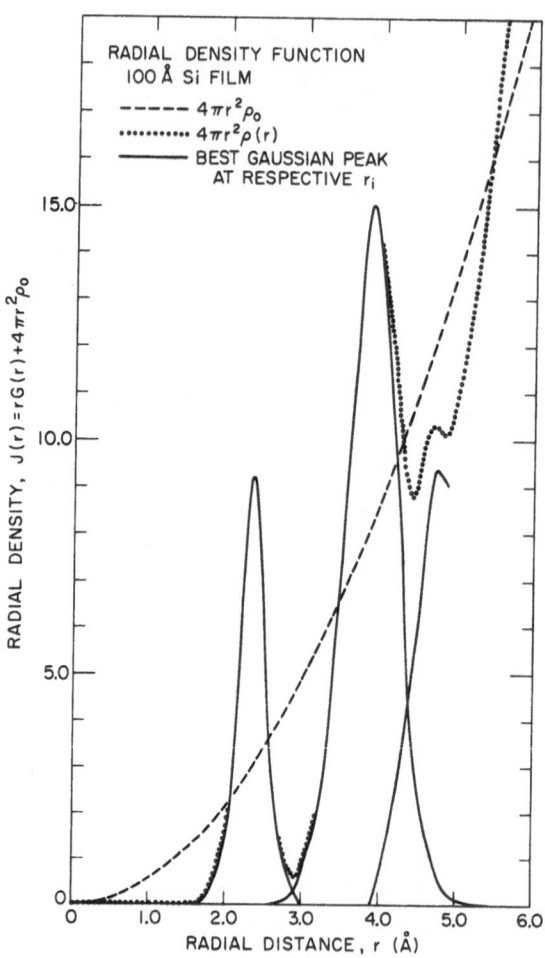

Fig. 7 Radial density function of the as deposited
100Å Si film with the near gaussian peaks drawn for calculating the C.N. values and assymmetry in the peaks.

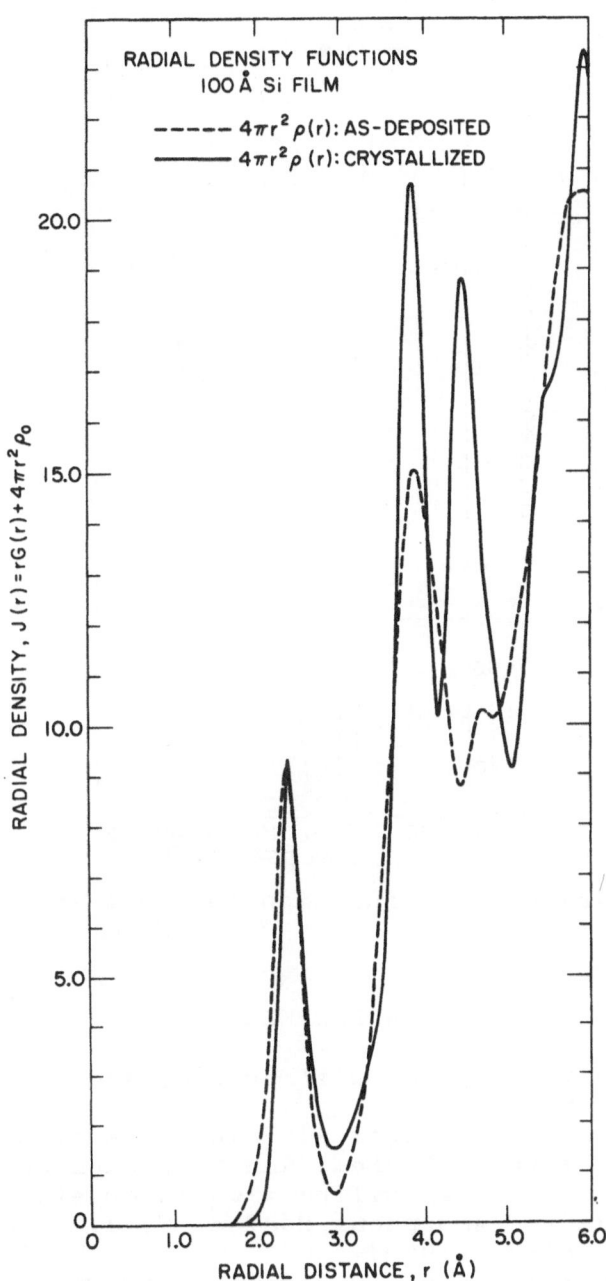

Fig. 8 A direct comparison between the radial density functions of the as deposited and crystallized films of Si.

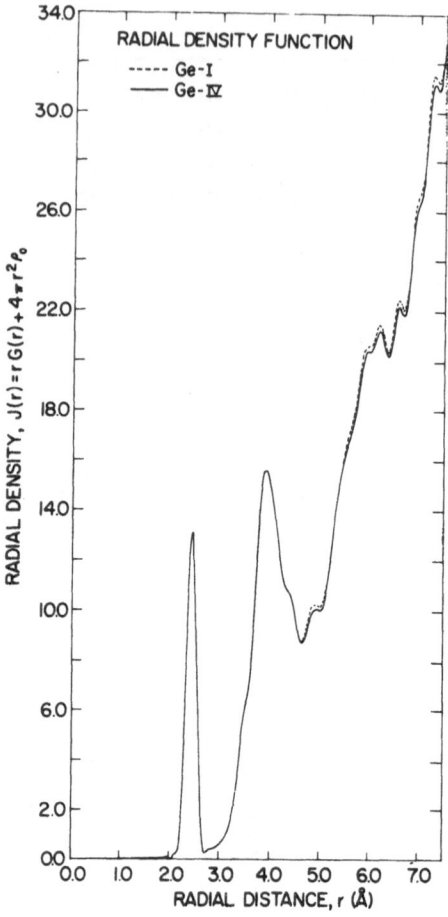

Fig. 9 Radial density function for Ge-I and Ge-IV.

REFERENCES

1. B. E. Warren, X-ray Diffraction, Addison-Wesley Pub-
    lishing Co. (1969).
2. R. W. James, The Optical Principles of the Diffrac-
    tion of X-rays, G. Bell and Sons, London (1954).
3. A. Guinier, X-ray Diffraction in Crystals, Imperfect
    Crystals and Amorphous Bodies, W. J. Freeman and Co.,
    San Francisco (1963).
4. R. D. Heidenreich, Fundamentals of Transmission
    Electron Microscopy, Interscience Publishers (1964).
5. N. F. Mott, H. S. Massey, The Theory of Atomic Col-
    lisions, Claredon Press, Oxford (1965).
6. N. F. Mott, Proc. Roy. Soc., London A127, 658 (1930).

7. R. E. Burge and G. H. Smith, Proc. Phys. Soc., 79, 673 (1962).
8. J. A. Ibers, Acta Cryst., 11, 178 (1958).
9. V. Schomaker and R. Glauber, Nature 170, 290 (1952).
10. B. K. Vainshtein, J. Expt. Theoret. Phys. USSR, 25, 157 (1953).
11. R. McWeeny, Acta Cryst. 4, 513 (1951).
12. R. W. James, G. W. Brindley, Phil. Mag. 7, 12, 81 (1931).
13. J. A. Hoerni and J. A. Ibers, Acta Cryst. 7, 744, (1954).
14. J. A. Hoerni and J. A. Ibers, Phys. Rev. 91, 1182 (1953).
15. International Tables for X-ray Crystallography, Vol. III, Tables 3.3.3, A(1) and A(2), Kynoch Press, 1962.
16. P. A. Doyle and P. S. Turner, Acta Cryst. A24, 390 (1968).
17. H. Raith, Acta Cryst. A24, 85 (1968).
18. J. Haase, Z. Naturforsch. 25a, 936 (1970).
19. S. C. Moss and J. F. Graczyk, Phys. Rev. Lett. 23, 1167 (1969); also, S. C. Moss and J. F. Graczyk, Proc. 10th Internatl. Conf. Phys. Semiconductors (Cambridge), eds. S. P. Keller, J. C. Hensel, and F. Stern, USAEC Div. Tech. Info., (1970) p. 658.
20. J. F. Graczyk and P. Chaudhari, Phys. Stat. Sol. (b) 58, 163 and 501 (1973); also, J. F. Graczyk and P. Chaudhari, Journ. of Non-Cryst. Solids, in press (1975).
21. F. C. Weinstein and E. A. Davis, J. Noncryst. Solids 13, 153 (1973).
22. J. F. Graczyk and P. Chaudhari, IBM Research Report No. 4856 (1974).
23. D. E. Polk, J. Noncryst. Solids 5, 365 (1971).
24. P. Chaudhari and J. F. Graczyk, 5th Int. Conf. on Amorphous and Liquid Semiconductors, Garmisch-Partenkirchen, Sept. 3-8, 1973, p. 59.
25. F. Betts and A. Bienenstock, J. Appl. Phys. 43 (11) 4591 (1972).
26. P. Chaudhari, J. F. Graczyk and H. P. Charbnau, Phys. Rev. Lett. 29, 425 (1972).
27. P. Chaudhari and J. F. Graczyk, Bull. Am. Phys. Soc. 19 (3), 317 (1974).
28. D. Henderson, Bull Amer. Phys. Soc. 16, 348 (1971).
29. R. Grigorovici, Thin Solid Films 9, 1 (1971).
30. C. J. Pings and J. Waser, Jour. Chem. Phys. 48, 7, 3016 (1968).
31. D. T. Keating, J. Appl. Phys. 34, 4, 923 (1963).
32. B. E. Warren and R. L. Mozzi, J. Appl. Cryst. 3, 59 (1970).

33.   R. Kaplow, S. L. Strong and B. L. Averbach, Phys. Rev. 138, 1336 (1965).
34.   L. Gatineau, J. Appl. Cryst. 5, 255 (1972).
35.   W. Paul, R. J. Temkin, and G. A. N. Connell, Advances in Physics 22(5), I-531, II, 581, III-643 (1973).

36.   J. H. Konnert and J. Karle, Nature-Physical Science 236, 92 (1972).
37.   J. Kakinoki, K. Katada, T. Hanawa, and T. Ino, Acta Cryst. 13, 171 (1960).
38.   D. Henderson, Bull. Amer. Phys. Soc. 16.16, 348 (1971).
39.   D. Turnbull and D. E. Polk, J. Non-Cryst. Solids, 8-10, 19 (1972).
40.   D. E. Polk and S. Boudreaux, Phys. Rev. Lett., 31, 92 (1973).

ON THE STRUCTURE OF AMORPHOUS FILMS

P. Chaudhari, J. F. Graczyk and S. R. Herd

IBM T. J. Watson Research Center

Yorktown Heights, New York 10598

ABSTRACT

In this note we present a qualitative and brief review of our understanding of the atomic arrangement in films of amorphous solids. We review the information obtained by diffraction techniques and the more recent results obtained by transmission electron microscopy. We conclude that the atomic arrangement is best described in terms of random network models.

INTRODUCTION

In this note we shall briefly review and summarize our current understanding of atomic arrangement in amorphous metal and semiconductor films. Information about the structure of a solid can be obtained from a variety of measurements. Generally the most direct technique has been diffraction which works most successfully in crystal structure determination. In the case of amorphous solids this technique has provided us with accurate information about the average structural environment around an arbitrarily chosen origin. However, it has not been able to yield explicit information about local atomic arrangement.

The limitations of this technique has prompted the use of transmission electron microscopy, which has the resolution required to resolve local atomic arrangements. Although the use of this instrument to investigate amorphous solids is not new it received considerable impetus

as a result of two sets of studies. The first is dark
field microscopy in which coherently scattering regions
have been observed (Rudee 1971; Chaudhari, Graczyk and
Herd 1972) and the second is interference microscopy in
which fringe-like contrast from a film of amorphous Ge
was reported (Rudee and Howie 1972). These results were
interpreted both in terms of the microcrystalline model
(Howie, Krivanek and Rudee 1973), and the random network
models (Chaudhari, Graczyk and Charbnau 1972; Chaudhari
and Graczyk 1973).

DIFFRACTION TECHNIQUES

This technique has been reviewed by Graczyk (this
proceedings) and we shall therefore not discuss the details here. The two functions that are obtained from
this analysis are the reduced intensity function and its
Fourier transform which gives the radial distribution
function. These two functions are used in comparing with
model calculations. If the model contains a small number
of atoms the radial distribution function is useful.
Where a larger cluster exists and computer time is not a
particular limitation the interference function is preferred. This approach has the advantage of comparing a
model calculation with data rather than data which has
been processed to avoid termination errors during Fourier transformation. In this note we shall be using the
interference function as the models are sufficiently large
that effects due to finite cluster size are small.

Moss and Graczyk (1969) showed that the diamond cubic crystallites could not describe the interference function obtained from films of amorphous silicon. Similarly
Graczyk and Chaudhari (1973) have shown that the data on
films of amorphous Ge cannot be described by the diamond
cubic microcrystallites. They also found that the amorphon proposed by Grigorovici and his colleagues (Grigorovici (1968, 1971), Grigorovici and Mănăilă (1967)) did
not give a satisfactory fit. Best fits were obtained from
random network models such as those of Polk (Polk (1971);
Turnbull and Polk (1972), Henderson (1971, 1974), and
Steinhardt, Alben and Weaire (1974)). The six-membered
random network model of Connell and Temkin (1974) also
gave a good fit (Chaudhari et.al. 1975). An example of
a good fit that can be obtained between experimental data
and model calculations is shown in Fig. 1. The various
models that we have referred to here have been discussed
at some length by Paul (this proceedings). On the basis

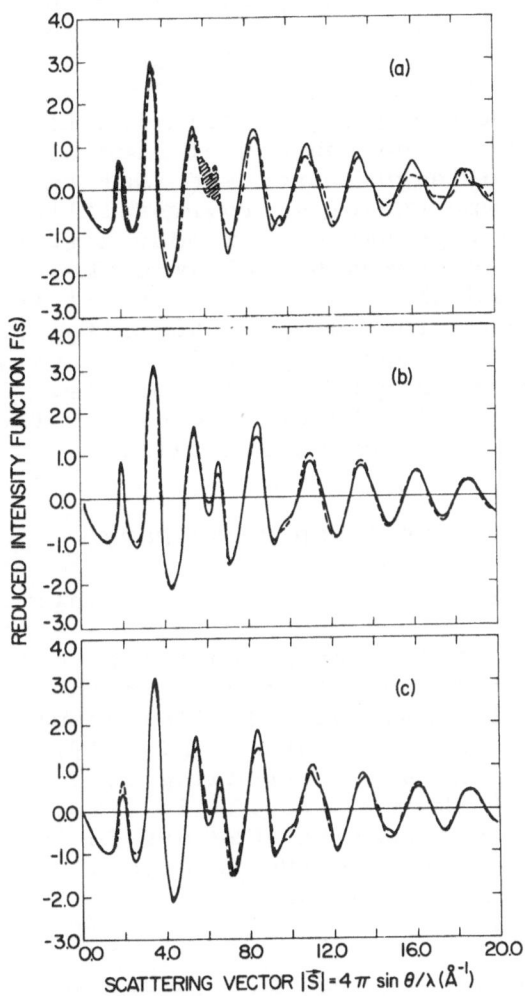

Fig. 1 Reduced intensity functions F(s) versus scatter-
ing vector |s⃗| for the following models: a) --- experi-
mental function with shaded regions showing the observed
spread in data for amorphous Ge, —— calculated function
for 519 atom unrelaxed Polk-Turnbull model. b) —— 519
atom model relaxed by Steinhardt, Alben and Weaire. c)
—— 201 atom model of Steinhardt, Alben and Weaire, relaxed as in b. For b) and c) --- is for the 519 unrelaxed model drawn in for reference. (Graczyk and
Chaudhari 1974).

of the diffraction techniques it is generally accepted
that the random networks describe the data best. However,
this does not preclude the possibility that some form of
a microcrystallite model can also describe the diffraction results. In an attempt to observe the microcrystallites Rudee (1971) and independently, Chaudhari and Herd
(1972) used dark field high resolution microscopy. On
the basis of these and subsequent studies of this kind
Rudee and his colleagues proposed that the structure of
amorphous Ge comprises of microcrystallites of Wurtzite
which are bound together by a random network, They argued that such an approach explained the electron microscope results and also gave a better fit to the interference function than that of pure Wurtzite, which is relatively poor. The composite fit is, however, still not
as good as for the random network models. In the next
two sections we shall therefore consider the electron
microscope results in greater detail.

DARK FIELD MICROSCOPY

In these experiments thin films ranging in thickness
from fifty to a few hundred angstroms are imaged in a
transmission electron microscope. The scattered electrons
which normally form the first halo in the diffraction pattern, are used for imaging. Initial results obtained by
Rudee (1971) and by Chaudhari et.al. (1972) indicated that
the dark field images corresponding to coherently scattering regions (CSR) were 10-15Å in size. Subsequent investigations showed that the CSR are in fact smaller and
more in the range of 5-7Å (Howie et.al. 1973; Chaudhari
et.al. 1972; Herd and Chaudhari 1974). An example is
shown in Fig. 2.

These observations prompted Rudee to propose the
Wurtzite structure whereas Chaudhari et.al. (1972) and
Shevchik (1972) pointed out that these observations were
not inconsistent with a random network. Cochran (1973)
has computed a dark field image using the Henderson model. Similarly, Chaudhari and Graczyk (1973) and Howie
and Krivanek (1975) have used the Polk-Turnbull model for
dark field calculations. An example of a dark field image
calculation carried out on the Henderson model and a diamond cubic crystallite is shown in Fig. 3(a,b,c). These
images correspond to the projections of atoms in the models shown in Fig. 3 (m,n,o). The dark field shown in
Fig. 3(a) corresponds to the calculations first reported
by Cochran (1973). The intensity trace corresponding to

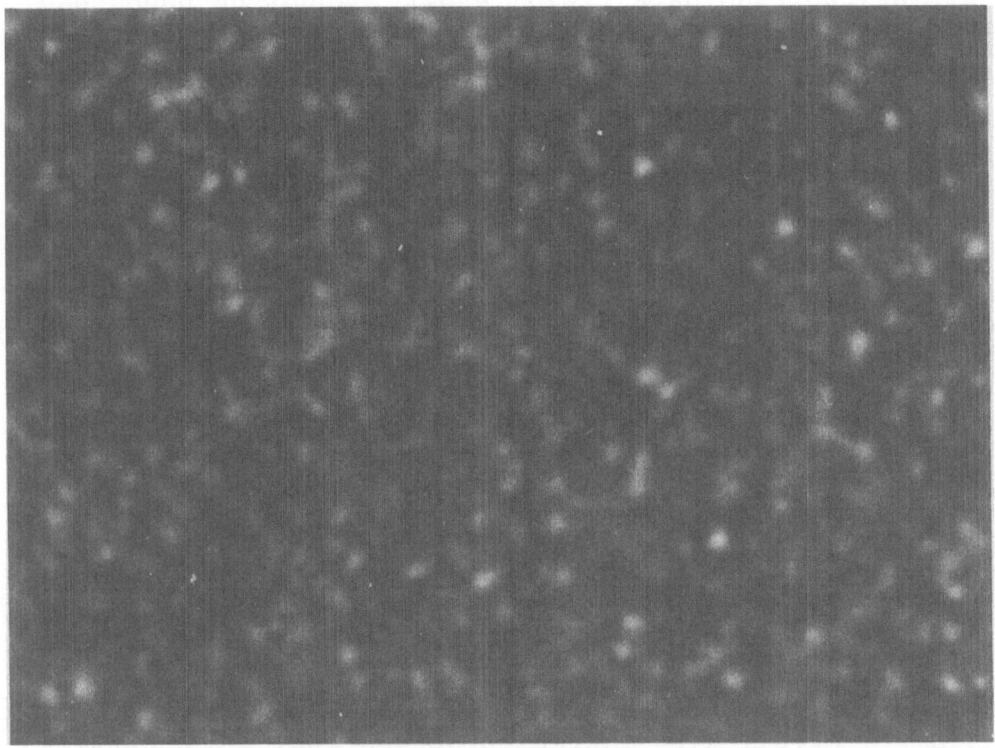

Fig. 2   High resolution dark field electron micrograph
of amorphous Ge.   MAG = 5 x 10⁶x

this image is shown in Fig.3(d) and taking the width of the
peak at half maximum we conclude that a CSR from this model is of the order of ~3Å.   If, however, we orient the
model for maximum scattered intensity (Chaudhari et.al.
1972) (H-H*) the image broadens and there is an increase
in intensity.   This can be seen in Fig. 3(b and e).   The
microcrystallite gives the broadest image.   The range of
image sizes that are computed for a variety of random networks are comparable with experimental observations.   The
isolated microcrystallites yield a broader image and do
not compare favorably with experimental data.   More recently, Howie and Krivanek (1975) have computed images
from microcrystallite models where interference effects
due to the presence of neighboring crystallites were taken
into account.   They showed that for typical film thickness used in dark field work, the interference effect

cannot be neglected in image calculations. They suggest
that the reported dark field results cannot be used to
differentiate between the random network and microcrystallite models. This finding negates their earlier conclusion based on calculations carried out on a single microcrystallite (Howie et.al. 1973).

OBSERVATIONS ON FRINGES

Although the dark field microscopy results had already reintroduced the microcrystallite model into structural considerations for amorphous solids the interference microscopy results reported by Rudee and Howie (1972)
appeared to provide striking evidence for the microcrystallite model. These authors reported seeing three to
four fringes in films of amorphous germanium. This cor-
responded to a crystallite size of 12-15Å.

There are several explanations for these observations. In the first these fringes, as proposed by Rudee
and Howie (1972), correspond to microcrystallites of
Wurtzite. The second is that fringes of this type are
also expected in random networks and therefore, as in
the case of dark field microscopy, we cannot use available results to distinguish between the two models. Berry
and Doyle (1973) have in fact proposed that any random
packing which gives rise to the observed diffraction pattern will produce interference fringes. According to
their calculations the first amorphous halo is expected
to yield 3-4 fringes and the second halo less than 2.
The third possibility is related to the experimental me-
thodology and suggests that the observed fringe contrast
is a combination of the way atoms are packed together and
the aberrations of the electron microscope imaging system.

Chaudhari and Graczyk (1973, 1974) have shown that
the Henderson, Polk, and Steinhardt et.al. random network
models for amorphous materials when correctly oriented
give rise to a fringe like contrast. An example taken
from the Henderson model is shown in Fig. 3 (g,h,i). A
trace of the fringe contrast is shown in Fig. 3 (j,k,l).
We note that the fringe contrast and spacing between the
correctly oriented Henderson model and the diamond cubic
crystallite is not dissimilar. As expected the crystallite yields the most regular fringes. The projections
of these models, Fig. 3(m,n,o) show the strong alignment
of atoms into crystalline like planes in the correctly
oriented Henderson model.

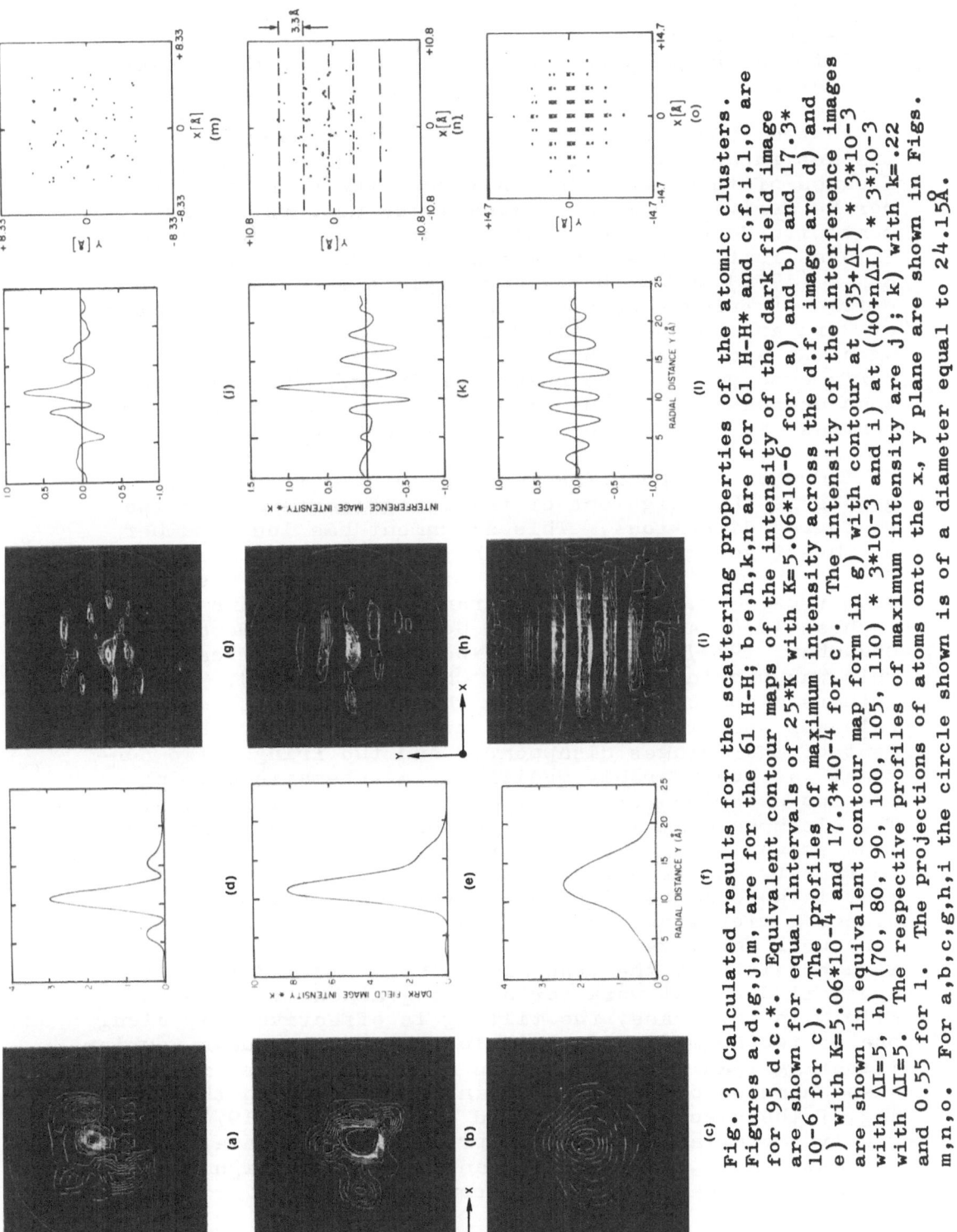

Fig. 3   Calculated results for the scattering properties of the atomic clusters. Figures a,d,g,j,m, are for the 61 H-H; b,e,h,k,n are for 61 H-H* and c,f,i,l,o are for 95 d.c.*. Equivalent contour maps of the intensity of the dark field image are shown for equal intervals of 25*K with K=5.06*10-6 for a) and b) and 17.3* 10-6 for c). The profiles of maximum intensity across the d.f. image are d) and e) with K=5.06*10-4 and 17.3*10-4 for c). The intensity of the interference images are shown in equivalent contour map form in g) with contour at (35+ΔI) * 3*10-3 with ΔI=5, h) (70, 80, 90, 100, 105, 110) * 3*10-3 and i) at (40+nΔI) * 3*1.0-3 with ΔI=5. The respective profiles of maximum intensity are j); k) with k=.22 and 0.55 for l. The projections of atoms onto the x, y plane are shown in Figs. m,n,o. For a,b,c,g,h,i the circle shown is of a diameter equal to 24.15Å.

Although these findings suggest that random networks
can also give rise to a fringe like contrast we still
need to calculate how often such contrast can be observed per unit area of a film when the thickness of a film
is specified. This question has not been satisfactorily
addressed. Cochran (1973) has used structure factor statistics in an attempt to obtain some idea of the significance of this question. Although the quantitative number he arrives at is questionable we believe, in agreement with Cochran, that the frequency of occurrence of
fringes is far less than the micrographs of Rudee and
Howie (1972) show. Berry and Doyle's explanation for the
origin of fringes does not agree with recent observations
of Herd and Chaudhari (1974). These authors showed that
when the second amorphous halo is used in forming the interference pattern the number of fringes increased rather
than decreased. We must therefore conclude that the fringe
contrast is not due to a random superposition of waves.

One of the puzzling features of the fringe micrographs is the alignment of the fringe pattern into two
preferred directions. This alignment has led a number
of investigators to suggest that the fringe contrast is
due to aberrations associated with tilting of the electron
beam (Howie et.al. 1973; Cochran 1974; McFarlane and
Cochran 1974; Ast, Krakow and Goldfarb 1974; Herd and
Chaudhari 1974). The first experimental verification of
this was provided by Herd and Chaudhari (1974). These
authors also showed that when the entire first amorphous
halo was used in untilted interference imaging the alignment of the fringes disappeared and the fringe like contrast was considerably modified. A schematic drawing of
the two experimental arrangements is shown in Fig. 4(a,b).
The electron beam is tilted in the experimental arrangement shown in Fig. 4(a) in order to minimize the phase
shift between the directly transmitted and scattered waves
that is associated with the spherical aberration of the
electron microscope lens. Although this approach works
very well for crystalline films where the diffraction pattern consists of the transmitted beam and a diffracting
spot, it does not work for a fine grained or amorphous
film. In this case, the tilting is effective only along
one or two segments of the halo. This introduces Fourier
filtering leading to preferred alignment. The effect is
also enhanced by astigmatism introduced during the tilting of the electron beam (Herd and Chaudhari 1974). Recently, McFarlane (1974) has proposed an explanation for
the astigmatic effect that leads to strong alignment. It
therefore appears that the strong and aligned fringes

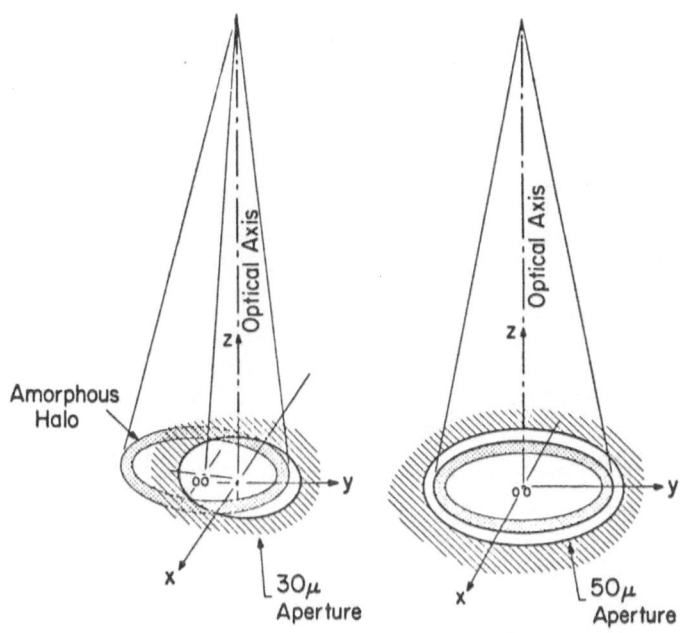

Fig. 4 Schematic representation of the experimental arrangement of the illumination and objective aperture relative to the electron optical axis for interference electron microscopy: a) tilted beam configuration, b) untilted symmetrical configuration.

observed first by Rudee and Howie (1972) are associated
with aberrations introduced during the tilting of the
electron beam.

    This conclusion suggests that the untilted beam in-
terference microscopy may be a better approach. Phase
contrast microscopy in this case is achieved by the phase
shift introduced in the scattered beam relative to the
directly transmitted beam. As in the preceding case this
phase shift is due to spherical aberration. However, in
this case it has the same value at all segments of the
halo and directional filtering is not present. The resultant image from the amorphous film does not show the
striking fringe contrast present in the tilted beam.
Using both a partially crystallized and a thin film Herd
and Chaudhari (1974) obtained contrast of the type shown
in Figs. 5 and 6. The crystallized area shows a well defined regular pattern whereas the amorphous area shows

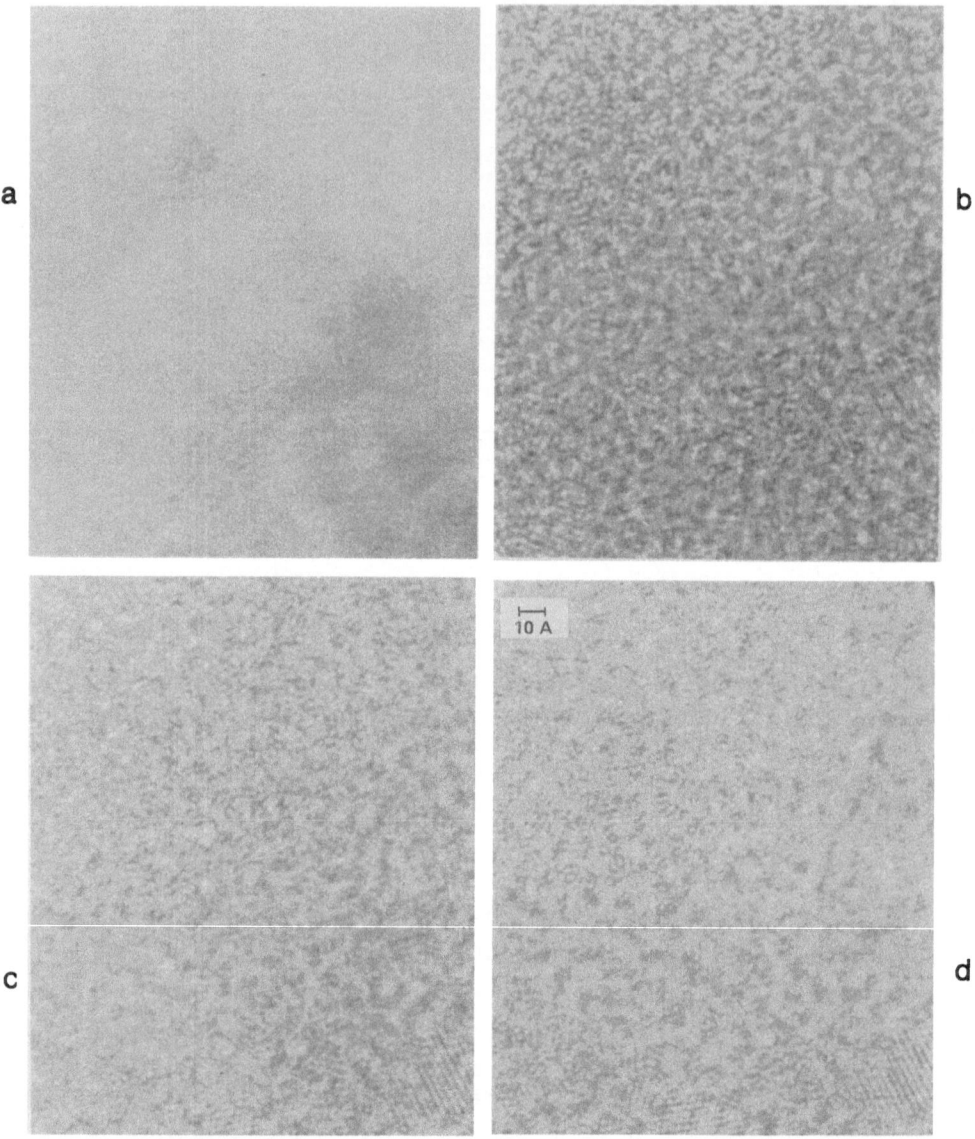

Fig. 5 Focussing series in the untilted interference image made of a partially crystallized Ge film. The central region which appears smoother than the amorphous area in the top left of the micrograph is also crystalline but is is not oriented for (111) interference (Herd and Chaudhari 1974). Thickness appr. 50Å)

50μ Obj. Apt.   Mag: 5x10^6

a. focus   b. focus - 1050Å   c. focus - 1450Å

d. focus - 1650Å

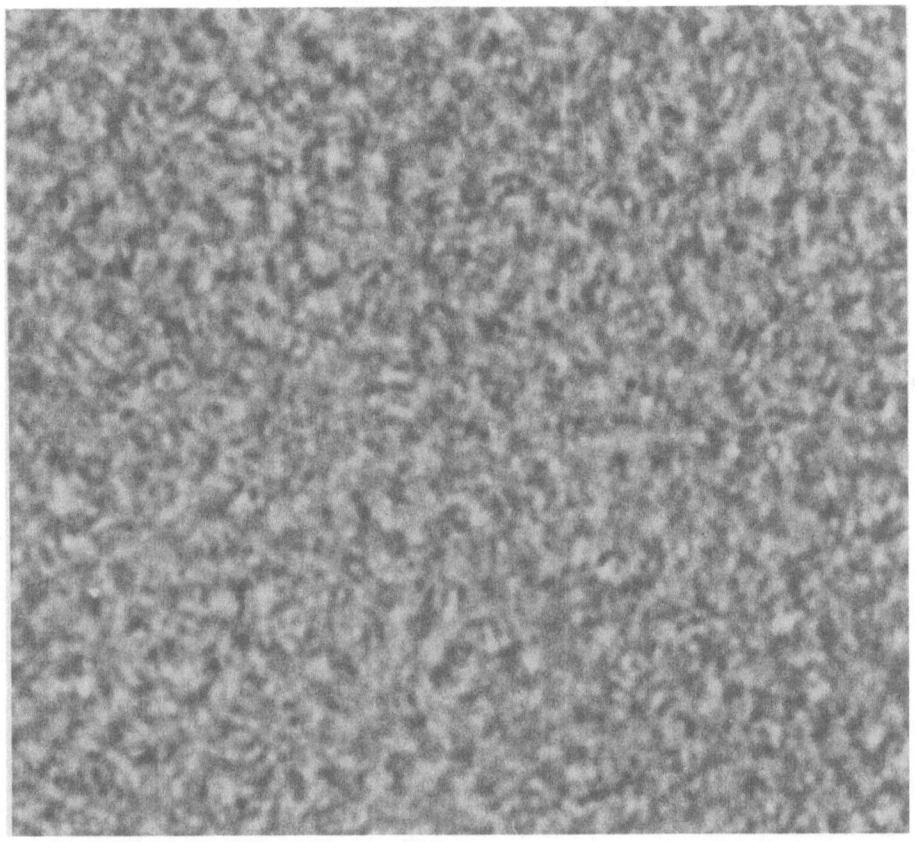

Fig. 6 Untilted interference electron microscopy micrograph for amorphous Ge.

Mag: 5×10^6x

no regularity which extends beyond the nearest neighbor
distance. Irregular fringe-like contrast which extends
over a larger area is only seen occasionally. These observations suggest that the amorphous films do not show
contrast which compares with results obtained on isolated
crystallites 10-15Å in diameter. We therefore conclude
that no evidence for microcrystallites has been obtained
by interference electron microscopy. Rather that contrast, to the extent that we understand it, is characteristic of random network models where strong nearest neighbor correlations are expected. The occasional strong
fringe contrast that is seen can also be explained within the framework of these models.

ACKNOWLEDGEMENTS

The authors are grateful to Professor Cochran and Dr. McFarlane for useful discussions and providing us with their results prior to publication. The authors thank Drs. A. Howie and O. Kirvanek and Professors M. L. Rudee and D. Ast for providing us with preprints prior to publication.

REFERENCES

Ast, D. G., Krakow, W., and Goldfarb, W., private communication.

Berry, M. V. and Doyle, P. A., 1973, J. Phys. $\underline{C6}$, L6.

Chaudhari, P., Graczyk, J. F. and Herd, S. R., 1972, Physica Status Solidi (b) $\underline{51}$(2), 801.

Chaudhari, P., Graczyk, J. F., and Charbnau, H. P., 1972, Phys. Rev. Lett. $\underline{29}$, 425.

Chaudhari, P., and Graczyk, J. F., 1973, 5th Intl Conf. on Amorphous and Liquid Semiconductors, Garmisch-Partenkirchen, p. 59.

Chaudhari, P., Graczyk, J. F., Henderson, D., and Steinhardt, P., 1974, IBM Res. Rep. 5027, also 1975, Philos. Mag. $\underline{31}$, 3, 727.

Cochran, W., 1973, Phys. Rev. $\underline{B8}$, 623.

Cochran, W., 1974, Proc. Int. Conf. on Tetrahedrally Bonded Semiconductors, eds. M. H. Brodsky, S. Kirkpatrick, and D. Weaire, AIP Conf. Proc., No. 20, p. 177.

Cochran, W., and McFarlane, S. C., 1974, Eighth Int. Congress on Electron Microscopy, pg. 228, also private communication.

Connell, G. A. N., and Temkin, R. J., 1974, Proc. Int. Conf. on Tetrahedrally Bonded Semiconductors, eds. M. H. Brodsky, S. Kirkpatrick and D. Weaire, AIP Conf. Proc., No. 20, p. 192.

Graczyk, J. F. and Chaudhari, P., 1973, Phys. Stat. Sol. (b) $\underline{58}$, 163 and 501.

Graczyk, J. F., and Chaudhari, P., 1974, IBM Res. Rep. 4856, J. Non-Cryst. Solids, in press.

Grigorovici, R., and Mǎnǎilǎ, R., 1967, Thin Solid Films $\underline{1}$, 343.

Grigorovici, R., 1968, Mat. Res. Bull. 3, 13; 1971,
Thin Solid Films, 9, 1.

Henderson, D., 1971, Bull. Amer. Phys. Soc. 16, 348.

Henderson, D., 1974, J. Non-Crystl. Solids 16, 317

Herd, S. R., and Chaudhari, P., 1974, Phys. Stat. Sol.
(a) 26, 627.

Howie, A., Krivanek, O., and Rudee, M. L., 1973, Phil.
Mag. 27, 235.

Krivanek, O.L. and Howie, A., 1975, J.Appl.Cryst. 8, 213.

McFarlane, S. C., private communication.

Moss, S. C., and Graczyk, J. F., 1969, Phys. Rev. Lett.
23, 1167.

Polk, D. E., 1971, J. Non-Cryst. Solids 5, 365;
Turnbull, D., and Polk, D.E., 1972, Proc. 4th Intl. Conf.
on Amorphous and Liquid Semiconductors, eds. M. H. Cohen
and G. Lucovsky, North Holland, Amsterdam, page 19.

Rudee, M. L., 1971, Phys. Stat. Sol. Short Notes (b)
46, K1.

Rudee, M. L. and Howie, A., 1972, Phil. Mag. 25, (4)
1001.

Shevchik, N. J., 1972, Phys. Stat. Sol. (b) 52, K121.

Steinhardt, P., Alben, R., and Weaire, D., 1974, J. Non-
Crystl. Solids 15, 199.

STRUCTURAL MODELLING OF DISORDERED SEMICONDUCTORS

William Paul and G. A. N. Connell

Division of Engineering and Applied Physics

Harvard University, Cambridge, Mass. 02138

I. INTRODUCTION

Fundamental understanding of the properties of any solid system requires that we know the atoms involved and their arrangement in space. Dr. Chaudhari's and Dr. Graczyk's lectures describe the extent to which this can be done through experimental measurements. Thus the gross, or average, chemical composition can be determined by standard chemical methods, and the local composition on a scale of microns by techniques such as the electron microprobe analyzer. Optical and electron microscopy may be used to study the macroscopic arrangement of phases and the material interconnecting them. X-ray, electron and neutron diffraction examine the microscopic atomic arrangement, expressing the result in terms of radial distribution functions and parameters deducible from them. Small-angle X-ray or electron scattering, electron spin resonance associated with uncompensated bonds at defects, nuclear magnetic resonance and its fine structure, infrared absorption and Raman scattering caused by the vibration of defect-impurity complexes are among the experimental methods invoked to yield indirect but often powerful evidence about local defects and therefore also about the underlying structure. Professor Turnbull's lectures, on the other hand, discuss thermodynamic and kinetic considerations which govern the atomic configuration and changes in it with temperature.

In this part we shall discuss what can be learnt

about the structure of disordered materials from the construction of three-dimensional models under appropriate
restrictive assumptions, and comparison of certain parameters deduced therefrom with those deduced from diffraction and other experiments. Most of the models (an exception is the computer-generated model for a-Ge of Shevchik[1] which is not fully coordinated) set out to describe
a low energy fully-coordinated metastable structure to
which it is supposed the actual disordered samples relax
without crystallization. Two separate questions arise
in comparing the model to the real substance. The first
is whether, for a given substance, there is only one such
metastable structure--structure being defined by <u>suffici-
ent</u> statistics regarding bond lengths, bond angles, connectivity, ring densities etc.--or whether there are several possible statistically-different metastable structures, depending on the preparation history. The second
is the extent to which the material will deviate from an
"ideal" structure despite the occurrence of local free
energy minimization (not leading to crystallization) during preparation and anneal. Experimental studies of radial distribution functions and associated parameters as
a function of parameters such as film deposition temperature do indeed show small but recognizable changes. However, there does not appear to be, at present, significant differences in the experimental radial distribution
functions of the materials discussed here, well correlated with preparation history, which would suggest the existence of more than one metastable disordered structure.
In the testing of models against experiment, therefore,
any of the most recent experimental radial distribution
functions may be employed. This is not to say, however,
that preparatory conditions could not be adjusted in any
given case, perhaps to the extent of the incorporation
of a low density of some vital stabilizing element, so
that a substance with a significant difference in its statistical parameters is produced.

II. MODELLING OF STRUCTURE

a. <u>Long range disorder and the maintenance of short range order</u>

Perfect crystalline order involves the periodic repetition in three dimensions of a fundamental molecular
unit which may, of course, be a single atom. Disorder
may be introduced without abandoning the basic lattice,
either by random changes in the atomic constitution of
the molecule (chemical, sometimes alloy, disorder), or

by a random arrangement of net spin on the molecule (spin
disorder), or by displacement of the atoms away from their
ideal lattice positions by an amount small compared to
their separation (vibrational disorder). These are examples of quantitative disorder. The disorder we are
most concerned with here is different, and involves a-
bandonment of the basic lattice. The arrangement of fundamental units is no longer isomorphous with any infinite periodic array. This topological disorder may, in
principle, occur alone but more likely occurs along with
quantitative disorder. The possibility of model construction and analysis rests on the plausible supposition (well buttressed by observation) that a certain a-
mount of short range order is maintained. For example,
we might suppose that the number of nearest neighbors
(first coordination number) of any atom is the same throughout the material, and that the configuration of neighbors deviates very little, if at all, from some mean value. In some cases we might envisage a mixture of a limited number of local configurations. The choice and
conservation of a limited amount of short range order
can be plausibly argued on the basis of the valence of
the atomic constituents and/or the geometry of the bonds
in crystalline polymorphs. Derived from it is an empirical rule, formulated by Joffe and Regel[2], that a material that is semiconducting in its crystalline state will
remain so in the disordered phase if the short range order remains the same.

b. Examples of short range order

Two-fold coordination is often displayed by elements
of group VI of the periodic table in both the elemental
and compound forms. Thus, for example, the O atoms in
a-SiO_2 are 2-fold coordinated with Si atoms. Amorphous
Se shows a narrow first peak in its radial distribution
function (RDF) at an interatomic distance intermediate
between those for crystalline trigonal and monoclinic Se,
but significantly, with a first coordination number of
exactly 2 as in the crystals[3]. Similarly the coordination numbers of S and Te in the elemental form, and of
the group VI elements generally in amorphous compound
form, appear to be 2. This suggests that modelling be
based on a first coordination number of 2, but leaves
open to decision in particular cases the choice of bond
angle. In principle, this may be guessed roughly from
the second or higher peaks in the RDF, or from consideration of molecular or crystalline bond angles. The former
is safer, if available, since the latter do not always
agree.

Three-fold coordination is displayed in diffraction data for group V elements such as As and Sb in both crystalline and amorphous forms[4]. Interpretation of the higher order peaks in the RDF of the amorphous materials to obtain coordination numbers, separations, and bond angles is not straightforward. Let us define an n-bond neighbor as one which can be reached in a minimum of n steps where each step is between nearest neighbors. The difficulty then is that there is overlap in the radial distributions of n-bond neighbors for n > 1.

We shall illustrate this problem with a brief description of As. There are two forms of crystalline As[5], in each of which there are layers with 3-fold covalent bonding within the layer and weaker bonding between the layers. In rhombohedral (α) As, neighbors with n = 1 are separated by 2.51 Å, are 3 in number, and lie within the layer, while neighbors with n = 2 are separated by 3.76Å, are 6 in number, and also lie within the layer. There are also, however, 3 neighbors at 3.15 Å for which n is undefined and which lie in the adjacent layer.

Amorphous As has 3 neighbors at 2.49 Å, 10 at 3.78 Å, and none near 3.15 Å. It is presumed that the layers have separated and that the 3 neighbors in the adjacent layer of the crystal contribute to the second neighbor distribution in the amorphous form. Modelling studies must take into account these differences in the RDF's.

Assuming, however, that the basic unit of short range order has non-planar 3-fold coordination, we face the problem that the bond angles are not uniquely defined in the way that, say, the tetrahedral bond angle of 109°28' is defined for crystalline group IV elements. Thus the bond angles are 97°, 96° and 94° in crystalline As, Sb and Bi, respectively[5]. The basic unit for modelling is like a triangular pyramid, which is anisotropic, and which has an uncertain mean bond angle.

An interesting question which arises here is whether the modelling of an amorphous 3-fold coordinated material such as As applies also to an amorphous IV-VI compound whose crystalline form is similar to that of the group V element. An example is GeTe whose structure is that of As. Should GeTe be modelled with each Ge and Te atom approximately 6-fold coordinated (three nearest neighbors plus three very slightly further away) or should the 8-N rule, which suggests 4-fold coordination of Ge and 2-fold coordination of Te, be followed? Both sides of this question have been argued recently; the present view appears to favor following the $(8-N)$ rule[6].

Four-fold coordination is displayed by Si and Ge in their amorphous forms, no matter how prepared, as well as in semiconducting compounds of the same family such as a-$A^{IV}B^{IV}$, a-$A^{III}B^V$, a-$A^{II}B^{VI}$, and a-$A^{II}B^{IV}C_2^V$. The RDF's show that the bonding is tetrahedral, as in the corresponding crystals, there are 4 nearest neighbors, that the bond lengths are approximately the same as in the crystals, but that the bond angles deviate on the order of 10°. Thus it is clear that a distorted tetrahedral unit is the appropriate modelling basis[7]. Apparently, in this case, the $(8-N)$ rule is not followed. Immediately a question arises regarding the nature of the 4-bonded atoms in $A^{III}B^V$ compounds[8]. Are the bonds always between unlike atoms, as in the crystal, or not? Maintenance of a high degree of short range order following the principles of bonding would suggest so. Evidence from the measured spectra of valence band[9] and core[10] electronic densities of states is confirmatory, but contrary evidence also exists from broadening in the first peak of the experimental amorphous RDF's[8], and in the energy of the infrared absorption edges and their pressure coefficients[11]. Modelling[12] must therefore apparently take into account both the tendency from energetic considerations to form a partially ionic bond and the difficulty from a kinetic viewpoint of avoiding the juxtaposition of like neighbors.

Other coordinations and mixed coordinations exist. Crystalline and amorphous[13] B both have a first coordination number of 6. Metallic elements especially tend to have high coordination numbers appropriate for dense packing, which for lack of space is deemphasized in this review. Among the materials showing mixed coordination is a-C.[14] The RDF of deposited films has a broadened nearest neighbor distribution suggestive of two merged peaks with an average distance between the values for diamond and graphite, strongly suggesting that the material is an intricate mixture of two coordinations. Similarly the area under the first peak gives a first coordination number between 3 and 4. Annealing of both a-C and vitreous C changes the structures and other properties toward those of graphite.

Other examples of mixed coordination are a-$(A^{III})_2(B^{VI})_3$ films such as a-In_2Se_3 where the coordination of each In atom is 4 Se atoms; of two-thirds of the

Se atoms 3 In atoms; and of the remaining one-third of
the Se atoms 2 In atoms[15]. This mixed coordination is
the same as in crystalline In_2Se_3. Other examples are
a-$(A^V)_2(B^{VI})_3$ films and glasses such as a-As_2Te_3, As_2Se_3
and a-As_2S_3.[16] They have short range order similar to
that of orpiment--a layer structure of puckered 12-member
rings with each B atom linked to two A atoms and each A
atom linked to two B atoms in one ring and to a third B
atom in a neighboring ring. Finally, as already observed, compounds and alloys of a-$(A^{IV})_x(B^{VI})_{1-x}$ usually
seem to have the coordination predicted by the (8-N) rule.
Thus in a-Ge_xTe_{1-x} each Ge atom is 4-coordinated and each
Te atom 2-coordinated, this being different from the crystals, which have distorted rock-salt (arsenic) structures with 6-fold coordination.

Although this short survey of bonding[17] has omitted
the details of many cases, it does show that amorphous
semiconductors usually have a nearest neighbor short
range order consistent with the bonding displayed in molecules or crystals. Care must be exercised, as is illustrated by the facts that the coordinations of As in c-GaAs
and a-GaAs are the same but not given by the (8-N) rule,
while the coordinations of Ge in c-GeTe and a-GeTe are
different, with the amorphous unit following the (8-N)
rule. Usually inspection of the RDF for the amorphous
form can decide the question, but sometimes the debate
is hard to close, as in the case of GeTe. Radial and
bond angle distortions will be present in the basic modelling unit if the next nearest neighbors are in a different position from the crystal. These facts are sufficient to permit modelling to proceed in the expectation
that a comparison of its results with experimental structural information will establish the detailed parameters.

c. Brief history of model-building

Before we embark on a more precise description of
the construction of atomic models for disordered systems
and a comparison of predictions from them with experiment,
we shall briefly recount the recent history of their study.
The models are of four kinds:

(1) Microcrystallite models

A conglomeration of very small crystals of one or
more polytypes of the molecule being studied is surely a
correct description of some real materials. However, when
the microcrystallite size extends only to three or four
unit cell dimensions, it is clear that a substantial

fraction of the atoms are on the cell boundaries and may
not be in exactly the same positions as if the crystallite were larger, and that furthermore, a large fraction
of the material must be contained in a disordered arrangement between the crystallites. Very few hand-built models of such a kind appear to have been built. An example reported by Rudee[18] is a model of wurtzite microcrystals constructed of ball and spoke units, which were connected at arbitrary angles with interfaces that were random networks with no unsatisfied bonds.

(2) Dense random packed models

The dense random packed structures[19] have been developed to describe monatomic liquids and amorphous solids and are appropriate for the high coordination numbers of metallic systems. In what follows, however, we
shall restrict consideration to networks suitable for 2,
3 and 4-coordinated atoms, which usually involve covalent, ionic or van der Waals forces.

(3) Random network models

The concept of a non-periodic random network was
first proposed by Zachariasen[20] in 1932 as a general model for glasses. Later it was adopted and developed by
Ordway[21], Bell and Dean[22], and Evans and King[23] to describe the structure of amorphous SiO_2. The work of Bell
and Dean marked the first quantitative success of the model. They were able to show that a ball and stick model
which arranged SiO_4 tetrahedra according to certain rules
yielded pair distribution functions in agreement with experiment. In 1969 Polk adopted the same general philosophy to build a 440-atom model of a-Ge and a-Si and showed
that it yielded a radial distribution function in reasonable agreement with experiment[24]. All later model-building of tetrahedrally-coordinated continuous random networks has been essentially a refinement of this work, as
will be discussed in detail below. Slightly later than
Polk's work came that of Shevchik[1] who built a 1000-atom
random network solely by a computer technique which constrained each added atom in position in much the same
way as the hand-built modelling of Polk.

(4) Amorphous cluster models

The amorphous cluster is defined to be a small group
of atoms (of order 100) arranged so that each atom has
exactly the same short range order, but in such a way
that it does not lead to a periodic arrangement capable
of filling all space. It is supposed that a very small

cluster of this type may have a lower free energy than a
cluster which is consistent with a crystallographic configuration. Thus, if the conditions of formation of a
sample are such that energy minimization occurs for groups
of atoms of this magnitude, the amorphous cluster may be
the preferred arrangement. The same difficulties of fitting the clusters together in a macroscopic sample occur
as in the case of the microcrystals. Such clusters may
be of several kinds. For example, Tilton[25] proposed a
model for silica glass, the basic unit of which was called the vitron, which contained SiO_4 tetrahedra arranged
to have the external symmetry of clusters of pentagonal
dodecahedra. A second well-known model due to Grigorovici[26] also used the pentagonal dodecahedron, called
by him the amorphon, as a basic structural unit in tetrahedrally-coordinated materials such as Ge. Again, if
the bond lengths and angles are slightly distorted, and
the dodecahedra possibly mixed with units containing hexagonal rings, a structure filling all space may be constructed.

We shall concentrate our attention on the topology
of the random network and amorphous cluster models and
on a comparison of predictions from them with experimental data.

d. Modelling of networks

A network must satisfy the following criteria in order to be accepted as a valid model:

(1) The statistical description of the structure by
certain distribution functions must correspond to those
deduced from diffraction experiments. One of these distribution functions is the radial distribution or density
function (RDF), defined by $J(r) = 4\pi r^2 \rho(r)$, where $\rho(r)$
is the atom density at distance r from any atom chosen
as origin. Associated functions comprising the RDF are
the partial distributions functions $J_n(r)$, which give the
distribution in r of n-bond neighbors. Also deducible
as quantities of particular interest are the average deviations in atomic separations and bond angles and the
statistical distribution of complete rings containing n
members. The corresponding distributions for the real
material may be obtained from Fourier inversion of the
experimental results of diffraction experiments:

$$J(r) = 4\pi r^2 \rho(r) = 4\pi r^2 \rho_0 + r\, G(r) \qquad (1)$$

$$G(r) = \frac{2}{\pi} \int_{0}^{\infty} F(s) \sin s r \, ds \qquad (2)$$

$$F(s) = s[I(s)/f^2(s) - 1] \qquad (3)$$

where $s = 2\pi \sin \Theta/\lambda$, the momentum transfer in elastic scattering, ρ_0 is the average atomic density, $f(s)$ is the atomic scattering factor, and $I(s)$ the normalized coherent intensity.

The experimental $J(r)$ may be in error not only because of inadequate correction of experimental errors in $I(s)$ (see Chaudhari's and Graczyk's review) but also because of termination effects associated with the finite range of s in the integral of equation (2). It therefore is a sensible check procedure to calculate $F(s)$ -- and so also $I(s)$ -- from Fourier inversion of the calculated $J(r)$ from the model and to compare those against the $F(s)$ and $I(s)$ derived from the experimental diffraction data.

Even if the model and experimental $F(s)$ and $J(r)$ fit perfectly, there is no guarantee that a uniquely appropriate network has been constructed, since these distributions are still only averages over the network. In order to approach a unique fit, the distribution of dihedral angles $P(\Theta_d)$, the ring statistics, and correlations of deviations of atomic separations and bond angles from the mean values should be added. In fact, only a complete accounting of all atom coordinates can fully describe the network. However, confidence in the usefulness of the statistical averages is improved if the model satisfies other tests discussed below.

(2) The atomic density of the network must correspond to the experimental density of material that is, or is close to being, fully coordinated. The area under the first or nearest-neighbor peak of the experimental $J(r)$ measures the average first coordination number of the real material.

(3) Indirect or secondary comparisons may, or may some day, be invoked to test the acceptability of the model. Thus, the heat of crystallization of the network should correspond to that measured experimentally. The experimental value is derived from measurements of the heat capacity as a function of temperature. The value for the constructed network is usually estimated from

approximate theories giving the extra elastic energy in
a distorted structure in terms of deviations of bond lengths and bond angles from their lowest energy (crystalline) values.

Also, the model should permit a plausible explanation of the effects of annealing, the kinetics of annealing and the process of recrystallization; this may involve the necessity of incorporation of defects into the
network. The model must also permit an explanation of
the results of any experiments such as nuclear magnetic
resonance, electron spin resonance, or Raman scattering
which can give the detailed local atomic configuration.
Finally, while taking into account the difficulties posed
by necessary approximations in the theory, the phonon energy and electron energy densities of states calculable
from the model coordinates and a plausible Hamiltonian
must correspond to those measured experimentally.

In the description to follow, we shall concentrate
on criteria (1) and (2). The estimated heat of crystallization, which is directly related to the expression for
the elastic distortion energy sometimes used in network
relaxation (see below) is a less sure criterion because
of uncertainties in that expression and the parameter values used in its evaluation. The other indirect tests are
more appropriately discussed along with the theory and
experiment for the properties involved.

An actual model may be constructed entirely by hand,
or constructed first by hand and its coordinates then adjusted by computer, or constructed entirely by computer.
In each case certain rules are consistently followed.
Thus, only a limited amount of radial or angular distortion may be permitted to the basic structural unit. A-
toms may be added to the model in such a way as not to
give crystalline periodicity, yet roughly simulating the
process of formation. How this is done varies from case
to case and is best illustrated with examples.

Hand-built models often use ball and spoke building
units with the number of spokes equal to the first coordination number (valence satisfaction) and the direction of the spokes reproducing the directions of the covalent bonds which may or may not be known at the outset.
Individual units may be easily connected. Very rigid
spokes give models which permit no bond compression or
bond angle variation, but do have possible bond extension
and relative unit rotation. Less rigid spokes allow

additional flexibility, and springs replacing the spokes
permit still more[27]. Some model-builders have used rigid Voronoi polyhedra (Wigner Seitz cells containing one
atom), bringing large area faces into contact to simulate
bonding[28].

III. MODELS FOR TETRAHEDRALLY-COORDINATED MATERIALS

a. Information from diffraction data

The most extensive modelling of a monatomic system
has been that for a-Si and a-Ge, which we shall adopt as
our case for detailed study. Electron and X-ray diffraction data have been reported for both elements and analyzed by methods discussed in Dr. Chaudhari's and Dr.
Graczyk's presentation. Selected graphical and tabular
data will be presented later. The information available
may be briefly summarized as

(1) the radial distribution function $J(r)$.

(2) the diffracted intensity $I(s)$, or $F(s)$.

(3) the first, second and perhaps higher coordination numbers.

(4) the corresponding average interatomic distances.

(5) the spread in these distances and its excess
over thermal broadening.

(6) the distortion of the tetrahedral bond angle,
θ_b, nominal value $109°28'$.

(7) The distribution of dihedral angles, θ_b, between
neighboring tetrahedra, where $\theta_d = 0$ corresponds
to the so-called "eclipsed" configuration, and
$\theta_d = 60°$ to the so-called "staggered" configuration.

(8) the density.

b. CRN models for amorphous tetrahedrally-coordinated materials

Modelling of this system was first reported by Grigorovici and Manaila in 1969.[28] Earlier, both Grigorovici[26], and Coleman and Thomas[29] had proposed that the
structure of a-Ge was a mixture of diamond-like structure
and a structure of pentagonal dodecahedra, since the RDF's
for a-Ge and a-Si showed that the basic tetrahedral arrangement of the diamond cubic structure was preserved,
there being four nearest neighbors at about the crystalline

distance and approximately twelve next nearest neighbors at $(8/3)^{1/2}$ times that distance. Pentagonal dodecahedra ("amorphons") are obtained when perfect tetrahedral ball and spoke units are assembled in the eclipsed configuration. Amorphons cannot be connected together to fill all space without distortion, but if small distortions are allowed, and also if the amorphons are mixed either with diamond structure, or framework hydrate structure (clathrate structure[30]),or a fraction of random network, the resultant structure can be extended indefinitely. Grigorovici's comparison of the experimental RDF with an RDF for the diamond-amorphon mixture is instructive. It is clear that the inclusion of pentagonal rings in the model leads to a less pronounced third-neighbor peak in the calculated RDF, in agreement with experiment. It is also clear that the disorder allowed is unnecessarily restrictive. For example, the two types of 6-member rings found in the wurtzite structure might be expected to show up in the amorphous structure, not to mention other rings that occur in other crystalline polytypes having the simple tetragonal and body centered cubic symmetries. In an attempt to secure a more complete mixing of the several types of ring, Grigorovici and Manaila used modified Voronoi polyhedra with tetrahedral symmetry arranged in either eclipsed or staggered configurations. The resultant model consisted of 6-fold puckered rings and 5-fold planar rings giving a continuous network capable of apparently indefinite extension. The agreement between the calculated RDF and experiment was again fair, but the bond length fluctuations were overestimated and the bond angle fluctuations underestimated, both attributable to the inflexibility of the building units employed.

Noting several apparent deficiencies in the model, especially the restriction to only eclipsed or staggered bonds, Polk[24] set out to test the possibility that a continuous random network might reproduce the main features of the experimental RDF. He used a flexible ball and spoke basic unit of tetrahedral symmetry to construct a model with the following characteristics (1) a core of five and six member rings which ensured the generation of a noncrystalline structure, (2) no unsatisfied bonds in the interior of the model, (3) bond length variations held to within 1% and bond angle variations (bending of the spokes) held to within $\pm 10°$, (4) choice of dihedral angle for a unit being added to the structure such that it connected to other units with a minimum of strain. The strain minimization was found to be operationally equivalent to requiring that the surface density of dangling bonds be kept constant as the model size was increased. The resultant model appeared capable of indefinite extension without incorporating dangling bonds and without producing volumes of increased strain. The final CRN consisted of 440 "atoms." Its RDF, determined by direct measurement of atomic separations from some 16 centrally-located atoms, and illustrated in Figure 1, was a better fit to experimental data for a-Si than the earlier work. The density was determined to be 97 ± 2°/o of that of the f.c.c. crystal, which is in reasonable agreement with experiment for the most void-free amorphous samples. It had 5, 6, and 7-member rings and a continuous distribution of dihedral angles from 0° to 60°.

The original Polk model was physically extended to 519 atoms and refined by Polk and Boudreaux[31]. The refinement consisted in first measuring, with an accurately positioned laser beam, the coordinates of all of the

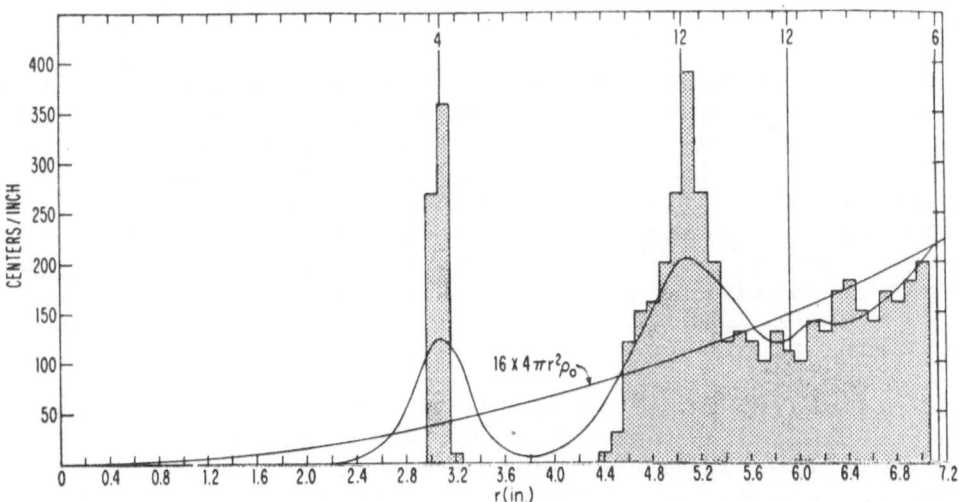

Fig. 1 Comparison of the RDF for the CRN model of Polk[24] (histogram) with experimental data from electron diffraction by a-Si by Moss and Graczyk[46]. The parabola represents the average density, and the vertical lines and corresponding numbers represent the position and number of neighbors in a crystal with a nearest neighbor separation equal to the mean separation in the model.

atoms, and then using computer techniques to determine
new atomic coordinates such that the variation in nearest neighbor bond lengths was essentially zero. This is
appropriate because it is known that the fluctuation in
nearest-neighbor bond length in the experimental RDF is
about equal to the thermal fluctuation in crystalline
germanium. First all of the nearest neighbor distances
were determined, then their mean, and then the pair of
atoms whose separation was farthest from the mean was i-
dentified. Their coordinates were then altered so as to
reduce the difference between their separation and the
mean by 50°/o. All of the atomic separations affected
by this change were then recomputed. This procedure was
iterated until the standard deviation of all interatomic
separations was found to be less than a predetermined
value (actually about 0.2°/o). Using the refined coordinates, an RDF was determined, as well as the mean tetrahedral bond angle and its standard deviation, the distribution of dihedral angles, and the density. The bond
length and bond angle fluctuations were found to be constant throughout the model; it is to be noted that the
latter had a standard deviation of 9.1° which happens to
be significantly smaller than the 10° determined in very
recent experimental determinations of the RDF. The mean
density was found to be 99°/o that of a diamond cubic
structure with a near-neighbor distance equal to the mean
of that of the random network. The distribution of dihedral angles was smooth and monatomic, with the staggered
configuration roughly twice as likely as the eclipsed.
Finally, the width of the second peak of the model was
similar to that of the experimental RDF after the effects
of thermal broadening are subtracted from the latter.
Figure 2 and this discussion, which is not intended to be
a full critical comparison of calculation and experiment,
show that the model meets at least the primary statistical tests applied to it.

Steinhardt et al.[32] and Duffy et al.[33] have recently
carried the refinement of the 519-atom model a stage further, starting from the coordinates of the hand-built models, as relaxed by computer in order to minimize bond
length fluctuations. The atom positions are systematically adjusted so as to minimize the Keating expression[34]
for the elastic energy of a tetrahedrally-bonded solid
in terms of the bond lengths and bond angles

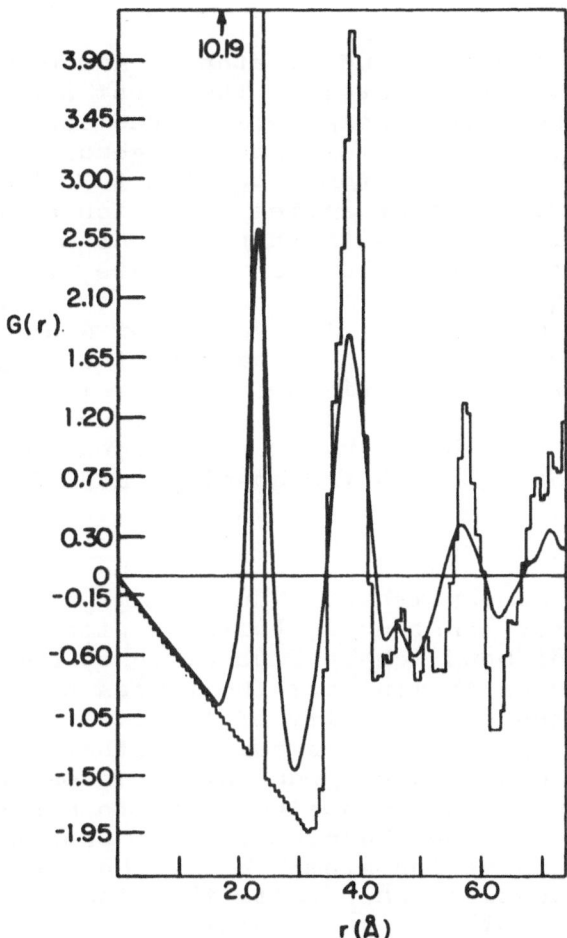

Fig. 2 Comparison of the $G(r) = 4\pi r \{\rho(r) - \rho_0\}$ derived from the CRN calculations of Polk and Boudreaux[31] (histogram) with the experimental function found for a-Si by Moss and Graczyk[46] (smooth curve). The calculated histogram is broadened to reduce statistical fluctuations.

$$E = \frac{3}{16} \frac{\alpha}{d^2} \sum_{\ell,i} (|r_{\ell i}|^2 - d^2)^2 + \frac{3}{8} \frac{\beta}{d^2} \sum_{\ell\{i,j\}} (r_{\ell i} \cdot r_{\ell j} - \frac{d^2}{3})^2$$

(4)

where α and β are the bond stretching and bond bending force constants respectively. The first sum is over all atoms ℓ and their four neighbors i; the second is over all atoms ℓ and all pairs of distinct neighbors i and j. $r_{\ell i}$ is the vector from atom ℓ to atom i. These two terms are the first in a power series expansion about the equilibrium positions and so the expressions may be incorrect for large distortions. Thus an estimate of the elastic energy from equation (4) and the model coordinates may be inaccurate. In addition, the values of α and β to be used are somewhat uncertain. Fortunately, the properties of the models ultimately derived do not seem to depend significantly on the choice of β/α, so that we may infer that the structure will be adjusted using equation (4) in the right sense, even though the absolute adjustment may be in some error.

There are several ways in which the energy minimization may be accomplished. Thus the model may be considered as first constructed (with or without any attempt at reduction of bond length fluctuations) and then the atoms moved so as to minimize the forces on them. In this process the connectivity of the atoms is not changed. Alternatively, as a model is built--either by computer or by hand--a newly added group of atoms may be considered and their positions adjusted so as to minimize the energy or the forces on the atoms. This process determines the connectivity of the atoms and is the analogue of a deposition process which permits the atoms a limited surface mobility. Clearly an attempt to model a low energy metastable continuous random network should ideally follow the latter procedure.

The method adopted by Steinhardt et al. is to move each atom separately and sequentially, while keeping the coordinates of the other atoms fixed, until in its altered position there is no force on it. Each atom is repositioned over 50-100 iterations until convergence is achieved. Steinhardt et al. used this method to find new coordinates for the 519-atom model of Polk and Boudreaux. They also built a new 201-atom model starting from a 21-atom seed and serially adding atoms. As a small group of atoms is added, their positions are relaxed by computer

to minimize the forces on them. Thus, the connectivity
of this network is determined by the relaxation procedure.
Apparently, however, Steinhardt et al. found no difference between the structural characteristics of the relaxed 519-atom model and the 201-atom model relaxed as
it was constructed. This appears to suggest that the instincts of the original model-builders led to topologies
in accord with local elastic energy minimization. Whether
these instincts would be sufficient if dihedral angle
terms and ionic forces contribute to the local free energy is a separate question. We shall return to this
point in discussing more automatic computer procedures
which seem to minimize the influence of the model-builder
in establishing the topology.

Steinhardt et al. carried out an extensive examination of the general characteristics of the 519 and 201-
atom networks. They found that the density of both was
within 1°/o of that of crystalline Ge. Energy minimization in both cases led to bond length fluctuations near
1°/o--six times larger than in the Polk-Boudreaux model
which sought to minimize these--and r.m.s. angular deviations near 7° -- smaller than the 9.25° of Polk and
Boudreaux, as expected. Correspondingly, the distortion
energies which depend largely on the square of the r.m.s.
angular deviation are much lower in the energy-relaxed
models. The elastic constants of the models were determined by applying different types of strains and allowing fully bonded atoms to relax so as to minimize the energy in equation (4). The functional dependence of energy on strain gives the elastic constants which were compared with those for the diamond cubic structure with
the same density and force constants, and for an unrelaxed
CRN model. The results for the bulk modulus of the models and c-Ge are found to be very close to each other.
It is notable that results like this for a-Ge have not
been verified experimentally.

The ratio of the densities of 5, 6, and 7-atom rings
was also determined. So also was the dihedral angle distribution $P(\Theta_d)$, which was found to be the same as in the
unrelaxed model of Polk and Boudreaux. Finally the radial distribution function $J(r)$, corrected for the finite
size of the models, was determined.

There were no significant differences between the
201 and the relaxed 519-atom models. As already implied,
the reduced bond angle distortion of the relaxed models
does sharpen their RDF with respect to the unrelaxed

RDF's. In addition, the $J(r)$ was decomposed into $J_n(r)$ corresponding to n-bond neighbors, and each $J_n(r)$ decomposed in turn into contributions from atoms forming parts of 5, 6, or 7-fold rings. This kind of analysis is instructive because--as we shall see later--the same RDF may correspond crudely to different ring statistics and so it is important to identify which features in it may be linked with certainty to the presence of a certain type of ring. This is especially important since it appears that the ring statistics have an important bearing on the physical properties.

These results for the RDF are displayed in Figure 3 and Table 1.

The procedure of Duffy et al. consisted of moving each atom of the already-constructed 519 atom model simultaneously, in the direction of the force on it, a dis-

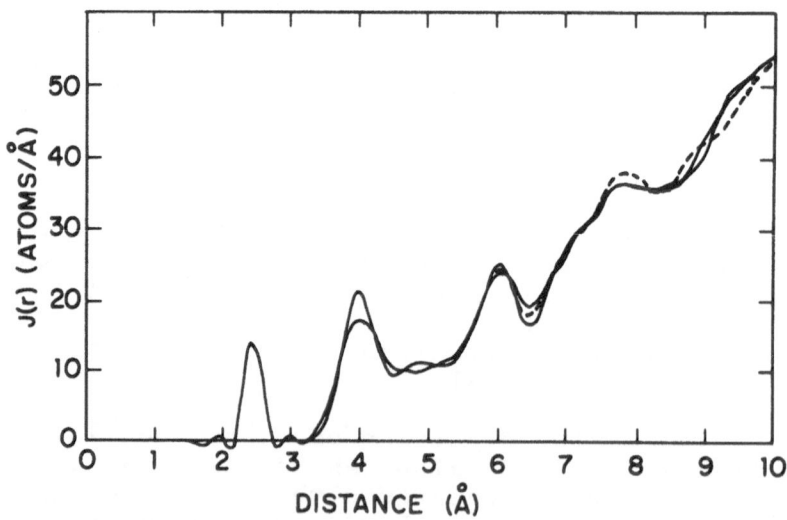

Fig. 3 Comparison of the RDF's derived from the energy-relaxed 519-atom model of Polk and Boudreaux[31] and the 201-atom model constructed by Steinhardt et al.[32] with experimental data for a-Ge by Temkin et al.[47]. The heavy line is the experimental result, the thin continuous line is for the 201-atom model and the dashed line is for the 519-atom model. The calculated curves are broadened to reduce statistical fluctuations.

TABLE I

Structural Parameter for Crystalline and Amorphous Germanium

	Shevchik and Paul (expt.)	Temkin, et al. (expt.)	Graczyk and Chaudhari (expt.)	Steinhardt et al.	Connell and Temkin	Diamond	Wurtzite
ρ_0 (atoms/Å³)	0.039	0.0428±0.0004	0.040 to 0.037	0.043	0.043	0.043	0.043
r_1^c (Å)		2.45±0.01				2.450	2.450
r_1^a (Å)	2.46±0.03	2.47±0.01	2.45±0.05	2.47	2.47		
$\Delta r_1^a / r_1^a$	0.015	0.015	0	0.014	0.012		
r_2^c (Å)		4.00±0.01				4.001	4.001
r_2^a (Å)	4.0±0.05	4.00±0.04	4.0±0.05	4.03	4.03		
θ_b (Deg)	109±2	108.1±1.7	109.47	109.49	109.21	109.47	109.47
$\Delta\theta_b$ (Deg)	10	9.5	6.5	6.45	10.71	0	0
N_1^c						4	4
N_1^a	3.85±0.1	3.91±0.10	3.95±0.1	4.044	4.029		
N_2^c						12	12
N_2^a	10.0±2	12.6±0.3	11.9±0.5	11.44	11.36		
N_3^c						24	25
N_3^a				25.25	23.77		
σ_1^c (exp.)		0.080±0.008				0.081 (theo.)	
σ_1^a (exp.)	0.085±0.01	0.089±0.008					
σ_1^s (exp.)		0.037+0.026-0.037					
σ_2^c (exp.)		0.12±0.02				0.104 (theo.)	
σ_2^a (exp.)	0.3±0.05	0.27±0.01					
σ_2^s (exp.)		0.25±0.02					
Eclipsed/Staggered				0.4/0.6	0.3/0.7	0/1.0	0.25/0.75
Rings/atom 5				0.38	0	0	0
6				0.91	2.3	2	2
7				1.04	0	0	0

tance proportional to that force. After each move the
forces are recalculated, the atoms moved again, and the
process iterated until the forces and displacements be-
come small. Thus the atom i is moved $\underline{\Delta x}_i = h \; \underline{\nabla}_i \underline{E}$.

The choice of h only matters in that a bad choice increases the number of iterations necessary, although it
need not lead to errors in the final coordinates. Duffy
et al. also carried out an extensive analysis of the statistical features for the relaxed network; there are no
significant differences for our purposes between their
conclusions and those of Steinhardt et al.

It seems clear that energy minimization as atoms are
added to a structure eliminates the possibility that the
connectivity is dependent on the whim of the model-builder.
On the other hand, it does rely to some extent on the cor-
rectness of the equation for the energy, since additional
terms in it might lead to a different choice of configuration in cases where two are in close competition. This
point is underscored by the work of Connell and Temkin[35]
who undertook to build a tetrahedrally coordinated CRN
model with no odd-membered rings. They were encouraged
to do so, partly because of a controversy involving the
presence of five-membered rings in amorphous 3-5 compounds, which would of course imply the occurrence of
bonds between like atoms, and partly because of suspicions
that models with quite different ring statistics could
lead to rather similar RDF's.

The Connell-Temkin model of 238 plastic and metal
tetrahedra was built following the same general procedure as employed by Polk. No difficulty was encountered
in avoiding the construction of a crystalline network or
of a random network containing odd-membered rings. The
coordinates of the fully-coordinated, even-membered ring
model were adjusted by computer so that the variance of
the bond lengths was equal to that of the static distortion measured experimentally in a-Ge, and the new coordinates used to compute the structural parameters of the
model. These are summarized in Table 1 and Figure 4.

c. <u>Computer-built CRN models for amorphous tetrahedrally-coordinated materials</u>

The first study of this type was apparently carried
out by Kaplow et al.[3] In an attempt to simulate the
structure of amorphous selenium, they started with crystalline arrays about 100 atoms in size and randomly perturbed the coordinates of each atom, only keeping those

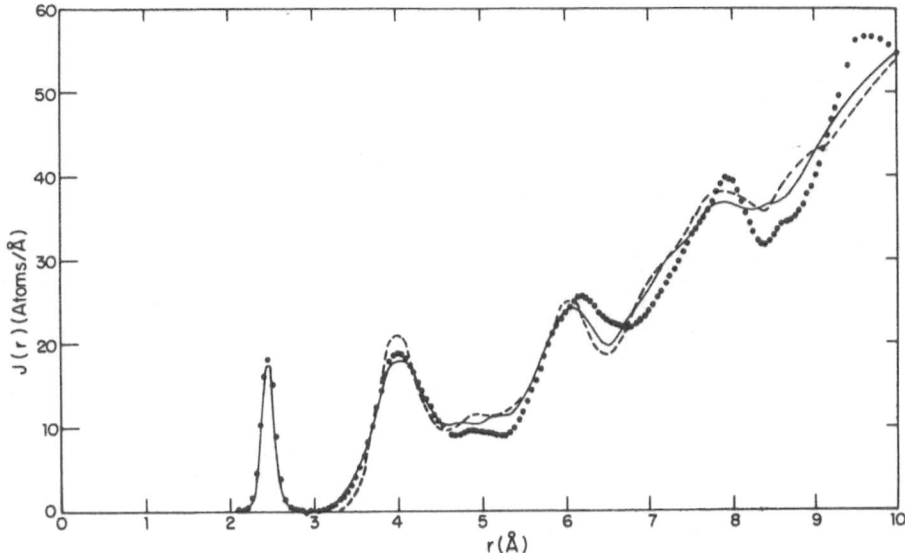

Fig. 4a Comparison of the RDF's derived from the even member ring model of Connell and Temkin[35] and the 519-atom model of Polk and Boudreaux[31] (not relaxed by energy minimization) with experimental data for a-Ge by Temkin et al.[47] The heavy line represents the experiment, the dotted line the calculations of Connell and Temkin and the dashed line the calculations of Polk and Boudreaux.

moves that improved the fit of a calculated RDF for the cluster with the experimental RDF. The resultant fit was reasonably good. However, it is hard to see how methods based on perturbations of a crystalline lattice, which hardly alter the topology at all, can be a true representation of the actual arrangement. Questions must arise regarding the metastability of atoms whose positions are found by simple displacements from crystalline positions that maintain the crystalline topology. The existence of non-crystallographic arrangements of low energy would seem to suggest that these should be permitted in the network that is built up. Therefore our view is that such fitting routines are unlikely to reproduce the details of the experimental RDF.

The most extensive computer modelling has been done for a-Ge by Shevchik[1], who built a 1000-atom cluster

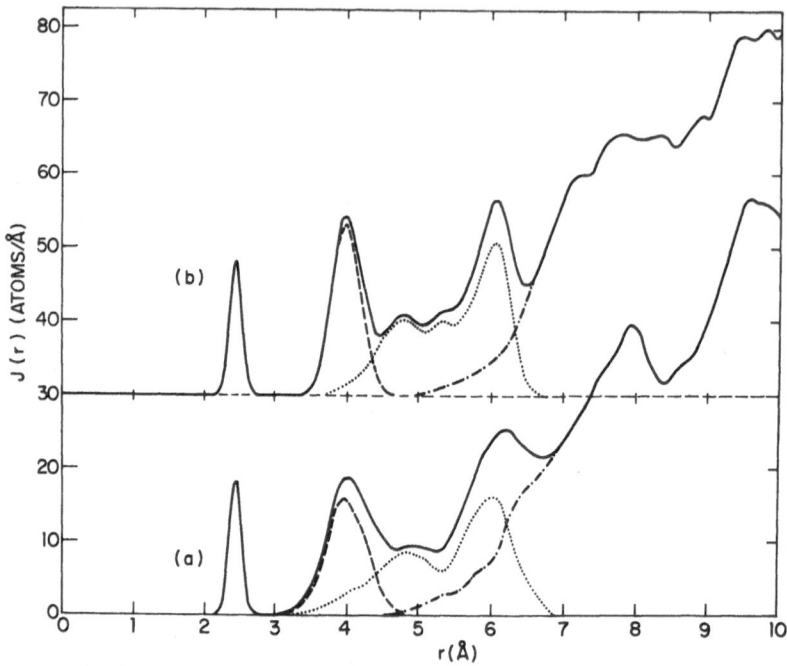

Fig. 4b RDF's for (a) the 238-atom even member ring model of Connell and Temkin[35] and (b) the 201-atom model
of Steinhardt et al.[32], and their analysis in terms of
the distributions of n-bond neighbors, $J_n(r)$; n = 2(---),
n = 3(····), n ⩾ 4(-· -·).

adatom by adatom, following certain rules suggested by
the known experimental RDF. These rules were that the
atoms were hard spheres 2.45Å in diameter (the first peak
in the RDF is sharp); no atoms could approach between
2.5Å and 3.2Å of each other (the RDF has almost no amplitude in this region), and each must have 4 or fewer nearest neighbors (the area under the first peak in the RDF
indicates tetrahedral coordination).

     In accordance with these rules, a seed of about 15
atoms was first generated by randomly distorting all tetrahedral bonds within a 20° limit. As each additional
atom was added, it was allowed to search its environment
so that it could find a location requiring a minimum of
bond distortion. First the adatom was allowed to search
for a site where it could make a bond to three existing

atoms within the stated restraints. This is supposedly
equivalent to allowing the atom to minimize its energy;
in practice, of course, a real adatom will not always
have sufficient surface mobility to allow it to search
very far from its point of impact. If there is no suitable 3-bond site, the adatom then searches for a site
where it can make two bonds satisfying the bond length
and angle constraints. If it makes either two or three
bonds, it is considered fixed in place, and no further
relaxation is permitted. We note immediately that the
search for a double or triple joining is essentially e-
quivalent to the hand-model builder's actions when he ad-
justs the dihedral angle rotation of an adatom unit so
as to join onto existing spokes of the model with a minimum of bond distortion.

In the event that an atom cannot find a 2-bond or
3-bond site, it attempts to find a place where it can
make a single bond that is consistent with the bond angle
fluctuation permitted to the atom to which it is joining.

The result was a continuous random network which had
a certain fraction of unsatisfied bonds, which are the
result of the rule against multiple atom rearrangements.
Thus although this procedure does not lead to an ideal
CRN, it may well be a very good representation of an actual deposition process, particularly one where the atoms
have low surface mobility (low temperature, moderate deposition rate).

The density, and bond angle and dihedral angle distributions, may be calculated from the model. An histogram RDF may also be calculated and smoothed for better
comparison with experimental data. It is shown in Figure 5.

The agreement is very good, and all of the experimental peaks are found. The first coordination number
is 3.8, indicating that about $5^o/o$ of the bonds are unsatisfied. The r.m.s. spread in the bond angle is 9.9^o,
and the dihedral angle distribution is flat from 0° to
60°.

Shevchik's model[1] has not been used in calculations
of the intrinsic vibrational and electronic properties
since the large fraction of broken bonds makes it less
suitable for that. It remains, however, as a very plausible representation of a real material. Moreover, it
is not unreasonable to suppose that an adaptation of the

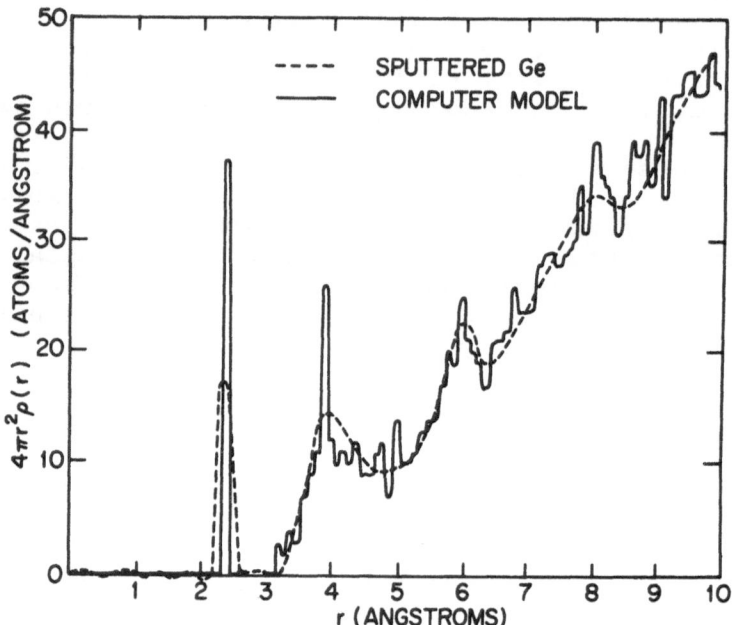

Fig. 5 Comparison of the RDF derived for the computer-generated CRN model of Shevchik[1] and experimental data from X-ray diffraction on a-Ge, also by Shevchik.

model which permitted multiple-atom local rearrangements could lead to a structure with far fewer broken bonds. Once a program is found to calculate this in reasonable computer time, the great advantages possessed by the computer method of easy parameter variation and easy accumulation of statistical data might make it a preferred approach to network modelling. Steps in this direction have recently been taken by Duffy et al.[33]

Another approach to computer modelling, which is less satisfactory in simulating the build-up of the real material, but which is more adapted to later calculation of vibrational and electronic properties, has been followed by Henderson and Herman[36]. In their method, the initial positions of a 64-61 atom cluster are taken to be those of the f.c.c. crystal, and each atom is then displaced randomly by about 10°/o of the distance to the centroid of its four nearest neighbors. After several hundred moves, the statistics of the assembly appear to converge. The resultant RDF is a fair fit to the experimental one, although not nearly as good as others considered in this review. It is, however, good enough to suggest that the coordinates may be used to calculate physical properties of amorphous tetrahedrally-bonded substances. A particular feature of the Henderson and Herman procedure which favors this is the fact that the structure has periodic boundary conditions.

Duffy, Boudreaux and Polk[33] have used several approaches in attempting to construct, solely by computer techniques, completely tetrahedrally-coordinated networks.

The first such technique is called the Vacancy Model. Spherical structures of diamond or wurtzite crystal are considered, a certain total number of atoms removed at random, the resulting dangling bonds connected in pairs, and the whole structure then relaxed. For example, 597-atom diamond structures had 23, 46 and 69 atoms removed and 370-atom wurtzite structures had 15, 30 and 45 atoms removed. The resulting structures had larger distortion energies than the relaxed CRN model. Also, if a sufficient number of atoms were removed to ensure that the third crystalline neighbor peak in the RDF was eliminated and the shape of the first two neighbor peaks had static broadening resembling the experimental broadening, the higher -r structure in the model RDF was much too washed out. It is concluded that this is not a good approach to computer modelling of tetrahedrally-coordinated materials.

The second technique is essentially a more elaborate version of the atom-by-atom build-up investigated by Shevchik. First, a cluster of 10 plastic tetrahedral units was constructed, and the atomic positions relaxed so as to minimize the elastic distortion energy. Atoms are then added one at a time to the model. The unsatisfied bond closest to the center of the model is first found in every case, so that no unsatisfied bonds are left in the interior as the model grows to an approximately spherical shape. If the added atom is doubly or triply connected, a local relaxation is performed to minimize the elastic distortion energy. After the addition of every five new atoms the whole structure is relaxed; this is reported to be necessary if the final network is to have low distortion energy.

The criteria used to decide whether an adatom should
be bonded, in addition, to a nearby atom are as follows.
If an atom in the core has an unsatisfied bond, and it
is within $1.4d_0$ of an atom just being added, a test is
made to find out if these two atoms should be made first
or second neighbors. First, they are joined, the local
configuration relaxed, and the increment in distortion
energy to the whole cluster determined. Second, a third
atom is installed between the first two, which then become second neighbors, this new assembly is relaxed to
minimize the strain energy, and the increment in strain
energy to the total for the assembly computed and compared with that for the first procedure.

By such procedures (the reader is referred to the
original paper for the full details) a 500-atom structure was assembled and statistically analyzed. The results are shown in Figure 6.

It is in order to comment at this point that this
computer model-building procedure relies on the accuracy
of the expression for the elastic distortion energy, which
is in fact known to be an approximation. For elemental
materials, additional terms enter which are concerned with
the relative rotation of neighboring tetrahedra. For binary materials ionic terms may have to be added. It seems
to be clear intuitively that such terms might influence
the network topology.

d. Microcrystallite and cluster models for amorphous tetrahedrally-coordinated materials

Although no physical models have been built in most
cases, it is appropriate to consider how space-filling
models of regular, but not necessarily crystalline units,
might arise.

The simplest conceivable microcrystallite model would
use microcrystals of the diamond form and some kind of
connecting tissue. This model was dismissed quite early
for a number of reasons, among them being the fact that
any reasonable crystallite size would probably display a
peak in the RDF at the third nearest neighbor separation.
This is absent in the diffraction results.

Rudee and Howie[37] found an improved fit to the RDF's
with a structure having the symmetry of wurtzite, which
has 25°/o of its bonds eclipsed. Wurtzite-Ge does not
exist in nature, but two other structures do. The first
is often called Si III or BC-8.[38] It has only 6-fold rings,

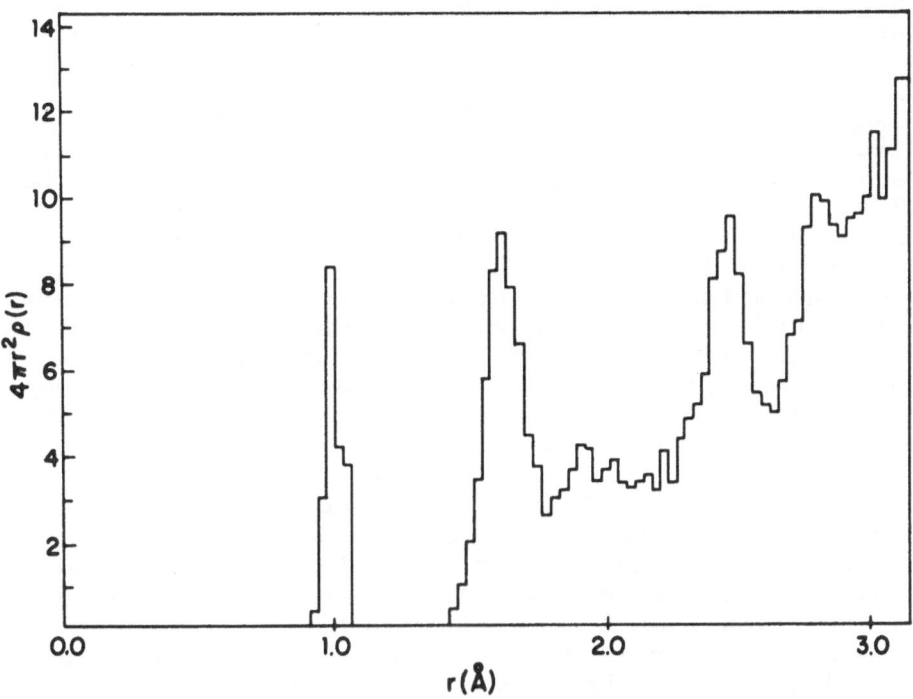

Fig. 6 RDF derived from the computer-generated 500-atom model of Duffy et al.[33]

but nearest neighbors have eclipsed, staggered and inter-mediate configurations. All the atoms have identical en-vironments. The second structure is often called Ge III or ST-12.[39] It has 5 and 7-fold rings as well as 6-fold; also there are two different kinds of atomic environment. These crystalline structures have been much investigated by theorists wishing to obtain some insight into electron-ic structure[40], since the lowered symmetry duplicates at least qualitatively (and perhaps even quantitatively in some respects) the situation in the amorphous materials.

Apparently it is also possible to fill space period-ically using only the pentagonal dodecahedra or amorphons (eclipsed bonds only) and larger units known as clathrates. There are two clathrate structures, both with cubic sym-metry, both with very large unit cells, and both with five and six-fold rings with nearly all eclipsed bonds

and severely distorted bond angles. The first clathrate
is composed of dodecahedra and tetrakaidecahedra, which
are units of 2 hexagons and 12 pentagons. The second
clathrate has dodecahedra and hexakaidecahedra, which are
units of 4 hexagons and 12 pentagons.

Weinstein and Davis[41] have recently calculated the
F(s) curves for randomly oriented microcrystallites of
different sizes having either the diamond, wurtzite, BC-8,
ST-12 or clathrate symmetry. They conclude that the fit
to experiment for a-Ge is poor, and presumably very in-
ferior to that for the CRN models. It is also reported
that a statistical mixture of microcrystallites of 60°/o
BC-8 and 40°/o clathrate I gives an acceptable fit to the
F(s). It is clear, however, that as statistical mixtures
are made, and also as the tissue connecting the microcrystallites is included in the calculation (it is probably
CRN), the fit to the experiment is very likely to improve.

Finally, for completeness, we mention a model, recently advanced by Betteridge and Heine,[42] which has perfect crystalline order in two directions but a random arrangement of either eclipsed or staggered bonds in the
third direction. Such a model can also, of course, fill
all space. However, contrary to the diffraction evidence,
it should display a sharp peak at the third nearest neighbor separation.

e. Comparison of the structural parameters of models
   with experimental diffraction data

The electron microscope studies of Rudee and coworkers[43] have been interpreted in terms of highly ordered domains with diameters about 5-15Å. Others have shown that
the effects observed by Rudee can arise from diffraction
from a random network structure.[44] This controversy has
been reviewed recently by Cochran[45] and will also be discussed in Dr. Chaudhari's and Dr. Graczyk's lectures.
Here we shall comment only on the fit of the experimental
J(r) and F(s) to the calculations of different microcrystallite models.

Microcrystallite models were originally discounted
by Moss and Graczyk[46] because their experimental RDF did
not possess a peak at the spacing for the third nearest
neighbor, and the second neighbor peak showed considerable dispersion in bond angle. Moss and Graczyk also calculated the diffraction functions F(s) for spherical microcrystallites of diamond structure of 9Å and 12Å diameter, and were unable to find a microcrystallite size

likely to give a good overall fit to experiment. Later
work has confirmed this analysis for different structural
polytypes.

Figures 7 and 8 show a comparison of F(s) data on
a-Ge from several laboratories. They are generally very
similar, although small, and possibly significant, differences do exist depending on the preparation history.
Figure 8a compares one of the experimental F(s) with F(s)
curves calculated for an array of randomly-oriented microcrystals of various polytypes of Ge. The fits are not
good. They should be compared with the fits in Figure
8b and 8c where the experimental data are compared with
calculations for a number of models of the CRN type. It
seems fair to conclude that most of the CRN models yield
F(s)'s in better agreement with experiment than the present microcrystallite models do.

It is to be noted particularly that part (c) of Figure 8b shows the F(s) of the random tetrahedral model[47],
which arises solely from the ordering of the first and
second nearest neighbors. The agreement between this F(s)
and experiment at large s is so good that it is clear
that differentiation of more sophisticated models must
be based on a detailed examination of the fits at all s.
The opposite side of this coin is that small differences
in experimental F(s) may presage significant differences
in the ring statistics of the corresponding networks.

The amorphous cluster models give better agreement
of their calculated RDF with the experimental distribution. This is illustrated in Figure 9. The agreement
of calculations of F(s) for amorphous clusters with experiment is shown in Figure 8c, curve b. The improved
agreement over microcrystallite models is connected with
the presence of eclipsed configurations, which reduces
the probability of occurrence of atomic separations corresponding to third neighbor spacings in the diamond cubic structure. However the small distortion of bond angle inherent in this model is not in accord with the static dispersion found experimentally.

Next we consider the fit of the RDF's of CRN models
and experiment. Figure 1 shows the fit of the original
440-atom model of Polk to the RDF for a-Si measured by
Moss and Graczyk, and Figure 5 a comparison of the RDF
from the computer model of Shevchik with experimental data
for a-Ge. Although the fits of the calculated RDF's are
quite good in both cases, they are surpassed by the results

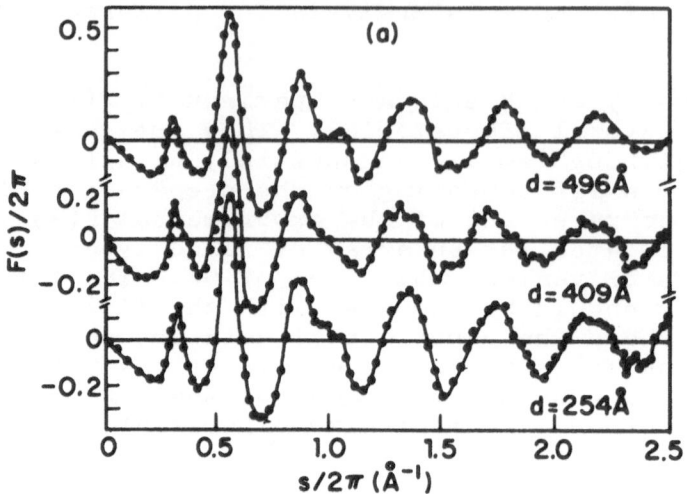

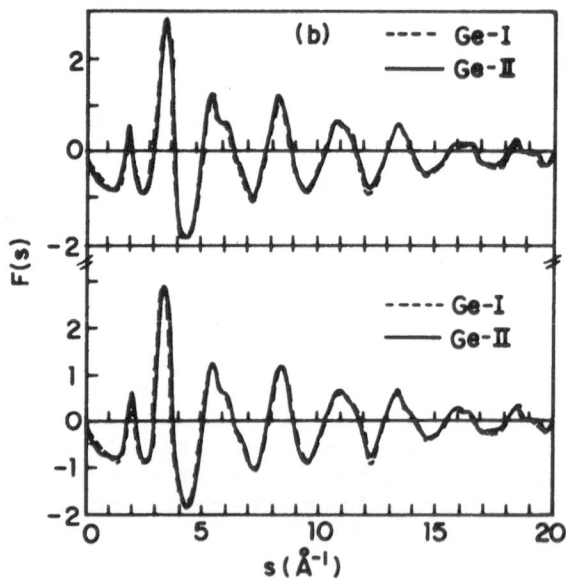

Fig. 7a Electron diffraction functions F(s) for three evaporated Ge films of different thickness, measured by Gandais et al.[48]

Fig. 7b Electron diffraction functions F(s) for sputtered (I) and electron-beam-gun evaporated (II) Ge films, measured by Graczyk and Chaudhari.[49]

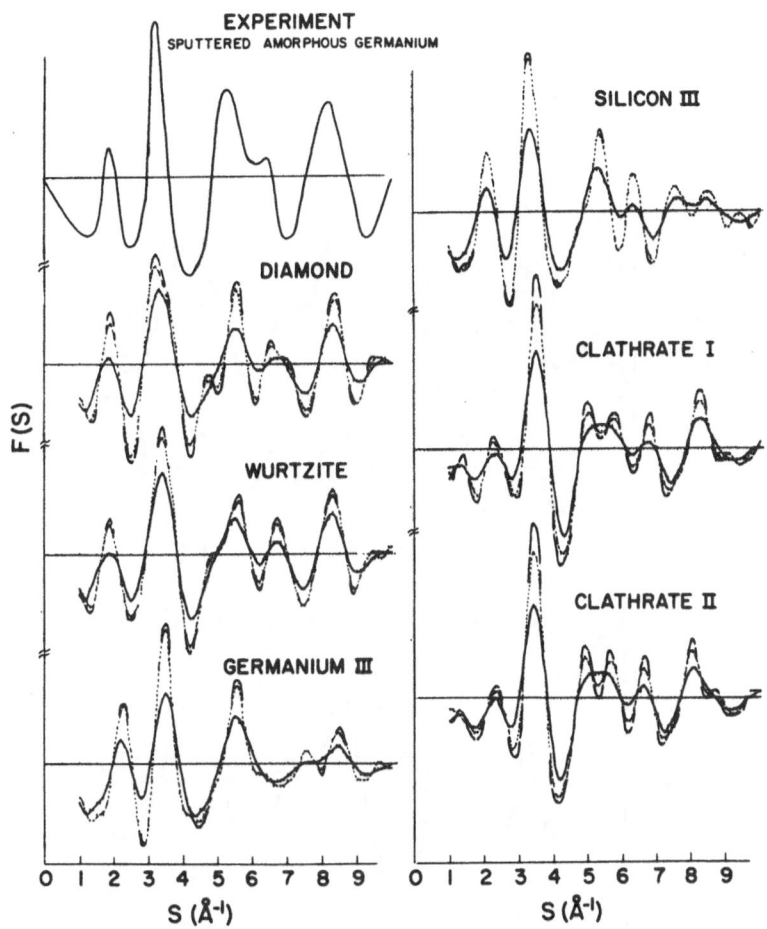

Fig. 8a Comparison of electron diffraction functions F(s) calculated by Weinstein and Davis[41] for microcrystallites of Ge, with experimental data. The latter data are for sputtered a-Ge, measured by Shevchik and Paul.[1] Three theoretical curves are plotted for each structure, corresponding to microcrystallites of (a) small (solid), (b) medium (dotted) and (c) large (dashed) radii. The small radii are about 4Å, and the large about 6.5Å.

of the recent work which refines the original procedure.

Thus Figure 2 compares the $G(r)$ [= $4\pi r\{\rho(r) - \rho_0\}$] derived from Polk and Boudreaux' calculations, which minimized bond length fluctuations, with that for a-Si measured by Moss and Graczyk. The agreement is good.

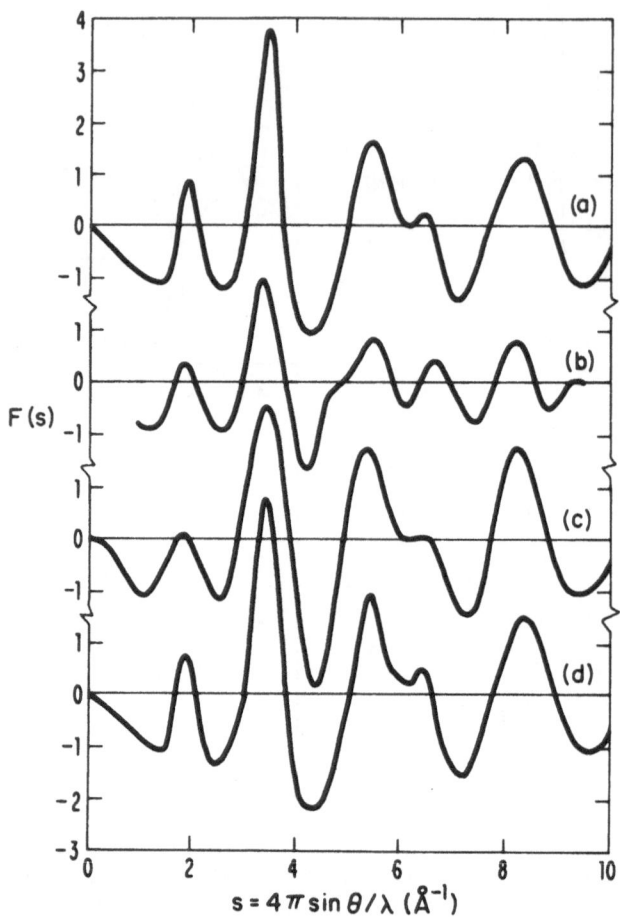

Fig. 8b Comparison of diffraction functions F(s) for
a-Ge, (a) experimental data for sputtered a-Ge, measured
by Temkin et al.[47] (b) calculated data for a wurtzite
microcrystallite model, by Howie et al.[50] (c) calculated
data for a random tetrahedral model, by Temkin et al.[47]
(d) data for the CRN model of Steinhardt et al.[32], calculated by Connell.[51]

Figure 4a includes a comparison of the RDF's of Polk and
Boudreaux and the even-membered ring CRN model of Connell
and Temkin with an experimental RDF for a-Ge of Temkin
et al. Figure 3 compares the same experimental RDF with
RDF's calculated by Steinhardt et al. for both the relaxed

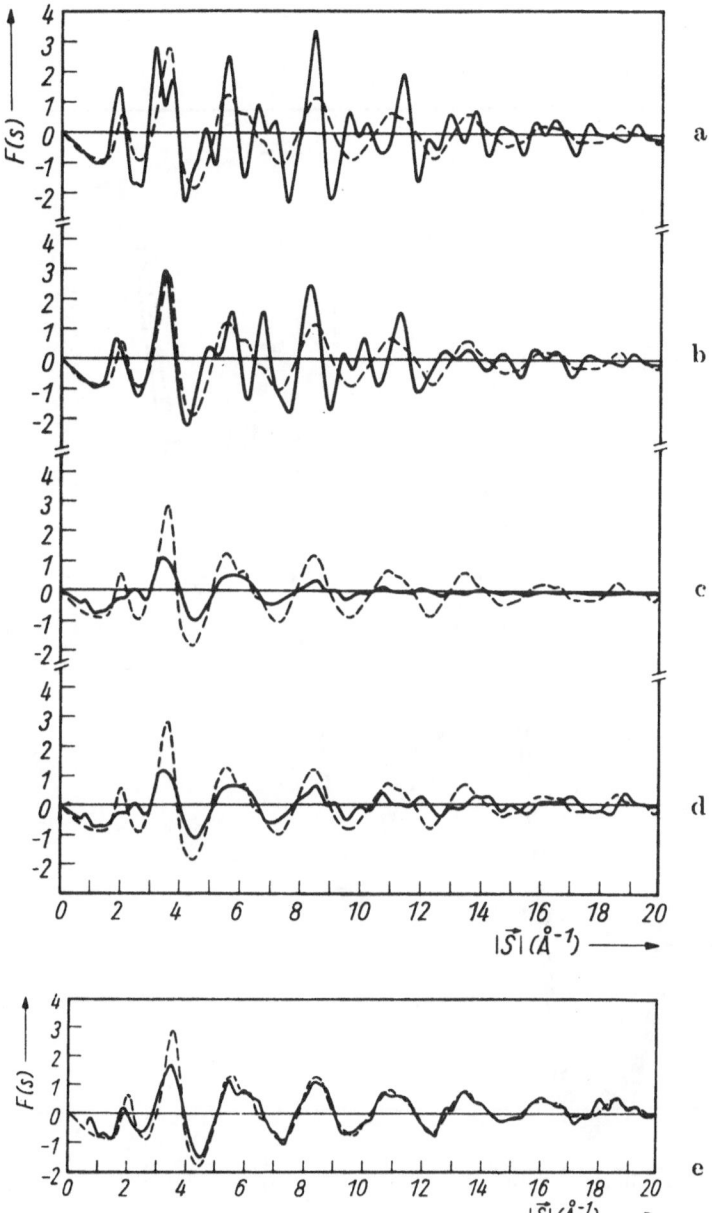

Fig. 8c Comparison of diffraction functions F(s) for
a-Ge[52] (a) 71-atom diamond cubic model. (b) Grigorovici[17]
68-atom amorphon model. (c) 64-atom model of Henderson
and Herman,[36] temperature effects included. (d) 64-atom model of Henderson and Herman, temperature effects excluded.
(e) 61-atom model of Henderson and Herman. The calculated
curves (solid)and experimental electron diffraction data
(dashed) are from Graczyk and Chaudhari.[49]

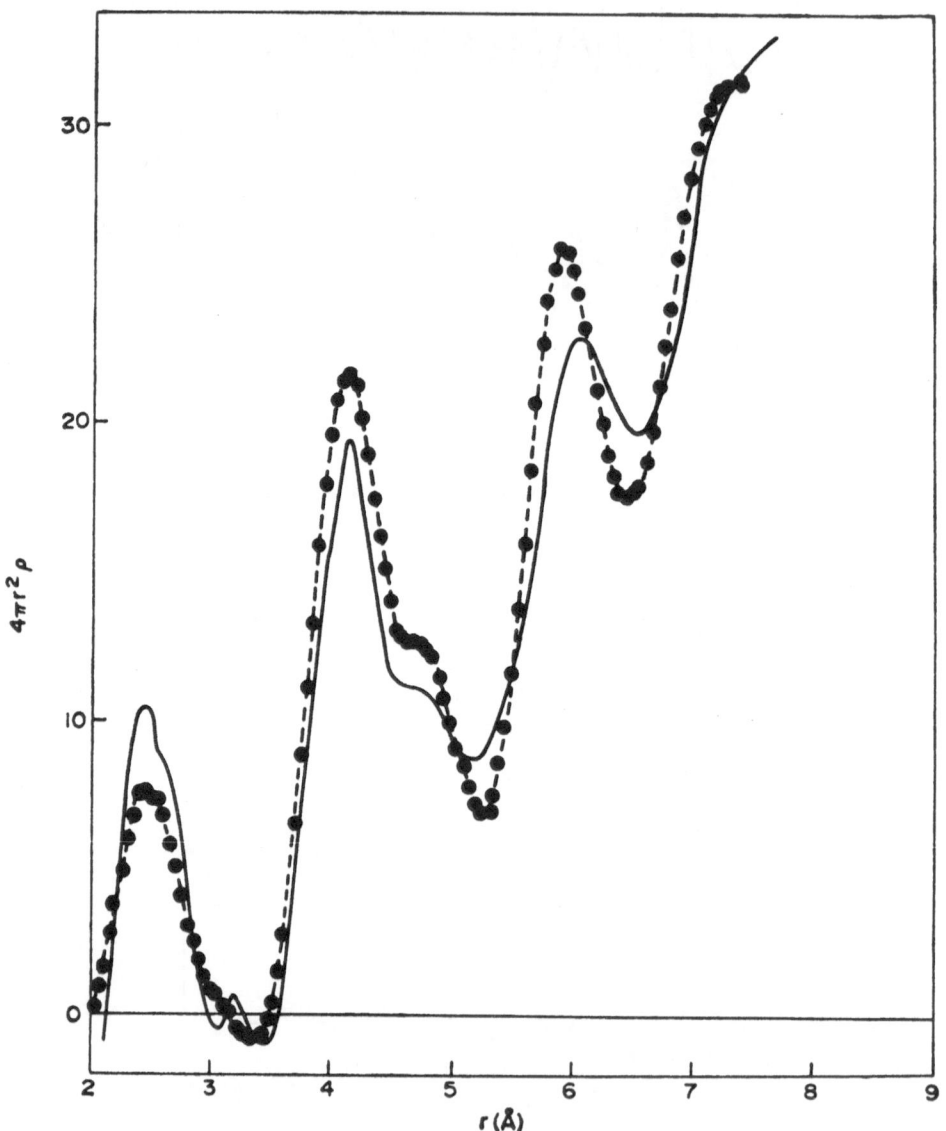

Fig. 9 Comparison of the RDF derived from the network
model of Grigorovici and Manaila[28] (full curve) and experimental data on a-Ge by Richter and Breitling[53] (dotted curve).

(energy minimization) 519 atom model of Polk and Boudreaux and for the relaxed 201-atom model. The fit of
all of these CRN models is very good, and may be improved
still further by combining the odd and even-ring models
i.e., by an adjustment that corresponds to somewhat different ring statistics. This is illustrated in Figure
10 by a 50:50 combination of the 201-atom model of Steinhardt et al. and the 238-atom model of Connell and Temkin.
It is evident that a model could be constructed which is
a fit to experiment within the errors of experiment and
errors stemming from smoothing of histogram RDF's. If
new calculations on microcrystallite models, perhaps with
different statistical averaging and a different treatment
of the disordered connective tissue, were to produce comparable fits to the experimental $J(r)$ and $F(s)$, then it
is likely that a different kind of test than modelling
would be required to differentiate the different structures.

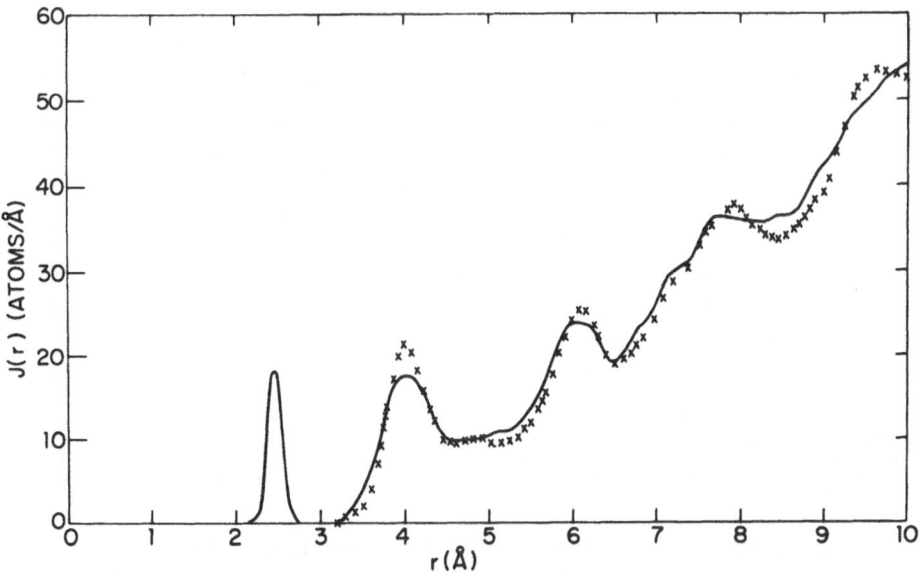

Fig. 10   Comparison of a 50:50 average of the RDF's for
the 201-atom energy relaxed model of Steinhardt et al.[32]
and the 238-atom even-member ring model of Connell and
Temkin[35] (xxx) with experimental data for a-Ge by Temkin
et al.[47] (————).

Figure 11 illustrates a comparison of the RDF for
the 201-atom model of Steinhardt et al. and experimental
data for a-Ge of Temkin et al., as well as a comparison
of the RDF for the 238-atom even-member-ring model of
Connell and Temkin and experimental data for a-GaAs of
Temkin.[54] The intention here is to test whether the calculated differences between the 201-atom model (which
has odd and even-member rings) and the 238-atom model
(which has no odd-member rings) are matched by the differences between the a-Ge and a-GaAs RDF's. From the
goodness of the two sets of fits this seems to be confirmed; in particular, the calculated and experimental

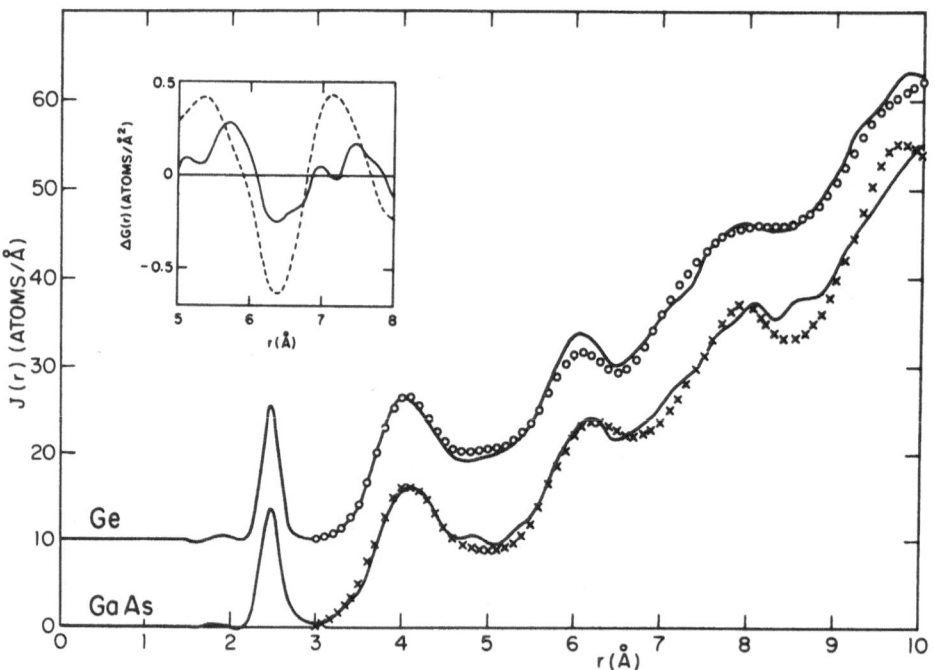

Fig. 11 Comparison of the RDF derived from the 201-atom
model of Steinhardt et al.[32] (ooo) with experimental data
on a-Ge by Temkin et al.[47]; also of the RDF derived from
the 238-atom model of Connell and Temkin[35] (xxx) with experimental data on a-GaAs by Temkin.[54] The insert shows
$\Delta G(r)$ for $\Delta J(r) = J_{Ge}(r) - J_{GaAs}(r)$ (solid) and
$\Delta J(r) = J_{201}(r) - J_{238}(r)$ (dotted).

peaks near 6Å fit quite well. The insert details the differences between the RDF's of the element and compound suggesting that there do exist differences in the topology that are probably important for calculation of vibrational and electronic properties. Since the data of Shevchik and Paul[8] on a wide variety of group III-V compounds also show the displacement to higher r of the third RDF peak, found in Temkin's GaAs data, it is possible that the above comments are generally applicable to the binary compounds.

Table 1 summarizes the experimental data from the studies of Shevchik and Paul,[1] Temkin et al.[47], and Graczyk and Chaudhari[49], and the calculated parameters from the modelling of Steinhardt et al.[32], and Connell and Temkin.[35] By and large, the data are self-consistent. Notable, however, is the fact that the ring statistics of the two CRN models are quite different, although the statistical parameters and RDF's are similar. It is also notable that the fluctuation of 10° in the tetrahedral bond angle, $\Delta\Theta_b$, deduced by Shevchik and Paul, and Temkin et al., is considerably smaller than that of 6.5° found by Graczyk and Chaudhari, and that the theoretical value of Steinhardt et al. of 6.45° is almost identical to that of Graczyk and Chaudhari. On the surface this evidence would seem to suggest a self-consistency between the experiments of Graczyk and Chaudhari and the calculations of Steinhardt et al., but, in fact, the situation is more complicated in two respects. The first of these is the fact that all of the experimental estimates of $\Delta\Theta_b$ in the first three columns of Table 1 are probably low; this is a result of neglect of the contribution of $J_3(r)$ to the second peak in the RDF, which Figure 4b suggests is quite considerable. Preliminary estimates indicate that the 9.5° quoted by Temkin et al. might be increased to about 12°.

The second problem is that the measured RDF's of Temkin et al., and of Graczyk and Chaudhari, which are shown in Figures 3 and 12 respectively, are in reasonable accord, and, if anything, the broadening of the second peak in the RDF of Graczyk and Chaudhari is greater. The smaller static angular distortion calculated by Graczyk and Chaudhari results from an apparently larger estimate of the thermal broadening of the RDF peaks. Graczyk and Chaudhari report only a breadth in excess of thermal vibrations giving $<\delta r_2^2> = 0.026$ Å^2, whereas Temkin et al. deduce $<\delta r_2^2> = 0.0625$ Å^2. Temkin et al.

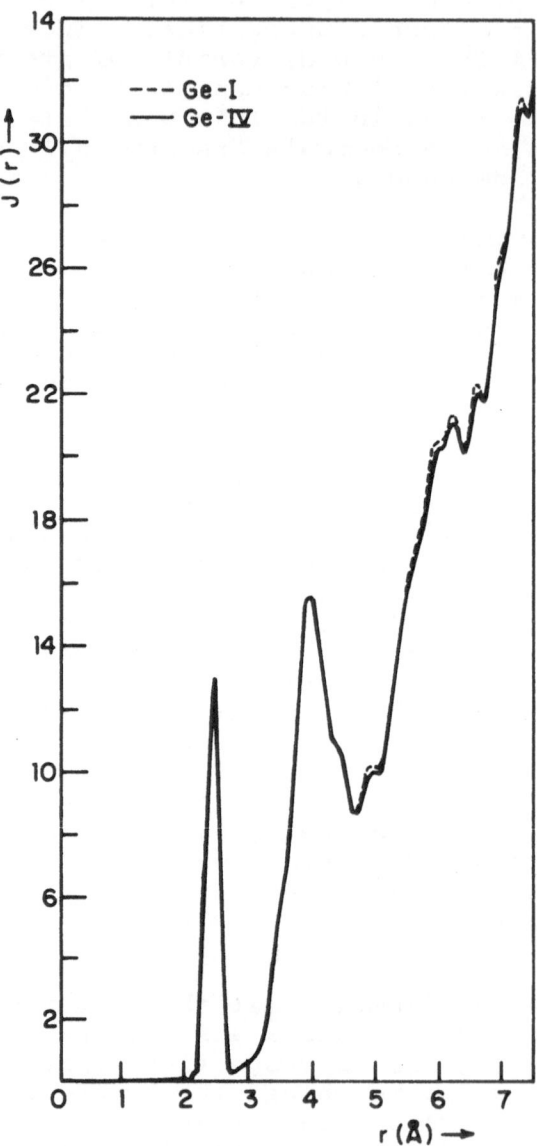

Fig. 12   RDF's deduced from electron diffraction data on two samples of a-Ge, by Graczyk and Chaudhari.[49]

assume that the thermal broadening in the crystalline and amorphous forms is the same and obtain $\delta r_2^2 = (\sigma_2^s)^2$

from $\sigma_2^s = [(\sigma_2^a)^2 - (\sigma_2^c)^2]^{1/2}$. The crystalline σ_2^c was

measured in X-ray diffraction studies of powdered crystalline germanium, and also estimated theoretically using the measured X-ray Debye temperature and coupling constants obtained from a model of Walker and Keating[55]. The experimental and calculated σ_2^c agree within $10^o/o$. Graczyk and Chaudhari do not give the details of their correction for thermal vibrations. Thus the reason for these different estimates of $\Delta\theta_b$ from experimental RDF curves that are in reasonable agreement is not at present understood.

IV. MODELLING OF OTHER CRN STRUCTURES

The modelling of other CRN structures will be summarized briefly in this part.

a. a-C

Kakinoki et al.[14] built a model of a-C using a mixture of two basic units, the first with the tetrahedral symmetry and bond length of the C atom in the diamond lattice, the second with the trigonal symmetry and shorter bond length of the C atom in graphite. Individual regions dominated by diamond-like or graphite-like structure were supposed to be no more than a few Angstroms across, in order to agree with electron diffraction data. The model, which contained on the order of 100 atoms, showed some stress in the regions connecting the two types of building unit. No detailed analysis seems to have been done, and it is not possible as a result to draw conclusions from the model not already inferred from the diffraction analysis.

b. a-As

Amorphous As has been modelled recently by Greaves and Davis[56] following the same general approach as already described for the amorphous group IV elements.

The model consisted of 533 units each with 3 bonds preset to an angle of 97°, which is the bond angle within the layers of the rhombohedral structure. The dihedral angles were chosen to minimize bond angle distortions. Although the bond lengths were constrained to be equal, no computer relaxation was carried out.

The RDF was obtained by Gaussian broadening the measured histogram RDF until its first peak had the same
height as the experimental RDF. Figure 13 shows a comparison of the result with experimental data. The agreement is quite good.

From the 2-bond neighbor distribution, $J_2(r)$, the
bond angle distribution was found to be $98.2° \pm 7.4°$.
The dihedral angle distribution was continuous but favored the staggered configuration, just as in the corresponding distribution for the tetrahedrally-bonded networks. There is a large variety of odd and even membered rings; in particular, an atom is about equally likely
to be found on a 5 or a 6-membered ring.

A notable difference between this model and those

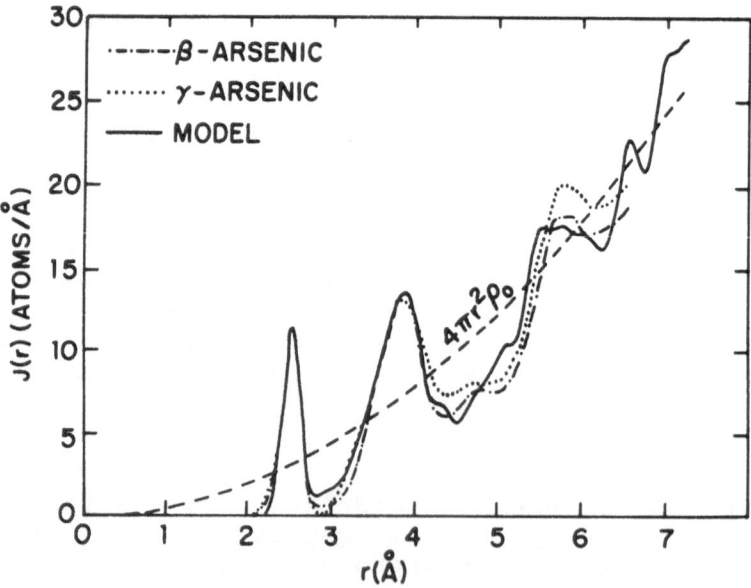

Fig. 13 Comparison of the RDF for a-As derived from the
model of Greaves and Davis[56] with experimental results
of Krebs and Steffen[4] for β- and γ-As.

for a-Ge and a-Si is that the model density is about 12°/o
less than that of the crystal but quite close to the value for the amorphous material. Thus the calculated density in gm-cm^{-3} is 4.93 ± .3, compared to 5.73 and 5.67
for the two crystalline forms, and 4.73, 5.01 and 5.11
for three forms of glassy material. The model seems to
be much more open than the crystal, with the layer-like
quality much diminished. It is notable that the semi-
metallic crystalline conductivity, usually associated
with overlap of orbitals in adjacent layers, has disappeared in the amorphous phase, which is reported to have
a band gap between 1.2 and 1.4 eV.

Greaves and Davis conclude that the minima in the
experimental RDF (and the model) at 2.9Å and 4.6Å prevent an interpretation in which either of the two crystalline phases is simply broadened, since these phases
both lead to interatomic distances of 4.5Å. Also, the
$J_n(r)$ for n = 2, 3, and 4 all have amplitudes at the position of the second peak in the experimental RDF. This
makes it unnecessary to assign the large amplitude of
this peak to remnants of adjacent layer structure, spread
out somewhat from the crystal, as had previously been
supposed.

Finally, using the X-ray scattering factors, $f(s)$,
and a $F(s)$ obtained by Fourier inversion of the RDF,
Greaves and Davis computed the X-ray scattered intensity
$I(s)$ for comparison with experiment. The fit is shown
in Figure 14; it is good in most respects, although there
are differences to be reconciled later.

Thus the essential features of the experimental $J(r)$
and $I(s)$ of a-As seem to emerge from a model based on a
CRN. This model is not two-dimensional, although the
basic unit used in constructing it was derived from the
double layers of the crystalline material. It seems very
likely that refinement of this model will bring the calculated and experimental structural data closer, and that
the calculated electronic and vibrational properties,
which will undoubtedly follow, will be very useful in the
interpretation of experimental data.

c. a-SiO$_2$

Investigation of the structure of a-SiO$_2$ has a long
history. Over many years Warren and his students analyzed X-ray diffraction results on the basis of the Zachariasen CRN model[20]. The 1969 paper by Mozzi and Warren[57]

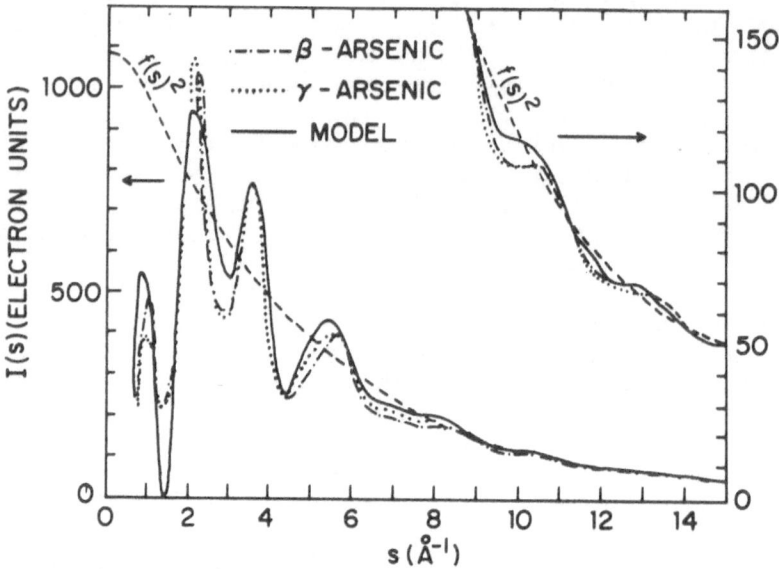

Fig. 14 Comparison of the diffracted intensity I(k) derived from the model of Greaves and Davis[56] for a-As with
experimental results of Krebs and Steffen[4] for β and
γ-As.

inferred that the structure involved perfect SiO_4 tetrahedra with a mean Si-O-Si inter-tetrahedral angle of 144°,
and a nearly random distribution of neighboring tetrahedra
about the common O atom.   Recently, Narten[58], and Konnert and Karle[59], have proposed different microcrystallite
models for SiO_2 and the similar materials GeO_2 and BeF_2.
We do not propose to analyze the pros and cons of this
revived debate here. Instead we shall concentrate on
the modelling of SiO_2 using a CRN network by Bell and
Dean[22] which fit very well the diffraction data of Mozzi
and Warren.

The Bell and Dean construction, which was itself pre-
ceded by other (but less successful) models for SiO_2 by
Ordway[21], and Evans and King[23], may be regarded as the
prototype for the extensive hand-built models of a-Ge and
a-Si. Bell and Dean used basic units consisting of a
central Si "atom" and four rigid wires in perfect tetrahedral arrangement. The O "atoms" were attached by

inserting the wires into pre-drilled holes in the O "atoms". The model was built so as to obtain a mean Si-O-Si bond angle near 160° but without any particular bond angle distribution, and to maintain full coordination in the interior of the model. It consisted of 614 atoms, whose coordinates were determined from stero-pair photographs.

From the coordinates, bond length and bond angle, histograms were constructed. The bond length variation was small, in agreement with the experimental diffraction results. The bond angle varied between 120° and 180° with a mean value of 153°. The pair distribution functions $\rho_{Si-Si}(r)$, $\rho_{Si-O}(r)$ and $\rho_{O-O}(r)$ were computed, properly corrected for finite model size, and combined with the appropriate atomic scattering factors for Si and O to give an RDF comparable with the RDF deduced from diffraction experiments. The results, shown for X-ray and neutron scattering in Figure 15, show startling agreement between calculation and experiment.

Finally, Bell and Dean analyzed the ring statistics of their model, finding finite numbers of all rings having 4 or more Si atoms. The presence of 19°/o of rings containing 4 Si atoms seemed to be necessary for a good fit to the RDF, while the presence of rings with 3 Si atoms worsened it, as did a very high precentage of rings with 5 Si atoms (the vitron unit).

V. CONCLUSION

In this section we hope to have demonstrated that hand- and computer-modelling of the structure of amorphous semiconductors is presently in a very active phase, and that comparison of such calculations with experimental diffraction data is leading to many useful conclusions about the structure of these materials.

REFERENCES

1. N. J. Shevchik and W. Paul, J. Non-Cryst. Solids 8-10, 381 (1972); N. J. Shevchik, Technical Report HP-29, Division of Engineering and Applied Physics, Harvard University (1972).
2. A. F. Joffe and A. R. Regel, in Progress in Semiconductors, (Heywood, London, A. F. Gibson, ed.) vol. 4, 237 (1960).

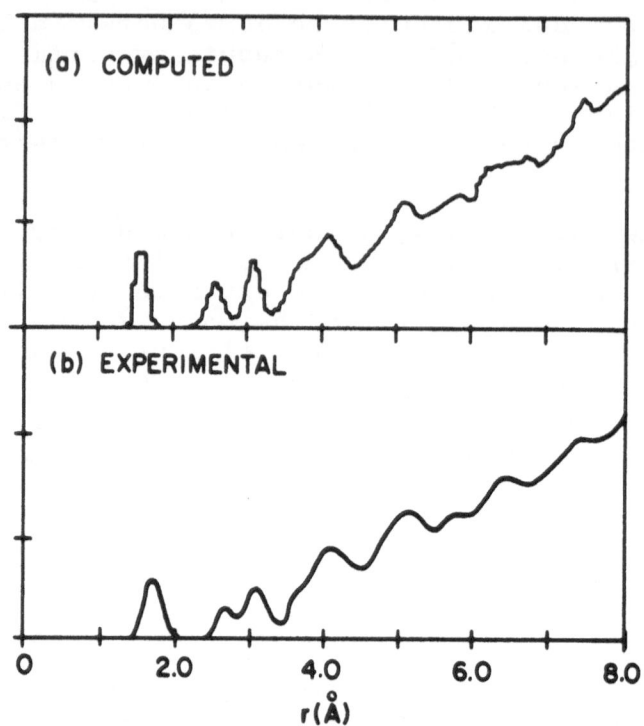

Fig. 15 Comparison of the RDF for SiO_2 derived from the
model of Bell and Dean[22] with an experimental curve based
on the results of Warren and Mozzi[57], Henninger et al.[60],
and Kitchens[61].

3.   H. Richter and F. Herre, Z. Naturforsch. 13a, 874
      (1958); R. Kaplow, T. A. Rowe, and B. L. Averbach,
      Phys. Rev. 168, 1068 (1968); H. Richter, J. Non-
      Cryst. Solids 8-10, 388 (1972).
4.   H. Krebs and F. Schultze-Gebhardt, Acta Cryst. 8,
      412 (1955); H. Krebs, F. Schultze-Gebhardt and R.
      Thees, Z. Anorg. Chem. 282, 177 (1955); H. Richter
      and G. Gommel, Z. Naturforsch. 12a, 996 (1957); H.
      Krebs and R. Steffen, Z. Anorg. Chem. 327, 224
      (1964); G. Breitling and H. Richter, Mat. Res. Bull.,
      4, 19 (1969); G. Breitling, J. Non-Cryst. Solids
      8-10, 395 (1972).
5.   H. Krebs, Angew. Chem. 70, 615 (1958); Fundamentals
      of Inorganic Chemistry (McGraw-Hill, London) 139
      (1968).

6. For recent reports and earlier references, see A. Bienenstock, Proceedings of the Fifth International Conference on Amorphous and Liquid Semiconductors, Garmisch-Partenkirchen, ed. J. Stuke and W. Brenig (Taylor and Francis, 1974), p. 49; also S. C. Moss, loc. cit.

7. Recent studies using either x-ray or electron diffraction have been reported by R. J. Temkin, W. Paul and G. A. N. Connell, Adv. in Phys. 22, 581 (1973); M. Gandais, M. L. Theye, S. Fisson and J. Boissonade, Phys. Stat. Sol. (b) 58, 601 (1973); and J. F. Graczyk and P. Chaudhari, Phys. Stat. Sol. (b) 58, 163 (1973) and 58, 501 (1973); References to and discussion of earlier papers are given in these three studies.

8. N. J. Shevchik and W. Paul, J. Non-Cryst. Solids, 13, 1 (1973/74).

9. G. A. N. Connell, Solid State Comm. 14, 377 (1974).

10. N. J. Shevchik, J. Tejeda and M. Cardona, Phys. Rev. B9, 2627 (1974).

11. G. A. N. Connell, Phys. Stat. Sol. (b) 53, 213 (1972); G. A. N. Connell and W. Paul, J. Non-Cryst. Solids, 8-10, 215 (1972).

12. G. A. N. Connell and R. J. Temkin, in Tetrahedrally Bonded Amorphous Semiconductors, ed. Brodsky, Kirkpatrick and Weaire (A.I.P. Conference Proceedings, No. 20, 1974), p. 192.

13. A. R. Badzian, Mat. Res. Bull, 2, 987 (1967).

14. J. Kakinoki, K. Katada, T. Hanawa and T. Ino, Acta Cryst. 13, 171 (1960); T. Moda and M. Inagaki, Bull. Chem. Soc. Japan 37, 1534 (1964).

15. A. I. Andreyevski, I. D. Nabitovitch and Ya. U. Voloshchuk, Kristallografiya, 7, 865 (1962).

16. A. I. Andreyevski, I. D. Nabitovitch and Ya. V. Voloshchuk, Kristallografiya 6, 662 (1961); A. R. Hilton, C. E. Jones and M. Brau, Phys. Chem. Glasses, 7, 105 (1966); E. A. Porai-Koshits and A. A. Vaipolin, Fiz. Tverd. Tela. 5, 246 (1963).

17. For further details and consideration of other systems the reader is referred to reviews by R. Grigorovici, in Amorphous and Liquid Semiconductors (Plenum, New York, J. Tauc, ed.) 45 (1974); and S. C. Moss, Proceedings of the Fifth International Conference on Amorphous and Liquid Semiconductors, Garmisch-Partenkirchen, ed. J. Stuke and W. Brenig (Taylor and Francis, 1974), p. 17

18. M. L. Rudee, Thin Solid Films, 12, 207 (1972).

19. J. D. Bernal, Proc. Roy. Soc., 280A, 299 (1964).

20.   W. H. Zachariasen, J. Amer. Chem. Soc. 54, 3841
       (1932).

21.   F. Ordway, Science 143, 800 (1964).

22.   R. J. Bell and P. Dean, Nature 212, 1354 (1966);
       R. J. Bell and P. Dean, Phil Mag. 25, 1381 (1972).

23.   D. L. Evans and S. V. King, Nature 212, 1353 (1966).

24.   D. E. Polk, J. Non-Cryst. Solids 5, 365 (1971). D.
       Turnbull and D. E. Polk, J. Non-Cryst. Solids 8-10,
       19 (1972).

25.   L. W. Tilton, J. Res. Nat. Bur. St., 59, 139 (1957).

26.   R. Grigorovici, Mat. Res. Bull. 3, 13 (1968).

27.   R. Grigorovici and A. Belu, Proc. 11th International
       Conference on Physics of Semiconductors, Warsaw,
       453 (1972).

28.   R. Grigorovici and R. Mănăilă, J. Non-Cryst. Solids
       1, 371 (1969).

29.   M. V. Coleman and D. J. D. Thomas, Phys. Stat. Sol.
       24, K111 (1967).

30.   A. F. Wells, Structural Inorganic Chemistry, 3rd
       edition, 578 (Oxford, New York)(1962); L. Pauling,
       The Nature of the Chemical Bond, 3rd edition, 469
       (Cornell, Ithaca)(1960); L. Pauling and R. E. Marsh,
       Proc. Nat. Acad. Sc. 38, 112 (1952); M. von Stackel-
       berg, J. Chem. Phys. 19, 1319 (1951).

31.   D. E. Polk and D. S. Boudreaux, Phys. Rev. Lett. 31,
       92 (1973).

32.   P. Steinhardt, R. Alben and D. Weaire, J. Non-Cryst.
       Solids 15, 199 (1974).

33.   M. G. Duffy, D. S. Boudreaux and D. E. Polk, J.
       Non-Cryst. Solids 15, 435 (1974).

34.   P. W. Keating, Phys. Rev. 145, 637 (1966); See also
       S. C. Moss, R. Alben, D. Adler and J. P. deNeu-
       fville, J. Non-Cryst. Solids, 13, 185 (1973/4).

35.   G. A. N. Connell and R. J. Temkin, Phys. Rev., B9,
       5323 (1974); G. A. N. Connell, and R. J. Temkin,
       in Tetrahedrally Bonded Amorphous Semiconductors,
       ed. Brodsky, Kirkpatrick and Weaire (A.I.P. Confer-
       ence Proceedings No. 20, 1974), p. 192.

36.   D. Henderson and F. Herman, J. Non-Cryst. Solids,
       8-10, 359 (1972); D. Henderson, in Computational
       Solid State Physics (Plenum, New York, F. Herman,
       ed.) 175 (1972).

37.   M. L. Rudee and A. Howie, Phil Mag. 25, 1001 (1972).

38.   R. H. Wentorf, Jr. and J. S. Kasper, Science 139,
       338 (1963); C. H. Bates, F. Dachille and R. Roy,
       Science, 147, 860 (1965).

39.   F. P. Bundy and J. S. Kasper, Science 139, 340
       (1963); J. S. Kasper and S. M. Richards, Acta Cryst.
       17, 752 (1964).

40. D. Henderson and I. B. Ortenburger, J. Phys. C6, 631 (1973). and references therein; J. D. Joannopoulos and M. L. Cohen, Phys. Rev. B8, 2733 (1973) and references therein.

41. F. C. Weinstein and E. A. Davis, J. Non-Cryst. Solids, 13, 153 (1973).

42. G. P. Betteridge and V. Heine, J.Phys.C6, L427 (1973)

43. M. L. Rudee and A. Howie, reference 37; M. L. Rudee, Thin Solid Films 12, 207 (1972); A. Howie, O. L. Krivanek and M. L. Rudee, Phil.Mag. 27, 235 (1973).

44. P. Chaudhari, J. F. Graczyk and H. P. Charbnau, Phys. Rev. Lett. 29, 425 (1972); N. J. Shevchik, Phys. Stat. Sol. (b) 52, K121 (1972); M. V. Berry, and P. A. Doyle, J. Phys. C6, L6 (1973).

45. W. Cochran, in Tetrahedrally Bonded Amorphous Semiconductors, ed. Brodsky, Kirkpatrick and Weaire (A.I.P. Conference Proceedings No. 20, 1974), p.177.

46. S. C. Moss and J. F. Graczyk, Phys. Rev. Lett. 23, 1167 (1969); S. C. Moss and J. F. Graczyk, Proc. 10th International Conference on the Physics of Semiconductors, Cambridge, Mass. p. 658 (1970).

47. R. J. Temkin, W. Paul, and G. A. N. Connell, Reference 7.

48. M. Gandais, et. al., reference 7.

49. J. F. Graczyk and P. Chaudhari, Phys. Stat. Sol. (b) 58, 163 (1973).

50. A. Howie, O. L. Krivanek and M. L. Rudee, Phil. Mag. 27, 235 (1973).

51. G. A. N. Connell, unpublished.

52. J. F. Graczyk and P. Chaudhari, Phys. Stat. Sol. (b) 58, 501 (1973).

53. H. Richter and G. Breitling, Z. Naturforsch. 13a, 988 (1958).

54. R. J. Temkin, Solid State Commun. 15, 1325-28 (1974).

55. C. B. Walker and D. T. Keating, Acta Cryst. 14, 1170 (1961).

56. G. N. Greaves and E. A. Davis, Phil. Mag. 29, 1201 (1974).

57. R. L. Mozzi and B. E. Warren, J. Appl. Cryst. 2, 164 (1969).

58. A. H. Narten, J. Chem. Phys. 56, 1905 (1972).

59. J. H. Konnert, and J. Karle, Nature-Physical Science 236, 92 (1972).

60. E. H. Henninger, R. C. Buschert and L. Heaton, J. Phys. Chem.Soc. 28, 423 (1967).

61. T. A. Kitchens, Thesis, Rice University (1963).

SOME THEOREMS RELATING TO THE STRUCTURE OF AMORPHOUS SOLIDS*

D. Weaire

Dept. of Physics, Heriot-Watt University,
Riccarton, Currie, Midlothian, Scotland
and
Dept. of Engineering and Applied Science
Yale University, New Haven, Conn., U.S.A.

I. STRUCTURES IN GENERAL

The theme of this lecture is the problem of structure-building with given constraints. Given the topic of this institute, it would not be appropriate to stress _periodic_ structures. We shall choose results and ideas which are not restricted to periodic structures, as opposed to group theoretical results related to symmetry.

This is an area in which obvious and simple results stand next to obvious and difficult ones. As an example of the latter, consider the widely accepted idea that the fcc packing density is the greatest possible for hard spheres of uniform size. There exists, to date, _no_ proof of this statement. The trouble is that the fcc structure does not achieve the maximum local density (as does the close-packed arrangement in _two_ dimensions) and it seems very difficult to prove that more is "lost on the swings than is gained on the roundabouts" in any rival structure.

However, there are some interesting results that can be proved or plausibly demonstrated, which are helpful in thinking about the structures of amorphous solids or crystalline solids having large unit cells. What follows is a selection of such results.

---

*Based in part on research supported by NSF.

Let us first make clear the distinction between <u>geo-metrical</u> and <u>topological</u> properties, the former having to do with the lengths and angles and the latter with the connectivity of a given structure. In covalent systems, the lengths and angles with which one is concerned are usually those of the bonds between nearest neighbors, and the connectivity properties are those associated with the network of such nearest neighbor bonds. In metallic structures the decomposition into Voronoi polyhedra (each of which contains all parts closest to a given atom) is often considered more meaningful. We begin with some topological properties.

II. EULER'S THEOREM

Euler's theorem (which may be generalized to any number of dimensions) states in the case of three dimensions that, in a connected array, the number of points minus the number of line segments plus the number of surfaces and minus the number of polyhedral cells is equal to one[1]

$$n_0 - n_1 + n_2 - n_3 = 1 \qquad (1)$$

If this is applied to an infinite two-dimensional network of finite cells, made up of three-fold vertices as in Fig. 1, it is easy to show that the average number of sides of a cell is <u>exactly</u> six.[1] Thus, even a random network must obey this severe constraint. This brings up the question of what we mean by a <u>random</u> topological network in the first place. I must confess to not having any neat definition. With such constraints as Euler's Theorem operating on any network one must be careful what one accepts as evidence of "non-randomness". For example, Aboav[2] observed that, in the apparently random two dimensional cellular structures obtained by sectioning polycrystals, there was a correlation between n, the number of sides of a grain and m, the average number of sides of neighboring grains, according to

$$m \approx 5 + \frac{8}{n} \qquad (2)$$

However, if the cells are considered as polygons, the angles at each vertex of the central cell must add to 2π (which is the essence of Euler's Theorem) and so one might

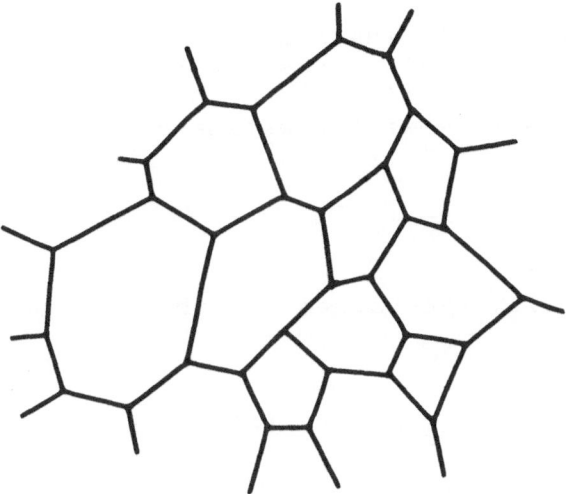

Fig. 1   Random network in two dimensions.

expect such a correlation even in a "random" system since
the internal angles can only be made smaller or larger
at the expense of the external ones. Indeed a simple argument[3], based on this, gives

$$m \approx 5 + \frac{6}{n} \tag{3}$$

For three dimensional networks, Euler's Theorem is
less useful since the bonds do not usually form the edges
of finite cells. There is, of course, a class of structures which satisfy this condition and it happens that
the so-called <u>clathrate</u> structures of Si and Ge do so.
For these structures, also mentioned in Lecture 4, Euler's
Theorem has some relevance, and it may indeed be used to
show that the average numbers of five-fold and six-fold
rings of bonds passing through an atom add to exactly six.

$$n_5 + n_6 = 6 \tag{4}$$

To this may be added the fact that the sum of the angles
at each four-fold vertex must be stationary with respect

to small distortions from the ideal tetrahedral configuration and hence the sum of the deviations from $\cos^{-1}(-1/3)$
is approximately zero. The small negative contributions
from the planar five-fold rings are to be balanced by the
large positive contributions from the planar six-fold rings.
In this way, we arrive at the estimates

$$n_5 = 5.25 \qquad n_6 = .75 \qquad\qquad (5)$$

to be compared with the values for the observed structures,

$$n_5 = 5.22 \qquad n_6 = .78 \quad (46 - \text{clathrate}) \qquad (6)$$

and

$$n_5 = 5.29 \qquad n_6 = .71 \quad (136 - \text{clathrate}) \qquad (7)$$

Thus a little simple geometry and topology suffices to
accurately predict the ring statistics of such structures!
However this is not true of tetrahedrally bonded structures in general, and the significance of the corresponding numbers for random networks is not so obvious. Nevertheless, different groups have reported rather similar
statistics for random networks and it is clear that they
are somewhat constrained if one adopts any "reasonable"
building procedure.

III. LABELLINGS

Connell and Temkin[4] have shown that random networks
can be built with only even rings. This implies that
such networks can be divided into interpenetrating sublattices (i.e. A atoms surrounded by B atoms and vice
versa). This is easily shown. One may label the structure in this manner, beginning at some arbitrary point
and expanding outwards in a series of steps, each consisting of the labelling of all atoms one step away. One
cannot arrive at any contradiction in so doing, since this
would imply the existence of an odd-membered ring. This
"bichromatic" or "alternating" property does not hold for
structures containing odd mumbered rings.

The AB labelling mentioned above may be visualized instead as a + - labelling, in which case we have constructed what amounts to a purely anti-bonding electronic s-state. The problem of electronic band structure prompts us to ask other "labelling questions". Can the bonds of any four-fold connected network be labelled + and - in such a way that each atom has two + and two - ? Can they be labelled with arrows in such a way that each atom has two inward, two outward pointing arrows? These questions were suggested by the study of purely p-like functions, bonding and antibonding, respectively and the answer in both cases is <u>yes</u>, although in fact the existence of (many) such functions is more obviously demonstrated in terms of functions localized on rings or chains[5], or by the type of argument given in the following section.

IV. HARD SPHERES

In the case of dense random packings of hard spheres, the division of (roughly twelve) near neighbors into "hard contacts" and "near contacts" loses physical significance when such packings are used as models for amorphous metals. Nevertheless, it is interesting that the number of hard contacts typically averages very close to six. Note that <u>twelve</u> is the obvious upper bound on this number, while <u>four</u> is a lower bound for a stable structure. Why <u>six</u> ? Bennett[6] has given a simple explanation. He visualizes the formation of a random packing as the coalescence of a random gas (not how such packings are usually made, it should be admitted). If hard contacts, once established, are thereafter maintained, the available degrees of freedom of a system of N atoms after the establishment of M hard contacts is expected to be 3N-M. After the establishment of M = 3N contacts, the system will "jam". This corresponds to <u>six</u> hard contacts per atom.

This style of argument, in which the number of (linear) constraints is subtracted from the number of degrees of freedom, is one which we have found very useful in a variety of contexts. Another example is the comparison of the behavior of the two polymorph structures BC8 and ST12 under pressure for reasonable forces[7]. In the latter case, there are many more degrees of freedom in the structure, more than the number of constraints which would be implied by keeping bond lengths constant while maintaining its symmetry. The ST12 structure can be expected to change markedly with pressure, since it can contract

at the expense of the weak bond bending forces rather
than the strong stretching forces. This effect, which
is confirmed by actual calculation using simple forces[7],
gives rise to an anomalously large predicted compressibility, among other things[7]. However for BC8, this is
not the case. (There are as yet, no experimental measurements in either case). Again, the number of pure
p-like bonding or antibonding eigenfunctions of a simple
tight binding Hamiltonian can be derived in such a manner
(or the number of TA or TO modes - see Chapter 12).

V. A THEOREM USEFUL IN MODEL BUILDING

Various groups (Connell and Temkin[4], Polk and
Boudreaux[8])have, after constructing random networks, used
computer procedures to alter the structure in such a way
as to make the bond lengths equal or nearly so. Others
have used energy minimization schemes which effectively
do the same thing and in addition minimize the variation
of bond angles from their ideal value. This raises the
following question: "Is it always possible, given a four-
fold connected network, to achieve a geometrical structure such that all nearest neighbor distances are exactly
equal?" The answer is, in general, <u>no</u>, as the simple
counter example shown in Fig. 2 clearly demonstrates.
Note that the kind of simple argument based on linear constraints which we have discussed above cannot directly
be applied to this problem since the required constraint
is a <u>non</u>-linear function of the coordinates. However,
one <u>can</u> prove the following. Suppose we define $\delta = \sum_{bonds} \Delta r^2$.
Since this is bounded below (by zero), there must exist
a structure which minimizes it (for any finite system).
If this structure is such that no atom has all of its
bonds lying in the same half-plane, then the minimum value
of δ is <u>zero</u>. To prove this, consider the structure in
question, and suppose δ is <u>not</u> zero for it. The bond length discrepancy Δr must take both positive and negative
values (otherwise a simple scaling would decrease δ).
There must therefore be two neighboring bonds somewhere
such that $\Delta r \geq 0$ on one and $\Delta r \leq 0$ on the other, with at
most one <u>inequality</u>. Let these two bonds have bond vectors
r_1, r_2, the other two belonging to the same atom being
r_3, r_4. Now consider the possibility of displacing the
atom along the direction $r_3 \times r_4$ by a small amount ε.
The lengths of the 3 and 4 bonds are not changed to first
order in ε. Bonds 1 and 2 are respectively increased and
decreased, or vice-versa, according to the sign of ε, by

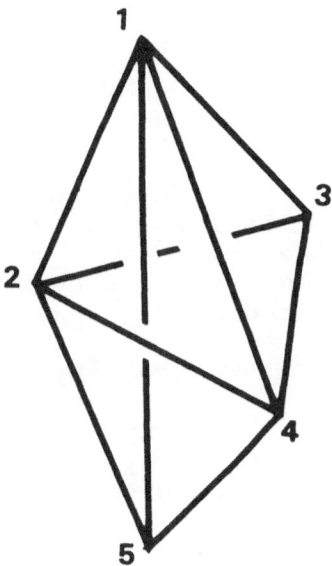

Fig. 2 Fourfold connected structure used as a counter-
example in the text.

amounts of <u>first</u> order in ε. It is therefore possible to
choose a small ε such that δ is decreased. This contra-
dicts our original assumption that δ was non-zero. Hence
it must be zero.

One can apply the same argument if <u>different</u> ideal
bond lengths are assigned, provided they are not too different.

The weakness of this result is, of course, the pro-
viso contained in its statement. (Note its correspondence
to the counter-example which we gave at the outset). We
cannot be sure that a structure which has no such "anomalous" atoms will not develop them as it is altered in the
process of refinement. Nevertheless, it does suggest that
most "reasonable" structures can indeed be reduced to
identical-bond-length structures, in accord with experience. Strickly speaking, one can only say that if and
when a minimally distorted structure is found, it <u>either</u>
has zero bond length distortions <u>or</u> has anomalous atoms as
described. (It will not be unique, since, as shown

previously there are as many modes which preserve given bond lengths as there are atoms in the system).

REFERENCES

1. C. S. Smith, Revs. Mod. Phys. <u>36</u>, 524 (1964).
2. D. A. Aboav, Metallography <u>4</u>, 425 (1971), and references cited therein.
3. D. Weaire, Metallography, to be published (1974).
4. G. A. N. Connell and R. J. Temkin, to be published, (1974).
5. M. F. Thorpe and D. Weaire, Phys. Rev. <u>4</u>, 3518 (1971).
6. C. H. Bennett, J. Appl. Phys. <u>43</u>, 2727 (1972).
7. R. Alben and D. Weaire, Phys. Rev. <u>9</u>, 1975 (1974).
8. D. E. Polk and D. S. Boudreaux, Phys. Rev. Letters <u>31</u>, 92 (1973).

POLYMORPHS*

D. Weaire

Dept. of Physics, Heriot-Watt University,
Riccarton, Currie, Midlothian, Scotland
and
Dept. of Engineering and Applied Science
Yale University, New Haven, Conn., U.S.A.

I. INTRODUCTION

Many of the systems which are widely studied in the amorphous form exhibit a variety of metastable crystalline forms (polymorphs). There is good reason for this, of course. If the bonding in a solid is such that a variety of local structures can be tolerated in the amorphous phase, then one might expect a variety of crystal structures to be possible also.

Polymorphs are quite fascinating in their own right, quite apart from any possible relevance to amorphous phases. However, because of their metastable nature, they usually cannot yet be prepared as good single crystals, so that the experimentalist is confronted with a considerable problem in attempting to investigate their properties. However, even when nothing is known of their properties, theorists have found them to be convenient models for the investigation of the effect of various structural parameters on electronic band structure and the like.

---

* Based in part on research supported by NSF.

II. Si III and Ge III

In the early sixties, F. P. Bundy and J. S. Kasper, who were at the time primarily engaged in trying to make diamonds, were studying the pressure-induced diamond cubic → white Sn structure transition in Si and Ge. The phase diagram was thought to be simple enough, with only these two solid phases within the range of attainable pressures. However, they noticed a change in the external dimensions of one of their Si samples recovered from one of these experiments. On checking, they found that it did not give the usual diamond cubic x-ray lines, but was an entirely new phase of Si! This they called Si III and its structure, which has come to be called BC8, is as follows.[1,2]

Si III   (BC8 Structure)

    Bravais lattice   bcc

    Number of atoms/Bravais cell   8

    Av. number of five fold
    rings through an atom   0

    Av. number of six fold
    rings through an atom   6

    Density, relative to
    diamond cubic   1.09

Ge also appears in a new form (Ge III or ST12) when recovered after going through the white Sn transition.[2,3]

Ge III   (ST12 Structure)

    Bravais lattice   simple tetragonal

    Number of atoms/
    Bravais cell   12

    Av. number of five fold
    rings through an atom   3.33

    Av. number of six fold
    rings through an atom   2

    Density, relative to
    diamond cubic   1.11

No one has yet explained why Si and Ge, so similar in many respects, should produce these two quite dissimilar structures when prepared in such a manner.

The ring statistics of two random networks are compared with those of these two polymorphs in Fig. 1.

It was further found that, upon annealing, the Si III structure returned not to the diamond cubic structure, but to the <u>wurtzite</u> structure[1]! Again, the reason for this is not apparent.*

Later work by Bates, et.al.[4] showed that the direct phase transitions to Si III and Ge III could be observed at pressures below the white Sn transition, but the transitions were exceedingly sluggish. These authors suggested that Ge III might even be the stable form of Ge at zero temperature and pressure. They also found the BC8 structure in Ge under some conditions.

Weaire and Williams[5] pointed out the usefulness of these phases as "halfway houses" of intermediate complexity between the diamond cubic and amorphous phases. They studied the stability of the structures under simple forces, consisting of central nearest neighbor forces together with forces resisting changes in angles. The principal conclusion of this work was a rather negative one, namely, that the use of such forces gave rather unreasonable results as regards the energies of these phases relative to diamond cubic (which may be estimated from experiment as $p_o \Delta V$). This, it was felt, suggested that some scepticism was in order if similar force schemes were applied to the amorphous form.

Extensive calculations have also been made by Cohen and Joannopoulos[6] and Ortenburger and Henderson[7] of the electronic band structure and related properties of these phases, although, to date, no directly relevant data exists.

---

* A small point of nomenclature. Wurtzite, being a stacking of hexagonal layers of Si-Si molecules is a poly<u>type</u> as well as a polymorph. The former term cannot properly be applied to Si III and Ge III.

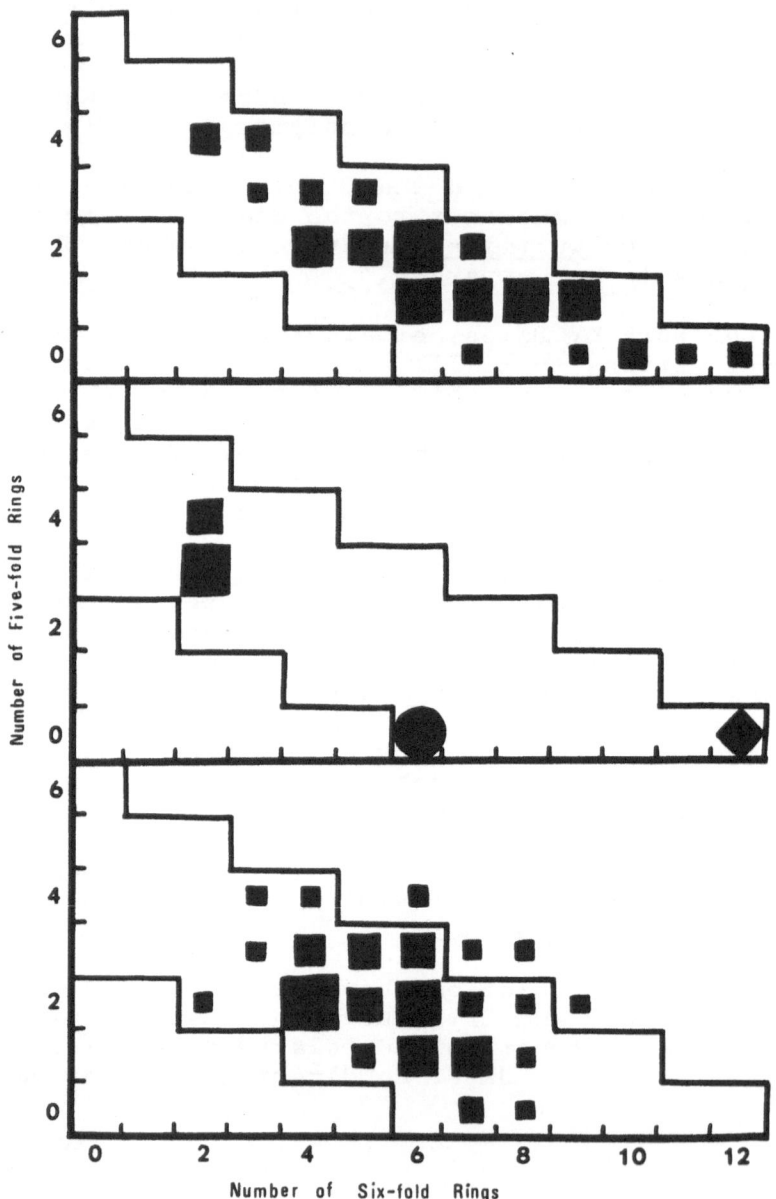

Fig. 1 Ring statistics of some structures. Each point
represents the proportion of atoms having a given number
of five and six-fold rings passing through it.   Top -
Polk random network. Middle - ◆ -diamond, ● -Si III,
■ -Ge III.   Bottom - Henderson model.

Figure 2 shows one such calculation. The following table makes it clear why such calculations are so help-ful in unravelling the dependence of properties on struc-ture.

Calculations of Raman and infrared spectra have also been performed for Si III and Ge III,[8] and some prelimi-nary Raman measurements have been made.[8] Raman data has also been reported for wurtzite Si[9]. In this case some preliminary experimental data exists, but there is room for considerably more work on both theory and experiment.

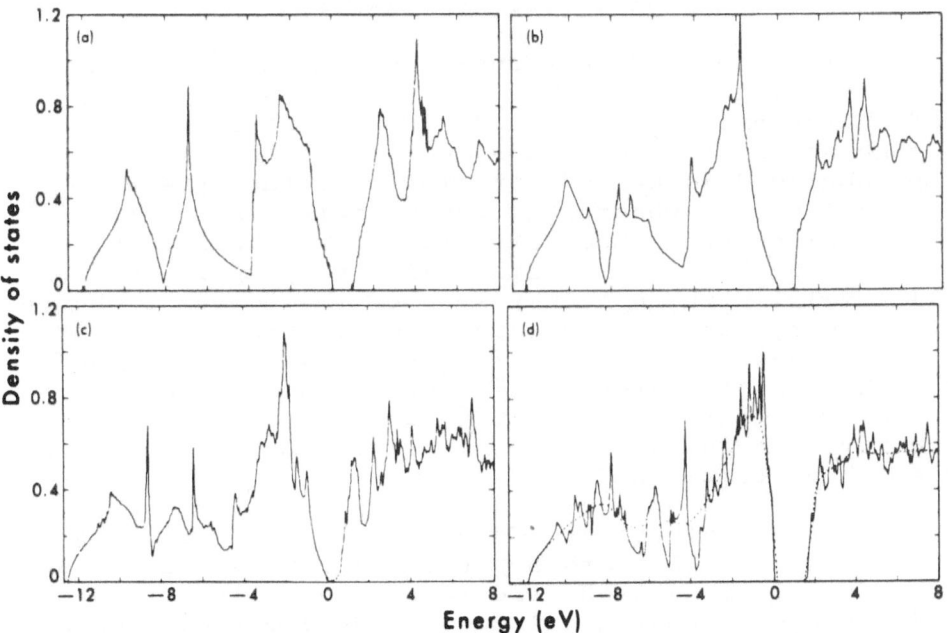

Fig. 2 The density of states calculated using the empirical pseudopotential method by Joannopoulos and Cohen (ref. 6), for Ge in the (a) diamond cubic, (b) wurtzite, (c) Si III and (d) Ge III structures.

TABLE I

	Diamond Cubic	Wurtzite	BC8	ST12
Tetrahedral Bonding?	√	√	√	√
Dihedral angles different from that in diamond cubic?		√	√	√
Significant distortions of bond angles?			√	√
Odd-numbered rings?				√

III. Si AND Ge CLATHRATES

All of the Si and Ge structures considered so far have had bonding arrangements of a predominantly "staggered" character. It seems that if one tries to make more than about 25°/o of bonds "eclipsed" or nearly so, one must build "clathrate" structures of the type already mentioned in Chapter 5, where their ring statistics were discussed. Such structures have been found for Si, Ge and Sn in combination with varying amounts of alkali metals,[10] which occupy some of the "cages". Hence these are not, strictly speaking, polymorphs of the pure elements. They are about 10°/o less dense than diamond cubic. We see, therefore, that the diamond cubic structure does not represent either a high or a low extreme of density when compared with other tetrahedral structures with small distortions.

Similar structures are found among rare gas hydrates. The clathrate with 136 atoms per cubic cell has an amusing and easily visualized structure. Note first that four "amorphons" (pentagonal dodecahedra) may be packed together around a common vertex in a tetrahedral arrangement. This immediately suggests placing one such amorphon on each bond of a diamond cubic lattice. This is indeed the structure of Na_xSi_{136}, etc! The "interstitial" regions consist of more complicated polyhedra containing some six membered rings.

This building procedure suggests that other clathrate structures might be built in the same way, starting
from other tetrahedral "super-structures", although eclipsed bonds in the latter present problems. [In particular there is a need to invent more dense clathrate
structures since denser forms do exist.[11]]

Again, these structures have been used for both electronic and vibrational studies[12,13] (see Fig. 3). Weinstein[14] has suggested that clathrate-like structures
might be found in amorphous Si and Ge with alkali metal
additions. It would indeed be fascinating to study the
structure of Si or Ge codeposited with, say, 1°/o of Na.

IV. OTHER SYSTEMS

Many other elements of interest here also have metastable polymorphs. For example, As can be prepared
with the black P structure which consists of layers in
which the bond lengths, angles and topology are similar
to the usual As structure, but dihedral angles are different.

Particularly worthy of mention are the polymorphs
of SiO_2 and H_2O, since they almost all have tetrahedral
bonding arrangements. The litany of polymorphs for these
is very long. For SiO_2, the list quartz, tridymite,
cristobalite, coesite, stishovite, keatite, melanophlogite-
is only a first approximation.[15] For H_2O, we have Ice
Ic, Ice Ih, and Ice II-IX.[16] There are some interesting
comparisons to be made between the tetrahedrally bonded

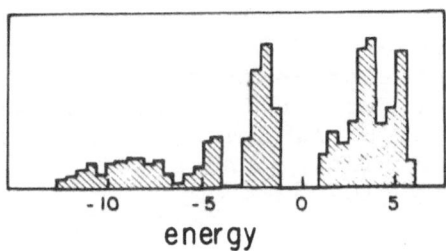

energy

Fig. 3 Density of states in the valence and conduction
bands for a silicon clathrate structure as calculated by
Henderson (ref. 9) using a tight binding Hamiltonian.

systems. For example, Ge III = keatite = Ice III. However, one looks in vain for the analog of the Si III
structure. This is all the more puzzling because of the
reasonableness of this structure - it is almost as dense
as the Ge III structure, but achieves this with much less
distortion from local tetrahedral symmetry. It is perhaps also puzzling that the Si III structure has not been
found for III-V or II-VI compounds (although it is found
in a distorted form in some II-V's[2]). One wonders if
pressure experiments of long duration, like those of
Bates et.al.[4] on Si and Ge, might uncover such phases.

Again, in exploring this potentially fascinating array of structures, one is confronted with the problems
of sample preparation. The exploration and comparison
of the properties of these phases should be most rewarding if such problems can be overcome wholly or partially.
Incidentally, the intriguing possibility of finding unexpected properties in new forms of familiar materials
has already been the subject of an excellent novel -
Cat's Cradle, by Kurt Vonnegut, Jr.!

REFERENCES

1.   R. H. Wentorf, Jr. and J. S. Kasper, Science 139,
     338 (1963).
2.   J. S. Kasper and S. M. Richards, Acta Cryst. 17,
     752 (1964).
3.   F. P. Bundy and J. S. Kasper, Science 139, 340
     (1963).
4.   C. H. Bates, F. Dachille and R. Roy, Science 147,
     860 (1965).
5.   D. Weaire and A. R. Williams, Physica Status Solidi
     49, 619 (1972).
6.   J. D. Joannopoulos and M. L. Cohen, Phys. Rev. 7,
     2644 (1973).
7.   J. B. Ortenburger and D. Henderson, Proc. 11th Int.
     Conf. on Physics of Semiconductors, Warsaw, PWN
     (Polish Scientific Publishers), 1972, p. 465.
8.   R. J. Kobliska, S. A. Solin, M. Selders, R. K. Chang,
     R. Alben, M. F. Thorpe and D. Weaire, Phys. Rev.
     Letts. 29, 725 (1972).
9.   R. J. Kobliska and S. A. Solin, Phys. Rev. B8, 3799
     (1973).
10. J. S. Kasper, P. Hagermuller, M. Pouchard, and C.
     Cros, Science 150, 1713 (1965).
11. F. P. Bundy and J. S. Kasper, High Temperatures -
     High Pressures 2, 429 (1970).

12.   D. Henderson, Physica Status Solidi (b) <u>57</u>, 661
      (1973).
13.   R. Alben, D. Weaire; J. E. Smith, Jr. and M. H.
      Brodsky, Phys. Rev. B <u>11</u>, 2271 (1975).
14.   F. C. Weinstein, to be published (1974).
15.   R. B. Sosman, "The Phases of Silica", Rutgers,
      New Brunswick (1965).
16.   N. H. Fletcher, "The Chemical Physics of Ice", CUP,
      Cambridge (1970).

OPTICAL PROPERTIES OF TETRAHEDRALLY BONDED AMORPHOUS

SEMICONDUCTORS: ABSORPTION SPECTRA AND ABSORPTION EDGE

M. L. Theye

Laboratoire d'Optique des Solides
Université de Paris VI
Paris, France

INTRODUCTION

The purpose of these lectures is to describe qualitatively the optical properties of amorphous semiconductors in a few simple cases, to discuss their variations
in relation to the sample structure and to compare them
with the predictions of the theoretical models which will
be presented in other lectures. We shall consider essentially the tetrahedrally bonded semiconductors: Ge,
Si and III-V compounds, which have been extensively investigated, both experimentally and theoretically, in
the crystalline and the amorphous states, and which are
particularly suited for studying effects of disorder because of their relative simplicity, viz., same coordination in crystalline and amorphous states and equivalence
of all bonds. The structural basis is a tetrahedron, the
bonds are purely covalent in Ge and Si, partly ionic in
the III-V coumpounds which may, in addition, present chemical inhomogeneity. In contrast to glasses, which will
be treated in other lectures, these materials can only
be obtained in the amorphous state as thin films, which
can be prepared by evaporation, sputtering, electrochemical deposition, glow-discharge decomposition, etc. These
films are known to present structural defects, like vacancies or voids of various sizes and shapes, intersti-
tials, dislocation loops, etc., and to contain impurities,
the influence of which may depend on the nature of their
bonding and on the film structure. The influence of these
defects and impurities, which is more or less important

depending on the property under consideration, must be
separated out from the intrinsic effects of topological
disorder and of quantitative or geometrical disorder
(atomic distance fluctuations and bond angle deviations
necessarily accompanying topological disorder), which
are accounted for in theoretical models. Therefore, the
more instructive studies are the studies performed on
well-characterized samples prepared under different conditions, e.g., deposited on substrates at higher and
higher temperatures or annealed at higher and higher temperatures below the onset of crystallization, and para-
lleled by thorough structural studies (density measurements, detection of voids by low-angle scattering experiments, determination of the dihedral angle distribution
function, etc.)

In Section 1, we shall give a few basic definitions
and general properties concerning the response of a solid,
either crystalline or amorphous, to an applied electromagnetic field. In Section 2, we shall describe various
experimental methods used to determine the optical properties, i.e., the complex dielectric constant of strongly or weakly absorbing films. Section 3 will be dedicated to the description and the discussion of the optical spectra at energies higher than the gap value. In
Section 4, the band edge properties will be deduced from
the optical absorption edge; special emphasis will be
put on the influence of defects and impurities, which in
certain cases almost dominate the optical properties in
this energy range.

1. Definitions - General Properties

1.1 Definitions

Let us consider the effect of a monochromatic plane
wave of angular frequency ω and wave vector $\vec{q}$ traveling
through an absorbing medium. The electric field $\vec{E}$ will
cause forced oscillations of the atomic particles. As
the wavelength is much longer than atomic distances, we
are dealing with an essentially macroscopic situation.
As the perturbing effect of the electromagnetic wave is
usually very small, the relation between the macroscopic
electric field $\vec{E}$ and the electric polarization $\vec{P}$ is linear.
Then:

$$\vec{D} = \vec{E} + 4\pi\vec{P} = \widetilde{\varepsilon}\vec{E}$$

The complex dielectric constant $\widetilde{\varepsilon}$, which is a second-rank tensor, represents the response of the medium to the electromagnetic field. It depends, not only on ω, but also on the photon wave-vector $\vec{q}$ ($q = 2\pi/\lambda$); however, this $\vec{q}$ dependence can be neglected in most cases, except when the relation between $\vec{D}$ and $\vec{E}$ becomes non-local (for example in case of strong excitonic effects, when the exciton radius is much larger than the lattice constant). In the following:

$$\tilde{\varepsilon}(\vec{q},\omega) \simeq \tilde{\varepsilon}(\omega).$$

If $\tilde{n}$ is the complex refractive index, then

$$\tilde{\varepsilon} = \varepsilon_1 + i\varepsilon_2 = \tilde{n}^2 = (n^2-k^2) + 2ink$$

where n is the index of refraction and k the extinction coefficient. The optical properties may also be expressed in terms of a complex optical conductivity; one then has:

$$\sigma = \sigma_1 + i\sigma_2; \quad \varepsilon_2 = \frac{4\pi\sigma_1}{\omega}; \quad \varepsilon_1 = 1 + \frac{4\pi\sigma_2}{\omega} \tag{1}$$

The electric field in the absorbing medium can be represented by the expression:

$$E = E_o \exp[-i(\omega t - \vec{q}\cdot\vec{z})]$$

where $\vec{q}$ is the complex propagation vector and $|\vec{z}|$ the distance traveled in the medium. This equation is the solution of the Maxwell's equations for the electromagnetic field, with magnetic permeability $\mu = 1$, if

$$q^2 = \frac{\omega^2}{c^2}\tilde{\varepsilon}$$

This represents a plane wave traveling in the $\vec{z}$ direction with velocity $\frac{c}{n}$, and attenuated as $\exp[-\frac{\omega k}{c}z] = [-\frac{2\pi k}{\lambda}z]$:

$$E_z = E_{oz}\cdot\exp[-i\omega(t - \frac{nz}{c})]\cdot\exp[-\frac{2\pi k}{\lambda}z]. \tag{2}$$

We shall speak of an absorption coefficient α, defined by $\alpha = \frac{4\pi k}{\lambda}$, which is related to the imaginary part of the dielectric constant ε_2 by $\alpha = \frac{\varepsilon_2\omega}{cn}$ (i.e. proportional to $\varepsilon_2\omega$ when n can be considered as constant, it is proportional to the fraction of energy absorbed by unit thickness.

1.2 Dispersion relations

$\tilde{\varepsilon}$ describes a linear relation between the amplitudes of $\vec{D}$ and $\vec{E}$, which implies a requirement of causality: this means that no displacement vector can exist before an electric field is applied. This is equivalent to the fact that $\tilde{\varepsilon}$ is an analytic function of the complex variable ω in the upper complex plane and leads to dispersion relations connecting the real and imaginary parts of $\tilde{\varepsilon}$, known as Kramers-Kronig relations[1]:

$$\varepsilon_1(\omega) = 1 + \frac{2}{\pi} P\int_0^\infty \frac{\omega' \cdot \varepsilon_2(\omega')}{\omega'^2 - \omega^2} \cdot d\omega'$$

$$\varepsilon_2(\omega) = -\frac{2}{\pi} P\int_0^\infty \frac{[\varepsilon_1(\omega)-1]\omega}{\omega'^2 - \omega^2} \cdot d\omega'$$

(3)

Therefore, ε_1 can be calculated at each frequency ω if ε_2 is known explicitly over the whole frequency range, and vice-versa.

Similar dispersion relations also hold for the complex refractive index, for the function $1/\tilde{\varepsilon}$ etc.; we shall see later that these relations for the complex reflection coefficient amplitude are of great practical use in computing the optical constants (Kramers-Kronig method) from reflectance measurements only.

If there is no absorption, i.e. if $\varepsilon_2(\omega) = 0$, then $\varepsilon_1(\omega) = 1$ everywhere and there is no dispersion.

By integrating by parts, the dispersion relations can also be written[2] as

$$\varepsilon_1(\omega) - 1 = \frac{1}{\pi} \int_0^\infty \frac{d[\varepsilon_2(\omega')]}{d\omega'} \cdot \ln \frac{1}{|\omega'^2 - \omega^2|} \cdot d\omega'$$

$$\varepsilon_2(\omega) = -\frac{1}{\pi} \int_0^\infty \frac{d[\varepsilon_1(\omega')]}{d\omega'} \cdot \ln \left|\frac{\omega' + \omega}{\omega' - \omega}\right| \cdot d\omega'$$

(4)

These local relations show that, because of the singularity of the logarithmic factor which weights heavily that value, $\varepsilon_2(\omega)$ is essentially determined by the value of the slope of ε_1 at this frequency ω, i.e., ε_2 will be maximum where ε_1 decreases strongly, etc. This property may be useful in various cases, e.g., for determining the general shape of the optical spectra, in the Kramers-Kronig method of computing optical constants, etc.

1.3 Sum rules

General expressions for the dielectric constant, known as sum rules, can be deduced from the dispersion relations. <u>For $\omega = 0$</u>

$$\varepsilon_1(0) = 1 + \frac{2}{\pi} \int_0^\infty \frac{\varepsilon_2(\omega')}{\omega'} \cdot d\omega' \qquad (5)$$

For semiconductors, the static dielectric constant $\varepsilon_1(0)$ is identical with the long wavelength optical dielectric constant n_0^2. The sum rule says that it is determined by the ε_2 values over the entire spectral range: in particular n_0^2 will be high if strong absorption occurs at low ω.

<u>For high ω</u>, $\omega \gg \omega_u$, ω_u being well above all absorption bands, if one assumes: $\omega' \leq \omega_u \ll \omega$, then:

$$\varepsilon_1(\omega) - 1 \simeq -\frac{2}{\pi\omega^2} \int_0^{\omega_u} \omega' \cdot \varepsilon_2(\omega') \cdot d\omega'$$

At sufficiently high frequencies, the electrons do not see any binding forces [in fact at frequencies higher than the corresponding plasma frequency $\omega_p = (\frac{4\pi N e^2}{m})^{1/2}$], and $\varepsilon_1(\omega)$ becomes the classical dielectric constant of a free electron gas of density N, given by the Drude formula: $\varepsilon_1(\omega) = 1 - \frac{\omega_p^2}{\omega^2}$; then:

$$\int_0^{\omega_u} \omega' \cdot \varepsilon_2(\omega') \cdot d\omega' = \frac{1}{2} \pi\omega_p^2 = \frac{2\pi^2 e^2}{m} N \qquad (6)$$

If $\omega_u \to \infty$, N is the total electron density.

In general, an uppermost group of valence states are well isolated in energy from all other filled states (core states). This is the situation in Ge, Si and the III-V compounds, where the four valence electrons occupy bands of about 10 eV width located well above the next d electron states (these miss in Si where 2p electron

states are situated about 100 eV below the valence band)
which are typically found 25 eV below the top of the
valence band. Then, if ω_u is such that the oscillator
strength of the valence electrons has been exhausted but
the core electrons have not yet come into play, the sum
rule must give the number of valence electrons. For
ω_u = 25 eV, one indeed finds N = 4 for Si; N is greater
than 4 for Ge and the III-V compounds because of a transfer of oscillator strength (especially for InSb) towards
lower energies, even below the frequency at which the
d-electrons begin to absorb[3]. If ω_u is high enough, the
effective number of core electrons involved in the optical transitions can be determined; it must be noticed
that they do not contribute very much to $\varepsilon_1(0)$.

Besides allowing a determination of the effective number of electrons, this sum rule provides another check on
the measured absorption spectrum and the presence of additional absorption bands above ω_u.

If, according to the dispersion theory, the response
of the medium can be accounted for by only one bound os-
cillator with frequency ωg, then $\varepsilon_2 \propto \delta(\omega-\omega_g)$ and the
two sum rules give[4]

$$\varepsilon_1(0) = 1 + \frac{\omega_p^2}{\omega_g^2} \tag{7}$$

If there is, as usually, a broad continuous distribution
of oscillator frequencies, equ. (7) can be used to define
an "average" oscillator frequency ω_g. If ω_p concerns with
only valence electrons, then ω_g will coincide with the
"average gap" defined in the Penn model[5] which will be discussed later on.

1.4 Microscopic formulation

Crystalline case

We shall essentially consider the contribution of
interband transitions of bound electrons. The contribution of free carriers should be added in the form of a
Drude expression:

$$\varepsilon_f = 1 - \frac{\omega_p^2}{\omega(\omega+i/\tau)} \ ,$$

τ being the carrier relaxation time and ω_p the plasma
frequency.

The perturbation of the crystal ground state by the time-dependent vector potential of the electromagnetic field induces electric dipole transitions from occupied valence states to empty conduction states across the Fermi level. The transition probability W_{vc} per unit time and unit volume between all states in a pair of bands can be computed by a time-dependent perturbation treatment[6]. As the energy loss suffered by the incident plane wave, $W_{vc} \cdot \hbar\omega$, must be equal to $\frac{1}{2} \sigma_1 E_o^2$

according to Maxwell's theory, then

$$\sigma_1 = \frac{\varepsilon_2 \omega}{4\pi} = \frac{2W_{vc} \cdot \hbar\omega}{E_o^2}$$

In the one-electron dipole approximation this yields the imaginary part of the dielectric constant (related to real interband transitions):

$$\varepsilon_2(\omega) = \frac{4\pi^2 e^2}{m^2 \omega^2} \sum_{v,c} |P_{vc}|^2 \cdot \delta(E_c - E_v - \hbar\omega) \qquad (8)$$

where the summation is performed over all valence (initial) and conduction (final) states in the unit volume.

The δ-function expresses the selection rule based on <u>energy conservation</u>. This means that only transitions between initial and final states for which the energy difference is equal to the photon energy $\hbar\omega$ are possible.

P_{vc} is the component along the direction of the electric field of the electron momentum matrix element between the valence and the conduction states Ψ_v and Ψ_c.

$$P_{vc} = i\hbar \frac{1}{\Omega} \int_{cell} \Psi_c^* \nabla \Psi_v \, d^3r$$

where the integration is performed over the unit cell of volume Ω. Because of the translational symmetry of the lattice, Ψ_v and Ψ_c are Bloch functions of the form $\exp(-i\vec{k}.\vec{r}).u(\vec{r})$, where u has the periodicity of the lattice. The matrix element will vanish unless $\vec{k}_c - \vec{q} = \vec{k}_v$. Since the photon wave-vector $\vec{q}$ is small compared to the Brillouin zone dimensions, it can be neglected. This yields a selection rule based on <u>momentum conservation</u>, i.e., only vertical, or <u>direct</u> transitions between states with the same wave vector are allowed.

The real part of the dielectric constant ε_1 can be deduced from ε_2 by the dispersion relation, it is related to virtual direct transitions. ε_2 can also be written as

$$\varepsilon_2 = \frac{4\pi^2 e^2}{m^2 \omega^2} \frac{2}{(2\pi)^3} \int_{BZ} d^3\vec{k} \cdot |P_{vc}|^2 \cdot \delta(E_c - E_v - \hbar\omega) \quad (9)$$

which gives, when integrated by parts:

$$\varepsilon_2 = \frac{e^2}{\pi m^2 \omega^2} \int_S \frac{|P_{vc}|^2 \cdot dS}{|\nabla_{\vec{k}}(E_c - E_v)|_{E_c - E_v = \hbar\omega}} \quad ,$$

the integration being performed over the equal energy-difference surface in $\vec{k}$-space defined by $E_c - E_v = \hbar\omega$.

In most cases, the matrix elements can be considered as slowly varying functions of $\vec{k}$, so that ε_2 is mainly determined by the joint density of states:

$$J_{vc}(\omega) = \int_S \frac{dS}{|\nabla_{\vec{k}}(E_c - E_v)|_{E_c - E_v = \hbar\omega}} \quad (10)$$

which measures the product density of full and empty states of equal energy difference. This quantity presents singularities for those frequencies at which:

$$\nabla_{\vec{k}}(E_c(\vec{k}) - E_v(\vec{k})) = 0$$

for some value of $\vec{k}$. Such critical points[7] may occur, either at high symmetry points of the Brillouin zone (where $\nabla_{\vec{k}}E_c = \nabla_{\vec{k}}E_v = 0$) or at any place where the valence and conduction bands are parallel.

Therefore in crystals, ε_2 will essentially be determined by the product of the joint density of states, which reflects the geometrical properties of the energy band structure in the Brillouin zone and combines the valence and the conduction band densities of states by matrix elements which are related to the wave-functions of both initial and final states. Therefore, ε_2 is not expected to yield directly an image of the density of states in

the valence or the conduction bands, unless the initial
or the final states are atomic like. This is the case
for absorption or emission in the X-ray range, which involves core levels and can give, respectively, a good picture of the conduction or the valence band density of
states.

When lattice perturbations like phonons are involved
in the interaction between the system and the electromagnetic field, non vertical or <u>indirect</u> transitions can
occur, provided the change in wave-vector is given by a
phonon participating in the process. These indirect allowed transtions, which come about only by second order
perturbation and are generally weaker than direct ones,
introduce for ε_2 a dependence on the square of energy,
e.g., in the vicinity of an indirect gap E_o, one has[8]

$$\varepsilon_2 \omega^2 \simeq \text{const.} \left(\hbar\omega - E_o \right)^2 \tag{11}$$

We shall just mention here another effect which may
be important in crystalline semiconductors: Creation of
an <u>exciton</u>. This effect results from electron-hole interaction and lies outside the one-electron approximation
used up to now. It usually gives rise to series of absorption lines in the vicinity of the absorption edges.

Amorphous case

In the absence of long-range order, i.e., without the
translational symmetry of the lattice, $\vec{k}$ is not a good
quantum number any more; one cannot define an energy band
structure $E(\vec{k})$ as in the crystalline case and the momentum conservation selection rule has to be relaxed. However, the valence and conduction bands retain their meaning in the amorphous case, and one can still speak of densities of states. Various theoretical treatments, based
on different approaches, have demonstrated the existence
of a gap, i.e., of a region of zero or small density of
states, separating a valence band from a conduction band.
The theoretical calculations of these densities of states,
as well as their determination by photoemission experiments, will be treated in other lectures. We shall concentrate here on $\varepsilon_2(\omega)$.

The expression (8) for $\varepsilon_2(\omega)$ still holds in the
amorphous case and the energy conservation selection rule
is still valid. However, the wave-functions of the initial and the final states entering the momentum matrix

elements are not Bloch functions anymore; they can be described as linear combinations of crystalline wavefunctions[9] as

$$|vK\rangle_a = \Sigma\Sigma_{\ell k} |\ell k\rangle\langle\ell k|vK\rangle_a$$

$$|cK'\rangle_a = \Sigma \Sigma_{\ell'k'} |\ell'k'\rangle\langle\ell'k'|cK'\rangle_a$$

K and K' are not quantum numbers but are introduced in order to count the states in the amorphous case (it is assumed that the number of states in the valence and conduction bands is the same in the amorphous and the crystalline cases, and that the number of unit cells per unit volume N is also the same). The summation in ℓ and ℓ' is made over all bands in the crystal, the summation in k and k' over the crystal Brillouin zone. Then $\varepsilon_2(\omega)$ can be written as

$$\varepsilon_2(\omega) = \left(\frac{2\pi e}{m\omega}\right)^2 \Sigma_{KK'\ell\ell'k} \left| \Sigma \Sigma P_{\ell\ell'}(\vec{k})_a \langle cK'|\ell'k\rangle\langle\ell k|vK\rangle_a \right|^2$$

$$\delta(E_c(K') - E_v(K) - \hbar\omega)$$

where $P_{\ell\ell'}(\vec{k})$ is the crystalline matrix element. With the usual assumption on $P_{\ell\ell'}$ and assuming that the perturbation leading to the amorphous case do not mix the wave functions from different bands

$$\varepsilon_2(\omega) = \left(\frac{2\pi e}{m\omega}\right)^2 \cdot \frac{2}{(2\pi)^3} |P_{vc}|^2 \frac{1}{B} \int d^3K \cdot d^3K' \cdot f(K,K') \cdot$$

$$\delta(E_c(K') - E_c(K) - \hbar\omega)$$

where B is the volume of the Brillouin zone, proportional to 1/N, and $f(K,K')$ is given by:

$$f(K,K') = N \cdot \Sigma_k \left|_a\langle cK'|ck\rangle\langle vk|vK\rangle_a\right|^2$$

In the crystal, $f(K,K') = N \cdot \delta(K;K')$, which immediately gives equation (9).

If the amorphous wave functions $|\ell K\rangle_a$ are linear combinations of crystalline $|\ell k>$ with k in the vicinity of K, then the δ-function can be replaced by a function of finite width: $f(K,K')$ can be replaced by a constant f_0 over B_0 of the Brillouin zone around $K = K'$. From the sum rule one has

$$\int_B d^3k \cdot d^3k' \cdot f(K,K') = B^2 \quad , \quad f_0 = B/B_0$$

then

$$\varepsilon_2(\omega) = (\frac{2\pi e}{m\omega})^2 \frac{2}{(2\pi)^3} \frac{1}{B_0} \cdot |P_{vc}|^2 \int d^3K \cdot d^3K'$$

$$\delta(E_c(K') - E_v(K) - \hbar\omega)$$

$$\varepsilon_2(\omega) = (\frac{2\pi e}{m\omega})^2 \frac{2}{(2\pi)^3} \frac{1}{B_0} \cdot |P_{vc}|^2 \int dE \int d^3K'$$

$$\qquad\qquad\qquad\qquad\qquad\qquad\qquad\qquad\qquad (12)$$

$$\delta(E_c(K')-E) \int d^3K \; \delta(E - E_v(K) - \hbar\omega)$$

$$\varepsilon_2(\omega) = (\frac{2\pi e}{m\omega})^2 \frac{2}{(2\pi)^3} \frac{1}{B_0} \cdot |P_{vc}|^2 \int dE \cdot g_v(E) \cdot g_c(\hbar\omega-E)$$

Therefore, $\varepsilon_2(\omega)$ is determined by the convolution of the density of states in the valence and conduction bands for which energy is conserved; this expression is what one intuitively expects for transitions in which only energy conservation is required. Of course, the matrix elements also play their role in $\varepsilon_2(\omega)$; we shall come back to that point later on.

If for example the density of states at the bottom of the conduction band is represented by $g_c \; \alpha \; (E-E_c)^s$ and at the top of the valence band by $g_v \; \alpha \; (E_v-E)^p$, then[10]

$$\varepsilon_2(\omega) \; \alpha \; [\; \int_{E_c-\hbar\omega}^{E_v} (E_v-E)^p \cdot (E-E_c + \hbar\omega)^s \cdot dE \; ;$$

setting $\quad y = \dfrac{E_c-E-\hbar\omega}{E_c-E_v-\hbar\omega} \quad$ and $E_c-E_v = E_0 \quad$ one has

$$\varepsilon_2(\omega) \; \alpha \; [\; \int_0^1 (1-y)^p \; y^s \; dy] \cdot (\hbar\omega-E_0)^{p+s+1}$$

The first term is a known integral $\dfrac{\Gamma_{(s+1)}\,\Gamma_{(p+1)}}{\Gamma_{(s+p+2)}}$.

If the band edges are parabolic, i.e., s = p = 1/2, then $\varepsilon_2\ \alpha(\hbar\omega-E_o)^2$, a behavior similar to indirect transitions in crystals. For the amorphous case as $\vec{k}$ is not well defined, it is termed a non-direct transition.

If the transitions involve states in band tails which are assumed to be linear in energy, i.e., if s=p=1, then $\varepsilon_2\ \alpha(\hbar\omega-E_o)^3$, which has indeed been observed in some cases. However, the case of optical transitions between localized states must be treated differently and the matrix elements are expected to be small[10,11].

There is no reason why excitonic states could not exist in disordered materials. Such states are reasonably well defined if the time needed for the completion of one orbit is shorter than the shorter of the lifetimes of the electron or the hole. Therefore, small radius excitons are more likely to be observed. Their effects, however, can be drastically modified by disorder.

2. Determination of Optical Constants

In all cases, one has to determine both the real and imaginary parts of the complex refractive index $\tilde{n}$ (or of the complex dielectric constant $\tilde{\varepsilon}$), i.e., two independent quantities. This requires at least two measurements; however, we shall see that the dispersion relations allow us to use only reflectance measurements. Optical measurements can be photometric measurements (reflected or transmitted intensity, at normal incidence), polarimetric measurements (the same with polarized light at oblique incidence), ellipsometric measurements (in which the phase changes in reflection are also considered). Taking into account the polarization of light, all these quantities can be expressed in terms of $\tilde{n}$, λ, the film thickness d and the angle of incidence ϕ_o. It may be recalled that the materials under consideration can only be obtained as films, with thicknesses ranging, according to the preparation method, between 100 A° and a few tens of microns. At low energies, the films are almost transparent and thick films are preferred in order to increase the accuracy; at high energies, the films are strongly absorbing and only reflection measurements can be performed, unless the films are thin enough to be semi-transparent (typically d <1000Å if $\alpha \simeq 10^6 cm^{-1}$).

2.1 Formulas

Films are considered as homogeneous, isotropic,

plane-parallel layers of thickness d comparable with the wavelength λ. The amplitude and the state of polarization of the complex, reflected or transmitted amplitude (for example $r = \sqrt{R} \cdot e^{i\delta R}$) can be computed either by summing the amplitudes of multiply reflected beams within the film, or by solving an electromagnetic boundary value problem; they can be expressed in terms of the Fresnel coefficients at the interface[12,13,14].

If d is of the order of λ, the path difference introduced by going through the film between successive reflections is small compared to the coherence length of monochromatic sources, and the total lateral displacement small compared to the beam breadth, the multiply reflected and transmitted beams combine coherently and can be summed up to infinity (the lateral dimensions can be taken as infinite except if total reflection or light amplification occurs). Then, if $\beta = \dfrac{2\pi\tilde{n}d}{\lambda}$, at normal incidence:

$$r = \frac{\tilde{n}(n_o - n_s)\cos\beta + i(n_o n_s - \tilde{n}^2)\sin\beta}{\tilde{n}(n_o + n_s)\cos\beta + i(n_o n_s + \tilde{n}^2)\sin\beta}$$

$$t = \frac{2n_o\tilde{n}}{\tilde{n}(n_o + n_s)\cos\beta + i(n_o n_s + \tilde{n}^2)\sin\beta}$$

n_o and n_s being the refractive indices of the first (air) and the third (glass) media.

The expressions for the reflected and transmitted intensities R and T are obtained by:

$$R = |r|^2 \quad \text{and} \quad T = \frac{n_s}{n_o}|t|^2$$

At oblique incidence ϕ_o, one has to differentiate between light polarized perpendicular (s) or parallel (p) to the plane of incidence. The expressions for r and t retain their validity, provided $\tilde{n}$ is replaced by $\tilde{n}\cos\phi$ and $\tilde{n}/\cos\phi$ for s and p polarizations respectively, and β by $\dfrac{2\pi d}{\lambda}\tilde{n}\cos\phi$ [ϕ is related to ϕ_o by $n_o\sin\phi_o = \tilde{n}\sin\phi = n_s\sin\phi_s$).

If the film is thick enough so that coherent multiple reflections within the film can be neglected (this is equivalent to neglecting $\exp(-\alpha d)$ with respect to 1), then:

$$T = \frac{16 n_o n_s (n^2 + k^2) \cdot \exp(-\frac{4\pi kd}{\lambda})}{[(n_o + n)^2 + k^2][(n_s + n)^2 + k^2]}$$

lnT is a linear function of d; the slope of this line gives k, its intercept with the axis, n.

For non-absorbing films, if the dispersion of the index of refraction can be neglected, R exhibits an oscillatory variation with d/λ; it reaches extreme values equal to

$$R_M = (\frac{n_o n_s - n^2}{n_o n_s + n^2})^2$$

for $\beta = (m + 1/2)\pi$, i.e., $d = \lambda_M(1+2m)/4n$. R_M is a maximum or a minimum depending on whether $n > n_s$ or $n < n_s$. It is only necessary that the optical thickness of the films is at least equal to $\lambda_M/4$, i.e., not too small. For $n^2 > n_o n_s$, which is practically always the case:

$$n^2 = n_o n_s \cdot \frac{1 + \sqrt{R_M}}{1 - \sqrt{R_M}}$$

more simply:

$$n = \frac{1+2m}{4} \cdot \frac{\lambda_M}{d}$$

In all cases, multiple incoherent reflections within the transparent substrate of finite thickness have to be taken into account[15]:

$$R_{meas} = R + \frac{T^2 R_o}{1 - R' R_o} \qquad T_{meas} = \frac{T(1-R_o)}{1 - R' R_o}$$

where R and T, as given earlier, represent the reflection and transmission coefficients of the air-film-substrate system, R' represents the reflection coefficient of the same system from the substrate side, and $R_o = (\frac{n_s - 1}{n_s + 1})^2$

represents the reflection coefficient of the air-substrate interface.

2.2 Methods

Absorbing films

These methods are the only ones which can be used
for opaque films, for which only measurements in reflection can be made; they can however be applied without
difficulty to semi-transparent films.

   a. reflected intensity measurements at normal
      incidence.

Only R is measured in this method, and an artifice
must be used. Since the complex reflected amplitude
$r = \sqrt{R} \cdot e^{i\delta_R}$ describes a linear relation between reflected and incident light, there is a dispersion rela-
tion similar to relation (3) connecting the real and imaginary parts of $\ln r = 1/2 \ln R + i\delta_R$:

$$\delta_R(\omega) = -\frac{1}{\pi} \int_0^\infty \frac{\omega \cdot \ln R(\omega')}{\omega'^2 - \omega^2} \cdot d\omega'$$

The phase shift δ_R at reflection can be calculated at
each ω if $R(\omega)$ is known over the whole spectral range.
The optical constants can then be determined from $R(\omega)$
and $\delta_R(\omega)$ by using the expressions (for an opaque film):

$$\begin{cases} R = \dfrac{(n-1)^2 + k^2}{(n+1)^2 + k^2} \\[4mm] \tan \delta_R = \dfrac{2k}{n^2 + k^2 - 1} \end{cases}$$

Generally, an expression similar to relation (4) is
used for the computation of δ_R

$$\delta_R(\omega) = \frac{1}{2\pi} \int_0^\infty \ln\left|\frac{\omega'-\omega}{\omega'+\omega}\right| \cdot \frac{d}{d\omega'} \left[\ln R(\omega')\right] \cdot d\omega'$$

This expression shows that $\delta_R(\omega)$ is principally determined by the values of the derivative of $\ln R$ near ω,
because of the logarithmic factor; however, regions where
this derivative is large may also influence the value of
δ_R.

Since $R(\omega)$ is known over a finite spectral range
only, the experimental curve must be extrapolated towards 0 and ∞. The extrapolation to ∞ is the more deli-
cate one[16], since only very few results are available a-
bove 10 eV; a common extrapolation, based on Drude be-
havior, is $R \propto \omega^{-4}$, or more generally $R \propto \omega^{-s}$, s being a
parameter which must be fitted at some ω where n and k
are known by another method. These extrapolations may
influence the results differently depending on the location of the spectral range of measurements with respect
to the main features of the optical spectrum. Usually,
this Kramers-Kronig method leads to correct positions
for the ε_2 peaks, but wrong intensities. However, it
may be the cause of non-negligible errors in the case of
amorphous semi-conductors, as illustrated[17] by Figure 1,
which may explain certain inconsistencies between different data. In conclusion, this widely used method is extremely useful for getting the overall shape of the optical spectrum, but not sufficient for the details and the
absolute values of $\varepsilon_2(\omega)$.

b. reflected intensity measurements at oblique incidence

A linearly polarized light is elliptically polarized
after reflection on an absorbing surface; both the amplitudes and the phases of the components perpendicular and
parallel to the plane of incidence are different. At
oblique incidence, using polarized light, various measurements are possible: R_s and R_p at one angle of incidence;
R_s or (and) R_p at several angles of incidence; $\sqrt{R_p/R_s}$ at
two angles of incidence, etc. One has to solve two or
more equations with two unknowns: n, k (three unknowns n,
k, d if the film is semi-transparent and three or more
quantities are measured); this is usually done on a computer by an iterative procedure. These methods, which
have the advantage to yield n and k directly at each ω,
are however difficult to perform; good polarizers are
difficult to find outside the visible range (but synchrotron radiation for example is naturally polarized) and
the results are strongly influenced by any surface defects
e.g., lack of flatness, roughness, contamination, etc.

c. ellipsometry

Instead of measuring only intensities, one measures
two quantities Ψ and Δ which are respectively related to
the ratio of the reflected intensities and the difference
in phase shifts at reflection corresponding to the

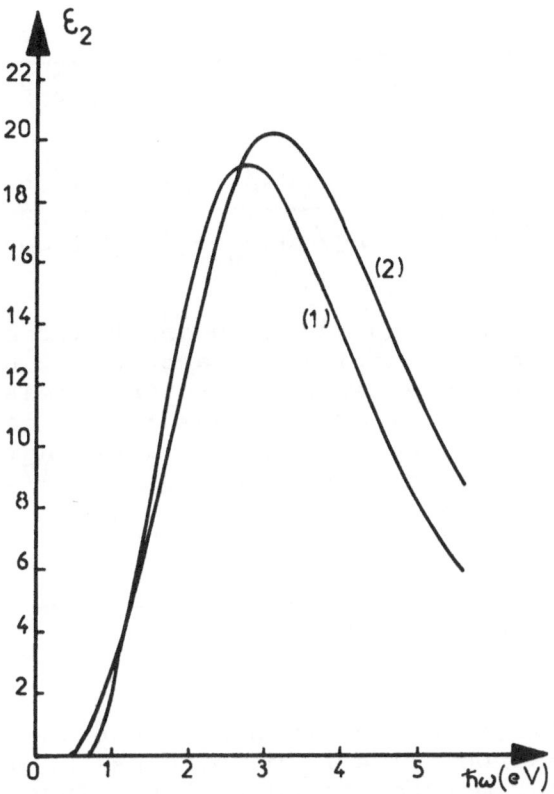

Fig. 1   Imaginary part of the dielectric constant ε_2 versus energy for a-Ge, calculated by Kramers-Kronig analysis from the combined reflectivity data of Bauer and Galeener (ref. 29) and Donovan et al. (ref. 28)- curve 1 or Theye (ref. 56)- curve 2. [Figure reproduced from reference 17].

components of the vibration parallel and perpendicular to the plane of incidence[18]

$$\tan \Psi = \sqrt{\frac{R_p}{R_s}} \qquad \Delta = \delta_{R_p} - \delta_{R_s}$$

These quantities are of course functions of ϕ_o, n and k (and d if the film is semi-transparent); for an opaque film for example

$$r_s = \sqrt{R_s} \quad e^{i\delta_{Rs}} = \frac{n_o \cos\phi_o - (n-ik)\cos\phi}{n_o \cos\phi_o + (n-ik)\cos\phi}$$

$$r_p = \sqrt{R_p} \quad e^{i\delta_{Rp}} = \frac{n_o/\cos\phi_o - (n-ik)/\cos\phi}{n_o/\cos\phi_o + (n-ik)/\cos\phi}$$

This method, which also yields n and k directly at each ω, proves to be very useful and accurate. However, the results are extremely sensitive to the presence of thin surface layers (for example oxides); they can be corrected for these effects when the layer characteristics are known, by using double layer formulas[17].

Slightly absorbing films

The films are then semi-transparent and suitable for transmission measurements.

a. (R,T)method

Both reflectance and transmittance at normal incidence are measured. The film thickness is measured independently, by methods based on X-ray interference in reflection at grazing incidence for small thicknesses (typically for Ge on glass: 100-1500 A°)[19] or on multiple beam optical interference for larger thicknesses[20]. n and k can then be computed at each ω by an iterative procedure, using the expressions for R and T given earlier, and correcting for reflections within the substrate as explained before. The accuracy is very good for large α values, except for some groups of values of n, k and d/λ for which the $R(n,k) = R_{ex}$ and $T(n,k) = T_{ex}$ curves in the (n,k) plane are almost tangent to each other; any error in R_{ex} or T_{ex}, even as small as 0.1°/o, may introduce large uncertainties in n and k, and even may cause that the system

$$\begin{cases} R_{ex} - R(n,k) = 0 \\ T_{ex} - T(n,k) = 0 \end{cases}$$

has no solution at all[21].

As the films must be thin enough for applying this method (typically ~1000 A°), the results are not very accurate for small α values, i.e., in the region of the absorption edge; typically for α values smaller than 10^4 cm^{-1}. High accuracy must be achieved in R and T, which means very uniform thickness, no substrate absorption, exact correction for reflections within the substrate etc. Even in this case, α is obtained with large uncertainties, as illustrated[22] by Figure 2; these curves correspond to a-Ge films of thicknesses 600-700 A^o prepared in different conditions, with estimated errors $\Delta R = \pm 0.001$ and $\Delta T = \pm 0.002$ (the sum (R+T) was never greater than 1.003). One can see that the exact shape and location of the edges are difficult to ascertain from these results and that the (R,T) method cannot differentiate, in most cases, between so-called "sharp" and "gradual" edges[23,24]. Measurements on thick films should be preferred.

b. transmittance measurements on thick films

When much thicker films (typically a few tens of microns) are available (prepared for example by cathode sputtering[17]), then the expression of T, valid in the limit of incoherent multiple reflections within the film, can be used in order to determine α directly; the results must be corrected for reflections inside the substrate[17]. One can use, either two samples of different thicknesses as explained earlier, provided they are prepared simultaneously and have the same structure and properties, or only one sample; n is then determined by another method, for example from the extrema of R for zero absorption (when k is negligible with respect to n), in which case the accuracy in n is only determined by the accuracy in d, and not in R. Moreover, these results are not affected by the presence of superficial layers. Such methods give good results for thick films provided these films are sufficiently well defined, homogeneous, without macroscopic defects and large stresses.

c. attenuated total reflection[14]

If a non-absorbing film is deposited on the base of a total reflection prism, then the reflection coefficient is equal to 1, whatever the values of d and the polarization. If the film is slightly absorbing $(k \neq 0)$, then attenuated total reflection takes place; R_s and R_p are different from 1 and present an oscillatory variation versus d (or λ) (Figure 3). Their expressions are computed without difficulty and allow the calculation of n and k directly. This is a rather difficult, but very sensitive method, and very small values of k can be determined with high accuracy.

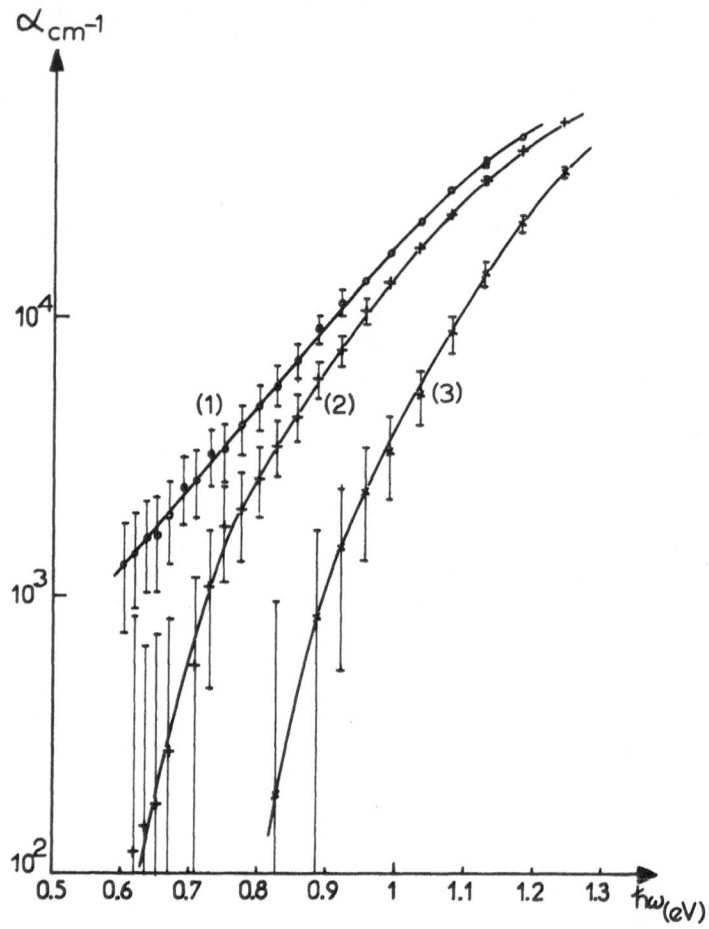

Fig. 2   Absorption coefficient α versus energy for evaporated a-Ge films (600-700 Å thick) deposited at high
rate at room temperature: as-deposited (1) and annealed
in the same vacuum up to 380°C (3); deposited at low
rate at room temperature (2) (α is computed by the (R,T)
method and the error bars correspond to ΔR = ± 0.001 and
ΔT = ± 0.002).

3. Absorption Spectrum

Different theoretical methods allowing to calculate the density of states of disordered materials will
be presented in other lectures. It will also be shown
how information about the role of local disorder in the

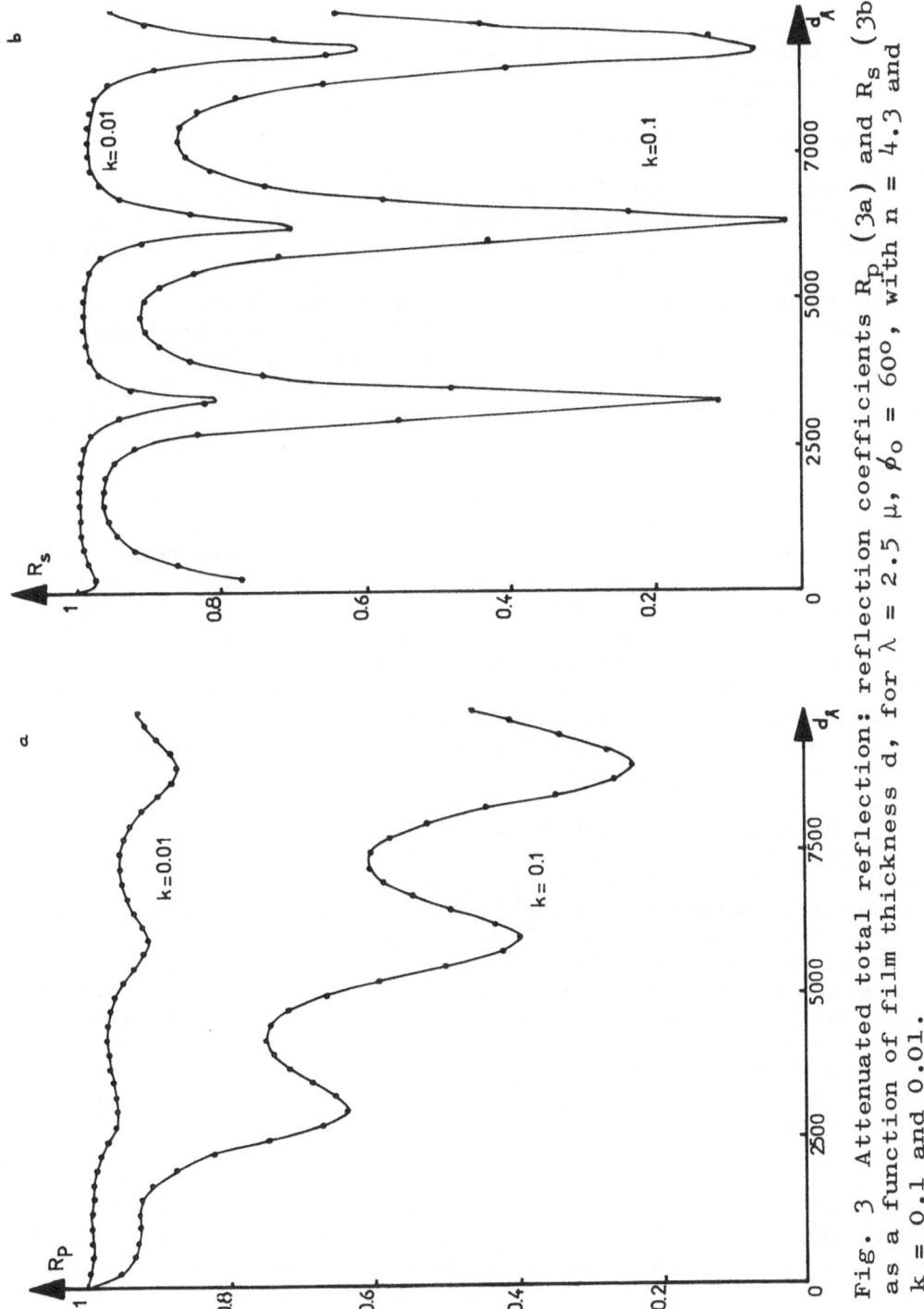

Fig. 3 Attenuated total reflection: reflection coefficients R_p (3a) and R_s (3b) as a function of film thickness d, for λ = 2.5 μ, ϕ_o = 60°, with n = 4.3 and k = 0.1 and 0.01.

properties of amorphous Ge or Si can be gained from investigations on their crystalline polytypes, for which
the atomic arrangement in the primitive cell becomes
more and more disordered and which may present (especially for ST 12) features similar to some characteristics
of the amorphous network.

Here, we shall essentially give a description of
the optical experimental results in a few representative
cases, compare these data to those for the corresponding
crystalline (diamond) phase, and discuss them in relation to the predictions of different theoretical models.
As emphasized earlier, we shall focuss our attention on
Ge, Si and the III-V compounds, which represent simple
cases particularly suited to a confrontation between experiment and theory, mainly because of the nature of
their bonding. It must however be emphasized again that,
for essentially the same reason, this apparent simplicity
is counterbalanced by the fact that some of their properties are strongly structure dependent, because of the
non-homogeneity of the as-deposited films and the presence of defects with varying nature and proportion with
the preparation conditions. This point is now widely
acknowledged for amorphous Ge and Si. Systematic work
on the a-III-V compounds has still to be done, in order
to test whether the partial ionicity of the bonds, which
makes them less directional, modifies this situation.
Consequently, the optical data will also be discussed
in relation to structural effects.

We have seen that optical absorption is determined
by the product of the convolution of the densities of
initial (valence) and final (conduction) states by matrix
elements; effects related to these two quantities will
have to be considered. Contrary to photoemission data,
optical absorption does not give a separate image of the
density of states in the valence band or in the conduction band, except if either the initial, or the final
state has a simple density. This is the case with measurements in the soft X-ray range. UPS and XPS experiments are presently developed and yield important information on the valence band density of states[25,26].
Very few X-ray absorption and emission measurements have
been done up to now. High resolution X-ray absorption
experiments near the L_{III} edge on amorphous Ge[27] show
that the threshold occurs exactly at the same energy as
in the crystal but that the peaks and the fine structure
characteristic of the crystalline phase are not seen any
more. This suggests that the conduction band density of

states is almost featureless.

3.1 Description of the absorption spectra.

We shall consider here the ε_2 spectra in a wide
spectral range, from the gap value (say 0.5-1.5 eV) to
the near ultra-violet (6-10 eV; as a matter of fact,
very few data are available between 6 and 10 eV). These
spectra present for all amorphous tetrahedrally bonded
semiconductors a typical shape with a single asymmetric
broad maximum, remarkably similar in all cases except
for variations in location, height and width, as illustrated by Figure 4[28,33]. When compared to the corresponding crystalline spectra, the amorphous spectra appear to be more or less shifted as a whole towards smaller energies; in particular, the maximum does not coincide with the main crystalline peak. This is illustrated by Figures 5a and 5b in two cases where the shift is
more (GaP) or less (GaAs) noticed[33]. As we shall see
later, these curves may present small variations with
the preparation conditions of the samples, therefore with
their structure.

3.2 Interpretation.

Crystalline spectrum

Let us first recall the relations between the crystalline ε_2 and the crystalline band structure. All semiconductors with diamond and zinc-blende structures have
qualitatively the same band structure (especially for
the valence band), and therefore the same optical properties. However, small variations near the fundamental edge
may be expected since, if the maximum of the valence band
is always at Γ, the minimum of the conduction band may
be either at Γ or at X. A now classical analysis[4,8] gives
the following assignments to the various features of the
absorption spectrum. For Ge, for example, (see Fig. 6)
the characteristic $(E-E_0)^{1/2}$ behavior observed at the
fundamental edge is related to a M_0 critical point at Γ;
it is clearly observed in Ge because it happens to be the
lowest critical point (the indirect edge occuring slightly below E_0 is too weak for obscuring this effect). The
peak at about 2.3 eV (E_1) is assigned to a M_1 saddle
point in the <111> direction (Λ). The two highest valence bands and the lowest conduction band are indeed parallel over most of the <111> lines, while they strongly
repel each other along directions perpendicular to <111>.
This is equivalent to a two-dimensional critical point,
since one of the effective masses is much larger than

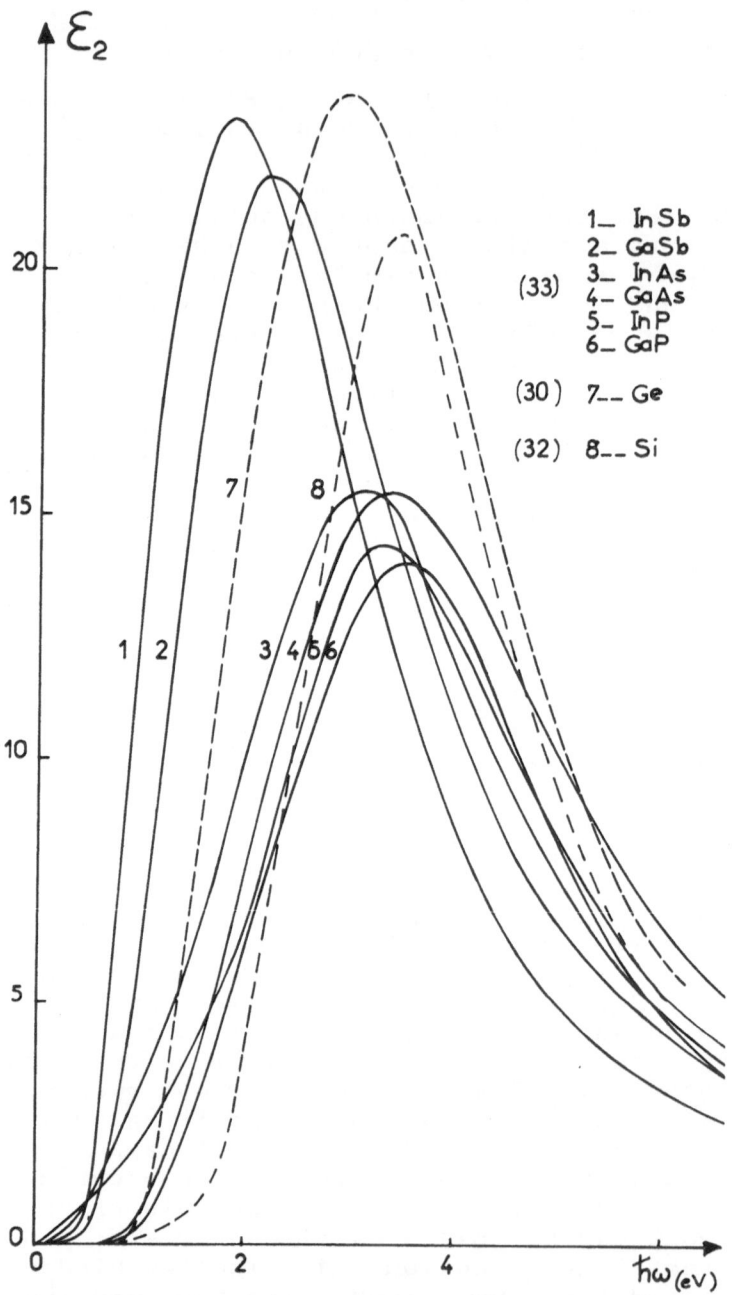

Fig. 4   Imaginary part of the dielectric constant ε_2
versus energy for a-Ge, a-Si and a-III-V compounds
(after Ref. 30, 32 and 33).

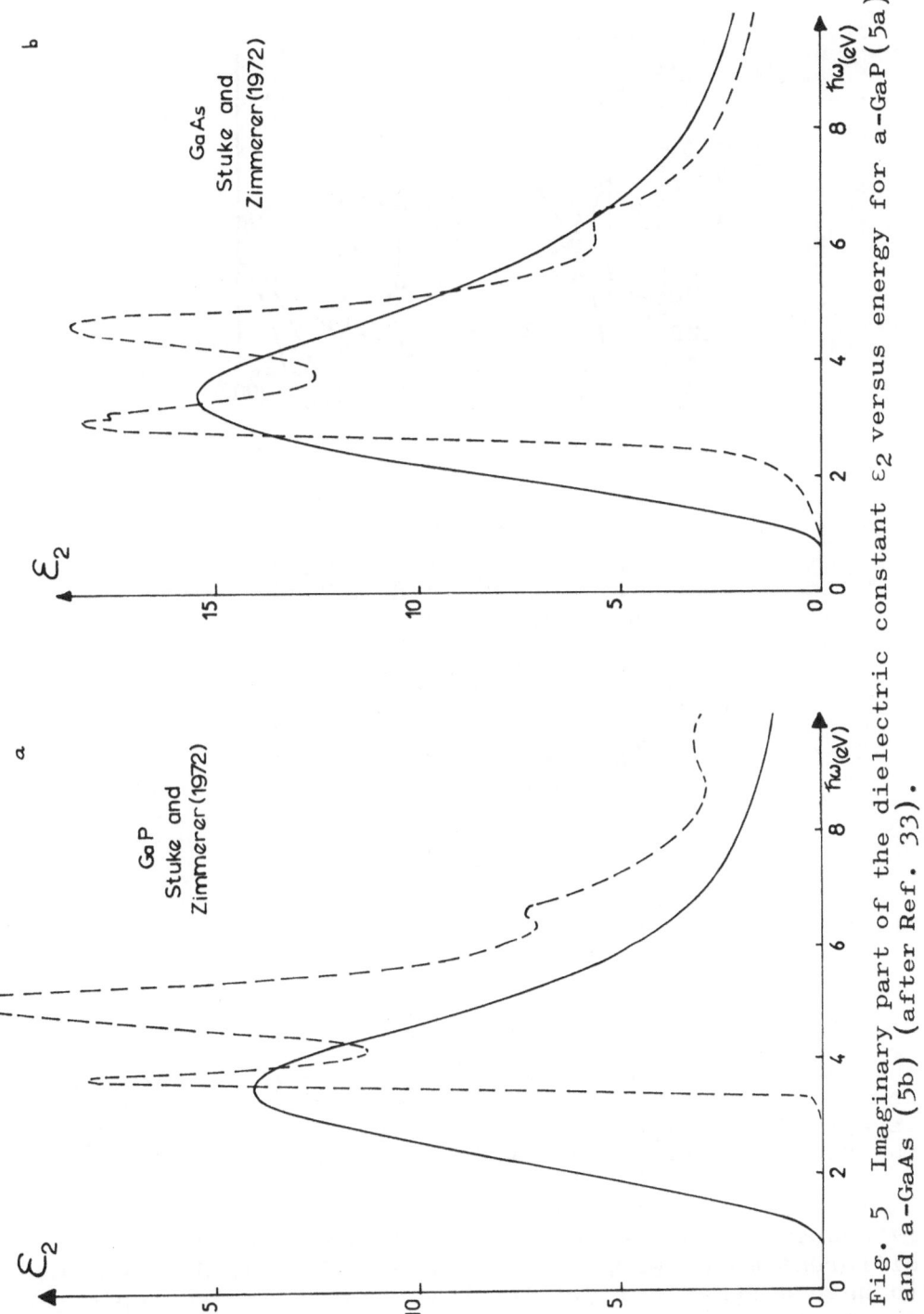

Fig. 5   Imaginary part of the dielectric constant ε_2 versus energy for a-GaP (5a) and a-GaAs (5b) (after Ref. 33).

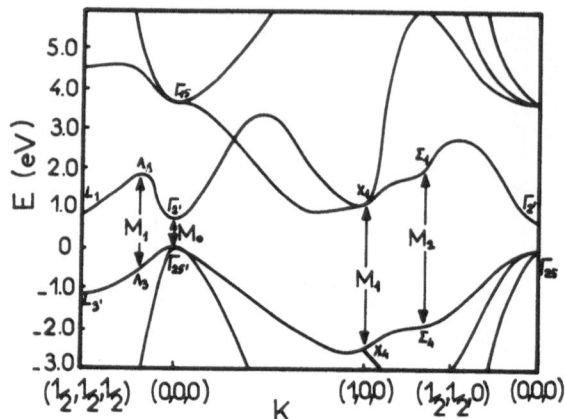

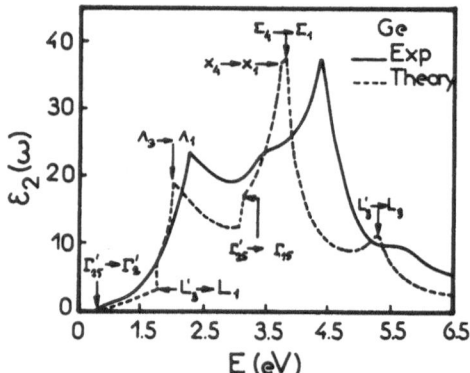

Fig. 6 Energy band structure and ε_2 spectrum for crystalline (diamond phase) Ge.

the other ones, and explains that the corresponding peak is quite intense. The highest peak at about 4.5 eV (E_2) is associated with two saddle points: M_1 at X and M_2 in the <110> direction (ε), which are almost degenerate.

This E_2 peak, in which most of the oscillator strength is concentrated, can also be accounted for in a quite different model proposed by Penn[5]. The solid is assumed to have a spherical Brillouin zone with an isotropic gap E_g at its boundary; the states are supposed to be free-

electron-like, except for the existence of this gap
which separates the valence from the conduction band.
Since the four valence electrons are in a single spin-
degenerate free-electron-like band, this Brillouin zone
must correspond to the Jones zone of the crystal. In
this model, there is a one-dimensional critical point
at the edge of the Brillouin zone. ε_2 is expected to
present a singularity in the form $(\hbar\omega - E_g)^{-1/2}$. The
static dielectric constant $\varepsilon(0)$ is related to the isotropic gap E_g and to the plasma frequency of the valence electrons ω_p by:

$$\varepsilon(0) = 1 + A \frac{(\hbar\omega_p)^2}{E_g^2} \tag{13}$$

The factor A accounts for matrix element effects; it can
be shown[4] that it is equal to $2/3$ if only transitions
at energies above E_g are considered; it is usually taken
as equal to 1, which must compensate for the fact that
real crystalline semiconductors have oscillator strength
below E_g.

The relationship between the Penn model and the actual energy bands has been discussed in detail[34]. The
real valence and conduction bands are indeed parallel
over large portions of the faces of the Jones zone; their
mean separation is roughly equal to the gap at the X
point of the crystal Brillouin zone. It has been shown
that:

$$E_g = 2 \left(V_{220} + \frac{V_{111}}{\Delta T}\right)^2$$

where V_{220} and V_{111} are the Fourier components of the
pseudopotential in the <220> and <111> directions. As
a matter of fact, E_g represents the average separation
of the valence and conduction bands, which depends essentially on the interaction between first neighbors[35]
and gives a measure of the strength of the bonds. The
value computed for Ge with the expression above, which
is in good agreement with the value deduced from $\varepsilon(0)$[4],
coincides with the position of the main peak (E_2) of the
absorption spectrum.

Amorphous spectrum

For interpreting the optical data relative to amorphous materials, two questions must be considered: (i)
why does the absorption spectrum present a single hump?

(ii) what determines its shape and its position in energy? Different theoretical approaches have been used, which must be considered as complementary rather than competing.

a. "Non-direct" transition model. It has already been seen that, in disordered systems, the selection rule based on $\vec{k}$-conservation which comes in the crystal from the presence of long-range order is relaxed and that ε_2 can be written [relation (12)] as

$$\varepsilon_2(\omega) \propto \frac{1}{\omega^2} \cdot |P_{vc}|^2 \cdot \int_0^{\hbar\omega} dE \cdot g_v(-E) \cdot g_c(\hbar\omega - E)$$

One has to take into account, not only the densities of states in the valence and conduction bands, which can be assumed to be the same as in the crystal except for some smearing out, but also the average transition probability $\overline{M}(\omega) = |P_{vc}|^2$. In crystalline materials, this quantity decreases smoothly with increasing frequency, except for sharp peaks associated with long-range order (umklapp-enhanced peaks); these peaks are expected to disappear due to disorder[36] but the overall frequency-dependence of $\overline{M}(\omega)$ is likely to be retained in the amorphous state. Indeed, the ε_2 curves computed in this model using the crystalline densities of states and constant matrix elements have a shape incompatible with the experimental data; on the contrary, using matrix elements which decrease with increasing frequency like the crystalline ones allows to enhance the low-energy part of the spectrum and leads to ε_2 curves with the correct shape[37]. The peak position is still slightly shifted towards higher energies with respect to the experimental one; this discrepancy would be removed if the exact amorphous densities of states[38] were used (Figure 7).

Unfortunately, it is not possible to really calculate the matrix elements for disordered systems. However, the similarity between the averaged matrix elements in the amorphous and the crystalline cases has been demonstrated by two different approaches[36,39]. Their conclusions are confirmed by the results obtained when computing the matrix elements for the crystalline polytypes of Ge and Si[40,41], especially for ST 12; it is emphasized that the averaged matrix elements must contain most of the information about ε_2, since the remaining part $J_{vc}(\omega)/\omega^2$ looks very much like a step function.

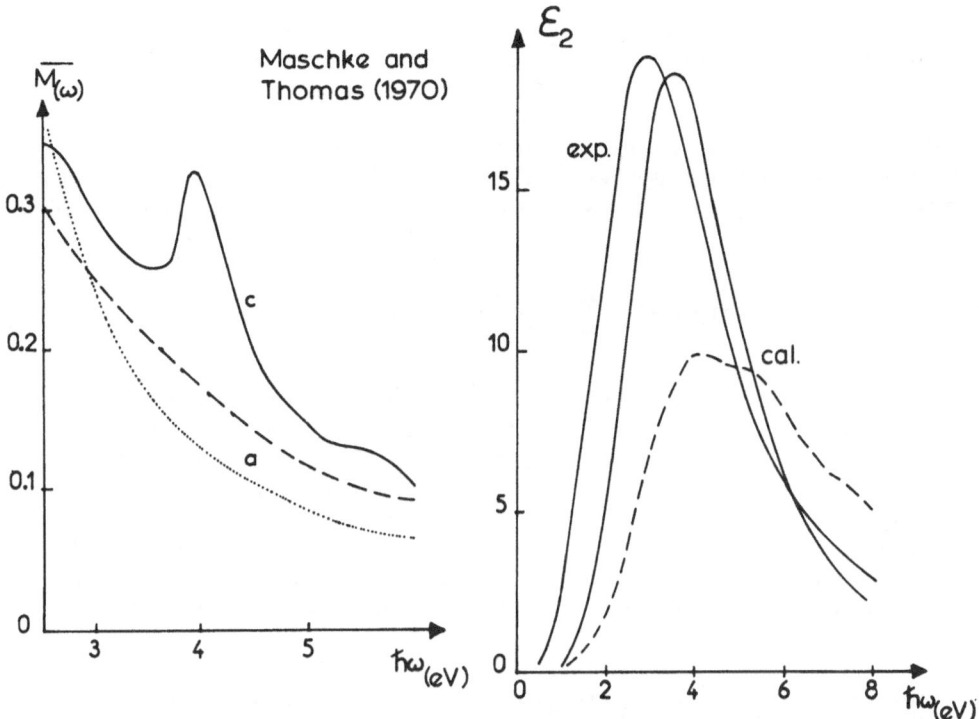

Fig. 7 (a) Average matrix element $\bar{M}(\omega)$ versus energy for Ge crystalline (C) and amorphous (a) (determined with the crystalline (...) and amorphous (---) densities of states). (b) ε_2 spectrum of a-Ge: experimental and calculated with constant (---) and energy dependent (———) matrix elements (from Ref. 36).

In conclusion, one can say that the remarkable narrowness of the ε_2 peak is essentially a matrix element effect, while its location is influenced by a density of states effect. The experimental data for a-Ge have been analyzed in this non-direct transition model under certain assumptions in order to yield the valence band density of states[39]. $\bar{M}(\omega)$ was taken to vary as ω^{-2} for $\hbar\omega > 3.5$ eV, and as ω for $\hbar\omega < 3.5$ eV (3.5 eV being the Penn gap value). It was assumed that the conduction band density of states is constant except in the lower edge region (about 0.25 eV wide at half-maximum); in view of the high energy data, this assumption is reasonable.

Then:

$$\omega^2 \cdot \varepsilon_2(\omega) = \text{const.} \ \bar{M}(\omega) \ \cdot \ g_c(E_a) \int_0^{\hbar\omega - E_a} g_v(-E) \cdot dE.$$

E_a being the separation between the top of the valence band and the energy at which the conduction band density becomes constant; E_a is expected to be close to the gap E_o.

$$g_v(E_o - \hbar\omega) = \text{const.} \ \frac{d}{d\omega} \ [\omega^2 \cdot \varepsilon_2(\omega)/\bar{M}(\omega)]$$

Figures 8a and 8b show that the results agree reasonably well with those of XPS measurements, which are believed to give a faithful picture of the valence band density of states[25]. There is only some discrepancy at the top of the band, where the assumptions involved in the analysis introduce uncertainties in the density of states computed from optical data. The same method has also been applied successfully to some of the a-III-V compounds[42].

b. Complex-band-structure calculations. The model used here implies partial $\vec{k}$-conservation but takes into account short-range order explicitly. As it is impossible to treat correctly multiple scattering at several atoms, which would be necessary because of atomic correlations due to short-range order, an approximate solution starting from the crystal is proposed. The method uses a Green's function technique and is based on a generalized pseudo-potential approach. This leads to a complex band structure, in which the disorder effects appear as imaginary values for the energy $E(\vec{k})$; the main effects are expected near critical points of the crystal[43]. A perturbation treatment proposed by other authors[44] would give essentially the same conclusions.

When looking at the results (Figure 9), it can be seen that the conduction bands, especially the highest ones, are more affected by disorder than the valence bands, and that the upper valence band itself is more affected near X than along the <111> directions. As a consequence, the valence band density of states is roughly preserved, while the conduction band density of states is completely smeared out. The fact that the lowest portion of the conduction band and the highest portion of the valence band, which both happen to be in the <111>

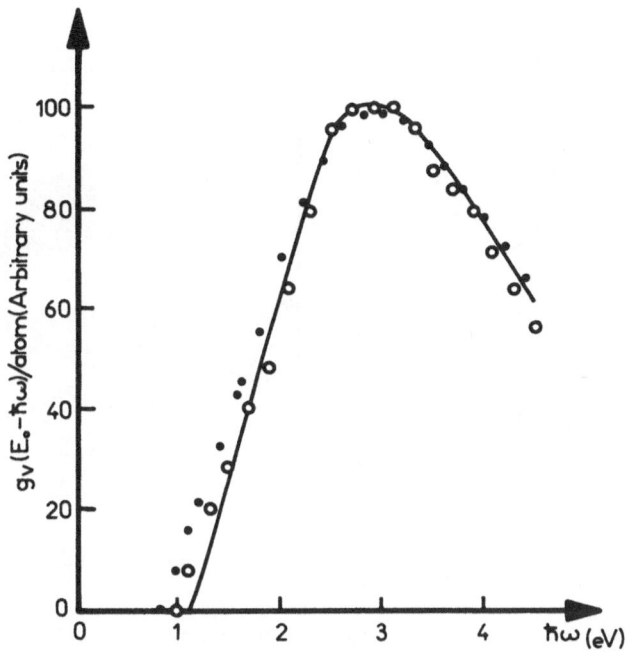

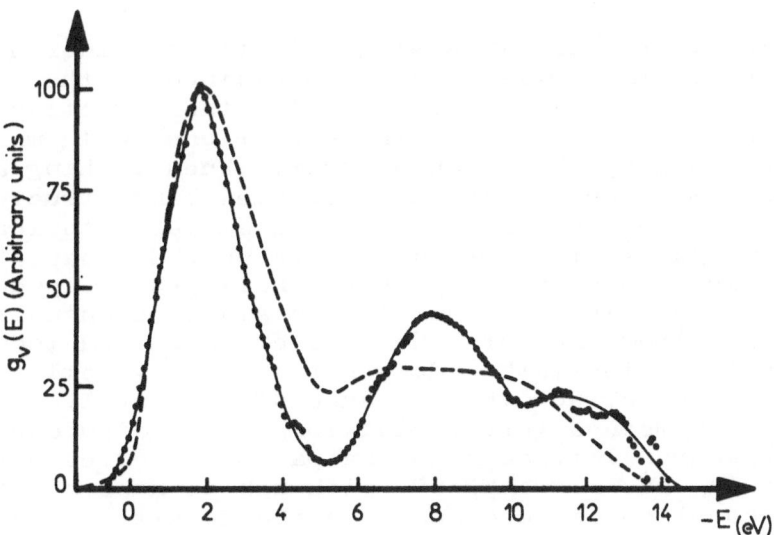

Fig. 8   Comparison between the experimental valence band
density of states of a-Ge (from XPS measurements of Ref.
25) and the one deduced from optical data as explained
in the text (low (.) and high (x) temperature films)
(from Ref. 39).

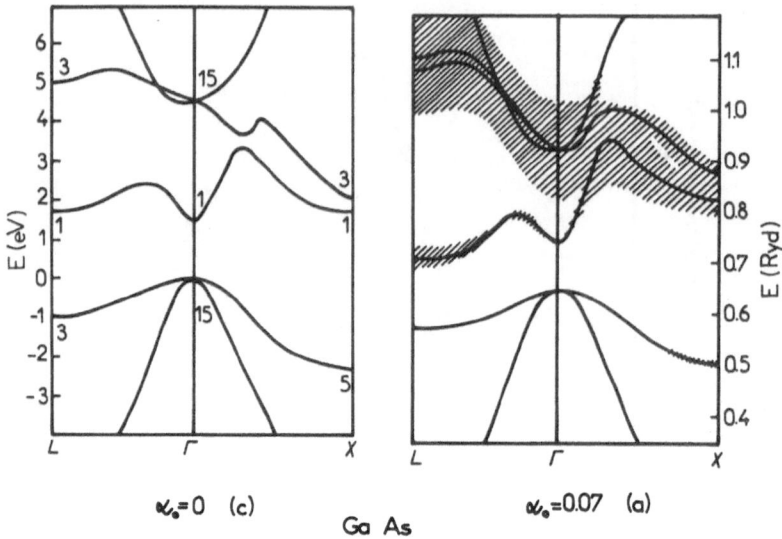

Fig. 9   Complex energy band structure for GaAs: crystalline $(\alpha = 0)$ and amorphous $(\alpha = 0.007)$(from Ref. 43).

directions, are more insensitive to the considered perturbation of the crystal can be explained by the coincidence of the <111> directions with the directions of
the bonds in the crystal. These portions would mainly
be determined by short range order. When looking at the
ε_2 spectrum computed from this complex band structure,
it appears that, as the disorder parameter entering the
model increases, the crystalline E_2 peak decreases much
more rapidly than the E_1 peak which is precisely due to
optical transitions in the <111> region. Eventually
the ε_2 spectrum presents the correct shape and location
(Figure 10). The method does not work very well for the
low energy side, however, because of the involved approximations; a method introducing long range disorder through an electron lattice coupling parameter gives better
results in this region[45,46], although it does not take
into account specific changes in the density of states[38].

This selective effect of disorder on the crystalline
ε_2 spectrum can therefore be interpreted as due to the
different origins of the E_1 and E_2 peaks; E_1, which is
associated with the tetrahedral bonding, is expected to
be preserved in all phases presenting the same short-

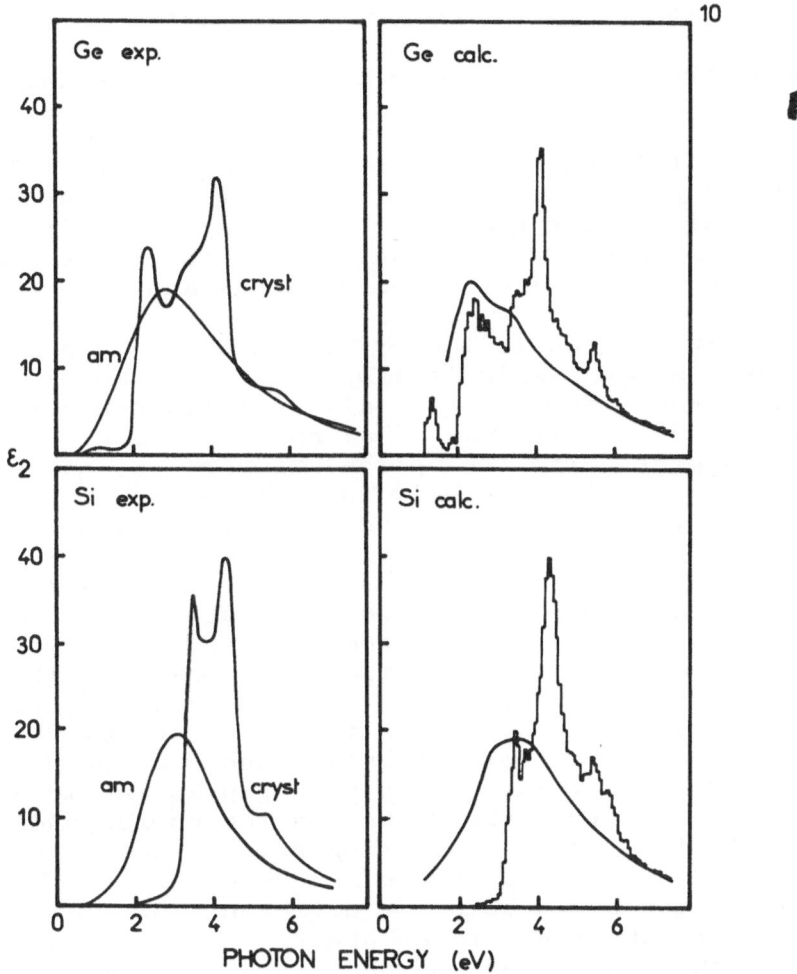

Fig. 10 Comparison between the ε_2 spectra of crystalline and amorphous Ge and Si: experimental and computed
from the energy band structure (from Ref. 43).

range order while E_2, which is more sensitive to the details of the structure and to long-range order, will be
more or less affected when the structure deviates from
the diamond structure. This is indeed confirmed when
computing ε_2 for the crystalline polytypes of Ge and Si;
for all polytypes the E_2 peak is selectively lowered
with respect to the E_1 peak as local disorder increases,
until for ST 12 the ε_2 spectrum presents a single peak
shape similar to the amorphous one[47,48,41]. In view of

the information provided by the approaches towards the
amorphous state via these polytypes, it can be asked
whether treatments based on short-range disorder models,
accounting for deviations in bond lengths and bond angles, presence of odd-membered rings etc., must not be
preferred to treatments starting from the crystalline
(diamond) case and introducing long-range disorder models. This is for example the approach used in cluster
models, in which multiple scattering theory is applied
to clusters of various configurations and sizes[49-51].

    c. Spectroscopic theory of bonding. This theory[52]
is based on the simple isotropic two-band model introduced by Penn[5] for the crystalline semiconductors (see
paragraph 3.1). It shows that many electronic properties, in particular the optical properties, can be explained by considering only the interaction of nearest
neighbors (and not the periodicity of the lattice).
Such an approach is expected to retain its validity in
the amorphous case since short-range order still exists,
and to be even more suited to this case because long-
range order effects do not come into play any more. Indeed, the amorphous ε_2 spectra look very much like a
broadened singularity of the type $(\hbar\omega-E_g)^{-1/2}$, with a
typical asymmetric shape, as first noticed by Phillips[53].
In the case of the tetrahedrally bonded amorphous semiconductors, the value of the Penn gap E_g should not be
very different from the value for the corresponding
crystal, since the coordination remains the same, at
least in "ideal" cases where all bonds are satisfied.
The situation would be quite different for example in
GeTe, where the short-range order is known to be drastically changed in the amorphous state; it might be different in the III-V compounds, as we shall see later,
if odd-membered rings, yielding chemical inhomogeneity
because of "wrong bonds", exist in the amorphous state.
In the case of Ge and Si anyway, small variations of the
Penn gap value from the crystalline to the amorphous
state are only expected due to quantitative or geometrical disorder, i.e., changes in bond lengths and bond angles and also in real samples where all bonds are not
satisfied and defects are present, due to changes in coordination number, density fluctuations etc. This last
point will be considered later on.

    The Penn gap value can be deduced (see relation 13)
from the static dielectric constant, i.e., from the value
of the refractive index n at low energies. It is found
experimentally that the amorphous n(0) is always slightly

(within 5 to 10°/o) greater than the crystalline value
55-59 [n(0) depending somewhat on the preparation con-
ditions], which leads to a Penn gap value slightly smaller than the crystalline one. As bond bending is not
expected to alter significantly the bond strength[39],
this decrease must principally be attributed to a variation in bond length (which is experimentally found to
be very small) since E_g was shown[60] to depend on
bond length d as $d^{-2.5}$, and also to a reduction of the
average coordination number, even in "nearly-ideal" samples, as we shall see later[39].

It is worth noting that, if the Penn gap value in
the crystal is taken to coincide with the location in
energy of the E_2 peak, i.e., about 4.3 eV, which is also
the value deduced from $\varepsilon(0)$ through

$$\varepsilon(0) = 1 + \frac{(\hbar\omega_p)^2}{E_g^2}$$

[taking A = 1 in relation (13)], then this value is
quite higher than the location in energy of the single
maximum of the amorphous ε_2 spectrum which is about
2.9 eV. However, this last value is close to the one
which would be deduced from $\varepsilon(0)$ when using A = 2/3 instead of 1. It is recalled that A = 2/3 results from a
computation of matrix elements with the assumption that
only transitions above the Penn gap contribute[4], an assumption which is certainly more valid in the amorphous
case than in the crystalline one. Therefore, this apparent discrepancy in ε_2 would be only a matrix element
effect, and the results would not be in contradiction
with the results deduced from n(0).

Another evidence for the similarity of the origin
of the Penn gap in the amorphous and crystalline states
was provided by measurements of the pressure coefficient
of n(0)[58]; contrarily to the pressure coefficient of ε_2
at the absorption edge, which does not resemble the pressure coefficient of any of the crystalline gaps, the
n(0) pressure coefficient is the same in both cases.

As pointed out earlier, significant modifications
of the Penn gap with respect to the crystalline value
are expected in the case of the amorphous III-V compounds if, because of the presence of odd-membered rings,
they contain bonds between like-atoms (for example Ga-
Ga or As-As bonds instead of only Ga-As bonds as in

crystalline GaAs). A systematic search for such effects
by an accurate determination of n at low frequencies,
in relation to the ionicity of the bonding, would be extremely interesting.

3.3 Influence of defects and impurities

It has already been mentioned that small variations
of the ε_2 spectra for a-Ge and a-Si were observed depending on the deposition conditions and the subsequent
heat treatments of the samples; these variations concern, not only the location and steepness of the low-
energy edge of the curve, but also, for some samples,
the location and height of the maximum. The first effect will be discussed later on, in relation to modifications of the band edges. Here we shall concentrate
on the second effect, which is not observed in all cases.
Figures 11a and 11b for example show the results obtained, on one hand for sputtered films deposited on substrates at 25°C and 350°C[17] (compared to the results relative to evaporated films deposited on substrates at
room temperature[28]), on the other hand for evaporated
films deposited at room temperature and high rate (200
Å/sec) before and after annealing at 200 and 300°C (as
well as crystallized)[30]. Effects similar to the ones
presented in Figure 11a have also been observed by other
authors[29,61]. These discrepancies may be explained, as
we shall see, by the different nature of the defects present in the films, depending on their preparation conditions.

Interpretation in terms of voids. The variations
in intensity of the ε_2 maximum with substrate temperature during deposition was first related to the presence
of macroscopic voids in the films under consideration
and attributed to changes in proportion to these voids[29,
62,63]. The model was based on a light-scattering or
void-resonance theory[64] and would also account for the
bump observed at about 8 eV for evaporated films with
small values of the ε_2 maximum[28]. Large voids were indeed seen directly by high resolution electron microscopy
in some thin evaporated films deposited on KCl substrates[65], as a network of cracks which would have a shape
consistent with the depolarization factor of 0.8 (corresponding to disk-like voids) deduced from the light-
scattering analysis of the optical data. However, such
large voids are probably due to the particular nature
of the substrate used for these electron microscopy experiments and are not necessarily present in films deposited on glass or silica for optical studies. It must

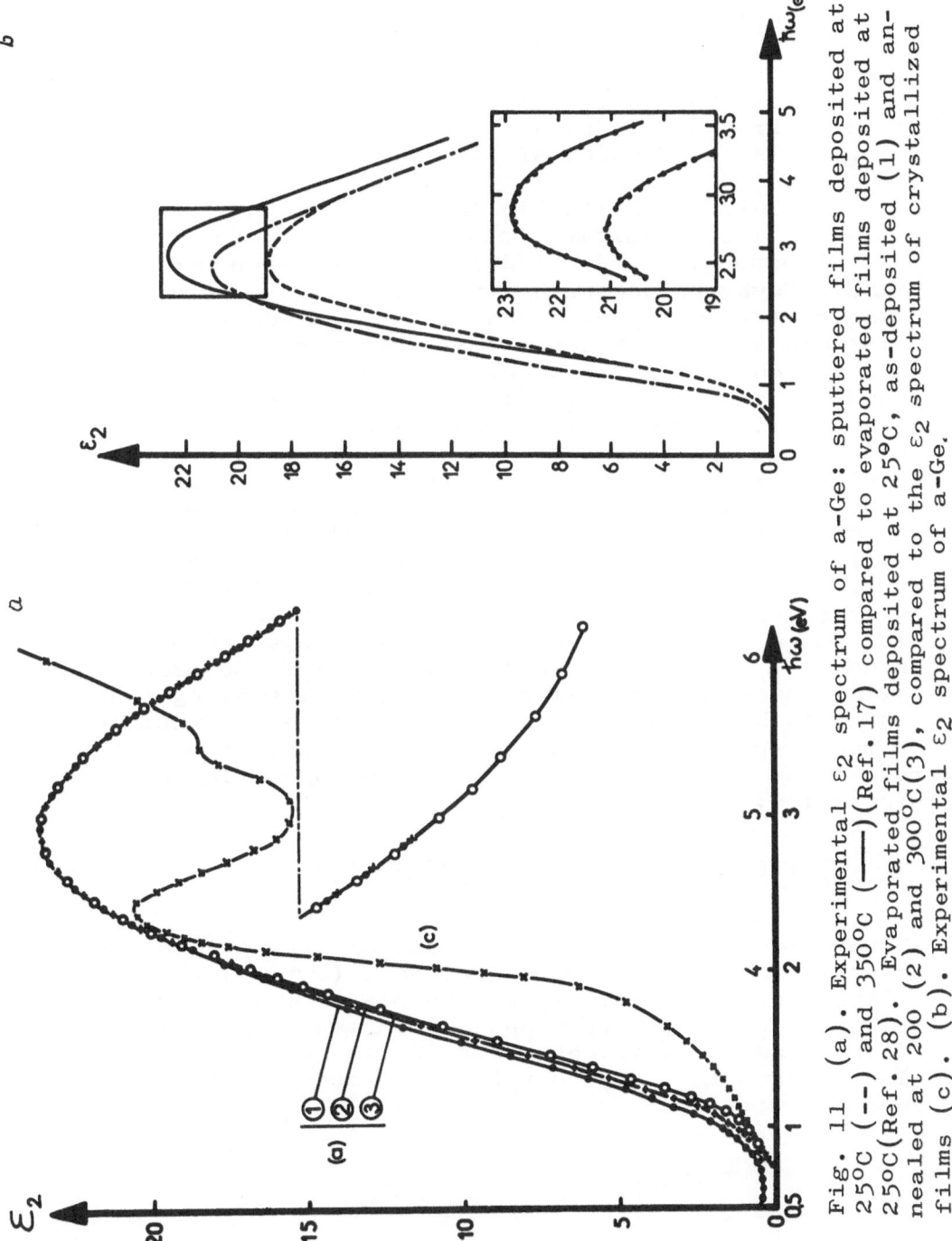

Fig. 11 (a). Experimental ε_2 spectrum of a-Ge: sputtered films deposited at 25°C (--) and 350°C (——)(Ref.17) compared to evaporated films deposited at 25°C(Ref.28). Evaporated films deposited at 25°C, as-deposited (1) and annealed at 200 (2) and 300°C(3), compared to the ε_2 spectrum of crystallized films (c). (b). Experimental ε_2 spectrum of a-Ge.

also be emphasized that, as pointed out earlier, the absolute values of ε_2 deduced by a Kramers-Kronig treatment of reflectance data only must be considered with prudence. If the above interpretation is nevertheless correct for the films under consideration, then it would also explain why these films have anomalously low values of the refractive index at low frequencies. The same model was also used successfully by other authors[66].

Not only variations in intensity, but shifts in energy of the ε_2 maximum with substrate temperature were also observed (see Figure 11a) for sputtered amorphous Ge films. However, thorough structural studies (in particular by low-angle X-ray scattering) showed that, if voids did exist in these films, they were not larger than 7 Å in diameter; the proportion of these voids was found to decrease significantly when increasing the substrate temperature during deposition. In this case, the observed optical effects on the ε_2 maximum were interpreted in the framework of the Penn-Phillips model[39]. It has already been seen that the Penn gap value E_g must depend on the bond length as $d^{-2.5}$[60]; it was also indicated[53] that a reduction of the average coordination number C_1 as a result of vacancies or small voids would decrease E_g as C_1^2. Such ideas, which had already been used in an attempt to interpret the ε_2 spectra by other authors[67], were developed by Paul et al[39] who made a detailed quantitative analysis of their data along these lines, both relative to the crystal and as a function of the preparation conditions, taking into account the results of their structure investigations. They used the expression:

$$E_g^a = E_g^c \cdot \left(\frac{d_a}{d_c}\right)^{-2.5} \cdot \left(\frac{C_{1a}}{C_{1c}}\right)^2 \cdot f(\sigma)$$

where c and a mean crystalline and amorphous, and $f(\sigma)$ accounts for bond distortion effects ($\sigma \simeq 10^o$) which can be considered as negligible; the plasma frequency of the valence electrons also necessary in the computation of E_g^a from the $n^a(0)$ values according to relation (13), was also changed in order to take into account the changes in density ρ:

$$(\omega_{pa})^2 = (\omega_{pc})^2 \cdot \frac{\rho_a}{\rho_c}$$

The correlation between the measured changes in $n(0)$ and the measured changes in nearest-neighbor distance and

coordination number or density was very good. As the
substrate temperature increases, the density increases,
which is attributed to the elimination of the small voids,
and the average coordination number increases; the consequence is an increase of the Penn gap value and of the
transition oscillator strength. On the other hand, the
determination of the valence band density of states as
described earlier in the non-direct transition model
showed that the upper bump of that band seems to narrow
with the densification of the network towards "ideal"
a-Ge.

Unfortunately, no density or low-angle scattering
measurements were performed on the evaporated films corresponding to Figure 11b, for which no variation of the
intensity or location of the ε_2 maximum with deposition
temperature or with annealing was observed. The only
indication was that the refractive index in the X-ray
range (deduced from interferences in reflection), which
is related to the film density, did not vary with the
preparation conditions. It is hard to believe that these
films do not contain any void or vacancy and that the
structural rearrangements which are responsible for the
changes of the absorption edge are only bond distortion
modifications. However, it must be noticed that the
curves of Figure 11b are identical to the one widely acknowledged now as characteristic of "ideal" a-Ge, i.e.,
the maximum of ε_2 is located at about 2.9 eV with a value of 23.

Contamination effects. Recent experiments on the
effect of air exposure on the reflectivity in the ultraviolet range (4-12 eV) of a-Ge films prepared in ultrahigh vacuum seem to show that this effect is much more
important than for crystalline Ge and does not saturate
with time[68]; it would be even more drastic when the films
are deposited at an angle, which is expected to lead to
more porous films. This was interpreted as due to the
fact that gaseous impurities like O_2, H_2O etc. are able
to diffuse in a-Ge because of the existence of voids;
the optical effect would then be the consequence of a
diffusion-limited oxidation in the bulk. Such experiments should be repeated under controled atmosphere on
differently prepared films.

3.4 Sum Rules

It is interesting to consider the sum rule giving
the effective number of electrons participating in optical transitions up to a frequency ω_μ:

$$Nn_{eff} = \frac{m}{2\pi^2 e^2} \int_0^{\omega_\mu} \omega' \cdot \varepsilon_2(\omega') \cdot d\omega'$$

(N is here the number of atoms per unit volume). n_{eff} has been computed from the ε_2 experimental values for amorphous Ge up to 5 eV only by Connell et al[17], up to 18 eV (i.e., above the plasma frequency ω_p) by Bauer[69,70]. At low energies, the results of these two determinations are the same: n_{eff} is first larger in the amorphous than in the crystalline case, then it becomes smaller at about 2.5 eV above the gap (Figure 12). This result is in a-greement with what we know about the amorphous density of states, especially about the shift of the hump at the top of the valence band towards the gap.

However, at high energies, the discrepancy between the amorphous and crystalline values of n_{eff} becomes too large, if it is recalled that the plasma frequency for valence electrons is about the same in the amorphous as

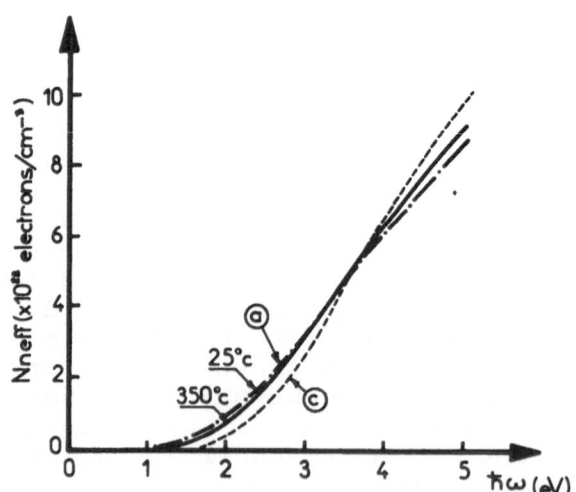

Fig. 12 Effective number of electrons involved in opti-cal transitions up to energy $\hbar\omega$ for sputtered a-Ge films deposited at 25°C (-.-) and 350°C (——) and for crystal-line Ge (...) (from Ref. 17).

in the crystalline case: 16.2 eV compared to 16.6 eV[71,72], the small variation being probably related to change in density (therefore in N). The proposed interpretation based on a matrix element argument is controversial[70], since it seems to discard the universal validity of the sum rule. New experimental data in the ultra-violet range are necessary to clear up this point.

4. Absorption Edge

4.1 General Remarks

It has been seen in Chapter 3 that the optical spectrum of a-Ge or a-Si at high enough energies above the gap (i.e., when the absorption coefficient is of the order of 10^5-10^6 cm^{-1}) is slightly sensitive to the sample structure and that the variations observed in some cases do not modify the overall shape of the curve. This is not surprising since the optical absorption is then due to transitions between "extended" states of the valence and the conduction bands, which are expected to be little affected by the details of the structure.

The situation is completely different when the absorption edge is considered. For energies close to the gap value, the absorption comes from transitions involving states at the band edges and in the band tails. The separation in energy of the valence and conduction bands, as well as the extension, intensity and shape of the tails and the localization of the states in these tails will certainly depend on the nature of the disorder (fluctuations of interatomic distances, bond distortions, local configurations of neighboring tetrahedrons etc.) and will be influenced by the presence of defects (dangling bonds, vacancies, voids etc.) and of impurities (depending on their nature and their bonding).

Different theoretical treatments, valid for "ideal" amorphous materials in which all bonds are satisfied, for example, treatments based on tight-binding simple models[74] or using multiple scattering theory on small clusters with appropriate short-range order[49,51], predict the existence of a (pseudo) gap in the amorphous state but are unable to give a quantitative description of the band edges. The investigations on the crystalline polytypes of Ge and Si provide some information about the influence of the local atomic arrangements on the gap value and the shape of the band edges[75,76]; according to these investigations, bond distortions would affect the top of the valence band and the bottom of the

conduction band (in particular cause an apparent shift
of the hump at the top of the valence band towards the
gap) while the presence of odd-membered rings would modify (increase) the gap value. But it is still impossible to obtain theoretically a quantitative estimation
of the density of states in the vicinity of the gap, as
well as of the matrix elements for optical transitions,
which must depend strongly on the localization of the
involved electron states. It has been suggested[53,54,77]
that band tails of localized states should not necessarily
exist in all disordered materials. Amorphous networks,
if not in internal equilibrium at their formation, would
tend to relax by slight atomic rearrangements in order
to reduce their total energy. This relaxation process,
possible in a-Ge and a-Si, would have the effect of e-
jecting states from the gap.

In addition to these theoretical difficulties, the
confrontation between experimental data and theoretical
models is complicated by the fact that the density of
states at the band edges is susceptible to vary with the
sample structure. The presence of defects should increase
the density of states in the tails and even introduce
bands of localized states in the gap; the elimination of
these defects, either when changing the deposition conditions, or when annealing, will affect the optical data.
In the same way, impurities might, depending on their
size, their chemical activity etc., modify the sample
structure and either introduce, or eliminate (if they
take up dangling bonds) states in the gap. Eventually,
it must be recalled that the band edges may be distorted by strains (which are probably present in all films)
and that internal electric fields associated with inhomogeneities may also play a role in the shape of the absorption edge by interacting with exciton effects.

4.2 Description of the Absorption Edge.

In order to obtain significant optical data in this
domain, it is more than ever necessary to control very
carefully the conditions of preparation and to characterize the films as completely as possible by structure
studies, density measurements, check of impurity content
etc. Comparative investigations on films prepared under
the same conditions except for one significant parameter
proved to be extremely useful for clearing up various
problems. On the other hand, the optical measurements
in this low absorption coefficient range ($\alpha < 10^4$ cm^{-1})
are difficult, as explained in Section 2; either sophisticated methods needing high accuracy must be used, or

thick films have to be preferred, provided they do not present cracks, non-homogeneity, large strains, rough surfaces etc.

Figure 13 shows the low-energy edge of the ε_2 spectrum for evaporated a-Ge films deposited at high rate (200A°/sec) on room temperature substrate, as-deposited and after annealing at increasing temperatures (the edge for a polycrystalline film is also shown for comparison). These edges can in fact be separated into two regions: (i) the higher energy region ($\hbar\omega > 1.2$ eV, which corres-

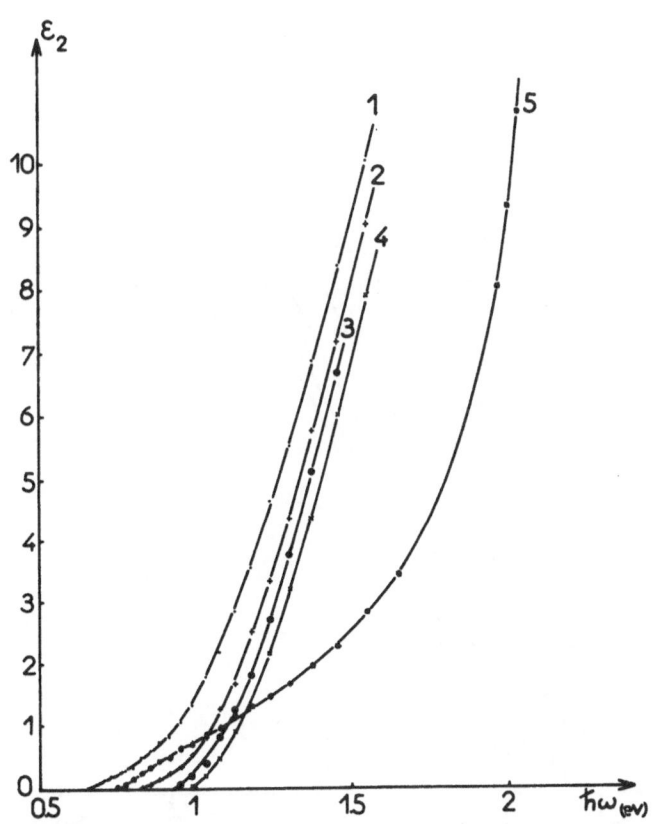

Fig. 13 Low-energy edge of the ε_2 spectrum for evaporated a-Ge films deposited at high rate at 25°C: as-deposited (1), after annealing at 200 (2), 300 (3) and 400°C (4), and for a recrystallized film (5).

ponds to $\alpha > 5.10^4$ cm^{-1}) where the experimental ε_2 follows the behavior (Figure 14)

$$\omega^2 \cdot \varepsilon_2 \propto (\hbar\omega - E_o)^2$$

The same empirical law, pointed out in early measurements

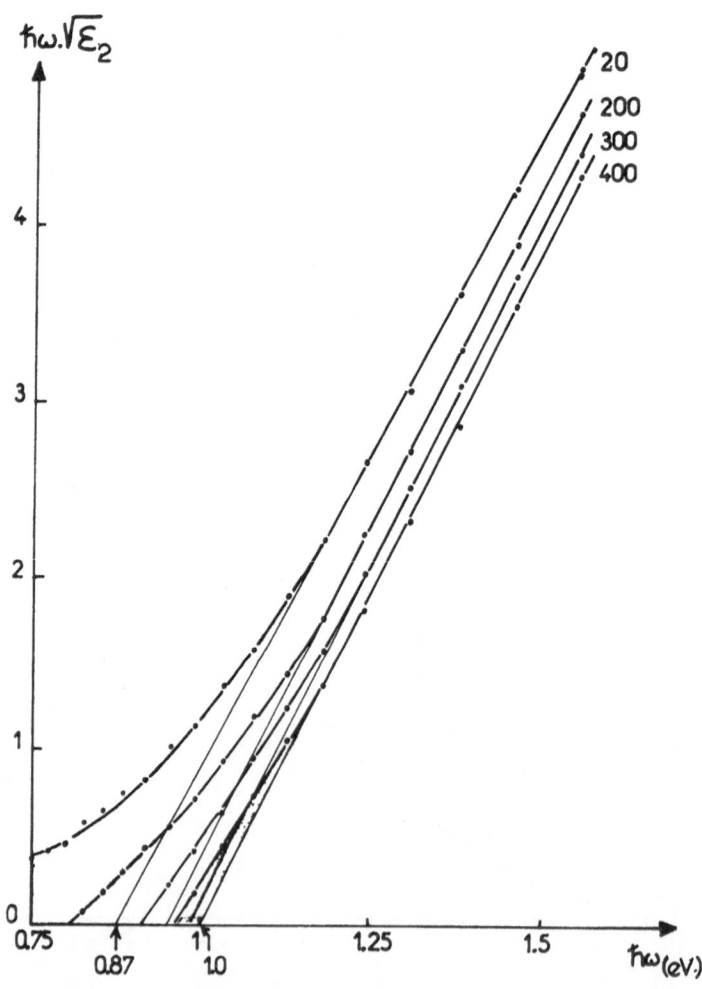

Fig. 14 Plot of $\hbar\omega \cdot \sqrt{\varepsilon_2}$ versus energy for the same films as in Figure 13 (the straight lines are extrapolated towards the optical gap value).

on a-Ge and a-Si[78,79], has been observed quite generally for all samples. It bears some resemblance to the expression for indirect (phonon-assisted) transitions in
crystalline semiconductors but cannot have the same
meaning here since k is not a good quantum number any
more. It has been seen earlier that it can be interpreted in the "non-direct" transition model as due to transitions between extended states in parabolic valence and
conduction bands; therefore it defines an <u>optical gap</u>
E_O equal to the separation in energy of these extended
states. As the accuracy of the data is still good in
this domain, E_O, which characterizes the material under
consideration, is usually well-determined.

(ii) The lower energy region ($\alpha < 10^4$ cm^{-1}), where
ε_2 decreases less rapidly with energy. This appears as
an extra-absorption below the optical gap E_O in the plots
of Figure 14. The exact shape of the absorption curves
is then difficult to ascertain because of experimental
uncertainties; besides, we shall see that it varies considerably with the sample structure. Early experiments
yielded almost exponential absorption edges, which was
currently interpreted as typical of transitions between
exponential tails of the densities of states[80,82]; these
conclusions were questioned by other data showing absorption edges almost as sharp as the crystalline ones, which
seemed to deny the existence of such tails[83,28]. Depending on the preparation conditions and the method of
measurement, different results showing more or less gradual edges have been found later on. The controversy
between the supporters of the exponential edges and of
the sharp edges seems now to be over. It is acknowledged that the absorption edge is quite gradual for very
imperfect films, and that it sharpens while moving towards higher energies (i.e., that the absorption at energies below the optical gap decreases strongly) when
the samples approach the "ideal" amorphous material.
Figures 15a and 15b illustrate this point in two different cases: for sputtered films deposited at 25 and
350°C[17] (for comparison are shown the results of early
measurements on evaporated films deposited at 25°C[81] and
the data relative to crystalline Ge[84])and for evaporated films deposited at 25°C, as-deposited and after annealing[57]. In the first case, the absorption coefficient
is deduced from transmittance measurements (at low energies) and ellipsometric measurements (at higher energies)
on thick films ($d \simeq 30\mu$); in the second case from reflectance and transmittance measurements on thin films, with
larger experimental uncertainties ($d \simeq 1000$ Å).

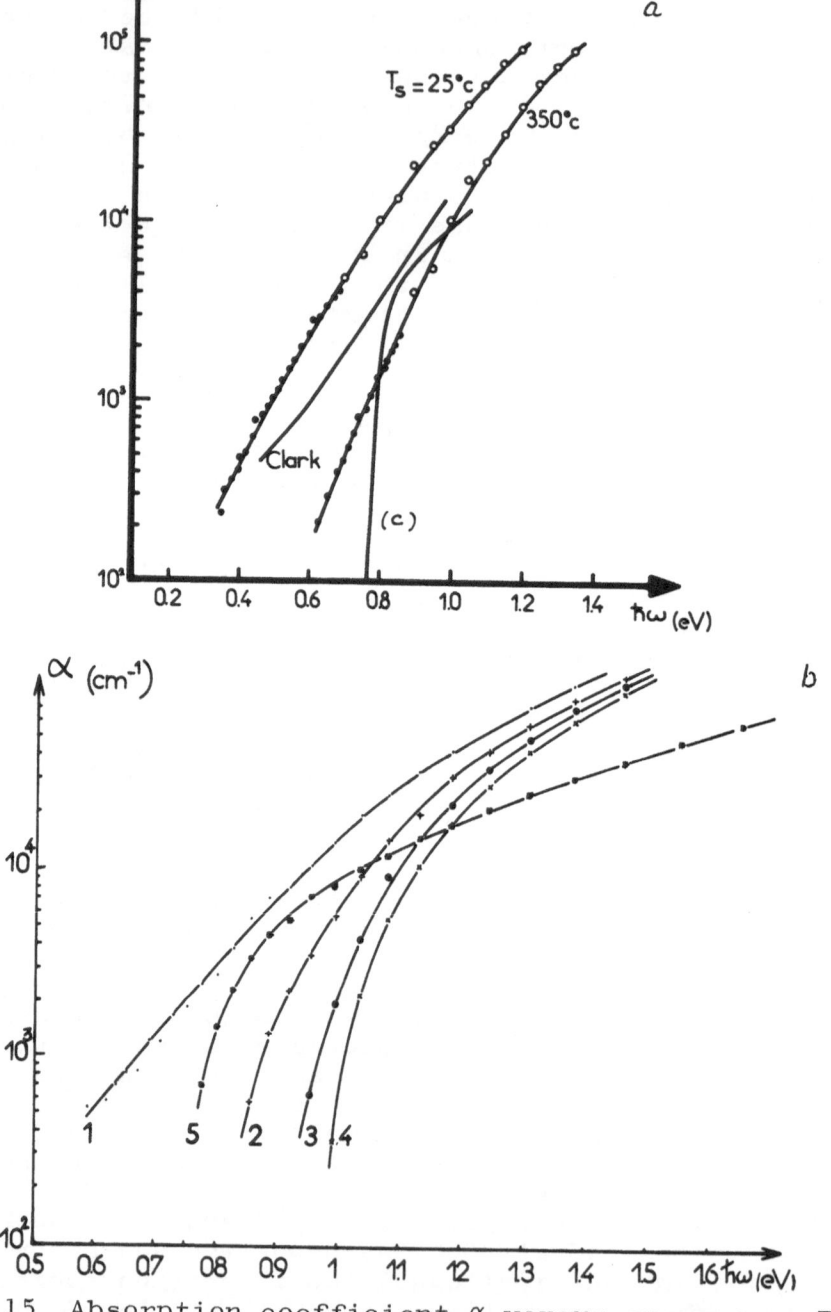

Fig. 15   Absorption coefficient α versus energy: a. For sputtered a-Ge films deposited at 25°C and 350°C (Ref. 17), for evaporated a-Ge films deposited at 25°C (Clark) (Ref. 81) and for crystalline Ge (Ref. 84).   b.   for the same films as in Figures 13 and 14.

4.3 Influence of defects and impurities

Influence of preparation conditions. For a given method of deposition, two main parameters influence the formation of the as-deposited films: the deposition rate and the substrate temperature. Varying the first one modifies the nature of the vapour species and their kinetic energy, the supersaturation of the vapour in the vicinity of the substrate, etc.; varying the second one essentially modifies the density of nucleation centers and the mobility of the ad-atoms. It has been shown by various studies on evaporated, sputtered and glow-discharge films that decreasing the deposition rate[88,89] and increasing the substrate temperature[56,85,87,17] produce a sharpening and a shift towards higher energies of the absorption edge. Effects in the same direction have also been observed with annealing films prepared under the same conditions at higher and higher temperatures (still below the crystallization onset)[57,58,90-92]. It must be noticed that these effects are irreversible and that the absorption edges tend towards a limit edge, which can be approached by any film if the deposition conditions are correctly modified or if this film is sufficiently annealed whatever its properties after deposition, and which cannot be overpassed unless the films begin to crystallize. This limit edge would correspond to the "ideal" metastable amorphous material; for example, the optical gap E_O is found to be about 1 eV for "nearly ideal" a-Ge. However, nearly ideal samples of different origin still present significant differences in the low absorption part of the edge, which must be related to the presence of residual defects of different nature and distribution.

On the contrary, ion bombardment produces effects in the opposite direction, i.e., a strong increase of the optical absorption below the gap, which can be cured by suitable annealing[93]. It must be recalled that implantation of energetic heavy ions in crystalline Ge and Si produces first clusters of highly disordered material, which eventually coalesce to form a continuous amorphous phase, the properties of which are similar to those of amorphous films prepared by vapour deposition[94,95].

Only a few comparative studies have been made on the influence of the preparation method[58,86,96]. These methods essentially differ by the nature of the species impinging the substrate surface and their kinetic energies, but the deposition rate as well as the impurity

condition are also different, which makes any comparison
difficult. As far as optical properties are concerned
(the discrepancies between transport properties may be
more important), it seems that evaporated or sputtered
films must be distinguished from electrolytically or
glow discharge deposited films, which in general present sharper absorption edges located at higher energies. This would indicate that deposition processes in
which the atoms arrive individually at the substrate
surface and not as clusters (like in vapour beams) are
more likely to avoid the formation of voids and other
defects, and to lead to less imperfect samples.

Discussion. The experimental results described in
the preceding paragraph must be interpreted in relation
to the structure of the amorphous network and the presence of extrinsic defects. The increase of the optical gap and the sharpening of the absorption edge (reflecting a decrease of the absorption below the gap),
observed when decreasing the substrate temperature during deposition or when annealing, can only be due to
modifications of the band edges and to a decrease of the
band tails. As pointed out earlier, this must be the
consequence, on the one hand, of slight changes in the
local atomic arrangements, tetrahedron configurations,
bond distortions etc.; on the other hand, of variations
of the nature and decrease of the proportion of defects,
dangling bonds, vacancies or voids.

It was attempted to correlate the optical results
with the results of simultaneous experiments sensitive
to structural details or to the presence of defects.
The most widely used experiments are electron spin resonance experiments, which can detect and measure the
density of unpaired spins which may be associated with
dangling bonds (for example at the internal surface of
voids). These dangling bonds are expected to create
states giving rise to optical absorption. In an early
work on a-Si[90,97] it was shown that the ESR signal, com-
ing from the bulk and not from the surfaces of the films,
decreased by annealing while the optical gap was
increasing. More systematic and careful studies have
been performed later on[98]. They seemed to confirm that
ESR experiments give a correct measurement of the internal void surface area, since the density measured by
nuclear back-scattering decreased with decreasing spin
concentration[99]. The observed dependence of the absorption edge on the spin concentration allowed to extrapo-
late the optical results to "ideal" a-Si free of effects

from voids and broken bonds. This leads to an optical
gap of 1.8 eV, which is much higher than any gap measured
directly for "nearly ideal" samples (about 1.5 eV). The
validity of such an extrapolation procedure is, however,
not certain because of the complexity of the problem and
because the coincidence between states giving an ESR signal and states giving rise to optical absorption has been
questioned.

Recently, optical data obtained from sputtered films
deposited at different substrate temperatures have been
interpreted in relation to the results of careful X-
ray diffraction experiments[39]. The shift of the absorption edge towards higher energies when increasing the
substrate temperature has been correlated with the densification of the films because of the inhibition of the
formation of small voids (smaller than 7 A° in diameter)
present in the low-temperature films. The increase of
the optical gap has been correlated with changes in the
dihedral angle distribution function, best approximated
by an increase in the proportion of staggered configurations for neighboring tetrahedrons. Using the model
described above, with parabolic band edges and exponential tails (assuming that the densities of states in the
valence and conduction bands are symmetric and that the
Fermi level is pinned at the center of the gap), it was
possible to estimate the density of states in the gap.
If

$$g(|\Delta E|) = g(0) \cdot \exp \frac{|\Delta E|}{E_s}$$

with E_s as deduced from the slope of the exponential absorption coefficient $\alpha \simeq \exp(\hbar\omega/E_s)$ equal to 0.13 and
0.10 respectively, it was found that $g(0) = 3.10^{19} cm^{-3}$
eV^{-1} for the 25°C film and $3.10^{18} cm^{-3} eV^{-1}$ for the
350°C film. These numbers provide a quantitative evidence for the decrease of the density of states in the
gap when considering "nearly ideal" samples; they are
in agreement with the density of $2.10^{18} cm^{-3} eV^{-1}$ at the
Fermi level deduced from transport measurements at low
temperatures on evaporated films (which, however, were
claimed to present much sharper optical edges than the
sputtered films considered here)[100].

Up to now no systematic study has been made on the
influence of the preparation conditions on the amorphous
III-V compounds[58,101]. It would however be extremely
interesting to check whether the absorption edge is as
structure dependent in partly-ionic materials as in

purely covalent ones, and whether the formation of extrinsic defects, which in a-Ge and a-Si seems to be a
consequence of the rigidity and high localization of the
bonds, is as easy.

Influence of contamination. More and more studies
are dedicated to the role of impurities on the properties of a-Ge and a-Si, which had previously been neglected under the assumption that an amorphous network
is flexible enough to accommodate foreign atoms. For
example, some observations on the absorption edge, following annealing, have been attributed, probably wrong-
ly, to contamination effects. The optical properties of
films prepared and studied in ultrahigh vacuum[87] do not
seem so different from the properties of films prepared
in standard vacuum and studied in air. However, impurity effects deserve consideration from different points
of view.

It was first deduced from optical measurements on
films sputtered in the presence of oxygen that O_2 produced a sharpening of the edge[102]. Some of the conclusions were difficult to reconcile with the usual observation of exponential edges for films deposited at 10^{-5}
or 10^{-6} Torr. The results of a comparison of the absorption edges for films annealed either without breaking the vacuum or after exposure to air were not very
clear[100] because other parameters than air contamination
might have come into play. We found that exposure to
air before annealing did not bring any significant difference in the optical (and electrical) behavior of evaporated films upon annealing (Figure 16).

Recently, a-Si films prepared by a number of methods
were exposed to oxygen and hydrogen either during or
after deposition and the consequences determined by different experiments[103]. It was found that the presence
of oxygen during deposition reduced the spin signal, decreased the conductivity and shifted the optical edge.
However, this shift was different from the shift due to
annealing under clean conditions. For films deposited
in oxygen, the edge tended to move towards that of amorphous SiO. These results suggested that O_2 interacts
with Si during deposition, forming Si-O bonds. This in-
hibits the vacancy formation but creates states which
give a gap (corresponding to SiO) larger than that expected for a-Si. Exposure to O_2 and H_2 after deposition
completely quenched the ESR signal but decreased only
slightly the conductivity and had negligible effect on

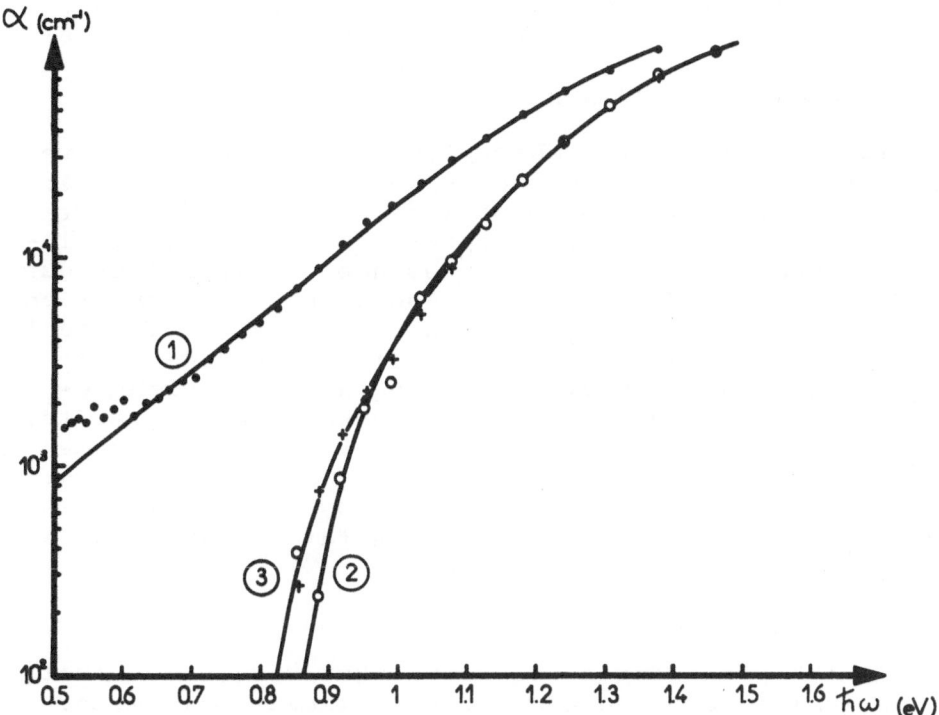

Fig. 16   Absorption coefficient α versus energy for eva-porated a-Ge films deposited at high rate at $25^{\circ}C$: as-deposited (1), annealed up to $380^{\circ}C$ after air exposure (2) and without breaking the vacuum (3).

the absorption edge.   Therefore the impurities must take up the unpaired spins but not the states giving rise to optical absorption.   It must be concluded that the states responsible for ESR signal are only a small fraction of the total localized state distribution, the majority of which appears insensitive to surface contamination.

Another approach was developed recently[104] for a-Ge sputtered films; it was attempted to use a "simple" im-purity like hydrogen(with a single electron and a small covalent radius) to compensate individual dangling bonds present in the films.   Structure studies seemed to show that the presence of hydrogen during deposition did not modify the sample structure; it was assumed that the bonded hydrogen atoms were located at the void internal

surfaces, probably with two configurations as suggested
by infra-red absorption measurements. The effects on
the absorption edge were similar to those from annealing or increasing the substrate temperature during deposition. The impurity produced a decrease of the density of states in the gap.

In conclusion, the absorption edge, and therefore
the density of states in the region of the gap, appear
only a little sensitive to surface contamination after
deposition(contrary to the properties in the ultra-violet) but are affected by the presence of impurities during deposition. These effects would depend strongly on
the nature of the bonding and on the presence of defects.
It must be pointed out that some impurities stabilize
the amorphous network by inhibiting atomic rearrangements, and allow the films to be annealed at higher temperatures without crystallization. Paradoxally, some
impurities might be a help in investigating the properties of "ideal" amorphous materials by compensating some
of the effects of extrinsic defects (at least as far as
states in the gap are concerned).

4.4 Models for the absorption edge

In the preceding sections, it has been noted that
the optical absorption edge is essentially determined
by the tailing of the valence band and conduction band
densities of states into the gap. This is strongly suggested by the variations in the slope and the location
of this edge with the preparation conditions etc. It
must however be recalled that roughly exponential edges
might be the result of edge broadening due to strains,
which are known to be quite large in most of the amorphous films[105]; such effects, although probably real,
must play a minor role compared to density of states effects. On the other hand, exponential absorption edges,
known as Urbach tails, are often observed in crystalline
semiconductors; they are attributed to the broadening
of excitonic transitions at the edge by internal electric fields[106]. The relation between the Urbach tails
and the effects of disorder (potential fluctuations) upon
excitonic absorption was demonstrated both on crystals
(PbI_2) and on glasses (As_2Se_3) by investigating the influence of heavy ion bombardment on the optical properties[107]. Such an interpretation, probably valid in
the case of glasses which all present similar typical
behavior, must be inadequate for a-Ge and a-Si. For example, the temperature dependence of the absorption edge
does not behave according to this model (its slope is

temperature independent, the edge being only shifted parallel to itself)[108,109]. This can be due to the fact that the exciton binding energy is small because of the high dielectric constant, making exciton effects negligible, and that random internal electric fields, if they exist, result from deformation effects (near defects) rather than from electrostatic ones because of the absence of substantial chemical inhomogeneities.

It has been pointed out that the high energy part of the absorption edge must be associated with transitions between extended states in the valence and conduction bands. The optical gap E_0 is therefore a pseudo-gap since localized states can exist in each band beyond the limits of the extended states; the localization of these states probably increases when going towards the center of the gap. It is hard to decide if the optical gap is identical to the mobility gap introduced in order to account for transport properties, although their values are in general close to each other. On the other hand, it seems meaningless to try to compare the optical gap to the smallest gap in the corresponding crystalline (diamond) phase. It has been shown that the pressure coefficient of E_0 bears no resemblance with the pressure coefficient of any of the crystalline gaps, which would suggest that disorder mixes states in different sub-bands[58]. If a comparison is needed, it must be made for the Penn gap, defined in a more chemical model and retaining the same meaning and validity in the amorphous as well as in the crystalline cases.

As far as the low-energy part of the absorption edge is concerned, it is not clear whether it is associated only with transitions between extended states in one band and localized states in the other one, or also with transitions between localized states in both bands. The matrix elements corresponding to these last transitions are however expected to be very small, because of the small overlap of the wavefunctions of the initial and final states of the transitions. Because of these ambiguities and theoretical difficulties, optical absorption measurements are not able to provide accurate quantitative information about localized states. For example, it is probable that the exact densities of states in the pseudo-gap are more complicated (in particular do not decrease monotonically with energy) than the model proposed in reference 39. Non-equilibrium experiments like photoconductivity or luminescence measurements are more suited to the elucidation of these problems.

However, optical results can help, as we have seen, in determining the effects of the different elements of disorder, both intrinsic and extrinsic, on the density of states. Theoretically, one expects that: (i) Topological disorder does not prevent sharp band edges and a well-defined gap which is probably slightly larger than the crystalline (diamond) gap. (ii) Geometrical disorder, i.e., fluctuations in interatomic distances and bond distortions, modifies the gap value, distorts the band edges and introduces restricted tailing in the gap. The tailing would be indeed very small for "ideal" amorphous materials in internal equilibrium (completely relaxed) with all bonds satisfied; (iii) defects like dangling bonds, non-satisfaction of tetrahedral coordination, vacancies and voids, long-range potential fluctuations etc., produce more or less important tailing at the band edges and introduce states in the gap. The variations in the optical gap E_0 would then reflect changes in the intrinsic properties of the amorphous network (characterized by the mean neighbor distances and the fluctuations of these distances, the mean distortion of the bonds, the dihedral angle distribution function and the proportion of five-fold rings etc.) while the low-energy absorption edge would essentially be associated with defects and impurities. It must be pointed out that these effects are not independent, since the network and its defects are correlated. The elimination of voids for example leads to a redistribution of bond distortions and interatomic distances, to variations in dihedral angle distribution function and ring statistics, etc.

REFERENCES

1. L. D. Landau and E. M. Lifshitz, "Electrodynamics of Continuous Media" (Pergamon Press, New York, 1960).
2. B. Velicky, Czech. J. Phys. B 11, 787 (1961).
3. H. Ehrenreich and H. R. Phillip, in Proceedings of the Int. Conf. on Physics of Semiconductors, Exeter 1962 (Inst. of Phys. Soc., London), p. 367.
4. M. Cardona, in Proceedings of the Int. School of Physics "Enrico Fermi", Varenna, Course 52 (Academic Press, New York, 1972), p. 514
5. D. Penn, Phys. Rev. 128, 2093 (1962).
6. F. Bassani, in Proceedings of the Int. School of Physics "Enrico Fermi", Varenna, Course 34 (Academic Press, New York, 1966), p. 33
7. L. Van Hove, Phys. Rev. 89, 1189 (1953); J. C. Phillips, Phys. Rev. 104, 1263 (1956).

8.   G. Harbeke in "Optical Properties of Solids", ed.
      F. Abeles (North-Holland, Amsterdam, 1972), p. 21.

9.   J. Tauc in "Optical Properties of Solids", ed. F.
      Abeles (North-Holland, Amsterdam, 1972), p. 277.

10.   E. A. Davis in Proceedings of the Aberdeen Summer
      School (Academic Press, New York, 1973), p. 425.

11.   N. F. Mott and E. A. Davis, "Electronic Processes
      in Non-crystalline Materials" (Clarendon Press,
      Oxford, 1971).

12.   O. S. Heavens in "Physics of Thin Films", vol. 2
      (Academic Press, New York, 1964), p. 193.

13.   F. Abeles in, "Progress in Optics", vol. 2, ed. E.
      Wolff (North-Holland, Amsterdam, 1962), p. 250.

14.   F. Abeles in "Advanced Optical Techniques", ed. A.
      C. S. Van Heel (North-Holland, Amsterdam, 1967),
      p. 144.

15.   M. L. Theye, thesis, Paris, 1968, unpublished.

16.   F. Stern in "Solid State Physics", vol. 15, eds.
      F. Seitz and D. Turnbull (Academic Press, New York,
      1963), p. 299.

17.   G. A. N. Connell, R. J. Temkin and W. Paul, Adv.
      in Physics 22, 643 (1973).

18.   Proceeding of the Symposium on "Recent Developments
      in Ellipsometry",eds. N. M. Bashara, A. B. Buckman
      and A. C. Haul, Surface Science 16 (1969).

19.   H. Kiessig, Ann. Physik 10, 769 (1931); W. Umrath,
      Z. Angew. Physik 22, 406 (1967); P. Croce, G.
      Devant, M. G. Sere and M. F. Verhaeghe, Surface
      Science 22, 173 (1970); A. Segmuller, Thin Solid
      Films 18, 287 (1973).

20.   S. Tolansky in "Multiple Beam Interferometry"
      (Clarendon Press, Oxford, 1948); H. Dupoisot, Ann.
      Phys. 3, 369 (1968).

21.   P. M. Grant, Tech. Report HP-14, Harvard University
      (1965); P. M. Grant and W. Paul, J. Appl. Phys. 37,
      3110 (1966); F. Abeles and M. L. Theye, Surface
      Science 5, 325 (1966); P. O. Nilsson, Appl. Optics
      7, 435 (1968).

22.   M. L. Theye, Proceedings of the 5th Intl. Conf. on
      Amorphous and Liquid Semiconductors, Garmisch,
      1973, (Taylor and Francis, London, 1974), p. 479.

23.   G. A. N. Connell and A. Lewis, Phys. Stat. Sol. b
      60, 291 (1973).

24.   T. M. Donovan in Proceedings of the Intl. Conf. on
      Tetrahedrally Bonded Amorphous Semiconductors,
      Yorktown Heights, (American Institute of Physics,
      New York, 1974), p. 1.

25.   L. Ley, S. Kowalczyk, R. Pollak and D. A. Shirley,
      Phys. Rev. Letters 29, 1088 (1972).

26.   D. E. Eastman and W. D. Grobman, Proceedings of the
       11th Intl. Conf. on Physics of Semiconductors,
       Warsaw (Polish Scientific Publishers, Warsaw,
       1972), p. 889.
27.   F. C. Brown and O. P. Rustgi, Phys. Rev. Letters
       28, 497 (1972).
28.   T. M. Donovan, W. E. Spicer, J. M. Bennett and
       E. J. Ashley, Phys. Rev. B2, 397 (1970).
29.   R. S. Bauer and F. L. Galeener, Solid State Comm.
       10, 1171 (1972).
30.   M. Gandais, J. Rivory and M. L. Theye, Thin Solid
       Films 12, 201 (1972).
31.   D. Beaglehole and M. Zavetova, J. Non-Cryst. So-
       lids, 4, 272 (1970).
32.   D. T. Pierce and W. E. Spicer, Phys. Rev. B5,
       3017 (1972).
33.   J. Stuke and G. Zimmerer, Phys. Stat. Sol. b 49,
       513 (1972).
34.   V. Heine and R. O. Jones, J. Phys. C 2, 719 (1969).
35.   J. C. Phillips in "Bands and Bonds in Semiconduc-
       tors" (Academic Press, New York, 1973).
36.   K. Maschke and P. Thomas, Phys. Stat. Sol. b 41,
       743 (1970).
37.   B. Kramer, K. Maschke and P. Thomas, Phys. Stat.
       Sol. b 49, 525 (1972).
38.   W. E. Spicer and T. M. Donovan, Phys. Rev. Letters
       24, 595 (1970).
39.   W. Paul, G. A. N. Connell and R. J. Temkin, Adv.
       in Physics 22, 531 (1973).
40.   J. D. Joannopoulos and M. L. Cohen, Solid State
       Comm. 13, 1115 (1973).
41.   J. D. Joannopoulos and M. L. Cohen, Phys. Rev. B
       8, 2733 (1973).
42.   G. A. N. Connell, Solid State Comm., 14, 377
       (1974).
43.   B. Kramer, Phys. Stat. Sol. b 47, 501 (1971).
44.   S. F. Edwards, Adv. in Phys. 16, 359 (1967).
45.   D. Brust, Phys. Rev. Letters 23, 1232 (1969).
46.   D. Brust, Phys. Rev. 186, 768 (1969).
47.   I. B. Ortenburger and D. Henderson, Phys. Rev. Let-
       ters 30, 1047 (1973).
48.   D. Henderson and I. Ortenburger, J. Phys. C 6,
       631 (1973).
49.   J. Keller, J. Phys. C 4, 3143 (1971).
50.   T. C. McGill and J. Klima, Phys. Rev. B5, 1517
       (1972).
51.   J. Keller and J. Ziman, J. Non-Cryst. Solids 8-10,
       111 (1972).
52.   J. C. Phillips, Rev. Mod. Phys. 42, 317 (1970).

53. J. C. Phillips, Phys. Stat. Sol. b 44, K1 (1971).
54. J. C. Phillips, Comments on Sol. State Phys. 4, 9 (1971).
55. J. Wales, G. J. Lowitt and R. A. Hill, Thin Solid Films I, 137 (1967).
56. M. L. Theye, Optics Comm. 2, 329 (1970).
57. M. L. Theye, Mat. Res. Bull. 6, 103 (1971).
58. G. A. N. Connell and W. Paul, J. Non-Cryst. Solids 8-10, 215 (1972).
59. M. H. Brodsky, R. S. Title, K. Weiser and G. D. Pettit, Phys. Rev. B I, 2632 (1970).
60. J. Van Vechten, Phys. Rev. 182, 891 (1969).
61. G. Jungk, Phys. Stat. Sol. b 44, 299 (1971); Phys. Stat. Sol. b 46, 603 (1971).
62. F. L. Galeener, Phys. Rev. Letters 27, 1716 (1971).
63. R. S. Bauer, F. L. Galeener and W. E. Spicer, J. Non-Cryst. Solids 8-10, 196 (1972).
64. F. L. Galeener, Phys. Rev. Letters 27, 421 (1971).
65. T. M. Donovan and K. Heinemann, Phys. Rev. Letters, 27, 1794 (1971).
66. G. Jungk, Phys. Stat. Sol. b 55, 579 (1973).
67. M. H. Brodsky and P. J. Stiles, Phys. Rev. Letters 25, 798 (1970).
68. C. R. Helms, W. E. Spicer and V. Pereskokov, Appl. Phys. Letters 24, 318 (1974).
69. R. S. Bauer in Proceedings of the 5th Intl. Conf. on Amorphous and Liquid Semiconductors, Garmisch, 1973, (Taylor and Francis, London, 1974), p. 595.
70. R. S. Bauer in Proceedings of the Intl. Conf. on Tetrahedrally Bonded Amorphous Semiconductors, Yorktown Hgts., (Amer. Inst. of Physics, N.Y.,1974),p.126.
71. K. Zeppenfeld and H. Raether, Zeits. für Phys. 193, 471 (1966).
72. D. L. Misell and R. A. Crick, J. Phys. C (Solid State Phys.) 4, 1591 (1971).
73. D. Weaire, Phys. Rev. Letters 26, 1541 (1971).
74. D. Weaire and M. F. Thorpe, Phys. Rev. B4, 2508 (1971).
75. J. D. Joannopoulos and M. L. Cohen, Phys. Rev. B 7, 2644 (1973).
76. R. Alben, S. Goldstein, M. F. Thorpe and D. Weaire, Phys. Stat. Sol. b 53, 545 (1972).
77. J. A. Van Vechten, Solid State Comm. 11, 7 (1972).
78. J. Tauc, R. Grigorovici and A. Vancu, Phys. Stat. Sol. 15, 627 (1966).
79. R. Grigorovici and A. Vancu, Thin Solid Films 2, 105 (1968).
80. A. M. Glass, Canad. J. Phys. 43, 1068 (1965).
81. A. H. Clark, Phys. Rev. 154, 750 (1967).

82.   R. C. Chittick, J. Non-Cryst. Solids 3, 255 (1970).
83.   T. M. Donovan, W. E. Spicer and J. M. Bennett,
      Phys. Rev. Letters 22, 1058 (1969).
84.   W. C. Dash and R. Newman, Phys. Rev. 99, 1151
      (1955).
85.   T. M. Donovan, E. J. Ashley and W. E. Spicer, Phys.
      Letters 32A, 85 (1970).
86.   R. J. Loveland, W. E. Spear and A. Al-Sharbaty,
      J. Non-Cryst. Solids 13, 55 (1973/74).
87.   P. J. Elliott, A. D. Yoffe and E. A. Davis, in
      Proceedings of the Intl. Conf. on Tetrahedrally
      Bonded Amorphous Semiconductors, Yorktown Heights,
      (Amer. Inst. of Physics, New York, 1974), p. 311.
88.   S. Koc, O. Renner, M. Zavetova and J. Zemek,
      Czech. J. Phys. B 22, 1296 (1972).
89.   S. K. Bahl, S. M. Bhagat and R. Glosser, Solid
      State Comm. 13, 1159 (1973).
90.   M. H. Brodsky, R. S. Title, K. Weiser and G. D.
      Pettit, Phys. Rev. B 1, 2632 (1970).
91.   S. Koc, M. Zavetova and J. Zemek, Thin Solid Films,
      10, 165 (1972).
92.   A. Lewis, Phys. Rev. Letters 29, 1555 (1972).
93.   J. A. Olley, Solid State Comm. 13, 1441 (1973).
94.   S. Kurtin, G. A. Shifrin and T. C. McGill, Appl.
      Phys. Letters 14, 223 (1969).
95.   B. L. Crowder, R. S. Title, M. H. Brodsky and G.
      D. Pettit, Appl. Phys. Letters 16, 205 (1970).
96.   S. Koc, I. Kubelik, M. Rozsival, L. Stourac, A.
      Triska, V. Vorlicek, M. Zavetova and J. Zemek in
      Proceedings of the Intl. Conf. on Physics and
      Chemistry of Semiconductors Heterojunctions and
      Layer Structures, Budapest, 1970.
97.   M. H. Brodsky and R. S. Title, Phys. Rev. Letters
      23, 581 (1969).
98.   M. H. Brodsky, D. Kaplan and J. F. Ziegler, in
      Proceedings of the 9th Intl. Conf. on the Physics
      of Semiconductors, Warsaw, 1972 (Polish Scientific
      Publishers, Warsaw).
99.   M. H. Brodsky, D. Kaplan and J. F. Ziegler, Appl.
      Phys. Letters 21, 305 (1972).
100. M. L. Knotek and T. M. Donovan, Phys. Rev. Letters
      30, 652 (1973); M. L. Knotek, M. Pollak and T. M.
      Donovan, Phys. Rev. Letters 30, 853 (1973); M.
      Pollak, M. L. Knotek, H. Kurtzman and H. Glick,
      Phys. Rev. Letters 30, 856 (1973).
101. W. Eckenbach, W Fuhs and J. Stuke, J. Non-Cryst.
      Solids 5, 264 (1971).
102. S. Koc, M. Zavetova and J. Zemek, Czech. J. Phys.
      22, 429 (1972).

103.   P. G. LeComber, R. J. Loveland, W. E. Spear and
        R. A. Vaughan, in Proceedings of the 5th Intl.
        Conf. on Amorphous and Liquid Semiconductors,
        Garmisch, 1973 (Taylor and Francis, London, 1974),
        p. 245.

104.   A. J. Lewis, G. A. N. Connell, W. Paul, J. R.
        Pawlik and R. J. Temkin, in Proceedings of the
        Intl. Conf. on Tetrahedrally Bonded Amorphous Semi-
        conductors, Yorktown Heights, 1974 (American Insti-
        tute of Physics, New York, 1974), p. 27.

105.   M. A. Paesler, S. C. Agarwal, S. J. Hudgens and H.
        Fritsche, in Proceedings of the Intl. Conf. on
        Tetrahedrally Bonded Amorphous Semiconductors,
        Yorktown Heights, 1974 (American Institute of
        Physics, New York, 1974), p. 37.

106.   J. D. Dow and D. Redfield, Phys. Rev. B1, 3358
        (1970); J. D. Dow and D. Redfield, Phys. Rev. Let-
        ters 26, 762 (1971); J. D. Dow and D. Redfield,
        Phys. Rev. B5, 594 (1972).

107.   J. A. Olley, Solid State Comm. 13, 1437 (1973).

108.   M. Zavetova and V. Vorlicek, Phys. Stat. Sol. b
        48, 113 (1971).

109.   G. A. N. Connell, Phys. Stat. Sol. b 53, 213 (1972).

PHOTOLUMINESCENCE

William Paul

Division of Engineering and Applied Physics

Harvard University, Cambridge, Mass. 02138

Photoluminescence is the term used to describe the emission of radiation when thermal equilibrium is upset by an external perturbation consisting of photons. It is the result of recombination, either of free electrons with free holes, or of one free carrier with the opposite sign carrier localized on a defect, or of two bound carriers of opposite sign on the same or different defects. Luminescence when the external perturbation is field injection at a contact or junction (electroluminescence), or particle bombardment, is also possible in principle for amorphous semiconductors, but neither has yet been reported.

It is worthwhile to recall briefly the phenomenon of photoluminescence in a crystal. This is a play in three acts:

(1) excitation, which distributes free electrons and free holes over the conduction and valence band states, and possibly also bound carriers in impurity states, with a density which depends on the spectrum of the incident radiation and the absorption spectrum and geometry of the crystal.

(2) relaxation, by carrier scattering which preserves the density of excited electrons and holes, but redistributes each type of carrier according to a Fermi-Dirac distribution within its own band.

(3) recombination, with single photon emission. This process comprises the first two out of at least six possibilities for transitions.

(a) band to band, with single photon emission
(b) band to recombination center, with single pho-
    ton emission
(c) band to band, with multiphonon emission
(d) band to recombination center, with multiphonon
    emission
(e) band to band, with energy release to a third
    carrier (Auger)
(f) band to recombination center, with energy re-
    lease to a third carrier (Auger).

In thermal equilibrium, the several processes of
carrier generation and recombination exactly balance.
Thus the six recombinative processes referred to above
are exactly balanced one by one by six inverse processes
of generation of pairs by ambient photons, ambient phonons and ambient carriers of high kinetic energy. For
example, the number of carrier pairs created by absorption of photons of a given energy in the Planck distribution for ambient temperature is equal to the number
which disappear by radiation of that photon energy during
recombination. This permits us to estimate the radiative
lifetime - crudely speaking the mean time between creation
of a pair and their recombination with photon emission -
without having to know the matrix element for the recombinative transition, but substituting for that a knowledge of the experimental absorption spectrum. It does
not tell us whether the recombination lifetimes for the
other processes in (3) above are shorter, nor how they
and relaxation processes inside a band affect the spectrum of recombination radiation. Usually, the relaxation
within bands is so rapid that the spectrum for band to
band radiative transitions has a band width proportional
to the sum of the widths of the steady state thermal distributions at the edges of the valence and conduction
bands, and the spectrum for band-recombination center processes is even narrower, provided the recombination energy
level is discrete.

What differences in the photoluminescence spectrum
might we expect when the semiconductor is disordered?
In considering this question, we shall suppose that states
of high energy in both the conduction and valence bands
remain delocalized, that states near the center of the
energy gap are highly localized, and that states whose
energy is near the band edges are intermediate in localization. Theory suggests that there may be a sharp transition in localization (often called a mobility edge*)

(* see bottom of next page).

at some critical energy in each band[1]; although this is
a simplifying feature for the analysis of the photoluminescence, we must remember that it has never been proved necessary to explain experimental results.

(1) Excitation

In the crystal, both electron and hole are in delocalized states after band to band excitation by photons
of high energy, and one or the other is in a delocalized
state after excitation from a defect level. There is a
qualitative difference between the wavefunctions of the
delocalized and the localized carriers, even though the
spatial extent of the latter is a sensitive function of
the energy of the defect in the band gap. In the disordered material, a similar qualitative distinction between
the two classes of delocalized and localized carriers applies if a sharp mobility edge exists. However, if it
doesn't, and we envisage a gradual decrease in localization of the wavefunction as its energy into the band increases, then we may well have, after excitation of the
material, a whole spectrum of localized and progressively more delocalized carriers.

(2) Relaxation

In the crystal, the relaxation time for scattering
of electrons and holes by phonons or impurities is of
the order of 10^{-12} sec., far shorter than the majority
of recombination times. Consequently, the electrons and
holes may be considered to relax to a distribution in
thermal equilibrium inside a single band, and so to be
concentrated in band widths of the order of thermal energies at the extrema of the valence and conduction bands
(of course, if one band is degenerate in thermal equilibrium, this is not strictly accurate and also, if carriers
are excited into a high lying extremum with a low matrix
element for scattering into other regions of the Brillouin zone, the statement above must also be modified.
However, it is applicable to the majority of cases). In
the amorphous material, the relaxation rate is at least
of the same order of magnitude for the truly extended
states, but very much slower as the states become more

---

* For clarity, we shall prefix this useful term by the
adjective "sharp" to describe an abrupt change in wavefunction localization with energy, thus differentiating
this case from the case of a more diffuse edge corresponding to a slower change in wave-function localization.

localized. The situation is represented schematically
in Figure 1 for the two differentiable cases of a sharply defined and an undefined mobility edge. We have plotted the relative probability of relaxation or recombination of a single carrier out of a state of any energy

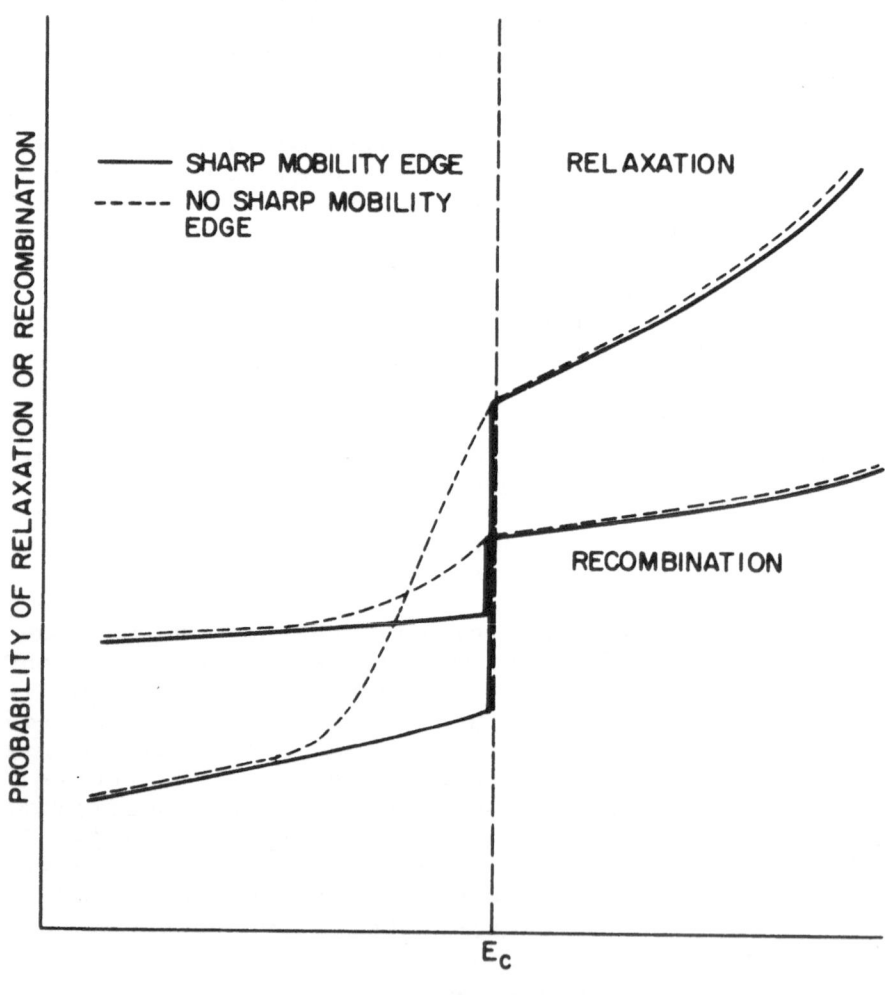

Fig. 1 Schematic representation of the relative probability of relaxation and recombination, as a function of
carrier energy above the Fermi level, for the two cases
(a) when there exists a sharp mobility edge (b) when
there does not.

above the Fermi level. In both cases, electrons and
holes are very likely to relax without recombination until they reach the mobility edges. The decrease in relaxation probability with decreasing energy is a consequence of decreasing final state density for the transition with phonon emission. Just below the sharp mobility edge, the space localization of the carrier sharply
decreases the probability of further relaxation with phonon emission, since the energetically available state
may be localized in a different region of space. Below
the mobility edge, relaxation proceeds at a progressively slower rate as the number of energetically-close, spa-
tially-overlapping localized states decrease. When the
mobility edge is diffuse, the curve of relaxation probability is smeared out as shown. In summary, if sharp mobility edges exist, there will be a concentration of carriers near them just after excitation. If the mobility
edges are diffuse, there will be few, if any, carriers
in the truly extended states, and a distribution over a
possibly wide range of localized states.[2]

(3) Recombination

In the crystal, recombination can occur by photon
emission, phonon emission and with Auger excitation.
The luminescence spectrum consists of a band to band
part, of energy close to the forbidden gap, and width
dependent on kT if the sample is statistically non-de-
generate, and on the position of the Fermi level in one
of the bands if it is degenerate. It may also have additional peaks if radiative recombination proceeds by recombination centers, either of discrete energy or in
bands. In the amorphous material, recombination can still
proceed by photon emission, phonon emission and with
Auger excitation. Of course the distinction between band
recombination and that via a recombination center with a
discrete energy will probably require modification.

The relative probability of recombination is also
plotted in Figure 1. For carrier energies above the mobility edge it is relatively constant, but may decrease
with decreasing energy as the density of final states for
a transition decreases. At the sharp mobility edge, the
probability shows a sharp decrease, because some of the
available transitions are between two localized states,
which are generally assumed to have lower probability,
when their overlap is small, than transitions between extended states and extended states and between extended
states and localized states. If the mobility edge is
diffuse, the curve is smeared out.

The net result is that, if a sharp mobility edge
exists, recombination becomes relatively more probable
than relaxation just below it; if the edge is diffuse,
the change-over in preference occurs at a lower energy.
Note that the luminescence spectrum is not defined by
this, since the recombination process has not been specified.

If sharp mobility edges exist, we can argue that the
luminescence spectrum will rise abruptly at energies just
below the mobility gap. We can also argue that radiative
recombination will be unlikely at low photon energies,
even though there is a continuous distribution of localized states in the energy gap, simply because the relative probability of recombination with multiphonon emission increases as the energy difference between initial
and final states decreases. Thus, unless the scenario
is changed by the presence of charged defects which have
a large cross-section for capture of either electrons or
holes, we would predict a luminescence spectrum increasing gradually from low photon energies, peaking at an energy below the mobility gap, and falling off rather sharply to a very small value at an energy equal to the mobility gap.

If sharp mobility edges do not exist, the spectrum
will be displaced to lower energies.

Temperature dependence

We can see in a qualitative way what will happen to
the curves in Figure 1, and to the luminescence spectrum,
when the temperature is lowered. Crudely speaking, all
of the curves of Figure 1 will be lowered in magnitude.
The relaxation probability is proportional to $(n_1 + 1)$
where n_1 is the Bose-Einstein excitation for an average
phonon. Similarly, the recombination probability is proportional to $(n_2 + 1)$ where n_2 is the Bose-Einstein excitation for an average photon emitted. Since $n_1 \gg n_2$,
the relaxation variations are changed more in magnitude.
The relaxation-recombination cross-over energy is unchanged when there exists a sharp mobility edge, but
moves to higher energy when the edge is diffuse.

The balance between radiative and non-radiative recombination changes in favor of the former as the temperature is lowered, so that the overall magnitude of the
luminescence increases greatly. If there is a sharp mobility edge, the shape of the spectrum on the high energy
side will be unchanged, unless there is a shift in the

mobility gap with temperature (some shift is quite probable); the spectrum may actually be extended on the low
energy side, as multiphonon emission is disfavored at
the low temperature, so that the band width of the luminescence spectrum is increased. If the mobility edge
is diffuse, the luminescence spectrum will be increased
in intensity, shifted to higher energy, and increased in
width at low temperatures.

We shall next examine how some of these speculations
compare with experimental results.

Experimental observations

Most of the experimental observations of photoluminescence in amorphous semiconductors have been on the
arsenic chalcogenides, first reported by Kolomiets et
al in 1968[3]. More recently, Engemann and Fischer[4] have
reported results for a-Si produced from silane, and we
shall take up an account of these first.

Photoluminescence in a-Si

Figure 2 shows the luminescence spectrum as a result of excitation with a Kr laser of photon energy
1.91 eV. There are peaks at 0.92, 1.1 and 1.25 eV, with
a total half-width of 0.3 eV. The integrated intensity
is roughly constant below about 80° K but falls rapidly
with increasing temperature thereafter. The 1.25 eV peak,
which is dominant at low temperature, becomes lower than
the 1.1 eV peak at 80 K, and lower than the 0.92 eV peak
at 170 K. The total half-width is almost constant with
changing temperature.

Pertinent to a discussion of these results is our
present knowledge of a-Si from other types of experiment.
Samples produced from silane are known to be orders of
magnitude more photoconducting than samples produced by
evaporation or sputtering. Engemann and Fischer found
no photoluminescence from the latter type of sample, even
when the sample was carefully annealed and showed the
same T dependence of the dark conductivity as the silane
type ones*. It is also known from measurements of the
dark conductivity and the photoconductivity that the mobility gap is about 1.5 eV and that there are localized
electron traps about 0.15 eV below the conduction band
edge.[5] From field effect measurements it is known that
the distribution of states in the gap is non-monotonic,

(* see bottom of next page).

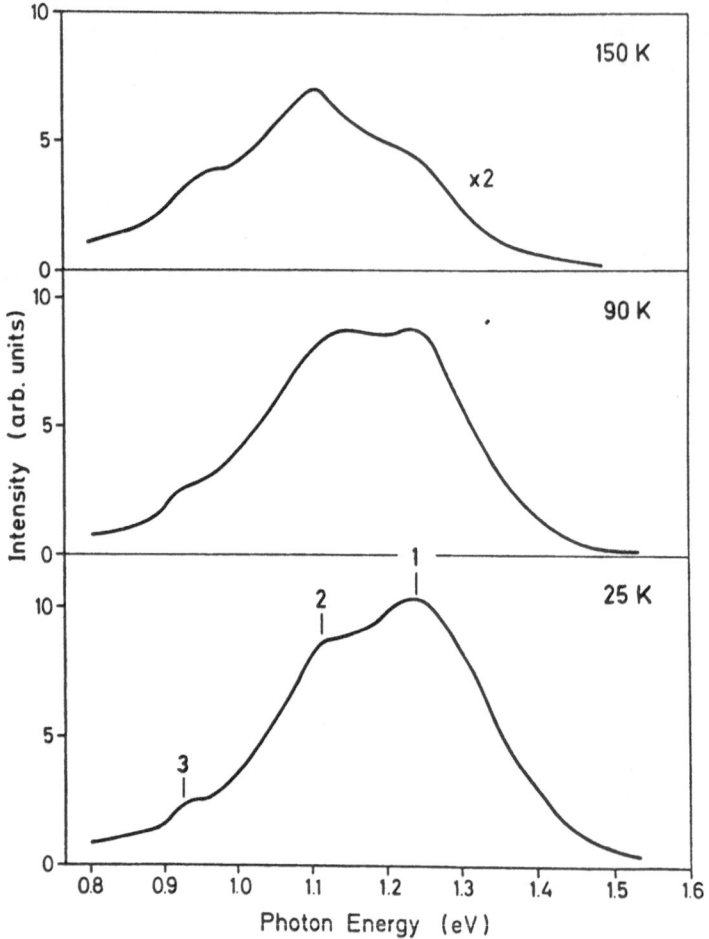

Fig. 2 Luminescence spectra of a-Si at 25K, 90K, and 150K, as measured by Engemann and Fischer.[4]

*This strongly suggest (although Engemann and Fischer do not state this point of view)that the nature of the states in the gap is quite different in the two kinds of sample. It may also mean that the silane type samples contain hydrogen, which, by compensating dangling bonds, modifies the number and nature of the states in the gap and reduces the non-radiative recombination. The fact that evaporated a-Si, annealed to give the same dependence of conductivity on temperature, shows no photoluminescence may mean that the photoluminescence depends on special properties of states associated with the hydrogen atoms.

with a maximum at 0.4 eV below the edge of the conduction band, a minimum near gap center, and a second maximum near the valence band.

The displacement of the 1.25 eV peak away from the 1.5 eV band gap suggests that the mobility edge is not sharp. Engemann and Fischer suggest that the electron traps (Loveland et al.) 0.15 eV below the edge of the conduction band are the initial states for the transition giving the 1.25 eV peak. The final state must then be 0.1 eV above the valence band edge, implying that the holes do not relax far into the band gap before some recombination sets in. The spectrum is asymmetric in the fashion predicted, although the spectra do not extend to as low energies as we might have expected. The occurrence of two subsidiary peaks requires that we add to our simple model local maxima in the density of states vs. energy function. Engemann and Fischer's model is shown in Figure 3; the two additional peaks shown are consistent with deductions from the field effect experiments[6], although they are not uniquely predicted by the PL data.

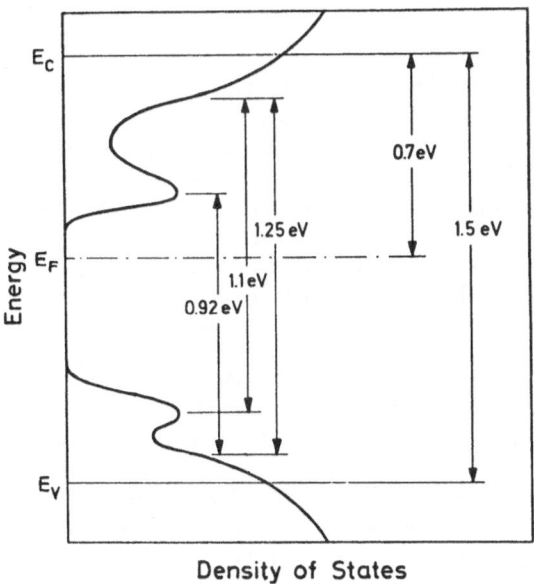

Fig. 3 Density of states model for a-Si proposed by Engemann and Fischer[4] to explain the luminescence spectra of Figure 2.

The overall decrease in luminescent intensity with increase of temperature is not unexpected, and increased multiphonon emission is one possible cause. The shift in relative intensity into the lower energy peaks is consistent with the notion that increased relaxation populates states further from the band edges before recombination occurs, as envisaged in Figure 1.

We conclude that there are no major mysteries in the photoluminescence spectra of a-Si, and that they provide very useful information on state density distributions and relaxation and recombination kinetics to be correlated with other data.

Photoluminescence in the arsenic chalcogenides

Photoluminescence in the arsenic chalcogenides (As_2S_3, As_2Se_3, As_2Te_3, and their alloys) was first reported by Kolomiets, Mamontova and Negreskul[3] in 1968. They excited a sample of As_2Se_3-As_2Te_3 with ultraviolet radiation at 77 K and found a luminescence spectrum with a peak near 0.67 eV and a much smaller peak near 1.16 eV. The energy gap from transport studies is 1.21 eV. Kolomiets et al interpreted the higher energy peak as band to band recombination and ascribed the 0.67 eV peak to transitions from a band edge to a band of localized states near the center of the gap.

In a later study, Kolomiets, Mamontova and Babaev[7] found similar spectra for As_2S_3 and As_2Se_3. The Kolomiets' group[8] then studied any differences in the intensity, energy and energy half-width of the spectra between crystalline and amorphous As_2Se_3, and in both crystal and amorphous material with various additions of impurity. The peak energy was found to be only slightly (20°/o) lower and the width of the principal luminescence peak was found to be nearly the same in undoped disordered As_2Se_3. The energy was roughly half of the energy of the forbidden gap. This fact suggested to the authors that, in addition to any continuous distribution of localized states in the pseudogap of the disordered material, there existed localized states, which were not very different from those in the crystal. Presumably the discrete energies of the states in the crystal are broadened into bands in the amorphous material. Kolomiets et al concluded that the changes in the luminescence spectra with stoichiometry and impurity incorporation were caused principally by changes in a band of localized states associated with structural defects. Impurity additions up to 10^{19} cm^{-3} had little or no effect on the spectra, but larger ad-

ditions were presumed to alter the number and nature of
the structural defects. From the changes in luminescent
energy and intensity with changes in stoichiometry, it
was inferred that the structural defect involved in both
crystal and amorphous material was the termination of
chains, especially when a Se atom was the last atom of
the chain.

The Kolomiets interpretation does not envisage much
difference in the interpretation of photoluminescence
for a disordered and a crystalline semiconductor. It is
consistent with relaxation in both valence and conduction
bands to a density of states or a mobility edge, followed by recombination without further relaxation. It is
not, however, a unique interpretation. For example, the
peak in the luminescence spectrum could be caused by
transitions between two narrow bands of localized states,
both associated with the same defect. This would not be
inconsistent with the existence of a continuous localized state density associated with disorder, and it is not
affected by the presence of mobility edges in the disordered material or indeed density of states edges in the
crystal.

We have already seen that it is possible to have a
luminescence spectrum with a maximum at an energy much
below the gap energy even in the presence of a continuum of quasi-localized states in the gap. Investigations
by Fischer and coworkers[9], Street and coworkers[10], and
by Bishop and Mitchell[11], have amplified and extended
the Russian work, without changing the basic experimental
result. However, the later studies demonstrate very significant differences in the details of their interpretation.

Fischer et al measured the luminescence spectra of
evaporated amorphous films of 2 As_2Te_3 - As_2Se_3 from 2
to 150 K for various intensities of 1.2 eV laser excitation. The luminescence spectrum showed a single peak
near 0.6 eV and a half-width of about 0.15 eV. Its shape
was independent of excitation intensity for low intensities, and independent of temperature from 2 to 110 K.
The integrated luminescent intensity increased linearly
with excitation intensity over a range of a factor of 50;
it was constant with T up to about 40 K and decreased
with increased T thereafter. The temperature dependence
was analyzed in terms of competing non-radiative processes whose nature is open to speculation, but which may
involve diffusion to the surface and/or increased recombination with phonon emission.

For the interpretation of their results, Fischer
et al used a quantitative version of the relaxation-
recombination model already described. This had previously been applied to studies of the photoconductivity
of $2As_2Te_3$ - As_2Se_3 by Weiser, Fischer and Brodsky[12].
There the notion of "recombination edges", about 0.2 eV
below the mobility edges, was introduced: essentially
these occur at the energy where the relative probabilities of relaxation and recombination depicted in Figure
1 cross*. In their quantitative model, Fischer et al
assumed that relaxation took place by emission of a single phonon, that the recombination rate was independent
of carrier energy, and that the conduction and valence
bands had identical properties with exponential tails of
quasi-localized states. Such a model permits the prediction of a luminescence spectrum in terms of parameters
such as the phonon energy, the exponential rate of increase of localized state density, the relaxation and recombination rate coefficients per carrier and the level of
pair excitation. Unfortunately, although judicious choice
of the parameters leads to a spectrum in reasonable a-
greement with experiment, the theory predicts that the
energy of the luminescence peak shift, when the excitation level is altered, at a rate roughly five times that
seen in practice. Fischer et al speculate that the cause
of the reduced peak shift may be local maxima in the density of states in the gap about 0.2 eV from the mobility
edges. This plausible speculation is consistent with an
internal transition at a structural defect, which, on
the one hand, does not require the assumption of a transition from band edges into a single localized band of
the Kolomiets' interpretation and, on the other, very
likely violates the assumption of a monotonically varying recombination rate with energy of the relaxation-
recombination model.

Street, Searle and Austin studied the photoluminescence and excitation spectra of a-As_2S_3 prepared both by
melting bulk glass and by vacuum evaporation. In the
former, the change in luminescence intensity with energy
is found for a fixed exciting photon energy. In the latter, the change in luminescence intensity at some chosen
energy is determined as the exciting photon energy is
varied. The shapes of the luminescence and excitation
spectra are found to be independent of the discrete ex-

---

*Figure 1, case (b) should be used here, since these au-
thors (see Weiser, reference 2, p. 924) believe that
there is no abrupt change in relaxation rate at a sharp
mobility edge.

citation and luminescent photon energies employed.

The shape of the luminescence spectrum, illustrated in Figure 4, was found to be independent of temperature between 8 and 69 K, although its intensity decreased by a factor of 10. The excitation spectrum, normalized for number of _incident_ photons, is illustrated in Figure 5. It shows that very little luminescence was produced at any wavelength for excitation energies below about 2.2 eV (absorption coefficient 1 cm^{-1}) or about about 2.8 eV (absorption coefficient 10^4 cm^{-1}, still in the exponential tail region of the absorption spectrum). The fall-off at low energies is ascribable to insufficient absorption in a sample of finite thickness; correction for this suggests that an excitation spectrum normalized to number of photons _absorbed_ would increase monotonically at least down to absorption coefficients of 1 cm^{-1}.

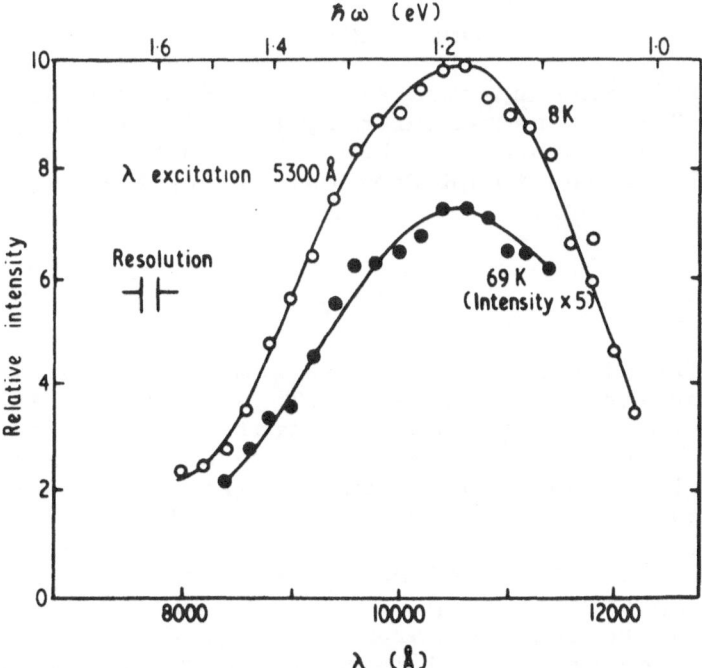

Fig. 4 Luminescence spectra at 8K and 69K of glassy As_2S_3, thickness 2.2 mm, as measured by Street, Searle and Austin[10].

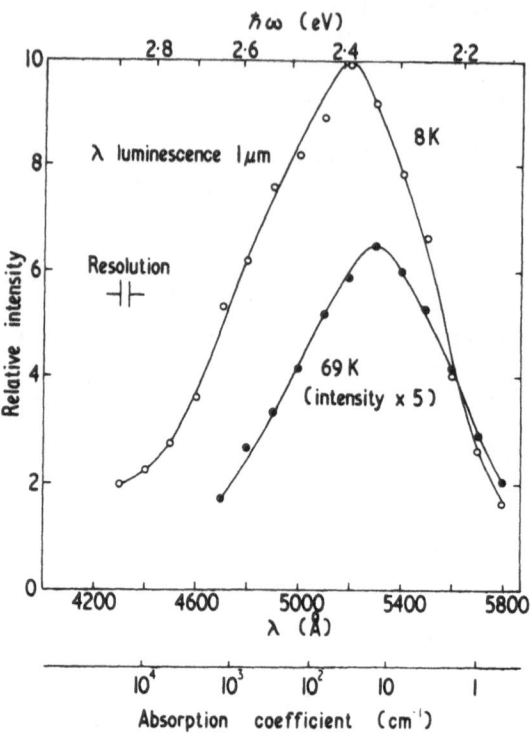

Fig. 5 Excitation spectra at 8K and 69K of glassy As$_2$S$_3$, thickness 2.2 mm, as measured by Street, Searle and Austin[10]. Also shown are absorption coefficients of As$_2$S$_3$ extrapolated to 8K from data given by Tauc and Menthe[22].

The fall-off at high energies is reminiscent of the effect of surface recombination. This, however, is hard to justify, since the light penetration depth is roughly 1 micron, which might be expected to be much larger than the diffusion length for carrier pairs (see below and under discussion of Bishop and Mitchell's work).

The effect of annealing on the excitation and luminescence spectra of evaporated films was also studied. Annealing decreased the intensity of the excitation spectra at high energies, and left it unaffected at low energies. This is illustrated in Figure 6. The change in the luminescence spectrum for samples prepared by different methods and annealed was studied; this is illustrated

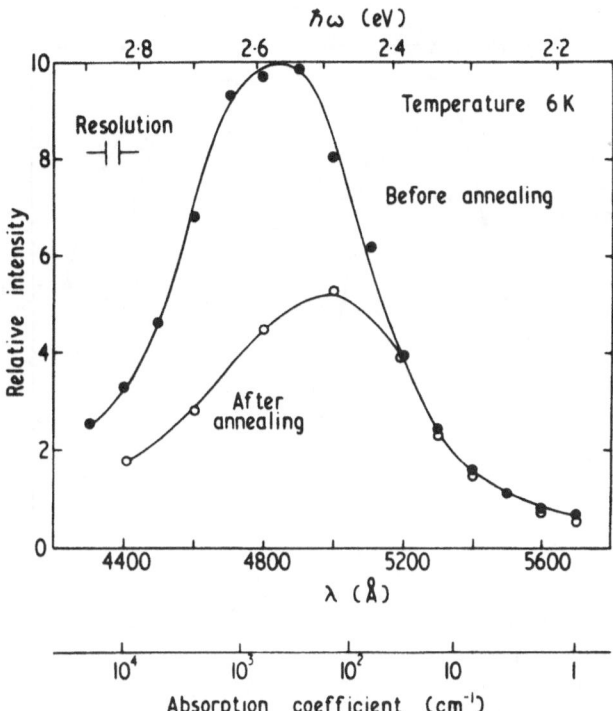

Fig. 6 Excitation spectra of an evaporated film of
As$_2$S$_3$, thickness 29 microns, showing the effect of an-
nealing at 200°C for 4 hours in an argon atmosphere.
From Street, Searle and Austin[10].

in Figure 7. Finally the intensity of the luminescence
was found to fall rapidly with rising temperature over
the full temperature range.

The lack of dependence of the luminescence spectrum
on excitation energy is consistent with the supposition
of rapid relaxation after excitation and before significant recombination. The similarity in the luminescence
spectra of amorphous and crystalline samples confirms
Kolomiets' general conclusion, even though the peaks in
Figure 7 are separated by 0.4 eV, rather than the 0.1 eV
found by Kolomiets. It is stressed that the similarity
in the spectra for the crystalline and amorphous phases
strongly suggests that the analysis of Fischer et al,
based on the existence of long band tails of states and

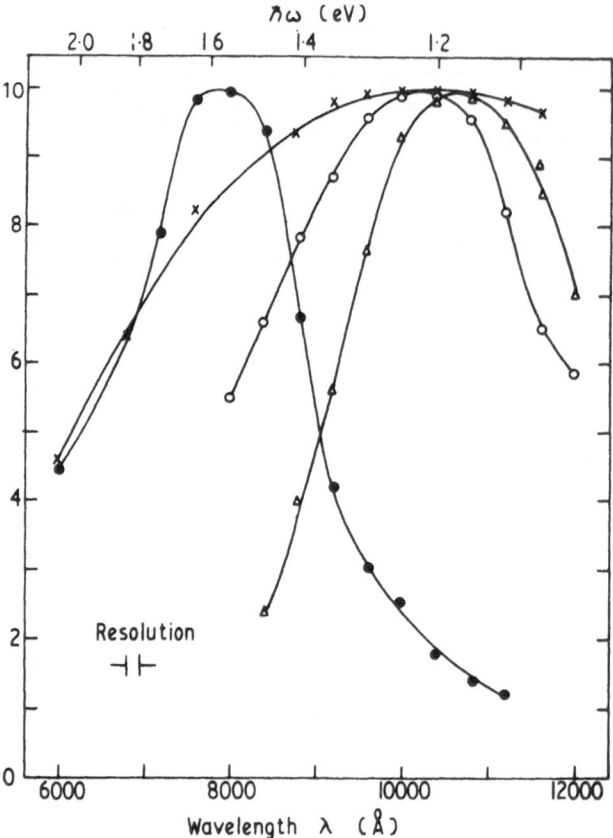

Fig. 7 Luminescence spectra of As$_2$S$_3$ prepared by different methods: $\triangle$, melted glass; x, freshly evaporated film; O, annealed evaporated film; $\bullet$, natural single crystal. From Street, Searle and Austin[10].

recombination edges, has omitted an important common element in the explanation.

The decrease of luminescence intensity with annealing supports the view that the luminescence in As$_2$S$_3$ involves recombination centers at broken bonds, whose number would be reduced by annealing. This agrees with the conclusions of Kolomiets et al on As$_2$Se$_3$. The broken bonds may be supposed to produce a band of recombination centers near midgap so that recombination involves transitions into them from the edges of the valence and con-

duction bands, or they may produce at least two bound
states between which the radiative transition takes place.
Street et al note that the former possibility is in accord with the Mott-Davis model[13] of amorphous semiconductors.

Street et al discard the possibility that the decrease in efficiency of luminescence as the excitation
energy is increased is caused by diffusion to and non-
radiative recombination at the surface. They argue that
the diffusion length required (of the order of 1 micron)
is unreasonably large and that, besides, the recombination at the surface is not significantly more rapid than
in the bulk[14]. They present instead a model for an energy-dependent quantum efficiency. The absorption in
the exponential tail at low excitation energies is supposed to be caused by transitions in the electric fields[15]
surrounding charged defects, so that the electron-hole
pairs are then conveniently in the vicinity of these defects which can act as efficient recombination centers.
At higher excitation energies, the electron-hole pair
can be produced by interband transitions without tunneling in an electric field, and the carriers can more
easily separate. Then, it is argued, competing non-radiative processes may dominate the recombination. If
one supposes that the mobility in the extended states is
1 cm^2/volt-sec at 10 K then a carrier would take about
10 μsec to diffuse 100Å, the separation between defects
that are present in a density of $10^{18}cm^{-3}$. It seems very
possible as suggested by Street et al, that the pairs
could recombine non-radiatively in times shorter than
this, or fall into deep traps from which they could neither escape at low temperature or recombine.

Street, Searle and Austin[16] have also investigated
the decay time of the photoluminescence in As_2S_3. The
experiments measured the variation of the photoluminescence at a fixed energy in the luminescence spectrum as
a function of the time after a brief burst (10 μsec) of
excitation by monochromatic light in the excitation spectrum. Typical results are illustrated in Figure 8; they
show no unique time constant. The variation of the decay rate with emission energy was also measured; the rate
of decay accelerated at higher energies at 7K, and was
almost independent of energy (although more rapid) at 49K.
An arbitrarily defined decay time constant was relatively constant up to 20K and decreased thereafter by a factor of five by 70K. The steady state response using 20
m sec pulses of light was also measured as a function of

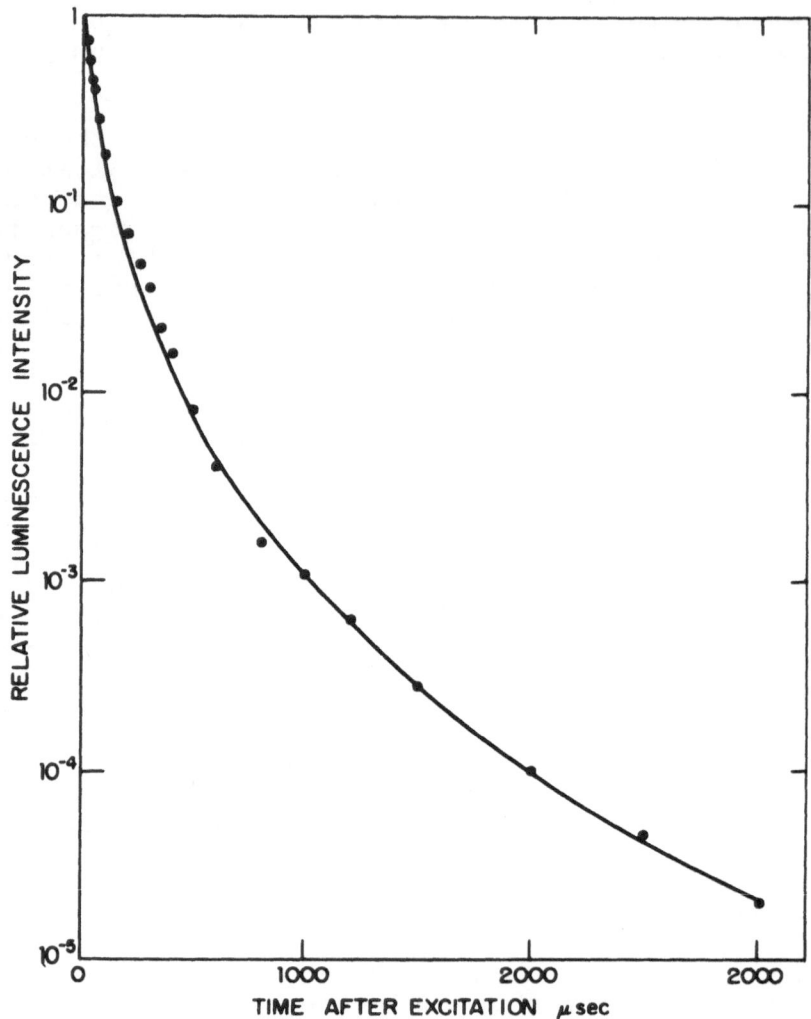

Fig. 8 Decay of luminescence at 7K of a glassy sample
of As_2S_3. Excitation energy 2.34 eV, emission energy
1.24 eV. From Street, Searle and Austin[16].

temperature. The intensity at a fixed energy of the luminescence spectrum decreased exponentially as $\exp(-T/26)$
up to at least 240K in melted glass samples, and up to
130K in as-evaporated films. Steady illumination of the
samples for 15 minutes at 5K with the same exciting photon energy, prior to measuring the temperature dependence

of the luminescence with pulses reduced the magnitude of
the luminescence signal. For the melted glass, the intensity, reduced by about a factor of two, still followed an $\exp(-T/26)$ law. For the evaporated sample, the
intensity was reduced by an order of magnitude at 5K by
the steady illumination and the subsequent decay of intensity with temperature occurred at a slower rate. Similar "fatiguing" of luminescence with a steady excitation has also been reported for a-As_2Se_3 by Cernogora,
Mollot and Benoit a la Guillaume[17].

Street et al have interpreted two main features from
this wealth of experimental data, namely, the decrease
in lifetime for electron-hole pairs of increasing energy
(increasing energy in the luminescence spectrum), and
the exponential decay of photoluminescence with temperature, on the basis of a model of fluctuations of electrostatic potential on a scale of 50Å . It is argued that
the radiative lifetime $\tau \sim \exp(2\,\alpha\,R)$ where R is the separation of the initial and final states localized in
real space and α is the state localization parameter.
Thus, in Figure 9, transition 1 (for a carrier relaxed
into a low energy in its band) is faster than transition
2 for two reasons: (a) R is smaller, (b) α is smaller, since
it is expected that further into the tail of states adjacent to each band edge, the localization parameter will
increase with penetration into the gap* [18].

The decrease of the luminescence efficiency with
temperature, which occurs at the same rate as the lifetime, is also explicable on a model of fluctuating electrostatic potential. Essentially, the competing non-radiative process involves first the tunneling of one carrier between neighboring minima in the fluctuating band.
If this tunneling is assumed to be a combination of excitation to a higher energy followed by tunneling through
a reduced potential barrier, as illustrated in Figure
9, it will clearly increase with temperature increase,
and the probability of radiative recombination will correspondingly decrease. Moreover, the efficiency of radiative recombination can be shown to depend simply on

---

*Street et al reject fluctuations in potential caused by
local strains, which would lead to variations in the energy gap, on the basis that the energy of states in midgap would be unchanged. It is not evident that this restriction is necessary or tenable, especially since the
strain dependence of the energy of defect levels is not
a simple function of their energy in the band gap.

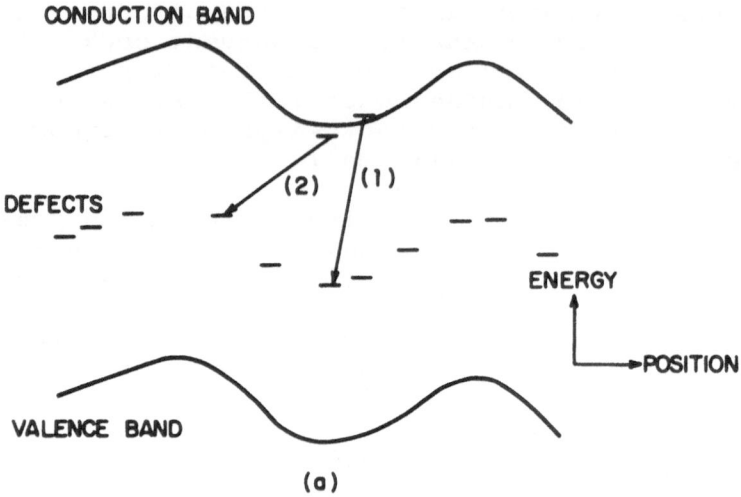

(a)

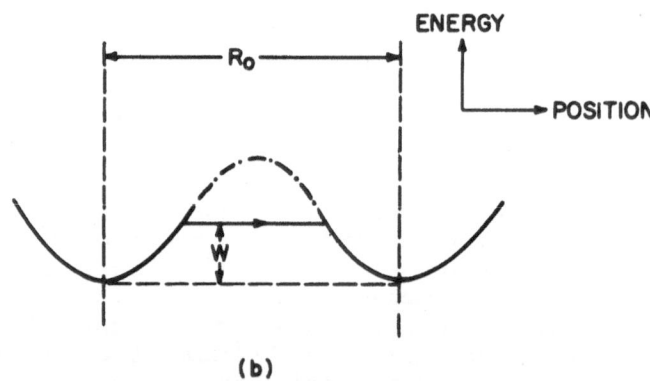

(b)

Fig. 9 (a) Schematic diagram of energy bands showing
typical radiative transitions. Transition 1 is faster
than transition 2 because of a smaller tunneling distance and a smaller localization parameter. (b) Idealized model of adjacent band minima showing a tunneling
path through a reduced potential barrier. From Street,
Searle and Austin[16].

the spatial separation of the extrema in potential, and
the localization parameter; a separation of 50Å for a
localization parameter α of 5Å^{-1} is found.

Street et al are then able to elaborate the model
for luminescence formulated earlier on the basis of the
luminescence and excitation spectra. They postulate that
an electron hole pair is formed by absorption of a photon in the exponential tail of the absorption spectrum,
and that this is de facto in the electric field of a
charged defect. One of the carriers then recombines rapidly by successive phonon emission through the hydro-
genic excited state levels of the system of charged carrier and oppositely charged defect. The carrier which
has thus recombined then stays in the defect, (even though its energy may not correspond to the minimum energy
in the fluctuating band to which it is close), because
of the Coulomb attraction. Subsequently the carrier that
is left, after thermalizing in a potential minimum of
the fluctuating potential, recombines radiatively with
the neutral defect; in the absence of a Coulomb attraction the intermediate energy levels are absent and the
competing process of simultaneous emission of some 30
phonons has relatively low probability. This model appears to be capable of explaining the observed features
of luminescence in As_2S_3.

Bishop and Mitchell have also studied the photoluminescence spectra, the excitation spectra and the photo-
luminescent decay time for glassy As_2S_3, As_2Se_3 and
As_2Se_3-As_2Te_3 at 6K and 80K.

The photoluminescence spectrum for As_2Se_3-As_2Te_3 at
6K showed a peak at 0.67 eV in good agreement with the
original report of Kolomiets et al. for a sample at 77K.
The excitation spectra for several sample thicknesses
are shown in Figure 10, normalized to the number of incident photons and to equal peak intensity. The optical
absorption spectrum is also shown. Clearly the integrated intensity of luminescence follows the interband absorption coefficient at low energies, with the thinner
samples showing less response at a given energy because
of incomplete absorption of the exciting radiation. All
of the excitation spectra decrease monotonically for energies above about 1.3 eV, reminiscent of the similar effect found for As_2S_3 by Street et al. In contrast to
these workers, however, Bishop and Mitchell interpret the
decrease in efficiency at high excitation energies as
due to diffusion of carriers to the sample surface

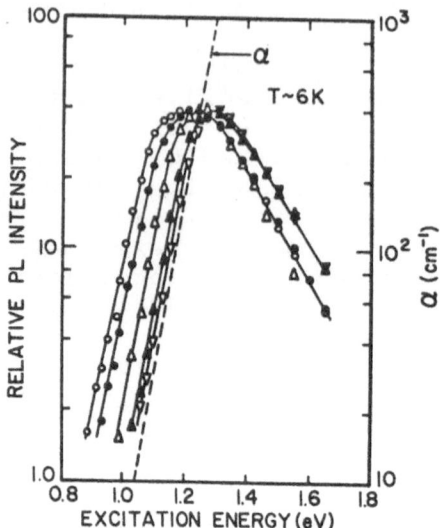

Fig. 10 Photoluminescence - excitation spectra for several samples of As_2Se_3-As_2Te_3 glass of varying thickness.
Sample thicknesses in cm are: o, 0.36; ●, 0.1; Δ, 3.4x10^{-2};
▲, 1.5x10^{-3}; and ▽, 1.0x10^{-3}. All spectra are normalized to the number of incident photons and to equal peak
intensity. The optical absorption spectrum is also shown.
From Bishop and Mitchell[11].

followed by non-radiative recombination. Since the maxima in the excitation spectra correspond to absorption
coefficients of the order of 10^2 cm^{-1}, the diffusion lengths must be large. Applying a theory by DeVore[19], whose
details we cannot discuss here, but which involves as
parameters the volume lifetime, the surface recombination
velocity, the diffusion coefficient, the diffusion length and the sample thickness, Bishop and Mitchell claim
to produce a reasonable fit to the part of the excitation
spectrum between 1.2 and 1.5 eV for which the theory is
applicable. The diffusion length deduced is about 1.5
microns.

Similar results found for glassy As_2Se_3 were in
agreement with the earlier work of Kolomiets et al, and
similar data for glassy As_2S_3 in agreement with Kolomiets
et al. and Street et al. The DeVore theory again appeared

to be suitable in its spectral range of applicability,
and yielded diffusion lengths of about 3 microns. The
data for As_2S_3 is illustrated in Figure 11.

Bishop and Mitchell also measured the photoluminescence intensity as a function of chopping frequency
and found that it decreased as the frequency increased.
Assuming monomolecular recombination kinetics, they were
able to deduce time constants for the decay of the luminescence of 0.1 m sec for As_2Se_3-As_2Te_3, 0.46 m sec for
As_2Se_3 and 0.84 m sec for As_2S_3.

Bishop and Mitchell emphasize that the shape of the
high energy decrease in excitation efficiency and its
dependence on thickness cannot be explained on the basis
of a single energy-dependent quantum-efficiency curve
and the absorption curves. An energy dependent quantum
efficiency which changes as a function of sample thickness, such as non-radiative surface-recombination produces, is their preference, and in this their interpretation is in direct contradiction to that of Street,
Searle and Austin. As part of their argument, Bishop
and Mitchell attack the conclusion of Mott and Davis that
the rates of recombination at the surface and in the bulk
of glasses are about the same. This conclusion was based

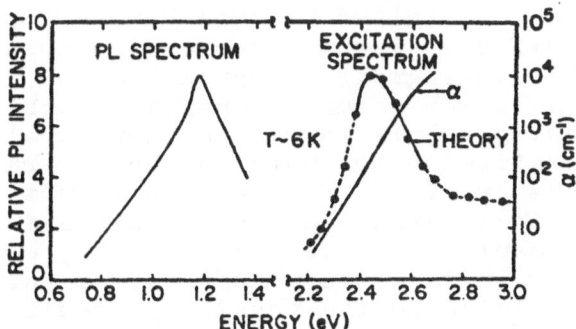

Fig. 11 Photoluminescence spectrum, excitation spectrum,
and absorption spectrum for glassy As_2S_3. The closed
circles and connecting dashed line represent the measured excitation spectrum for a 2.5×10^{-2}cm thick sample.
The theoretical solid curve segment represents the fit
of DeVore's theory[19] for the best choice of parameters.

on an observed constancy of photoresponse at high photon
energies, but Bishop and Mitchell point out that the samples on which experiments were reported were all less
than one micron thick, in which case no fall-off in photoresponse from strong surface recombination is predicted by theory. They also point out that their measured
millisecond lifetimes - which appear to be large at first
glance - are 6K radiative lifetimes, that are appropriate for the experiment, but may be much longer than the
high temperature non-radiative lifetimes estimated from
photoconductivity data. This is correct, and the two
experiments, which measure different kinds of lifetime,
are not necessarily inconsistent. What _is_ unlikely is
that non-radiative lifetimes of 10^{-8} sec at, say, 300K
will lengthen to 10^{-4} sec at 6K. Yet Bishop and Mitchell
require that the carriers not be in localized states just
before recombination, if they are to have diffusion lengths of the order of 1 micron. The appropriate lifetime
for carrier-transport in delocalized states is the shortest (probably non-radiative) one, which may $\sim 10^{-7}$ sec at
6K. This, along with a 1 micron diffusion length, suggests an extended state mobility of ~ 1000 cm^2/volt-sec,
which is wholly unreasonable for disordered materials.
Bishop and Mitchell, on the other hand, use $\tau \sim 10^{-4}$ sec
and $L \sim 10^{-4}$ cm to deduce a mobility of ~ 0.5 cm^2/volt-sec;
it is the contention of this author that this calcula-
tion is not defensible.

    In summary, many of the data on photoluminescence
in the arsenic chalcogenides are mutually consistent.
The luminescence is excited by interband optical absorption, with an efficiency which decreases rapidly for absorption coefficients greater than $10^4 cm^{-1}$. The luminescence spectra, which occur at energies much less than
the optical gap, probably involve a structural defect
with characteristics common both to the several kinds of
glass and to the crystal. However, it is not clear whether the fall-off in efficiency with increasing photon
energy is the result of diffusion of a very mobile carrier followed by surface recombination, or whether it is
a bulk effect related to the physics of radiative recombination in the electric field of a charged defect. More
work is apparently needed to finally resolve this question.

CONCLUSIONS

    Experience with amorphous silicon and the amorphous
arsenic chalcogenides suggests that photoluminescence

should be considered as a valuable additional method of
studying the distribution of delocalized and localized
states in amorphous semiconductors. It is preferable
to carry out such investigations in parallel with measurements of the absorption and photoconductivity spectra on the same samples, and to correlate the results
with other measurements (field effect, other transport)
in order to obtain information on the distribution of
states and on the carrier kinetics in non-equilibrium
situations.

The data on a-Si, although sparse at the moment of
writing, suggest, on the one hand, that a sharp mobility
edge is unlikely, and on the other, that a-Si has structure in the density of states in its forbidden gap. These
density of states maxima are presumably caused by a relatively simple structural defect, which, in view of the
similarity in the short range order of the disordered
substance to that of the crystal, probably occurs in the
crystalline phase as well. One plausible candidate is
the divacancy (Spear, reference 6).

In the arsenic chalcogenides, there is similarly no
positive evidence in favor of the existence of a sharp
mobility edge. The early model of Cohen, Fritzsche and
Ovshinsky[20], which envisioned a continuous, monotonically varying density of states tail between the mobility
edges and a Fermi level near gap center, requires elaboration in order to explain the photoluminescence data.
The CFO model assumes that the tail states are the result of potential fluctuations in completely coordinated
material. The PL data suggest that there exists structure in the density of states in the gap between the extended states of the conduction and valence bands, and
that this is probably caused by recombination centers
which are structural defects. The work of Kolomiets et
al provides clues regarding the nature of the structural
defects, but this is a difficult problem which must be
regarded as incompletely resolved. Structure in the local state density of the chalcogenides is also suggested
by a variety of current evidence from transport and photo-
transport investigations[21]. Finally, there remains interesting unresolved questions regarding the decrease
of photoluminescence efficiency at high exciting photon
energies and the fatiguing of photoluminescence by a
steady intensity of exciting light[22].

ACKNOWLEDGMENTS

I would like to thank Dr. G. A. N. Connell and Dr.
T. Moustakas for helpful comments on the manuscript.

REFERENCES

1.   See N. F. Mott and E. A. Davis, _Electronic Processes_
     in Non-Crystalline Semiconductors (Oxford, 1971)p.43.
2.   The model described here is similar to the one a-
     dopted by Fischer et al., reference 9, and discussed
     in detail by K. Weiser, J. Non-Cryst. Sol. 8-10,
     922 (1972). It is different, however, in several
     respects, the most important being the assertion of
     a discontinuous change in relaxation and recombina-
     tion rates near a sharp mobility edge. In this,
     our view seems to be closer to that of M. H. Cohen
     and D. L. Johnson, J. Non-Cryst. Sol., 3, 271 (1970).
3.   B. T. Kolomiets, T. N. Mamontova and V. V. Negreskul,
     Phys. Stat. Sol., 27, K 15 (1968).
4.   D. Engemann and R. Fischer, Proceedings of the 5th
     International Conference on Amorphous and Liquid
     Semiconductors, Garmisch-Partenkirchen, ed. J.Stuke
     and W. Brenig (Taylor and Francis, 1974) pp.947.
5.   R. J. Loveland, W. E. Spear, A. Al-Sharbati, J.
     Non-Cryst. Sol., 13, 55 (1973/74).
6.   W. E. Spear, Proceedings of the 5th International
     Conference on Amorphous and Liquid Semiconductors,
     Garmisch-Partenkirchen, ed. J. Stuke and W. Brenig
     (Taylor and Francis, 1974) p. 1.
7.   B. T. Kolomiets, T. N. Mamontova and A. A. Babaev,
     J. Non-Cryst. Sol., 4, 289 (1970).
8.   B. T. Kolomiets, J. Non-Cryst. Sol., 8-10, 1004
     (1972).
9.   R. Fischer, V. Heim, F. Stern and K. Weiser, Phys.
     Rev. Lett., 26, 1182 (1971).
10. R. A. Street, T. M. Searle and I. G. Austin, J.
     Phys. C 6, 1830 (1973).
11. S. G. Bishop and D. L. Mitchell, Phys. Rev. B, 8,
     5696 (1973).
12. K. Weiser, R. Fischer and M. H. Brodsky, Proceed-
     ings of the 10th International Conference on the
     Physics of Semiconductors, Cambridge, Mass. (1970)
     p. 667.
13. E. A. Davis and N. F. Mott, Phil Mag. 22 903 (1970).
14. See N. F. Mott and E. A. Davis, _Electronic Processes_
     in Non-Crystalline Semiconductors (Oxford, 1971),
     p. 230.

15.   See J. D. Dow and D. Redfield, Phys. Rev. B$\underline{1}$, 3358
        (1970); Phys. Rev. B$\underline{5}$, 594 (1972); N. F. Mott and
        E. A. Davis, loc.cit., p. 237.
16.   R. A. Street, T. M. Searle and I. G. Austin, Pro-
        ceedings of the 5th International Conference on
        Amorphous and Liquid Semiconductors, Garmisch-
        Partenkirchen, ed. J. Stuke and W. Brenig (Taylor
        and Francis, 1974), p. 953.
17.   J. Cernogora, F. Mollot and C. Benoit a la Guillaume,
        Phys. Stat. Sol., (a) $\underline{15}$, 401 (1973).
18.   N. F. Mott and E. A. Davis, <u>Electronic Processes in
        Non-Crystalline Semiconductors</u>, (Oxford, 1971),
        p. 17.
19.   H. B. DeVore, Phys. Rev., $\underline{102}$, 86 (1956).
20.   M. H. Cohen, H. Fritzsche and S. R. Ovshinsky, Phys.
        Rev. Letters $\underline{22}$, 1065 (1969).
21.   J. M. Marshall and A. E. Owen, Phil. Mag. $\underline{24}$, 1281
        (1971). J. M. Marshall, C. Main and A. E. Owen,
        J. Non-Cryst. Sol. $\underline{8\text{-}10}$, 760 (1972). C. Main and
        A. E. Owen, Proceedings of the 5th International
        Conference on Amorphous and Liquid Semiconductors,
        Garmisch-Partenkirchen, ed. J. Stuke and W. Brenig
        (Taylor and Francis, 1974), p. 953.
22.   J. Tauc and A. Menthe, J. Non-Cryst. Sol., $\underline{8\text{-}10}$,
        569 (1972).

INTRODUCTION TO PHOTOEMISSION

D. Weaire

Dept. of Physics, Heriot-Watt University
Riccarton, Currie, Midlothian, Scotland
and
Dept. of Engineering and Applied Science
Yale University, New Haven, Conn., U.S.A.

I. INTRODUCTION

This lecture is intended to be an introduction to the interpretation of photoemission energy distributions. It is not a comprehensive guide to the rapid developments in theory and experiment which have occurred in the last few years, but by clarifying the terminology and general ideas, may serve to clear the ground for later lectures.

The technique is emerging from an initial period of much uncertainty, particularly concerning the relative weighting of surface and bulk contributions and of k-conserving and non-k-conserving transitions in observed spectra. Inasmuch as the theoretical analysis is still in a process of refinement by testing against results for single crystals, it is not inappropriate to devote some time to ideas (critical points and lines) which in fact have relevance only to crystals.

Let us begin with a reminder of some of the various spectroscopic techniques which may be brought to bear on the problem of the electronic structure of a solid. Fig.1 illustrates some of these. Photoemission has much in common with optical or x-ray absorption. It has the advantage that it yields "two-dimensional" data, i.e., a function of incident and final energies, but the added complication that the excited electron must escape from the solid before it is observed. This makes interpretation

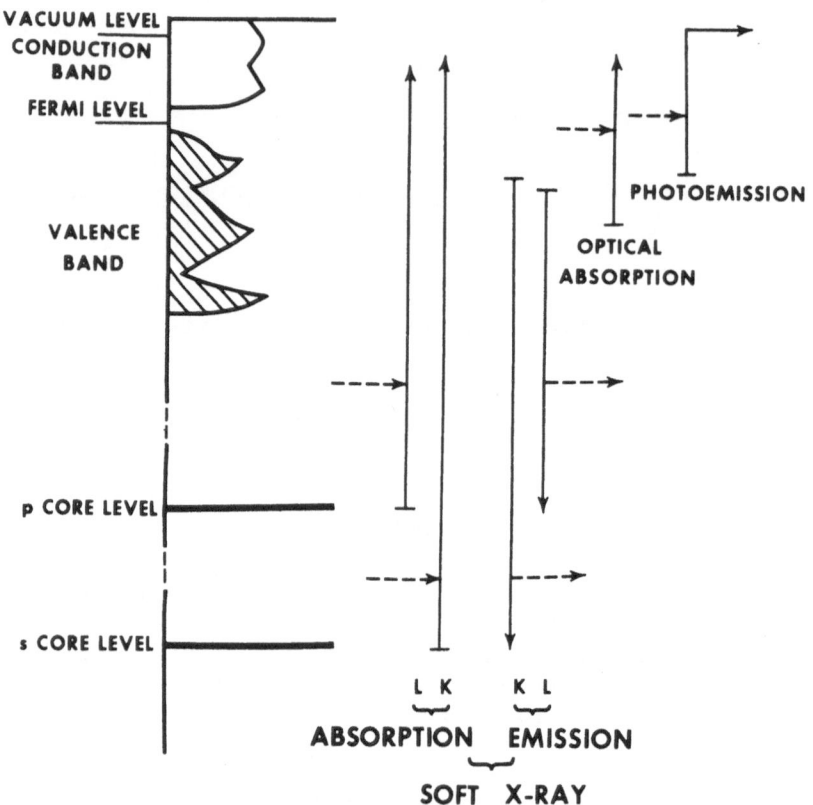

Fig. 1   Various spectroscopic techniques.
————electrons   - - - - photons

more difficult but, as we shall see, may be used to advantage in the study of surface properties.

II. NAIVE VIEW OF PHOTOEMISSION

Naively, the photoemission spectrum associated with
a valence band may be visualized as follows. Suppose we
neglect any structure in the conduction band. Apart from
a sharp cut-off to be expected when the energy of the excited electron falls below the vacuum level, the spectrum
of emitted electrons will be that of the initial states.
Let us immediately list some of the many things which are
wrong with such a picture.

(a) The conduction band is not structureless and matrix elements between valence band and conduction band states may vary considerably also. The relevent matrix element is the dipole matrix element < initial $|\underline{p}|$ final >.

(b) This is a one-electron picture. Many-body effects can shift and modify photoemission spectra.

(c) Because of the finite escape depth, the spectrum may contain significant contributions from the surface region where the valence band density of states is different.

(d) Secondaries produced by the inelastic scattering responsible for the finite escape depth will overlap the (unscattered) spectrum.

(e) The cut-off is certainly not sharp and the escape probability may not be smooth above the threshold.

(f) Other processes (e.g. Auger) may produce overlapping spectra.

No doubt there are further difficulties not so familiar to an outsider such as myself!

III. THE X-RAY LIMIT

In this rather difficult situation, which is not necessarily eased by the protests of theorists that descriptions such as the above are inappropriate and should be replaced by Green's functions[1], it is natural to seek means of eliminating as many of the complicating factors as possible by the choice of appropriate experimental parameters. In this respect, the recent rapid development of UV and x-ray photoemission techniques has been a great boon. As the photon energy is increased into the UV, and above, problem (a) is eliminated as one achieves what Eastman has called the "x-ray limit", typically at about 40 eV or so. At such energies and higher, the conduction bands become smooth and, at least in some average sense, free-electron-like, and it would appear that the spectrum of excited electrons indeed reflects the density of initial states. One must be careful here - the possibility of resonances (peaks) in the conduction band does exist and should be checked for by measuring at several energies.

The role of secondaries is to add a steeply rising but smooth contribution to the spectrum on the low energy side. Attempts have been made to calculate this[3] for semiconductors, using a theory due to Kane[4] but since it appears to be quite smooth, it can be subtracted out by an unsophisticated extrapolation without much difficulty.

IV. PHOTOEMISSION IN THE OPTICAL REGION

We have first discussed photoemission in the x-ray region of the spectrum because it is such a direct probe of the density of states of the valence band.

At lower energies, (in which regime most measurements were made until recently - the important cut-off is that of the LiF window at about 11 eV) the question of the proper calculation of matrix elements must be faced. In view of the difficulty of this, many experiments were interpreted on the following basis:[2] If a feature, let us say a peak, in the spectrum moves to higher energies when the exciting energy is increased, then it must be associated with the initial states. If it stays fixed, then it is in the final density of states. However the spectrum is not just a convolution of the initial and final densities of states. It involves the matrix element between the states, implying k-selection rules in crystals as well as a weighting depending on the character of the states (e.g. largely s-like states will have rather small matrix elements with largely p-like states, because of the dipole selection rule). Such complications made "rule of the thumb" interpretations difficult.

Kane[5] has provided some guiding principles for the interpretation of spectra of crystalline solids based on the identification of critical points (singularities) in the spectrum. In practice, the spectrum has so many singularities that a very careful theoretical calculation is still necessary if such interpretations are to be fully convincing. A brief description of critical points follows.

V. CRITICAL POINTS

Let us first recall the form of the usual singularities found in the density of states integral associated with a function $E(k)$ in one, two and three dimensions. Since the density of states may be written

$$\rho(E) = \frac{1}{(2\pi)^3} \int \frac{dS}{|\nabla E(k)|} \, , \tag{1}$$

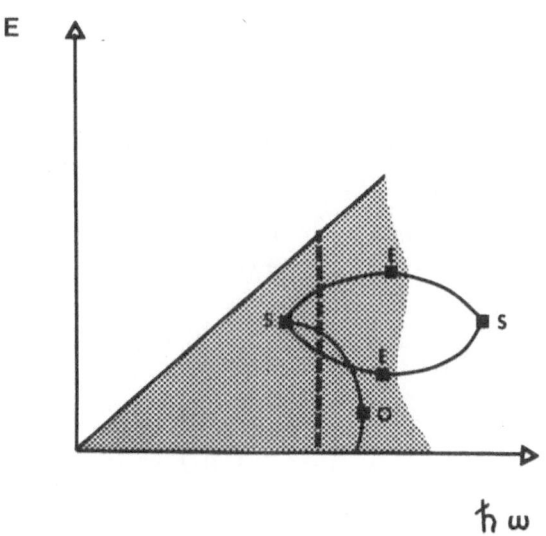

Fig. 2 The E,ℏω diagram.   E is the final state energy
(relative to the top of the valence band), ℏω is the energy of incident photons. A spectrum at fixed ω corresponds to the dashed line. Also shown are possible images
of critical lines, as suggested by Kane (ref. 5).
S = Symmetry critical point, E = (final state) energy
critical point, O = optical critical point.

Such singularities occur whenever

$$\nabla E(\underset{\sim}{k}) = 0 \qquad\qquad\qquad (2)$$

(In the above integral S is the equipotential surface corresponding to energy E.) In the neighborhood of a point
$\underset{\sim}{k}_o$ satisfying (2), we may write

$$E(\underset{\sim}{k}_o + \underset{\sim}{\delta k}) = E(\underset{\sim}{k}_o) + \frac{1}{2} \sum_{ij} \delta k_i \, \delta k_j \, \partial_i \partial_j \, E\Big|_{\underset{\sim}{k}_o} \qquad (3)$$

We shall use a,b,c to denote principal values of the inverse effective mass tensor $\partial_i \partial_j$ E.

The forms of the appropriate singularities are as

follows, where ΔE denotes a small increment of E at $E(\underset{\sim}{k}_o)$.

TABLE I

Van Hove Singularities

Dimension	Signs of a, b, c	Form of singularity	Sketch
1	+	$(\Delta E)^{-1/2}$	
	−	$(\Delta E)^{-1/2}$	
2	− +	$H(\Delta E)$	
	+ −	$-\ln|\Delta E|$	
	− −	$H(-\Delta)$	
3	+ + +	$\sqrt{\Delta}$	
	+ + −	$\sqrt{-\Delta}$	
	+ − −	$-\sqrt{\Delta}$	
	− − −	$\sqrt{-\Delta}$	

The electron density of states is given by a surface integral in three-dimensional $\underset{\sim}{k}$-space. Hence singularities are generally of the third (square-root) type. The same is true of the optical absorption (if we neglect the effects of matrix elements except for $\underset{\sim}{k}$-selection). This is defined by[5]

$$\varepsilon_2(\omega) = A \sum_{i,f} \int_{BZ} d\underset{\sim}{k} \left|P_{if}\right|^2 \delta(\hbar\omega - E_f(\underset{\sim}{k}) + E_i(\underset{\sim}{k}))$$

$$\tag{4}$$

$$A = e^2/6\pi \, m_o^2 \, \omega^2 \, V$$

Here E_i is an initial (occupied) state and E_f is a final (unoccupied) state.

Now consider the theoretical expression for (bulk) photoemission. Leaving aside any factors necessary to account for transport to the surface and emission, this is given by[5]

$$n(E,\omega) = A \sum_{i,j} \int_{BZ} d\underset{\sim}{k} \left|P_{if}\right|^2 \delta(\hbar\omega - E_f(\underset{\sim}{k}) - E_i(\underset{\sim}{k}))$$

$$\delta(E - E_f(\underset{\sim}{k})) \tag{5}$$

This is the same as (4), apart from the absence of an integration over E, i.e.

$$\varepsilon_2(\omega) = \int n(E,\omega) \, dE \tag{6}$$

For given E, Eq. (5) may be regarded as an integral over a surface (the equipotential surface for that energy) which contains all possible final states. However, there is a second condition to be met, which may be expressed in terms of either the initial energy or the difference $\hbar\omega$. Following Kane, we choose the latter. The locus in $\underset{\sim}{k}$-space of all points such that

$$E_f - E_i = \hbar\omega \tag{7}$$

is also a surface which, if it intersects the first surface at all, will generally do so on a <u>line</u>. Therefore, as we vary ω for fixed E, or vice-versa, we are performing a density of states integral on a surface (i.e. two dimensions). Singularities which occur must be of the two-dimensional type, i.e. step functions or logarithmic singularities.

   In terms of the two surfaces mentioned above, what happens (in the case of the step function singularities) is that the two surfaces fail to intersect after a certain point, which is clearly given by the condition that they <u>touch</u>, and indeed this is the condition for both kinds of singularities. Mathematically, if at a given point in k-space,

$$\nabla E_f \; // \; \nabla E_i \tag{8}$$

then the energies $E = E_f$, $\hbar\omega = E_f - E_i$ define a point on the $(E, \hbar\omega)$ plot such that a singularity occurs if we scan through that point (usually by varying E for constant ω). Suppose we now change E slightly and hence deform the final energy surface. Clearly $\hbar\omega$ can, in general, be changed slightly to deform the optical energy surface so as to maintain the condition (8) for touching. In this way we can trace out a <u>critical line</u> in k-space, where the two surfaces touch and also a corresponding line in the E, $\hbar\omega$ plane. The latter is what one hopes to observe - to interpret it in terms of the details of band structure it must be associated with the correct critical line in k-space. Kane[5] has enumerated various topological rules to help in this procedure. It is argued that critical lines will only cross, bifurcate etc., at <u>symmetry points</u> in k <u>space</u>, where

$$\nabla E \, (k) = 0 \; , \tag{9}$$

by symmetry, for both the functions involved, so the critical lines can be thought of as lines joining such critical points. Any symmetry line along which the direction of $E(k)$ is dictated by symmetry alone, such that (8) is automatically obeyed, <u>must</u> be a critical line. Any point which is a critical point of either the final energy or the optical energy, so that one of the gradients in (8) vanishes, must lie on a critical line (see Fig. 2).

   These rules relate to critical lines in k-space and one must be careful in applying them to the "images" of such lines in the (E, ω) plane. In this plane, lines can cross at general points. However, a bifurcation, or the confluence of many lines at one point is indicative of a symmetry critical point and is hence an important guide to the interpretation of the data.

The strategy outlined by Kane has been pursued with
mixed success. Perhaps the most convincing analysis made
so far has been that of Grobman et al who have recently
been able to obtain good qualitative agreement between
theory (based on a pseudopotential band structure calculation) and data for Ge in the range 6.5 - 23 eV [6]. The
theory contains a small but apparently crucial refinement
in the treatment of the factors which account for transport to the surface and escape to the vacuum. In earlier
work, isotropic averages[7] had been used here. Instead,
Grobman et.al. incorporate into the integral (5) two
factors $T(\underline{k})$ and $D(\underline{k})$ representing respectively the probability of reaching the surface without inelastic scat-
tering and the probability of transmission. This selectively weights various features and so has a significant
effect, without invalidating the picture based on critical lines.

VI. SURFACE CONTRIBUTIONS

Brundle[8] has recently summarized our present experimental knowledge of escape depths. Fig. 3 shows the approximate range of escape depths for a large number of
solids. We might expect "surface effects" in electronic
band structure to extend only over the first few Angstroms.
Hence these effects are negligible in the low and high
energy limits, but in the UV region are appreciable. Indeed there has been much recent effort to use this region
to extract a "surface density of states".

VII. APPLICATIONS TO AMORPHOUS SOLIDS

Shevchik[9] has reviewed a number of recent applications
of photoelectron spectroscopy to amorphous solids. We
will give just one example here - the data of Ley et al.
on Si and Ge, which is in the x-ray range. Shown in Fig. 4
is the raw data. Note the large contribution of secondaries. When these are subtracted, what remains is a direct picture of the density of states of the valence band
apart from a smoothly varying factor due to matrix element
effects associated with the s-like or p-like character of
the states. This data has been the primary basis for arguments regarding the washing-out of structure at the bottom of the valence band. UV photoemission[11] gives rather
similar results. For a review of low energy photoemission

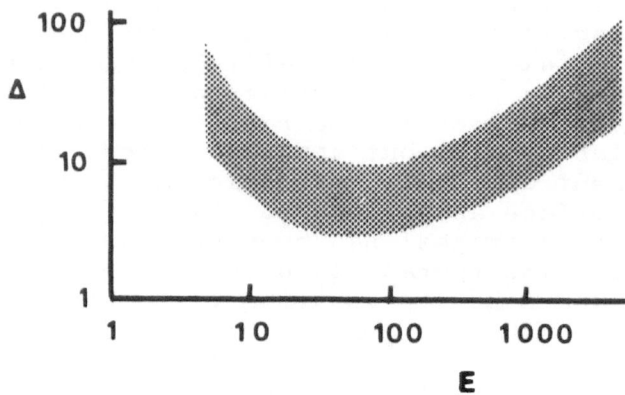

Fig. 3 Range of escape depths for solids surveyed by
Brundle (ref. 8). Units are Å and eV.

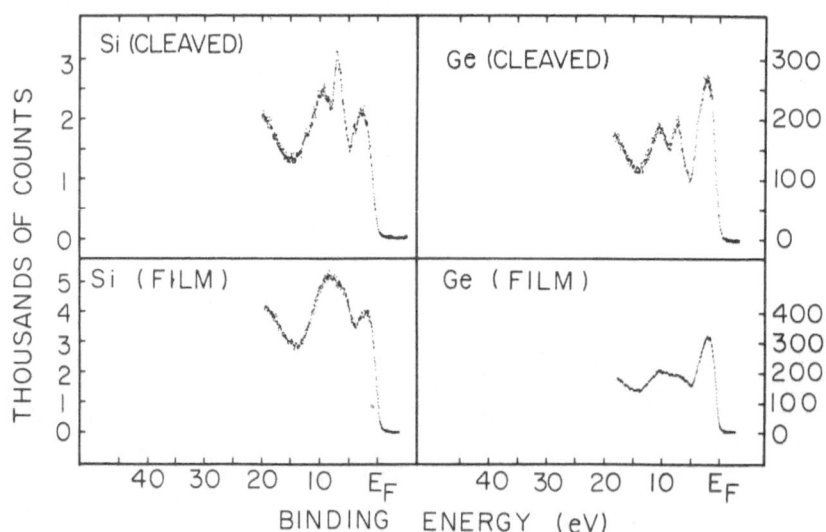

Fig. 4 X-ray photoelectron spectra of amorphous and crystalline Si and Ge, as reported by Ley et.al. (ref. 10).

as it applies to amorphous semiconductors, see the suc-
ceeding lectures by L. Laude.

REFERENCES

1. G. D. Mahan, Phys. Rev. <u>2</u>, 4334 (1970).
2. See, e.g., T. M. Donovan and W. E. Spicer, Phys.
    Rev. Lett. <u>21</u>, 1572 (1968).
3. D. E. Eastman, W. D. Grobman, J. L. Freeouf and
    M. Erbudak, Phys. Rev. <u>9</u>, 3473 (1974).
4. E. O. Kane, Phys. Rev. <u>159</u>, 624 (1967).
5. E. O. Kane, Phys. Rev. <u>175</u>, 1039 (1968).
6. W. D. Grobman, D. E. Eastman, J. L. Freeouf and J.
    Shaw, Proceedings of the 12th Intl. Conf. on the
    Physics of Semiconductors, ed. M. H. Pilkuhn,
    Teubner, Stuttgart, W.Germany, 1974, p. 1275.
7. C. N. Berglund and W. E. Spicer, Phys. Rev. <u>136</u>,
    A1030, 1044 (1964).
8. C. R. Brundle, J. Vac. Sci. Tech. <u>11</u>, 212 (1974);
    Surface Sci. <u>48</u>, 99 (1975).
9. N. Shevchik, Proceedings of International Conf.
    on Tetrahedrally Bonded Amorphous Semiconductors,
    ed. M. H. Brodsky, S. Kirkpatrick and D. Weaire,
    AIP, 1974, p. 72.
10. L. Ley, S. Kowalczyk, R. Pollak and D. A. Shirley,
    Phys. Rev. Lett. <u>29</u>, 1088 (1972).
11. D. E. Eastman, J. L. Freeouf and M. Erbudak, Pro-
    ceedings of International Topical Conference on
    Tetrahedrally Bonded Amorphous Semiconductors, ed.
    M. H. Brodsky, S. Kirkpatrick and D. Weaire, AIP,
    1974, p. 95.

PHOTOEMISSION OF DISORDERED ELEMENTAL SEMICONDUCTORS

L. D. Laude

Université de l'Etat à Mons
Faculte des Sciences
Mons, Belgium

PREAMBLE

Photoemission is one of the very basic tools of investigation in solid state physics, the one which is in fact responsible for the latest development in the elaboration of the electronic structure of solids. It is the experiment in which a monochromatic beam of photons (energy $h\nu$) is shined onto a solid surface which eventually emits a flux of electrons. The kinetic energy distribution of these emitted electrons (or photoelectrons) constitutes the photoemission spectrum of the solid at that photon energy $h\nu$. The structure of this spectrum can be expected to be representative of the history of all photoelectrons, between the moment when they absorb the energy of the photon and become conduction electrons, and the moment when they are emitted in the vacuum after surpassing the potential barrier present at the surface and which equals the ionization energy of the solid (or work function, in the case of a metal).

It is clear that all photoelectrons will not have the same history. In fact, the interpretation of the photoemission spectrum is bound to be a formidable one if one wants to take effectively into account, for instance, all possible mechanisms of inelastic scattering which an electron may be subjected to before being emitted.

Other factors will tend to complicate even further the problem. Therefore, it is not in the scope of this

211

series of lectures on photoemission to present an elaborate theory but, rather, we will try to optimize the
use of photoemission according to the kind of information we are looking for: basically, we want to obtain
not only the features of the bulk electronic structure
of a disordered material, but also details of its evolution upon annealing for instance, i.e., information on
the reordering processes occurring in a film under such
circumstances (configurational changes, activation energies, etc). To this end, we will introduce a few parameters and approximations which will enable us to delineate the best conditions under which we have to work.

The first section will be, therefore, devoted to
theoretical considerations regarding optical absorption,
inelastic scattering, and the elastic electron energy
distribution. In the second section, the photoemission
experiment will be described, stressing which of the instrumental parameters have to be evaluated correctly.
This being done, both theoretical and experimental arguments will be applied to the study of tetrahedrally and
non-tetrahedrally coordinated semiconductors in section
three. In this latter, we will restrict ourselves to two
materials, Ge and Te, which will be treated in detail.
These examples will be aimed at demonstrating the wealth
of information provided by photoemission and the important contribution of this technique to our understanding
of the properties of disordered solids.

I. THEORETICAL CONSIDERATIONS ON PHOTOEMISSION

1. Introduction

As said in the preamble, the photoemission effect
occurs when a solid emits electrons (or photoelectrons)
upon illumination with a beam of photons. What we are
concerned with is, then, two fold:
(1) the kinetic energy (E_{kin}) distribution of the
photoelectrons in the presence of an incident monochromatic photon beam of energy $h\nu$: $N(E_{kin}, h\nu)$;
(2) the intensity of the electron emission, or number of photoelectrons per incoming photon, otherwise called the quantum yield $Y(h\nu)$. This quantity is obviously
related to the preceeding via the relation:

$$Y(h\nu) = \int_0^{+\infty} N(E_{kin}, h\nu) dE_{kin},\qquad(1)$$

so that the integral of the experimentally measured energy distribution curve (EDC) at $h\nu$ must equal the value of the quantum yield at $h\nu$, $Y(h\nu)$. The measures of these two quantities, $N(E_{kin}, h\nu)$ and $Y(h\nu)$ constitute the essence of the photoemission experiment, and we must interpret their outcome.

On one hand, we inject into the solid a certain amount of energy (photons) and on the other, a certain number of electrons are emitted which have acquired from these photons an energy higher than the so-called ionization energy of the solid, E_i. This latter energy is the one necessary to pull an electron lying about (or just beneath) the Fermi level, out of the solid. Its magnitude is, in general, of the order of 5 eV for metals and semiconductors, i.e., slightly less than the free-atom ionization energies (E_i^f) (for Ge, $E_i = 4.8$ eV and $E_i^f = 7.9$ eV). The ionization energy E_i is materialized by a potential barrier at the surface. The height of this barrier measures E_i, i.e., the energy difference between a so-called "surface potential" and the uppermost limit of the distribution of valence electrons (Fermi level in metals, or valence band edge in semiconductors). As a consequence, using a monochromatic beam of energy $h\nu$, photoemission will occur only if $h\nu$ is higher than a threshold energy E_T which should, in principle, equal the ionization energy. This threshold E_T has the advantage of being reasonably well delineated by the abrupt onset of the curve giving $Y(h\nu)$. As a matter of definition, the work function E_F is equal to the energy difference between surface potential and Fermi level E_f, so that it equals the ionization energy only in the case of a metal.

Suppose now that an electron fulfilling the above condition is effectively emitted out of a metal, for instance, with a kinetic energy E_{kin}. The kinetic energy zero being the one corresponding to a hypothetical photoelectron leaving the surface with zero energy (i.e. occupying before emission a surface energy level practically aligned with the surface potential), one can refer the energy of every photoelectron to the Fermi level since the surface potential is independently defined versus this level by E_T. We have therefore:

$$\varepsilon = E_T + E_{kin},\qquad\qquad(2)$$

where ε is the <u>absolute</u> energy of the photoelectron referred to Fermi level. If we recall that only the

relative position of the bulk energy levels at some
points of the Brillouin zone can be obtained by reflectivity, this characteristic of photoemission could be
considered as a definite advantage over pure optical measurements. However, before such a conclusion can be drawn,
several points have to be secured which will be discussed now.

First, we want to measure the kinetic energy E_{kin}
of the photoelectron and, to this end, a collector (or
anode) is positioned in front of the emitting surface
and is negatively biased at a variable potential V. The
photoelectron is collected by the anode when: $-eV = E_{kin}$,
where e is the electron charge. For this measured energy to be the effective kinetic energy of the photoelectron at the time it leaves the surface, one has to ensure
that it will not loose energy by colliding with gas molecules upon travelling between the emitter surface and the
collector, i.e., its mean free-path in vacuum will have
to be larger than the emitter-collector distance. The
latter is, in general, of the order of a few cms and the
mean free path of an electron travelling in a 10^{-4} torr
vacuum is of that order of magnitude. It will be therefore sufficient to place the system emitter-collector in
a chamber evacuated at pressures $\leq 10^{-5}$ torr to avoid
energy (and electron) losses from gas molecules outside
the emitter. Suppose now that the latter is a freshly
cleaved surface. The average time necessary to measure
an EDC runs between a few seconds and a few minutes.
Allowing for several measurements to be performed at
various photon energies (only, say, 0.2 eV apart) over
a 5 eV range, the time necessary to complete a photoemission experiment may be estimated to be of the order
of 2-3 hours. We want to make sure that the sample we
are investigating remains the same all along this period
of time, i.e., its surface does not change upon contamination which would a priori affect the surface potential.
The time required for a monolayer of H_2, for instance,
to grow on a metal surface is about 2-3 secs. (with a
sticking coefficient = 1) in a 10^{-6} torr vacuum. It
drops down to 7-8 hrs. in a 10^{-10} torr vacuum and to 3
days in 10^{-11} torr. Therefore, the pressure in the
vacuum chamber should not exceed 10^{-10} torr in order to
preserve clean surface conditions all along an experiment.

We assume now that the above conditions are fulfilled, i.e., an electron is emitted from a clean surface
in a 10^{-10} torr vacuum, its kinetic energy is measured
to be E_{kin}, which can be considered to be its true value

just after emission from the surface of the solid. Other
electrons are emitted with different kinetic energies in
the presence of the same photon beam of energy $h\nu$. The
total spectrum measured (the EDC mentioned before) looks
like the one shown in Fig. 1, where the low-energy onset,
at $E_{kin} = 0$, is equalled to E_T and the high-energy onset
delineates the uppermost limit of photoemission which
can only be reached by electrons lying about E_f in their
steady state, absorbing $h\nu$ and being emitted in vacuum
with an energy, $(h\nu - E_T)$. The width of the EDC is
therefore equal to $h\nu$, which can be easily verified by
varying $h\nu$. We will not explain here how we measure
these EDC's, which will be done in the second section.
We simply want to point at the rich structure which is,
in general, exhibited by these curves obtained from clean
surfaces. Such a structure is observed to vary in a complex manner upon varying $h\nu$. It is the interpretation
of this structure and of its evolution that we will try
now to clarify. Before getting to that point, we will
however, have to visualize the photoemission effect

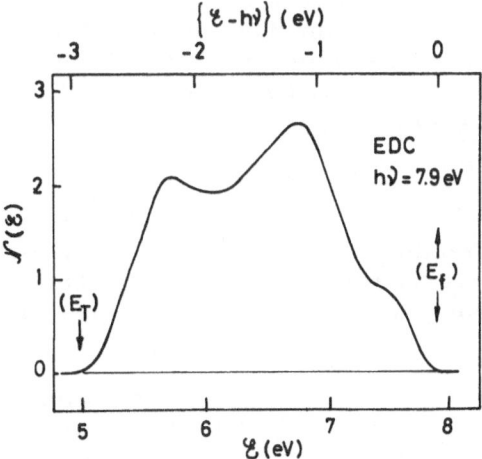

Fig. 1 Conventional profile of a photoelectron energy
distribution curve (EDC) obtained from a material, at
$h\nu = 7.9$ eV. Structure in the EDC can be referred to a
"final" state energy scale E (bottom) or to an "initial"
state energy scale $(E - h\nu)$, (top). The low-energy onset is delineated by E_T on the E scale. The high-energy
onset locates emission associated with electron states
lying about E_f (zero on $(E - h\nu)$ scale).

within the solid and this will be the subject of the
following subsection.

2. A Model

A number of models have been proposed to describe
photoemission. We can roughly classify them into two
categories, depending on whether we consider it as being
the outcome of (i) the interaction of an incident light
wave with the eigenstates of the electrons at and near
the surface of the solid, or (ii) the photo-electron interaction, as a two particle interaction. Although the
former approach[1] would seem to be the most realistic, it
must be realized that it may lead to an unsolvable problem since it assumes a perfect knowledge of the surface
potential distribution, i.e., of the microscopic (Å
scale) surface atomic configurations. This being not the
case at the present stage of our knowledge in this field,
one is forced to introduce model configurations which
may be misleading in general. In view of this major difficulty, it seems preferable to visualize photoemission
within the two-particle approach[2] which might be described as follows:

(i) a photon of energy $h\nu$ penetrates the surface of
a solid and diffuses into it until colliding with a
steady (valence or core) electron of energy $E_v(\vec{k})$; it
transfers to this electron its energy $h\nu$ and wave vector
$\vec{k}_{ph}$; the electron then undergoes a "direct" optical transition to a conduction state level $E_c(\vec{k}')$, such that
$\vec{k}' = \vec{k} + \vec{k}_{ph} \approx \vec{k}$, and

$$E_c(\vec{k}) - E_v(\vec{k}) = h\nu; \tag{3}$$

(ii) this "conduction" electron now propagates into
the solid until it reaches eventually the surface and is
emitted if its energy when reaching the surface is higher that E_T.

The first step is purely optical and could a priori
be investigated within the band structure framework of
the solid, in terms of a calculated "optical" energy distribution which we shall define later. If photoemission
could be limited to this optical absorption process it
is clear that this technique would be far more powerful
than reflectivity in elaborating a band structure, since
one could assign pieces of structure appearing in the
EDC's to well defined transitions in the Brillouin Zone.

Unfortunately, the second step disturbs, in general, this idealized picture. It involves a number of secondary effects which are associated with the transport of the optically excited electrons to and through the surface. Essentially, these electrons may inelastically scatter with other electrons, with the lattice and with impurities or defects both in the bulk and at the surface, before being emitted. Each of these effects would, therefore, tend to alter the original energy of each excited electron, in a manner which depends on the characteristics of the scattering process involved (energy loss per collision, mean-free path). The problem is that we will analyze simultaneously the energy of photoelectrons having different histories, although emitted in the presence of the same incoming monochromatic photon beam. In other words, EDC's will be contributed by _elastic_ electrons (the ones which are emitted without suffering inelastic scattering in the solid) and by _inelastic_ electrons (those which have lost energy in various ways before their emission). The structure of the energy distribution of the former should be conveniently described with the "optical" energy distribution mentioned before, whilst the latter should be responsible for all observed departures from that calculated distribution. It is evident that a proper account of the scattering processes is a hazardous game, since one would have (i) to evaluate _separately_ all different scattering mechanisms, (ii) calculate a corresponding energy distribution which would be further "weighted" by some semi-empirical parameter, and (iii) admix it finally with the elastic electron distribution. One can _always_ reach a "good fit" to the measured EDC's using such a procedure. It seems clear that the resulting description of the experimental data would be widely open to criticism and, in general, would not prove much. On the other hand, the "optical" distribution calculation is the only one which may be checked _independently_ by considering the other optical means which are available to us in elaborating a band structure, namely: the imaginary part of the dielectric constant spectra (ε_2) and the optical absorption coefficient (K), which, in many respects, are the counterparts, in pure optical measurements, of the photoemission quantities $N(E_{kin}, h\nu)$ and $Y(h\nu)$.

The approach we will ultimately adopt in investigating the electronic structure of solids by photoemission becomes now clear: we will tend to minimize, as much as we can, the contributions of inelastic scattering processes to the EDC's, in such a way that one could more

reliably consider the "optical" distribution as a fair
approximation to the structure of these EDC's.

   As far as disordered solids are concerned, the a-
bove approach is essential for two main reasons:

   (i) other scattering mechanisms than the ones present in a crystal may exist in a disordered film. Some
others may be more or less important in such a film,
compared to the crystalline counterpart of this material;
   (ii) since we will try to obtain information on the
electronic structure of these disordered solids by iden-
tifying the structure of their EDC's, we will have to
compare systematically these EDC's to EDC's measured from
their crystalline counterparts at the same photon energies. The latter being optimized in terms of elastic
electron contribution, the former will have also to be
optimized the same way.

   What we will do essentially is the following. Before studying the photoemission properties of a disordered solid, we will collect as much information as we
can on the photoemission properties of the crystalline
phase of this solid, stressing the elastic electron contributions to the spectra. Then, we will study the disordered phases of this solid under exactly the same experimental conditions. By comparison of the two sets of
data (position and intensity of peaks in the respective
EDC's), one should be able, in principle, to outline the
main properties of the electronic structure of these disordered phases. However, one may expect, in general,
that the number of disordered phases of the same material will be large, and to each disordered phase will correspond one set of structural properties. As a consequence, these properties should vary widely depending on
the preparation conditions of the disordered phases.
Therefore, one will try essentially to outline the _trends_
of the evolution of the electronic structure in the presence of configurational _and_ quantitative disorder.

   The optimization process which we mentioned earlier
will be developed using relevant parameters. We shall
then focus our attention onto the optical absorption step
of photoemission and relate it to the theory of the optical properties of solids.

3. Elastic and Inelastic Electron   Contributions
    to the Photoemission Spectra

We will base our reasoning on an attempt at evaluating the relative contributions of what we have defined
as elastic and inelastic electrons to the photoemission
spectra (the EDC's). To this end, we have first to answer two questions, namely: (i) how many photons (of
energy hν) are absorbed at a depth x behind a solid surface on which the photon beam is focussed?, and (ii) how
many of the optically excited electrons will reach the
surface and will be emitted in the vacuum without having
suffered inelastic scattering in the solid? Answers to
these questions will be obtained by considering the photon
penetration depth, (i.e., in fact, the optical absorption
coefficient K) and the electron scattering processes
(mean-free path and energy loss).

Optical Absorption. If I_o is the value of the light
intensity at the surface of the solid, its value I after
passing through a thickness x of the material is given
by the Lambert's law:

$$I = I_o \exp(-Kx), \tag{4}$$

where K is the optical absorption coefficient of the solid. The dependence on photon energy of the latter, in
crystalline Si, is given in Fig. 2 [3]. Note the decrease
of K above hν = 10 eV. Its strength loses two orders
of magnitude between 10 and 100 eV, at which energy optical transitions initiating at deep 2p levels causes K to
increase sharply again.

The density of photons absorbed between the surface
and a plane parallel to it at x is proportional to:

$$\mathcal{J} = \frac{I_o - I}{I_o} = 1 - \exp(-Kx) \tag{5}$$

This function is represented in Fig. 3, for different
values of hν. At low photon energies (hν < 5 eV), a
great majority of photons are absorbed very deep in the
solid. At hν = 5 eV, absorption takes place close to the
surface. Above hν = 5 eV, it occurs again mostly deep
in the solid. This means that an increasing majority of
valence electrons will be optically excited deep behind
the surface of the solid as we increase the photon energy
above 5 eV. Another way to look at these data is represented in Fig. 4 where we plot $\mathcal{J}$ as a function of hν at

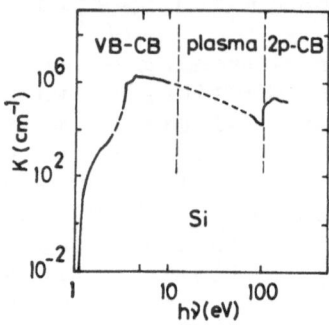

Fig. 2   Optical absorption coefficient K as a function of hν for Si, from Ref. 3.   Three regions are distinguished: valence-to-conduction band transitions up to hν = 12 eV, plasma region between 12 and 100 eV, 2p-to-conduction band transitions above 100 eV.

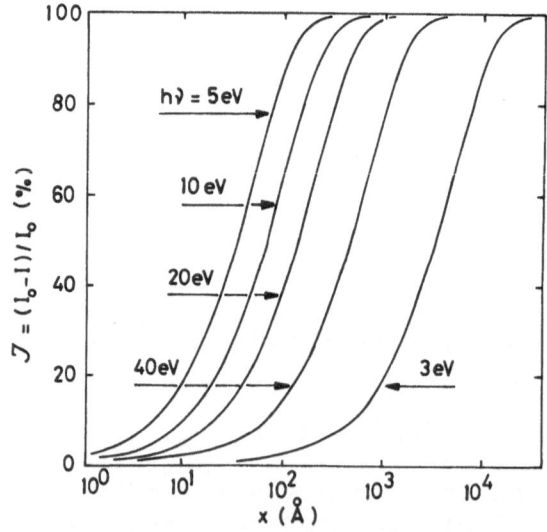

Fig. 3   Relative density $\mathcal{I}(x)$ of photons (energy ranging from 3 to 40 eV) absorbed between the surface and depth x.

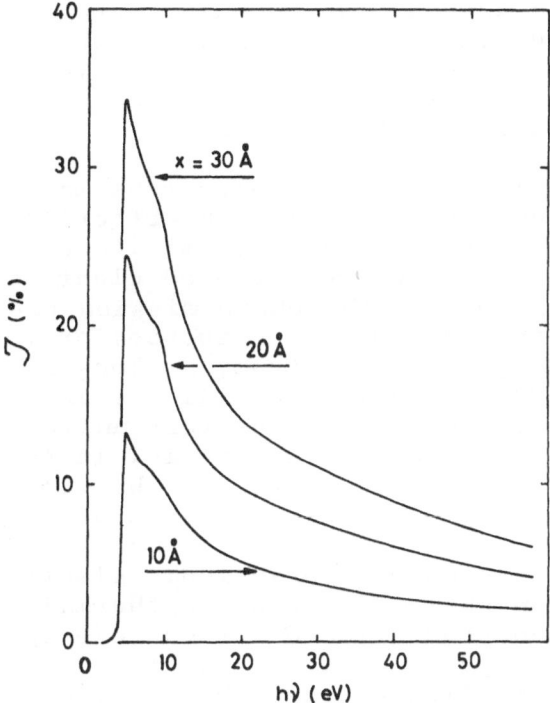

Fig. 4 $\mathcal{J}$ (hν) at depths between 10 to 30 Å.

different depths x (measurements leading to the values
of K used here, Fig. 2, were performed at near normal incidence). We see that absorption is maximum around
hν - 5 eV and is a factor of 5 smaller at hν = 40 eV, at
depths up to a few hundred Å. Optimizing optical absorption will mean working at photon energies around 5 eV.
This value corresponding in general to most values of E_T
experienced on solids (metals and semiconductors), the
yield Y(hν) should be already maximum just above its onset (E_T). We will see that this is not the case for reasons associated with the transport of the photoexcited
electrons to the surface.

        Inelastic Scattering. Suppose that upon absorption
of a photon, a conduction electron is created at a depth
x. To be emitted, such an electron has to reach the surface sometime during its propagation through the solid
and to have at that time an energy larger that or equal
to the ionization energy of the solid. Even so, this
electron may be reflected back into the solid depending

on its kinetic energy and propagation axis, just like an electromagnetic wave at the boundary between two media having different indices. Along its path, the electron may interact with the medium (defects, vibrations) and/ or with the other conduction electrons. We briefly review these scattering possibilities.

a. Defects: Propagating electrons may be trapped on impurity atoms or in lattice imperfections. Electron states associated with these defects lie within the forbidden band gap. Therefore, trapped electrons could in principle contribute to the photoemission process upon absorption of a photon and give evidence for the trapping levels which they originate from, as long as these levels lie below the Fermi level. We will see in section two that this possibility is however very unlikely. The main perturbation introduced by defects is, in fact, to lower considerably the electron mobility. We will see later how this effect may be viewed in terms of photoemission.

b. Electron-phonon interaction: Electron waves may couple with phonons to give rise to thermalization processes[4] in which conduction electrons eventually lose considerable energy and pile up at high density of states regions of the Brillouin Zone. However, on the one hand, this process necessitates a very large number of electron-phonon collisions, since each of them introduces an electron energy loss of the order of the phonon energy (20-40 meV). On the other hand, the electron-phonon mean free path, Z_{ph}, is of the order of 30 Å at room temperature and is assumed to increase slowly with the energy of the electron. Therefore, such a large number of electron-phonon collisions could be experienced by a conduction electron before reaching the surface only if this conduction electron is created very deep into the solid. Such a possibility may have a high probability to occur essentially for energies below $h\nu = 4$ eV as it has been demonstrated experimentally[4] (see Fig. 3). The electron-phonon interaction could also play an important role directly into the optical absorption step giving rise to phonon-assisted (indirect) transitions, as will be seen later on.

c. Electron-electron interaction: This is, by far, the most important inelastic scattering event which an electron may encounter on its way to the surface. It is characterized by large energy losses and a mean-free path, ℓ_e, which varies widely when going from the visible to the X-ray region. The behavior of ℓ_e with $h\nu$ is shown

in Fig. 5 for a wide variety of materials[5]. It decreases
from a few hundreds Å at hν < 5 eV to some 5 Å at
20 < hν < 100 eV in most materials. The solid line is
an approximate average of all the values reported in the
literature, but is not meant to be an universal curve.
For simplicity, we adopt this curve to represent the behavior of ℓ_e with hν, in silicon. Compared to ℓ_{ph}, it
is seen that the most effective scattering mechanism is
due to (i) the electron-phonon interaction at hν < 5 eV
and (ii) the electron-electron interaction at hν > 5 eV.
Around 5 eV, both types of interaction should have the
same probability to occur ($\ell_e \simeq \ell_{ph} \simeq$ 30 to 40 Å).

As an example, $\ell_e \simeq$ 15 Å at hν = 10 eV. Only these
electrons which are excited at sites x between 0 and 15 Å
from the surface may reach it, in the first approximation,
without colliding with other conduction electrons, i.e.,
have a chance to be emitted with a kinetic energy E_{kin},
corresponding to the energy level E they have originally
populated upon absorbing hν ($E_{kin} = E - E_T$). The proportion of these electrons is given in Fig. 4 to be around
15°/o at hν = 10 eV, i.e., 15°/o of the emitted electrons
(which contribute to the photoemission spectrum) are elastic. Knowing the total number of electrons emitted at
hν = 10 eV, which is given by the total yield curve of
Si, Y_{tot} in Fig. 6, one obtains, this way an estimate of
the elastic electron contribution to the yield at
hν = 10 eV. The same procedure repeated for other photon
energies yields the elastic yield Y_e (dashed line, Fig. 6),

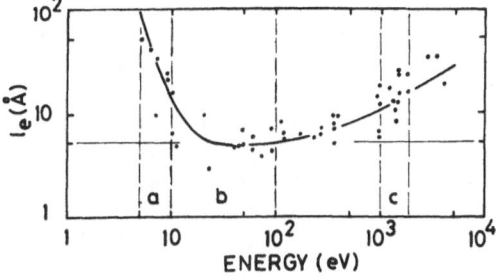

Fig. 5 Electron-electron mean free-path, ℓ_e, as a function of hν for various materials, from Ref. 5. The solid
line averages experimental data. Labels a, b, c, denote
the three basic photon energy ranges outlined in the
text (5 - 10 eV, 10 - 100 eV, 1 - 2 keV, respectively).

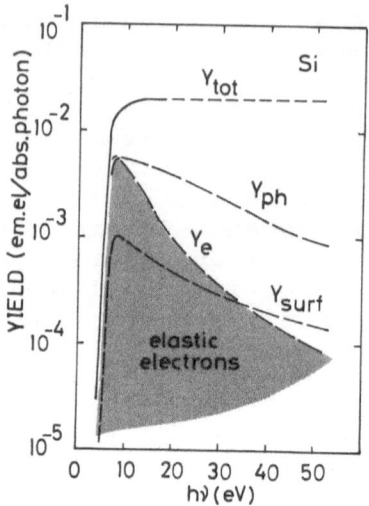

Fig. 6 Absolute quantum yield $Y_{tot}(h\nu)$ of crystalline
Si. Y_{ph} and Y_e are the partial elastic yields defined
in the text, from the point-of-view of electron-phonon
and electron-electron scatterings, respectively. Y_{surf}
is the surface yield, i.e., emission from the surface region arbitrarily defined as extending from x = 0 to
x = 10 Å. The shaded area represents the total elastic
electron emission of the solid, Y_e^{tot}, which is seen to
be optimum between $h\nu$ = 5 and 8 eV.

from the electron-electron interaction point of view.
An identical approach for the electron-phonon interaction (ℓ_{ph} given in Fig. 5) gives the electron-phonon
elastic yield Y_{ph} (dotted line, Fig. 6). The shaded
area represents the total elastic emission Y_e^{tot} which is
limited by (i) electron-phonon interactions at energies
$h\nu$ < 5 eV, and (ii) by electron-electron interactions
above 5 eV. Yield values above 12 eV are only estimates,
since no experimental data are available and the values
of the elastic yield Y_e^{tot} above 12 eV should be taken as
first approximations. The fact that Y_e^{tot} is maximum between 5 and 8 eV is a clear indication that, should we
be interested in the elastic part of the photoemission
spectra, we should work at such low energies to optimize
photoemission in that respect.

It is interesting to estimate also what we might
call the "surface emission", the one which originates in
the volume comprised between the surface and a depth of,

say, 10 Å. The profile of the corresponding "surface
absorption" is given in Fig. 4. We used it to obtain
the "surface yield" Y_s represented in Fig. 6. This yield
is also maximum at $h\nu \sim$ 5 to 7 eV. The interesting thing
to notice is the fact that, at photon energies in the
range 20 to 100 eV, the only elastic electrons emitted
originate in the vicinity of the surface. However, while
surface emission is totally elastic at $h\nu$ < 10 eV (at
least from the point of view of electron-electron and
electron-phonon interactions), one cannot be sure any
longer that this is the case in the 20 - 100 eV range,
since ℓ_e is of the order of 5 to 10 Å in that region.
Statistically, the proportions of elastic and inelastic
electrons, excited within the top 10 Å of the solid at
20 < $h\nu$ < 100 eV, would be nearly equivalent. Further,
these surface elastic electrons are expected to have an
energy distribution reflecting the electronic structure
of that surface, which is bound to the actual surface
atomic configuration. The latter being, at the best, a
distorted version of the bulk configuration (a cleaved
(111) Si surface is not necessarily equivalent to a (111)
plane in the bulk of the crystal), these surface <u>elastic</u>
electrons have every chance to yield energy distributions
associated with disorder effects "within" the surface region of a crystal. This argument should be kept in mind
when investigating the electronic structure of disordered
solids by photoemission.

We have estimated the contribution of inelastic electrons to the total yield $Y_{tot}(h\nu)$. Let us look now at
the energy distribution of inelastic electrons at a given
photon energy $h\nu$ (electron-electron scattering process).
In this respect, a simple model has been proposed for
metals by Berglund and Spicer[6]. We present it here in
order to describe the general profile of an inelastic
electron distribution, although a more exact description
of the experimental data is far more complicated. Assuming (i) valence and conduction densities-of-states
(DOS) to be constant, and (ii) the scattering probability
to be uniform for all conduction electron energies, this
model states that the inelastic electron energy distribution could be written, after one electron-electron scattering event, as:

$$N_s(E, h\nu) = N_{ns} + 4 N_{ns} \left[\frac{E}{h\nu} - 1 + \log \frac{h\nu}{E}\right] \qquad (6)$$

where N_{ns} is the density of the non-scattered (elastic)
electrons produced at the same photon energy. The profile

of $\dfrac{N_s - N_{ns}}{N_{ns}}$ is shown in Fig. 7 for different photon

energies. The essential characteristics of such a distribution is the large peak produced at the low-energy
edge of the distribution. This means that, upon scattering with other conduction electrons, an electron
exited into a conduction state of energy $E_c(\vec{k})$,
will tend to lose part of the acquired energy and populate a state lying closer to the bottom of the conduction
band. This low-energy peak does not, of course, contribute the photoemission spectra since its position lies
below the low-energy edge of the latter (at $E_T \simeq 5$ eV).
However, a tail of scattered electrons would always enhance the low-energy region of the EDC's. As we increase
$h\nu$, tail strengthens considerably and, depending on the
more or less sharp low-energy cut-off of the EDC's at
E_T, a "peak" is produced in these spectra just above E_T.
It is this feature which is usually referred to as the
"scattering peak". Its strength would, then, reflect
approximately, the density of the inelastic electrons contributing to the EDC's. According to Fig. 7 and as expected from the scattering yield $Y_s = Y_{tot} - Y_e^{tot}$, this peak
turns out to be the major feature of the EDC's above
$h\nu = 15$ eV. A remark should be made here. Since scattered electrons tend to pile up at the bottom of the conduction band, one might expect structure to be superimposed on the smooth profile of the scattered electrons
(Fig. 7), the structure which could be associated with
the conduction band density-of-states alone. Further.
these electrons being scattered deep into the bulk of the
solid, at $h\nu < 10$ eV, such a structure would reflect bulk
properties, whilst the elastic electron contribution to
the same spectra would reflect approximately surface properties, at least up to $h\nu = 100$ eV, as we have shown in
Fig. 6. The problem we are now facing is to deconvolute
the two types of structure (bulk in the inelastic profile,
and surface in the elastic one). In the 10-15 eV range,
the two overlap strongly and one should be extremely
careful in assigning actual spectral features to either
"surface" or "bulk" electron structure. Above 15 eV,
the elastic profile has a fixed width set by the valence
band width which now contributes to it totally. As $h\nu$
increases, this profile tends to be more and more separated from the scattering peak region.

In the X-ray range (1-2 keV), we are in the same
situation with the difference that ℓ_e is now some 10 to
20 Å large. The elastic distribution should therefore

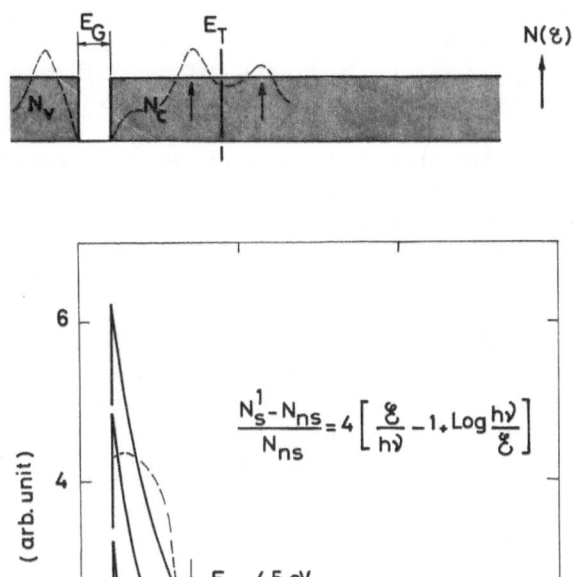

Fig. 7 Energy distribution of the relative density of once-scattered electrons N_s^1 versus non-scattered electrons N_{ns}, $\dfrac{N_s^1 - N_{ns}}{N_{ns}}$, at various photon energies in a solid. Valence and conduction band DOS $(N_v$ and $N_c)$ are taken to be constant on both sides of the band gap (width E_G). Structure in N_c may eventually show up in the scattering profile (dashed line). E_T is the photoelectric threshold value, taken here to be 4.5 eV.

reflect properties close to the bulk ones, as the inelastic distribution does. However, the equivalent De Broglie wave lengths of the energies used in that region are very small (~ 8 Å) and the bulk ordering we are now able to explore would not extend over that limit.

The conclusion we reach, after considering optical

absorption and inelastic scattering mechanisms, can be
stated now very simply: If we are concerned with the
electronic structure in the bulk of a disordered solid
and want to investigate it by photoemission under the
best conditions, we should restrict our study to low
photon energies (in the 5-10 eV range). However, since
such energies are not sufficient to explore the structure of the density-of-states of the lower half of the
valence band, higher photon energies have to be used for
that purpose, within the limits discussed above.

The elastic electron energy distribution: We now
concentrate on the energy distribution of the elastic
photoelectrons, those which (i) are optically excited into the conduction band of a three dimensional solid, and
(ii) are emitted without suffering inelastic scattering.

What we want to know here are the details of the
population of the conduction states in the presence of
a given photon energy.

a. Case of a Crystal: The energy distribution of
the excited electrons can here be described by convoluting valence and conduction densities-of-states and imposing energy and $\vec{k}$-conservations to the valence-conduction band transitions. It is given by the expression:

$$N(E, h\nu) = \sum_{i,j} \int_{\vec{k}_j} P_{ij} \, \delta[E - E_c(\vec{k}_j)] d^3\vec{k}_j \qquad (7)$$

where $\delta[E - E_c(\vec{k}_j)]$ represents the condition imposed on
the excited state to be a Bloch state (at energy level
$E_c(\vec{k}_j)$). P_{ij} is the probability (per unit time) for an
electron to be excited from the valence-state $\psi_v(\vec{k}_i)$ to
the conduction state $\psi_c(\vec{k}_j)$. This probability is proportional to:

$$P_{ij} \sim \sum_{i=j} |< \psi_v(\vec{k}_i) | \vec{e} \cdot \vec{p} | \psi_c(\vec{k}_j) >|^2 \, \delta[E_c(\vec{k}_j) - E_v(\vec{k}_i) - h\nu]$$

$$+ \sum_{i \neq j} (\text{ITME})^2 \, \delta[E_c(\vec{k}_j) - E_v(\vec{k}_i) - h\nu]. \qquad (8)$$

(ITME) are the indirect transition matrix elements.

The first term on the right of eq. (8) for i = j describes direct transitions. The second term for i ≠ j is
the probability of phonon-assisted (indirect) transitions
to occur (via intermediate states) between valence and
conduction states belonging to different $\vec{k}$'s but lying
at levels separated by hν. This latter probability is
limited by stringent conditions imposed by the symmetries
of the states and phonons involved. In general, one may
consider electron-phonon interaction to introduce structure broadening of the order of the phonon energies
(<0.1 eV). For the sake of simplicity, we will consider
only direct transitions ($\vec{k}_i = \vec{k}_j = \vec{k}$).

An expression of the density of the elastic photoelectrons, N(E, hν), is further obtained by convoluting
N(E, hν) with an elastic escape function P(E, $\vec{k}$) which
describes the probability of a conduction electron (at
state $\psi_c(\vec{k})$ and energy E = $E_c(\vec{k})$ to overpass the surface
potential barrier, without loosing energy before reach-
ing the surface. We obtain:

$$N(E,\ h\nu) = \sum_{v,c} \int P(E,\vec{k})\ |\ <\psi_v(\vec{k})\ |e\cdot p|\ \psi_c(\vec{k})\ >|^2$$

$$\times\ \delta[E_c(\vec{k}) - E_v(\vec{k}) - h\nu]\ \ \delta[E - E_c(\vec{k})]d^3\vec{k}$$

$$(9)$$

Using the property of the δ functions, the volume integral on the right hand side of eq. (9) reduces to a line
integral in k space:

$$N(E,h\nu) =$$

$$\sum_{v,c} \int_{C(\vec{k})} P(E,\vec{k})\ \frac{|<\psi_v(\vec{k})\ |\vec{e}\ \cdot\ \vec{p}|\ \psi_c(\vec{k})>|^2}{|\nabla_{\vec{k}}[E_c(\vec{k}) - E_v(\vec{k}) - h\nu]\ \times\ \nabla_{\vec{k}}E_c(\vec{k})|}\ dZ(\vec{k})$$

$$(10)$$

where the contour C($\vec{k}$) is the intersection of the equal
energy surfaces defined by:

$$E_c(\vec{k}) - E_v(\vec{k}) = h\nu$$

and

$$E_c(\vec{k}) = E,$$

$$(11)$$

dZ$(\vec{k})$ being an element of curve C$(\vec{k})$.

The energy distribution of the elastic photoelectrons given in eq. (10) exhibits Van Hove singularities
if $[\ \nabla_{\vec{k}}[E_c(\vec{k}) - E_v(\vec{k}) - h\nu] \times \nabla_{\vec{k}}E_c(\vec{k})]$ vanishes..
These singularities determine the structure of the ener-
gy distribution, as discussed by Kane[7], and one should
be able, in principle, to assign features of the elastic
EDC's to transitions at some critical points of BZ from
a line-shape analysis of these EDC's. On the other hand,
the intensity of the structure depends mostly on the
transition matrix elements $<\psi_v(\vec{k})|\vec{e} \cdot \vec{p}|\psi_c(\vec{k})>$ and on
the elastic escape function P$(E,\vec{k})$. We now discuss these
two terms of eq. (10).

At first glance, an evaluation of P$(E,\vec{k})$ would appear to be rather difficult since we do not know the details of the surface potential distribution on an atomic
scale (perturbation of the crystal atomic configuration
is very likely to take place at the surface but cannot
be explored pricisely). However, one may remark the following: one conduction electron reaching the surface from
the bulk side is emitted in the vacuum only if the wave
functions which describe its propagation inside (Bloch
functions) and outside the solid (free-electron plane
waves) do match at the surface, no matter how the matching occurs.

Since the final state energy $E_c(\vec{k})$ of the excited
electron in the solid is relatively large compared to the
band gap width E_G of the semiconductor we are studying,
let us assume here that the crystal field seen by the
propagating electron to be weak. In this approximation,
the vector $\vec{k}$ in the Bloch function of the electron, which
indicates the direction of propagation of the associated
Bloch wave, would also indicate approximately the direction of propagation of the nearly-free electron in the
solid. In that sense, the vector $\vec{k}$ of the electron might
be considered as a wave vector which we might now decom-
pose, in the solid, into two components parallel $(\vec{k}_p)$ and
normal $(\vec{k}_n)$ to the surface. Decomposing in the same way
the wave vector of the free-electron plane wave outside
the solid, the above condition imposes the conservation
of $\vec{k}_p$ during the transfer of the excited electron through
the surface. As a result, all such electrons which propagate in the solid towards the surface with a group vel-
ocity equal to $\frac{1}{\hbar} \nabla_{\vec{k}} E_c(\vec{k})$, are eventually emitted in

the vacuum if their energy $E = E_c(\vec{k})$ fulfills the condition[8]:

$$E_c(\vec{k}) - E_i \geqslant \frac{\hbar^2}{2m} k_p^2, \tag{12}$$

where E_i is the ionization energy of the crystal (i.e., E_T).

Since we are only dealing here with elastic electrons, we should impose also the total electron momemtum $\hbar\vec{k}$ to be preserved during the process. Thus, writing the energy $E_c(\vec{k})$ of the propagating electron in the solid as:

$$E_c(\vec{k}) = \frac{\hbar^2}{2m} (k_n^2 + k_p^2),$$

the zero of energy being then at the bottom of the valence band, eq. (12) reduces to:

$$\frac{h^2}{2m} k_n^2 \geqslant E_i. \tag{13}$$

This equation fixes the probability of a conduction electron in the state $\psi_c(\vec{k})$ and of energy $E_c(\vec{k})$ (with $E_c(\vec{k}) - E_F \gg E_G$) to be elastically emitted in the vacuum, i.e., the escape probability function $P(E,\vec{k})$.

Since $P(E,\vec{k})$ depends on $\vec{k}$, it cannot be explicitly taken out of the integral in eq. (10). However, one can estimate its behavior upon varying E and $\vec{k}$, and this is what we intend to do now.

Nearly-free electrons travelling towards the surface along the perpendicular to this surface "see" a surface potential barrier reduced to $E_i (\vec{k}_p = 0$ in this case). They are emitted simply if their energy $E_c(\vec{k})$ is higher or equal to E_i. Among these (elastic) electrons, those which are emitted with the smallest kinetic energy $(E_{kin} = E_c(\vec{k}) - E_i \simeq 0)$ have a final state energy $E_c (\vec{k} = \vec{K}_n) = E_i$. Their wave-vector $\vec{K}_n$ is then, defined by $E_i = \frac{\hbar^2}{2m} K_n^2$. In Fig. 8a, we represent $\vec{K}_n$ and the wave-vectors $\vec{K}_1$ and $\vec{K}_2$ of two off-normal electrons such that: $\vec{K}_{1,n} = \vec{K}_{2,n} = \vec{K}_n$. All three electrons fulfill the condition expressed by eq. (13) and are emitted. However,

$|K_2| > |K_1| > |K_n|$ and $E_2 > E_1 > E_c(\vec{K}_n) = E_i$, where E_2
and E_1 are the energies of the respective off-normal
electrons. The energies of these electrons are represented in the contruction of Fig. 8b, which allows us
to define a so-called "escape cone" associated with each
electron energy $E_c(\vec{k})$ (we remember that this is only possible because we are in the nearly-free electron approximation, emphasizing $\vec{k}$ as a wave vector rather than a vector in the reciprocal lattice). The half-angle of this
cone is given by:

$$\cos \Theta = \frac{E_i}{E_c(\vec{k})} = (\frac{K_n}{k})^2 \qquad (14)$$

If $E_c(\vec{k})$ increases, the cone opens so that one may
then expect the average escape probability of all the
electrons (i.e., whatever the details of the symmetry of
their conduction states $\psi_c(\vec{k})$ are, at large $\vec{k}$) to increase with the electron energy. In other words, elec-
trons would tend to see <u>less</u> the surface as their energy
increases. This goes together with the fact that, by so
doing, electrons tend also to see a more and more uniform crystal potential. Their propagation in the solid
could, then, be described more adequately by pure free-
electron plane waves, just like outside the solid, and
there would not be any matching difficulty at the surface

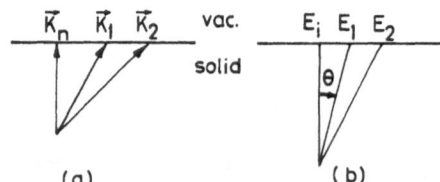

(a)   (b)

Fig. 8 (a) Wave vectors of normal incidence ($\vec{K}_n$) and
off-normal free-electrons ($\vec{K}_1$ and $\vec{K}_2$) propagating in a
solid towards the surface and such that all three components perpendicular to the surface are equal.

(b) $\vec{K}_n$ is taken to be such that $\frac{\hbar^2}{2m}K_n^2 = E_i = E_T$.
The minimum energy necessary for an off-normal electron
to be emitted is shown here (E_1 or E_2) to be larger than
the one required by a normal electron (E_i).

for these high-energy electrons. The average escape probability may be, then, expected to nearly saturate above
some energy and that energy would correspond to the electron plasma energy (16 to 17 eV in Si and Ge) as derived
from pure optical absorption data[9].

For energies decreasing towards E_i, the above description indicates that elastic emission would be more
and more restricted to electrons reaching the surface at
near normal angle of incidence. Although we do not expect the nearly free electron approximation to hold in
this case as well as at higher energies (the vector $\vec{k}$
should no longer indicate the direction of propagation
of the electron), we may remark that E_i is still large
in general ($\simeq$5 eV) compared to the forbidden band gap
width of a semiconductor ($\leqslant$ 1 eV). Within the limits of
this argument, one might consider (as a crude approximation) that the elastic electrons which are responsible
for the yield onset (measure of E_i) have an energy $E_c(\vec{K}_n)$
equal to E_i and a vector ($\vec{K}_n$) perpendicular to the actual
surface of the crystal. Since this surface is characterized by a reciprocal lattice vector $\vec{K}_s$ perpendicular to
it, the yield onset would be contributed by electrons in
states near $\vec{K}_n$ such that:

$$\vec{K}_n \cdot \vec{K}_s = |K_n K_s| \tag{15}$$

Would this mean that only electrons in states $\vec{K}_n$
may contribute to E_T? No, of course, since electrons
which have scattered in the solid and/or at its surface
(with or without losing energy, but in any case random-
izing their $\vec{k}$'s) may contribute also to the yield onset.
However, eq. (15) points at a quite promising eventuality,
namely, at photon energies just above E_T, elastic photoelectrons would be emitted approximately about the perpendicular to the surface. At higher energies, one might
then anticipate, from eqs. (13) to (15), some "angular
dependence" of photoemission: $N(E,h\nu)$ would vary with
the angle of emission. Altogether, the probability of
emission and the angle of emission of a given electron
in state $\psi_c(\vec{k}_j)$ would be set just by comparing its $\vec{k}_j$
with $\vec{K}_n$ defined earlier. How far may we go along this
line of thinking? Not much for low-energy electrons and
we have said why. However the subject is sufficiently
important to try some other way to understand it. Let
us write the vector $\vec{k}_j$ of the electron state as:

$$\vec{k}_j = (j - 1) \vec{K}_a + \vec{K}_j \tag{16}$$

where j is the band index, $\vec{K}_a$ is a vector the dimension
of which is the average dimension of the Brillouin Zone
in $\vec{k}$ space and $\vec{K}_j$ is the Brillouin Zone vector which has
the symmetry of the electron state $\psi_c(\vec{K}_j)$. In other
words, the behavior of the electron in the solid is here
described by the convolution of a pure plane wave (vector $\vec{K}_a$) and a Bloch wave (vector $\vec{K}_j$).

For large values of j, $\vec{k}_j \simeq (j - 1) \vec{K}_a$ and
$\vec{k}_j \cdot \vec{K}_s \sim \vec{K}_a \cdot \vec{K}_s$. The electron couples with the surface via a wave-vector the direction of which is at ran-
dom in the crystal. As a result, no angular dependence
of photoemission would be expected at large electron
energies.

For small values of j, $\vec{k}_j \cdot \vec{K}_s \approx \vec{K}_j \cdot \vec{K}_s$ and low-
energy photoemission could exhibit angular dependence.
However, we have seen that $\vec{K}_j$ does not correspond to the
direction of propagation of the electron and, therefore,
one cannot predict the angular distribution of the photoelectrons.

The only case where angular emission could be predicted reasonably well would be, in principle, surface
emission. At the surface, the star of $\vec{K}$ reduces to one
single $\vec{K}$ vector: $\vec{K}_s$, so that elastic conduction electrons
originating from the surface region would be preferential-
ly emitted along the normal to the surface. However,
surface reconstruction takes place, especially on open
lattices (like Si and Ge), which results into a mixing
of the one-electron Bloch states at and near the surface
and it is now the energy distribution of these surface
electrons which becomes difficult to predict. In turn,
this reconstruction process is likely to perturb locally
the image of the surface potential barrier we introduced
earlier and by extension, the so-called ionization energy
E_i of the material.

As a matter of fact, this energy E_i is the smallest
energy necessary to pull a valence electron out of the
crystal. This electron (i) should lie at an energy level
very close to the top of the valence band, and (ii) should
originate from the surface region (inelastic scattering
probability minimum, see Fig. 6). Now, since such an
electron participates to the bonding strength between
two adjacent atoms at or near the surface, the larger is
this strength, the larger is the (ionization) energy required to pull this electron out of its orbital. Namely,
a bonding orbital electron would require a larger energy

than a non-bonding orbital one. As a consequence, for such materials, which are characterized by both bonding and non-bonding valence energy levels, the yield onset might vary from one face to another depending on the nature of the bonds present on the actual surfaces. It is expected that disordered forms of these materials would then be characterized by a yield onset averaging all possible values of E_i in the crystal. On the other hand, in a material where the valence electrons belong to bonding orbitals only, it would not matter much which surface of the crystal is emitting and a nearly uniform value of E_i should be obtained for all crystal surfaces, except in the presence of important surface reconstructions which may perturb strongly the inter-atomic distances between neighboring surface atoms. Any disordered form of such a material would have a yield onset very close to the nearly uniform crystal one.

In order to illustrate the behavior of the escape probability with energy alone, we represent in Fig. 9 the average escape probability calculated by Saravia et al[8] in the cases of (111) Si and Ge. In this calculation, one takes:

$$P(E,\vec{k}) = \text{a constant A, for } E_c(\vec{k}) - E_i \geqslant \frac{\hbar^2}{2m} k_p^2$$

and $P(E,\vec{k}) = 0$ for $E_c(\vec{k}) - E_i < \frac{\hbar^2}{2m} k_p^2$.

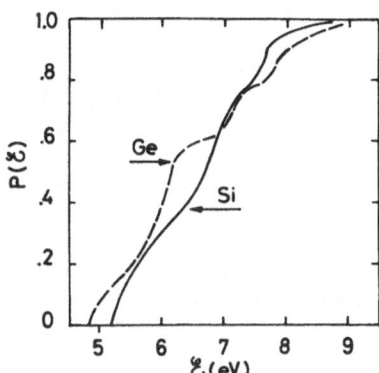

Fig. 9 Average escape probability function $P(E)$ for Si and Ge crystals, after Ref. 8.

$P(E,\vec{k})$ is further averaged over all the available states of energy E in the Brillouin Zone, and this yields the average escape probability factor $P(E)$, which is independent of $\vec{k}$:

$$P(E) = \int_{BZ} P(E,\vec{k}) \, d\vec{k} . \tag{17}$$

It can be seen, Fig. 9, that $P(E)$ saturates at $E \geqslant 8$ eV, i.e., some 3 eV above E_i, for both Si and Ge. Also, $P(E)$ does not exhibit any important structure. Once taken out of the integral in eq. (10), the only important effect of $P(E)$ will be to supress in the EDC's the density of the low energy elastic or inelastic electrons up to 3 eV above their low-energy onset. This, in turn, slows down the abrupt increase of $Y(h\nu)$ which we might expect from optical absorption data. In general, $Y(h\nu)$ saturates between $h\nu = 8$ and 10 eV (Fig. 6).

The intensity of the structure of $N(E,h\nu)$ also depends on the transition matrix elements [see eq. (10)]. An obvious simplification would be to consider them as constant which would reduce the integral of eq. (10) to a true joint density of states. However, (i) at low photon energies these matrix elements do vary widely depending on the relative symmetry of the states involved and on the polarization of the photon beam; (ii) at higher energies (above $h\nu = 10$ eV), the transition oscillator strength is affected by the presence of electron bands below the valence band[9]. This is the case for Ge and III-V compounds, in which d-band electrons lie some 30 to 20 eV, respectively, below the valence band upper limit. This causes the transition oscillator strength (and, further, the matrix elements) to increase for valence-conduction transitions occurring at photon energies between $h\nu \simeq 15$ eV (exhaustion of the valence band electrons and, approximately, the plasma frequency) and the onset of d-band transitions in these materials.

In silicon, this is not the case since the next 2p band starts 100 eV below valence-band edge. One may therefor consider matrix elements to be constant in Si at photon energies above 15 eV. This point explains also the profile of the optical absorption coefficient K in Fig. 2, which decreases above $h\nu = 10$ eV corresponding to the progressive exhaustion of the f-sum rule in Si.

b. Case of disordered solids: Since we have no

periodicity in such materials, we do not need to impose $\vec{k}$-conservation in calculating the energy distribution of their elastic photoelectrons. The expression of $N(E,h\nu)$ in eq. (10), reduces to :

$$N(E,h\nu) = \sum_{v,c} P(E) \left| M_{vc} \right|^2 n_v(E-h\nu) \, n_c(E), \qquad (18)$$

where $n_v(E-h\nu)$ and $n_c(E)$ are the values of the DOS in the actual valence and conduction bands of these solids, respectively; $\left| M_{vc} \right|^2$ is an average matrix element and $P(E)$ could be the average escape function calculated in eq. (17) and represented in Fig. 9 for Si and Ge crystals. A discussion of the respective profiles of n_v and n_c is presented elsewhere in this book. Essentially, the principal effect of disorder is to nearly obliterate the structure of the conduction band DOS so that the only structure present in the EDC's of disordered solids is related to features in the valence band DOS of these solids. (Experimental evidence for these effects is given in section three). This being the case, and assuming a similar electron-electron scattering probability in crystalline and disordered solids, increasing the photon energy at between 10 and 100 eV would, in principle, yield information on the valence states closer and closer to the surface, adding surface disorder to bulk disorder effects. In the X-ray region, the situation would, however, be close to the one found in crystals.

4. Conclusions and Comparison Crystal-Disordered Solid

Altogether, we may distinguish three regions in the photon energy range:
   (i) 5 - 10 eV (low U.V. photoemission): Information obtained is mostly on the <u>bulk</u> electronic structure of both crystals and disordered phases. However, $N(E,h\nu)$ has to be calculated explicitly from eq. (10) in crystals and can only be approximated from eq. (18) in disordered phases. In crystals, structure is rich and combine valence and conduction DOS; uncertainty exists as to the evaluation of a proper escape function $P(E,\vec{k})$, which obviously affects the low-energy photoelectrons. Since various models predict a vanishing of the conduction band DOS structure in the presence of disorder, low U.V. photoemission is ideal to test this prediction.

   (ii) 10 - 100 eV (far U.V. photoemission): The elastic electron contribution to the EDC's is well above

the approximate saturation point of $P(E)$. Assuming the
matrix elements $|M_{VC}|^2$ to be slowly varying with E, in
the crystals and in the disordered solids, and since the
actual energy distribution of the conduction states involved should be that typical of free electrons ($\sqrt{E}$ and
structureless) in eq. (10) and (18), both reduce to:

$$N(E,h\nu) \sim n_V(E-h\nu). \tag{19}$$

However, since electron-electron scattering is crucial
in this energy range, $n_V(E-h\nu)$ is in fact here the surface valence DOS which is a combination of a true surface states distribution (vacant or "dangling bonds")
and a distorted version of the bulk valence DOS. Comparison between the elastic parts of EDC's of crystalline
and disordered forms of the same material should take account of these complications. On the other hand, practically all the inelastic scattering occurs in the bulk.
Consequently structure appearing in the inelastic (low-
energy) part of the crystalline spectra (Fig. 7) would
represent structure in the bulk conduction band DOS.
Therefore, it is expected that a comparison of the inelastic emission of a disordered solid with the inelastic
emission of the corresponding crystalline form would
clarify directly the question of remanent structure in
the conduction band DOS of the disordered solid.

(iii) 1 - 2 keV: Both the elastic and inelastic
parts of the X-ray photoemission spectra would yield information mostly on the bulk electronic structure: Valence band DOS in the elastic region and conduction band
DOS in the inelastic one. One limitation in this photon
energy range comes from the wavelengths of the
photons used which are only 6 to 12 Å, thus restricting
this bulk information to the electronic structure features associated mostly with the short-range periodicity
even in the crystal. As a result, effects of long-range
order, if present in the disordered solids, could not be
evidenced here.

All these conclusions are summarized in Fig. 10,
where the upper edges of all the EDC's have been aligned
so that structure is delineated on an initial state energy scale $(E-h\nu)$. The ratio Y_s/Y_e gives the proportion of
scattered over elastic electrons. Below 10-15 eV, the structure of the EDC's follows a behavior set by eq. (10), or
optical joint density of states (OJDOS) behavior, while
at higher photon energies, valence and conduction band

DOS tend to contribute separately the elastic and inelastic EDC profiles, respectively, as shown in Fig. 11.
On the latter, local order only is effective between
1000 and 2000 eV. Above 2000 eV (or DeBroglie's limit),
the wavelengths associated to the photons drop below the
unit cell dimension of the crystal.

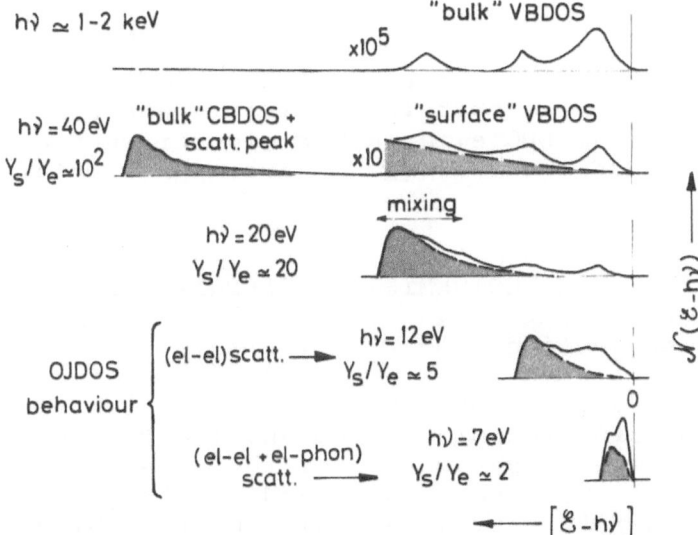

Fig. 10   General profiles of EDC's at various photon
energies. Up to hν ≃ 10 to 15 eV, the EDC structure has
an optical joint density-of-states behavior (eq. (1)).
At hν ≃ 20 eV, structure in the scattering peak (see text)
mixes with elastic emission originating near the bottom
of the valence band. At hν ≃ 40 eV, the elastic (surface
valence band DOS) and inelastic profiles (bulk conduction
band DOS) are well separated. At hν ≃ 1-2 keV, the elastic profile represents to some extent (see text) the bulk
valence band DOS; the inelastic part of the spectrum(in
particular plasmon losses) is not shown. Upon increasing
hν, the proportion of scattered over elastic photoelectrons (Y_s/Y_e) increases according to Fig.6.

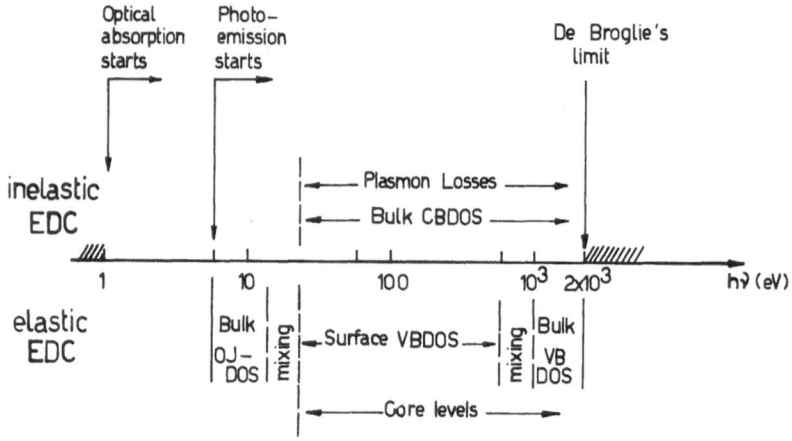

Fig. 11   Diagram giving trends of the qualitative information obtained from both the inelastic and elastic contributions to the EDC's, as a function of photon energy.
Limits of the delineated regions are approximate.

II. THE PHOTOEMISSION EXPERIMENT

1. Introduction

In this section, we intend to give a description of
the photoemission technique. We shall optimize the relevant instrumental parameters of the experiment, i.e.,
sample definition, light source and electron energy analyzer, in order to obtain reliable spectra which could
be analyzed using the procedure described in section one.

We show in Fig. 12 a very simplified schematic diagram of the photoemission set-up based on the so-called
retarding potential method[10]. A monochromatic photon
beam (energy $h\nu$) is focussed onto the material under study
(M) which eventually emits photoelectrons. These electrons are collected on an anode (A). Sample and anode
are immerged into a vacuum chamber and, thus form a diode. Its external circuit allows us to bias A (potential
V_R) and to measure the diode current (I) on M. The "retarding" potential V_R is varied from large negative values to positive ones with a ramp voltage and $I(V_R)$ is recorded simultaneously. We represent, in Fig. 13, the
$I(V_R)$ characteristic curve of the diode, measured in the

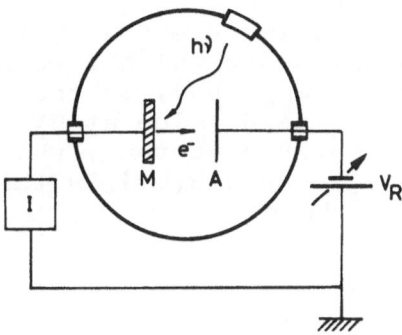

Fig. 12 Schematic diagram of a photoemission set-up, using the retarding potential method. M: sample; A: anode or collector; V_R: variable retarding potential; I is the photocurrent or diode current measured in the presence of photons of energy $h\nu$.

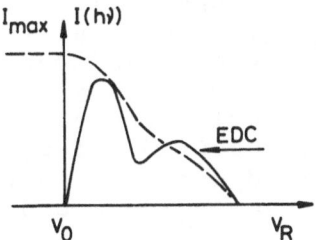

Fig. 13 Characteristic $I(V_R)$ and its first-derivative (EDC).

presence of a photon beam of energy $h\nu$. For large negative values of V_R, all the photoelectrons are repelled
by A and $I = 0$. Upon increasing V_R, the fastest photoelectrons start to be collected at some value of V_R,
thus marking the onset of I. As V_R increases again, more
and more photoelectrons can be collected in A. This progressively increases I which eventually reaches a saturation point at $V_R = V_0$, above which I remains constant at
$I_{max}(h\nu)$. This value, $I_{max}(h\nu)$, is the one which allows
us to set the quantum yield value at $h\nu$, $Y(h\nu)$. Once we
know the number of absorbed photons, which is proportional to $[1 - R(h\nu)]$, where $R(h\nu)$ is the reflection coefficient of M, then we have:

$$Y(h\nu) = A \frac{I_{max}(h\nu)}{I - R(h\nu)} ,$$

where A is a normalizing constant.

When $V_R = V_0$, all emitted electrons are collected,
including those which have the smallest kinetic energy
$E_{kin} \simeq 0$. Therefore, V_0 is taken to be the kinetic energy zero and we change the V_R scale into a E_{kin} scale.
Next, we consider the onset of I which occurs at some
value of E_{kin} and we note that it can only be contributed
by elastic electrons originally excited from states lying
very close to the upper valence band edge. Thus, such
electrons would reach conduction states at an energy
$E = h\nu$ above valence band edge. These electrons being
elastically emitted, we may then assign the energy $h\nu$ to
the onset of I and adopt now a "final state" energy scale
E (on that scale, energy zero is at the valence band edge).
V_0 corresponds to electrons which populate states around
the surface potential before being emitted with a kinetic
energy $E_{kin} \approx 0$. V_0 should then correspond to the photoelectric threshold E_T measured independently from the onset of the quantum yield curve as corresponding to the
energy difference between valence band edge and the surface potential.

On the other hand, we can write:

$$I(h\nu) \sim e \int_0^\infty N(E_{kin},h\nu) \, d\, E_{kin} \sim e \int_0^\infty N(E,h\nu)dE$$

$$(20)$$

where e is the electron charge. $N(E,h\nu)$ is the number
of electrons emitted with kinetic energy E_{kin}, per second.

From eq. (20) it follows that:

$$\frac{dI(h\nu)}{dE} = N(E,h\nu) \tag{21}$$

The derivative of the I(E) curve, which should therefore represent the total energy distribution of the photo-electrons (EDC), is shown in Fig. 13.

As far as only the elastic electrons are concerned [eq. (10)], one can take the energy zero at the upper onset of the EDC and change the final state energy (E) scale into an initial state energy (E - hν) scale. The low and high energy onsets are delineated by E_T (E scale) and $E_F = 0$ ([E - hν] scale), respectively. For the width of the EDC to be exactly equal to (hν - E_T) will, however, depend on the sharpness of its edges both of which are associated with the definition of E_T in a semiconductor.

We have seen that, by nature, the ionization energy of a material may not be uniform on an atomic scale owing to the possible anisotropy of the bond nature at the surface (surface states associated with dangling bonds, inhomogeneity in valence orbitals, reconstruction). In the case of a semiconductor, the situation is further complicated by the existence of a certain density of impurity states below E_f which may contribute to the photoemission process. The charge compensation which takes place near the surface results into a positive (n-type) or negative (p-type) band bending, which extends into the solid to a depth x of the order of, say 20 Å for 10^{17} cm^{-3} impurity concentration and increases with increasing doping.

The energy diagrams of four different situations are represented schematically in Fig. 14, from strong p-type to strong n-type. The band edges in the bulk are kept fixed while the Fermi level is moved in the band gap. Elastic electrons excited at depths of 50 and 10 Å have an energy distribution represented by curves 1 and 2, respectively. In drawing these curves, we assume the electronic structure not to be perturbed, in the first approximation, by fluctuations of the atomic configuration near the surface, but simply shifted following an arbitrarily uniform band bending. Relative intensities of curves 1 and 2 should be in the ratio of 5 to 1, respectively, according to Fig. 14.

For strong p-type, E_f is nearly at the top (E_{v_o}) of the valence band. Emission originating at 10 Å behind

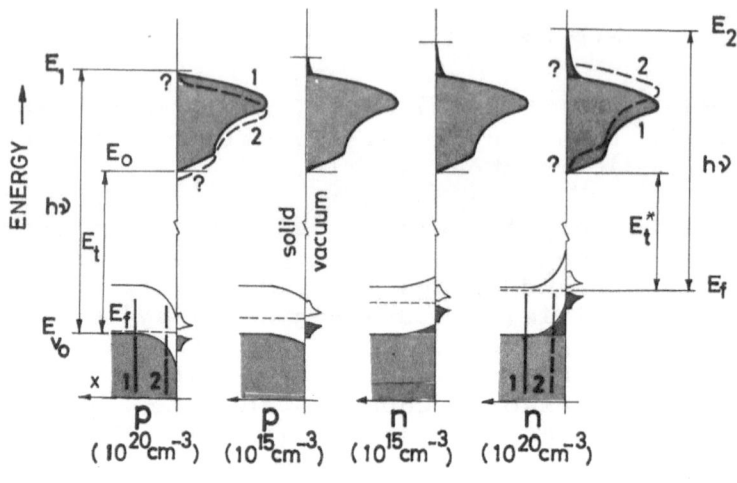

in 1 :   x ≃ 50 Å
in 2 :   x ≃ 10 Å

Fig. 14 Energy diagrams near the surface of four different semiconductors ranging from strongly p doped to
strongly n doped $(10^{20}$ cm$^{-3})$; x is the depth behind the
surface $(x = 0)$.

the surface (curve 2) is seen to overlap the low energy
onset (E_O) of the main distribution which is due to bulk
electrons (curve 1). This results in an apparent perturbation of E_T, originally set to equal the energy difference $(E_O - E_{VO})$. The surface state distribution which
is represented in the band gap region may also contribute to the spectrum but cannot be evidenced since its
contribution would show up at energies below the upper
edge of the EDC (E_1), which is associated to photoexci-
tation initiated at the top of the valence band (E_{VO}).

For strong n-type, emission from the band bending
region $(x = 10$ Å, dashed line) now overlaps the upper
edge of the bulk spectrum. In addition, some surface
states lying below E_f in the band gap may now contribute
to the spectrum above E_1. Altogether, the result is a
large EDC tailing which extends up to E_2 over a range
nearly equal to E_G. The low-energy onset of the EDC, E_O,
remains here unperturbed by the space charge region effect. Since emission from the band gap region is now

strong, it is possible to evaluate the work function of
this semiconductor $(E_o - E_f)$ which should be identified
in the yield curve as a secondary photoelectric threshold $E_T^* < E_T$.

At intermediate doping levels (either p- or n-type),
band bending is negligible in intensity, although extending deeper in the solid. Also E_f is about at the
center of the band gap. Surface states emission may contribute a high-energy tail. Fluctuations of the yield
onset which we have just mentioned are exemplified in
Fig. 15 where the yield of crystalline Si is shown to
vary with the nature and density of impurities[11]. The
uncertainty in defining the edges of the EDC (Fig. 13)
should not allow one to evaluate its width $(E_1 - E_o)$
with a precision better that $\pm$ 0.1 eV. The particular
difficulty encountered in delineating emission originating just below E_f in a relatively pure semiconductor is
usually lifted (i) by replacing the semiconductor with
a metal which is characterized by a high DOS value about
its Fermi level, and (ii) by keeping the rest of the set-
up unchanged. Just like in the semiconductor case, the
Fermi level of a metal aligns with the Fermi level of
the (metallic) sample holder. Therefore, the high-energy
onset of the metal EDC (measured with the same anode A)
delineates emission originating about E_f in the metal at
the energy $E = E_f + h\nu$ (= $h\nu$, with origin at E_f). The
position of this onset on the final state energy scale
associated with E_f does not depend on the material studied as long as sample holder and anode are not changed.
Going back to the semiconductor sample, a comparison of
the metal high-energy onset with the semiconductor one
gives (Fig. 14) the actual energy difference between E_f
and the top of the valence band in the semiconductor.

One word should be said about the position of the
low-energy onset of the EDC on the retarding potential
scale (V_o). Anode and sample (cathode) do not have, in
general, the same ionization energy. This results in
an offset voltage between them which shifts the EDC's by
an amount equal to the energy difference between the ionization energies of the anode and the sample. As a result, the low-energy onset of the EDC's does not coincide,
in general, with $V_R = 0$. The important thing is to observe the EDC width at some given photon energy $h\nu$ to remain constant along the experiment: a uniform energy
shift of the total EDC could only be attributed to a variation of the offset voltage, in fact, of the anode ionization energy, but a shift of the low-energy onset alone

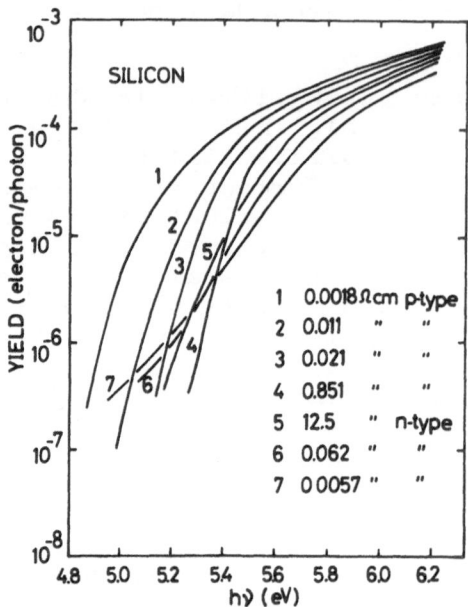

Fig. 15   Yield curves obtained for differently doped Si
crystals cleaved along (111), from Ref. 11.

should be attributable to a change of the ionization energy of the sample during the experiment (upon contamination
or surface reconstruction, for instance).

Finally, we shall make the following remark concerning the photocurrent $I(h\nu)$ measured on a semiconductor.
An electron emitted from M, (Fig. 12) contributes effectively the photocurrent I not only if it is collected on
A (depending on V_R) but also if its image charge left on
the emitting surface can travel through the sample towards its back surface and, from there, to the rest of the
circuit.   Until now we only considered the first switch
(V_R), implicitly assuming that all electrons (and holes)
travel the same way in the solid.   The transport of the
image-charge (the second switch is the sample itself) is,
however, an important parameter which deserves consideration in two instances:

(i) Suppose the sample is a moderately doped (for
instance, p-type) semiconductor (Fig. 14).   Most of the

tailing above the high-energy onset of the EDC's is due
to electrons excited from localized states lying between
E_f and E_{vo}, i.e., states within the mobility gap of the
semiconductor. Once one of the electrons trapped on these
acceptor states is optically excited into the conduction
band and further emitted, another electron is trapped
which eventually initiates the transport of a hole towards
the back face of the sample. This transport is described
by a mobility which is characteristic of the type of localized state involved. In any case, such a mobility is
several orders of magnitude smaller than the one of a
truly propagating electron (i.e., excited from extended
states lying below the valence band mobility edge to conduction states lying well above the conduction band mobility edge).

In this respect we may write the actual photocurrent
$I(h\nu)$, measured as a matter of fact between the front
(emitting) and back faces of the sample, as:

$$I(h\nu) = e\int [N_{ext}(E,h\nu) \mu_{ext}(E) + n_{loc}(E,h\nu \mu_{loc}(E)]dE$$

$$= I_{ext}(h\nu) + I_{loc}(h\nu), \qquad (22)$$

where $N_{ext}(E,h\nu)$ is the density of propagating (elastic
and inelastic) electrons and $\mu_{ext}(E)$ the mobility of their
image charges; $n_{loc}(E,h\nu)$ is the density of the localized states involved and $\mu_{loc}(E)$ the corresponding mobility of the image charges. Since $\mu_{loc}(E) \ll \mu_{ext}(E)$ and
$N(E,h\nu) \gg n_{loc}(E,h\nu)$, the second term in the integral
of eq. (22), $[I_{loc}(h\nu)]$ can be neglected. Further,
$\mu_{ext}(E)$ is constant for all propagating electrons so that
$I(h\nu) \approx I(h\nu)_{ext}$ is effectively proportional to
$\int N(E,h\nu)dE$, [eq. (20)]. As a crude approximation, the
ratio between $I_{loc}(h\nu)$ and $I_{ext}(h\nu)$ can be estimated to
be around

$$10^{-6} \left(\frac{n_{loc}}{n_{ext}} \frac{\mu_{loc}}{\mu_{ext}} = 10^{-4} \times 10^{-2}\right),$$

which means that it would be impossible, in general, to
measure an EDC contributed by localized electrons alone
($I_{ext}(E) = 0$ in this case). Even when this is possible,
the details of the energy distribution of the localized
states and the corresponding mobility remain almost un-
predictable and cannot be deconvoluted from the data.

(ii) Assume that an electron is emitted and collected on A (first switch closed), but that its image
charge cannot reach the sample holder (second switch open)
due to the low mobility of the material $[\mu_{ext}(E)$ is very
small in eq. (22)]. Although collected, the photoelectron does not contribute directly to $I(h\nu)$. What happens
in such a case is that we obtain an accumulation of positive charges at the surface which may travel towards the
holder only via the surface of the material, but with a
much lower velocity than the one they would have in the
bulk of a normally doped semiconductor. This surface
electrostatic charging effect eventually perturbs locally the surface potential. As a result, the photocurrent
does not saturate until V_R reaches relatively large positive values and the EDC width is much larger than the
value set by the difference $(h\nu - E_T)$. Such a situation
occurs in insulators and very low mobility, high purity
semiconductors at low temperatures.

2. The Diode

We have mentioned in Section one the importance of
operating at pressures of the order of 10^{-10} torr or less
if one desires to minimize or control the contamination
of the emitting surfaces. Whatever the photoemission
set-up adopted, this pressure requirement is mandatory.
Next, this set-up has to fulfill three basic conditions
if we want it to be accurate, efficient and reliable,
i.e,:

(i) Only the sample under study should emit electrons; (ii) all electrons emitted by the sample should
be collected; (iii) all electrons should see the same
retarding field whatever the angle of emission is. The
importance of these three conditions will now be discussed with regard to the arguments presented in this section before and in section one.

The first point in (i) is obvious. Suppose that
part of the light beam strikes the metallic anode A which
is held at a potential $V_R < 0$. Photoelectrons are produced by the anode which are immediately repelled by the
anode itself. Since the sample M is at a positive potential compared to A, most of these anode photoelectrons
are collected by M and contribute to the measured current $I(h\nu)$. Their energy distribution being characteristic of the anode metal, it will overlap with the semiconductor EDC especially about the upper edge where it
will show up as an additional tailing between the Fermi
level emission limit and E_1 (Fig. 14).

The second point (ii) should guarantee that the
quantum yield measured from M integrates all emitted
electrons. The third point (iii) deals with the spatial
distribution of these electrons: All photoelectrons
should look at the same retarding field distribution between M (the emitter) and A (the collector), at any angle
of emission, in order to ensure a uniform resolution of
the kinetic energy for all of them.

    The emitter-collector assembly: The only collector
which fulfills the above conditions is a sphere in which
the sample M is positioned at its center. Since M has
finite dimensions, perturbation of the spherical equipotentials in the vicinity of M is likely to occur. To
minimize such perturbation requires small dimensions of
M compared to A. The energy resolution $\Delta E_{kin}/E_{kin}$ of
such a spherical collector has been defined to be [12]

$$\frac{\Delta E_{kin}}{E_{kin}} \leqslant \left(\frac{a}{b}\right)^2 ,$$

where a is the diameter of the smallest spherical equipotential and b is the collector diameter. Since we want
to prevent perturbation of the electron propagation in
the vacuum associated with residual magnetic field, the
emitter-collector assembly is enclosed in a μ-metal can,
immerged in the vacuum chamber. For this reason, b is
usually of the order of 5 to 10 cm. A one percent energy
resolution is then obtained with small sample (a $\leqslant$ 5 mm).

    The position and dimension of the light ports on the
spherical collector are also important. If we have a
normal incidence beam, there will be only one port for
incoming and outgoing light. However, practically all
electrons emitted perpendicularly to the surface of the
sample will be lost through this port ($|V_R|$ is only of
the order of a few volts). It should be more convenient,
therefore, to operate with an oblique incidence although
this requires two ports. If the photon beam has only a
few mm^2 cross-section, the area of the ports can be reduced to about one percent of the spherical annulus which
covers the ports (average angle: twice the angle of incidence). On the other hand, this angle should not be too
large in order to minimize reflection losses at increasing angles. For nearly all the semiconductors, this incidence angle can be as large as 60° with still near-
normal reflectivity conditions[13]. It is noticeable that
under such incidence conditions, polarization of light
becomes an important parameter since the polarization

vector may be set to have a component perpendicular to
the surface. This can be used to study surface versus
bulk photoemission.

    Such a spherical collector[14] is shown in Fig. 16a.
The conical shape of the collector around the sample
holder is designed (i) to allow the positioning of the
sample at the center of A, (ii) to avoid emission from
the sample holder, and (iii) in such a way that the spherical symmetry of the equipotentials in front of the emitting surface of M is still preserved. This is shown in
Fig. 16b which represents the distribution of these equipotentials for the adopted geometry.

    In some cases, it is not possible to reduce the di-
mension of the sample and the energy resolution $\Delta E/E$ deteriorates. In such cases $\Delta E/E$ is improved by surrounding the non-ideal sample geometry with a field-free region[12]. The sample is screened by a fine spherical mesh
which restores the spherical symmetry of the retarding
field between the mesh and the collector itself. Such
a system (similar to a triode) may present difficulties
in handling and data acquisition.

    A detailed review is given in Ref. 12 of various
collector geometries which integrate all the photoelectrons. For these geometries, conditions (i) and (ii) are
easily satisfied but not condition (iii). The latter is
important since, although probably present, the angular
dependence of photoemission cannot be predicted in general
(see section one). As a consequence, any perturbation
of the spherical geometry of the retarding field may be
expected to perturb the energy distribution of photoelectrons. Since most disordered solids are studied by photoemission in the form of films prepared in situ, it is
worth pointing out the fact that one should avoid depositing the material to be studied onto the collector surface, thus perturbing its surface potential. The deposition being non-uniform in general, the perturbation introduced is also non-uniform and, as mentioned before,
this would result in a bluring of the EDC edges as well
as of the EDC structure.

    Other types of analyzers (which are not integral col-
lectors) allow to energy-analyze selectively the photoelectrons with respect to their angle of emission. This
is for instance the case of the 90 or 127° sector analyzers[15] in which electrons are retarded between two 90°
or 127° cylindrical deflection plates biased at + and $-V_R$,

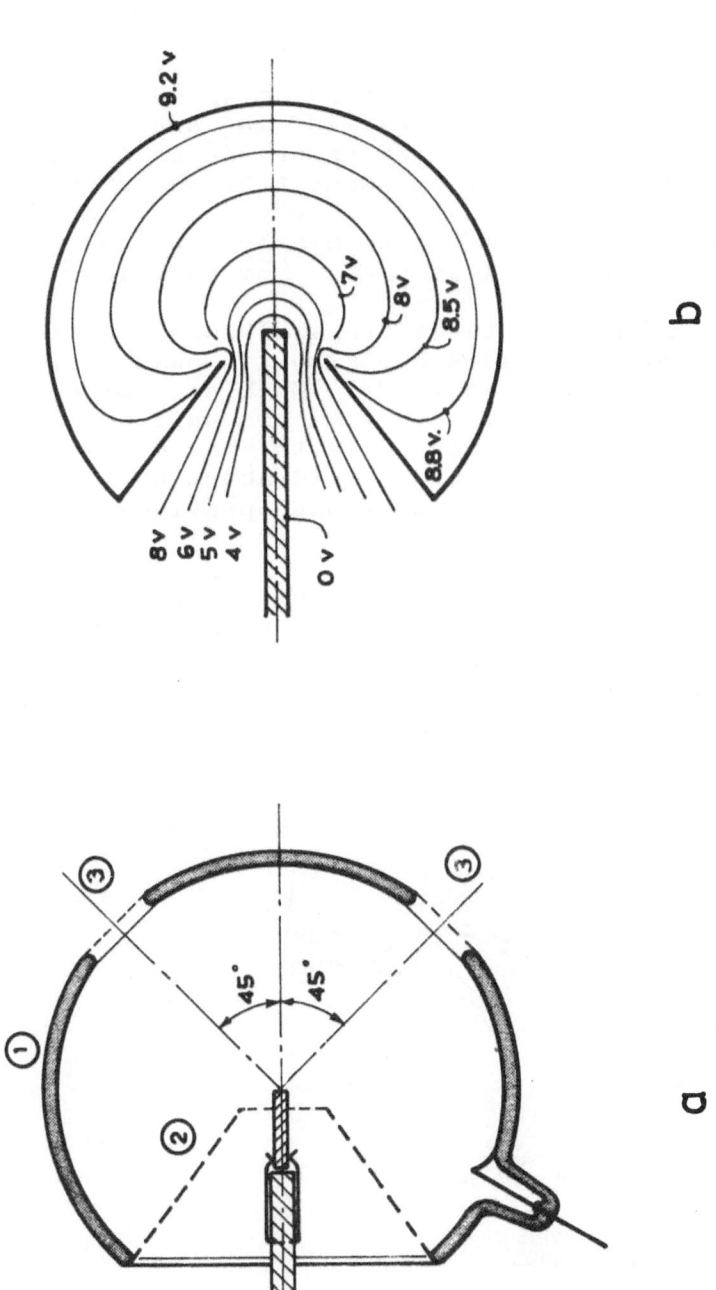

Fig. 16 Schematic view of a spherical analyzer (from Ref. 14) - a: sectioned top view with (1) glass sphere (gold coated inside surface), (2) electrical feedthrough, (3) light ports - b: distribution of equipotentials in a vertical plane.

respectively, and energy-analyzed using a channeltron electron multiplier at the exit slit of the analyzer (Fig. 17). Energy resolution is typically of the order of one per cent. Since the acceptance angle of such analyzers is reduced to 10-15°, only a small proportion of the overall photoelectrons are analyzed [condition (ii) is not fulfilled here]. However, by rotating the sample surface in front of the entrance slit of the analyzer one may select the photoelectrons which enter the analyzer with respect to their angle of emission from the sample surface. This makes sector analyzers very useful in investigating angular emissions, particularly the ones which are associated with surface states and chemisorption[16].

The detection limit of the above analyzers lies in the 10^{-14} amp range, which does not allow to measure the EDC's at photon energies close to E_T. Much smaller currents can be detected directly by usual channeltrons (down to 10^{-17} amp). Therefore the latter may be used to evaluate details of the quantum yield about E_T, regarding the gap states emission mentioned previously in this section.

The light source: If we want to follow accurately the evolution of the EDC's of a material with photon energy, we need a light source which is altogether rich and intense for all wavelengths below the one of the

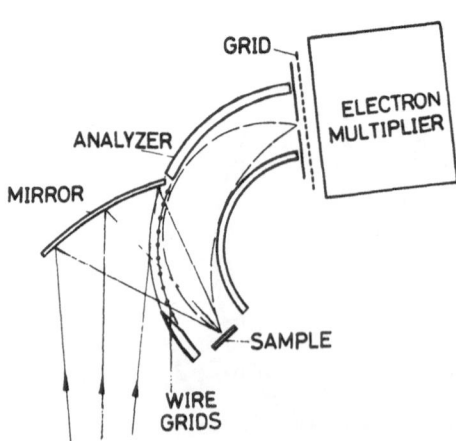

Fig. 17 Schematic view of a 127° sector analyzer (from Ref. 15).

photoelectric threshold. The lamp source which is the
best adapted to the 4 to 6 eV range is the Hg lamp.
Above 6 eV, the hydrogen lamp takes over up to the Lyman-
α region (around 12 eV). At higher energies, a very limited number of line-sources is available: Neon I
(16.8 eV), He I (21.2 eV), Ne II (26.9 eV) and He II
(40.8 eV). At even higher energies, hard or soft X-ray
sources are used (for instances, Al Kα at 1486.6 eV).

    While lamp sources provide photons at all energies
up to 12 eV, allowing a precise tracing of the EDC structure, say, every 100 meV, such a possibility vanishes
above 12 eV. We remember that around this energy, the
electron mean-free path Z_e (Fig. 5) decreases rapidly at
below 10 Å, i.e., the elastic EDC is rapidly confined to
electrons originally excited within the surface region
of the sample. It is unfortunate that at the point where
surface interaction becomes very effective, one cannot
trace correctly the elastic EDC structure anymore. One
could argue that above 15 eV the elastic EDC more or
less represents an initial state density $n_v(E-h\nu)$ at the
surface and, therefore, there is no reason to expect variations in this elastic EDC structure by varying hν above
15 eV. This might be about correct above hν = 20 eV, but
not in the 10 - 15 eV range. In this latter range, it
should be possible, for instance, to evaluate the progressive distortions introduced in the "elastic" electronic
structure of the solid as the escape depth of the elastic
electrons decreases with progressively increasing hν.
This progressive probing of the electronic structure when
approaching from the surface could reveal details on the
atomic configuration at, or very close to, the surface.

    Such a study is not possible using the above lamp
sources. It can only be investigated using the output
of a synchrotron radiation storage ring. The latter source
is characterized by: (i) a continuum extending between a
few eV and a few keV; (ii) its intensity which can be
several orders of magnitude larger than the ones of the
few lamp sources mentioned before; (iii) an effectively
negligible noise level and (iv) the polarization of the
beam. These four arguments clearly demonstrate the tre-
mendous advantages of a storage ring output over conventional light sources, at all energies. The only problem
with it is that, so far, there are not so many rings
available, reducing the number of its users to a happy
few dozens of experimentalists in the world. Yet, many
problems can be investigated in an ordinary laboratory,
at energies below 12 eV.

The emitter. What we are mainly concerned with is
now the contamination problem of the surface of the emitter. We already said why we needed to work at pressures
$\leqslant 10^{-10}$ torr. Up to $h\nu$ = 12 eV, the light beam can go
in and out of the chamber through LiF windows which remain transparent up to that energy. At $h\nu$ = 12 eV, we
have to operate with "open" chambers provided with differential pumping systems at each open port. Usually,
pressure in such open systems can be maintained in the
upper 10^{-10} torr range or higher, so that surface contamination may become critical after one hour of operation.
In such instances, the actual sample surface has to be
renewed every hour either by ion bombardment or by cleav-
ing, or by preparing a new film.

As far as films are concerned, the problem is even
more serious. It is not the surface contamination which
is the most critical, but the bulk contamination which
takes place during the growth of the films. During evaporation of the material, the temperature of the evaporation is high enough to heat up parts of the assembly.
The pressure in the chamber may then increase into the
10^{-8} torr range (or more) by the end of the evaporation.
Typically, if the latter last for 2 min., and if pressure
is in the 10^{-7} range for most of that time, the bulk contamination rate of a, say, 1000 Å layer could be already
of the order of two to three percent. In this respect,
pressure should be kept at below 10^{-9} torr all along the
preparation of films.

Many other parameters may influence the characteristics of a film. Among these, the most important seem
to be the temperature of substrate at the time of deposition, its nature, the annealing temperature, the annealing and cooling rates. The impurity content of the
evaporated material may also play an important role, which
can be minimized by evaporating the same material which
is used for the comparative crystalline study.

The importance of the substrate should be stressed.
As a matter of fact, ordering processes take place in a
film at the very early stage of its growth, i.e., as soon
as the first monolayers are formed on the surface. These
atoms do see the substrate atomic configuration and its
defects and, consequently, will acquire a certain mobility which depends on the respective temperatures of the
substrate and of the impinging particle and also on these
defects (surface dislocations, etc...). The next atom
layers will not see anymore the substrate surface but

the film surface and their mobility will then be totally
different to the one of the first atoms reaching the surface. The choice of the substrate is therefore critical.
In particular, one should try to avoid imposing any ordering from the substrate into the film by choosing some
stable and disordered form of a material (glass, for instance).

All such interactions between film growth and substrate usually show up as distinct evolutions of the film
electronic structure which are materialized, in the photoemission spectra, by the appearance or disappearance of
peaks at some given energy E, by a shift of a given piece
of structure or by fluctuation in the emission associated
with the gap states or space charge region behind the
surface.

3. Measuring the Yield and the EDC.

Yield. Use of a nearly closed collector like the
one shown in Fig. 16 ensures that all emitted electrons
can be effectively collected when the collector potential V_R is set to a positive value (say, + 10 eV). The
saturation value of $I(h\nu)$, $I_{max}(h\nu)$, gives the total number of electrons photoemitted. The total number of absorbed photons (energy $h\nu$) is obtained by a proper cali-
bration of the source against a calibrated thermopile,
taking into account eventually the transmission of the
LiF windows, and dividing the resulting number of incident photon by $[1 - R(h\nu)]$ (where $R(h\nu)$ is the reflectivity of the sample at $h\nu$).

Usual resolution is of the order of twenty to thirty
percent although resolution is sometimes claimed at better that ten percent. Below 10^{-6} electron/abs. photon,
resolution deteriorates ($>50°/o$). Usually, E_T is taken
at $Y(h\nu) \simeq 10^{-6}$ although emission may be still detected
at energies just below the adopted value of E_T.

What we may reasonably expect is that yield curves
would not exhibit strong structure. This can be directly seen from the fact that $\varepsilon_2(h\nu)$ curves, which are the
optical equivalents of yield curves, exhibit structure
only up to $h\nu \simeq 6$ eV. Above this energy, $\varepsilon_2(h\nu)$ usually becomes monotonic. Therefore, structure in $Y(h\nu)$
should be restricted to the energy range just next of
E_T, i.e., in a region where yield resolution is not at
its best ($\sim 50°/o$). As can be seen, the quantum yield
$Y(h\nu)$ is not a very convincing quantity in photoemission.

Its nearly only use is to help in locating the surface potential versus E_f. Even so, resolution is not sufficient and definition of the surface potential leaves many
questions open.

   EDC. The EDC is obtained by differentiating the $I(V_R)$
characteristic. This differentiation can be done elec-
tronically by superimposing a small a.c. voltage v on
the ramp sweep d.c. voltage V_R on the collector and de-
tecting its first derivative via lock-in techniques.
With this technique, the capacitive effect between collector and emitter has to be compensated. Differentiation
can also be performed by a computer which has the advantage of a much simpler detection system limited to the
measure of $I(V_R)$; also capacitance compensation is avoided.

   The overall resolution of the photoemission set-up
is determined by the analyzer resolution (set by its geometry), the monochromator operation (slit width and stray
light) and temperature broadening of the electron states
($\sim kT$ or 30 meV at $300^{\circ}K$). Using the analyzer described
in Fig. 16, the kinetic energy of a 1 eV photoelectron
(at $h\nu = 8$ eV) is resolved at $\pm$ 100 meV with 0.5 mm slits
and at $300^{\circ}K$. For this resolution to be effective in the
experimental EDC's one has, in addition, to use peak-to-
peak modulation amplitudes of v smaller than the overall
resolution mentioned above: if v is 0.2 V_{pp}, features
of the $I(V_R)$ curve would be resolved independently if
they are separated at least by 0.2 eV, whatever the resolution of the set-up is.

   In the far-UV range (20 - 100 eV), ΔE_{kin} increases
but photon energies are better resolved, so that overall
resolution should be still about 0.2 - 0.3 eV.

   In the X-ray region, we no longer have to care about
the spatial geometry of the collector, due to the very
large kinetic energy of the elastic photoelectrons. Instrumental resolution, although relatively excellent
($\sim 5 \times 10^{-4}$), is limited to a constant but poor level compared to U.V. photoemission ($\pm$ 0.7 eV). In addition,
plasmon losses affect the intensity of the elastic spectrum in the manner described in Fig. 18. The dashed line
is the inelastic electron-electron scattering profile
(Fig. 7). After sufficient magnification, the elastic
EDC (solid line) appears accompanied by characteristic
one-plasmon, two-plasmon etc... loss spectra. The elastic EDC has then to be corrected for these losses which
introduces additional uncertainty in locating the

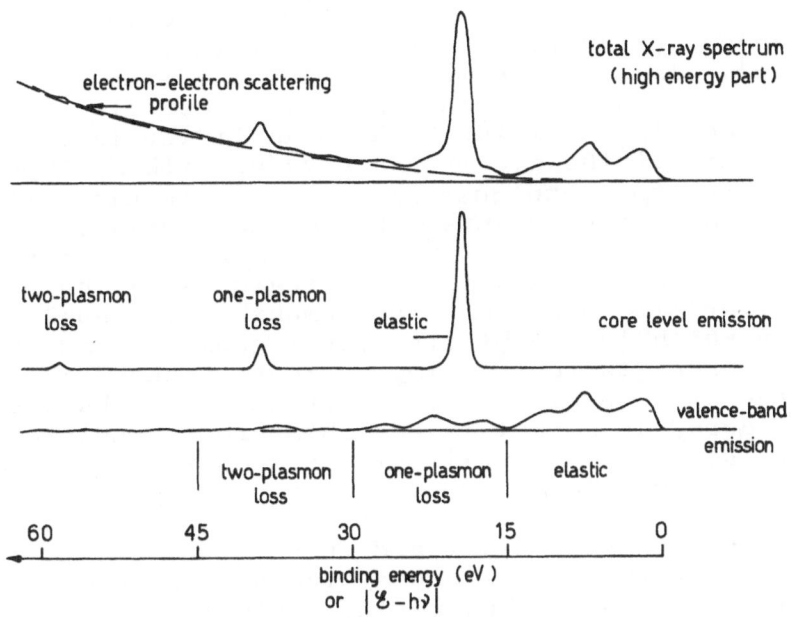

Fig. 18 Schematic X-ray EDC. The inelastic electron-electron scattering peak appears at the low-energy edge. Upon magnification, elastic emission from deep core levels appears first which is accompanied at lower energies by satellite structure associated with plasmon losses ($h\nu_p$ = plasmon energy). With higher magnification, elastic emission originating in the valence band itself is evidenced, again with plasmon-loss inelastic electron distributions.

structure of the elastic EDC. This is particularly important when trying to identify elastic emission from the bottom of the valence band. The width of the latter is measured at ± 1 to 1.5 eV after considering these corrections as well as the elastic signal-to-noise ratio. This would render a comparison between X-ray and far-U.V. elastic EDC's rather difficult since both sets of spectra may or may not agree within experimental errors which are, however, too large (± 1.0 and ± 0.3 eV, respectively) to allow conclusions to be drawn. We already noted the difference in the nature of the elastic distributions in far-U.V. and X-ray photoemission which is associated with the inelastic electron-electron scattering probability. The above instrumental argument would probably not help clarifying the ambiguity.

It may appear useful to enhance structure which exists in the EDC but is either hard to locate(shoulders,
for instance), or intricate (especially in low U.V. photoemission). This enhancement is done by twice differentiating [17] the EDC as shown in Fig. 19, either electronical-
ly (reference signal of the lock-in amplifier is the
third harmonic of the a.c. voltage v) or with a computer.
Both differention techniques give similar results. Fig.
19 shows clearly the advantage of second-derivatives over
usual EDC's. Pronounced structure is more precisely located on the energy scale and weak features may be resolved. However, it is not an improvement in the analyzer
resolution which is set by.its actual geometry and the
monochromator slits, but an improvement in the "visibility" of the structure. Higher-derivatives may be therefore considered as a tool in searching for structure within the EDC's. When uncertainty exists as to whether a
piece of structure in an EDC second derivative is attributable to true EDC structure or to an artifact of the differentiation, one can vary v and observe the evolution of
the relevant feature: its energy location should not
vary with decreasing v (increasing visibility) if it is
associated with a feature in the EDC.

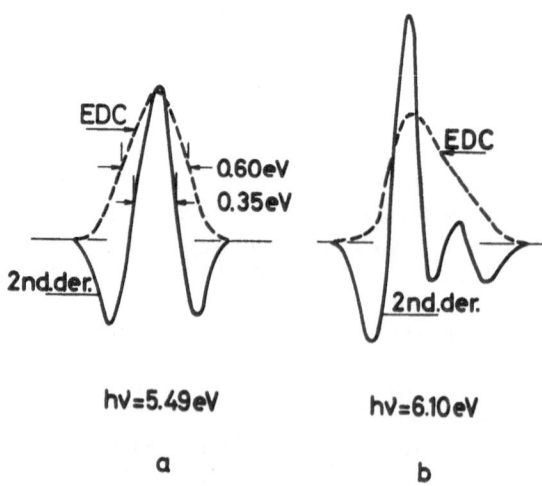

Fig. 19 EDC's and their second derivatives obtained at
(a) hν = 5.49 eV and (b) hν = 6.10 eV, from a vacuum
cleaved (1010) Te face.

III. PHOTOEMISSION OF DISORDERED SEMICONDUCTORS

1. Introduction

In sections one and two we evaluated the wealth of information one could expect from photoemission as well as some of its instrumental aspects. We will now try to investigate thoroughly the photoemission spectra of crystalline and disordered forms of semiconductors, in the light of the theoretical and instrumental considerations presented earlier.

Following the procedure used in the preceding sections, the data obtained from both crystals and disordered films are analyzed in parallel in each of the three basic energy ranges (5-10 eV; 10-100 eV; 1-2 KeV). Where applicable, a comparison of these data with those obtained by other techniques is made to help resolve ambiguities in the interpretation.

2. Tetrahedrally-coordinated semiconductors

The material which has attracted most interest until recently has been germanium. The following is mostly devoted to this elemental semiconductor. Attention is also paid to Si and III-V compounds when data are available.

Low-energy UV photoemission: (5-10 eV). We have seen, in Fig. 10, that at such energies the structure of the crystalline EDC's is essentially described by a combination of valence and conduction band DOS, in which energy- and $\vec{k}$-conservations are imposed. As expected from the absence of long-range periodicity in the disordered configurations, EDC's from disordered films should no longer retain such fine structure arising, in the crystal EDC's, from the $\vec{k}$-conservation rule. As a consequence, photon energies in the 5-10 eV range should be essential in evaluating and characterizing the lack of long-range order in the films, from a comparison between the EDC's obtained from these films and those from the corresponding crystal.

In order to give some confidence in this kind of comparison, let us start with the photoemissive properties of Ge (111) surfaces. Fig. 20 represents an EDC and its second-derivative obtained from vacuum-cleaved (111) Ge[18], at $h\nu = 7.72$ eV. As shown in section two, structure appearing in the EDC is emphasized in its second derivative. This structure can be followed with photon energy. It may then be useful to plot the evolution

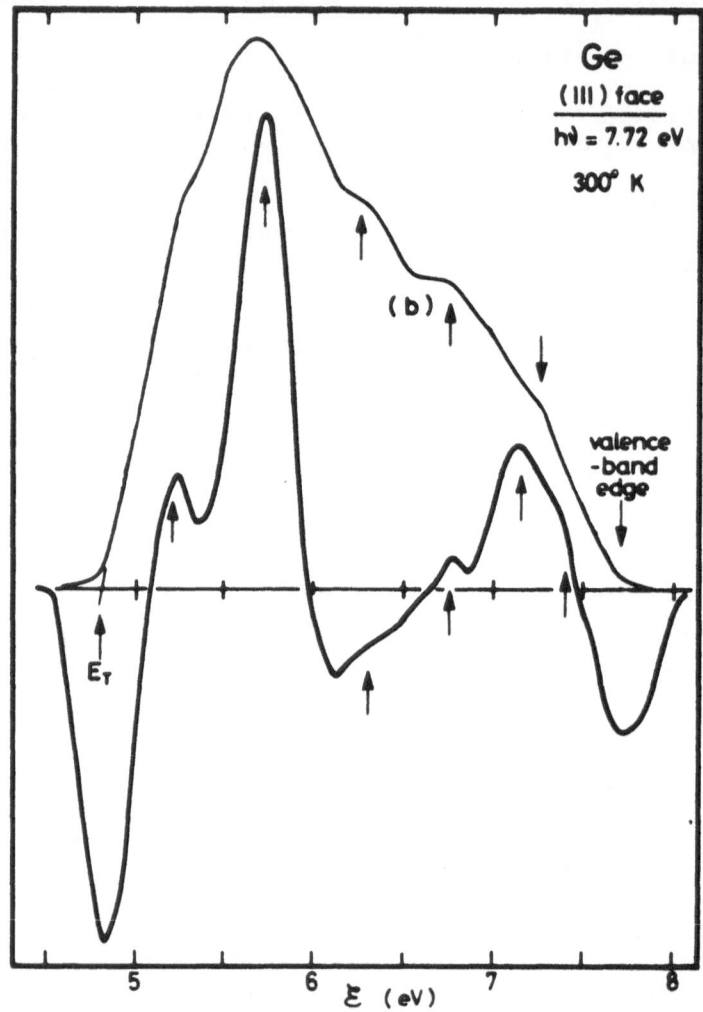

Fig. 20   EDC and its second derivative obtained from
(111) Ge at hν = 7.72 eV.   Final state energy scale ℰ
is indicated.

of this structure on a diagram as shown in Fig 21 (from
ref. 19), where the position of spectral features is
quoted on the final state energy scale (ℰ) on ordinate
at each photon energy hν (abcissa).   The solid lines are
the direct-transition lines obtained directly from the
ℰ(k⃗) diagram of Ge, by coupling vertically valence and
conduction states separated by the photon energy hν.
This procedure corresponds to the follow-up of the peak

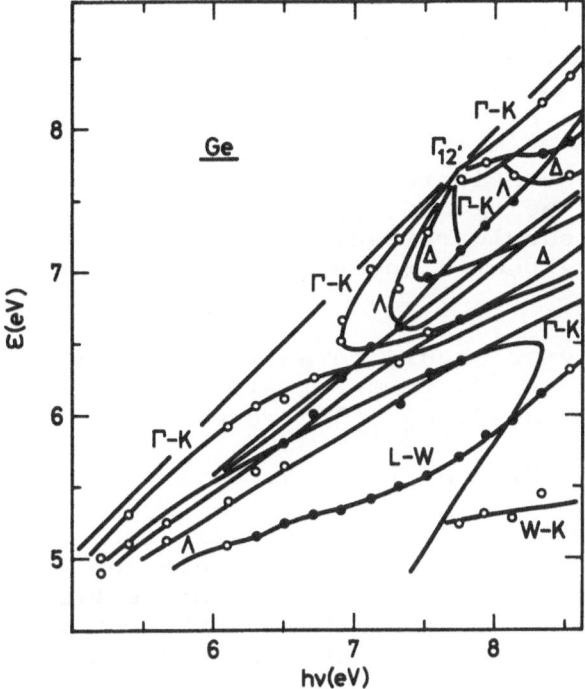

Fig. 21   Energy diagram $\mathcal{E}(h\nu)$ for (111) Ge. Points refer to the position of spectral features observed at
various photon energies (solid or open circles for strong
or weaker spectral features, respectively). Lines are
the direct-transition lines obtained directly from the
band structure of Ge, Ref. 8.

positions(and not intensity) in the EDC's calculated from
eq. (10). Among others, the main peak in the low-energy
part of the spectra (Fig. 20) is well described by direct
transitions occuring in the L-W region of the Brillouin
zone. Generally speaking, one can say that the EDC structure in this energy range evolves in a complex manner
which may be predicted from the calculated elastic electron distribution of eq. (10). Indirect transitions
would, in general, introduce structure broadening.

     In the case of disordered films, we have seen that
the structure in EDC's is dominated by the consequences
of the lack of long-range periodicity (the Brillouin
zone scheme itself becomes irrelevant to some extent).
The latter introduces two specific effects: (i) $\vec{k}$-con-
servation becomes pointless, and (ii) the conduction band

DOS is more or less monotonic.   The first point does
not mean that the vector $\vec{k}'$ of the conduction state is
different from the vector $\vec{k}$ of the valence state, but this
vector $\vec{k}$ could no longer be decomposed into reciprocal
lattice vectors as is done in the crystal case (extended
zone scheme). From that point of view, conduction electrons would be considered as free electrons, the "crystal" potential being nearly constant. Thus, the conduction band DOS should reflect this situation and its profile would be expected to be featureless (the second
point).

This behavior should be easy to verify in photoemission spectra. In terms of structure (peak positions, for instance), eq. (18) reduces to:

$$N(\mathcal{E},\ h\nu) \sim P(\mathcal{E})\ n_v\ (\mathcal{E}-h\nu),\quad\quad\quad\quad\quad (23)$$

i.e., EDC's would reflect essentially the valence band
DOS. In the energy range 5-10 eV, only the top 5 eV of
the valence band can be explored. Since $P(\mathcal{E})$ tends to
suppress the lower 3 eV of the spectra, only the upper
2 eV range of a 10 eV EDC would effectively follow the
valence band DOS profile.

An example of low energy EDC's measured from some
amorphous Ge film[20] (deposited at room temperature onto
silica, and not annealed) is shown in Fig. 22. The upper
edges of these EDC's have been aligned on an initial
state scale $[\mathcal{E}-h\nu]$. Apparently, their structure behaves
as predicted in eq. (23). A peak appears on the high
energy side at somewhere between -1.0 and -1.5 eV, thus
apparently indicating a high DOS at 1 to 1.5 eV below
valence-band edge in this film. Whether this would be a
consequence of the escape function $P(\mathcal{E})$ which would saturate approximately at $\mathcal{E} = 8$ eV (or -2 eV on the $[\mathcal{E}-h\nu]$
scale) cannot be yet decided. No conduction band DOS fea-
tures seem to show up. Had such features existed, they
would remain fixed on final state energy scale $\mathcal{E}$, according to eq. (18):

$$N(\mathcal{E},\ h\nu) \sim P(\mathcal{E})\ n_c\ (\mathcal{E}),\quad\quad\quad\quad\quad (24)$$

where we have taken $n_v(\mathcal{E}-h\nu)$ nearly constant over the
range of energy where such a conduction band feature
would manifest itself. In fact, spectral features may
overlap and lose their specific character (either valence
or conduction band DOS) and EDC structure has to be

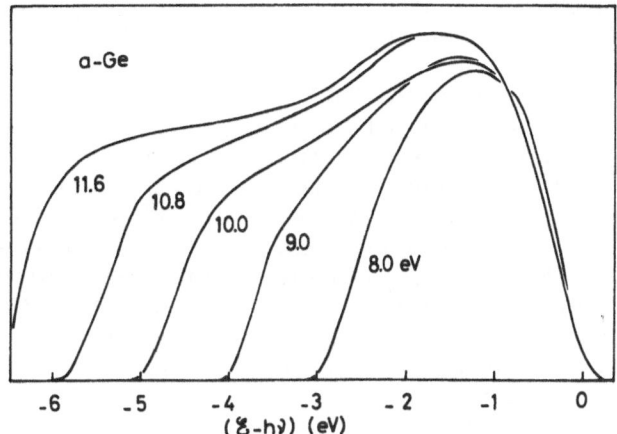

Fig. 22   EDC's obtained from a disordered Ge film,
Ref. 20.

searched in detail using higher derivatives.

The next point is that because of $P(\mathcal{E})$, no information is obtained on conduction band states lying below
the surface potential ($\sim$5 eV). The result of these various limitations is shown in Fig. 23, where the solid line
indicates the DOS estimated profile obtained from the
EDC's. The rest of the DOS remains undetermined by photoemission means, as seen from the alternative profiles
(dashed curves shown). This is particularly critical a-
round the bottom of the conduction band. Soft X-ray absorption data[21] would support a DOS maximum just above
the conduction band edge.

Upon annealing, EDC's of such Ge films do evolve[20,22]
towards the crystalline spectrum above 300°C. However,
the emergence of crystalline spectral features is not,
in general, well identified (energy location) which pro-
hibits the use of these curves to explore qualitatively
and quantitatively the amorphous to crystalline transition. Higher derivatives of EDC's can be used, in this
respect, as will be demonstrated below.

A comparison of Figs. 20 and 22 already indicates
that EDC's are very different in (111) Ge and in the disordered Ge film mentioned before. At first glance, one

would say that the EDC maximum has moved from the low-
energy (crystal) to the high-energy side ("amorphous"
film) of the EDC's (widths are nearly equal here,
E_T = 4.7 to 4.8 eV in both materials). The EDC second
derivative of (111) Ge, Fig. 20, is compared in Fig. 24
with the second derivative EDC of a disordered Ge film
(a-Ge). The integrals of the two curves measured under
the same conditions have been normalized to the same
value, taking the quantum yield to be the same for both
materials. Peak labelled A(L-W transitions) in the crystal spectrum is seen to disappear totally in the disordered case. Peak B (transitions about the top of the valence band, $\Gamma_{25'}$) apparently strengthens in a-Ge. The
absence of Peak A is probably the most striking effect
of disorder in the a-Ge electronic structure, whilst no
conclusion can be drawn out of peak B behavior (bands are
nearly parabolic about $\Gamma_{25'}$ in the crystal). As can be
seen in Fig. 24, we have here a very accurate way to test
the nature and extent of the ordering present in films
prepared in various ways. Indeed, since the position
and strength of peak A at $\mathcal{E}$ = 5.7 eV in the 7.72 eV crystal spectrum are not fortuitous (see Fig. 21), they may
be taken to be representative of the bulk crystal atomic
configuration. For instance,should this peak appear in
the 7.72 eV spectrum of a disordered film precisely at
$\mathcal{E}$ = 5.7 eV, we would then have evidence for diamond-
like ordering in such a film. Now, the extent of this
ordering, i.e., the proportion of atoms which are diamond-like assembled in that volume of the film which

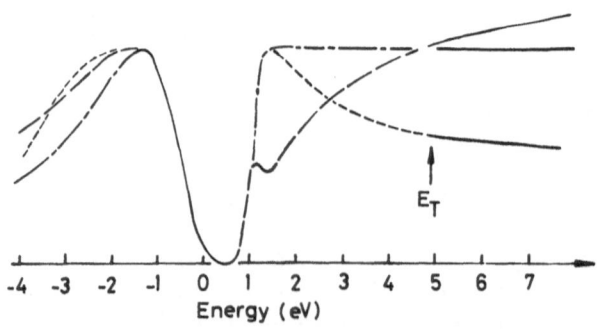

Fig. 23 Estimated profile of a-Ge DOS from Fig. 21. Uncertainties in experimental and theoretical parameters
are represented by alternative profiles (dashed lines).

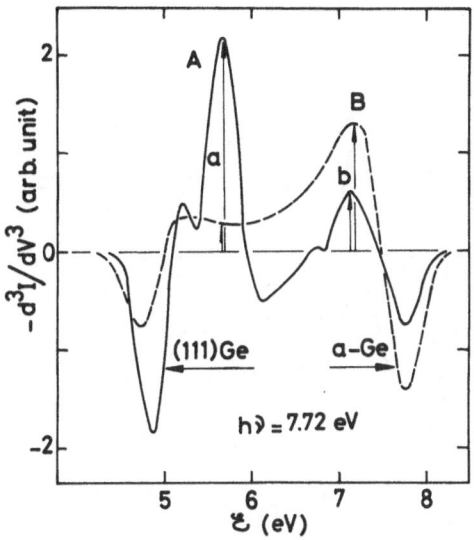

Fig. 24 Comparison of EDC second derivatives from (111)
Ge and a disordered Ge film defined in the text. Intensities of regions A and B of the spectra are evaluated approximately by the quantities a and b.

contributes to the elastic photoemission spectrum at such a
photon energy(l_e is about 20 Å at $h\nu$ = 7.72 eV, see Fig.
5), should be somehow related to the relative strengths
of peaks A and B in both crystalline and disordered spectra. In contrast, <u>non-diamond-like</u> ordering would result in structure which could not be compared with the
crystal spectrum, both in position and strength.

We shall now illustrate the above considerations
with examples on Ge films. In view of the extreme complexity of film preparation, it is essential to explore
the largest possible variety of films in order to evaluate their most fundamental properties. This is what we
will try to do in the following.

As demonstrated in Figs. 20 and 24, great advantage is obtained from curves of higher derivatives and
we will mostly restrict ourselves to the analysis of such
curves. In Fig. 25 is shown a series of second derivatives of EDC obtained at $h\nu$ = 7.72 eV from Ge films

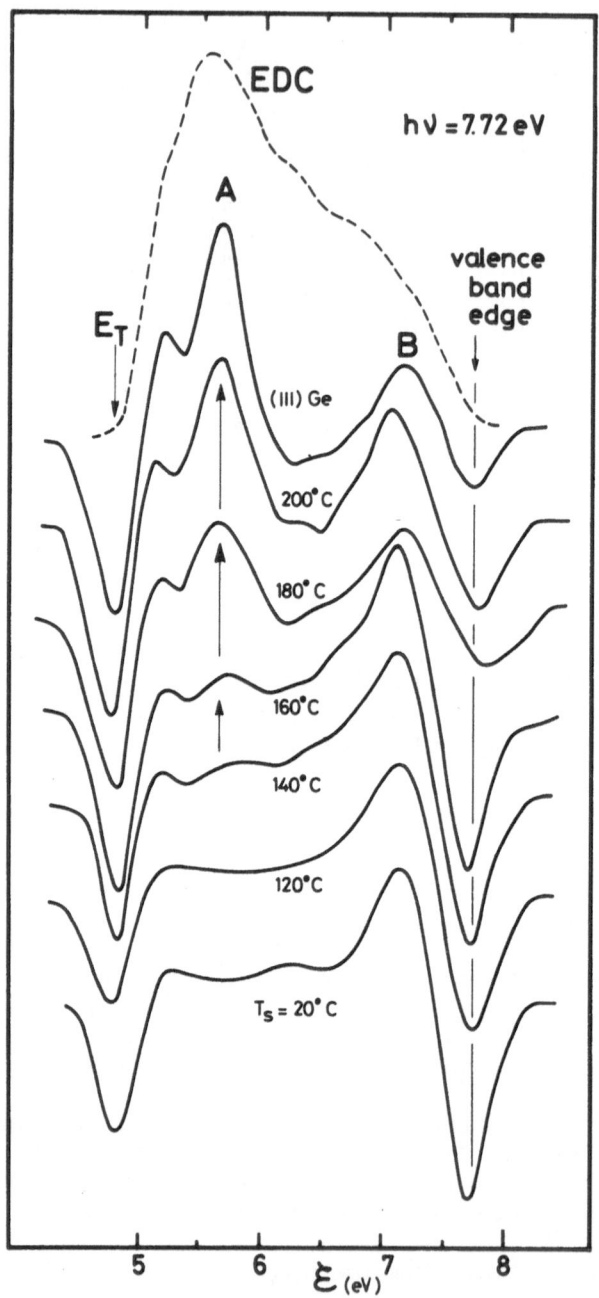

Fig. 25 EDC second derivatives obtained at $h\nu = 7.72$ eV
from strain-free Ge films prepared by deposition onto
vitreous carbon substrates held at temperatures T_S from
20°C to 200°C, and slow-cooled to 20°C before measurement.
The crystal EDC (dashed line) and EDC second derivative
(solid line, (111) Ge) are shown for comparison.

deposited in 10^{-10} torr vacuum onto polished, vacuum-
clean vitreous carbon (vc) substrates held at various
temperatures, T_S, between 20°C and 200°C during deposition[23]. Data were measured after slow-cooling the films
(10^{-2} oC/sec) to room temperature and without any further
annealing. Since EDC's are usually normalized to the
total photocurrents measured, the integrated areas of
EDC second derivative were here also systematically normalized to photocurrent. The crystalline spectrum of
Fig. 20 (EDC and second derivative) is shown at the top
of the figure for comparison.

The film deposited at 20°C on vc substrate shows no
evidence of peak A at $\mathcal{E}$ = 5.7 eV. Instead, a small peak
appears around $\mathcal{E}$ = 6.2 eV, with no counterpart in the
crystal spectrum. Upon varying $h\nu$, this latter peak does
not move on the $\mathcal{E}$ scale which might indicate that it
could delineate a relatively high DOS in the conduction
band at 6.2 eV above valence band edge. At T_S = 120°C,
this peak disappears totally. Tentatively, one might
think of some ordering induced in the film after low-
temperature deposition onto vitreous carbon. It has been
observed that the strength of this 6.2 eV peak increases
when depositing onto silica, always at $\mathcal{E}$ = 6.2 ± 0.2 eV.
This, by itself, would be an interesting subject to explore, particularly with regard to the various models
proposed to date to describe residual ordering in Ge random matrices[24]. At T_S = 120°C and 140°C, peak A is not
yet present. It becomes prominent at T_S = 160°C, exactly
at $\mathcal{E}$ = 5.7 eV as in the crystal, and simultaneously, peak
B region weakens. At higher deposition temperatures,
peak A strength increases markedly and at $T_S \geq 200^\circ$C, the
film spectrum is now very close to the crystal one. At
T_S = 180°C, this spectrum is just halfway between the
crystal and the 120°C spectrum (let us call the latter
the "amorphous" spectrum). What happens if we fast-cool
(say, at a 1°C/sec rate) such a mixed film? This is
shown in Fig. 26 which illustrates the evolution of the
T_S = 180°C spectrum[18] of Fig. 25, with time. Some two
hours after cooling starts (film is at T = 20°C after
~ 30 min.), the characteristic peak A has disappeared almost completely and three hours after cooling, the spectrum is then identical to the spectrum obtained from a
film deposited at T_S = 120°C and slowly cooled. Does
this mean that the electronic structures of the quenched
film and of the slow-cooled, T_S = 120°C films are the
same? No, of course, since one has been heavily strained from the substrate-film interface and the other has
not, i.e., their atomic configurations should not be

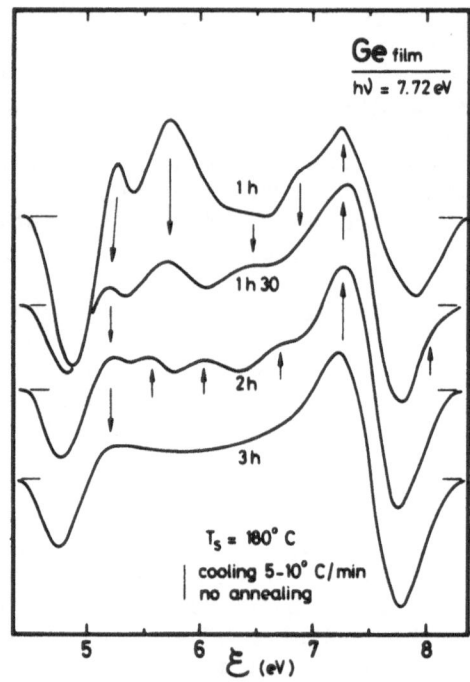

Fig. 26   Quenching a Ge film deposited at T_S = 180°C
and, then fast-cooled to 20°C. Spectra were recorded
between one and three hours after deposition.

similar. We reach here one of the limits of photoemission; its signature might well be the same in the presence of two different disordered atomic networks.
Photoemission indicates here that probably, disorder introduced by the strain into the partially diamond-like
ordered film suppresses totally this partial order and
"amorphisizes" the film completely (second-,third-,near-
est neighbors being at random). Incidentally, it is
shown (Fig.26) that such a strain effect may be studied
(propagation in the solid, energies involved, etc.) by
photoemission. A further remark is in order. This effect observed on a quenched film deposited at T_S = 180°C
(50°/o of atoms are diamond-like coordinated) tends to
disappear in films deposited at higher (200°C) or lower
temperature (160°C), i.e., in films which (from the
photoemission point of view) are closer to one or the
other phase available; (stable) crystalline or

(metastable) "amorphous".

     How could we lift the ambiguity evidenced in Fig.
26? Answers to this question can be found in the behavior of the films upon annealing. Fig. 27 shows second derivatives of EDC measured from the quenched film
(T_S = 180°C) of Fig. 26. As we increase the annealing
temperature (T_A) above T_S, peak A progressively emerges
at $\mathcal{E}$ = 5.7 eV. At T_A = 200°C, the indication is very
faint but is more pronounced at 250°C and further, well
established at 300°C and above. Obviously, we would say
that the film somehow "crystallizes". How this occurs
is revealed by the evolution of the high energy region
of the curves upon annealing. As soon as peak A appears
on the low-energy part of the spectrum, the Fermi level
E_f moves to higher energies in the band gap (see Fig. 14).
This allows emission from gap states to contribute photoemission in terms of a tail above $\mathcal{E}$ = 7.7 eV, which gives
a definite structure centered at about $\mathcal{E}$ = 8.1 eV (or
0.4 eV above the uppermost limit of the extended valence
states elastic emission) after annealing at T_A = 250°C.
As T_A increases, this gap emission evolves in two distinct ways: (i) It weakens as peak A strengthens; and
(ii) features within this gap structure evolve in energy
location. At T_A = 450°C the overall profile of the film
spectrum is identical to the crystal one, with the exception of a remanent, faint structure centered at
$\mathcal{E}$ = 7.9 eV which will disappear totally upon annealing
at T_A = 500°C. Considering this tail emission as being
characteristic of the existence of a certain density of
localized (defect) states in the film, one would quite
naturally interpret the above behavior by saying that
the film does crystallize, but "the hard way". That is,
as the temperature T_A increases, the strain present in
the film would be progressively removed allowing diamond-
like order to appear in the strain-free regions of the
films and to form diamond-like clusters. This process
is accompanied by the appearance and evolution of defects about these regions. Such defects might tentatively be described as broken bonds (i.e., free standing,
non-saturated orbitals giving rise to localized states),
which are evidently annealed out when the average dimension of the diamond-like clusters increases progressively.

     What happens upon annealing an unstrained film is
presented in Fig. 28. We have taken here the case of an
"amorphous" film, deposited at T_S = 120°C (Fig. 25) and
progressively annealed up to T_A > 300°C. The spectrum

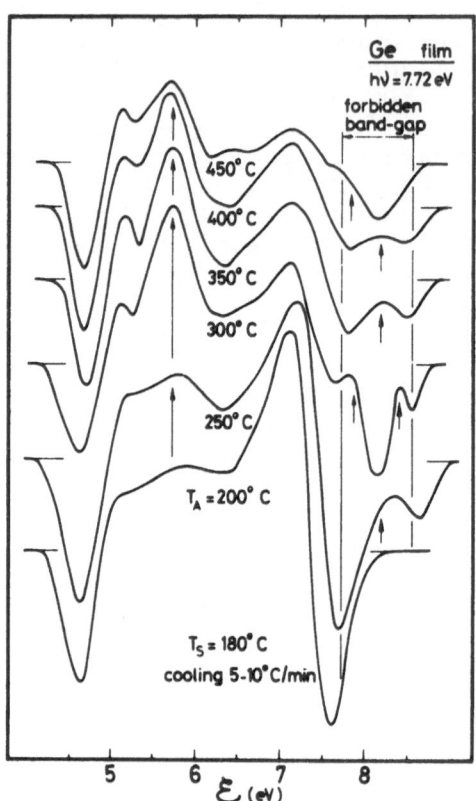

Fig. 27 Effect of annealing at T_A between 200°C and
450°C on strained, fast-cooled films deposited at
$T_S = 180^\circ$C. Peak A, characteristic of the crystal, progressively increases in intensity at $\mathcal{E} = 5.7$ eV, when
T_A increases. Simultaneously, structure appears at
$\mathcal{E} \simeq 7.8$ to 8.4 eV, corresponding to emission from localized states located above valence-band edge emission, in
the region of the crystalline forbidden bandgap (indicated at the top of the figure).

remains perfectly unaffected by annealing up to $T_A = 240^\circ$C,
at which temperature peak A suddenly emerges. The spectrum is then already similar to the one obtained from a
slow-cooled film deposited directly at $T_S = 200^\circ$C, Fig.
25. Here, the "crystallization" is very sharp. Finer
measurements, performed every 5°C, have shown that the
above amorphous-to-crystalline transition occurs over a

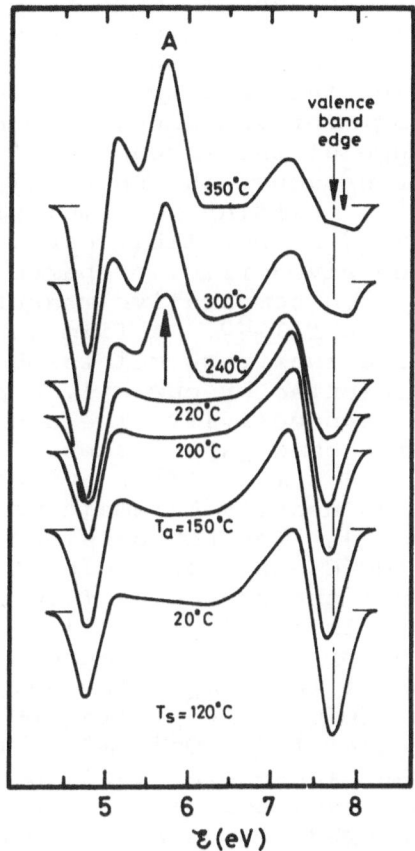

Fig. 28 Effect of annealing to T_A between $150°C$ and $350°C$ a film deposited at $T_S = 120°C$ and slow-cooled to $20°C$ after deposition. Note the sudden appearance of peak A in the spectra at $T_A = 240°C$, indicative of a sharp disordered-crystalline phase transition.

2 to $4°C$ temperature range centered around $T_A = 240°C$ in the case of a vitreous carbon substrate, and around $T_A = 255°C$ in the case of a flame-polished silica substrate. This abrupt evolution of the films electronic structure would indicate nearly phase-to-phase transition conditions.

One will notice also that gap states emission is absent in such films. Does this mean that their density

is negligible, and can we put a lower limit for such a
density to be detected by photoemission? We mentioned
in section two, that we cannot hope to extract an absolute density of localized states from a photoemission
spectrum. This point is now demonstrated experimentally.
If this was possible, then a large variation of this density (upon quenching, for instance) should correspond to
a proportional change of the intensity of the associated tailing appearing above the sharp high-energy onset
of the EDC (Fig. 14). Lifting the ambiguity can only be
achieved by measuring, in parallel, conductivity and
photoemission of one given film, prepared under well defined conditions. Such comparative measurements have
been performed on a Ge film[25]. A film deposited on a
flame-polished silica substrate maintained at 142°C
yields the EDC represented by curve 1, Fig. 29. From
the behavior of its conductivity, the density of the localized states in the band gap of these films is estimated[26] at N = 2 x 10^{17} cm^{-3}. After quenching this film,
its EDC (curve 1) and density N remain unchanged (within
experimental accuracy). The EDC obtained from a film
deposited at $T_S = 108^{\circ}$C and slow-cooled to 20°C is identical to the one from the preceding film and
N = 2 x 10^{18}cm^{-3} for this film (this larger value could
be attributed to the presence of vacancies or "voids" in
the films deposited at low temperatures). Upon quench-
ing this film from 250°C to 20°C, its EDC evolves into
curve 2 and N increases to $\sim 10^{21}$ cm^{-3}. Let us examine
the evolution of the high-energy tail in the EDC of the
second film. Fermi level emission is delineated by the
onset of an EDC measured from a gold target. All emission above E_f in gold is due to instrumental and temperature broadenings of the edge. We then normalize
the valence band edge (E_v) emission of a-Ge to that of
the Fermi level in Au. This is shown 10 times magnified
on the right of the figure. Upon quenching the film deposited at $T_S = 108^{\circ}$C, (E_f-E_v) increases from 0.35 to
0.45 eV and the intensity of the extra tailing (above
the gold-profile shaded area) increases by a factor of
2 to 3. Meanwhile the actual density N of the gap states
(which should somehow contribute to the extra tailing)
increases by a factor of nearly three orders of magnitude.
This can obviously not be related to the small increase
of the tail intensity. Therefore, the latter would be
mostly associated with the 0.1 eV shift of E_f upon annealing.

From what we have just said, the gap states evidenced in Fig. 27 would have a very high density which

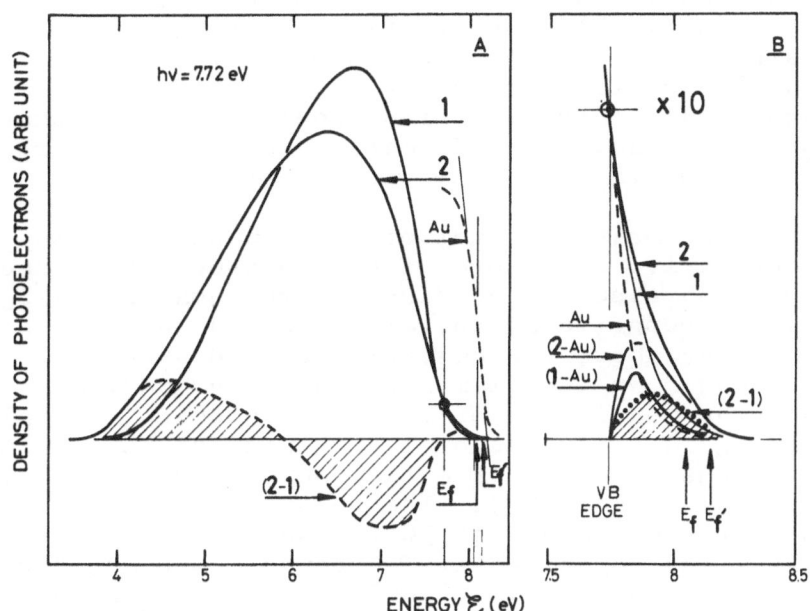

Fig. 29 In (a): EDC's measured from a-Ge films deposited at T_S = 142°C before and after fast-cooling to 20°C
(curve 1), and at T_S = 108°C before (curve 1 and after
(curve 2) fast-cooling to 20°C. E_f delineates Fermi
level emission for curve 1, which is displaced to E_f for
curve 2. The high energy tailing is magnified in (b) to
evidence extra emission associated with the Fermi level
shift upon fast-cooling the T_S = 108°C film. Dashed
line in (a) represents the upper edge of a gold EDC, used
to delineate E_f.

is, however, impossible to figure out for reasons given
in section two. Worth noticing is the fact that for
films, either quenched or not, the forbidden band gap
width remains constant at E_G = 0.6 eV, obtained by adding to $E_f - E_{vo}$ = 0.35 eV (measured by photoemission)
the activation energy 0.25 eV measured[24] from the conductivity above 20°C. This is valid for any deposition
temperature between 100 and 200°C. Therefore, E_G should
not be considered as a decisive factor in characterizing
Ge films. Rather, it is the nature and density of the
gap states which would be characteristic of the stage of
development of the crystallization process in these films.

From the above description, it would appear that
only films deposited at T_S = 140 to 150°C could be considered as ideally amorphous, in the sense that, (i) diamond-like ordering is absent; (ii) the density of localized states in the gap is the minimum obtainable using
classical evaporation techniques; (iii) only these films
withstand strain induced by fast-cooling without under-
going any detectable transformation in their electron
properties from the photoemission point-of-view.

A comprehensive account of the evolution of the
second-derivative structure of EDC in a-Ge, upon annealing, can be obtained by considering the quantity I/I_0
where: I is proportional to $(a/a_0 + b_0/b)$, a and b being
the magnitudes above the base-line of the spectral
regions A and B in the film spectrum (Fig. 24), and
a_0 and b_0 the magnitudes of the corresponding crystalline
features; I_0 is the value of I obtained for the featureless amorphous spectrum of a film deposited at T_S = 120°C
and slow-cooled (Fig. 25). The behavior of I/I_0 with
$1/T_a$ is represented in Fig. 30 for several slow-cooled
films (Fig. 25) deposited at T_S between 20°C and 200°C.
Data from fast-cooled, T_S = 180°C film are also shown
(Fig. 27). The highest slope of the curves gives the activation energy E_a of the main ordering process involved
in each film. The value of ϑ = Log I/I_0 found for the
crystal spectrum (ϑ = 1.35) is taken to represent 100°/o
diamond-like ordering. Comparatively, the right-hand
vertical scale gives an estimate (in °/o) of the density
of the diamond-like clusters in the films. The highest
activation energy E_a^{max} is obtained for films deposited
at T_S = 140 - 150°C and is equal to 1.4 ± 0.4 eV. Such
a high value would suggest an ordering mechanism involving vacancy formation in these films, since E_a^{max} is
of the order of magnitude of the energy required to ac-
tivate vacancy formation in Ge crystals. All other films
are characterized by much lower values of E_a (E_a<0.4 eV)
which indicates that ordering in these films occurs in
a complex manner involving probably vacancy and interstitial formation and migration.

Finally, we plot in Fig. 31 the activation energies
of the slow-cooled films, obtained from Fig. 29, as a
function of the parameter T_S. Together on this figure,
we plot the density N_F of localized states about the
Fermi level derived from the $T^{-1/4}$ behavior[26] of the film
conductivity measured in parallel with photoemission.
E_A is seen to be maximum, and N_F to be minimum, for films
condensed around T_S = 150°C.   Did an "ideal" amorphous

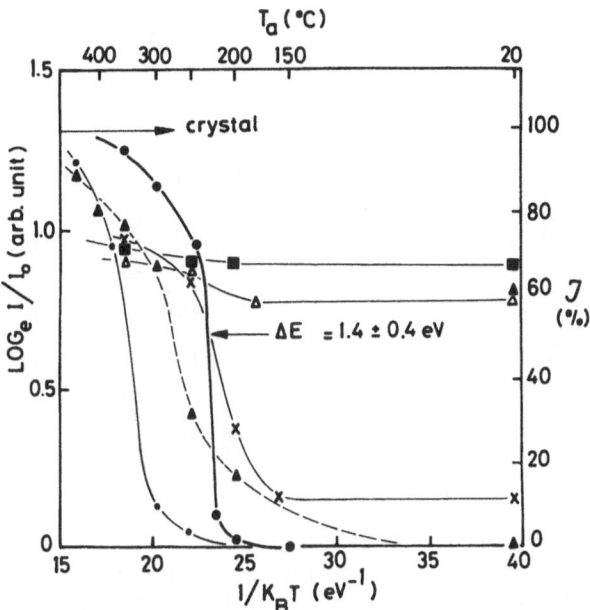

Fig. 30 Evolution of a-Ge EDC structure upon annealing (T_A). Deposition temperatures are: $T_S = 20°C$ (·), $140°C$ (●), $160°C$ (X), $180°C$ (△: slow-cooled, ▲: fast-cooled); $200°C$ (■). Thermal activation energies are obtained by measuring the slope of the $\mathcal{J} = Log\ I/I_0$ curve (see text) versus $1/k_B T_a$. On the right vertical scale, $\mathcal{J}$ is given in °/o (taking $\mathcal{J}_{crystal} = 100°/o$ as a reference). Accuracy is of the order of ±2°/o on the position of each point. Highest slope ($E_A = 1.4 ± 0.4$ eV) is obtained about $T_A = 240 - 250°C$ for film deposited at $T_S = 140°C$.

Ge film exist, Fig. 31 strongly indicates that it could be prepared by condensation onto a vitreous carbon or silica substrate held at $T_S ≃ 150°C$.

The above study does not necessarily limit the conditions for preparing ideal amorphous Ge to the ones just outlined. A very large number of film preparation parameters should be reviewed and the importance of each of them reasonably estimated in order to optimize even further the preparation technique. As far as deposition and annealing temperatures are concerned, it has been shown by photoemission, that films deposited at $T_S ≃ 150°C$ on glassy substrates remain in a somewhat stable phase

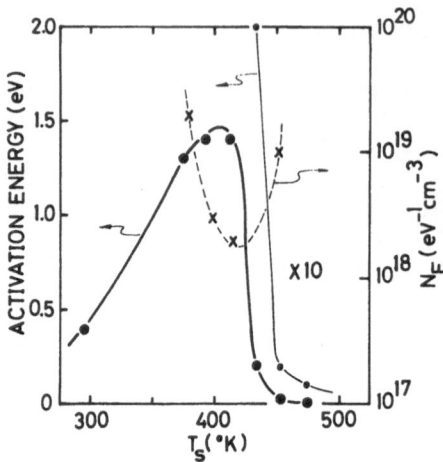

Fig. 31   Thermal activation energies (left vertical scale) and density of localized states about E_f (right vertical scale) versus T_S, for films deposited at T_S between $20°C$ and $200°C$ and slow-cooled to $20°C$.

up to $T_A \simeq 240$ to $250°C$ where <u>abrupt</u> and <u>bulk</u> crystallization occurs accompanied with a high (1 to 1.5 eV) thermal activation energy.

In summary, low-energy UV photomission appears to be of prime importance in investigating the ordering mechanisms which take place in films upon annealing. A number of effects can be thoroughly investigated in this energy range, viz., film-substrate interface problems, strain-induced distortions of the atomic network, partial ordering persisting in some films etc. All these effects contribute to the definition of a high-density, continuous network "glassy" state containing a minimum of localized states and which we might identify as "amorphous" Ge.

Such a study has not been performed on Si films. Partial results obtained by Pierce et al[27] would indicate that most of the above conclusions on a-Ge would still about hold for a-Si, with all characteristic temperatures shifted to higher values by some 200 to $300°C$.

No equivalent study exists for III-V compounds. One may remark, in this case, that compositional disorder is

likely to add on topological disorder which should render an evaluation of the crystallization process in these
III-V films rather difficult. Now, assuming the conclusions reached for a-Ge film to still hold in III-V's as
far as topological disorder is concerned, there might be
here a way to evaluate the effect of compositional disorder __alone__ onto the crystalization process of such compound films.

In a very general sense, far-UV and X-ray data tend
to "mimic" each other in a rather puzzling way (see section one). After deconvolution of all inelastic electrons, both elastic spectra are expected to follow eq.
(19), i.e., an initial density of states profile $n_v(\mathcal{E}-h\nu)$,
which should be characteristic either of the surface
(far UV) or somehow of the bulk (X-ray) properties.
Since this results into some uncertainty in interpreting
crystalline spectra, one would not feel very safe in comparing elastic EDC's obtained from a-Ge at, say, 40 and
1486.6 eV, respectively.

Fig. 32 is a demonstration of this ambiguity. Here,
are represented elastic EDC's measured from (111) Ge and
from some disordered Ge films at $h\nu$ = 25 eV[28] and
1486.6 eV[29]. Let us start with the crystal case. Despite the experimental uncertainties (± 0.3 and ±1.0 eV,
respectively) it is clear that the two curves do not fit

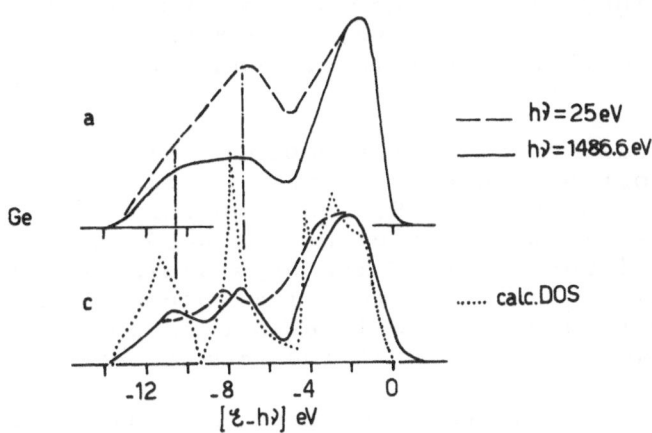

Fig. 32 Far-UV and X-ray elastic EDC's from (111)
Ge[c] and a-Ge[a], from Refs. 27 and 28.

convincingly. This is particularly true for the two
upper peaks which seem to be shifted by ~1 eV to lower
energies in the 25 eV spectrum compared to the X-ray one.
Tentatively, this would indicate differences in short-
range atomic configuration present at the crystal .sur-
face (far-UV) and up to some 15 Å (ℓ_e) from this surface
(X-ray).

In the case of a disordered film, comparison with a
crystal in the X-ray region shows that short-range order
should be about the same at about 10 Å from the surface
of the film and of the crystal. The two low-energy peaks
at $\mathcal{E}$ - hν =-6 and-9 eV, respectively, in the crystal
spectrum correspond reasonably well with two distinct
shoulders at the same energies in the a-Ge film spectrum.
In the far-UV spectrum of a-Ge, these two shoulders are
replaced by one broad peak centered at $\mathcal{E}$ - hν =-7 eV.
Although one can only speculate about such a difference
(due either to matrix elements or photoionization cross
sections of s vs. p electrons), a best guess would be to
interpret it as due to differences in short-range order
present at the surface of the film (far-UV) and at, say,
10 to 15 Å behind it. The only positive information is,
here, the agreement observed in the position of the high-
energy peak at $\mathcal{E}$-hν = -1 to -2 eV below valence-band edge
emission limit. This would concur to assign the peak
obtained in the low-energy UV spectra (Fig. 23) effectively to a high-density of valence states. Yet, such
limited information does not seem to be very useful in
evaluating precisely the nature, extent and evolution
(upon annealing) of any ordering present in a-Ge films.
As anticipated in section one, far-UV and X-ray photoemission would then appear to be unadapted to this kind
of study.

In the case of Si, X-ray photoemission data are also
available[29] but not vacuum UV ones. X-ray data are shown
in Fig. 33 for both (111) Si and some form of disordered
Si. Just like in the case of Ge, it would seem here too
that the two lower peaks of the crystalline spectrum are
approximately averaged (in the film spectrum) into a
broad peak retaining two well-delineated edges corresponding to the two lower peaks of the crystal spectrum.

Far-UV and X-ray photoemission data have been obtained from III-V compounds, by Shevchik et al[30]. We
show in Fig. 34 the GaSb data, which are then compared
to GaAs ones in Fig. 35, after deconvoluting the elastic
(shown on Fig. 35) and inelastic EDC's. Fig. 34 is a

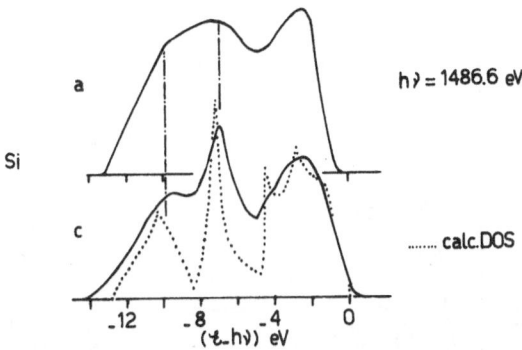

Fig. 33 X-ray elastic EDC's from (111) Si[c] and
a-Si[a] (Ref. 28).

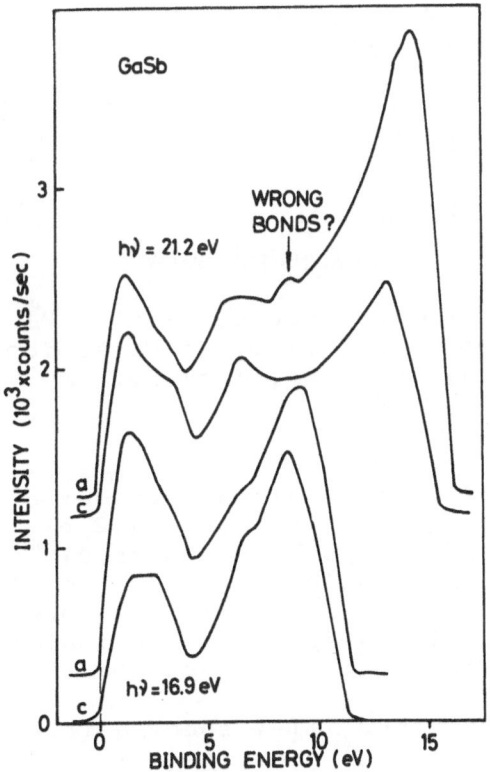

Fig. 34 Far-UV EDC's measured at hν = 16.9 and 21.2 eV
from crystalline (c) and "amorphous" GaSb films. The
abcissa refers to the binding energy of the photoelec-
trons, which is equivalent to $|\mathcal{E}-h\nu|$, from Ref. 29.

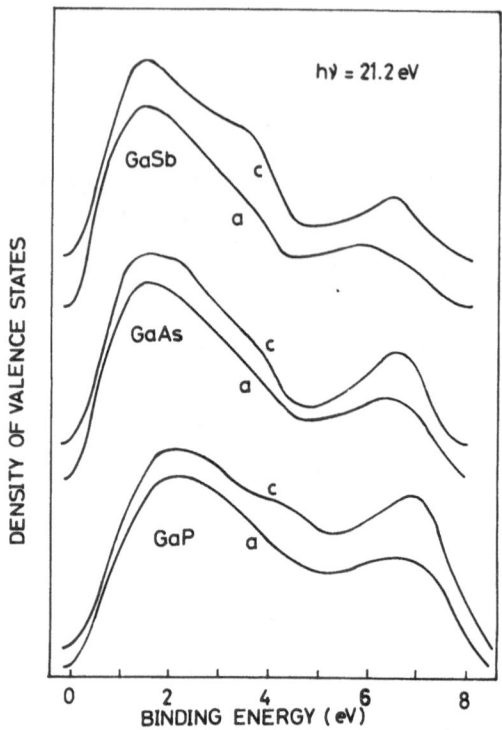

Fig. 35   Far-UV EDC's measured at hν = 21.2 eV from crystalline and "amorphous" films of Ga semiconducting compounds, GaSb, GaAs and GaP, from Ref. 29.

demonstration of what we already expected (section one, Fig. 10). For both hν = 16.9 and 21.2 eV, the low-energy region of the spectra is hardly dissociated into elastic and inelastic contributions in the crystalline case (curve labelled c) and, a fortiori, in the amorphous case (a). Some features in the a-spectrum are tentatively attributed to surface emission of "wrong bonds"

   Compared to other III-V spectra (Fig. 35), two points are worth noticing: (i) there is not much difference between the amorphous and crystalline spectra of the same compound; and (ii) all the spectra (either c- or a-) of these Ga compounds are very close to each other. The obvious conclusion is that, just like in the case of Ge, far-UV photoemission spectra do not seem to be very effective in revealing structural differences which are otherwise well identified in low-UV photoemission spectra.

3. Non-Tetrahedrally-Coordinated Semiconductors

For such materials, one would tend to follow essentially the same procedure as the one used for Ge. If this was effectively the case, then there would be nothing to add to what has been said earlier. However, the fundamental characteristic of the materials we are dealing with now is the following: bonding is no longer uniform (i.e., covalent via bonding orbitals) as in column IV elements, but varies, within the same material, from a purely covalent character to Van der Waals-type cohesion (via non-bonding orbitals). For instance in tellurium, an atom is covalently bound to its two first-nearest neighbors (along a chain), but Van der Waals-like bound to its four second-nearest neighbors which belong to adjacent chains. Therefore, one may expect, for instance, the presence of strain in a Te film to be far more effective than in a Ge film, by introducing large fluctuation of local order both in extent and nature. As a consequence, this complicates its crystallization process which would be more difficult to initiate experimentally and, conversely, to characterize. For this reason, one cannot expect photoemission to yield, in the case of non-tetrahedrally bonded semiconductors, results similar to the ones obtainable from column IV elements. The following discussion refers to the Te case, which can be easily extended to the Se one. Emphasis is put on those photoemissive properties which reveal difference between tetrahedrally and non-tetrahedrally bound materials, both in their disordered configurations and upon annealing.

Low-energy UV photoemission: We have seen that, in the 5-10 eV energy range, information on the bulk electronic structure is obtained. As far as trigonal Te is concerned, photoemission spectral features are extremely sensitive to photon-energy as shown in Fig. 36 where second derivatives of EDC's measured from trigonal Te between $h\nu = 5.4$ and 8.7 eV are plotted[31]. From their behavior, DOS maxima can be traced either in the valence band (1v to 5v) or in the conduction band (1c), owing to the relative narrowness of the bands in Te. We already pointed out that the conduction band DOS of a disordered solid would tend to be structureless, with the consequence that EDC's would essentially reflect its valence band DOS (Eq. 23). Therefore, a simple way to decide whether

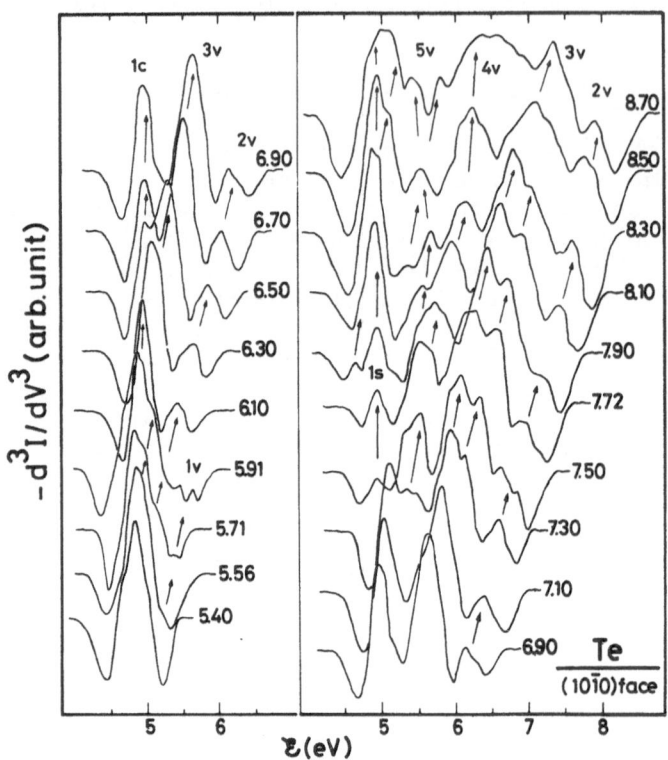

Fig. 36 EDC second-derivatives measured from trigonal
Te at energies between hν = 5.4 and 8.7 eV. (Ref. 31).

a Te film is somewhat disordered is to compare one of its
photoemission spectra measured at hν with the trigonal
Te spectrum also measured at the same hν. This energy
hν is chosen so that the crystalline spectrum presents
an extremely characteristic piece of structure. Peak 1c,
Fig. 36, has been observed to depend strongly on light
polarization (electric field E perpendicular to c axis
only). This behavior together with its energy position
in the spectra between hν = 6.5 and 7.1 eV allows us to
assign it to direct optical transitions occuring in the
L-M region of the Brillouin zone, between states belonging to the first (bonding) p valence band (denoted p_1)
and the second (d-like) conduction band (SCB). The second (non-bonding) p valence band, (denoted p_2), covers
the upper 2 eV of the valence band. In fact, the onset
of peak 1c in the spectra marks the onset of optical absorption between p_1 and SCB. It is this peak 1c which

we decide to take as a reference in evaluating disorder
in Te films, just like peak A has been chosen in the case
of Ge (Fig. 24). Note that peak 1c has practically no
contribution from inelastically scattered electrons, as
deduced from its polarization dependence. The so-called
"scattering" peak (1s) appears at $h\nu$ = 7.5 eV and increases in apparent strength with increasing $h\nu$. However,
its profile and, therefore its real strength also are continuously disturbed by elastic electrons since, at such
low energies, the valence band is not yet completely explored.

Spectra measured at $h\nu$ = 7.1 eV from a $(10\bar{1}0)$ face
of Te and a variety of Te films[31] are compared in Figs.
37 and 38. A film deposited at $100°$K onto silica and
warmed up to $300°$K before measurement yields a spectrum
where peak 1c is absent, as well as the peak centered
at $\mathcal{E} - h\nu$ = -0.6 eV. This indicates: (i) the absence of
structure in the conduction band DOS (at least between
$\mathcal{E} = E_T = 4.7$ eV and 7.1 eV), and (ii) a weakening of
structure present in the top 0.8 eV of the crystalline
valence band DOS (upper part of p_2). Upon annealing the
same film to $430°$K (limit of sublimation of Te in 10^{-10}
torr vacuum), the low-energy part of the film spectrum
evolves towards the crystalline profile, while no change
is noticed in the upper region of the spectrum. Simultaneously, X-ray diffraction indicates a reordering process which might be considered as a very imperfect "crystallization" developing in the film[32]. However, it is
important to notice that peak 1c does not show up yet,
or, more precisely, elastic electrons emitted with low
kinetic energies are characterized by a distribution peaking up at an energy approaching that of 1c closer and closer
as the annealing temperature, T_A, increases. This behavior is totally different from the one observed in Ge
films (Figs. 27 and 28). In the latter, the strength of
the characteristic crystalline spectral feature, peak A,
was either strengthening (Fig. 27) or suddenly emerging
in the Ge film spectrum (Fig. 28), upon increasing T_A,
<u>exactly</u> at $\mathcal{E}$ = 5.7 eV for $h\nu$ = 7.72 eV, i.e., just like
in the crystal. This was considered to be indicative of
diamond-like ordering developing into the film. In the
case of Te, both the strength and energy location of the
spectral features on the low-energy region of the film
spectrum do change upon annealing (Fig. 37). A more complete study of the ordering process in Te films by photoemission is unfortunately limited by the Te sublimation
occuring at T_A > $430°$K in 10^{-10} torr vacuum. Despite
this, it seems clear that one cannot consider trigonal

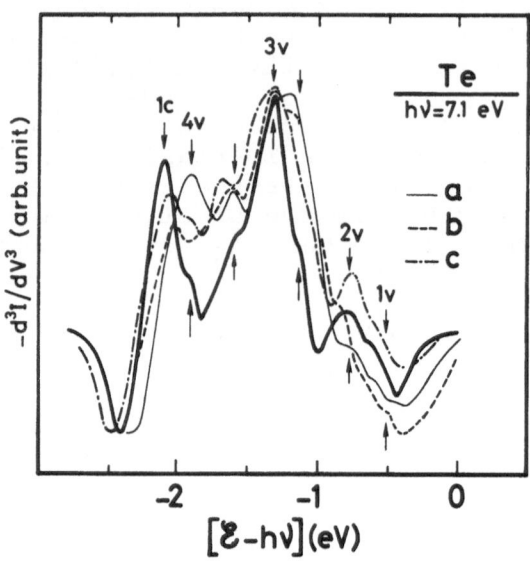

Fig. 37 EDC second derivatives measured from a film deposited on silica at $T_D = 100^\circ$K and warmed up to $T_A =$
300°K (curve a) and $T_D = 430^\circ$K (curve b), and from a film
deposited on Au substrate maintained at $T_D = 350^\circ$K (curve
c). Films spectra are compared to the trigonal Te spectrum (thick line) (Ref. 31).

ordering to develop into the Te films the way diamond-
like ordering progresses in Ge films. In that sense, it
seems practically impossible to characterize the crystallization of Te films in terms of activation energy as is
done in the case of Ge. Furthermore, owing also to the
anisotropy of the bonding character in trigonal Te, film-
substrate interface problems are critical and render the
preparation conditions of reasonably good polycrystalline Te films difficult to achieve and maintain. Otherwise said, Te films present, in general, a large number
of defects and at the limit, may be considered as disordered in various ways. Changing one parameter in the
film preparation may have therefore, important consequences on local order, i.e., on the actual electronic
structure of the films. Consequently, photoemission
spectra are very sensitive to film preparation as shown
in Figs. 37 and 38.

The spectrum of a Te film condensed at 350°K onto

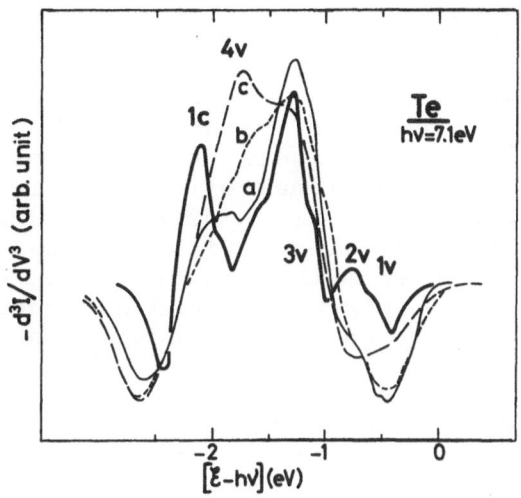

Fig. 38   EDC second derivatives measured from a film
prepared on silica at $T_D = 100^\circ K$, warmed up to $T_A = 300^\circ K$
(curve a), quenched to $130^\circ K$ and warmed up to $300^\circ K$ (curve b), and further annealed to $T_A = 430^\circ K$ (curve c).
They are compared to the crystal spectrum (thick line)
(Ref. 31).

a polycrystalline gold substrate (Fig. 37, curve c) is
shown to exhibit peak 1c about completely restored in
position and strength, as well as the high energy peak
at $\mathcal{E} - h\nu = -0.7$ eV. Curve a, Fig. 38, is the spectrum
obtained from a film prepared under conditions similar
to film a, Fig. 37. Note their dissimilarity about
$\mathcal{E} - h\nu = -1.5$ to -2 eV. Such a film, after quenching
to $130^\circ K$, yields curve b. If one anneals this quenched
film to $430^\circ K$, the spectrum evolves into curve c which
now presents a maximum at -1.8 eV. Although we would expect a certain evolution towards trigonal Te (as in film
b, Fig. 37, which has also been annealed to $430^\circ K$, but
not previously quenched), it seems that we have here a
totally different material in which short- and middle-
range order would be no longer trigonal but could, tentatively, be approached via a simple cubic configuration[33].

To summarize, effects of disorder on the electronic
structure of Te films can be stated briefly: (i) the
conduction band DOS seems to be totally devoid of structure, which could be attributable to the loss of long-

range order as in a-Ge; (ii) perturbation of the valence
band DOS are limited to the upper part of p_2 (contributed by point H in the trigonal Te Brillouin zone) in
room temperature Te films, but may affect the total width
of p_2 if such (metastable) films are further disturbed
(strained by quenching, for instance); since p_2 corresponds to p non-bonding orbitals pointing towards adjacent chains[34], this additional effect of disorder could
be associated to a perturbation of short-range order,
i.e., to a loosening of binding forces between adjacent
chains. It does not seem that the p_1 valence band DOS
is very different in trigonal Te and disordered Te films,
as deduced from higher energy photoemission data.

Te and Se having basically the same atomic configuration, the above conclusions can be applied to Se as
well. Model calculations have been tested in order to
describe the above effects. Qualitative agreement is
found with experiment using the method introduced by
Kramer[35] both for Te and Se. The molecular orbital approach of Chen[34] has also been proven to describe reasonably well the electronic structure of disordered Se
films, at least as far as p valence bands are concerned.

High-energy UV and X-ray photoemission: With such
energies, the bottom of the valence band is explored.
Spectra obtained at $h\nu$ = 21.2 eV from trigonal and disordered Te and Se films[36] are shown in Fig. 39. The
total valence band width is seen to be unaffected by disorder in both materials.

X-ray photoemission data are compared to far-UV
ones[37] in the case of Se, in Fig. 40. One notices the
important structure which appears in the X-ray spectrum
of trigonal Se, between ~8 and 18 eV. At $h\nu$ = 21.1 eV,
this structure is totally absent. It is due to emission
initiating in the s-band, which lies just below the p
valence band. Photoionization cross-section differences
between s and p electrons at far-UV and X-ray energies
are probably responsible for the absence of the s-band
emission in Se far-UV spectra. Taking account of the
better resolution at low energies, a composite DOS is
constructed using far-UV (p-valence band) and X-ray data
(s-band). One should remember, however, that by so doing
surface (far-UV) and bulk (X-ray) data are not differenti-
ated. Comparing amorphous Se to trigonal Se, it seems
reasonable to conclude that the two types of materials
have close s and p "valence" bands, except for the fine
structure in p_2 which disappears in a-Se. Here again,

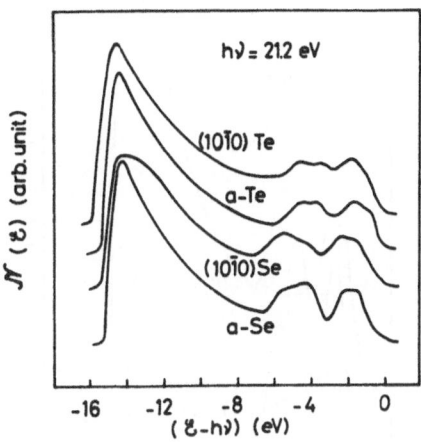

Fig. 39   EDC's measured at hν = 21.2 eV from trigonal
(10$\bar{1}$0) and disordered films (a) of Te and Se (Ref. 36).

characterization of the stage of disorder present in a
film seems to be impossible to achieve at high energies,
either in the far-UV or in the X-ray range.

IV. CONCLUSIONS

Generally speaking, the number of independently collected data are still very limited. Considering the various parameters which may affect the preparation and properties of disordered semiconducting films, proper evaluation of the importance of each experimental parameter
is difficult to achieve. However, the following conclusions seem to be reasonably well supported. Low-energy
UV photoemission has been shown, in the case of UHV-clean
Ge films, to provide an efficient and reliable tool
in the exploration of strain effects originating at the
film-substrate interface. Once such effects are minimized, it is possible to define an ideal amorphous network of Ge atoms the properties of which are: high density, minimum concentration of localized states (and
associated dangling bonds, see high density), a structureless conduction band DOS (at least between $\mathcal{E}$ = 5 and
10 eV). Such a material withstands strain induced by
fast-cooling contrary to all other films studied. Most

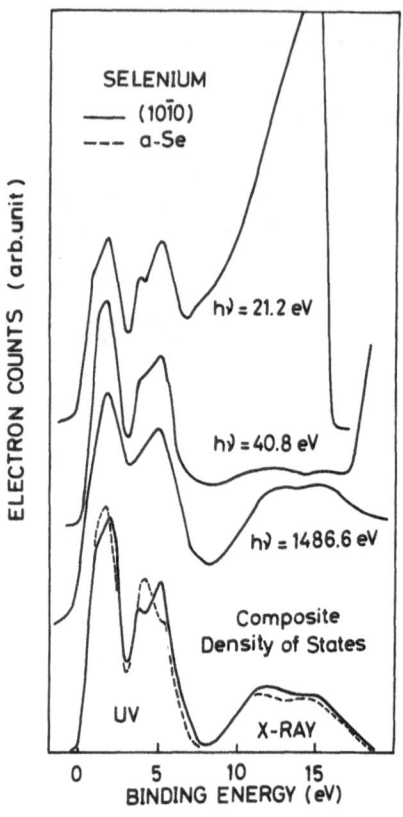

Fig. 40 Composite DOS of trigonal and amorphous Se, as
derived from X-ray (hν = 1486.6 eV) and far-UV (hν =
21.2 and 40.8 eV) spectra. Binding energy is equivalent
to (ε-hν) (Ref. 37).

of all, it undergoes an abrupt transition to diamond-like
ordering characterized by the highest, by far, thermal
activation energy. Similar results should be obtainable
in the case of Si.

In the high energy range, data are even fewer. Difficulties in the interpretation of the crystalline spectra renders the use of these high energies rather hazardous in exploring the bulk electronic structure of disordered solids. Consequently, extensive studies by far-UV
and X-ray photoemission, although useful in exploring
deep valence states, are not expected to yield important

information which could help characterizing accurately the disordered films. Altogether, low-energy photoemission is far more promising and should attract interest in the future.

REFERENCES

1. W.L. Schaich and N.W.Ashcroft, Sol. State Commun. 8, 1959 (1970) and Phys. Rev. B 3, 2452 (1971); G. D. Mahan, Phys. Rev. B 2, 4334 (1970); C. Caroli et al, Phys. Rev. B 8, 4552 (1973).

2. W. E. Spicer, Phys. Rev. 112, 114 (1958).

3. J. Tauc in Proc. E. Fermi International School of Physics, Varenna, Italy (ed. by J. Tauc, Acad. Press, N.Y., 1966), p. 63.

4. L.W. James and J.L.Moll, Phys. Rev. 183, 740 (1969).

5. C. R. Brundle, J. Vac. Sci Tech. 11, 212 (1974).

6. C. N. Berglund, W. E. Spicer, Phys. Rev. 136, A 1030 (1964).

7. E. O. Kane, Phys. Rev. 147, 335 (1966).

8. L.R. Saravia and L.Casamayou, J. Phys. Chem. Solids 32, 1075 (1971); ibid, 32, 1541 (1971).

9. H. Ehrenreich, in Proc. E. Fermi International School of Physics, Varenna, Italy (ed. by J. Tauc, Acad. Press, N. Y., 1966), p. 106.

10. R. C. Eden, Rev. Sci Instrum. 41 252 (1970).

11. G. W. Gobeli and F. G. Allen, Phys. Rev. 127, 141 (1962).

12. T. H. DiStefano and D. T. Pierce, Rev. Sci. Instrum. 41, 180 (1970); D. T. Pierce and T. H. DiStefano, Rev. Sci. Instrum. 41, 1740 (1970).

13. H. B. Holl, J. Opt. Soc. Am. 57, 683 (1967).

14. M. R. Barnes and L. D. Laude, Rev. Sci. Instrum. 42, 1191 (1971).

15. J. K. Cashion et al., Rev. Sci. Instrum. 42, 1670 (1971); B. Feuerbacher and B. Fitton, Rev. Sci. Instrum. 42, 1172 (1971).

16. D. E. Eastman and J. K. Cashion, Phys. Rev. Letters 24, 310 (1970); B. Feuerbacher and B. Fitton, Phys. Rev. Letters 29, 786 (1972).

17. N. V. Smith and M. M. Traum, Phys. Rev. Letters 25, 1017 (1970); L. D. Laude, B. Fitton and M. Anderegg, Phys. Rev. Letters 26, 637 (1971).

18. L.D. Laude, R.F.Willis and B. Fitton, in Proc. of 11th Internatl. Conf. on Physics of Semiconductors, Warsaw (Polish Sci. Publ., 1973), p. 561.

19. L. D. Laude, to be published.

20.   T. M. Donovan, Tech. Rep. No. 5221-2, Stanford
       Electronics Lab., Stanford Univ. (1970, unpublished).
21.   F. C. Brown and O. P. Rustgi, Phys. Rev. Letters
       $\underline{28}$, 497 (1972).
22.   C. G. Ribbing, D. T. Pierce and W. E. Spicer, Phys.
       Rev. B $\underline{4}$, 4417 (1971).
23.   L. D. Laude, R. F. Willis and B. Fitton, Solid
       State Commun. $\underline{12}$, 1007 (1973).
24.   T. C. McGill and J. K. Klima, Phys. Rev.B $\underline{5}$, 1517
       (1972).
25.   L. D. Laude, R. F. Willis and B. Fitton, Proc. 5th
       Internatl. Conf. on Amorphous and Liquid Semicon-
       ductors, Garmisch, (Taylor and Francis, London),
       p. 277.
26.   T. Pollak, et al., Phys. Rev. Letters, $\underline{30}$, 856
       (1973).
27.   D. T. Pierce and W. E. Spicer, Phys. Rev. B $\underline{5}$, 3017
       (1972).
28.   D. E. Eastman and W. D. Grobman, Phys. Rev. Letters
       $\underline{28}$, 1378 (1972).
29.   L. Ley, et al., Phys. Rev. Letters, $\underline{29}$, 1088 (1972).
30.   N. J. Shevchik, J. Tejeda and M. Cardona, in Proc.
       5th Internatl. Conf. on Amorphous and Liquid Semi-
       conductors, Garmisch, (Taylor and Francis,
       London), p. 609.
31.   L. D. Laude, B. Kramer and K. Maschke, Phys. Rev.
       B $\underline{8}$, 5794 (1973).
32.   M. J. Capers and M. White, Thin Solid Films $\underline{8}$,
       353 (1971).
33.   B. Kramer, K. Maschke and L. D. Laude, Phys. Stat.
       Sol. (b) $\underline{59}$, 219 (1973).
34.   I. Chen, Phys. Rev. B $\underline{7}$, 3672 (1973).
35.   B. Kramer, Phys. Stat. Sol. $\underline{41}$, 649 (1970); ibid
       $\underline{41}$, 725 (1970); L. D. Laude, B. Fitton, B. Kramer
       and K. Maschke, Phys. Rev. Letters $\underline{27}$, 1053 (1971).
36.   P. Nielsen, in Proceedings 5th Internatl. Conf. on
       Amorphous and Liquid Semiconductors, Garmisch,
       (Taylor and Francis, London, 1974), p. 639.

37.   N. J. Shevchik, J. Tejeda and M. Cardona, in Proc.
       5th Internatl. Conf. on Amorphous and Liquid Semi-
       conductors, Garmisch, (Taylor and Francis,
       London, 1974), p. 609.

MATHEMATICAL METHODS FOR CALCULATING THE ELECTRON SPECTRUM

B. Kramer

University of Dortmund, 4600 Dortmund 50

Federal Republic of Germany

I. FORMULATION OF THE PROBLEM

I-1. Qualitative Remarks on the Electron Spectrum of Solids

Before going into details of the formal theory used in the calculation of the electronic properties, we will discuss qualitatively some features of the electronic density of states of disordered systems and their consequences for the electronic properties[1,2].

Ideal Crystal

First consider an ideal crystalline system of infinite size. It is described by a Hamiltonian, which is invariant upon lattice translations. Therefore, it is possible to classify the about 10^{23} energy eigenvalues and the respective eigenstates by a wave-vector $\vec{k}$, (Fig. 1). The band structure $E_n(\vec{k})$ is a quasi continuous function of the wave vector $\vec{k}$ within each band. The corresponding density of states $n(E)$ (Fig. 2) consists of "bands", where $n(E) \neq 0$ and "band-gaps", where $n(E) = 0$. The characteristic shape of the bands is determined by the number, form, and the energetic distance of the van Hove singularities, which are connected with the property of long-range order of the system.

The electron states are Bloch states, i.e. of the form

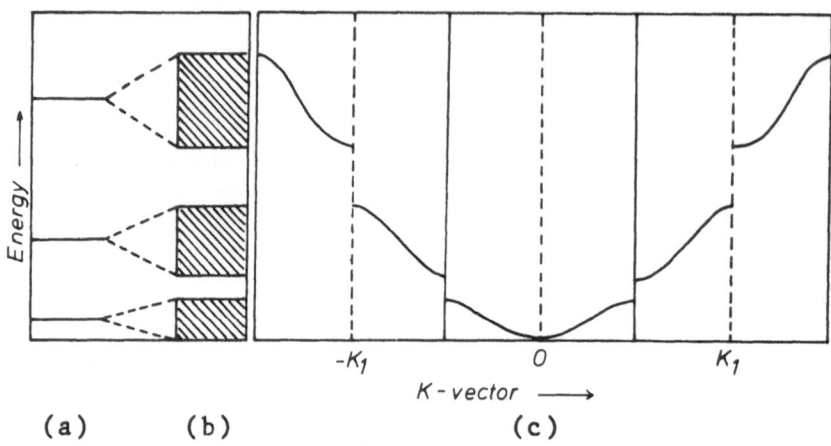

Fig. 1   Qualitative energy spectrum of a crystalline electron. (a) Isolated atoms: discrete energy levels, (b)
Crystal: bands and bandgaps, (c) Crystal: energy band
structure $E(\vec{k})$.

$$\Psi_{n\vec{k}}(\vec{r}) = e^{i\vec{k}\cdot\vec{r}}\, u_{n\vec{k}}(\vec{r}) \tag{1}$$

where $u_{n\vec{k}}(\vec{r})$ has lattice periodicity, and therefore does
not vanish for $|\vec{r}|\to\infty$ (Fig. 3(a)).

The Bloch wave functions $\Psi_{n\vec{k}}(\vec{r})$ are typical examples
for extended states, these states are the only ones which
can occur in long range ordered systems. As a consequence,
the electrical conductivity of an ideal crystal at zero
temperature is either infinite (metal, Fermi energy within a band) or zero (insulator, Fermi energy within a band
gap).

Impurity System

In "real life", no ideal crystalline system exists.
A real crystal is limited by surfaces and will always contain some impurity atoms, structural defects, etc. If
the number of deviations from the crystalline order is
very small (compared to the number of atoms in the crystal), we can use the ideal crystal as a starting point
and calculate the energy spectrum of the impurity or
structural defect relative to that of the crystal. The

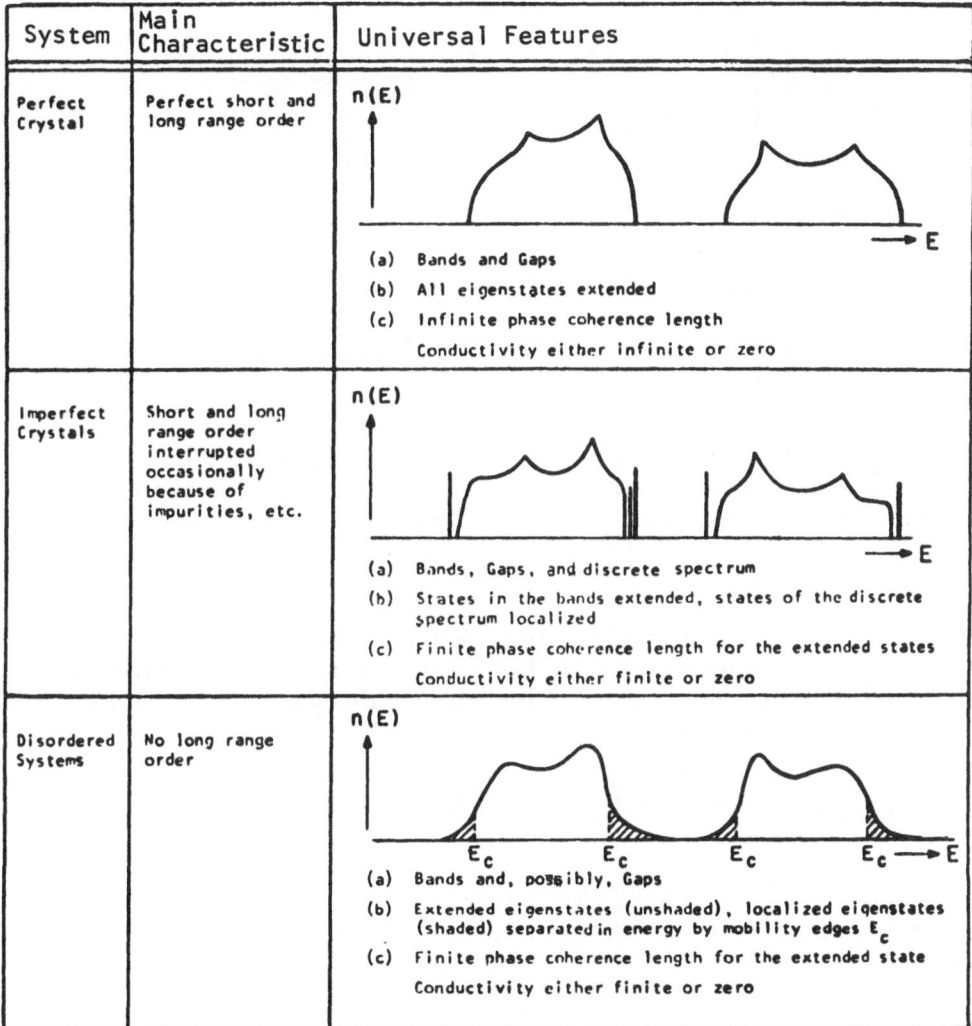

System	Main Characteristic	Universal Features
Perfect Crystal	Perfect short and long range order	n(E) ↑ → E (a) Bands and Gaps (b) All eigenstates extended (c) Infinite phase coherence length Conductivity either infinite or zero
Imperfect Crystals	Short and long range order interrupted occasionally because of impurities, etc.	n(E) ↑ → E (a) Bands, Gaps, and discrete spectrum (b) States in the bands extended, states of the discrete spectrum localized (c) Finite phase coherence length for the extended states Conductivity either finite or zero
Disordered Systems	No long range order	n(E) ↑ E_c E_c E_c E_c → E (a) Bands and, possibly, Gaps (b) Extended eigenstates (unshaded), localized eigenstates (shaded) separated in energy by mobility edges E_c (c) Finite phase coherence length for the extended state Conductivity either finite or zero

Fig. 2   Qualitative dependence of the density of states and the electronic properties on structural changes (from ref. 5).

result is a density of states, which consists of the (slightly changed) energy bands of the ideal crystal plus some characteristic discrete energy levels, which are associated with the nature of the defects (Fig. 2). In

Fig. 3 Qualitative shape of the electron wave function
in a one-dimensional system: (a) Crystal: extended
state, (b) Disordered system: localized state.

effective-mass approximation, the problem of calculating
the impurity spectrum can be reduced to the eigenvalue
problem of an atom, i.e. the envelope wave functions behave as Fig. 3(b))

$$X_I(\vec{r}) \sim e^{-\alpha|\vec{r} - \vec{R}_I|} \qquad |\vec{r}| \rightarrow \infty \qquad (2)$$

This is a typical example of a localized electron state.
The contribution of these states to the electrical conductivity vanishes for T→0 because the electron is bounded to the impurity at $\vec{R}_I$.

Disordered System

     If we increase the number of the deviations from the
ideal crystalline structure, the number and density of
the discrete impurity levels will increase and the shape
of the continuous energy bands will deviate more and more
from that of the crystalline bands, (Fig. 2). For a disordered system, which may be considered as an impurity
system with about 10^{23} different impurities, we expect
to have about 10^{23} "impurity energy levels" everywhere
at the energy axis. Many of the respective energy states
may be localized as in the low density limit. Many of

the energy levels will be situated within the continuous energy bands of the fictituous host crystal and the respective wave functions become delocalized.

Hence, in general, we expect the density of states of a disordered system to consist of a distribution of localized and extended states. The localized states will be associated with <u>finite</u> atom clusters, the extended states with clusters <u>extending to infinity</u> along at least one path. Therefore, electrons in localized states are restricted to one atom cluster, whereas electrons in extended states can escape to infinity. It is reasonable to assume, that there is no coexistence of localized and delocalized states at the same energy. Therefore, the separation between the localized state energy regions and the extended state regions will be discontinuous. Since the localized states do not contribute to the electronic transport at T = 0, the energies separating the localized and extended state regions are denoted as mobility edges.

Questions

There are four main theoretical questions arising from the qualitative picture of the density of states we have developed so far:
(i) What is a disordered system?
(ii) Is it possible to prove the existence of localized states in an infinite, three dimensional disordered system?
(iii) What is the ratio of localized versus extended state numbers for a given kind of disorder, and how does the density of localized states depend on the disorder?
(iv) How does the density of extended states depend on the disorder?

The answer to the first question forms the basis of any quantitative theory of the electron spectrum. We shall discuss it in the subsequent chapters of this lecture.

The second question is of a fundamental nature and has been subject to very elaborate theoretical studies[3,4,5]. As far as we know, no rigorous proof for the existence of localized states in a general disordered system has been given. Therefore, we refer the reader to the literature for the details of these theories.

The answer to the third question will be useful for the explanation of the electronic properties associated with states in energy regions, where, in the corresponding

crystal the density of states vanishes. These properties
are, for instance, the electronic conductivity near T=0
and optical absorption below and about the crystalline
fundamental absorption edge. Since the localized states
are associated with <u>finite</u> atom clusters, the answer to
this question is mathematically complicated and will depend strongly on the <u>microscopic details</u> of the structure
of the particular system in consideration.

The answer to the fourth question will be useful for
the explanation of properties as, for instance, the band
to band optical absorption and the photoconductivity.
It will not depend too much on the microscopic structural
details as, for instance, special <u>atom-cluster</u> configurations and <u>dangling bonds</u>. Instead of this, it is expected
to be possible to relate the electronic density of states
to <u>average structural properties</u> as the "degree of disorder", i.e. the <u>average</u> diameter of atom clusters, or the
<u>average</u> diameter of a short range order region. This
problem is strongly connected with the question of the
existence of an "ideal amorphous state" as represented
for example by a random network of atoms or by a microcrystalline model[6,7,8]. We shall restrict the discussion of the following lectures to this case.

I-2 Quantum Mechanical Description of the Electron

The One-Electron Model

In the one-electron approximation one can reduce the
calculation of the electronic properties of a solid to
the problem of calculating the properties of one electron
moving in the potential of N atomic scatterers situated
at sites $\vec{R}_1 \ldots \vec{R}_N$ within a volume Ω. Surface properties
are eliminated by taking the thermodynamic limit $N, \Omega \to \infty$,
N/Ω = constant, at the end of the calculation. The
Hamiltonian of the system in one-electron approximation
is

$$H_{\vec{R}_1 \ldots \vec{R}_N} = P^2 + \sum_{i=1}^{N} v_i(\vec{R}_i) \tag{3}$$

P^2 is the kinetic energy of the electron. $v_i(\vec{R}_i)$ is the
potential of the ion at the site $\vec{R}_i$. The atom sites
$\vec{R}_1 \ldots \vec{R}_N$ are assumed to be at rest, T = 0.

The one-electron model is the simplest model one
can find for the investigation of the electronic properties of a solid. Nevertheless, it explains many of the

experimental data obtained from "ideal" single crystals. This encourages us to assume that it is also a reasonable first approximation for the description of the electronic properties of non-crystalline materials.

Since the quantum mechanical one-electron problem is exactly solvable for the case of ideal order, it is reasonable to recall the two main approaches used in crystalline band structure theory. This discussion leads then automatically to forms of the Hamiltonian, which are most suitable for the description of the properties of structurally and compositionally disordered systems, respectively.

The Tight-Binding Hamiltonian

The first approach is obtained by regarding the properties of the crystal electron to be determined mainly by the properties of the constituent atoms. Then it provides a good starting point to expand an electron state into a linear combination of basis states $\varphi_{i\nu}$ associated with the atoms at $\vec{R}_i$. ν are some "atomic quantum numbers.

$$|\Psi> = \sum_{i\nu} c_{i\nu} |\varphi_{i\nu}> \qquad (4)$$

Using the basis $\{\varphi_{i\nu}\}$ the spectral representation of the Hamiltonian is

$$H_{\vec{R}_1\ldots\vec{R}_N} = \varepsilon_{i\nu} |\varphi_{i\nu}><\varphi_{i\nu}| + \sum_{\substack{i\neq j \\ \nu\mu}} W_{ij}^{\nu\mu} |\varphi_{i\nu}><\varphi_{j\mu}| \qquad (5)$$

where

$$\varepsilon_{i\nu} = <\varphi_{i\nu}|H|\varphi_{i\nu}>$$

$$\qquad (6)$$

$$W_{ij}^{\nu\mu} = <\varphi_{i\nu}|H|\varphi_{j\mu}>$$

In first approximation, $\varepsilon_{i\nu}$ can be interpreted as the νth energy level of the atom at the site $\vec{R}_i$. $W_{ij}^{\nu\mu}$ can be interpreted as the probability amplitude for an electron transition from the state $|\varphi_{i\nu}>$ to the state $|\varphi_{j\mu}>$.

One should recall two points:

(i) Writing the expansions (4) and (5), we assumed an orthonormal set of basis states. The atomic orbitals usually used in LCAO-theory are generally not linearly independent for i≠j. This leads to computational difficulties.

(ii) In practical calculations, one is automatically restricted to a finite number of basis states. Hence, the validity of the results is restricted to the valence bands and (in some cases) to the lowest conduction bands. However, the Hamiltonian (5) may be useful as a model Hamiltonian to derive general properties of certain classes of systems[9].

The Nearly-Free Electron Hamiltonian

The second approach is obtained by taking the superposition of the core potentials as a weakly varying background for the motion of a free electron. Then it is reasonable to expand an electron state into plane wave states $|\vec{k}>$

$$|\Psi> = \sum_{\vec{k}} C_{\vec{k}} |\vec{k}>; \quad <\vec{r}|\vec{k}> = \frac{1}{\sqrt{\Omega}} \exp(i\vec{k}\vec{r}) \tag{7}$$

The spectral representation of the Hamiltonian is

$$H_{\vec{R}_1 \ldots \vec{R}_N} = \sum_{\vec{k}} k^2 |\vec{k}><\vec{k}| + \sum_{\vec{k},\vec{k}'} W_{\vec{k}\vec{k}'} |\vec{k}><\vec{k}'| \tag{8}$$

where

$$k^2 = <\vec{k}|p^2|\vec{k}'>$$

$$W_{\vec{k}\vec{k}'} = <\vec{k}|\sum_{j=1}^{N} v_j|\vec{k}'> = \sum_{j=1}^{N} \exp(-i(\vec{k}-\vec{k}')\vec{R}_j)v_{\vec{k}\vec{k}'}^{(j)} \tag{9}$$

$$v_{\vec{k}\vec{k}'}^{(j)} = <\vec{k}|v_j|\vec{k}'>$$

k^2 is the free-electron energy, $v_{\vec{k}\vec{k}'}^{(j)}$ the atomic potential scattering form factor of the atom at site $\vec{R}_j$, and $\sum_{j=1}^{N} \exp\{-i(\vec{k}-\vec{k}')\vec{R}_j\}$ is the structure factor.

In the thermodynamic limit, the $\vec{k}$-space sums in eqs. (7) and (8) have to be taken as integrals. Therefore, the solution of the eigenvalue problem of the Hamiltonian (8), in general, requires the solution of an integral equation instead of a matrix equation as in the tight-binding case of eq. (5). Also the free-electron approach is only reasonable for sufficiently smooth atomic potentials, in order to ensure a sufficiently short ranged form factor $v_{kk'}^{(j)}$. This condition is usually assumed to be fulfilled for conduction band electron states and (in some cases) for the highest valence bands. However, as we shall see immediately, in spite of some computational difficulties, the "nearly-free electron" Hamiltonian (8) has some advantages in calculating the electronic spectrum of structurally disordered materials.

I-3. Structural and Compositional Disorder

Definition of Disorder

Disordered systems can be defined by the lack of long range order. This means that in the case of <u>structural</u> (or positional) disorder, the atom positions $\vec{R}_1, \vec{R}_2$ are statistically independent if the difference $|\vec{R}_1 - \vec{R}_2|$ is much larger than the mean distance between two atoms, (Fig. 4(a)). In the case of compositional disorder, the lack of long range order can be achieved by the requirement, that the occupation of two lattice sites $\vec{R}_1, \vec{R}_2$ with the various constituent atoms is statistically independent if $|\vec{R}_1 - \vec{R}_2|$ is much larger than the lattice constant, (Fig. 4(b)). In general, a disordered system exhibits both compositional and structural disorder. However, the two limiting cases can be considered in a certain sense as "ideal" disordered models, and can be used as a basis for a theoretical treatment of the problem of calculating the electronic properties of random systems.

Electron Hamiltonian for Compositional Disorder

Until now, we have considered the positions of the atomic scatterers, $\vec{R}_1 \ldots \vec{R}_N$, and the "atomic" potentials, v_i as the given parameters characterizing the system. For further specification we distinguish two limiting cases: The first is the case of <u>compositional disorder</u>, which is obtained by assuming different kinds of atoms distributed somehow at the points of a crystal lattice. Examples are the non-perfect crystal with impurities distributed randomly through the crystal lattice, or the

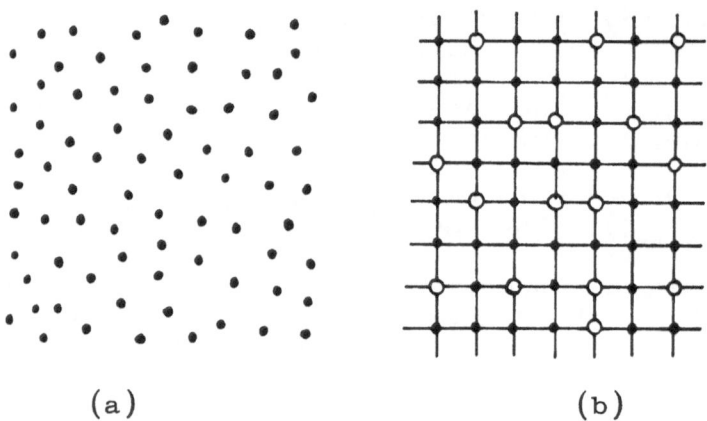

(a)   (b)

Fig. 4 Two-dimensional examples of (a) structural and
(b) compositional disorder.

two component alloy consisting of two kinds of atoms distributed randomly at the lattice sites. Taking now the
simplest form of the Hamiltonian (5), which is obtained by taking only one state per atom,

$$H = \sum_{i} \varepsilon_i |\varphi_i\rangle\langle\varphi_i| + \sum_{i \neq j} W_{ij} |\varphi_i\rangle\langle\varphi_j| \qquad (10)$$

we observe that we obtain a particularly convenient
Hamiltonian for studying the electron spectrum by assuming W_{ij} to depend only on the difference between $\vec{R}_i$ and
$\vec{R}_j$ (lattice vector!) and <u>not</u> on the nature of the atoms
at $\vec{R}_i$ and $\vec{R}_j$, respectively. In this case, disorder occurs only in the diagonal part of H, ε_i, which varies
statistically from site to site (Fig. 5(a)).

Electron Hamiltonian for Structural Disorder

The second case is that of <u>structural disorder</u>,
which is obtained by assuming only one kind of atom situated at sites $\vec{R}_1 \ldots \vec{R}_N$, which are distributed somehow
within the system volume Ω. Examples are structurally
distorted crystals or random networks of atoms which are
assumed to be realized in amorphous Ge films for instance.
Since now in the Hamiltonian (10), disorder occurs also
in the nondiagonal elements W_{ij} (Fig. 5(b)), it would appear to be no longer a convenient choice. However, the

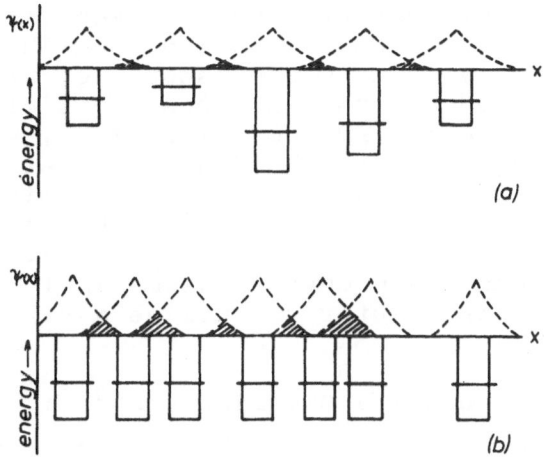

Fig. 5 One-dimensional model for (a) compositional and
(b) structural disorder. The dashed lines are schematic
electron wave functions associated with the atoms represented by the square well potentials each containing one
energy level denoted by a horizontal line. The overlap
between two wave functions being a measure of the magnitude of the respective hopping matrix element is shown
as a shaded area. In the case of compositional disorder
(fluctuating positions of the atomic energy levels) the
overlap is almost constant. In the case of structural
disorder (constant positions of the energy levels) the
overlap is strongly fluctuating.

Hamiltonian of eq. (8) now looks particularly convenient
for the investigation of the electronic properties, because the disorder occurs in the off-diagonal elements
only via the structure factor. The form factor is independent of $\vec{R}_i$.

    Since we are concerned with structurally disordered
solids, we shall use in the following lectures the Hamiltonian (8). Most of the results we shall derive can also
be obtained by starting from a Hamiltonian of the type
(5) [10-14].

    I-4. Statistical Description of the Structure

    Since we are dealing with a macroscopic system, it

is clear from the beginning that we will not be able to
know the density of states for one specific configuration because we are not even able to know all the posi-
tions $\vec{R}_1 \ldots \vec{R}_N$ for the configuration. On the other hand,
the density of states as a macroscopic quantity is not
expected to depend on the exact positions of any finite
number of atoms in the system. Therefore, what we shall
try to do is to calculate the density of states as an
average over certain <u>classes</u> of configurations having
certain common features. These shall be classified by
some parameters, the number of which certainly has to be
much smaller than the number of atoms in the system. It
is the aim of this chapter to outline a structural classification.

Probability Density Distribution[15]

We consider a macroscopic homogeneous system of
$N(\approx 10^{23})$ atoms at $\vec{R}_1 \ldots \vec{R}_N$ in the volume Ω. We define
the probability for the atoms to be situated within the
infinitesimal volumes $d^3R_1 \ldots d^3R_N$ at $\vec{R}_1 \ldots \vec{R}_N$ as

$$P(\vec{R}_1 \ldots \vec{R}_N) \; d^3\vec{R}_1 \ldots d^3\vec{R}_N \tag{11}$$

$P(\vec{R}_1 \ldots \vec{R}_N)$ is the probability distribution function normalized to unity.

$$\int \ldots \int \; P(\vec{R}_1 \ldots \vec{R}_N) \; d^3\vec{R}_1 \ldots d^3\vec{R}_N = 1 \tag{12}$$

because of the macroscopic homogeneity we have

$$P(\vec{R}_1 + \vec{a}, \ldots \vec{R}_N + \vec{a}) = P(\vec{R}_1 \ldots \vec{R}_N) \tag{13}$$

for arbitrary translations $\vec{a}$. Probability distributions
for single atoms, pairs of atoms etc., are obtained from
P by

$$P(\vec{R}_i) = \int \ldots \int \; P(\vec{R}_1 \ldots \vec{R}_N) \; \underset{i \neq j}{\pi} \; d^3\vec{R}_j \tag{14}$$

$$P(\vec{R}_i, \vec{R}_j) = \int \ldots \int \; P(\vec{R}_1 \ldots \vec{R}_N) \; \underset{k \neq i, j}{\pi} \; d^3\vec{R}_k \tag{15}$$

The configurational average of any configuration dependent quantity $A(\vec{R}_1 \ldots \vec{R}_N)$ can be calculated by

$$<A>: = \int \ldots \int \ A(\vec{R}_1 \ldots \vec{R}_N) \ P(\vec{R}_1 \ldots \vec{R}_N) \ \underset{i}{\pi} \ d^3R_i \qquad (16)$$

Conditional Probabilities

For applications in the subsequent lectures we need the following useful relation: If $P_1(\vec{R}_{\ell_1})$ is the probability distribution for an atom to be at the site $\vec{R}_{\ell_1}$, $P(\vec{R}_{\ell_1} \ldots \vec{R}_{\ell_N}$, can be written as

$$P(\vec{R}_{\ell_1} \ldots \vec{R}_{\ell_N}) = P_1(\vec{R}_{\ell_1})P(\vec{R}_{\ell_1} ; \vec{R}_{\ell_2} \ldots \vec{R}_{\ell_N}) \qquad (17)$$

Eq. (17) defines the conditional probability distribution for the (N-1) atoms $\ell_2 \ldots \ell_N$ to be at $\vec{R}_{\ell_2} \ldots \vec{R}_{\ell_N}$ assuming the atom ℓ_1 to be fixed at $\vec{R}_{\ell_1}$. Taking now $P_2(\vec{R}_{\ell_1} ; \vec{R}_{\ell_2})$ as the probability distribution for having the atom ℓ_2 at $\vec{R}_{\ell_2}$ with the atom ℓ_1 at $\vec{R}_{\ell_1}$ fixed, $P_3(\vec{R}_{\ell_1}, \vec{R}_{\ell_2} ; \vec{R}_{\ell_3})$ for finding the atom ℓ_3 at site $\vec{R}_{\ell_3}$ provided atoms ℓ_1 ℓ_2 are fixed at $\vec{R}_{\ell_1}$ and $\vec{R}_{\ell_2}$ etc., we continue the procedure of eq. (17) and define

$$P(\vec{R}_{\ell_1} ; \vec{R}_{\ell_2} \ldots \vec{R}_{\ell_N}) = P_2(\vec{R}_{\ell_1} ; \vec{R}_{\ell_2})P(\vec{R}_{\ell_1} \vec{R}_{\ell_2} ; \vec{R}_{\ell_3} \ldots \vec{R}_{\ell_N})$$

$$P(\vec{R}_{\ell_1} \vec{R}_{\ell_2} ; \vec{R}_{\ell_3} \ldots \vec{R}_{\ell_N}) = P_3(\vec{R}_{\ell_1} \vec{R}_{\ell_2} ; \vec{R}_{\ell_3})P(\vec{R}_{\ell_1} \vec{R}_{\ell_2} \vec{R}_{\ell_3} ; \vec{R}_{\ell_4} \ldots \vec{R}_{\ell_N})$$

$$\qquad (18)$$

$$P(\vec{R}_{\ell_1} \ldots \vec{R}_{\ell_{N-2}} ; \vec{R}_{\ell_{N-1}}) = P_{N-1}(\vec{R}_{\ell_1} \ldots \vec{R}_{\ell_{N-2}} ; \vec{R}_{\ell_{N-1}}) \times$$

$$P(\vec{R}_{\ell_1} \ldots \vec{R}_{\ell_{N-1}} ; \vec{R}_{\ell_N})$$

$$P(\vec{R}_{\ell_1} \ldots \vec{R}_{\ell_{N-1}} ; \vec{R}_{\ell_N}) = P_N(\vec{R}_{\ell_1} \ldots \vec{R}_{\ell_{N-1}} ; \vec{R}_{\ell_N})$$

$P(\vec{R}_{\ell_1} \ldots \vec{R}_{\ell_{j-1}} ; \vec{R}_{\ell_j} \ldots \vec{R}_{\ell_N})$ are the obvious generalizations of $P(\vec{R}_{\ell_1};\vec{R}_{\ell_2}...\vec{R}_{\ell_N})$. Obviously the factorization of eq. (18) holds for every sequence of atom positions $\vec{R}_{\ell_1}..\vec{R}_{\ell_N}$.

The conditional probabilities can be used to calculate restricted averages of any configuration dependent quantities

$$<A>_{\ell_1...\ell_j} := \int \prod_{\nu=j+1}^{N} d^2\vec{R}_{\ell_\nu} A(\vec{R}_1...\vec{R}_N) P(\vec{R}_{\ell_1}..\vec{R}_{\ell_j};\vec{R}_{\ell_{j+1}}..\vec{R}_{\ell_N})$$

$$(19)$$

Eqs. (17) and (18) establish a factorization of $P(\vec{R}_1..\vec{R}_N)$

$$P(\vec{R}_{\ell_1}..\vec{R}_{\ell_N}) = P_1(\vec{R}_{\ell_1})P_2(\vec{R}_{\ell_1};\vec{R}_{\ell_2})....P_N(\vec{R}_{\ell_1}..\vec{R}_{\ell_{N-1}};\vec{R}_{\ell_N})$$

$$(20)$$

As we shall see later, the factorization of $P(\vec{R}_1...\vec{R}_N)$ into the conditional probabilities P_j gives the possibility of a decomposition of the integral equation for the scattering matrix of the system containing the positions of all the atomic scatterers ($N\to\infty$ in the thermodynamic limit) into an infinite set of integral equations, each containing explicitly only a <u>finite</u> number of atom sites. Each member of this set of integral equations will contain the structure of the system explicitly only via the conditional probabilities $P_j = P_j(\vec{R}_1...\vec{R}_{j-1};\vec{R}_j)$. The definition of the structural model requests the specification of the P_j.

Examples

In the subsequent lectures we shall not need the conditional probabilities $P_j(\vec{R}_1...\vec{R}_{j-1};\vec{R}_j)$ for the atom j to be at $\vec{R}_j$ provided the atoms $1...j-1$ are fixed at $R_1...R_{j-1}$ but instead of this

$$g_j(\vec{R}_1...\vec{R}_{j-1};\vec{R}_j) = (N-j+1)P_j(\vec{R}_1...\vec{R}_{j-1};\vec{R}_j) \qquad (21)$$

This is the probability for <u>any</u> atom to be at the site $\vec{R}_j$ provided $j-1$ atoms are fixed at $\vec{R}_1...\vec{R}_{j-1}$.

For g_1 we have

$$g_1(\vec{R}) = N\, P_1(\vec{R}) = \frac{N}{\Omega} = d \tag{22}$$

because of homogeneity.

$g_1(\vec{R})$ does not depend on the position of the atom and therefore contains no information about the microscopic atomic arrangement, i.e. first, second, third, etc. neighbors coordination.

Information about the short range order is contained in the higher order probability distributions. For the two-atom distribution $g_2(\vec{R}_1,\vec{R}_2)$ we have

$$g_2(\vec{R}_1;\vec{R}_2) = (N-1)P_2(\vec{R}_1;\vec{R}_2) \equiv g_2(\,|\vec{R}_1-\vec{R}_2|\,) \tag{23}$$

because of macroscopic homogeneity and isotropy. For a completely random system without any short range order $g_2(x)$ is constant (Fig. 6(a))

$$g_2(x) = \frac{N-1}{\Omega} \tag{24}$$

In the case of ideal order $g_2(x)$ exhibits δ-function peaks at the distances between first, second, third, etc. neighbors, (Fig. 6(b)).

$$g_2(x) = \sum_{i=1}^{\infty} z_i\, \delta(x - a_i)\, \frac{1}{4\pi x^2} \tag{25}$$

z_i is the number of ith neighbors.

For a disordered system with short range order $g_2(x)$ exhibits maxima at the mean distances between the first, second, etc. neighbors and becomes a constant for $x\to\infty$ because then the two atoms are statistically independent due to the lack of long range order (Fig. 6(c)). We write

$$g_2(x) = \sum_{i=1}^{\infty} f_i\, (x - a_i) \tag{26}$$

where $f_i(x-a_i)$ are chosen to be localized near a_i with the width λ_i and

$$\int f_i\, (x-a_i)d^3x = z_i \tag{27}$$

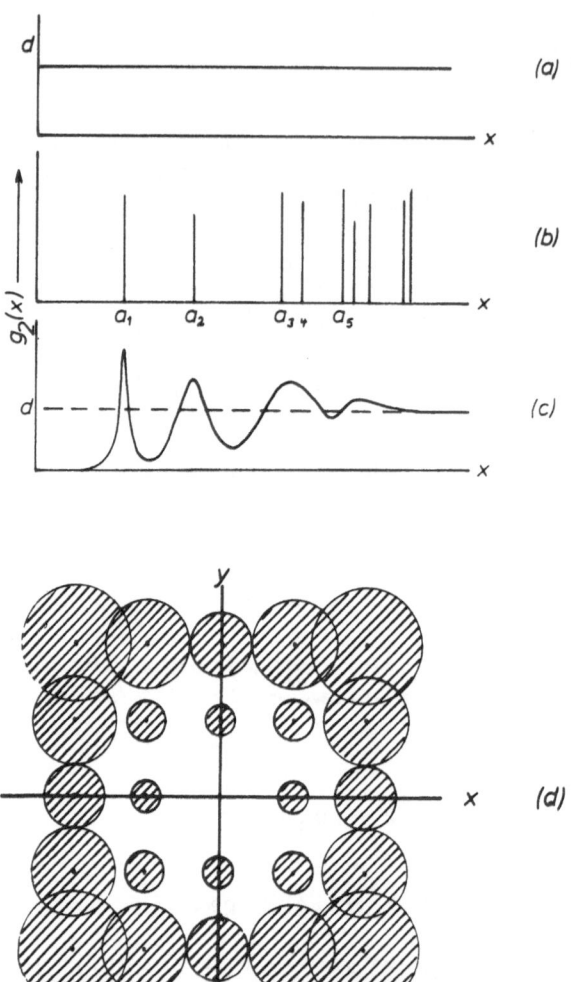

Fig. 6 Qualitative picture of the two atom distributions for (a) complete disorder; (b) complete order; (c) short-range ordered system; (d) nonisotropic short-range ordered system in two dimensions. In (d) the shaded circles denote the half-widths of the partial contributions to g_2 associated with the lattice points.

As $g_2(x)$ contains only information about the <u>distance</u> between two atoms, it yields no information about the angular distribution. Such information is obtained from the higher order probability distributions. However, these are not available experimentally for disordered systems in contrast to g_2, which can be obtained in principle by Fourier transformation of x-ray or electron scattering spectra. One could try to construct the higher order atom distributions by using some suitable model. In general, this is very complicated[16]. The only case in which all the g_j are known is that of complete disorder.

$$g_j(\vec{R}_1 \ldots \vec{R}_{j-1}; \vec{R}_j) = \frac{N-j+1}{\Omega} \qquad (28)$$

We conclude this section by giving an expression for the two-atom distribution which is very useful for some applications[17] (Fig. 6(d)).

$$g_2(\vec{X}) = \sum_{i=1}^{\infty} f_i \ (\vec{x} - \vec{\ell}_i) \qquad (29)$$

This expression looks quite similar to that of eq. (26). However, f_i are now taken to be localized near the points of a <u>Bravais lattice</u>, ℓ_i, and do not only depend on the distance between two atoms, but also on the <u>direction</u>. Hence, (29) describes a system with <u>no</u> macroscopic isotropy. If the widths λ_i of the $f_i(\vec{x} - \vec{a}_i)$ is chosen appropriately, in order to have $g_2(\vec{x}) \to$ constant for $|x| \to \infty$ the expression (29) can be used to study the effect of the lack of long range order on the density of states of a system with a given short range order, which is represented by the Bravais lattice $\{\vec{\ell}_i\}$[18].

The expression (29) for the two-atom distribution function can easily be generalized to include the case of a lattice with a unit cell consisting of more than one atom. Instead of $\vec{\ell}_i$ one has to insert simply $\vec{\ell}_i + \vec{b}_j$, where $\vec{b}_j$ are the basis vectors of the atoms in the unit cell. The sum over i has to be replaced by a sum over i and j.

II. FORMAL THEORY

II-1. The Electronic Density of States

Configurational Averaged Density of States

The quantity of interest is the number of electron states per unit energy and atom. This density of electron states is configuration dependent. For a system described by a Hamilton $H_{\vec{R}_1 \ldots \vec{R}_N}$ it is defined as

$$n(E)_{\vec{R}_1 \ldots \vec{R}_N} = \frac{1}{N} \, \mathrm{Tr}\, \delta(E - H_{\vec{R}_1 \ldots \vec{R}_N}) \qquad (30)$$

since $n(E)_{\vec{R}_1 \ldots \vec{R}_N}$ is a macroscopic quantity, it should not depend on the microscopic details of the structure of the system, i.e. the detailed positions of one, two, or any finite number of atoms. Therefore, the density of states has to be calculated as a configurationally averaged quantity.

$$n(E) = \langle n(E)_{\vec{R}_1 \ldots \vec{R}_N} \rangle = \int \prod_i d^3 R_i \; P(\vec{R}_1 \ldots \vec{R}_N) n(E)_{\vec{R}_1 \ldots \vec{R}_N})$$

$$\qquad (31)$$

Using eq. (30), we obtain

$$n(E) = \frac{1}{N}\mathrm{Tr} \int \prod_i \pi d^3 R_i \; P(\vec{R}_1 \ldots \vec{R}_N) \delta(E - H_{\vec{R}_1 \ldots \vec{R}_N}) \qquad (32)$$

We notice that in order to calculate the configurationally averaged density of states, we have to calculate the density operator

$$\rho(E) = \int d^3 R_i \; P(\vec{R}_1 \ldots \vec{R}_N) \; \delta(E - H_{\vec{R}_1 \ldots \vec{R}_N}) \qquad (33)$$

Relation to the Green's Function

A great variety of methods for the calculation of the density of states are based on Green's function and multiple scattering techniques [10,19,20,21]. The relation between the density of states and the one-electron Green's function is easily obtained from eq. (30) or, alternatively, for the averaged density of states from the eq. (33) for the density operator by inserting the well known definition of the δ-function

$$\delta(x) = \lim_{\eta \to o} \frac{i}{2\pi} \left(\frac{1}{x+i\eta} - \frac{1}{x-i\eta}\right) \; ; \; \eta > o \tag{34}$$

we obtain

$$\rho(E) = -\frac{1}{\pi} \text{Im} < G(E^+)_{\vec{R}_1 \ldots \vec{R}_N} > = -\frac{1}{\pi} \text{Im} G(E^+) \tag{35}$$

where E^+ in $G_{\vec{R}_1 \ldots \vec{R}_N}$ and G, respectively, means that the limit $\eta \to +o$ has been taken. $G(Z)_{\vec{R}_1 \ldots \vec{R}_N}$ and $G(Z)$ are the configuration dependent and the configurationally averaged one-electron resolvent, respectively.

$$G(Z)_{\vec{R}_1 \ldots \vec{R}_N} = \frac{1}{Z - H_{\vec{R}_1 \ldots \vec{R}_N}} \tag{36}$$

$$g(Z) = < G(Z)_{\vec{R}_1 \ldots \vec{R}_N} > \tag{37}$$

Z is the complex energy parameter. In the scattering theory, $G(Z)$ relates the scattered electron states $\psi^s_{\vec{k}}$ to the unscattered states $\varphi_{\vec{k}}$ via the Lipmann-Schwinger equations.

$$\psi^s_{\vec{k}} = \varphi_{\vec{k}} + \frac{1}{E - s\eta - H} V\varphi_k \; , \; s = \pm \tag{38}$$

where H is decomposed into an unperturbed Hamiltonian H_O and a perturbation, the scattering potential V. Taking $G(Z)$ in some representation, for instance, momentum representation, we have the Green's function of the system.

$$G(\vec{kk}';Z) = < \vec{k} | G(Z) | \vec{k}' > \tag{39}$$

Spectral Representation of the Green's Function

Particularly for a free electron described by the Hamiltonian $H_O = p^2 = -\Delta$ we obtain the spectral representation

$$G_o(Z) = \sum_{\vec{k}} \frac{|\vec{k}> < \vec{k}|}{Z - k^2} \tag{40}$$

In general, if we use the eigenrepresentation of H, we have

$$G(Z) = \sum_n \frac{|\Psi_n \rangle\langle \Psi_n|}{Z - E_n} \tag{41}$$

E_n, Ψ_n are the eigenvalues and eigenstates of H, respectively. We observe, that G has poles at the discrete eigenvalues of H and a cut in the region of the continuous energy spectrum

Since $G(Z)$ is related to the scattering problem, we expect that calculating the one-electron Green's function we should gain much benefit from using multiple scattering expansions of $G(Z)$[19,20,21]. However, in the following chapters, we shall use a slightly different method in deriving expressions for the Green's function, which avoids the sometimes rather puzzling diagrammatic notation. It is based on the factorization of $P(\vec{R}_1 \ldots \vec{R}_N)$ of the preceeding lecture and on the concept of a mean, effective medium describing the propagation of the electron between two scattering events. The results, however, are equivalent to those obtained by the multiple scattering expansion method.

II-2. The Configurational Averaged Green's Function

Integral Equation and Relation to the Scattering Operator

In the following, we shall assume the Hamiltonian to consist of a configuration independent part H_0 and a configuration dependent part $V_{\vec{R}_1 \ldots \vec{R}_N}$. As a special example, we may remember the Hamiltonian of eq. (8). However, we are not necessarily restricted to H_0 being the free electron Hamiltonian. H_0 may be considered as a Hamiltonian describing the properties of an electron moving in some effective medium and $V_{\vec{R}_1 \ldots \vec{R}_N}$ to be the scattering potential of the atomic scatters relative to the medium. In general, H_0 and V may depend on Z.

$$H_{\vec{R}_1 \ldots \vec{R}_N} = H_0(Z) + V_{\vec{R}_1 \ldots \vec{R}_N}(Z) \tag{42}$$

Taking

$$G_0(Z) = \frac{1}{Z - H_0} \tag{43}$$

as the resolvent corresponding to the "free-electron", we obtain an equation for $G(Z)$ by applying the resolvent identity

$$\frac{1}{Z-H_o-V} = \frac{1}{Z-H_o} + \frac{1}{Z-H_o} \; V \; \frac{1}{Z-H_o-V} \tag{44}$$

$$G(Z) = G_o(Z) + G_o(Z) \; V \; G(Z) \tag{45}$$

In momentum representation, this corresponds to an integral equation

$$G(\vec{k},\vec{k}';Z) = G_o(\vec{k}\vec{k}';Z) + \sum_{\vec{k}''\vec{k}'''} G_o(\vec{k}\vec{k}'';Z)V(\vec{k}'',\vec{k}''')G(\vec{k}'''\vec{k}';Z) \tag{46}$$

This integral equation takes a particular simple form if we choose H_o to be diagonal in momentum representation

$$G(\vec{k}\vec{k}';Z) = G_o(\vec{k};Z)\delta_{\vec{k}\vec{k}'} + G_o(\vec{k};Z) \sum_{\vec{k}''} V(\vec{k},\vec{k}'')G(\vec{k}''\vec{k}';Z) \tag{47}$$

Relating the resolvent $G(Z)$ to the total scattering matrix of the system, $T(Z)$

$$VG(Z) \equiv T(Z)G_o(Z) \tag{48}$$

we obtain from eq. (45)

$$G(Z) = G_o(Z) + G_o(Z)T(Z)G_o(Z) \tag{49}$$

and $T(Z)$ satisfies

$$T(Z) = V + VG_o(Z)T(Z) \tag{50}$$

$T(Z)$ yields the total scattering amplitude of the scattering potential V. Iterating eq. (50) we obtain the Born-series

$$T = V + VG_oV + VG_oVG_oV + --- \tag{51}$$

Introducing now V as a superposition of atomic potentials

$$V = \sum_{i=1}^{N} v_i(\vec{R}_i) \tag{52}$$

and the total scattering matrix T as a sum of partial <u>a-tomic scattering contributions</u> associated with the atomic scatterers

$$T(Z) = \sum_{i=1}^{N} Q_i(Z) \tag{53}$$

we obtain from eq. (50)

$$T(Z) = \sum_{i=1}^{N} v_i + \sum_{i=1}^{N} v_i \, G_o(Z) \sum_{j=1}^{N} Q_i(Z) \tag{54}$$

$$Q_i(Z) = v_i + v_i G_o(Z) \sum_{j=1}^{N} Q_j(Z) \; ; \; i=1\ldots N \tag{55}$$

Eq. (55) shows, that the partial contribution of the i-th atom to the total scattering matrix is related to all the other scatterers in the system. This becomes still more evident if we relate $Q_i(Z)$ to $t_i(Z)$, the scattering matrix of an isolated atomic scatterer, defined in analogy to eq. (50) by

$$t_i(Z) = v_i + v_i G_o(Z) \, t_i(Z) \tag{56}$$

$$Q_i(Z) = t_i(Z) + t_i(Z) \, G_o \sum_{j \neq i} Q_i(Z) \tag{57}$$

We notice that eqs. (55) and (57) are formally equivalent. v_i is replaced by $t_i(Z)$ and the sum at the right hand side of eq. (57) is restricted to $j \neq i$. Eq. (57) indicates, that the partial scattering contribution of the i-th atom described by $Q_i(Z)$ can be considered to consist of the scattering contribution of an isolated scatterer represented by $t_i(Z)$ plus a contribution, which depends on the other scatterers in the medium[10,15]. Eqs. (55) and (57), being equivalent in describing the scattering properties of the system, give rise to different approaches to the problem. If we have an assembly of weak scatterers, each describable within the Born approximation , it will be more convenient to start from eq. (55). However, if we are dealing with an assembly of strong scatterers, it is expected to be more suitable to start from eq. (57), which contains the atomic scattering matrix explicitly.

Configurational Averaging[15,22]

Next, we calculate the configuration average of the

electron propagator, i.e. the scattering matrix,

$$<G(z)> = G_o(z) + G_o(z) <T(z)> G_o(z) \tag{58}$$

because H_o, hence $G_o(z)$, is assumed to be configuration independent. From eqs. (53), (55) and (57) we have

$$<T(z)> = \sum_{i=1}^{N} <Q_i(z)> \tag{59}$$

and, alternatively,

$$<Q_i(z)> = <v_i> + \sum_{j=1}^{N} <v_i G_o(z) Q_j(z)> \tag{60}$$

$$<Q_i(z)> = <t_i> + \sum_{j \neq 1} <t_i G_o(z) Q_j(z)> \tag{61}$$

Performing the configurational average, we consider a structurally disordered system with $v_i(\vec{r}-\vec{R}_i) = v(\vec{r}-\vec{R}_i)$. $Q_i(z)$, the partial scattering contribution of the i-th atom, depends on all atom positions $\vec{R}_1 \ldots \vec{R}_N$. We denote this dependence as

$$Q_i(z) = Q(\vec{R}_i; \{\vec{R}_j\}_{j \neq i}) \tag{62}$$

where we have omitted the variable Z, which is not important for the configurational averaging process.

Using the decomposition of the probability distribution function into conditional probabilities (eqs. (17) and (18), we have

$$<Q_i> = \int \prod_k d^3 R_k \, P(\vec{R}_1 \ldots \vec{R}_N) \, Q(\vec{R}_i; \{\vec{R}_j\}_{j \neq i}$$

$$= \int d^3 R_i \, P_1(\vec{R}_i) \, \bar{Q}(\vec{R}_i) \tag{63}$$

where

$$\bar{Q}(\vec{R}_i) = \int \prod_{k \neq i} d^3 R_k \, P(\vec{R}_i; \vec{R}_1 \ldots \vec{R}_{i-1} \vec{R}_{i+1} \ldots \vec{R}_N) Q(\vec{R}_i; \{\vec{R}_k\}_{k \neq i}) \tag{64}$$

is the conditionally averaged partial scattering operator of the atom at $\vec{R}_i$ fixed. Conditionally averaging eq. (55) yields an equation for $\bar{Q}(\vec{R}_i)$

$$\bar{Q}(\vec{R}_i) = v(\vec{R}_i) + v(\vec{R}_i)G_o\,\bar{Q}(\vec{R}_i)$$

$$+ \int d^3R_j v(\vec{R}_i)\ G_o\bar{Q}(\vec{R}_j;\vec{R}_i)g_2(\vec{R}_i;\vec{R}_j) \tag{65}$$

where we have treated the term $i = j$ separately, and $g_2 = (N-1)P_2$. $\bar{Q}(\vec{R}_j;\vec{R}_i)$ is the conditionally averaged partial scattering operator of the atom at $\vec{R}_j$ with $\vec{R}_j$ and $\vec{R}_i$ fixed. Continuing the procedure, we obtain the following relation between the conditionally averaged partial scattering contributions with n and $n+1$ atoms respectively, fixed

$$\bar{Q}(\vec{R}_{\ell_1};\vec{R}_{\ell_2}\ldots\vec{R}_{\ell_n}) = v(\vec{R}_{\ell_1}) + v(\vec{R}_{\ell_1})\ G_o\bar{Q}(\vec{R}_{\ell_1};\vec{R}_{\ell_2}\ldots\vec{R}_{\ell_n})$$

$$+ v(\vec{R}_{\ell_1})G_o\bar{Q}(\vec{R}_{\ell_2};\vec{R}_{\ell_1}\vec{R}_{\ell_3}\ldots\vec{R}_{\ell_n})$$

$$\vdots \tag{66}$$

$$+ v(\vec{R}_{\ell_1})G_o\bar{Q}(\vec{R}_{\ell_n};\vec{R}_{\ell_1}\ldots\vec{R}_{\ell_{n-1}})$$

$$+ \int d^3R_{\ell_{n+1}} v(\vec{R}_{\ell_1})G_o\bar{Q}(\vec{R}_{\ell_{n+1}};\vec{R}_{\ell_1}..\vec{R}_{\ell_n})g_{n+1}(\vec{R}_{\ell_1}..\vec{R}_{\ell_n};\vec{R}_{\ell_{n+1}})$$

with

$$\bar{Q}(\vec{R}_{\ell_1};\vec{R}_{\ell_2}..\vec{R}_{\ell_n}) = \int \prod_{k\neq\ell_1..\ell_n} d^3R_k\ P(\vec{R}_{\ell_1}..\vec{R}_{\ell_n};\{\vec{R}_k\}_{k\neq\ell_1..\ell_n}) \times$$

$$\times Q(R_{\ell_1};\ \{R_j\}_{j\neq\ell_1}) \tag{67}$$

and $g_{n+1} = (N-n)P_{n+1}$

The first $(n+1)$ terms at the right hand side of eq. (66) containing only the sites of the fixed atoms, relate the averaged partial scattering contribution of the atom at $\vec{R}_{\ell_1}$ so those of the remaining $(n-1)$ atoms at $R_{\ell_2}\ldots R_{\ell_n}$ of the cluster. The last term at the right hand side of eq. (66) describes the contribution of the remaining $(N-n)$ atoms. It is easy to reformulate eq.(66) in terms of the atomic scattering matrix

$$\bar{Q}(\vec{R}_{\ell_1};\vec{R}_{\ell_2}\ldots\vec{R}_{\ell_n}) = t(\vec{R}_{\ell_1})+t(\vec{R}_{\ell_1})G_o\bar{Q}(\vec{R}_{\ell_2};\vec{R}_{\ell_1}\vec{R}_{\ell_3}\ldots\vec{R}_{\ell_n})$$

$$+t(\vec{R}_{\ell_1})G_o\bar{Q}(\vec{R}_{\ell_3};\vec{R}_{\ell_1}\vec{R}_{\ell_2}\vec{R}_{\ell_4}\ldots\vec{R}_{\ell_n})$$

$$\tag{68}$$

$$t(\vec{R}_{\ell_1})G_o\bar{Q}(\vec{R}_{\ell_n};\vec{R}_{\ell_1}\ldots\vec{R}_{\ell_{n-1}})$$

$$+\int d^3R_{\ell_{n+1}}t(\vec{R}_{\ell_1})G_o\bar{Q}(\vec{R}_{\ell_{n+1}};\vec{R}_{\ell_1}\ldots\vec{R}_{\ell_n})g_{n+1}(\vec{R}_{\ell_1}\ldots\vec{R}_{\ell_n};\vec{R}_{\ell_{n+1}})$$

Eq. (68) shows quantitatively what we stated already qualitatively in the first section of this chapter: On the average, the total scattered wave corresponding to an atom of a cluster with fixed atom positions consists of a superposition of

(a) the scattered wave of the atom in the absence of the system, described by $t(\vec{R}_{\ell_1})$,

(b) waves, scattered at the atom at $\vec{R}_{\ell_1}$ when scattered before at one of the other atoms of the cluster,

(c) waves, scattered at the atom at $\vec{R}_{\ell_1}$ when scattered before at one of the atoms of the system, which is not a member of the cluster.

Eqs. (66) and (68) respectively, yield an (in the thermodynamic limit) infinite hierarchy of equations for the scattering matrix, for instance from eq. (66)[22]

$$\bar{Q}(\vec{R}_{\ell_1}) = v(\vec{R}_{\ell_1})$$

$$+ v(\vec{R}_{\ell_1})G_o\bar{Q}(\vec{R}_{\ell_1})$$

$$+ \int_{\ell_2} v(\vec{R}_{\ell_1})G_o\bar{Q}(\vec{R}_{\ell_2};\vec{R}_{\ell_1})g_2(\vec{R}_{\ell_1};\vec{R}_{\ell_2})$$

$$\bar{Q}(\vec{R}_{\ell_2};\vec{R}_{\ell_1}) = v(\vec{R}_{\ell_2})$$

$$+ v(\vec{R}_{\ell_2})G_o\bar{Q}(\vec{R}_{\ell_2};\vec{R}_{\ell_1})$$

$$+ v(\vec{R}_{\ell_2})G_o\bar{Q}(\vec{R}_{\ell_1};\vec{R}_{\ell_2})$$

$$+ \int_{\ell_3} v(\vec{R}_{\ell_2})G_o\bar{Q}(\vec{R}_{\ell_3};\vec{R}_{\ell_1}\vec{R}_{\ell_2})g_3(\vec{R}_{\ell_1}\vec{R}_{\ell_2};\vec{R}_{\ell_3})$$

(69)

$$\bar{Q}(\vec{R}_{\ell_3};\vec{R}_{\ell_1}\vec{R}_{\ell_2}) = v(\vec{R}_{\ell_3})$$

$$+ v(\vec{R}_{\ell_3})G_o\bar{Q}(\vec{R}_{\ell_3};\vec{R}_{\ell_1}\vec{R}_{\ell_2})$$

$$+ v(\vec{R}_{\ell_3})G_o\bar{Q}(\vec{R}_{\ell_2};\vec{R}_{\ell_1}\vec{R}_{\ell_3})$$

$$+ v(\vec{R}_{\ell_3})G_o\bar{Q}(\vec{R}_{\ell_1};\vec{R}_{\ell_2}\vec{R}_{\ell_3})$$

$$+ \int_{\ell_4} v(\vec{R}_{\ell_3})G_o\bar{Q}(\vec{R}_{\ell_4};\vec{R}_{\ell_1}..\vec{R}_{\ell_3})g_4(\vec{R}_{\ell_1}..\vec{R}_{\ell_3};\vec{R}_{\ell_4})$$

and so forth.

n-Center Approach

Each of these equations contains only a finite number of atom sites explicitly. However, in order to calculate the total averaged scattering matrix one has to solve the total infinite hierarchy, which is of course impossible to do exactly. The usual procedure for calculating $\bar{Q}(\vec{R})$ is therefore to cut the hierarchy at some finite n by some procedure, which consists mainly in a suitable choice of $\bar{Q}(\vec{R}_{\ell_{n+1}};\vec{R}_{\ell_1}...\vec{R}_{\ell_n}) \equiv \bar{Q}_{(n+1)}$. Cutting the hierarchy at some finite n is equivalent to taking into account only multiple scattering processes at atom clusters containing up to n atoms[11].

The simplest possibility of truncating the hierarchy is to assume the scattering contribution of an atom of a cluster $\vec{R}_1...\vec{R}_n$ to not be affected by any scattering from outside the cluster, i.e. neglecting the last term at the right hand side of the n-th of the eqs. (69). Writing the equation for $\bar{Q}_{(n)}$ for every cluster site and summing yields

$$\sum_{j=1}^{n}\bar{Q}(\vec{R}_j;\vec{R}_1...\vec{R}_{j-1}\vec{R}_{j+1}...\vec{R}_n) = \sum_{j=1}^{n} v(\vec{R}_j)$$

$$+\sum_{j=1}^{n} v(\vec{R}_j)G_o\sum_{k=1}^{n}\bar{Q}(\vec{R}_k;\vec{R}_1...\vec{R}_{k-1}\vec{R}_{k+1}...\vec{R}_n)$$

(70)

which is the equation for the scattering matrix of an atom cluster with atoms at $\vec{R}_1 \ldots \vec{R}_n$. Having calculated the scattering contribution of the n atoms in the cluster with fixed atom positions, one obtains the scattering contributions with n-1, n-2, etc. atom positions fixed by successive substitution into the respective equations of the hierarchy.

This truncation of the hierarchy neglects completely the effect of the inter-cluster interaction on the electron spectrum.

The next simple possibility of truncating the hierarchy, which takes into account the interactions with atoms outside the cluster, is to set[11,14,16,24]

$$\bar{Q}(\vec{R}_{\ell_{n+1}}; \vec{R}_{\ell_1} \ldots \vec{R}_{\ell_n}) = \bar{Q}(\vec{R}_{\ell_{n+1}}) \qquad (71)$$

i.e., the scattering contribution of the atom at $\vec{R}_{\ell_{n+1}}$, which is not one of the fixed atom positions $\vec{R}_{\ell_1} \ldots \vec{R}_{\ell_n}$ to be independent of $\vec{R}_{\ell_1} \ldots \vec{R}_{\ell_n}$, the particular cluster configuration. This truncating procedure is a generalization of the quasi-crystalline approximation of Lax[23] to which we shall come back in the following lecture. Assumption (71) yields a closed set of equations

$$\bar{Q}(\vec{R}_{\ell_1}) = v(\vec{R}_{\ell_1})$$

$$+ \; v(\vec{R}_{\ell_1}) G_o \bar{Q}(\vec{R}_{\ell_1})$$

$$+ \int u(\vec{R}_{\ell_1}) G_o \bar{Q}(\vec{R}_{\ell_2}; \vec{R}_{\ell_1}) g_2(\vec{R}_{\ell_1}; \vec{R}_{\ell_2}) d^3 R_{\ell_2}$$

$$\vdots \qquad\qquad (72)$$

$$\bar{Q}(\vec{R}_{\ell_n}; \vec{R}_{\ell_1} \ldots \vec{R}_{\ell_{n-1}}) = v(\vec{R}_{\ell_n})$$

$$+ \; v(\vec{R}_{\ell_n}) G_o \bar{Q}(\vec{R}_{\ell_n}; \vec{R}_{\ell_1} \ldots \vec{R}_{\ell_{n-1}})$$

$$\vdots$$

$$v(\vec{R}_{\ell_n}) G_o \bar{Q}(\vec{R}_{\ell_1}; \vec{R}_{\ell_2} \ldots \vec{R}_{\ell_n})$$

$$+ \int v(\vec{R}_{\ell_n}) G_o \bar{Q}(\vec{R}_{\ell_{n+1}}) g_{n+1}(\vec{R}_{\ell_1} .. \vec{R}_{\ell_n}; \vec{R}_{\ell_{n+1}}) d^3R_{\ell_{n+1}}$$

Subsequent substitution of the first $(n+1)$ equations (72) into the last one yields an equation containing only $\bar{Q}(R_{\ell_n}; R_{\ell_1} \cdots R_{\ell_{n-1}}) = \bar{Q}_{(n)}$ which can be solved in principle.

Having evaluated $\bar{Q}_{(n)}$, the remaining scattering contributions with $(n-1)$, $(n-2) \ldots 1$ atoms fixed can be obtained by inserting $\bar{Q}_{(n)}$ into the equation for $\bar{Q}_{(n-1)}$, $\bar{Q}_{(n-1)}$ into the equation for $\bar{Q}_{(n-2)}$, etc.

The various n-center approaches are expected, of course, to become exact for $n \to \infty$ (in the thermodynamic limit). However, in particular the set of equations (72), in spite of being finite, is still very complicated and therefore far from being applicable to any real system. There is an additional difficulty arising from the fact that the conditional probability distributions of higher order than two are not known experimentally. Therfore, one has to construct model probability distributions which is a difficult task in itself[16].

II-3. Self-Consistent Effective Medium Theory

The neglecting of the higher order scattering processes creates some mistakes, which cannot be estimated quantitatively, but which should be decreasing with increasing number of cluster atoms. Usually, it is believed, that this neglecting of multiple scattering terms can be partially counter-balanced by introducing an "effective medium" $\hat{H}(Z)$, which is configuration independent, by the following procedure[25]:

$$H = \hat{H} + H - \hat{H} =: \hat{H}(Z) + \hat{V}(Z)_{\vec{R}_1 \ldots \vec{R}_N} \tag{73}$$

Then

$$\hat{G}(Z) = \frac{1}{Z - \hat{H}(Z)} \tag{74}$$

is the resolvent corresponding to the electron moving in the effective medium. It is reasonable to choose $\hat{H}(Z)$ appropriately, in order to provide that on the one hand $\hat{G}$ takes a simple form and on the other hand $V_{R_1 \cdots R_N}$ can

again be written as a superposition of atom potentials

$$\hat{V}_{\vec{R}_1 \ldots \vec{R}_N} = \sum_{i=1}^{N} \hat{v}(\vec{R}_i) \tag{75}$$

Taking $\hat{H}$ as the "unperturbed" part of the total Hamiltonian we have in analogy to the equations (45) and (49)

$$\begin{aligned} G &= \hat{G} + \hat{G}\,\hat{V}\,G \\ &= \hat{G} + \hat{G}\,\hat{T}\,\hat{G} \end{aligned} \tag{76}$$

where $\hat{T}$ means the scattering matrix relative to the effective medium,

$$\hat{T}_{\vec{R}_1 \ldots \vec{R}_N} = \hat{V}_{\vec{R}_1 \ldots \vec{R}_N} + \hat{V}_{\vec{R}_1 \ldots \vec{R}_N} \hat{G}\, \hat{T}_{\vec{R}_1 \ldots \vec{R}_N} \tag{77}$$

and

$$\hat{Q}_i = \hat{v}(\vec{R}_i) + \hat{v}(\vec{R}_i) G_o \sum_{j=1}^{N} \hat{Q}_j; \quad i=1\ldots N \tag{78}$$

are the partial scattering contributions.

Now the configurational averaging process and the truncating of the hierarchy can be carried out analogously as in the previous sections of this chapter. It seems, that we gained nothing by introducing the effective Hamiltonian but the difficulty of choosing $\hat{H}$ appropriately. However, taking $\hat{H}$ to fulfill the condition

$$\langle \hat{T}\,(z)_{\vec{R}_1 \ldots \vec{R}_N} \rangle = \sum_{j=1}^{N} \langle \hat{Q}_j(z) \rangle = 0 \tag{79}$$

we obtain

$$\langle G(z) \rangle \equiv g(z) = \hat{G}(z) \tag{80}$$

i.e., $\hat{H}$ has the meaning of a self energy operator. Condition (79) means, that on the average the electron is assumed not to be scattered from the system relative to the effective medium.

In the case of a one-center approach (single-site approximation) this concept of a self-consistently determined effective Hamiltonian has been discussed extensively in the literature[10,15,20,21,22,25], and we shall come back to this in the subsequent lectures.

III. SINGLE SITE APPROXIMATION

III-1. Single-Site Decoupling

A particular simple, but nontrivial decoupling of the hierarchy of equations for the scattering contribution of a single atom, is obtained by taking n = 1 in eq. (71)

$$\bar{Q}(\vec{R}_{\ell_2};\vec{R}_{\ell_1}) = \bar{Q}(\vec{R}_{\ell_2}) \tag{81}$$

This is the single-site approximation[15,26] because it takes into account only multiple scattering at one atom. This decoupling procedure is equivalent to assuming the n-center probability distribution can be factored into a <u>product of two-center</u> probability distributions.

In the single-site approximation the closed system of eq. (72) reduces to only one equation

$$\bar{Q}(\vec{R}) = v(\vec{R}) + v(\vec{R})G_o \int d^3R' \bar{Q}(\vec{R}')D_2(\vec{R};\vec{R}') \tag{82}$$

where we defined the two-atom correlation function

$$D_2(\vec{R};\vec{R}') = g_2(\vec{R};\vec{R}') + \delta(\vec{R}-\vec{R}') \tag{83}$$

Carrying out the R'-integral for the δ-function part of D_2 we can reformulate eq. (82) in terms of the atomic scattering matrix. We obtain

$$\bar{Q}(\vec{R}) = t(\vec{R}) + t(\vec{R})G_o \int d^3R' \ \bar{Q}(\vec{R}') \ g_2(\vec{R};\vec{R}') \tag{84}$$

Having obtained $\bar{Q}(\vec{R})$ by solving eqs. (82) or (84) respectively, we calculate $<Q(z)>$ by

$$<Q(z)> = \int P_1(\vec{R}) \ \bar{Q}(\vec{R}) \ d^3R = \frac{1}{\Omega} \int \bar{Q}(\vec{R}) \ d^3R$$

$$= \frac{1}{\Omega} \ \bar{Q}(z) \tag{85}$$

and

$$g(z) = G_o(z) + G_o(z) \frac{N}{\Omega} \bar{Q}(z) \ G_o(z) \tag{86}$$

III-2.   Average t-Matrix Approximation

Consider an assembly of strongly scattering atoms and take $\hat{H}(z) = H_o$, the free electron Hamiltonian. Taking the matrix element of eq. (84) in momentum representation we obtain,

$$\bar{Q}_{\vec{k}\vec{k}'}(\vec{R}) = t_{\vec{k}\vec{k}'} \; e^{-i(\vec{k}-\vec{k}')\vec{R}} \tag{87}$$

$$+ \sum_{\vec{k}''} e^{-i(\vec{k}-\vec{k}'')} t_{\vec{k}\vec{k}''} \; G_o(\vec{k}'') \int d^3R' \, \bar{Q}_{\vec{k}''\vec{k}'}(\vec{R}') g_2(\vec{R};\vec{R}')$$

with

$$G_o(\vec{k}'') = \langle \vec{k}'' | G_o | \vec{k}'' \rangle \tag{88}$$

and

$$t_{\vec{k}\vec{k}'} e^{-i(\vec{k}-\vec{k}')R} = \langle \vec{k} | t(\vec{R}) | \vec{k}' \rangle \tag{89}$$

The decomposition of the matrix element of $t(\vec{R})$ into the position independent part $t_{\vec{k}\vec{k}'}$ and the phase factor $\exp\{-i(\vec{k}-\vec{k}')\vec{R}\}$ is obtained readily from eq. (56). Iteration of eq. (87) yields the analogous decomposition for $\bar{Q}(\vec{R})$

$$\bar{Q}_{\vec{k}\vec{k}'}(\vec{R}) = \bar{Q}_{\vec{k}\vec{k}'} \; e^{-i(\vec{k}-\vec{k}')\vec{R}} \tag{90}$$

Hence

$$\bar{Q}_{\vec{k}\vec{k}'} = t_{\vec{k}\vec{k}'} + \sum_{\vec{k}''} t_{\vec{k}\vec{k}''} \; G_o(\vec{k}'') \; \bar{Q}_{\vec{k}''\vec{k}'} h_2(\vec{k}''-\vec{k}') \tag{91}$$

with

$$h_2(\vec{q}) = \int d^3x \; e^{-i\vec{q}\vec{x}} \; g_2(\vec{x}) \tag{92}$$

the Fourier transformed two-atom distribution. For atomic scattering matrices, which can be factored as

$$t_{\vec{k}\vec{k}'} \sim \sum_{\mu} \phi_\mu^*(\vec{k}) \, \chi_\mu(\vec{k}') \tag{93}$$

eq. (91) can be solved exactly by iteration. A factorization as in eq. (93) is provided, for instance, by assuming the system to consist of an assembly of muffin-tin scatterers[27,28].

As an example, let us discuss the case of a completely random liquid.

$$g_2(\vec{x}) = \frac{N}{\Omega} = d \tag{94}$$

$$h_2(\vec{q}) = N \cdot \delta_{\vec{q},0} \tag{95}$$

Then from eq. (91), we have for the diagonal part
of $\bar{Q}$

$$\bar{Q}_{\vec{k}\vec{k}} = t_{\vec{k}\vec{k}} \frac{1}{1 - NG_o(\vec{k})t_{\vec{k}\vec{k}}} \tag{96}$$

Because

$$<Q_{\vec{k}\vec{k}'}(z)> = \frac{1}{\Omega} \int e^{-i(\vec{k}-\vec{k}')\vec{R}} \bar{Q}_{\vec{k}\vec{k}'}(z) \tag{97}$$

$$= \delta_{\vec{k}\vec{k}'} \bar{Q}_{\vec{k}\vec{k}}(z)$$

and

$$G_o(\vec{k}) = \frac{1}{z - k^2} \tag{98}$$

we have for the diagonal part of the averaged Green's
function

$$g(\vec{k};z) = \frac{1}{z-k^2 - Nt_{\vec{k}\vec{k}}(z)} \tag{99}$$

$t_{\vec{k}\vec{k}}(z)$ is the forward scattering amplitude of the
atom potential. This result is the average t-matrix approximation[29,30]. The poles of the averaged electron
Green's function are given by the equation

$$z - k^2 - Nt_{\vec{k}\vec{k}}(z) = 0 \tag{100}$$

Hence, $t_{\vec{k}\vec{k}}(z)$ plays the role of a self-energy giving
the corrections to the unperturbed electron energy
$E(\vec{k}) = k^2$. Eq. (91) is the generalization of the averaged
t-matrix approximation to the case of a system where the
correlation between the scatterers are taken into account

by the two-atom distribution function.

III-3. The Coherent Potential Approximation[22]

Self-consistent, Effective Medium

We have already introduced the concept of an "effective medium", which can be used to counter-balance the errors resulting from neglecting the multiple scattering processes at atom clusters (truncation of the hierarchy of equations for the averaged scattering operator). We shall now discuss the effect of the "effective medium" in the case of the single-site approximation.

We start from the one-electron Hamiltonian in momentum representation (eqs. (8) and (9))

$$H = \sum_{\vec{k}} k^2 \, |\vec{k}\rangle\langle\vec{k}| + \sum_{\vec{k}\vec{k}'} W_{\vec{k}\vec{k}'} \, |\vec{k}\rangle\langle\vec{k}'| \tag{101}$$

and assume the Hamiltonian of the "effective medium" to be of the form

$$\hat{H}(z) = \sum_{\vec{k}} \left(k^2 + \frac{N}{\Omega} M(\vec{k};z)\right) \, |\vec{k}\rangle\langle\vec{k}| \tag{102}$$

Using $\hat{H}(z)$ we write the electron Hamiltonian in the form

$$H = \sum_{\vec{k}} \left(k^2 + \frac{N}{\Omega} M(\vec{k};z)\right)|\vec{k}\rangle\langle\vec{k}| + \sum_{\vec{k}\vec{k}'} \sum_{j=1} e^{-i(\vec{k}-\vec{k}')\vec{R}_j} \hat{v}_{\vec{k}\vec{k}'}(z) \tag{103}$$

where

$$\hat{v}_{\vec{k}\vec{k}'}(z) = v_{\vec{k}\vec{k}'} - \frac{1}{\Omega} M(\vec{k};z)\delta_{\vec{k}\vec{k}'} \tag{104}$$

represents the potential of an atom relative to the "effective medium".

We use the diagonal part of H as the "unperturbed" Hamiltonian yielding for the "unperturbed" electron propagator

$$\hat{G}_{\vec{k}\vec{k}'}(z) = \frac{1}{z-k^2-dM(\vec{k};z)} \, \delta_{\vec{k}\vec{k}'} = \hat{G}(\vec{k};z) \, \delta_{\vec{k}\vec{k}'} \tag{105}$$

Taking now eq. (82) in momentum representation, yields

$$\bar{Q}_{\vec{k}\vec{k}'} = \hat{v}_{\vec{k}\vec{k}'} + \sum_{\vec{k}''} \hat{v}_{\vec{k}\vec{k}''} \hat{G}(\vec{k}'';z) \bar{Q}_{\vec{k}''\vec{k}'}(z) H_2(\vec{k}''-\vec{k}') \tag{106}$$

where

$$H_2(\vec{q}) = \int e^{-i\vec{q}\vec{x}} D_2(\vec{x}) d^3x \tag{107}$$

Because of homogeneity, the averaged partial scattering operator is diagonal in momentum representation. It depends on the function M

$$<Q_{\vec{k}\vec{k}}> \equiv <Q_{\vec{k}} (\{M\})> \tag{108}$$

The averaged one-electron Green's function depends on the choice of M also. Therefore, we can use M as an adjustable parameter to improve the single-site approximation by a suitable choice of the zeroth order Hamiltonian.

As we have learned in the preceding lecture, a more favorable choice of M is provided by assuming no scattering on the average from the system relative to the effective medium, i.e., $<T> = 0$, which means in single-site approximation

$$<Q_{\vec{k}}(\{M\})> \equiv \bar{Q}_{\vec{k}}(\{M\}) = 0 \tag{109}$$

This yields for the averaged electron Green's function

$$g(\vec{k};z) = \hat{G}(\vec{k};z) \tag{110}$$

and $dM(\vec{k};z)$ is the self-energy of the electron giving the corrections to the unperturbed electron energies.

Equation for the Self-Energy

We shall calculate now the self-energy from the self-consistency condition (109). First, we reintroduce the atom potential into eq. (106); this yields

$$S_{\vec{k}\vec{k}'} = v_{\vec{k}\vec{k}'} - \frac{1}{\Omega} M(\vec{k};z)\delta_{\vec{k}\vec{k}'}$$

$$+ \sum_{\vec{k}''} v_{\vec{k}\vec{k}''} \frac{H_2(\vec{k}''-\vec{k}')}{z-k''^2 - dM(\vec{k}'';z)(1 - \frac{1}{N} H_2(\vec{k}''-\vec{k}'))} S_{\vec{k}''\vec{k}'} \tag{111}$$

with

$$S_{\vec{k}\vec{k}'} = \bar{Q}_{\vec{k}\vec{k}'}(1 + \frac{1}{\Omega} M(\vec{k}) \hat{G}(\vec{k}) H_2(\vec{k}-\vec{k}')) \tag{112}$$

Eq. (111) can be reformulated into

$$R_{\vec{k}\vec{k}'} = v_{\vec{k}\vec{k}'} + \sum_{\vec{k}''} v_{\vec{k}\vec{k}''} G_{eff}(\vec{k}'',\vec{k}') H_2(\vec{k}''-\vec{k}') R_{\vec{k}''\vec{k}'} \tag{113}$$

where

$$R_{\vec{k}\vec{k}'}(1 - G_{eff}(\vec{k}',\vec{k}')M(\vec{k}')) = S_{\vec{k}\vec{k}'} + M(\vec{k}')\delta_{\vec{k}\vec{k}'} \tag{114}$$

and

$$G_{eff}(\vec{k}'',\vec{k}') = \frac{1}{z - k''^2 - dM(\vec{k}'')(1 - \frac{1}{N} H_2(\vec{k}''-\vec{k}'))} \tag{115}$$

Application of the self-consistency condition eq. (109) to eqs. (112) and (114) yields for the self-energy

$$\frac{1}{\Omega} M(\vec{k}) = \frac{R_{\vec{k}\vec{k}}}{1 + G_{eff}(\vec{k},\vec{k})H_2(0)R_{\vec{k}\vec{k}}} \tag{116}$$

This can be further simplified. Since

$$H_2(\vec{q}) = (N-1) \delta_{\vec{q},o} + a_2(\vec{q}) \tag{117}$$

we can rewrite eq. (113) into an integral equation for an operator $\sigma_{\vec{k}\vec{k}'}$

$$\sigma_{\vec{k}\vec{k}'} = v_{\vec{k}\vec{k}'} + \sum_{\vec{k}''\neq\vec{k}'} v_{\vec{k}\vec{k}''} \ G_{eff}(\vec{k}'',\vec{k}')a_2(\vec{k}''-\vec{k}')\sigma_{\vec{k}''\vec{k}'} \tag{118}$$

the diagonal part of which yields the self-energy

$$M(\vec{k};z) = \Omega \ \sigma_{\vec{k}\vec{k}}(z) \tag{119}$$

The final result for the self-energy is, of course, finite in the thermodynamic limit, because the plane wave matrix elements in eq. (118) have to be calculated with plane waves normalized to the volume Ω, i.e.,

$$v_{\vec{k}\vec{k}'} = \frac{1}{\Omega} \int d^3r \ e^{-(\vec{k}-\vec{k}')\vec{r}} v(\vec{r}) = \frac{1}{\Omega} \ v(\vec{k}-\vec{k}') \tag{120}$$

$v(\vec{k}-\vec{k}')$ is the usual atomic form factor.

III-4. Examples

Complete Order

In the case of complete order, we can take

$$H_2(\vec{q}) = N \sum_m \delta_{\vec{q},\vec{K}_m} \tag{121}$$

($\vec{K}_m$ reciprocal lattice vector)

and

$$a_2(\vec{q}) = N \sum_{m\neq o} \delta_{\vec{q},\vec{K}_m} \tag{122}$$

The equation for the self-energy then reads

$$\sigma_{\vec{k}\vec{k}'} = v_{\vec{k}\vec{k}'} + \sum_{m\neq o} v_{\vec{k}\vec{k}'+\vec{K}_m} \frac{N}{z-(\vec{k}'+\vec{K}_m)^2} \ \sigma_{\vec{k}'+\vec{K}_m,\vec{k}'} \tag{123}$$

This is exactly the equation for the self-energy of a crystalline electron. Note that at the right-hand side of eq. (123), instead of the renormalized electron propagator $G_{eff}(z)$, we have now the free electron propagator $(z -(\vec{k}' + \vec{K}_m)^2)^{-1}$, because $1 - \frac{1}{N} H_2(\vec{k}''-\vec{k}') = 0$ for

$\vec{k}''-\vec{k}' = \vec{K}_m \neq 0$. This complete <u>cancellation of the ef-fective medium</u> in the crystalline case is due to the fact that the multiple scattering processes at atom clusters neglected in the single-site approximation are reintro-duced by the particular form of the two-atom correlation function

$$D_2(\vec{x}) = \sum_{\ell} \delta(\vec{x}-\vec{\ell}) \tag{124}$$

($\vec{\ell}$ lattice points)

Complete Disorder

In the case of complete disorder, the two-atom cor-relation function is

$$D_2(\vec{x}) = \delta(\vec{x}) + \frac{N-1}{.2} \tag{125}$$

Hence

$$H_2(\vec{q}) = 1 + (N-1)\delta_{\vec{q},0} \tag{126}$$

Thus we obtain for the self-energy[20]

$$\sigma_{\vec{k}\vec{k}'} = v_{\vec{k}\vec{k}'} + \sum_{\vec{k}''\neq\vec{k}'} v_{\vec{k}\vec{k}''} \frac{1}{z-k''^2-N\sigma_{\vec{k}''\vec{k}''}} \sigma_{\vec{k}''\vec{k}'} \tag{127}$$

Note the <u>complete renormalization of the electron propagator</u> at the right-hand side of eq. (127) from $1 - \frac{1}{N} H_2(\vec{k}''-\vec{k}') = 1$ for $N \gg 1$ and $\vec{k}'' \neq \vec{k}'$. The occur-rence of the averaged electron Green's function in eq. (127) emphasizes, that on the average between two scat-tering processes, the electron is moving in the effect-ive medium.

Eq. (127) is a nonlinear integral equation and can-not be solved simply by iteration in general. However, taking as a zeroth approximation

$$\sigma^{(0)}(z) = 0 \tag{128}$$

$\sigma^{(1)}(z)$ fulfills the equation for the atomic scattering matrix

$$\sigma_{\vec{k}\vec{k}'}^{(1)} = v_{\vec{k}\vec{k}'} + \sum_{k''} v_{\vec{k}\vec{k}''} \; G_o(\vec{k}'') \; \sigma_{\vec{k}''\vec{k}'}^{(1)} \tag{129}$$

Therefore, eq. (127) together with eq. (110) can be considered to be the generalization of the averaged t-matrix approximation eq. (99).

Short Range Order

Returning now to the general eq. (118), we first state that for a system with a <u>finite region of short range order</u> $a_2(\vec{k}''-\vec{k}')$ is finite for all values of $(\vec{k}''-\vec{k}')$ in the thermodynamic limit. Therefore,

$$1 - \frac{1}{N} H_2(\vec{q}) = \begin{cases} 0 & q=o \\ 1 & q\neq o \end{cases} \quad N \gg 1 \tag{130}$$

and

$$\sigma_{\vec{k}\vec{k}'} = v_{\vec{k}\vec{k}'} + \sum_{\vec{k}''\neq \vec{k}'} v_{\vec{k}\vec{k}''} \; \frac{a_2(\vec{k}''-\vec{k}')}{z-k''^2-dM(\vec{k}'')} \; \sigma_{\vec{k}''\vec{k}'} \tag{131}$$

Note the <u>complete renormalization</u> of the electron propagator by the effective medium. The only difference in eq. (127) for the completely random distribution, is the appearance of the structure function $a_2(\vec{k}''-\vec{k}')$ describing the correlation between two subsequent scatterers. It is important to note that after having performed the thermodynamic limit providing a <u>finite region of short range order</u>, it is impossible to obtain the crystalline result eq. (123) simply by taking $a_2(\vec{q}) = N \sum_{m\neq o} \delta_{\vec{q},\vec{K}_m}$.

Therefore, because a reasonable approximation for the electron spectrum of a structurally disordered system should allow for the interpolation between the two limiting cases of complete order <u>and</u> complete disorder, eq. (131) cannot be considered to be correct. The reason why we cannot obtain the crystalline limit from eq. (131) is the occurrence of the full "effective medium" at the right-hand side. We know the crystalline case eq. (123) to be exact. On the other hand, we know that in deriving eq. (131), which contains of course eq. (127), the case of complete disorder, we have neglected multiple scattering at atom clusters. Therefore, we must conclude that the <u>complete renormalization</u> of the electron propagator in the equation for the self-energy is partly <u>an</u> artifact

of the truncating of the hierarchy of the equations for
the scattering matrix. This means, that for a system
with a _finite_ short range order region in the thermodynamic limit the self-consistently determined effective
medium will be over-estimated not only in the single-
site approximation, but also in the n-center approximation with any finite n. The results obtained from any
finite-cluster approximation can only be considered to
be qualitatively correct. The error can be expected to
be large in energy regions, where the density of states
is _zero_ in the crystalline case because clustering effects should play an important role in determining the
density of states in these "gap-regions". The error can
be expected to be negligible in energy regions, where
the density of states is _large_ in the crystalline case.
One can assume that the error decreases with increasing
number of scattering centers, n, treated correctly by
the approach. Having these restrictions in mind, we
shall give some examples of methods for the calculation
of the density of states which can be used at least for
the investigation of the effect of the lack of long range
order.

IV. PRACTICABLE METHODS

IV-1. Miscellaneous Approximations for Practical Use

The Muffin-Tin Model

Assuming the system to consist of non-overlapping
muffin-tin potentials (Fig. 7), each being spherically
symmetric, the equation for the partial scattering contribution (eq. (87)) can be solved. The atomic scattering matrix of a spherically symmetric potential can be
written as[15]

$$t_{\vec{k}\vec{k}'}(E) = \sum_L Y_L^*(\vec{k}')Y_L(\vec{k})\ \tau_\ell(K) \qquad (132)$$

where $Y_L(\vec{k}) = Y_{\ell m}(\vec{k})$ are the spherical harmonics of the
angles associated with $\vec{k}$, $K = \sqrt{E}$, and

$$\tau_\ell(K) = -\ \frac{1}{K}\ e^{i\delta_\ell}\ \sin\delta_\ell \qquad (133)$$

δ_ℓ are the phase shifts and can be calculated from the
logarithmic derivatives of the radial wave-functions at

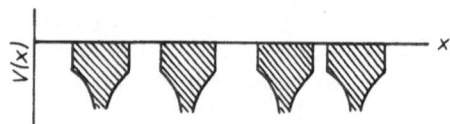

Fig. 7 One-dimensional disordered muffin-tin model.

the muffin-tin radius.

Iterating now eq. (87) and inserting $t_{\vec{k}\vec{k}'}$ of the above eq. (132), we obtain for $\vec{k} = \vec{k}'$

$$\bar{Q}_{kk} = \sum_{LL'} Y_L(\vec{k})\tau_\ell(k) \left[\delta_{LL'} + \tilde{G}_{LL'}(\vec{k})\tau_{\ell'}(k) \right.$$

$$+ \sum_{L_1} \tilde{G}_{LL_1}(\vec{k})\tau_{\ell_1}(k)\tilde{G}_{L_1L'}(\vec{k})\tau_{\ell'}(k)$$

$$\left. + \cdots \right] Y^*_{L'}(\vec{k}) \tag{134}$$

where we defined

$$\tilde{G}_{LL'}(\vec{k}) = \sum_q Y^*_L(\vec{q})G_o(\vec{q})\, h_2(\vec{q}-\vec{k})Y_{L'}(\vec{q}) \tag{135}$$

For the crystalline case, $g_2(\vec{x}) = \sum_{\ell\neq o} \delta(\vec{x}-\vec{\ell})$, $\tilde{G}_{LL'}(\vec{k};z)$ are the usual structure factors, well known from KKR-band structure theory. Eq. (134) represents a geometric series for the atomic scattering contribution and can be summed formally, yielding

$$\bar{Q}_{\vec{k}\vec{k}} = \sum_{LL'} Y_L(\vec{k})\tau_\ell(k)\{1 - \tilde{G}(\vec{k})\tau(k)\}^{-1}_{LL'} Y^*_{L'}(\vec{k}) \tag{136}$$

Inserting this into the equation for the averaged Green's function yields for the singularities of $G(\vec{k};z)$

$$\det \left| \left| \tilde{G}_{LL'}(\vec{k},E) - \tau_\ell^{-1}(k)\delta_{LL'} \right| \right| = 0 \tag{137}$$

In the crystal, the solution of eq. (137) yields the electron band structure $E_m(\vec{k})$. For a disordered system, where g_2 is no longer a sum of δ-functions the solutions of eq. (137) are no longer real. If z is chosen to be real, the secular determinant (137) yields complex wave numbers $k_0(E)$. If the imaginary part of k_0 is sufficiently small (weak incoherent scattering) the associated spectral distribution, $ImG(k,E)$, will be centered at Rek_0 and will have a width of the order of $2Imk_0$, (Fig. 8)[15],[17].

Self-Consistent Muffin-Tin Model

It is possible to generalize the above KKR-equations to be self-consistent in the CPA-sense. However, in order to preserve the formal simplicity of the secular equation (137) one has to assume the self-energy to be k-independent.

$$M(\vec{k};E) = M_c(E) \tag{138}$$

Using M_c, which has to be determined self-consistently from the actual CPA condition

$$\bar{Q}_{kk}(E - M_c(E)) = 0 \tag{139}$$

the complex poles of the averaged Green's function are situated at

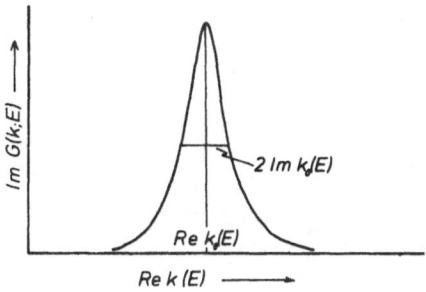

Fig. 8 Qualitative behavior of the imaginary part of the averaged Green's function for given real E as a function of the wave vector. $[Imk_0(E)]^{-1}$ can be interpreted as a coherence length.

$$k_o^2(E) = E - M_c(E) \tag{140}$$

Inspection of eq. (136) shows, that the self-consistency condition eq. (139) yields a rather complicated equation for the effective medium. Moreover, in general it will not be possible to choose M_c independent of the wave vector, because $\tilde{G}$ depends on $\vec{k}$. However, taking into account only the first term in the power series we have

$$\sum_\ell (2\ell+1) \, \tau_\ell \left(\sqrt{E-M_c(E)}\right) = 0 \tag{141}$$

This form of the self-consistency condition was first suggested and applied by Anderson and McMillan[31] and further developed to include higher order corrections for the single-site scattering contribution by Schwartz and Ehrenreich[15].

Generalized Pseudopotential Approach

In crystalline band structure theory great advantage has been taken of the use of "weak" model atom potentials, which are determined empirically by fitting to the gap between the valence and the conduction band. Using the OPW-method, it can be shown, by projecting out the deep core levels, that such "weak" potentials do exist and that an expansion of the electron wave function into plane waves yields a reasonable convergence of the respective secular equation for the band structure, $E_m(\vec{k})$, which is of the form:

$$\det \left|\left|(E - (\vec{k}+\vec{K}_m)^2)\delta_{mm'} - v_{mm'}\right|\right| = 0 \tag{142}$$

with the reciprocal lattice vectors $\vec{K}_m$ and $v_{mm'} = \langle\vec{K}_m|v|\vec{K}_{m'}\rangle$ the plane wave matrix elements of the atomic model potential. Regarding $\Omega v_{mm'} = v(|\vec{K}_m-\vec{K}_{m'}|)$ to be a continuous function of its argument it can be argued that it is slowly varying in q-space (Fig. 9(a)).

It is well known from experimental structure investigation, that in disordered semiconductors the short range order of the atomic configuration is almost the same as in the respective crystalline modification. Therefore, it is reasonable to assume a structure function

(Fig. 9(b))

$$a_2(\vec{q}) = \sum_{m \neq o} f_m(\vec{q} - \vec{K}_m)$$ (143)

where $f_m(\vec{q} - \vec{K}_m)$ are considered to be localized near $\vec{K}_m$ with a certain width λ_m. The width of the localization is a measure of the degree of the disorder. $\lambda_m \to 0$ for all m represents the case of complete order. λ_m as a

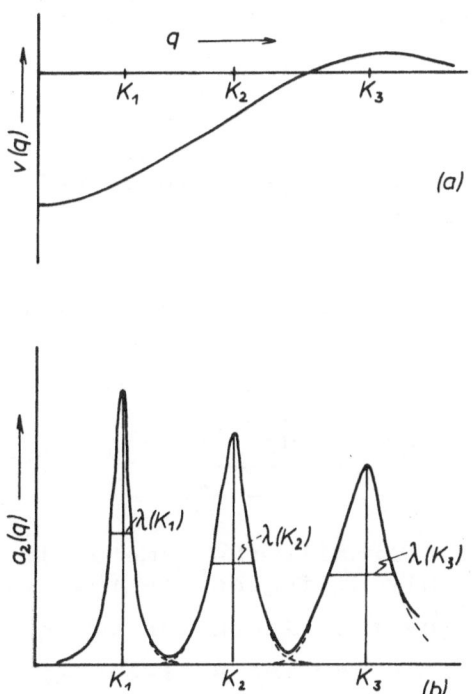

Fig. 9 (a) Schematic picture of the atom form factor as a function of q. (b) Shape of the structure function a_2 exhibiting typical short-range order peaks, K_1, K_2, K_3 ....$\lambda(K_i)$ is a measure of the degree of disorder.

function of $\vec{K}_m$ has to be chosen to guarantee $a_2(\vec{q}) \to 1$; $q \to \infty$.

Keeping in mind these two properties concerning the atom potential and the structure function we can find a first approximation to the self-energy from eq. (131) by taking in the zeroth approximation $M(\vec{k};z) = 0$. We obtain in the thermodynamic limit for $M(\vec{k}\vec{k}') = \Omega \sigma_{\vec{k}\vec{k}'}$

$$M(\vec{k}\vec{k}') = v(\vec{k}-\vec{k}') + \int d^3q \; v(\vec{k}-\vec{q}) \; \frac{a_2(\vec{q}-\vec{k}')}{z - q^2} \; M(\vec{q},\vec{k}')$$

$$(144)$$

Introducing the structure function of eq. (143) we have

$$M(\vec{k},\vec{k}') = v(\vec{k}-\vec{k}') + \sum_{m \neq o} \int d^3q \; v(\vec{k}-\vec{q}) \; \frac{f_m(\vec{q}-\vec{k}'-\vec{K}_m)}{z - q^2} \; M(\vec{q},\vec{k}')$$

$$(145)$$

Recalling that $v(\vec{k}-\vec{q})$ is assumed as slowly varying in $\vec{q}$-space and taking $f_m(\vec{q}-\vec{K}_m)$ to be well-localized near $\vec{K}_m$ we have approximately

$$M(\vec{k},\vec{k}') = v(\vec{k}-\vec{k}') + \sum_{m \neq o} v(\vec{k}-\vec{k}'-\vec{K}_m)\phi_m(\vec{k};z) \; M(\vec{k}'+\vec{K}_m,\vec{k}')$$

$$(146)$$

where

$$\phi_m(\vec{k}';z) = \int d^3q \; \frac{f_m(\vec{q}-\vec{k}'-\vec{K}_m)}{z - q^2}$$

$$(147)$$

Eq. (146) is formally the same as in the crystalline case. The only difference is that we have replaced $[z - (\vec{k}'+\vec{K}_m)^2]^{-1}$ by ϕ_m. Iteration of eq. (146) yields a geometric series for the self-energy

$$M(\vec{k};z) = v_{oo} + \sum_{m \neq o} v_{om} \; \phi_m \; v_{mo}$$

$$+ \sum_{m,m' \neq o} v_{om} \; \phi_m \; v_{mm'} \; \phi_{m'} \; v_{m'o} + \ldots$$

$$(148)$$

which can be summed in analogy to the crystalline case. It is straightforward to show that the poles of the averaged Green's function

$$z - k^2 - dM(\vec{k};z) = 0 \qquad (149)$$

are given by the zeros of the determinant[18]

$$det \ || \ \phi_m^{-1}(k;z) \ \delta_{mm'} - v_{mm'} \ || = 0 \qquad (150)$$

For the crystal we have $f_m(\vec{q}-\vec{k}) = \delta(\vec{q}-\vec{k}-\vec{K}_m)$, $\phi_m^{-1} = z - (k+K_m)^2$ and eq. (150) reduces to eq. (142), the secular equation of the crystal. The crystalline band structure $E_m(\vec{k})$ is real for real $\vec{k}$-vectors, because $E-(\vec{k}+\vec{K}_m)^2$ is real. For the disordered phase $\phi_m(\vec{k};E)$ is in general complex for real $\vec{k}$, and real E. Hence, if we take the wave-vector to be real the solutions of eq. (150), will be situated in the complex energy plane. For small degree of disorder ($\lambda_m << \frac{2\pi}{a}$) the real part of the energy, $R_e\varepsilon_m(\vec{k})$, will be approximately the same as in the crystalline case and the imaginary part of the energy, $Im\varepsilon_m(\vec{k})$, can be considered as a $\vec{k}$-dependent, reciprocal lifetime of the respective Bloch state.

Complex Band Structure

As we emphasized already, the averaged density of states is determined by the imaginary part of averaged Green's function. Therefore, in order to obtain a first insight into the behavior of the density of states when relaxing the long-range order it suffices to investigate the behavior of $ImG(\vec{k};E)$, the spectral function. It is useful to rewrite the averaged Green's function into a form more suitable for practical purposes using its poles in the complex energy plane as determined from eq. (150), $\varepsilon_m(\vec{k})$

$$G(k,E) = \sum_m \frac{Res \ G(\vec{k};z)_{z=\varepsilon_m(\vec{k})}}{E - \varepsilon_m(\vec{k})} \qquad (151)$$

For vanishing disorder, i.e., vanishing width of localization, we have

$$G(\vec{k};E)_{crystal} = \sum_m \frac{|\langle\vec{k}|\Psi_{m\vec{k}}\rangle|^2}{E - E_m(\vec{k})} \qquad (152)$$

and the spectral function is

$$\text{ImG}(\vec{k};E)_{crystal} \sim \sum_m |\langle\vec{k}|\Psi_{m\vec{k}}\rangle|^2 \, \delta(E-E_m(\vec{k})) \qquad (153)$$

because $E_m(\vec{k})$ is real.

For small degrees of disorder, i.e. small widths of localization of the Fourier transformed two-atom distribution function, the main effect of the relaxation of the long-range order is the shift of the poles of $G(\vec{k};z)$ into the complex energy plane. Now, the spectral function is approximately a superposition of Lorentzian curves centered near $E_m(\vec{k}) = R_e\varepsilon_m(\vec{k})$ with the width $2\Gamma_m(\vec{k}) := 2\text{Im}\varepsilon_m(\vec{k})$, (Fig. 10).

$$\text{ImG}(\vec{k};E) \sim \sum_m A_m(\vec{k}) \frac{\Gamma_m(\vec{k})}{(E-E_m(\vec{k}))^2 + (\Gamma_m(\vec{k}))^2} \qquad (154)$$

If in a region of the Brillouin zone $\Gamma_m(\vec{k})$ is only small for a given degree of disorder, then the corresponding structure in the density of states is only affected by the lack of long range order. However, if $\Gamma_m(\vec{k})$ is large, the corresponding structures in the density of states vanish. In this sense, the imaginary part

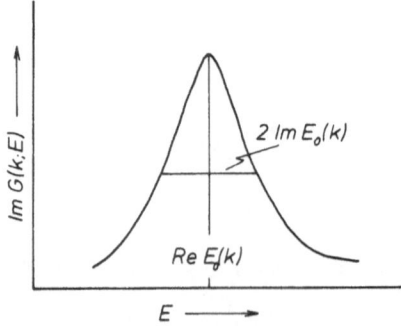

Fig. 10 Behavior of the spectral function for given real k as a function of the energy. $[\text{ImE}_0(k)]^{-1}$ can be interpreted as a lifetime.

of energy $\Gamma_m(\vec{k})$ can be regarded as to represent the disorder sensitivity of a crystalline energy level.

   The qualitative behavior of the energy spectrum and
the density of states can be studied by investigating a
nearly-free electron model with only one potential coefficient different from zero. The results are plotted in
Figs. 11 and 12. We observe the effect of the lack of
long-range order to be large in energy regions where the
free-electron parabolas originating in the various reciprocal lattice points intersect and therefore split in
the crystalline case. As a consequence the respective
singularities in the density of states are removed. These
results are essentially the same as already obtained
by Edwards[32], by application of perturbation theory and
evaluation of the self-energy in powers of the atom potential. The advantage of eqs (146) and (150) compared to
the work of Edwards lies in the fact that they can easily
be applied to realistic amorphous models by a suitable
choice of the atom distribution.

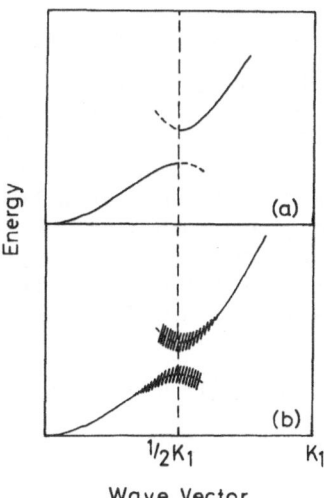

Fig. 11 Complex band structure for the nearly free-electron model. (a) $\alpha=0$: crystalline case (b) $\alpha \neq 0$: disordered case. The imaginary part of the energy is shown
as a broadening of the real part of energy (from Ref. 38).

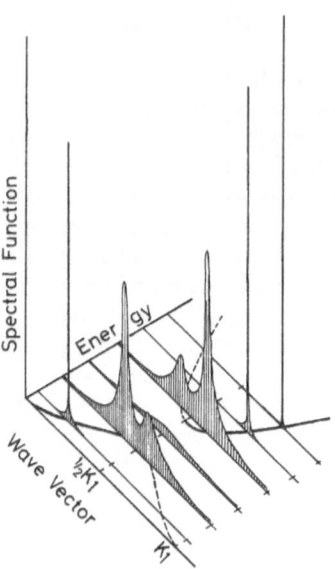

Fig. 12 Spectral function for the nearly free electron
model (from Ref. 38).

Self-Consistent Complex Band Structure Approach

The generalized pseudopotential approach described
in the section above has the disadvantage of not being
self-consistent in CPA sense. The result is a discon-
tinuity in the electron spectrum due to the poles of the
averaged Green's function being complex only for E>0, as
can be seen immediately by inspection of eq. (147) for
the averaged free-electron Green's function. This dis-
continuity can be removed in principle by determining
the self-energy self-consistently from

$$M(\vec{k},\vec{k}') = v(\vec{k}-\vec{k}'') + \sum_{m \neq o} v(\vec{k}-\vec{k}'-\vec{K}_m) \phi_m(\vec{k}';z-M) M(\vec{k}'+\vec{K}_m,\vec{k}')$$

$$(155)$$

with

$$\phi_m(\vec{k}';z-M) = \int d^3q \ \frac{f_m(\vec{q}-\vec{k}-\vec{K}_m)}{z - q^2 - dM(\vec{q},z)}$$

$$(156)$$

Eqs. (155) and (156) result from eq. (131) by using

the same approximating procedure used in obtaining eqs. (146) and (147). However, taking the crystalline limits of eqs. (155) and (156) we observe again, that it is not possible to obtain the correct equation for the crystalline self energy, because of the full renormalization of the free-electron propagator[33].

One could have the idea of removing the shortcoming of the <u>finite</u> short range order region in the <u>thermodynamic limit</u> by assuming[34]

$$\lim_{\Omega \to \infty} \frac{\Omega_s}{\Omega} = C \tag{157}$$

Ω_s = short-range order volume.

The result is only a slightly modified crystalline energy spectrum. However, such a calculation can be used to obtain additional information concerning the <u>degree of renormalization</u> in eq. (156).

Instead of using the full effective medium $M(\vec{k};z)$ one should multiply it by a factor $s(\vec{k},c)$ which depends on the ratio between the short-range order volume Ω_s and the total volume of the system. This can be seen to be reasonable also by inspection of our general single-site result of eqs. (118) and (119).

$$\phi_m(\vec{k}',z-M) = \int d^3q \; \frac{f_m(\vec{q}-\vec{k}'-\vec{K}_m)}{z - q^2 - dS(\vec{q},c)M(\vec{q},z)} \tag{158}$$

The renormalization fraction $S(\vec{k},c)$ removes the <u>overestimation of the renormalization</u> resulting from the <u>underestimation of the cluster multiple scattering</u> effects. If $c \to 1 (\Omega_s \to \Omega$ crystalline case$)$ $S(\vec{k},c)$ should vanish, for $c \to o$ $(\Omega_s \to o$, completely random distribution$)$ $S(\vec{k},c)$ should tend to unity. In practice, $S(\vec{k},c)$ can either be used as an adjustable quantity or it can be estimated by consideration of some simple models[34].

IV-2. Examples

KKR - Examples

In this last chapter we shall discuss some examples of application of the methods described in the preceeding chapter. First, we concentrate on the simplest structurally disordered system, the liquid. The first

self-consistent single-site calculation on a liquid has
been performed by Anderson and McMillan[31] for liquid Fe.
A similar calculation, but with an improved self-consistency condition has been carried out by Schwartz and
Ehrenreich[15] for liquid Cu. The numerical calculation
was based on the fact that the radial wave function (and
therefore their logarithmic derivatives) inside the muffin-tin sphere is not affected by the existence of the
complex medium outside the sphere. The result is shown
in Fig. 13 as the function $E(Rek_o(E))$. The imaginary
part of $k_o(E)$ is shown as a broadening of the curve.
The "complex band structure" exhibits the splitting associated with s-d hybridization induced by a resonant
d-level. This interaction of the s- and d-states leads
to a pronounced minimum in the density of states (Fig.
14) at 0.45 Ryd., because in the isotropic liquid, the
hybridization takes place at the same energy for all directions of $\vec{k}$-space. This is in contrast to the crystalline phase, where the d-bands have structure and the

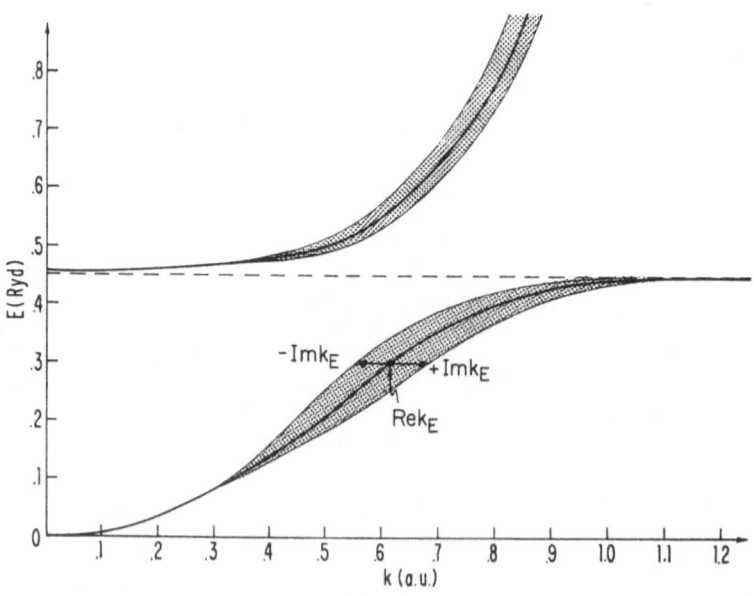

Fig. 13 Complex band structure for a simple model of
liquid Cu. The imaginary part of the wave vector is
shown as a broadening of $E(Re\ k)$ (from Ref. 15).

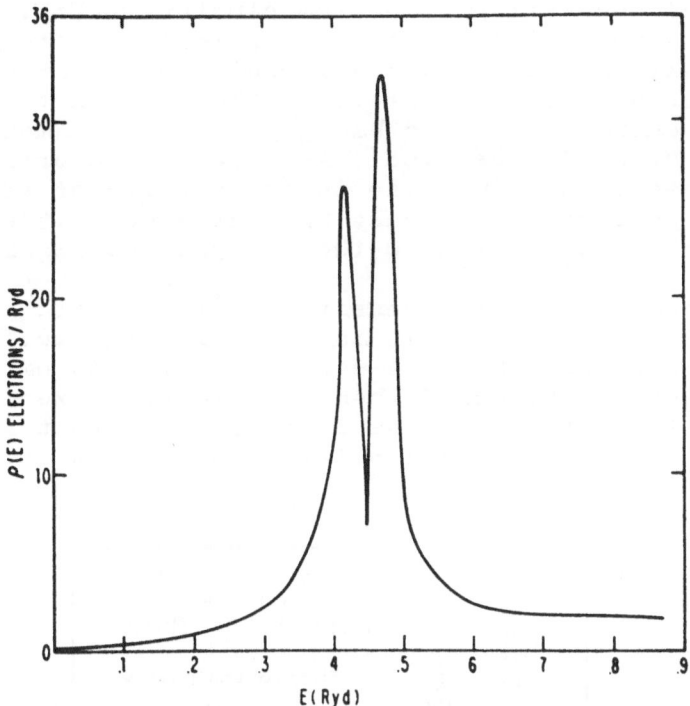

Fig. 14 Density of states of liquid Cu. (from Ref.15).

hybridization takes place at different energies for each $\vec{k}$-direction. In the liquid, the energetic position of the minimum in the density of states coincides with the energetic position of the d-level, which can be determined by investigating the scattering phase shift for $\ell=2$. It passes rapidly through $\pi/2$ at $E_d = 0.45$ Ryd. The most obvious deficiency of the above described results is the complete absence of structure in the d-band except for the hybridization gap. This is due to the complete neglect of correlations between the atom positions in the present model, which would, as we mentioned already, require also a correct treatment of cluster multiple scattering.

Such a treatment of the microscopic atom correlations is provided by the cluster KKR methods as developed and applicated by McGill and Klima[35] and Keller and Ziman[36]. They treated models for amorphous group-IV

materials using clusters with varying number of atoms in
the cluster. The first of these calculations dealt with
an Si-model consisting of 8-atom clusters. Fig. 15 shows
the results for the two characteristically different
clusters with staggered and eclipsed configuration. The
free electron density of states is lowered in the same
region of energy where the bandgap for the diamond cubic
structure of Si is observed. As a second important result we observe the increase in the density of states of
the lower part of the valence band at about 1.5 eV, when
going from the staggered to the eclipsed configuration.

The behavior of the density of states with increasing number of cluster atoms is shown in Fig. 16. It
clearly indicates, that the density of states within the
"energy gap" is decreased when the cluster size is increased. Thus it is reasonable to assume, that at least
part of the gap states are related to surface atoms, the

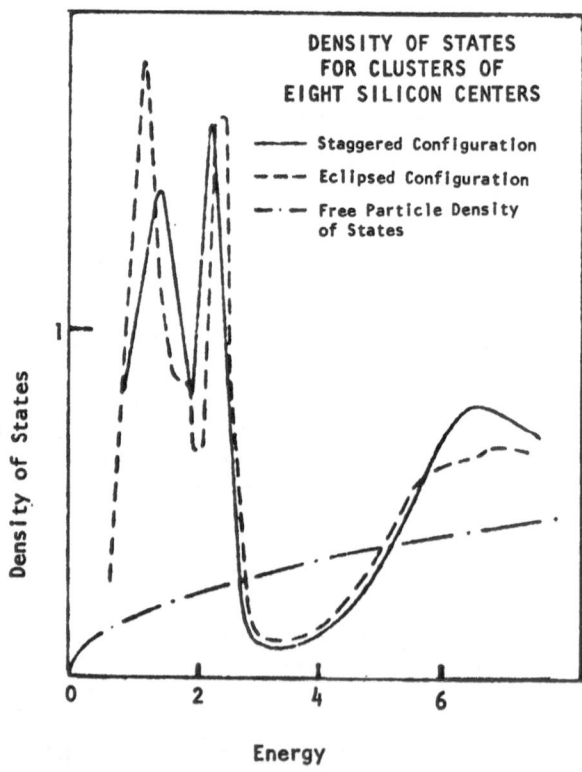

Fig. 15 Density of states for 8-atom clusters of Si and
the free-electron density of states (from Ref. 35).

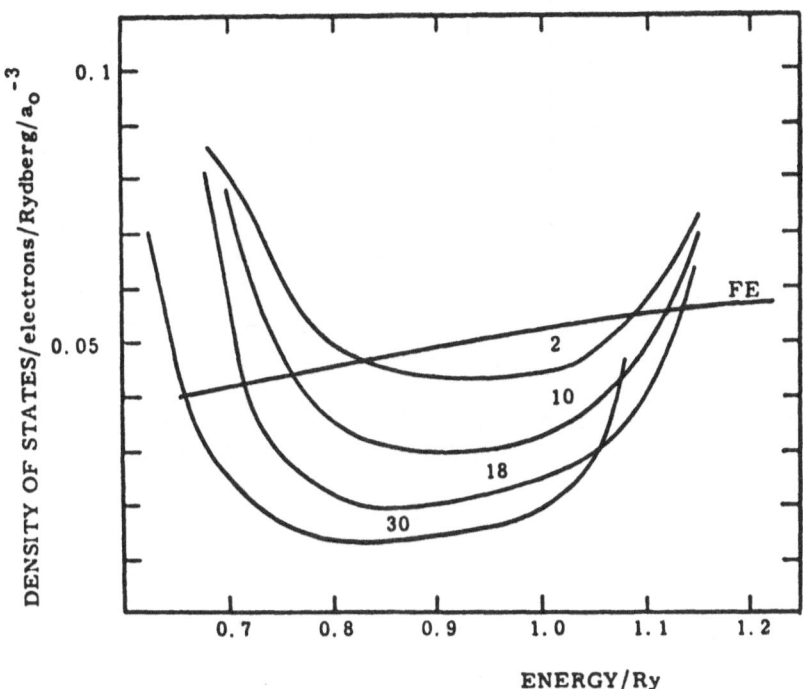

Fig. 16 Density of states in the gap-region for clusters of 2, 10, 18, 30 C-atoms compared to the free-electron density of states (FE) (from Ref.36).

number of which is still considerably high even for the 30 atom cluster.

A self-consistent version of the KKR-cluster method has recently been presented by S. Yoshimo et al[37] . The results of their calculation of the density of states is shown in Fig. 17. It admits again the same tendency as indicated already by the calculation of Keller and Ziman. Interesting are their results on the self-consistent effective medium, which are shown in Fig. 18. The total amount of the effective medium $\sigma = \sigma_1 + i\sigma_2$ decreases with increasing cluster size. This emphasizes quantitatively, what we obtained already qualitatively at the end of part III, namely that the role of the <u>effective medium is overestimated when underestimating the effect of cluster multiple scattering</u>.

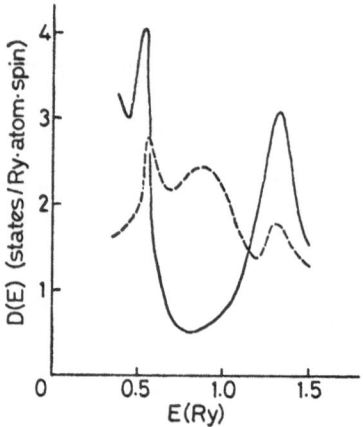

Fig. 17 Density of states of 2-atom (dashed line) and
8-atom (full line) clusters of C. (from Ref. 37).

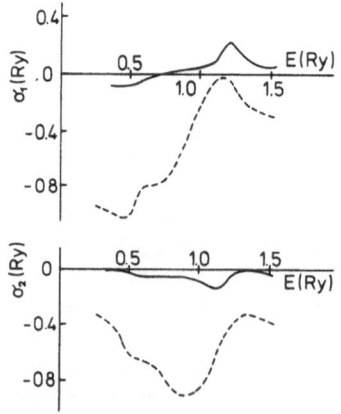

Fig. 18 Effective medium $\sigma = \sigma_1 + i\sigma_2$ for 2-atom
(dashed line) and 8-atom (full line) clusters of C.
(from Ref. 37).

Pseudopotential Examples

The generalized pseudopotential formalism as described in the preceeding chapter has been applied to a great variety of structurally disordered systems, namely the disordered group-IV semiconductors and III-V compounds and disordered Te and Se systems[38]. We shall discuss in this section only the example of Ge.

The application of eq. (150) was based on the assumption that the radial distribution curve of amorphous Ge can be interpreted as indicating about the same short-range order of the disordered Ge phase as the cubic crystal[39]. Hence, the cubic crystal was taken as the model crystalline structure underlying the atom correlation function $a_2(\vec{q})$. The single contributions to a_2 were assumed to be Gaussians and a_2 was taken to represent distributions of "unit cells" consisting of two atoms as in the crystal.

$$a_2(\vec{q}) = \sum_{o \neq m < N_o} \frac{1}{\pi 3/2 \; \alpha^3 K_m^3} \; e^{\frac{(\vec{q}-\vec{K}_m)^2}{\alpha^2 K_m^2}} \qquad (159)$$

$N_o >$ dimension of the determinant.

The widths of the Gaussians were assumed to be proportional to $|\vec{K}_m|$; the proportionality constant, α, being a measure of the degree of disorder. The above Fourier transformed two-atom distribution function does not agree very well with the experimental curve[38] but since we are mainly interested in the effect of the lack of the long-range order on the electronic density of states eq. (159) may serve as a model. The pseudopotential matrix elements in eq. (150) were taken to be the same as in the crystal.

The complex roots of the secular equation (150) are plotted in Fig. 19 versus the real $\vec{k}$-vector for the axis Γ-X and Γ-L of the crystalline Brillouin zone. (The occurrence of the crystalline notation Γ, X, L, should not confuse the reader. Of course, the complex band structure is not periodic in $\vec{k}$-space. However, the whole $\vec{k}$-space does not contribute to the density of states, since the residues of the electron Green's function are non-zero only near the free electron parabola).

The main features of the electron spectrum can be compiled as follows:

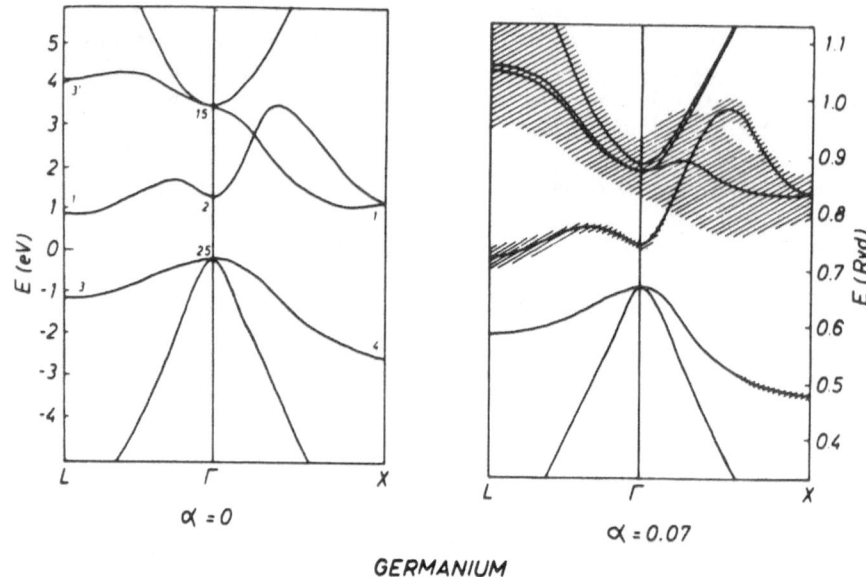

GERMANIUM

Fig. 19 Complex band structure of a disordered Ge model from the generalized pseudopotential method. (From
Ref. 38) α = 0: crystal; α = 0.07: amorphous.

    a. The imaginary parts of energies are smaller for
the valence band than for the conduction band.
    b. The imaginary parts of energies at the valence
band edge near the axis Γ - L are smaller than near point
X. The same holds for the conduction band edge.
    c. The imaginary parts of energies in the higher
conduction bands near Γ - L are large, compared with
those from the conduction band edge.

    As we mentioned already, the imaginary part of the
energy can be considered as a measure for the "disorder
sensitivity" of a crystalline energy level. Hence, the
rather small imaginary parts of the energies in the valence band indicate, that the corresponding density of
states is almost conserved when transforming the crystalline into the amorphous phase. The conduction band density of states is smoothed out because of the rather
large imaginary parts of energies. Due to the compara-

tively small imaginary parts of the energy near the
Γ - L axis, the density of states of the higher energetic parts of the valence band and the lower energetic
parts of the conduction band should not be seriously affected by increasing the disorder, whereas the energetically deeper (valence band) respectively higher (conduction band) structures should be decreased. Inspection
of the behavior of the calculated density of states in
Fig. 20 confirms this behavior. We may interpret the
upper valence band edge and the lowest conduction band
edge to be determined mainly by the short-range order.

The above discussion shows again, that in order to
obtain the qualitative behavior of the density of states
of a given material upon disorder, it is only necessary
to investigate the behavior of the complex energy spectrum, i.e. the behavior of $\Gamma_m(\vec{k})$ as a function of the
disorder parameter.

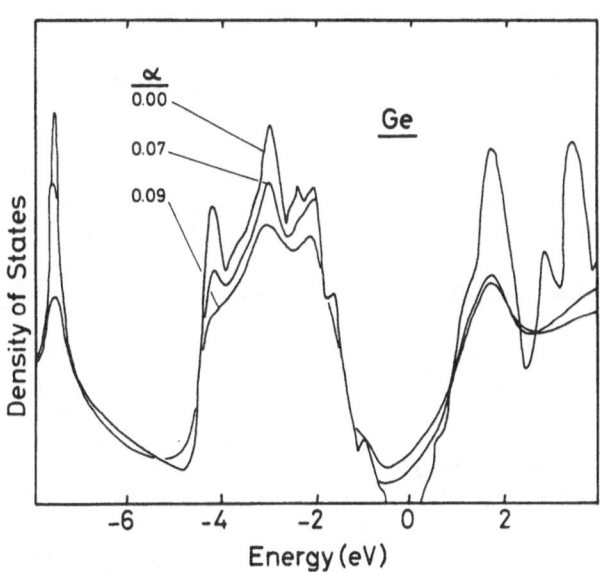

Fig. 20 Density of states of amorphous Ge models for
various degrees of disorder. (various values of the disorder parameter α) (from Ref. 38).

REFERENCES

1.   N. F. Mott and E. A. Davis, Electronic Processes in Non-Crystalline Materials, Oxford University Press (1971); J. Tauc, ed., Amorphous and Liquid Semiconductors, Plenum Press, London and New York, (1974).
2.   M. H. Cohen, J. Non-Crystalline Solids $\underline{4}$, 391 (1970).
3.   P. W. Anderson, Phys. Rev. $\underline{109}$, 1492 (1958).
4.   D. J. Thouless, J. Non-Crystalline Solids $\underline{8-10}$, 461 (1972).
5.   E.N. Economou, Journal de Physique $\underline{33}$, C3-157 (1972).
6.   D. E. Polk, J. Non-Crystalline Solids $\underline{5}$, 365 (1971).
7.   D. Turnbull, D. E. Polk, J. Non-Crystalline Solids $\underline{8-10}$, 19 (1972).
8.   G.A.N. Connell, R. J. Temkin, Proceedings of the International Conference on Tetrahedrally Bonded Amorphous Semiconductors, Yorktown Heights (1974) to be published.
9.   See lectures of D. Weaire.
10.   B. Velicky, S. Kirkpatrick, H. Ehrenreich, Phys. Rev. B $\underline{175}$, 747 (1968).
11.   B. G. Nickel, J. A. Krumhansl, Phys. Rev. B $\underline{4}$, 4354 (1971).
12.   P. L. Leath, J. Phys. C $\underline{6}$, 1559 (1973).
13.   T. Morita, C. C. Chen, J. Phys. Soc. Japan, $\underline{34}$, 1136 (1973).
14.   J. Zittartz, Z. Physik $\underline{267}$, 243 (1974).
15.   L. Schwartz and H. Ehrenreich, Annals of Physics $\underline{64}$, 100 (1971).
16.   A. Bansil, L. Schwartz, Proceedings of the International Conference on Tetrahedrally Bonded Amorphous Semiconductors, Yorktown Heights (1974), to be published.
17.   K. Maschke, P. Thomas, Phys. Stat. Sol. $\underline{39}$, 453 (1970).
18.   B. Kramer, Phys. Stat. Sol. $\underline{41}$, 649 (1970).
19.   P. Lloyd, Proc. Phys. Soc. $\underline{90}$, 207 and 217 (1967)
20.   F. Yonezawa, Progr. Theor. Phys. $\underline{31}$, 357 (1964).
21.   F. Yonezawa, T. Matsubara, Progr. Theor. Phys. $\underline{35}$, 357 and 759 (1966).
22.   B. L. Gyorffy, Phys. Rev. B $\underline{1}$, 3290 (1970).
23.   M. Lax, Phys. Rev. $\underline{85}$, 621 (1952).
24.   H. J. Fischbeck, Phys. Stat. Sol. $\underline{62}$, 425 (1974).
25.   P. Soven, Phys. Rev. $\underline{156}$, 809 (1967); Phys. Rev. $\underline{178}$, 1136 (1969).
26.   W. Jones and N. H. March, Theoretical Solid State Physics, J. Wiley and Sons, London (1973).
27.   J. Korringa, Physica $\underline{13}$, 392 (1974).
28.   W. Kohn, N. Rostoker, Phys. Rev. $\underline{94}$, 1111 (1954).

29.   J.Korringa, Phys. and Chem. of Solids $\underline{7}$, 252-8 (1958).
30.   J. L. Beeby, Proc. Phys. Soc. A $\underline{279}$, 82 (1964).
31.   P. W. Anderson and W. L. McMillan, Proc. Int. Sch.
        Phys. "Enrico Fermi" $\underline{37}$, 50 (1967).
32.   S. F. Edwards, Proc. Roy. Soc., $\underline{267}$, 518 (1962).
33.   See Lecture III on Mathematical Methods.
34.   B. Kramer, to be published.
35.   T. C. McGill and J. Klima, J. Phys. C. $\underline{3}$, L 163
        (1972) and Phys. Rev. B $\underline{5}$, 1517 (1972).
36.   J. Keller and J. M. Ziman, J. Noncrystalline Solids
        $\underline{8-10}$, 111 (1972).
37.   S. Yoshimo, M. Okasaki, M. Inoue, Solid State
        Commun. $\underline{15}$, 683 (1974).
38.   B. Kramer, Adv. in Solid State Physics/Festkörper-
        probleme XII, Pergamon/Vieweg (1972).
39.   N. J. Shevchik and W. Paul, J. Non-Crystalline
        Solids $\underline{8-10}$, 381 (1972).

SOME THEOREMS RELATING TO DENSITIES OF STATES*

D. Weaire

Dept. of Physics, Heriot-Watt University,
Riccarton, Currie, Midlothian, Scotland
and
Dept. of Engineering and Applied Science
Yale University, New Haven, Conn., U.S.A.

I. BLOCH'S THEOREM

Let us recall the significance of Bloch's Theorem for periodic systems. In the case of the one-electron Schrödinger equation, it states that the wave functions can be chosen to be of the form

$$\psi_{n\underset{\sim}{k}}(\underset{\sim}{r}) = e^{i\underset{\sim}{k}\cdot\underset{\sim}{r}} u_{n\underset{\sim}{k}}(\underset{\sim}{r}) \tag{1}$$

where n is a band index, $\underset{\sim}{k}$ is the "crystal momentum vector" and u has the periodicity of the Hamiltonian.

For small departures from exact periodicity, it would clearly be appropriate to use Bloch's Theorem at least as a zeroth approximation - thus it is still very useful for alloys with similar constituents, dilute alloys or even liquid metals (in which case the structure is highly disordered but the potential departs only slightly from a uniform potential). However, for typical amorphous solids, such an approach is highly questionable, and there is much to be said for not trying to define a $\underset{\sim}{k}$-vector even in some approximate sense. This is the attitude that we adopt here. We will emphasise general results which have

*Based in part on research supported by N.S.F.

relevance to non-periodic as well as periodic systems.

Mathematically speaking, the problem of the determination of electronic or vibrational densities of states
is just the eigenvalue problem for Hermitian matrices.
There are, of course, plenty of useful theorems which
place computable constraints on the spectra of such matrices, regardless of their detailed structure, which
may be used to make rigorous statements about densities
of states of amorphous solids. These include variational (inner) bounds on the range of the spectrum. As for
outer bounds, a powerful theorem is that of Gerschgorin[1,2],
which states that the characteristic roots of a matrix A
must lie within the (closed) disks centered on its dia-
gonal elements a_{ii} and having radii $\sum_{j \neq i} |a_{ij}|$.

Consider, for example, a matrix of the kind which
we shall deal with later, with zeros on the diagonal and
each row or column adding to exactly four. The application of the above theorem yields the result

$$-4 \leq \lambda \leq +4 \qquad\qquad\qquad (5)$$

giving outer bounds on the spectrum. This may be proved
by a variety of other theorems, but this theorem has the
advantage of great generality, as pointed out by Thorpe[3],
and has a remarkably simple proof[2].

II. MOMENTS OF SPECTRUM

The nth moment of the eigenvalue spectrum, defined
by the sum over all eigenvalues,

$$\mu_n = \sum_i \lambda_i^n , \qquad\qquad\qquad (6)$$

is given by

$$\mu_n = \text{Tr } M^n \qquad\qquad\qquad (7)$$

This theorem is trivial, given the invariance of the trace
under similarity transformation. Note that the trace can
be written out as a sum of products of n elements

$$\mu_n = \sum_{i,j,k,\ell \ldots q} m_{ij}\, m_{jk}\, m_{k\ell} \cdots m_{qi} \qquad\qquad (8)$$

In applications, a non-zero value of m_{ij} corresponds to a "bond" or "interaction" of some kind. We see that the moments μ_n are associated with closed chains or paths of such interactions. It is often convenient to evaluate them in just this way.

If the spectrum is weighted with the squared i^{th} component of the eigenvector, a similar result holds for what is, in practice, termed a "local density of states", i.e.,

$$\mu_n^{(i)} = \sum_{j,k,\ell,q} m_{ij} \, m_{jk} \, m_{k\ell} \cdots m_{qi} \tag{9}$$

which is the same as (8), without the summation on i.

III. GREEN'S FUNCTION

If the density of states is bounded, the Green's function $(E-H)^{-1}$ may be expanded, for large $|E|$, as

$$G = (E - H)^{-1} = E^{-1} \sum_{n=0}^{\infty} H^n E^{-n} \tag{10}$$

In this way, matrix elements of the Green's function can be expressed in terms of matrix elements of H. For example, the diagonal element G_{ii} will be given by a series

$$G_{ii} = E^{-1} \sum_{n=0}^{\infty} (H^n)_{ii} \, E^{-n}$$
$$= E^{-1} \sum_{n=0}^{\infty} \mu_n^{(i)} \, E^{-n} \tag{11}$$

Thus, for large $|E|$, G may be expressed in terms of moments and hence paths of bonds. (Results of such treatment are, however, not necessarily restricted to large $|E|$ since analytic continuation arguments may be used).

IV. TYPES OF DISORDER

It is useful to distinguish between different kinds of disorder in Hamiltonians of the kind used in simple tight binding theory. If the matrix elements between the localized basis functions fluctuate from site to site, we term this <u>quantitative</u> disorder. On the other hand, if the network of neighbor relationships defined by the Hamiltonian is (topologically) non-periodic, we term this <u>topological</u> disorder.

V. QUANTITATIVE DISORDER

This case is represented by the <u>alloy</u> problem, in which the structure is periodic but matrix elements vary from site to site. In its simplest idealized form, the problem is that of the determination of the spectrum of the simple Hamiltonian

$$H = \sum_i \varepsilon_i |\phi_i\rangle\langle\phi_i| + \sum_{i \neq j} t_{ij} |\phi_i\rangle\langle\phi_j| \tag{12}$$

Here the basis functions i are localized at the atomic sites (one per site) and only the diagonal elements ε_i vary from site to site. The off-diagonal terms, by themselves, define a periodic system. Usually only two values $\pm \delta$ of ε_i are allowed, corresponding to A and B atoms in a binary alloy. Kirkpatrick et al.[4] have given rigorous bonds, as shown in Fig. 1, for the density of states in this case. The A and B sub-bands must lie within limits given simply by shifting the $\delta = 0$ band limits by $\pm \delta$. This was proved using Green's function techniques[4] due to Kato. In cases (e.g. bcc with nearest neighbor interactions) in which the limits of the $\delta = 0$ spectrum coincide with the Gerschgorin bounds, the same result follows from Gerschgorin's theorem, but the results used by Kirkpatrick, et al are more general.

The most widely studied technique for obtaining an approximate shape spectrum is the Coherent Potential Approximation[4], in which an effective medium is set up, such that the scattering produced by replacing one atom by an A or B atom vanishes when appropriately weighted (with the proportions of A and B). We shall not go into any more detail than this here. (See Kramer's lectures).

VI. TOPOLOGICAL DISORDER

The simplest problem of this kind is that of a single s-band Hamiltonian of the form

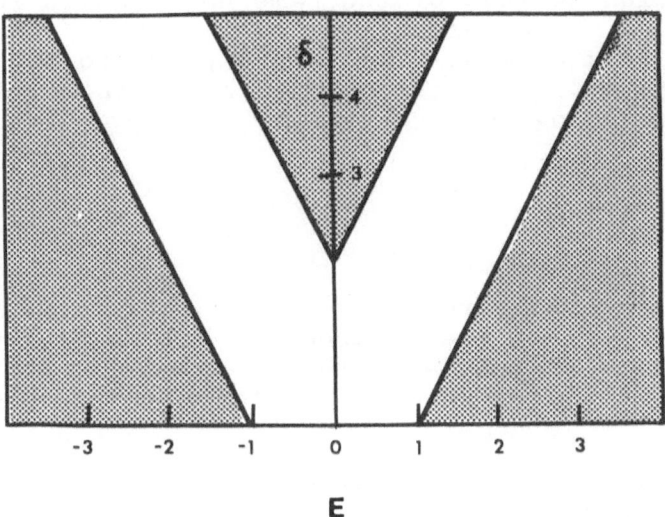

Fig. 1 Bounds on the density of states of a simple alloy
Hamiltonian. The band width at δ = 0 is taken to be 2
(from ref. 4).

$$H = V \sum_{ij} |i> <j| \tag{13}$$

where the sum is over all pairs of nearest neighbors ij
in a topologically disordered structure. Suppose, in addition, that there is a constant coordination number N.
Then, various of the theorems noted in the mathematical
introduction above give the result that the spectrum is
bounded by ±N V. Note also that there must be an eigenstate at + N V, which is just that state which has weight
+1 at each site. The opposite bound -N V is only attained if the structure is of the "alternating" type (see
Chapter 5). In the latter case, we can also assert that
the spectrum is symmetric about E = 0, as may be shown,
for example, by considering the relationship between moments and paths given above. (All odd moments vanish).
Note, in addition, that when H is squared, the two sublattices decouple so that the spectrum of H^2 (and hence
that of H) is related to that of some appropriate

Hamiltonian for the sublattices[5,6]. To be specific, for the case of diamond cubic structure, the above Hamiltonian gives a spectrum (shown in Fig. 2) related by a simple transformation to that of the same Hamiltonian for the fcc structure! Each eigenvalue of ε of the fcc Hamiltonian with matrix element $V = 1$ corresponds to two eigenvalues E of the fcc Hamiltonian (again with matrix element 1), according to

$$E^2 = \varepsilon + 4 \tag{14}$$

Note that while the symmetry of the spectrum of Fig. 2 follows from the bichromatic nature of the structure, the fact that it goes to zero at $E = 0$ does not. Fo example, consider the case of the simple cubic structure (Fig. 3).

Another such transformation has been derived by Thorpe and Weaire[6]. It relates to the spectrum of the Hamiltonian

$$H = V_1 \sum_{i,j \neq j'} |\phi_{ij}> <\phi_{ij'}| + V_2 \sum_{j,i \neq i'} |\phi_{ij}> <\phi_{i'j}| \tag{15}$$

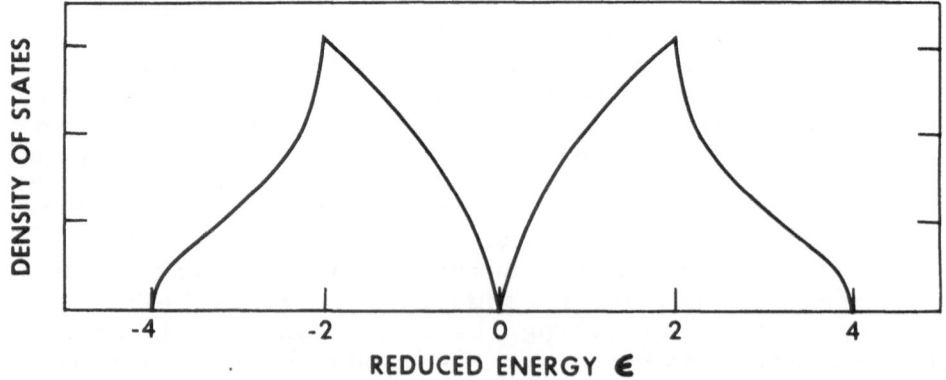

Fig. 2   Density of States of the simple Hamiltonian (13) for the diamond cubic structure.

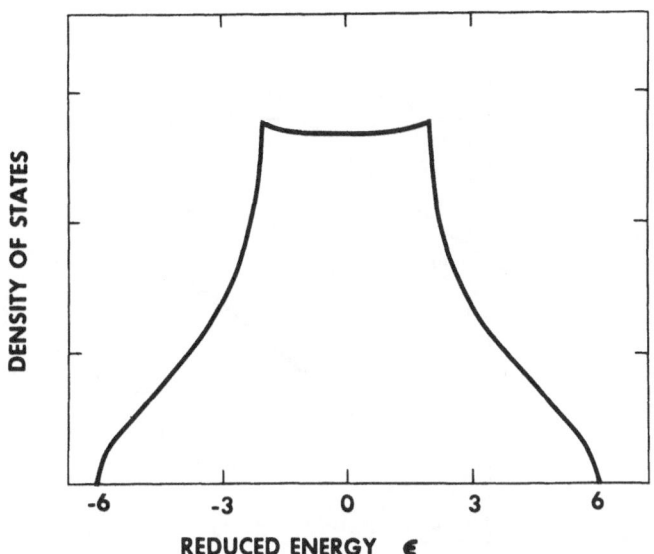

Fig. 3 Density of states for the simple cubic structure
(from ref. 7).

in which matrix elements V_1 connect basis functions associated with the same atom and different bonds, while
others V_2 connect those associated with the same bond and
different atoms. According to this theorem, the spectrum of (15) can be obtained by transforming that of (13)
and adding certain <u>delta functions</u>. For the usual case
($N = 4$, or "sp^3" orbitals) the transformation is given
by

$$E = V_1 \pm (V_2^2 + 4V_1^2 + V_1 V_2 \varepsilon)^{1/2} \tag{16}$$

where ε is an eigenvalue of (13) and E is an eigenvalue
of (15). Note that here the delta functions are, respectively, pure p-like bonding and antibonding functions.
Note also that the bounds (5) now give bounds on the continuous parts of the valence and conduction bonds, shown
in Fig. 4. The form of the bands for three values of
V_1/V_2 is shown in Fig. 5. A comparison with the density
of states for a much more realistic Hamiltonian[8] is shown
in Fig. 6.

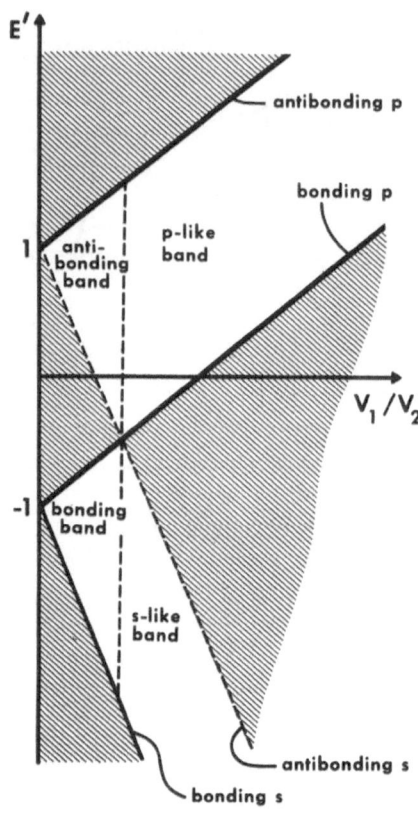

Fig. 4 Bounds for the density of states of the Hamiltonian (15). $E' = E/V_2$.

        There are two interesting limits of (15), in which
V_1 and V_2 go to infinity. The latter possibility simply
gives two infinitely separated bands of pure bonding and
antibonding character, in each of which the continuous
portion is just the familiar spectrum of (13) again.
The bonding bands could be obtained by using a Hamiltonian
with only bonding basis functions. This is the "Hall
Hamiltonian".

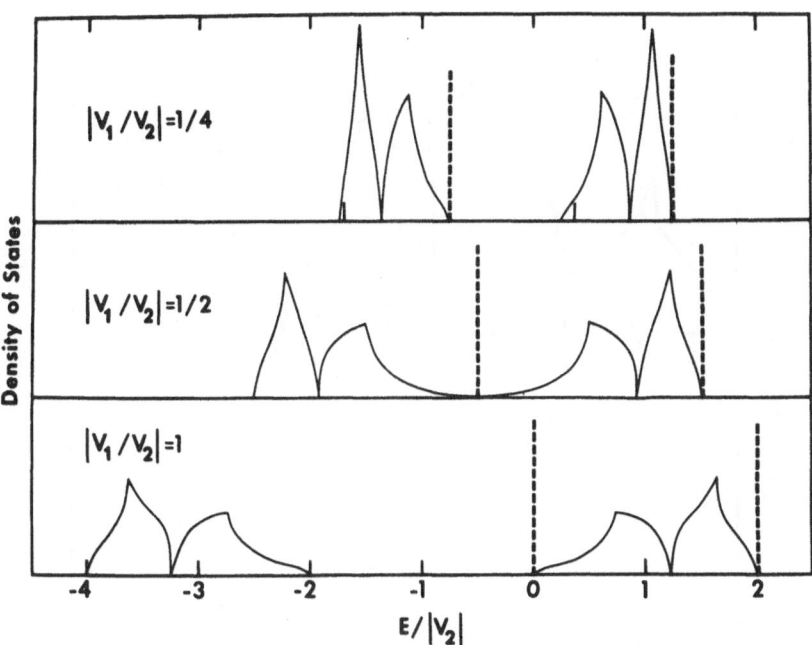

Fig. 5 Density of states of the Hamiltonian (15) for the diamond cubic structure. Vertical lines denote delta functions.

When V_1 goes to infinity we instead obtain bands of purely s and p-like character. Again, the continuous part of each reduces to the seemingly inescapable spectrum of (13). This result has surprising significance, because the p-like bands may be related to the vibrational density of states[9]. To see this, note that if we assume perfect tetrahedral coordination of nearest neighbors, the nearest neighbor bond vectors r_i satisfy

$$\sum_{i}^{4} r_i = 0 \qquad (17)$$

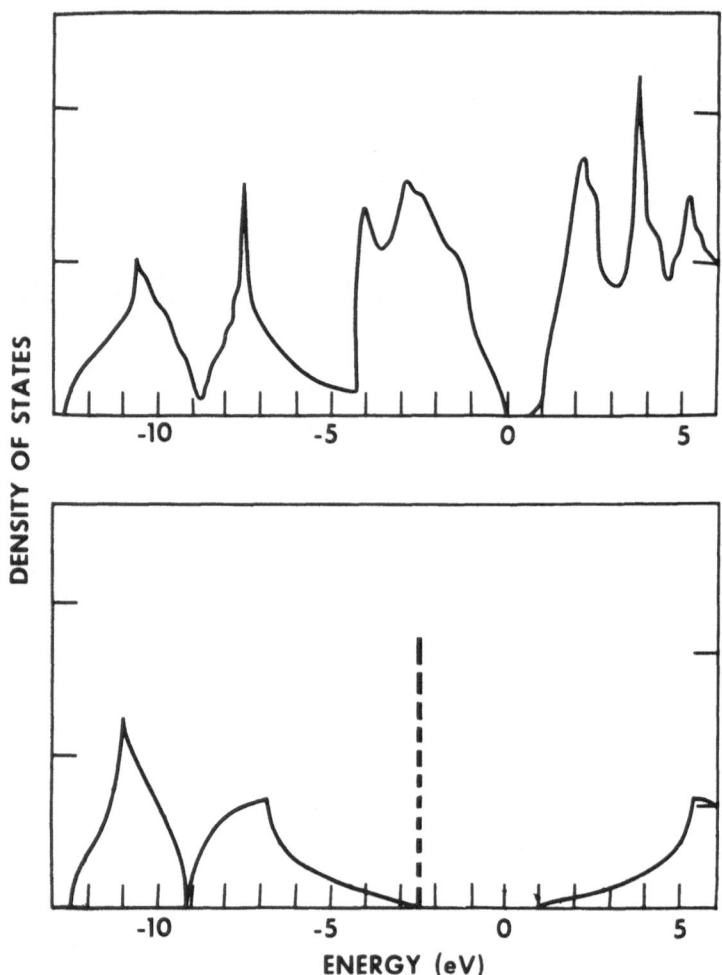

Fig. 6 Comparison with a pseudopotential calculation
(from ref. 8).

where the sum is over the four bonds associated with each
atom. If we use as coordinates not the displacement vectors $\underset{\sim}{u}$ of the atoms but rather their projections
$a_i = \underset{\sim}{u} \cdot \underset{\sim}{r}_i$ on to the bonds then

$$\sum_i^4 a_i = 0 \qquad\qquad (18)$$

follows from (17). If we think of a_i as the coefficients
in a tight binding wave function of the kind considered
above, this is just the condition that the wave function
is p-like (i.e. orthogonal to the s-like combination on
each atom). What then is the significance of the V_2
term? This is just a <u>central</u> <u>force</u> term. Hence, again
appealing to the transformation mentioned above, one has
the simple (but not obvious) result that for tetrahedral
bonding and central forces, the vibrational density of
states consists of two delta functions, at $\omega = 0$ and at
the top of the spectrum respectively, with the remaining one third of the spectrum being simply that of the
s-band Hamiltonian (13), when expressed in terms of
ω^2. Figs. 7 and 8 illustrate this result and its relationship to more realistic models. All of the relationships mentioned above are summarized in Table I.

VII. THE DELTA FUNCTIONS

Although there are neater ways of deriving the existence of the delta functions in the density of states,
the necessity of their existence is most easily seen by
noting that many wave functions of the kind shown in
Fig. 9 can be constructed. As for their <u>weight</u>, its significance is most easily seen by using arguments of the
type ("degrees of freedom minus constraints") mentioned
in Chapter 5. The wave functions in the delta functions
have purely p-like bonding or antibonding character.
The system has 3N linearly independent p wave functions
and the bonding (or antibonding) condition is represented by 2N constraints.

VIII. AVERAGE LOCAL CHARACTER OF EIGENFUNCTIONS

Even within the continuous bands, precise statements
can be made about average local character of eigenstates[11].
These have been proved in a rather clumsy fashion by
Weaire and Thorpe[11], but doubtless could be derived more
elegantly in the manner in which Straley[12] has studied
this Hamiltonian (see Fig. 10).

Such statements are useful in interpreting experiments in which <u>matrix</u> <u>elements</u> are important. For example, in the case of the infrared absorption spectrum,
the upper delta function has a local character such that
a factor in the simplest model expression for the matrix
element involved vanishes, while it does not for most of

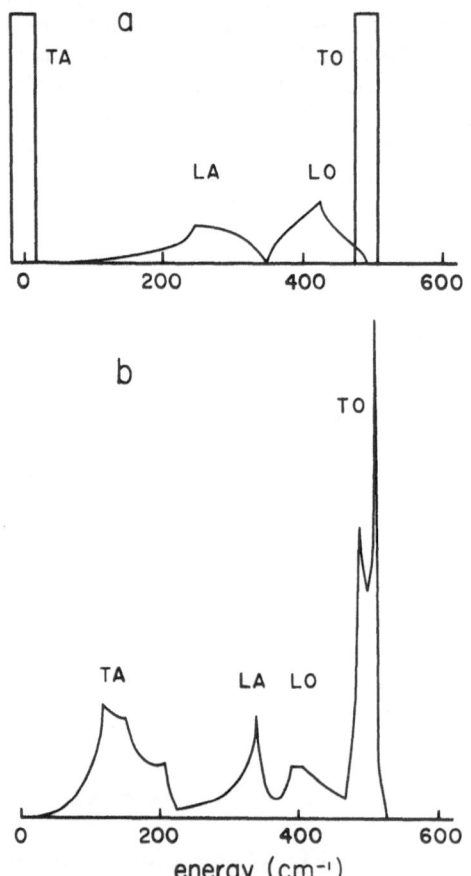

Fig. 7 Vibrational spectrum of diamond for central
forces between nearest neighbors (top).   Shell model
calculation (ref. 10) for Ge (bottom).

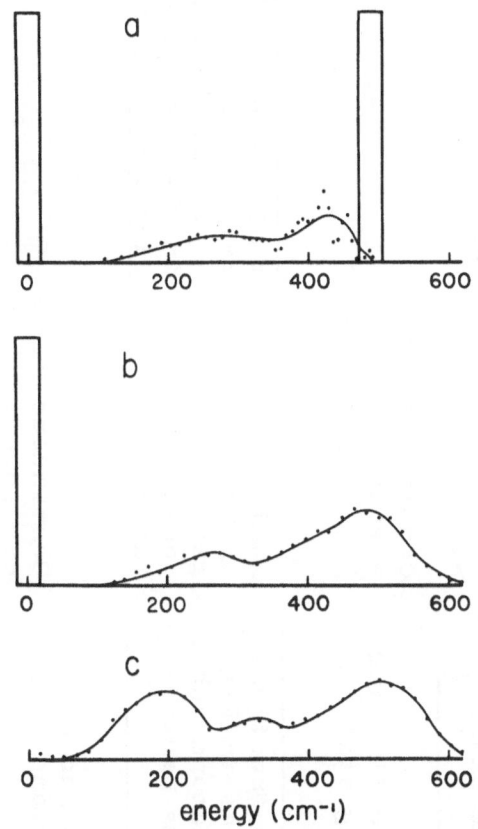

Fig. 8 (a) Vibrational density of states for a random
network, as given by the transformation discussed in the
text, for central forces. (b) Same, directly calculated. (c) Same, with bond bending forces added. For more
details, see Ref. 9.

TABLE I

Relationships between the spectra of various simple Hamiltonians

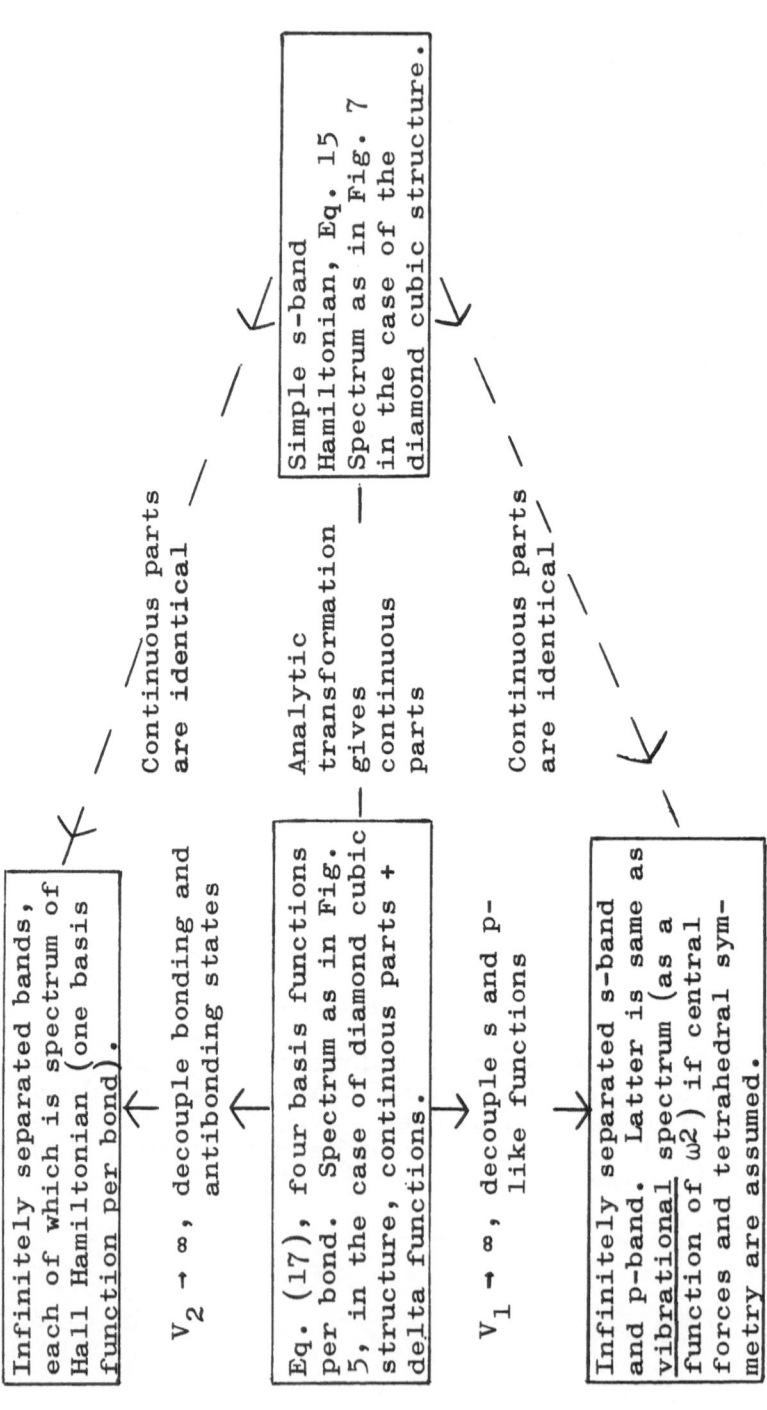

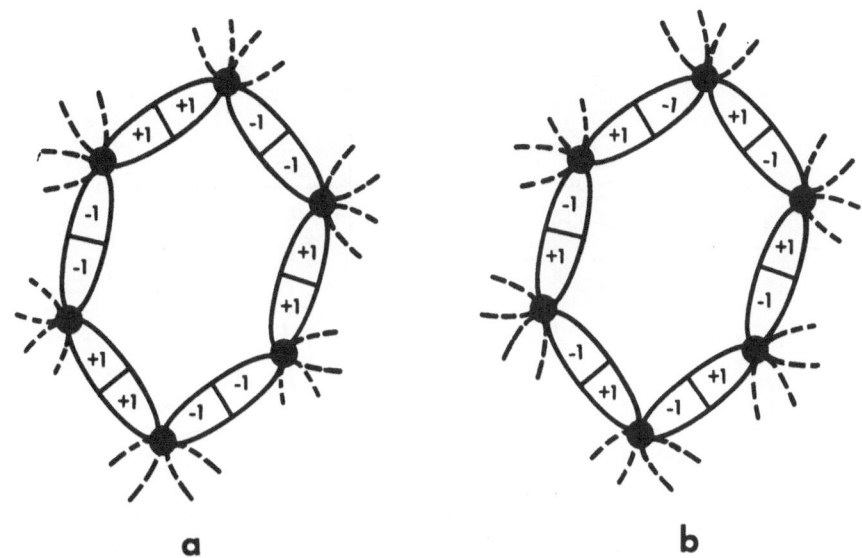

a   b

Fig. 9 Examples of pure p-like bonding and antibonding modes, localized on rings. For the vibrational problem, these correspond to "TO" and "TA" respectively.

the rest of the spectrum. This is useful in understanding the overall shape of the observed spectrum[9].

IX. GENERALIZATIONS

Some of the results of the previous sections are highly specialized. However as we emphasized at the outset, at least the results on bounds can be generalized to more complicated Hamiltonians. Gerschgorin's Theorem or the theorems used by Kirkpatrick, et al. clearly offer an opportunity to do so. Thorpe[3] has used Gerschgorin's Theorem to derive results for the case in which matrix elements fluctuate but are strongly correlated.

X. SUMMARY

We have seen how some interesting results may be derived rigorously for sufficiently simple periodic or

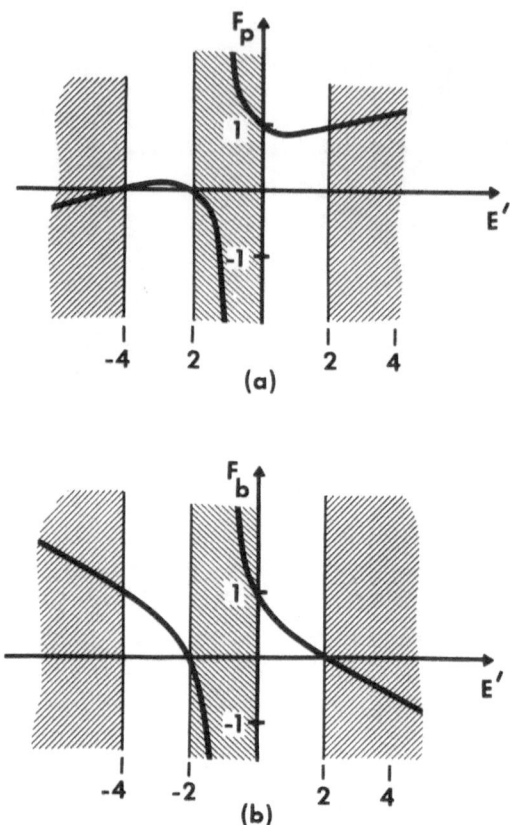

Fig. 10   Variation with E of the fractional p bonding
character of eigenfunctions of (15) for any structure.
This is for the case $|V_1/V_2|<1/2$.

non-periodic Hamiltonians alike, using (primarily) familiar matrix theorems. Many other such theorems might be
brought to bear, not to mention the wealth of theorems
relating spectra to their moments. Some progress in the
latter direction has lately been made by the Cambridge
group and others.

We have also noted that matrix element effects a-
rising from the average local character of eigenfunctions

can be studied. In crystals, one may distinguish two types of matrix element effects or selection rules - those relating to the local character and those relating to more distant correlations, which are given in the k-vector. The former remain in the amorphous case while the latter are (in practice) almost entirely lost.

REFERENCES

1. R. Bellman, "Introduction to Matrix Analysis", McGraw-Hill, New York (1960), p. 106.
2. E. Kreyszig, "Advanced Engineering Mathematics", Wiley, New York (1972), p. 685.
3. M. F. Thorpe, J. Phys. C, L75 (1973).
4. B. Velicky, S. Kirkpatrick, and H. Ehrenreich, Phys. Rev. 175, 747 (1968); S. Kirkpatrick, B. Velicky, and H. Ehrenreich, Phys. Rev. B1, 3250 (1970); L. Schwartz and H. Ehrenreich, Phys. Rev. B6, 4088 (1972).
5. R. J. Bell and P. Dean, Proc. Camb. Phil. Soc. 67, 97 (1968).
6. M. F. Thorpe and D. Weaire, Phys. Rev. B4, 3518 (1971).
7. T. Wolfram and J. Callaway, Phys. Rev. 130, 2207 (1963).
8. F. Herman, R. L. Kortum, C. D. Kuglin, and J. L. Shay, in Proc. Intern. Conf. on II-VI Semiconducting Compounds, Providence 1967, ed. D. G. Thomas (Benjamin, New York, 1967), p. 503.
9. D. Weaire and R. Alben, Phys. Rev. Lett. 29, 1505 (1972); R. Alben, D. Weaire, J. E. Smith, Jr. and M. H. Brodsky, Phys. Rev. B 11, 2271 (1975).
10. G. Dolling and R. A. Cowley, Proc. Phys. Soc. (London) 88, 463 (1966).
11. D. Weaire and M. F. Thorpe, Phys. Rev. B4, 2508 (1971).
12. J. P. Straley, Phys. Rev. B6, 4086 (1972).

THEORY OF OPTICAL ABSORPTION OF DISORDERED SOLIDS

B. Kramer

University of Dortmund, 4600 Dortmund 50

Federal Republic of Germany

1. DIELECTRIC CONSTANT

One of the simplest experimental methods for obtaining information about the electronic spectrum of solids is the measurement of optical absorption and/or reflection spectra[1,2]. The absorption coefficient is related to the imaginary part of the <u>dielectric</u> constant, $\varepsilon_2(\omega)$

$$\alpha(\omega) \sim \omega \varepsilon_2(\omega) \tag{1}$$

The real part of the dielectric constant, $\varepsilon_1(\omega)$, can be calculated from the imaginary part via <u>Kramers-Kronig relations</u>

$$\varepsilon_1(\omega) = 1 + \frac{2}{\pi} \, \mathcal{P} \int_0^\infty \frac{\omega' \varepsilon_2(\omega')}{\omega'^2 - \omega^2} \, d\omega' \tag{2}$$

$$\varepsilon_2(\omega) = -\frac{2\omega}{\pi} \, \mathcal{P} \int_0^\infty \frac{(\varepsilon_1(\omega') - 1)}{\omega'^2 - \omega^2} \, d\omega' \tag{3}$$

From the Kramers-Kronig relations, one can find the following structural correlations between ε_1 and ε_2. ε_1 has a maximum, where the derivative of ε_2 is large and positive. Correspondingly there will be a minimum in ε_1 for large and negative values of the derivative of ε_2 (see Fig. 1).

There are <u>sum-rules</u> on ε. One of them is obtained by taking in eq. (2) the upper limit of the integration as $\bar{\omega}$, $\omega \gg \bar{\omega}$, and using the Drude formula for $\varepsilon_1(\omega)$.

$$\int_0^{\bar{\omega}} \omega' \varepsilon_2(\omega')d\omega' = \frac{\pi}{2}\omega_p^2 = \frac{\pi}{2}\frac{e^2}{\varepsilon_o}\frac{N_{eff}(\bar{\omega})}{m} \qquad (4)$$

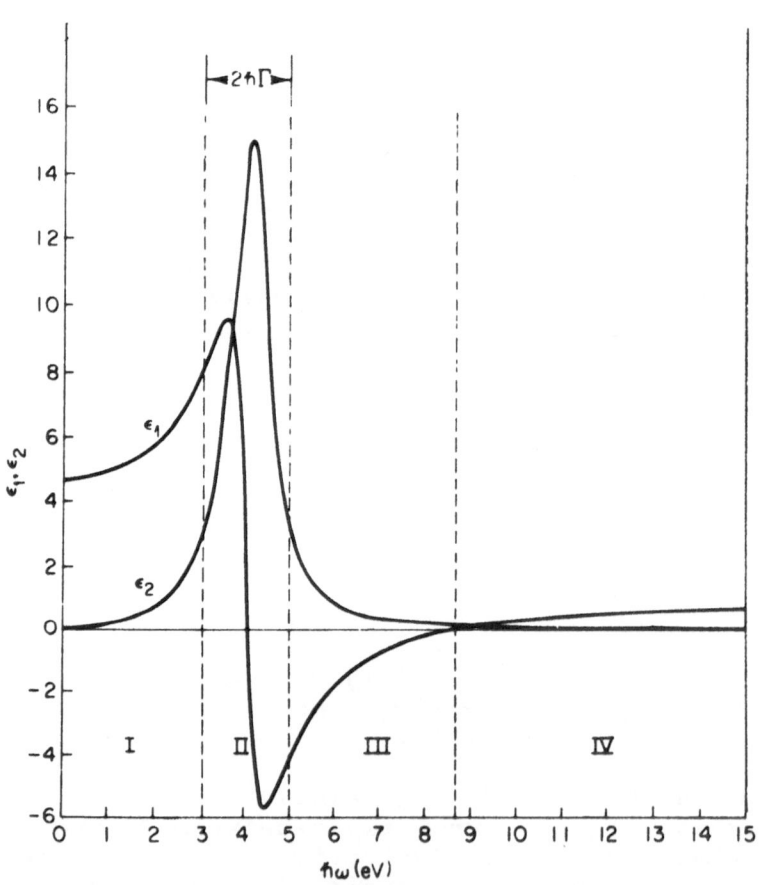

Fig. 1 Shape of the real part, ε_1, and the imaginary part, ε_2, of the dielectric constant as a function of the photon energy $\hbar\omega$ as calculated for the Lorentz-oscillator model (from ref. 1).

where m is the free electron mass and N_{eff} the effective
valence electron density. This sum-rule provides a check
on the presence of further absorption bands above $\bar{\omega}$.
For $\bar{\omega} \rightarrow \infty$ the sum-rule (4) is a measure of the energy absorption for all frequencies.

The optical process consists of the excitation of
an electron from an occupied (valence) state into an unoccupied (conduction) state. The absorption coefficient
and therefore ε_2 is proportional to the transition probability for the electron transition and the probabilities to find an occupied initial and an unoccupied final
state, the latter are given by the Fermi-distribution of
the electrons. In linear approximation one obtains for
ε_2 Fermi's golden rule

$$\varepsilon_2(\omega) \sim \frac{1}{\omega^2} \sum_{if} \left| <i|\vec{e}\cdot\vec{p}|f> \right|^2 \delta(E_f - E_i - \omega) n_T(E_i)(1 - n_T(E_f))$$

(5)

where $\vec{e}$ is the polarization vector of the light, $\vec{p}$ the
momentum operator, $<i|$ the initial state, $<f|$ the final
state, E_i and E_f the respective energies. $n_T(E)$, the
Fermi distribution function, is for $T = o$ a step-func-
tion.

$$n_o(E) = \{ \begin{matrix} 1 & E<E_F \\ 0 & E>E_F \end{matrix}$$

(6)

E_F is the Fermi energy. $\delta(E_f - E_i - \omega)$ is the energy conservation law.

2. SELECTION RULES

General Considerations

In physical systems with certain symmetry properties,
the electron wave functions and energies are classified
by quantum numbers related to the symmetry properties.
In these cases, there exist selection rules for the optical transitions due to vanishing transition probabili-
ties $|<i|e\cdot p|f>|^2$ for certain combinations of quantum
numbers. In the case of a spherically symmetric atom potential, the electron states can be classified by the angular momentum numbers ℓ and m yielding, for instance,
the angular momentum selection rule $\Delta\ell = \ell_i - \ell_f = \pm 1$.

In the case of a crystalline solid, the electron

states and energies can be classified by the momentum quantum number $\vec{k}$: $\Psi_{m\vec{k}}$, $E_{m\vec{k}}$. Calculating now the transition probability one obtains the so called k-selection rule

$$\vec{k}_i - \vec{k}_f = 0 \tag{7}$$

Therefore, the crystalline dielectric constant at T = 0 reads

$$\varepsilon_2(\omega) \sim \frac{1}{\omega^2} \sum_{n,m} \int_{\text{Brillouin zone}} d^3k \, |\langle \Psi_{n\vec{k}} | \vec{e} \cdot \vec{p} | \Psi_{m\vec{k}} \rangle|^2 \, \delta(E_{m\vec{k}} - E_{n\vec{k}} - \omega) \tag{8}$$

where the sums over n and m run over the occupied and unoccupied states, respectively. Due to $\omega_{mn}(\vec{k}) := E_{m\vec{k}} - E_{n\vec{k}}$ having some extrema in $\vec{k}$-space, the ε_2 - spectrum of the crystal exhibits pronounced peak structure. The peaks are related with the critical points defined by

$$\nabla_{\vec{k}} \, \omega_{mn}(\vec{k}) = 0 \tag{9}$$

yielding van Hove singularities in the absorption spectrum. The strengths of the van Hove singularities are given by the transition probabilities.

Nondirect-Transition Model

In the case of a disordered system we do not have distinct symmetry properties. Therefore, in general the transition probability will be non-zero for every combination of initial and final states

$$M_{if} := |\langle i | \vec{e} \cdot \vec{p} | f \rangle|^2 = M(E_i, E_f) \neq 0 \tag{10}$$

We rewrite eq. (5) by introducing an additional δ-function

$$\varepsilon_2(\omega) \sim \frac{1}{\omega^2} \sum_{if} \int dE \, M(E_i E_f) \delta(E - E_i) \delta(E_f - E - \omega) F_T(E, \omega) \tag{11}$$

Taking now $M(E_i E_f)$ to depend only on the difference $\omega = E_f - E_i$ we obtain by using the definition of the valence and conduction band density of states

$$n_v(E) = \sum_{\text{valence bands}} \delta(E-E_i)$$

$$n_c(E) = \sum_{\text{conduction bands}} \delta(E-E_f)$$

(12)

$$\varepsilon_2(\omega) \sim \frac{1}{\omega^2} M(\omega) \int_{E_F-\omega}^{E_F} dE\, n_v(E) n_c(E+\omega)$$

(13)

(at T = 0)

the convoluted density of states multiplied by an average transition probability[3,4]. Expression (13) can be used to calculate the ε_2-spectrum by starting from the crystalline density of states and using some suitable function for M (Fig. 2).

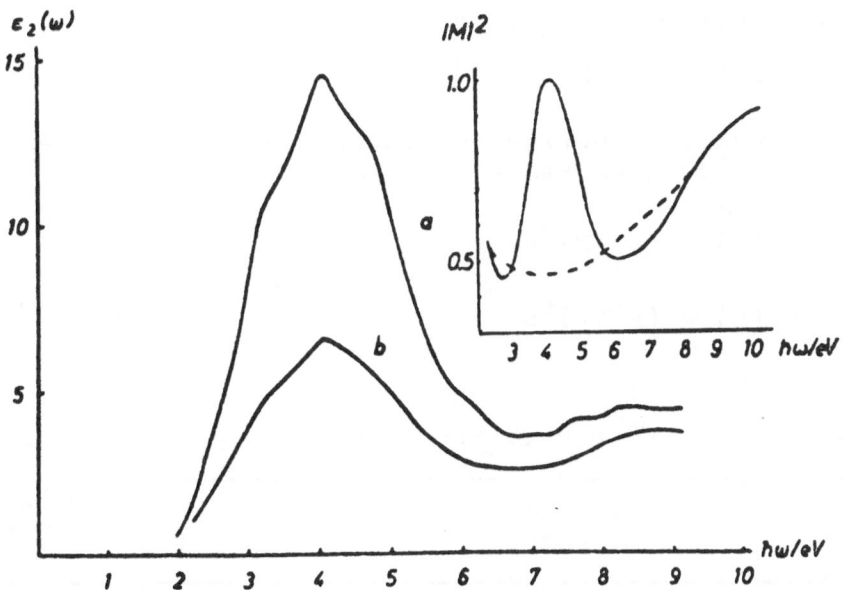

Fig. 2 ε_2-spectrum of Se calculated by using the non-direct transition model (eq. (13)). (a) with constant matrix elements, (b) with energy dependent matrix elements as shown in the insert (dashed line). The full line in the insert is $|M(\omega)|^2$ for the crystal (from ref. 4).

3. EMPIRICAL BEHAVIOR OF THE ABSORPTION SPECTRUM

In disordered solids we can distinguish three characteristically different regions of absorption (Table 1, Fig. 3 and 4).

The transitions from localized into localized states are important in explaining the shape of the ε_2-curve for small photon energies. This region is very difficult to treat theoretically and experimentally[5]. Using exponential valence and conduction band tails one can estimate the absorption coefficient

$$\omega\alpha(\omega) \sim e^{\omega/E_1} \tag{14}$$

where E_1 is some characteristic, temperature dependent parameter, ($E_1 \approx 0.1$ eV). For transitions between localized and delocalized states (and vice versa) one can also derive an exponential law, by using an exponential density of states for the localized states and a free-electron density of states for the extended states.

$$\omega\alpha(\omega) \sim e^{\omega/E_2} \tag{15}$$

For transitions between extended valence states and extended conduction states we can assume a power law to hold, i.e.

$$n_v(E) \sim (E_v - E)^{r_v}$$

$$n_c(E) \sim (E - E_c)^{r_c} \tag{16}$$

Table 1   Optical transitions in disordered semiconductors

Notation in fig.4	Notation in fig.3	Ψ_i	Ψ_f	$\omega^2\varepsilon_2(\omega)$
C	1	Localized	Localized	$\exp\{\omega/E_1\}$
B	2	Extended	Localized	$\exp\{\omega/E_2\}$
	3	Localized	Extended	
A	4	Extended	Extended	$(\omega - E_g)^r$ $r > 0$

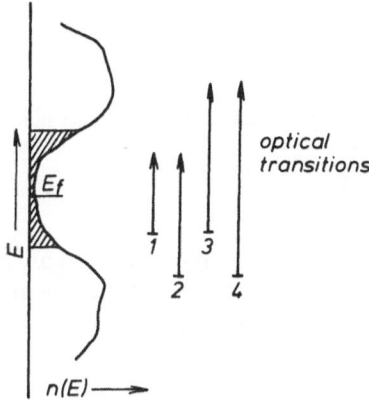

Fig. 3 Optical transitions in disordered materials (see Table 1).

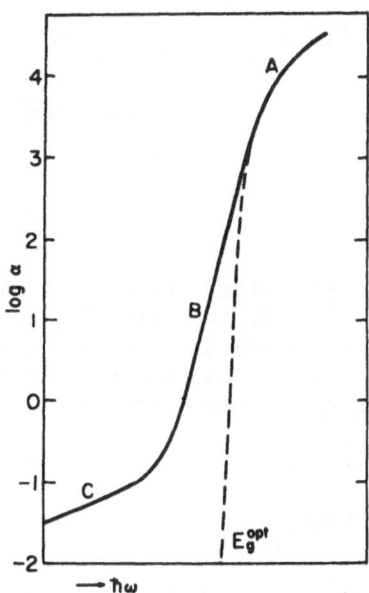

Fig. 4 Shape of the optical absorption edge of a disordered material shown schematically (from ref. 5). (absorption regions A, B, C, see Table 1).

Then it is straightforward to derive (with constant matrix elements)

$$\omega\alpha(\omega) \sim (E-E_g)^{r_v+r_c+1} \tag{17}$$

with a phenomenological <u>optical energy</u> gap E_g. For $r_v=r_c=1/2$ we obtain

$$\omega\alpha(\omega) \sim (E-E_g)^2 \tag{18}$$

the so-called <u>nondirect-transition</u> law, because it was derived originally by relaxing the $\vec{k}$-selection rule in the crystalline ε_2-formula eq. (8)[3].

4. RELATION TO ELECTRON GREEN'S FUNCTION

To obtain a quantitative description of the optical properties of disordered materials, it is useful to write our starting equation for the dielectric constant in terms of electron Green's functions

$$\omega^2\varepsilon_2(\omega) \sim \sum_{if}\int dE <i|\vec{e}\cdot\vec{p}|f><f|\vec{e}\cdot\vec{p}|i>$$

$$\times \delta(E-E_i)\delta(E_f-\omega-E)F_T(E,\omega) \tag{19}$$

$$\sim Tr\{\int dE F_T(E,\omega)(e\cdot p)ImG(E^+)(e\cdot p)ImG(E+\omega)\}$$

where

$$G(z) = \frac{1}{z-H} \tag{20}$$

is the resolvent of the Hamiltonian H and z is the complex energy variable[6]. H, as the Hamiltonian describing the motion of the electron in the fluctuating potential of the system, is configuration dependent. $\varepsilon_2(\omega)$, as being a macroscopic quantity, has to be averaged over all configurations.

$$\omega^2\varepsilon_e(\omega) = <\omega^2\varepsilon_2(\omega)_{\vec{R}_1\ldots\vec{R}_N}> \tag{21}$$

As the Fermi-distribution, $F_T(E,\omega)$, is configuration independent, we have to average the product of the electron Green's functions

$$< \text{ImG}(E^+)_{\vec{R}_1 \ldots \vec{R}_n} (\vec{e} \cdot \vec{p}) \ \text{ImG}(E+\omega)_{\vec{R}_1 \ldots \vec{R}_N} > \qquad (22)$$

Recalling

$$\text{ImG} = \frac{1}{2i} (G - G^*) \qquad (23)$$

we note that we have to calculate operators of the form

$$G_2(Z_1, Z_2) = <G(Z_1)_{\vec{R}_1 \ldots \vec{R}_N} G(Z_2)_{\vec{R}_1 \ldots \vec{R}_N}> \qquad (24)$$

which are <u>two-electron propagators</u>. Note, that we obtain correlations between the electron and the hole in
spite of starting with an equation for ε_2 which is based
on one-electron theory, neglecting the electron-hole interaction. This <u>effective electron-hole interaction</u> is induced by the averaging process, i.e. the disorder. This
is among one of the fundamental facts obtained from the
theory of optical absorption of disordered systems.

5. FORMAL THEORY

Bethe-Salpeter Equation

For the averaged one-electron Green's function in
momentum representation we have the integral equation

$$g_1(\vec{k}; Z) \delta_{\vec{k}\vec{k}'} = G_o(\vec{k}; Z) \delta_{\vec{k}\vec{k}'}$$
$$+ G_o(\vec{k}; Z) <\sum_{\vec{k}''} V_{\vec{k}\vec{k}''} \ G(\vec{k}'', \vec{k}'; Z)> \qquad (25)$$

where

$$G_o(\vec{k}; Z) = \frac{1}{Z - k^2} \qquad (26)$$

is the free-electron Green's function and $V_{\vec{k}\vec{k}'}$ the total
potential of the solid. Introducing the configuration
independent self-energy $M_1(\vec{k}; Z)$ we can write the integral
equation for the <u>configuration dependent</u> one-electron
Green's function as

$$G(\vec{k}, \vec{k}; Z) = g_1(\vec{k}; Z) \delta_{\vec{k}\vec{k}'}$$
$$+ g_1(\vec{k}; Z) \sum_{\vec{k}''} \hat{V}_{\vec{k}\vec{k}''} \ G(\vec{k}'', \vec{k}'; Z) \qquad (27)$$

where

$$\hat{V}_{\vec{k}\vec{k}''} = V_{\vec{k}\vec{k}''} - M_1(\vec{k};z)\delta_{\vec{k}\vec{k}''} \qquad (28)$$

M_1 fulfills the condition

$$\sum_{\vec{k}''} <(M_k(\vec{k};z)\delta_{\vec{k}\vec{k}''} - V_{\vec{k}\vec{k}''})G(\vec{k}''\vec{k}';z)> = 0 \qquad (29)$$

which is of course equivalent to[6]

$$<T(z)> = 0 \qquad (30)$$

Using eq. (27) we derive an equation for the averaged two-electron Green's function (omitting the variables $z_1 z_2$ for simplicity)

$$g_2(\vec{k}\vec{k}'\vec{k}''\vec{k}''') = g_1(\vec{k})_1(\vec{k}'')\delta_{\vec{k}\vec{k}'}\delta_{\vec{k}''\vec{k}'''}$$

$$+ g_1(\vec{k})g_1(\vec{k}'')\sum_{\vec{q},\vec{p}} M_2(\vec{k}\vec{q}\vec{k}''\vec{p})\ g_2(\vec{q}\vec{k}'\vec{p}\vec{k}''') \qquad (31)$$

The two electron self-energy M_2 (or vertex part) in analogy to M_1 is defined by

$$\sum_{qp} M(\vec{k}\vec{q}\vec{k}''\vec{p})\ g_2(\vec{q}\vec{k}'\vec{p}\vec{k}''') =$$

$$\sum_{\vec{q}} (V_{\vec{k}\vec{q}} - M_1(\vec{k})\delta_{\vec{k}\vec{q}})G(\vec{q},\vec{k}') \times \qquad (32)$$

$$\times \sum_{\vec{p}} (V_{\vec{k}''\vec{p}} - M_1(\vec{k}'')\delta_{\vec{k}''\vec{p}})G(\vec{p},\vec{k}''') >$$

Eq. (31) is a Bethe-Salpeter equation.

Approximation for Weak Potentials

As a first approximation, one can neglect the disorder induced electron-hole interaction by taking

$$g_2 = g_1 g_2 \qquad (33)$$

Expansion of the self-energy M_2 of eq. (32) into powers of the atom potential yields in lowest non-vanishing order

$$M_2(...) = ...a_2(\vec{k}-\vec{k}',\vec{k}''-\vec{k}''')v_{\vec{k}\vec{k}'}v_{\vec{k}''\vec{k}'''} \qquad (34)$$

This is a repulsive electron-hole interaction yielding a disorder induced screening of excitonic effects at the absorption edge[7]. It is acting __against__ the broadening of the absorption edge due to the broadening of the valence and conduction band edges.

6. APPLICATION

Lifetime Broadened Absorption Spectrum

Let us calculate the dielectric constant of a crystalline system in terms of Green's functions. For complete order the n-atom correlation function can be factored into a product of two-atom correlation functions

$$D_n(\vec{R}_1 \ldots \vec{R}_n) = \sum_{\ell_1} \delta(\vec{R}_1 - \vec{R}_2 - \vec{\ell}_1) \sum_{\ell_2} \delta(\vec{R}_2 - \vec{R}_3 - \vec{\ell}_3) \ldots$$

$$\ldots \sum_{\ell_{n-1}} \delta(\vec{R}_{n-1} - \vec{R}_n - \vec{\ell}_{n-1}) \tag{35}$$

Using the expansion of the configuration dependent one-electron Green's function into a power series in the potential

$$G = G_o + G_o \sum_{\ell_1} v_{\ell_1} G_o + G_o \sum_{\ell_1} v_{\ell_1} G_o \sum_{\ell_2} v_{\ell_2} G_o + \ldots \tag{36}$$

we observe that the product

$$(\vec{e} \cdot \vec{p}) \, G(z_1)(\vec{e} \cdot \vec{p}) G(z_2) \equiv \pi(z_1 z_2) \tag{37}$$

can be written as a sum of terms of the form[8]

$$(\vec{e} \cdot \vec{p}) \, G_o \, v_{\ell_1} G_o \ldots v_{\ell_m} G_o (\vec{e} \cdot \vec{p}) \, G_o v_{\ell_{m+1}} G_o \ldots v_{\ell_{n+m}} G_o \tag{38}$$

Taking this in momentum representation

$$\vec{p} \, |\vec{k}> = \vec{k} \, |\vec{k}> \tag{39}$$

$$<\vec{k}|v_\ell|\vec{k}'> = e^{-i(\vec{k} - \vec{k}')\vec{R}_\ell} v_{\vec{k}\vec{k}'}$$

and averaging by using $D_{m+n}(\vec{R}_1 \ldots \vec{R}_m, \vec{R}_{m+1} \ldots \vec{R}_{m+n})$ of eq. (35) we obtain for the diagonal part of π

$$\pi(\vec{k};Z_1 Z_2) = (\vec{e} \cdot \vec{k}) \sum_m G_c(\vec{k}, \vec{k}-\vec{K}_m; Z_1) \times$$
$$\times (\vec{e} \cdot (\vec{k}-\vec{K}_m)) \; G_c(\vec{k}-\vec{K}_m, \vec{k}; Z_2) \tag{40}$$

G_C is the Green's function of a crystalline electron. Using eq. (23), (22) and performing the trace in eq. (19) for $T = 0$

$$\omega^2 \epsilon_2(\omega) \sim \int_{E_F - \omega}^{E_F} dE \int d^3k \sum_m (\vec{e} \cdot \vec{k}) \; \text{Im} G_c(\vec{k}, \vec{k}-\vec{K}_m; E)$$
$$\times (\vec{e}(\vec{k}-\vec{K}_m)) \; \text{Im} G_c(\vec{k}-\vec{K}_m, \vec{k}; E+\omega) \tag{41}$$

and with the residues and poles of G_c

$$G_c(\vec{k}, \vec{k}-\vec{K}_m; Z) = \sum_n \frac{\text{Res} \; G_c(\vec{k}, \vec{k}-\vec{K}_m; Z)\big|_{Z=E_n(\vec{k})}}{Z - E_n(\vec{k})} \tag{42}$$

we get the expression (8) for the dielectric constant if we remember

$$\text{Res} \; G_c\big|_{Z=E_n(\vec{k})} = \langle \vec{k} | \Psi_{n\vec{k}} \rangle \langle \Psi_{n\vec{k}} | \vec{k}-\vec{K}_m \rangle \tag{43}$$

Thus, the residues of G_C determine the optical transition probabilities, whereas the poles in the energy plane determine the shape and postion of the van Hove singularities in the spectrum.

We know that in the case of a short range ordered (but long-range disordered) solid the one-electron Green's function can be written in a form quite analoguous to eq. (42). However, the poles of g are now complex numbers

$$\epsilon_n(\vec{k}) = E_n(\vec{k}) + i\Gamma_n(\vec{k}) \tag{44}$$

For "small disorder", the real parts of energy are about the same as in the crystal. If we assume in addition the residues of G_C to be not strongly affected by the lack of long-range order we can take also the transition probabilities to be about unchanged. Then, we obtain for the dielectric constant

$$\omega^2 \varepsilon_2(\omega) \sim \int dE \int d^3k \; M_{mn}(\vec{k}) \times$$

$$\times L(\Gamma_m(\vec{k}); E - E_m(\vec{k})) \qquad (45)$$

$$\times L(\Gamma_n(\vec{k}); E + \omega - E_n(\vec{k}))$$

which involves a convolution of the Lorentzians

$$L(\Gamma; E) = \quad \text{m} \; \frac{1}{E + i\Gamma} \qquad (46)$$

Eq. (45) can also be derived by performing the configurational average on the expression (38) by using a factorization of the n-atom correlation function analoguously to the crystal (eq. (35)) and assuming the atomic form factor $v_{\vec{k}\vec{k}'}$ to be slowly varying in $\vec{k}$-space[8]. For small disorder, i.e. large short-range order volume, eq. (45) represents essentially a lifetime broadened "crystalline" ε_2-spectrum. $\Gamma_m(\vec{k})$, a $\vec{k}$-dependent reciprocal lifetime, reflects the effect of the lack of long-range order. If $\Gamma_m(\vec{k}) + \Gamma_n(\vec{k})$ is only small, the corresponding structure in the ε_2-spectrum will be preserved in the disordered phase. If $\Gamma_m(\vec{k}) + \Gamma_n(\vec{k})$ is large, the respective structure will vanish.

Examples

The calculation of the ε_2-spectrum of an amorphous solid by using eq. (45) is still a very complicated procedure, in spite of the formal simplicity of the formula. It requires the knowledge of at least $\Gamma_m(\vec{k})$ and a "crystalline" ε_2-spectrum subdivided into contributions from the different regions of the $\vec{k}$-space, where $\Gamma_m(\vec{k})$ can be taken as a constant. Calculations have been performed for the group-IV and III-V semiconductors as well as for selenium[9]. They were based on the pseudopotential band structure calculation and the corresponding ε_2-spectra of the crystalline phases. The complex band structures were calculated using the generalized pseudopotential method[9]. As an example, we discuss the ε_2-spectrum[8] of Se.

Fig. 5 shows the complex energy band structure. The real parts of energy are about the same as for the crystal. The imaginary parts of energy in the valence bands (bands up to about 8 eV) are smaller than those of the conduction bands. Within the two valence-band triplets

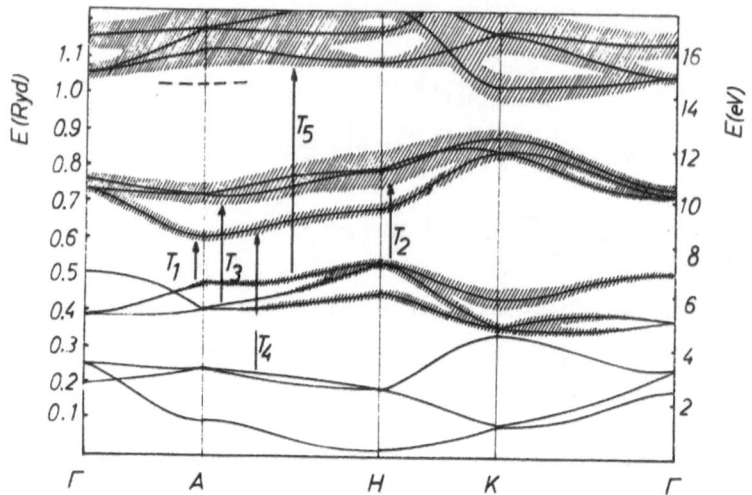

Fig. 5 Complex energy band structure for amorphous Se
(α = 0.075). The arrows T_1....T_5 denote optical transitions important for the explanation of the ε_2-spectrum
(see also Fig. 7). The imaginary part of energy is shown
as a broadening of the real part[8].

we observe relatively small imaginary parts of energy to
occur along the $\vec{k}$-space Γ- A, which corresponds to the
trigonal axis of the crystal. Note also the clear separation of the lowest band in the first conduction band triplet (at about 9 to 10 eV) from the two higher subbands
in the $\vec{k}$-space region A-H.

In Fig. 6 we show that part of the ε_2-spectrum which
corresponds to transitions between the highest valence
band and the lowest conduction band triplet as a function
of the disorder parameter[6]. α can be related to the diameter, R_s, of a short range order region

$$R_s \sim \frac{a}{\alpha} \tag{47}$$

where a is the mean interatomic distance.

We observe, that the energetically higher part of
the spectrum, between hω ≈ 3.5 eV to 5 eV (transitions T_2,
T_3 in Fig. 5) is more strongly decreased by disorder (increasing α, decreasing diameter of the short range order
region) than the peak at about 3 eV. This peak corres-

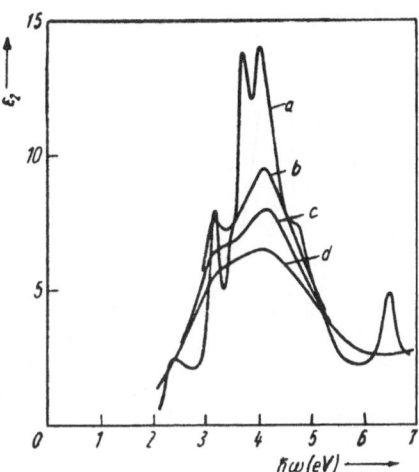

Fig. 6 ε_2-spectrum of Se with relaxed long-range order.
(a) $\alpha=0$ crystalline spectrum (isotropic average), (b)
$\alpha = 0.05$, (c) $\alpha = 0.075$, (d) $\alpha = 0.1$ (from ref. 8).

ponds to transitions from the valence band edge to the
conduction band edge near $\vec{k}$-point A (T_1 in Fig. 5), where
the imaginary parts of energy are rather small. Therefore, a considerable amount of T_1 is determined by short
range order, whereas T_2, and T_3 are more influenced by
long-range order.

The conservation of the peak T_1 is confirmed by a
comparison with experiment in Fig. 7, which shows also a
good overall agreement between experimental and calculated ε_2-spectrum in the whole energy range up to 10 eV.
The conclusions from this comparison are twofold:
(i) It is possible to obtain a good overall agree-
ment between experimental and theoretical ε_2-spectrum by
using the crystal as a starting point (short-range order)
and introducing the disorder by only one structural broadening parameter α.
(ii) The disorder behavior of the various structures
in the crystalline ε_2-spectrum can be estimated by calculating the complex energy spectrum and investigating the
behavior of the imaginary parts of energy in the different $\vec{k}$-space regions as a function of the disorder parameters.

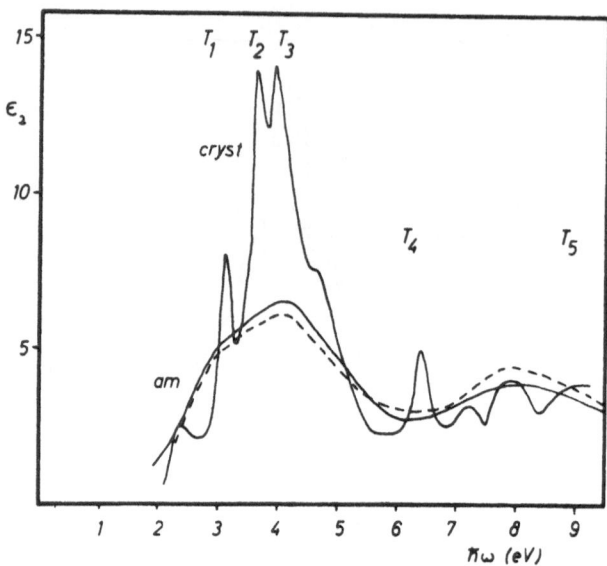

Fig. 7 ε_2-spectrum of Se. Crystalline spectrum: isotropic average based on pseudopotential calculations[11].
Amorphous spectra: Full line as calculated using the complex band structure. Dashed Line: experimental curve[10].

REFERENCES

1.   F. Wooten, Optical Properties of Solids, Academic
     Press (1972).
2.   D. L. Greenaway, G. Harbeke, Optical Properties and
     Bandstructure of Semiconductors, Pergamon (1968).
3.   J. Tauc, Optical Properties of Non-Crystalline So-
     lids in: F. Abeles, Optical Properties of Solids,
     North Holland, Amsterdam (1972).
4.   K. Maschke, Thesis, Marburg (1969); K. Maschke, P.
     Thomas, Phys.Stat.Sol. 41, 743 (1970).
5.   J. Tauc, A. Menth, J. Non-Crystalline Solids 8-10,
     569 (1973).
6.   see lectures on Mathematical Methods.
7.   J.D. Dow, J.J. Hopfield, J. Non-Crystalline Solids
     8-10, 664 (1972).
8.   B. Kramer, Phys. Stat. Sol. 41, 725 (1970).
9.   B. Kramer, Advances in Solid State Physics/ Fest-
     körperprobleme XII, Pergamon/Vieweg (1972).
10. A. G. Leiga, J. Appl. Phys. 39, 2149 (1968); J. Opt.
     Soc. Amer. 58, 880 (1968).
11. R. Sandrock, Phys. Rev. 169, 642 (1968).

COMMENTS ON THE THEORY OF LOCALIZED STATES IN SEMI-CONDUCTING NONCRYSTALLINE SOLIDS*

David Emin

Sandia Laboratories

Albuquerque, New Mexico 87115

ABSTRACT

After presenting examples of so-called localized states in crystals and commenting on the nature of electronic states in disordered solids, the question of the relationship between the nature of the eigenstates of an electron in a rigid noncrystalline solid and its transport properties is addressed. It is pointed out that the absence or presence of a finite-temperature thermally activated mobility does not necessarily follow from the electronic states being "extended" or "localized", since electron-lattice interaction often plays a major role in determining the transport properties of a solid. The effect of the electron-lattice interaction on the electronic eigenstates of a deformable solid, characterized by a short-range electron lattice interaction, is then discussed at length. In particular, the question of when it is a reasonable first approximation to neglect the electron-lattice interaction in studying the eigenstates of the system is considered. When this approximation fails, the charge carriers are often best viewed as "localized" small polarons. Finally, the situations which characterize small-polaron formation in disordered insulators are outlined.

---

*This work supported by the
U. S. Atomic Energy Commission.

In considerations of the electronic properties of
noncrystalline semiconductors there has been an ever-
present and growing interest in so-called "localized
states". The present lecture will consider several fundamental aspects of what may loosely be termed the theory of localized states. This talk is meant to complement other discussions of this subject which dwell on
topics related to Anderson localization[1]. In particular,
the primary thrust of this discussion will be to address
an <u>operational</u> definition of localized states and then
to elaborate the role played by the electron-lattice interaction in the theory.

When we think of an electron in a localized state
we envision the electron as being contained within a
finite region of space. Thus, for example as illustrated in Fig. 1, the potential well associated with a single donor atom, such as phosphorus, placed in a crystal,
such as germanium, may provide a bound state for the
"donor" (chemically unpaired) electron. An electron occupying this bound state will then be unable to move a-
way from the potential well associated with the donor
atom. Considering the donor atom to be stationary for
all time, the bound electron may be said to be occupying a localized state. An occupied donor state results
in an electrically neutral unit. Similarly, defects such
as vacancies, can provide localized states in crystalline
solids. While donor and acceptor states are electrically neutral when "occupied" by an electron or hole, respectively, other types of localized states (trap states)
are electrically neutral when unoccupied, and charged
when occupied. One particular type of defect which has
frequently been mentioned in connection with noncrystalline solids is the "dangling bond."[2] In this situation,
depicted in Fig. 2, due to a failure of all local bonding requirements to be fulfilled, a normally paired electron is left unpaired. The atom with the unpaired electron may serve as a center which attracts yet a second
electron, or the first electron may itself escape. In
other words an atom associated with a dangling bond may
be regarded as a defect which can accomodate zero, one,
or two electrons. This defect is electrically neutral
when occupied by a single electron. Defects which can
accomodate more than one electron are known in crystals
as well. For instance, a negative-ion vacancy in an alkali halide crystal can trap one or two electrons; in
these circumstances it is known respectively as an F-
center or an F'-center.[3] Similarly, deep-lying impurity
levels, such as those attributed to Cu in Ge, may accept

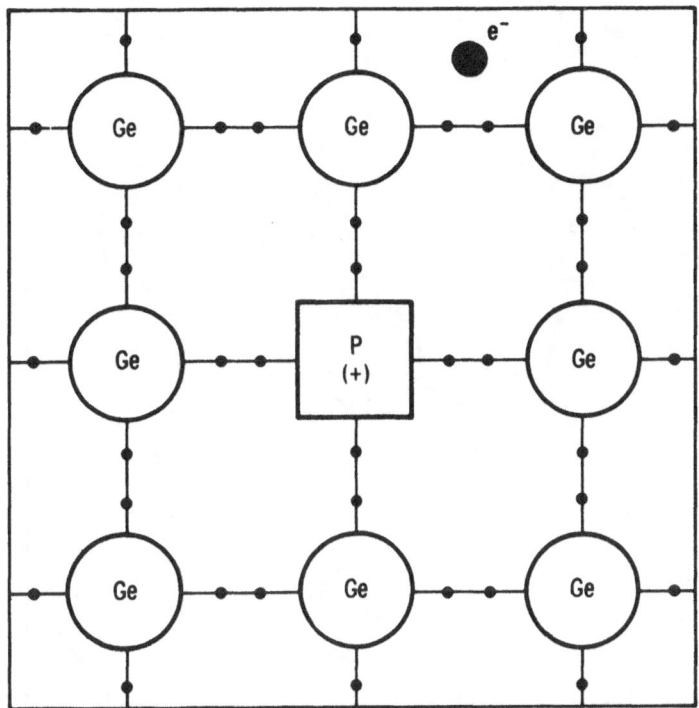

Fig. 1 A phosphorus donor atom and "donor" electron in
germanium. Four of the five 2p electrons of phosphorus
are used to fulfill the bonding requirements appropriate
to a germanium lattice, while the fifth electron of the
phosphorus (large dot) is loosely bound by the coulombic
field of the phosphorus atom with four bonding electrons.
The electrons involved in bonds with germanium are depicted as small dots on the bond line connecting the involved atoms.

as many as three electrons.[4] Numerous types of localized defect states are known to exist; these are just a
few examples.

    Barring such defect-related "localized states", non-
crystalline solids may possess spatial fluctuations in
the one-electron potential which arise simply from the
topological disorder of the material. These fluctuations may be sufficiently great so as to give rise to
electronic states which are said to be "localized".
When one speaks of a localized state of energy E one

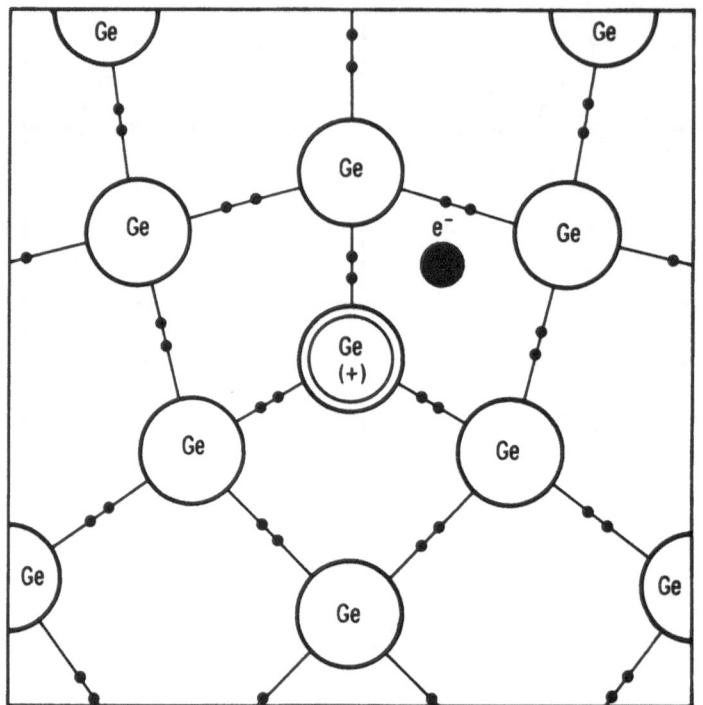

Fig. 2 A "dangling bond". The dangling-bond germanium
atom has formed chemical bonds with only three rather
than four germanium atoms. The extra electron is weakly
attracted to the dangling-bond germanium atom. This defect is also hypothesized to be able to act as a trap
for a second electron.

thinks of it as being associated with a wavefunction of
finite spatial extent in the sense that it is simply nor-
malizable:[4]

$$\int |\Psi_E(\underset{\sim}{r})|^2 \, d\underset{\sim}{r} = \text{finite constant.} \qquad (1)$$

However, such localized wavefunctions may not be as simple as those which one visualizes when one thinks of an
electron bound to a defect or impurity. This is because
the potential which is experienced by an electron in a
noncrystalline solid is typically pictured as being much
more involved than that associated with isolated defects.

For example, the potential well attributed to a phosphorus donor in Ge, schematically shown in Fig. 3a, may
in some approximation be viewed as just a somewhat modified coulombic well, while, as illustrated in Fig. 3b,
the potential which would be associated with localized
states in a noncrystalline solid is typically viewed as
highly convoluted. One must always be cognizant that
the electronic potential and hence the electronic eigenfunctions are much more complex in a noncrystalline solid

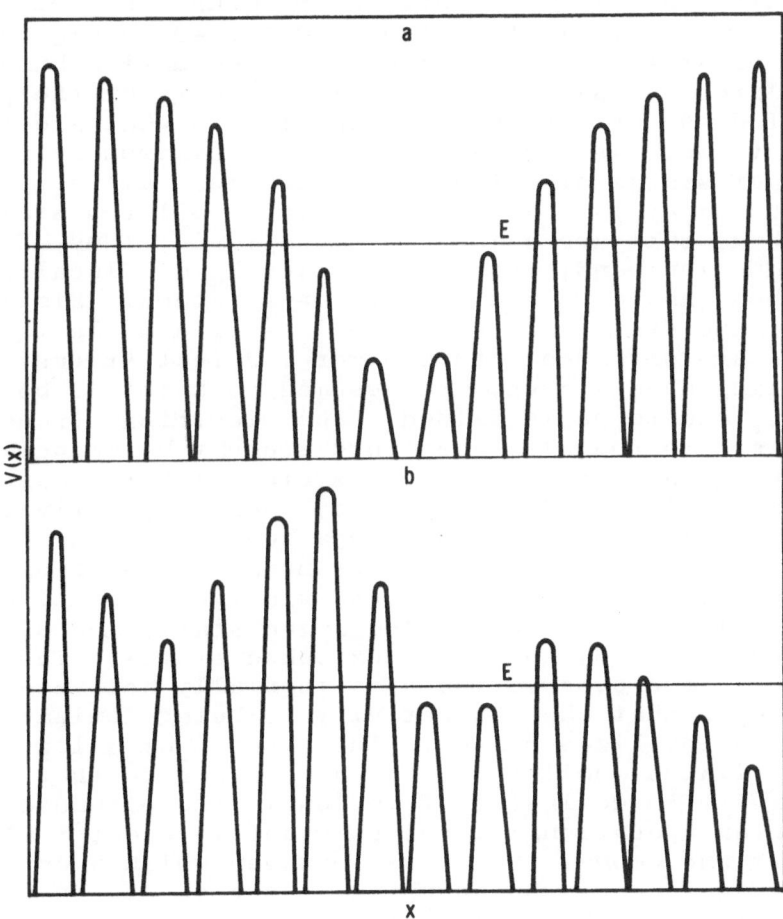

Fig. 3 The potential seen by a donor electron in a crystal is illustrated in the top portion of this figure.
In the lower portion, the highly convoluted nature of
the electronic potential hypothesized for some noncrystalline solids is schematically depicted.

than in other better defined situations, such as in
lightly doped semiconductors. (This observation will be
returned to later in subsequent lectures.)

One is quickly led to ask how the presence of localized states will manifest itself experimentally. In
particular, attention will now be restricted to their relationship to the electrical-transport properties of a
material. The initial attempt to relate the formation
of localized states to electrical-transport properties
in noncrystalline semiconductors is found in the hypo-
theses which are termed the Mott-CFO model, named after
its architects:[6,7] N. F. Mott, M. H. Cohen, H. Fritzsche,
and S. Ovshinsky. In this model, depicted in Fig. 4, the
density of states of the valence and conduction bands of
an insulator is assumed to be "blurred out" by the presence of disorder, thereby giving rise to what are termed band tails. As one passes from the maximum in the
density of states of either band into the tail of the
band, an energy is reached beyond which all the states
in the band tail are said to be localized. Thus, for
the conduction band, the states below E_c are localized
while those above E_c are said to be extended. Similarly,
valence band states above E_v are localized while those
below E_v are extended. Furthermore, defect states, such
as dangling bond states, are assumed both to lie between
E_c and E_v and to be localized. The electrical transport
properties are roughly taken into account by associating
the occupation of the localized states with the carriers
possessing a low ($\lesssim 10^{-2}$ cm^2/V-sec) thermally activated
mobility, while a carrier's occupation of an extended
state is associated with a higher nonactivated mobility.[8]
The transport is taken to proceed via two parallel chan-
nels, hopping conduction in localized states and band-
like (nonactivated) motion in extended states. Thus, in
this scheme the observation of a thermally activated mobility means that the conductivity is being dominated by
hopping in localized states. The absence of a low thermally activated mobility is taken as evidence that the
conductivity is dominated by motion in the extended states.
Thus, on an operational level relevant to electrical-
transport phenomena, the states between which a carrier
is believed to hop, when hopping is thought to be observed, are taken to be localized states.

In thinking about this classification scheme it
should be pointed out that the measurement of transport
quantities in effect involves measuring the transition
rates, which govern the motion of a carrier between a-
tomic sites. Determining these rates involves considering

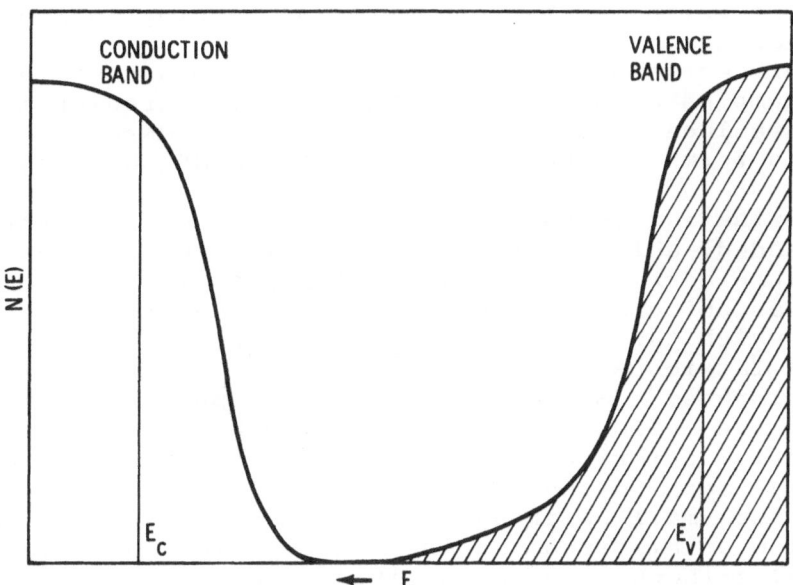

Fig. 4 An hypothesized model of the density of states
in an "ideal" noncrystalline solid. The valence and conduction band edges are smeared out, giving rise to band
tails. Below some energy E_c the conduction-band states
are taken to be localized. Above E_c, deeper into the
conduction band, the states are not localized (extended).
Similarly, E_v denotes the energy above which valence
band states are localized. The hatched area denotes the
levels which are occupied at absolute zero.

the system to which the electronic state of the carrier
is coupled. For instance, the phonon-assisted hopping
of an electron from one site to another involves consideration of the coupling of the carrier to the vibrational motion; in the absence of an electron-lattice interaction hopping motion cannot take place. Furthermore,
the notion of scattering of a "band-like" carrier involves interaction of the carrier, at least implicitly, with
a heat bath with which it can exchange energy. However,
the attributing of, say, thermally activated hopping motion uniquely to hopping between localized states is in
essence stating that this mobility is a unique signature
of the localized nature of the eigenstates of the purely
electronic system. Since the mobility is not solely

determined by electronic eigenstates of the system, one
might suspect that the strict correspondence between an
electron existing in localized states and its mobility
being that which is characteristic of hopping motion may
not be valid.   In fact, there are examples of the hopping
motion of carriers whose eigenstates are Bloch states.
Also, examples exist of carriers whose electronic eigenstates are localized but whose mobility is described most
easily in terms of the scattering of "band" carriers.

A situation in which the eigenstates of the system
are extended, actually Bloch-like, while the finite-
temperature carrier motion is most appropriately characterized as being hopping motion, is encountered in crystals in which the electronic charge carriers interact
with the lattice sufficiently strongly so as to form
small polarons.[9]   Some examples of this phenomenon are
found in the motion of holes in some alkali halides such
as potassium chloride $(KC\ell)$[10] or in the transition-metal
oxide, manganese oxide (MnO)[11].   In the case of the alkali
halides the small polarons are often called V_K centers.
The motion of electrons in the chalcogenide molecular
crystals, orthorhombic sulfur (S_8)[12] and realgar (As_4S_4)[13],
is also observed to be small-polaron hopping motion.
While both the concept of a small polaron and small-
polaron motion will be discussed in detail later in these
talks, mention is only made of this topic at this point
to provide a few examples of periodic systems in which an
extended-state carrier's motion is best characterized as
hopping motion.   In these examples the carriers are sometimes spoken of as "localized" or the transport properties are said to be describable in terms of a "localized
picture."   Here the term "localized" refers merely to the
fact that a convenient starting point for treating the
phonon-assisted motion of a particle in these cases is
to consider the carrier as initially placed (or "localized")
on a site and then calculate the rate with which it moves
to another site.   The localized electronic sites which
are considered as the starting points in this procedure
(the Wannier states) are not, however, eigenstates of the
system.

An example of a situation in which localized electronic states are associated with so-called band-like motion arises in what is sometimes called the Stark-ladder
problem.[14]   In this problem one recognizes that the application of a finite electric field over a periodic array of sites partially lifts the energetic degeneracy associated with the zero-field occupancy of any site and
yields field-dependent eigenstates which are localized

in that they possess a finite spatial extent in the direction of the applied field. In other words, for finite values of the applied electric field the eigenstates of this system are no longer Bloch states but are states termed Stark-ladder states.[15] Thus, in the absence of any "scattering," due to electron-lattice interaction say, a carrier cannot move. However, with interaction the mobility associated with the motion of a particle between these states is not necessarily characterized as hopping motion; in fact, one typically speaks of the electron as moving with a finite mean free path associated with its interaction with lattice vibrations. The resolution of what at first sight may appear to be a contradiction is that in the mobility calculation the electron-lattice interaction typically provides sufficient "scattering" so that one cannot utilize the Stark-ladder states as zeroth order states while one treats the electron-lattice interaction simply as a small perturbation on the system. In more physical terms, when the mean free path associated with the carrier's interaction with the lattice vibrations is very much smaller than the characteristic radius of the Stark states, the Stark states are no longer meaningful zeroth order states in a simple perturbative approach to charge transport.[16] In this typical situation it is the applied electric field which is considered the smallest perturbation on the system; this is the domain of so-called "linear response theory." Thus, the Stark-ladder problem illustrates that just because electronic states of a system are spatially localized does not necessarily imply that the carrier transport associated with such states is what would generally be referred to as hopping motion.

These examples illustrate the failure of a one-to-one correspondence between whether the eigenstates of a system are "localized" or "extended" and whether hopping does or does not characterize the carrier's finite-temperature motion. If, however, one adopts an operational point of view relevant, say, to transport measurements, one is then only concerned with understanding the circumstances under which hopping-type motion is manifested. While this is a complicated issue, one major aspect of the problem is that of understanding the effect of the electron-lattice interaction on the eigenstates of the system. The remainder of this talk will be restricted to addressing this question. In particular, attention will be directed toward elucidating the conditions under which a carrier can exist in a solid with and without producing a substantial displacement of the atoms of the material from their carrier-free equilibrium positions.

In other words, the situations in which a carrier does
and does not form a small polaron will now be considered.

    To begin, let us define what is meant by a polaron
in general, and a small polaron in particular. Consider
the two-dimensional array of atoms illustrated in Fig.
5a. As schematically depicted in Fig. 5b, if a stationary electron is added to this system the atoms will generally be displaced in response to its presence. These
atomic displacements will be such as to produce an additional potential well for the added electron. If this
carrier-induced potential well is sufficiently deep, the
carrier may occupy a bound state, being unable to move
without an alteration of the positions of the surrounding
atoms. The unit comprising the bound carrier and its induced lattice deformation is termed a <u>polaron</u>. Alternatively, since the potential well results from the carrier it is often referred to as being "<u>self-trapped</u>". In
addition, the adjective "small" in the term <u>small</u> <u>polaron</u>

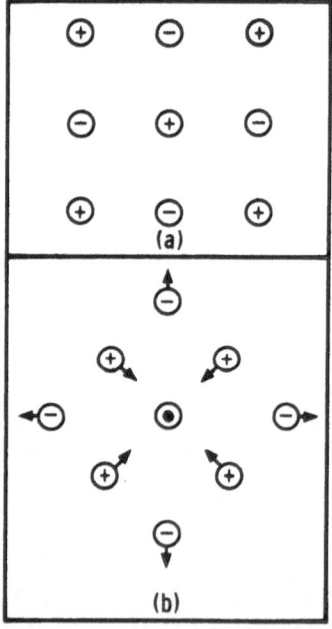

Fig. 5 Subfigure (a) depicts a portion of a lattice of
ionic charges in which no excess charge carriers are present, while (b) illustrates the readjustment of the positions of the ions in response to an excess electron being placed on the central positive ion.

indicates that the "self-trapped" carrier is essentially confined to a "small region" which is typically associated with a single atomic site.

Unfortunately, the term polaron is somewhat misleading. The name "polaron" is derived from early consideration of a model in which the carrier-induced potential well arises solely from the classical (electrostatic) interaction of a carrier with the optical-mode dipoles of a polar material. However, the argument that a stationary carrier will induce a lattice deformation, and hence potential well, which will tend to hold it in place, is quite general and is not at all restricted to a polar system. In fact, the best experimental evidence for the existence of small polarons is found in situations in which the carrier interacts with the atoms via an electron-lattice interaction that is short-range rather than via the long-range (electrostatic) interaction characteristic of the early theoretical discussions. For instance, small polarons even exist in non-polar crystals such as orthorhombic sulfur[12] and realgar, As_4S_4.[13] Thus, one should not restrict consideration of small-polaron formation to polar materials in which the classical coupling of an electron to the optical modes of an ionic lattice may be large; in this sense the term polaron is a misnomer. It should be remembered that "self trapping" occurs in both polar and nonpolar materials and may be associated with the carrier's interaction with both acoustic and optical-mode atomic displacements.

Let us now consider the energetic situation which prevails in the case of an excess electron forming a small polaron. Following the usual procedure of assuming that the energy of the electron is a linear function of the displacements of the atoms from their equilibrium positions, one finds that as a result of a carrier displacing the atoms surrounding it, its energy is reduced by an amount $2E_b$ while the strain energy of the lattice associated with producing this distortion is increased by an amount E_b.[17] As depicted in Fig. 6, this yields a net reduction of the energy of the system comprising a stationary electron and a deformed lattice, relative to that of a stationary electron in an undeformed lattice (in the tight-binding limit), by the amount E_b; this energy is termed the small-polaron binding energy. Furthermore, since the small polaron may equally well reside on any one of the geometrically equivalent sites in the crystal, we may expect the formation of a small-polaron band analogous to an electronic band of a rigid lattice. Pursuing the analogy with an electronic band, we may

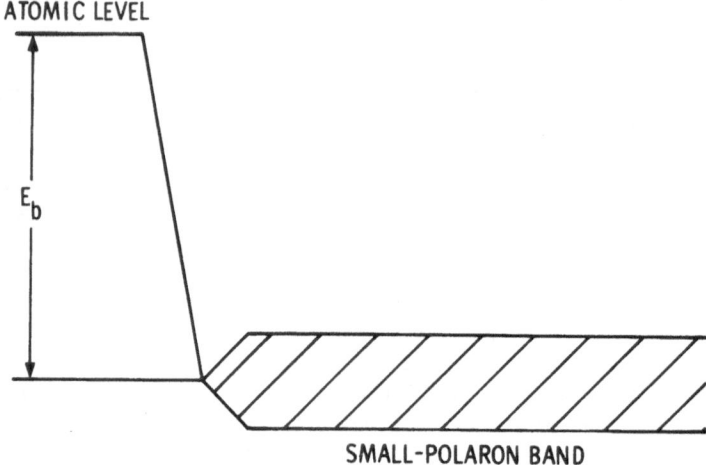

Fig. 6 As a result of allowing the atoms surrounding a
charge carrier at a site in a crystal to adjust fully to
its presence, the energy of the system is reduced [from
its value in the tight-binding (zero-overlap) limit] by
an amount E_b, the small-polaron binding energy. If one
then takes cognizance of the fact that the degeneracy
associated with the geometric equivalence of a regular
lattice is lifted when finite electronic overlap exists,
a typically very narrow small-polaron band is formed.

proceed as in the standard tight-binding approach to
view the wavefunction of the small polaron as the sum of
local wavefunctions; in particular, in the case of the
small polaron it is appropriate to take the local wavefunction to be a product of the <u>local electronic</u> wavefunction and the vibrational wavefunction associated with
the concomitant local deformation.[9] The width of the
small-polaron band calculated in this modified tight-
binding scheme is proportional to the product of an electronic transfer integral and a vibrational overlap integral. Each of the two vibrational wavefunctions in the
vibrational overlap integral represents a localized lattice deformation, the two deformations and concomitant
wavefunctions being related by a simple lattice translation of one lattice spacing. This vibrational overlap
factor, related to atomic tunneling, is typically very
small. Thus, in a crystal the small-polaron band is

usually extremely narrow, its width being very small compared with even vibrational energies. This is because
the motion of the carrier requires the concomitant transport of both an electron and its associated atomic displacement pattern. Indeed, due to the extreme narrowness of the small-polaron band one might guess that any
non-trival amount of energetic disorder ($\gtrsim 10^{-4}$ eV) will
yield Anderson localized small-polaron states.

Thus at least two distinct possible approaches to
the eigenstates of a carrier in a deformable material
present themselves. First, one may adopt the rigid-
solid approximation and assume that an electron can exist
and move through a solid without substantially displacing
the atoms surrounding it. Second, one may view the carrier as existing and moving with the atoms adjacent to it
being substantially displaced in response to its presence;
that is, it forms a small polaron. The issue is to define the conditions under which each of these points of
view is appropriate.

In order to gain insight into this matter a tight-
binding variational determination of the eigenstates of
an excess electron in a deformable molecular cyrstal has
been obtained.[18,19] In this calculation the electron's
energy is assumed to only be a function of the displacement of the atoms of the molecule of the molecular crystal upon which the electron resides. Furthermore, for
simplicity the electron is assumed to be coupled to only
the optical-mode vibrations of the crystal.

The principal results of this calculation are illustrated in Fig. 7, in which the energy spectrum corresponding to self-consistent (nearly) rigid-lattice
states, the conduction-band states, and to small-polaron
states is plotted versus a measure of the electron-lattice coupling strength, $E_b/\hbar\omega_o$. Here E_b is the small-
polaron binding energy, ω_o is the optical-mode frequency
and J is the rigid-lattice electronic transfer integral
(in a cubic lattice the conduction band has the width 12J).
Weak-coupling solutions only exist for values of $E_b/\hbar\omega_o$
less than some temperature-dependent maximum value
$C_{weak}(T)$. Small-polaron states only exist for values of
$E_b/\hbar\omega_o$ greater than the temperature-dependent minimum
value $C_{small}(T)$. Between these two limiting values both
types of states may coexist. With a decrease in the
adiabaticity parameter, $6J/\hbar\omega_o$, the coexistence regime is
decreased until for extremely narrow conduction bands,
$6J \lesssim \hbar\omega_o$, only one solution occurs; in this case the

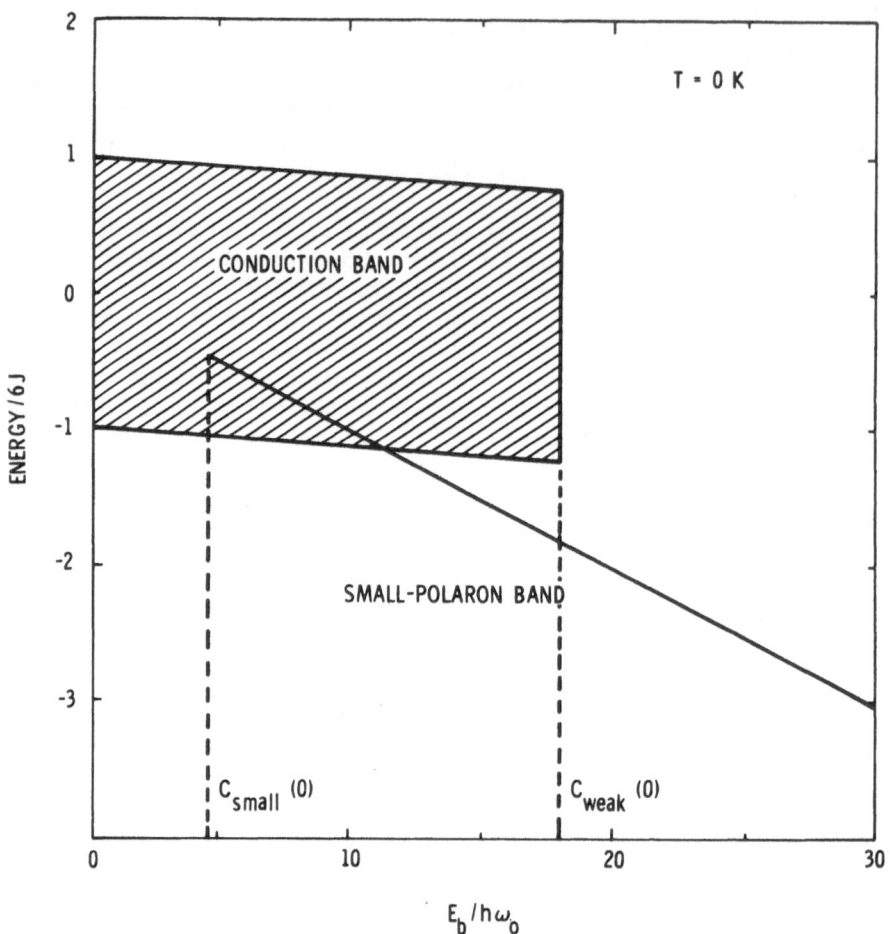

Fig. 7 The one-electron energy spectrum is plotted as a function of the electron-lattice coupling strength, defined as $E_b/\hbar\omega_o$, for $6J/\hbar\omega_o = 10$ at $T = 0$ K. The energy of the coupled system minus the zero-point vibrational energy, $n\hbar\omega_o/2$, is measured in units of the rigid-lattice band half-width, $6J$.

solution changes its character continuously from weak-coupling to small-polaron as the electron-lattice coupling strength is increased. In the opposite limit, the adiabatic limit, $\hbar\omega_o \to 0$ with $M\omega_o^2$ finite, the coexistence regime expands to include all nonzero values of the electron-lattice coupling strength.

Focusing attention on the regime in which the co-existence region exists and extends over a finite range of the coupling strength, $E_b/\hbar\omega_o$, one can obtain formulae for the two existence conditions. The weak-coupling existence condition may, apart from a minor numerical factor, be obtained from the following argument. To begin, it is noted that a carrier associated with an energy band of halfwidth 6J can only be confined to a single site for a time $\approx \hbar/6J$. It is this amount of time that is available for the carrier to force a substantial alteration of the positions of the atomic constituents associated with the occupied site. The force that the carrier exerts on the atoms which are adjacent to the occupied site is simply the electron-lattice interaction constant A. [The constant A is related to the small-polaron binding energy E_b by the relation $E_b = A^2/2M\omega_o^2$; $M\omega_o^2$ is the lattice stiffness.] Thus, during the time that a site is occupied the atoms surrounding the carrier are subjected to the carrier-induced acceleration A/M. The electron-lattice interaction will only be ineffective in producing a deviation of the motion of the surrounding atoms from the motion which characterizes them in the absence of a carrier, if the typical carrier-induced change in the "velocity" of these atoms is small compared with their carrier-free "velocity":

$$\left(\frac{A}{M}\right)\left(\frac{\hbar}{6J}\right) < \sqrt{\frac{\hbar\omega_o(N+1/2)}{M}} \ , \qquad (2)$$

where N is the Bose factor, $[\exp(\hbar\omega_o/kT)-1]^{-1}$. Thus, one may conclude that for a carrier to exist in a material characterized by a short-range electron-lattice interaction without being self-trapped, it must be able to move between sites sufficiently rapidly so as to preclude the atomic constituents from being able to respond to its presence. As the temperature is raised the atomic velocities increase and the presence of an electron on a site poses less of a perturbation to the atomic vibratory motion. Concomitantly, as illustrated in Fig. 8, if the temperature is raised the weak-coupling states remain dynamically stable to an increased value of the electron-lattice coupling strength.

Thus, it is stressed that above some (typically strongly temperature-dependent) value of the electron-lattice coupling strength the rigid-atom conduction-band

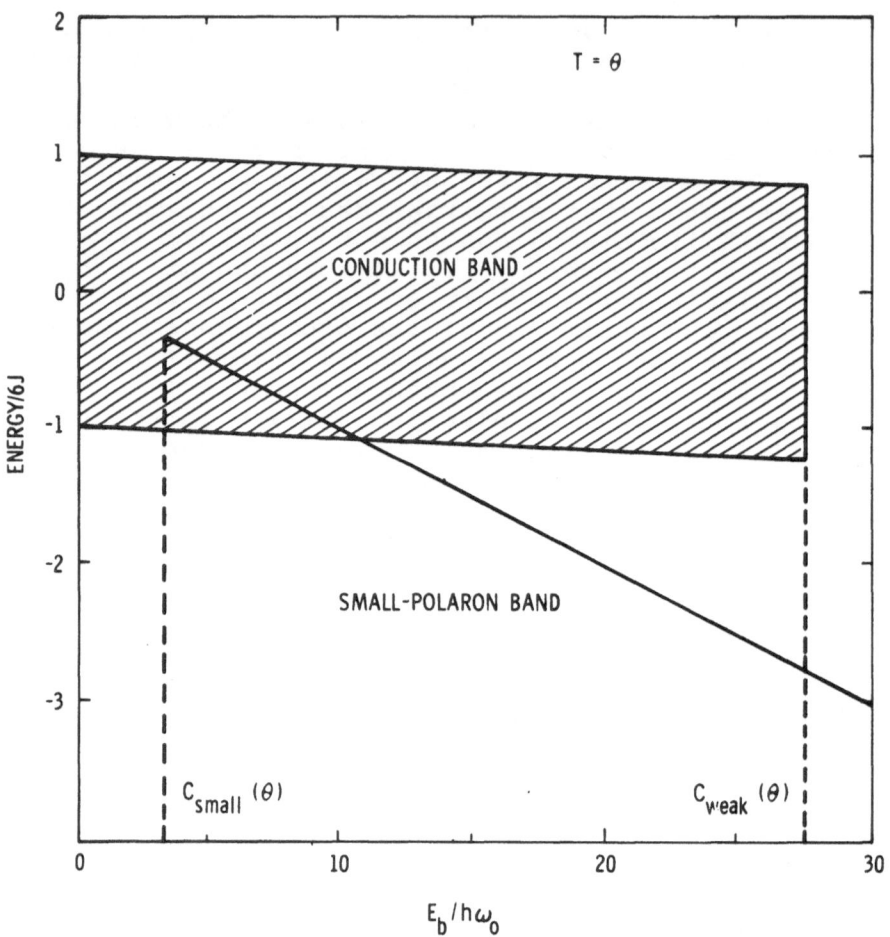

Fig. 8 The one-electron spectrum is plotted as a func-
tion of $E_b/\hbar\omega_o$ at the temperature $T = \hbar\omega_o/k \equiv \Theta$ with
$6J/\hbar\omega_o$ taken, as in Fig. 7, to be 10. The energy of the
coupled electron-lattice system minus the purely vibrational contribution to the energy, $n\hbar\omega_o(N + 1/2)$, is measured in units of the rigid-lattice conduction-band half-
width, $6J$.

states no longer are a reasonable representation of the
<u>dynamically</u> stable eigenstates of the coupled electron-
lattice system. Since the interaction between a stationary electron and the atoms surrounding it is typically

great enough to produce a large displacement of the atoms surrounding the electron,[17,20] it is only the ability of an electron to move between sites sufficiently rapidly so as to preclude the atomic displacements from occurring which justifies the rigid-atom approximation. If a carrier were to linger in the vicinity of a particular site too long the rigid-lattice approach would fail and the carrier would form a small polaron. It is speculated that one role which disorder may play in these considerations is to act to impede the rigid-atom motion of a carrier through a material[21]. Thus, a carrier placed at a site in a noncrystalline solid may not be able to diffuse away before the surrounding atoms adjust to its presence; as a result of the tendency of the presence of disorder to localize an electron, the rigid-atom approach may break down more readily (at smaller values of $E_b/\hbar\omega_o$) in a disordered material than in its ordered counterpart. The extent to which such an effect occurs in any particular system is hard to estimate <u>a priori</u>. However, there is experimental evidence of small-polaron formation in a variety of chalcogenide and vanadate glass systems. Thus one should at least be aware of the possibility that the rigid-atom picture of electrons in an insulating solid may fail in some of the frequently studied noncrystalline insulators.[22]

A further aid to understanding the role played by the electron-lattice interaction in determining the eigenstates of an electron in a deformable material is found by employing the adiabatic approximation to study the eigenstates of an electron in a deformable continuum (an elastic medium).[19,23,24] In particular, one can study the effect of the electron-lattice interaction on an electron bound in a (coulombic) impurity potential in this deformable continuum as well as its effect on an electron in a uniform deformable continuum. Within the adiabatic approach the motion of the atoms making up the elastic continuum is considered to be sufficiently slow compared with the motion of an electron so that the electron may be assumed to adjust to the instantaneous state of the atoms (continuum). In this scheme, the Born-Oppenheimer method, the energy of the (vibrationless) groundstate is the minimum of the sum of the electronic energy, itself a function of atomic positions, and the potential energy associated with the atomic vibratory motion. The resulting groundstate energy, E may be written as a sum of terms:

$$E = T_e - (V_{imp} + V_{int}) + E_{strain} , \qquad (3)$$

where the contributions to the total energy are, respectively, the expectation values of the electron kinetic energy operator, its potential energy, and the strain
energy of the continuum; the electron's potential energy
operator is expressed as a sum of contributions arising
from its interaction with a defect or impurity and from
its interaction with strains in the deformable continuum,
V_{imp} and V_{int}, respectively.[25] To proceed further, consider a model in which: 1) the electron-lattice interaction is represented by a linear dependence of the electron's potential energy at any point in the continuum on
the dilation of the continuum at that point (analogous
to the deformation potential in semiconductors); 2) the
impurity potential is taken to be a spherically symmetric
(r^{-1}) coulombic well; and 3) the strain energy is the sum
of contributions from the entire continuum, each of which
is quadratic in the dilation of the continuum at each
point. To find the essential features of the adiabatic
eigenstates of this model one may exploit the fact that
with a change of scale [the electron's position vector
$\underset{\sim}{r}$ being replaced by $\underset{\sim}{r}/R$] the ground state energy associated with a finite-radius eigenstate must be at a minimum
at the scale corresponding to the actual eigenstate, R=1.
In other words, writing out the energy as a function of
R for the above-described (three-dimensional) model, one
can associate the minima in the E(R)-versus-R curve with
the eigenstates of the system. Specifically, one finds
in a three-dimensional system that

$$E(R) = \frac{T_e}{R^2} - \frac{V_{imp}}{R} - \frac{V_{int}}{2R^3} , \qquad (4)$$

where use has been made of the relation $E_{strain} = \frac{1}{2}V_{int}$.
Looking for the minimum in the E(R)-versus-R curve in
the limit of a vanishing electron-continuum interaction,
$V_{int} = 0$, one finds, as illustrated in Fig. 9, a single
minimum at $R = 2T_e/V_{int} \equiv 1$. This confirms what is well-
known, namely that a coulombic well in a nondeformable
system supports a finite-radius bound state in which the
kinetic energy contribution to the total energy is half
of the magnitude of the potential energy contribution.
On the other hand, as shown in Fig. 10, for a finite electron-continuum interaction but a vanishing defect potential (an electron in a homogeneous deformable continuum)
there is no finite radius minimum in the E(R)-versus-R
curve. The only minima which occur are at $R = \infty$, corresponding to an unbound electron in an unstrained

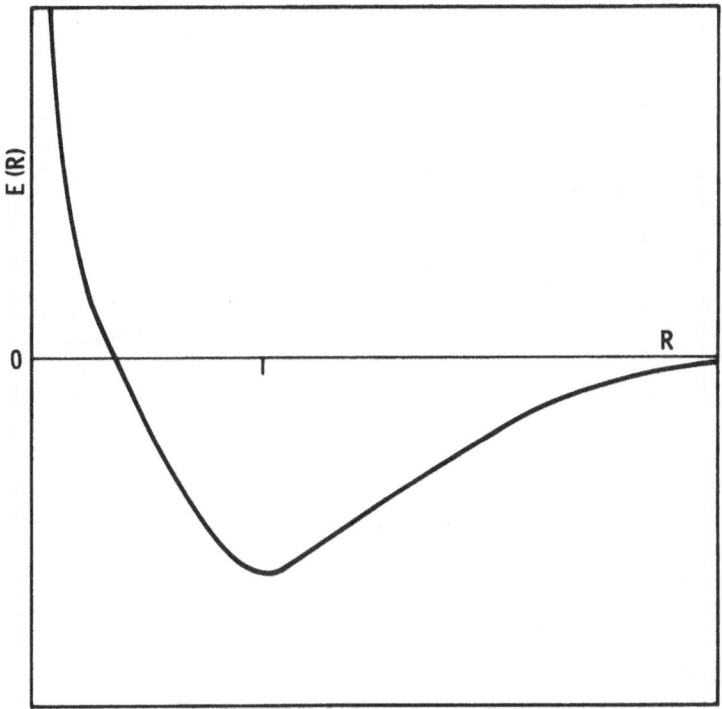

Fig. 9 The adiabatic groundstate energy of the system comprising an electron bound to a donor-like impurity in a three-dimensional deformable continuum is plotted against the scaling factor R, related to the spatial extent of the electron's wavefunction, from Eq. (4), in the case of the vanishing electron-lattice interaction, $V_{int} = 0$.

continuum, and at R = 0, corresponding to an electron self-trapped in an infinitely deep and infinitesimally narrow potential well associated with an infinite local strain. These two states are, respectively, the continuum analogies of the previously-discussed conduction-band and small-polaron eigenstates of an excess electron in a deformable crystal.

In the situation of primary interest, that in which both V_{int} and V_{imp} are finite, one finds that either of two situations can arise: 1) for $T_e^2 > \frac{3}{2} V_{int} V_{imp}$

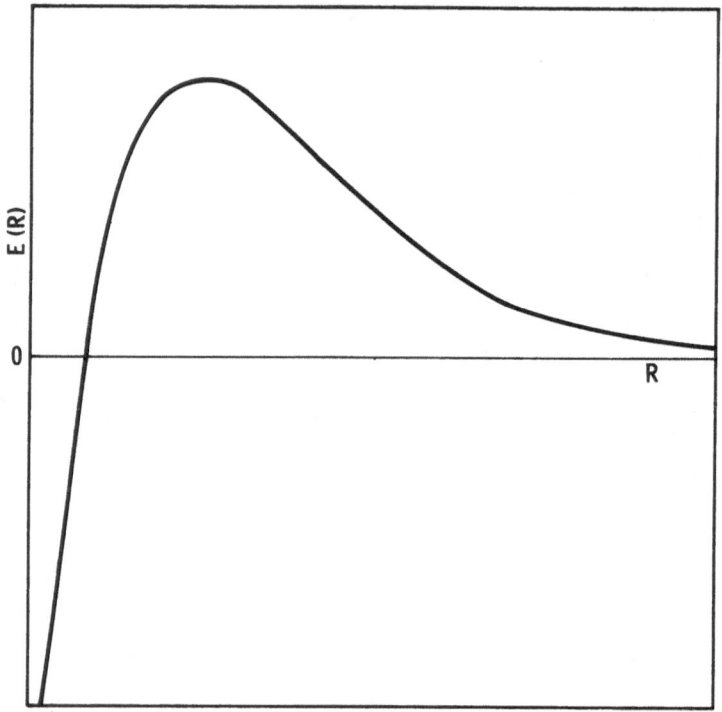

Fig. 10   The adiabatic ground state energy of the system
composed of an electron in a three-dimensional deformable continuum is plotted against the scaling factor R.
This corresponds to $V_{imp} = 0$ in Eq. (4).

(Fig. 11a), in addition to a finite-radius bound state
a small-polaron solution will exist, and 2) for
$T_e^2 < \frac{3}{2} V_{int} V_{imp}$ (Fig. 11b), only the small-polaron solution exists. In other words, for a sufficiently strong
electron-continuum interaction finite-radius states (non-
small-polaron states) will not exist even in the adiabatic limit. In a real material, characterized by discrete atoms as distinct from a deformable continuum, the
displacement of the atoms surrounding a localized electron saturates at a maximum value when the characteristic radius of the electronic wavefunction becomes smaller
than some distance comparable to an interatomic spacing.
Thus, while an electron in a continuum can lower the energy of the coupled electron-continuum system without
limit simply by shrinking its spatial extent indefinitely,

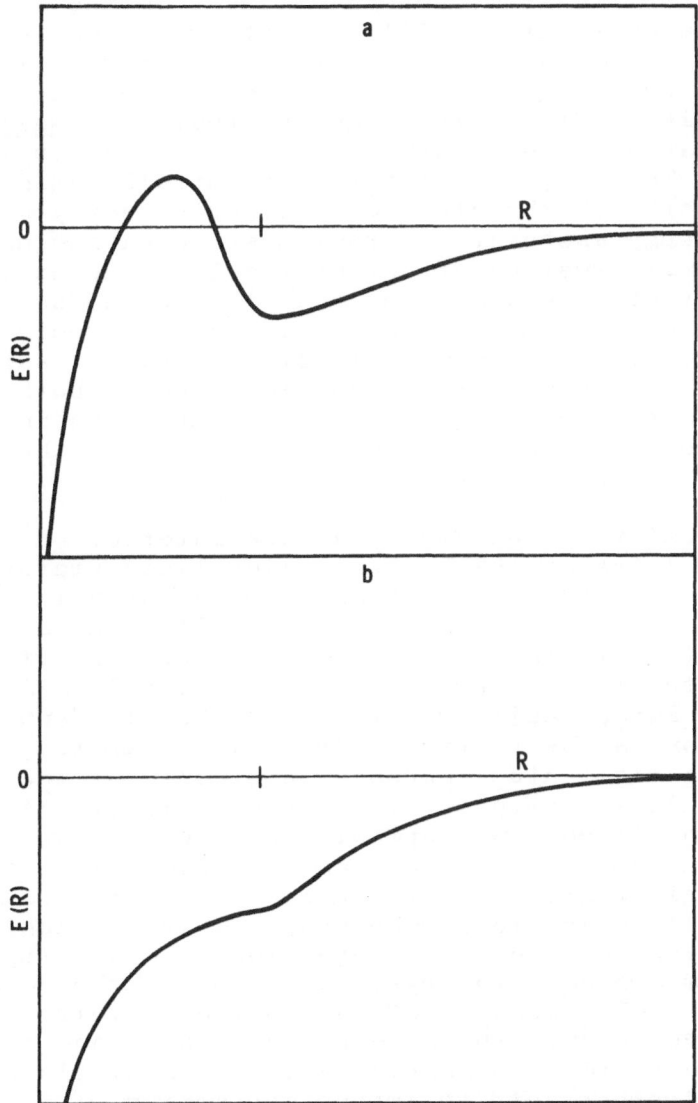

Fig. 11 The $E(R)$-versus-R curve of Eq. (4) is plotted in the general case when neither V_{int} nor V_{imp} vanishes. In Fig. 11a one has $T_e^2 > \frac{3}{2} V_{int} V_{imp}$ and a finite-radius minimum exists. In Fig. 11b one has $T_e^2 = \frac{3}{2} V_{int} V_{imp}$. When $T_e^2 < \frac{3}{2} V_{int} V_{imp}$, no finite minimum exists.

in a real material this is not the case.[26]

However, in a discrete atomic system the localized small-polaron solution will exist (carrier and concomitant atomic displacement pattern restricted to the vicinity of a single site). If the electron-lattice coupling strength is sufficiently strong so that the <u>finite-radius</u> small-polaron state is dynamically stable against spreading out. That is, to exist (be dynamically stable) the lowering of both the electronic kinetic energy and the strain energy associated with the state expanding infinitesimally must be offset by the increase in the electron's potential energy. On the other hand, as illustrated in Fig. 11, the large-radius nonpolaronic localized state will be dynamically stable against shrinking to a small-polaron state if the increase in both kinetic energy and strain energy upon shrinking by an infinitesimal amount is greater than the decrease in the electron's potential energy.

Typically the electron-lattice interaction is sufficiently large so that severely localized states (radii ~ lattice constant) are accompanied by substantial lattice deformations; they form localized small-polarons. For example, this is the situation characteristic of the groundstate of F-centers.[27] However, for localized states of rather large radii the effect of the electron-lattice interaction may be sufficiently small so as to justify adopting a rigid-atom approach to studies of the groundstates of these systems. The large-radii ($\sim 10^2$A) shallow dopants in silicon and germanium are well-known examples of this circumstance. Thus, as a carrier finds itself increasingly confined, the likelihood of its groundstate being small-polaronic is enhanced. This conclusion complements that obtained from studying the validity of the rigid-atom approach in crystals. Namely, the more time that a particle spends confined in the vicinity of a particular atomic site the greater the likelihood of a failure of the rigid-atom approximation. Thus, since the presence of disorder acts to impede the motion of a carrier through a noncrystalline semiconductor both by creating regions in which an electron is confined (localized) and by reducing the rate that characterizes the real intersite motion of a particle, it is reasonable to anticipate that at least some of the states which a carrier can occupy will be small-polaronic.

To gain some feeling for the ramifications of this occurence consider Fig. 12a, in which a simple version of the density of states of the rigid-atom conduction band of a disordered solid with some defect states lying in the gap is illustrated. Permitting the atoms to adjust to the presence of a charge carrier, making the system deformable, may yield (among others) one of the two stable-state situations illustrated in Figs. 12b and 12c. In the first instance, the stable states associated with the defect states and some deep-lying localized states are small-polaronic.[28] The energy spread of the polaronic levels is assumed to be relatively narrow, consistent with the assumption of the presence of substantial short-range order in this hypothetical disordered insulator. In the second instance, all of the states of the rigid system are taken to have stable small-polaron counterparts.[24] The deep levels in a-Ge are presumably a defect-related illustration of the first instance[29] if not the second. Transport experiments suggest that some chalcogenide glasses[30,31] as well as some vanadate glasses[32] are examples of the second case. Finally it should be noted that the rigid-atom scheme should be appropriate to optical absorption experiments since, via the Franck-Condon principle, the atoms are assumed to be unable to adjust to the optical excitation of a carrier.[33] Radiationless transport experiments, however, presumably can sample the non-rigid small-polaron levels.

While this discussion of "localized states" has suggested that small-polaron formation may occur in disordered insulators, it is well beyond the present state of the theory to predict their absence or presence in any particular noncrystalline, or even crystalline, system. The existence of small polarons in any system must therefore be decided on the basis of experimental results which will or will not manifest those properties which are hallmarks of the presence of small polarons. In particular, it is measurements of electrical transport properties that presently appear to be most likely to provide the desired information.

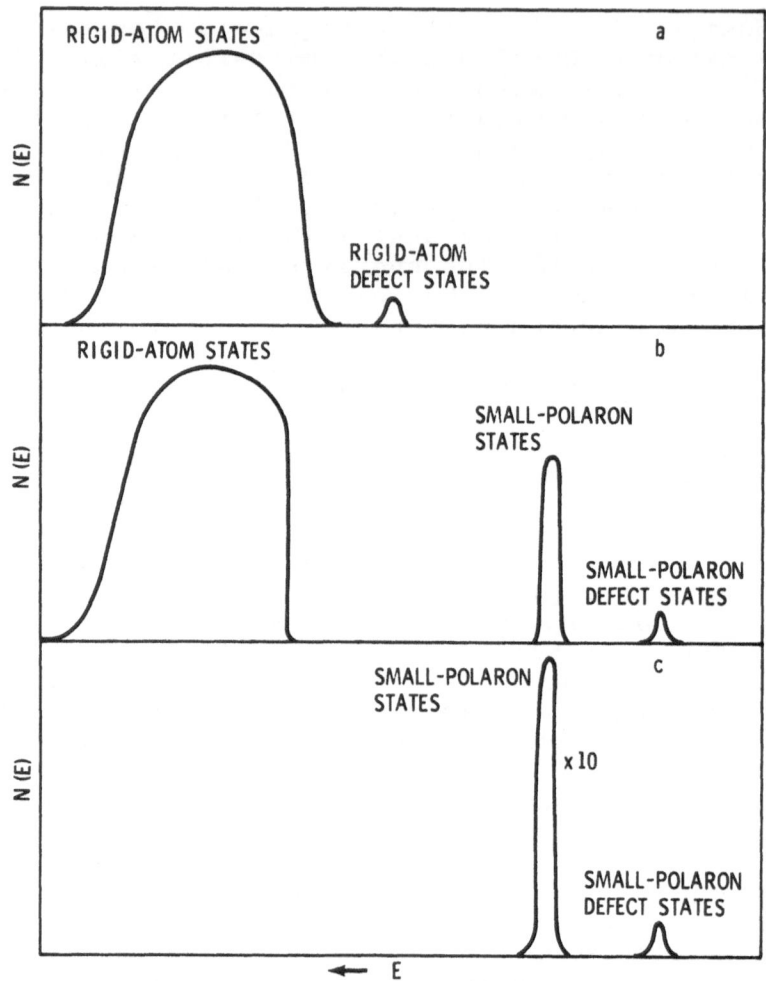

Fig. 12 In subfigure 12a, a hypothetical rigid-atom density of states for a conduction band and set of defect
levels in a noncrystalline solid is illustrated. With a
finite electron-lattice interaction the deep-lying bandtail and defect states may be energetically unstable with
respect to small-polaron formation. Subfigure 12b illustrates the case when only some low-lying states have low-
er-lying small-polaron counterparts. The question of
whether any of the energetically unstable rigid-atom states
associated with these small-polaron states can coexist
with these is not addressed in this figure since only the
energetically stable states are shown. In subfigure 12c
the case when all the rigid-atom states are unstable with
respect to small-polaron formation is depicted. Note
that the presumably narrow band of small-polaron states
is associated with a very high density of states.

REFERENCES

1. P. W. Anderson, Phys. Rev. 109, 1492 (1958).
2. N. F. Mott, Phil. Mag. 24, 911 (1971).
3. C. Kittel, Introduction to Solid State Physics (Wiley, New York, 1956), pp. 493-497.
4. H. H. Woodbury and W. W. Tyler, Phys. Rev. 105, 84 (1957).
5. Specifically, the normalization constant of the localized wavefunction should be independent of the size of the sample.
6. N. F. Mott, Adv. Phys. 16, 49 (1967).
7. M. H. Cohen, H. Fritzsche, and S. R. Ovshinsky, Phys. Rev. Lett. 22, 1065 (1969).
8. M. H. Cohen, Amorphous and Liquid Semiconductors, edited by N. F. Mott (North-Holland, Amsterdam, 1970), p. 406.
9. T. Holstein, Ann. Phys. (N. Y.) 8, 343 (1959).
10. F. J. Keller and R. G. Murray, Phys. Rev. 150, 670 (1966).
11. C. Crevecoeur and H. J. DeWit, J. Phys. Chem. Solids 31, 783 (1970).
12. D. J. Gibbins and W. E. Spear, J. Phys. Chem. Solids 27, 1917 (1966).
13. G. B. Street and W. D. Gill, Phys. Status Solidi 18, 601 (1966).
14. G. Wannier, Elements of Solid State Theory (Cambridge, Cambridge, 1960), p. 190-192.
15. The phrase Stark-ladder denotes the fact that the energy spacing between adjacent eigenstates is just the intersite potential energy difference eEa, where e is the electron's charge, E is the strength of the electric field, and a is the lattice constant; the energy spectrum therefore resembles a ladder with the interrung spacing eEa
16. In the case of an electron hopping between large-radius ($\sim$50-100 A) shallow impurity states at low temperatures, the free carrier's mean free path is typically much greater than the impurity-state radii.
17. Both model calculations and experiments yield estimates of the energy associated with such displacements (the small-polaron binding energy, E_b) which are substantial: Typically E_b lies between several tenths of an electron volt and several electron volts.
18. D. Emin, Phys. Rev. Lett. 28, 604 (1972).
19. D. Emin, Adv. Phys. 22, 57 (1973).
20. Typically $E_b/\hbar\omega_o$ is estimated to range from 10 to 100.
21. In terms of tight-binding theory this arises from disorder of the local-site energies (Anderson disorder)

and perhaps from disorder of intersite transfer integrals. For example, while the time required for
a particle to diffuse from a site in a crystal is
$\sim \hbar/6|J|$, if the site in question is surrounded by
sites which differ in energy from the occupied site
by Δ, with $|\Delta| \gg |J|$, then the time to diffuse to de-
generate (next-nearest-neighbor) sites is much long-
er, $\sim (\hbar/6J)(|\Delta/J|)$.

22. Some of the experimental ramifications of such an
occurrence will be pursued in subsequent lectures.
Suffice it to comment here that room temperature
small-polaron motion is typically characterized as
thermally-activated hopping motion; hence what has
frequently been dubbed "hopping in localized states"
may be alternatively described as small-polaron motion.

23. A fuller discussion of this calculation will be published. Aspects of the calculation are discussed in
greater detail in Sec. 7.3 of Ref. (24) and in Appendix D of Ref. (19).

24. D. Emin, Electronic and Structural Properties of
Amorphous Semiconductors, edited by P. G. LeComber
and J. Mort (Academic, New York, 1973), p. 268.

25. These expectation values are calculated with respect to the exact adiabatic electronic wavefunction.
That is, any change in the electronic potential produces a change in the electronic wavefunction which
in turn alters all of the individual contributions
to the adiabatic ground-state energy. This is all
implicitly taken into account in this procedure.

26. This result is contingent on the electron-continuum
interaction possessing a short-range component and
the continuum being three-dimensional: See Appendix
D of Ref. (19).

27. W. Harrison, Solid State Theory (McGraw Hill, New
York, 1970), p. 333.

28. P. W. Anderson, Nature (Lond.) 235, 163 (1972).
Anderson also mentions the possibility that the conduction-band small-polaron levels will fall into the
valence band.

29. N. F. Mott, private communication.

30. D. Emin, C.H. Seager, and R. K. Quinn, Phys. Rev.
Lett. 28, 813 (1972).

31. C. H. Seager, D. Emin and R. K. Quinn, Phys. Rev.
B8, 4746 (1973).

32. I. G. Austin and N. F. Mott, Adv. Phys. 18, 41 (1967).

33. See page 323 of Ref. (24) and Ref. (27).

TRANSPORT PROPERTIES OF AMORPHOUS SEMICONDUCTORS

W. Fuhs

Fachbereich Physik

University of Marburg, Germany

I. INTRODUCTION

The investigation of transport properties of amorphous semiconductors in the past has been much stimulated by the interest in possible technical applications. So far these expectations have only partly been fulfilled and at present mainly chalcogenide glasses are of some technical importance as photoreceptors, switching devices, memories etc. In the recent years, the tetrahedrally bonded amorphous semiconductors Ge and Si have probably become the most intensively studied disordered materials. This is a remarkable trend which can hardly be explained by technological reasons. These materials are the simplest disordered systems exhibiting only positional disorder and in the crystalline form these semiconductors are the best understood. The philosophy, therefore, is that particularly the investigation of amorphous Ge and Si offers a chance to find a profound understanding of the physics of the disordered state in general. In the following I will also follow this trend and therefore most of the presented experimental results will be on amorphous Ge and Si. It is, however, not my aim to give a complete review of all the interesting work done in this field. For this the reader is refered to the various review articles in the literature[1,2,3,4]. I will rather concentrate in these lectures on those methods which, in my opinion, have been the most useful ones for the investigation of amorphous semiconductors and therefore have strongly influenced our present ideas

about the transport mechanism. Of particular interest
will be those methods which yield information about the
distribution of localized gap states. In the first section, I will deal with the various models, which have
been suggested for the density of gap states and their
experimental evidence in field effect measurements. A
brief discussion of the carrier mobility in these energy
states will follow. In this context the Hall effect
and drift experiments will be covered. The investigation
of the conductivity and the thermoelectric power has contributed considerably to our present knowledge about the
transport in amorphous semiconductors. That is why these
measurements and the kind of information they provide,
will be treated in some detail. A type of conduction
which is unique for amorphous semiconductors is variable
range hopping in states near the Fermi level. The experimental information about this type of conduction is
critically discussed. Then some aspects of photoconductivity in amorphous photoconductors will be treated.
Here, I will confine myself on those features which are
typical for the amorphous state and originate from the
high density of localized gap states.

II. DENSITY OF LOCALIZED STATES

A. Models

Various models have been discussed for the density
of states of amorphous semiconductors in order to explain one or the other experimental result. The one proposed for certain glasses by Cohen, Fitzsche and Ovshinsky[5] is shown in Fig. 1a. Tails of localized states are
assumed to extend from the valence and conductions bands
far into the gap. Near the center of the gap these
states overlap. Since the states, which tail from the
valence band are donor like i.e. neutral when occupied
and those tailing from the conduction band are acceptor
like, charged states result near the gap center which
effectively pin the Fermi level. Inside the bands at
energies E_C and E_v a sharp transition between localized
and extended states is assumed to occur.

Another model suggested originally by Davis and
Mott[6] and later improved by Mott[7] is illustrated in
Fig. 1b. Due to the loss of long range order the electron states near the band edges E_A and E_B are localized.
Near the middle of the gap two overlapping bands of localized states are assumed to exist, which may originate
from dangling bonds which can either act as donors D or

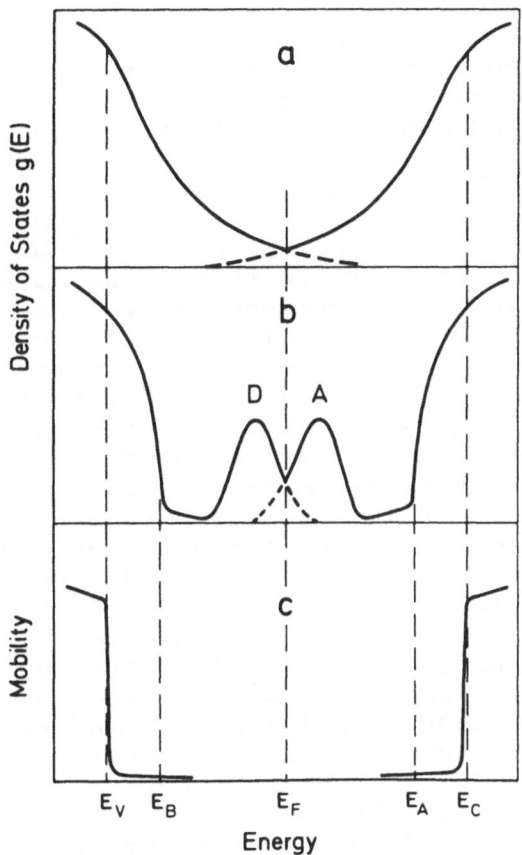

Fig. 1 Models for density of states proposed by (a) Cohen, Fitzsche and Ovshinsky (Ref. 5) and (b) Mott and Davis (Ref. 6,7). (c) Energy dependence of the mobility.

acceptors A. These states pin the Fermi level near the gap center like in the CFO-model.

A common feature of all these density of states models is, that there is a finite density of localized states at the Fermi energy. Due to the different character of the electron states above and below E_C and E_V, different conduction mechanisms have to be expected. Whereas in extended states charge carriers are assumed

to move with an almost normal mobility of about 1 to 10
cm^2/Vs, transport in localized states can only occur by
phonon assisted tunnelling. That is why at the transition from extended to localized states at E_C and E_V the
mobility drops more or less sharply by some orders of
magnitude giving rise to mobility edges. As a consequence a so called mobility gap E_C-E_V arises as is indicated in Fig. 1c.

The above models assume that the amorphous material be homogeneous. There are, however, numerous experimental results which indicate that these materials are
inhomogeneous due to the presence of vacancies, voids of
microscopic and macroscopic dimensions. The only model
that takes heterogeneity into account is the one originally proposed by Heywang and Haberland[8] for some chalcogenide glasses. The application of this model to tetrahedrally bonded amorphous semiconductors has been discussed in detail by Fritzsche[1,9]. This model is illustrated in Fig. 2. It is assumed that as a consequence
of disorder the energetic position of the band edges
fluctuates in space (Fig. 2a). In addition also the
size of the gap is assumed to vary due to fluctuations
in the density. The density of states distribution (Fig.
2b) obtained by projection of all states on the energy
scale has some similarity with the CFO-model in Fig. 1a.
However, one can expect additional energy states appearing inside the gap due to certain specific defects in
the structure. Two types of such gap states have been
included in Fig. 2c. That leads to a density of states
distribution (Fig. 2d) which resembles the one deduced
from field effect for amorphous Si (see B). Also in
this model a mobility edge is to be expected, which can
be explained in terms of classical percolation theory.
If E_C is the height of the lowest path in this potential
mountain, electrons with $E > E_C$ will be free to move throughout the semiconductor whereas those with $E < E_C$ will be
more or less localized. These percolation thresholds
E_C and E_V are the analogs of the mobility edges in the
homogeneous models.

B.   Experimental Evidence

The knowledge about details of the density of states
distribution is relatively poor. Experimental evidence
for a high density of localized states near E_F is obtained from ac- and dc-conductivity, thermoelectric power, contact properties and field effect. However,
there are remarkable differences between the values determined by different methods.

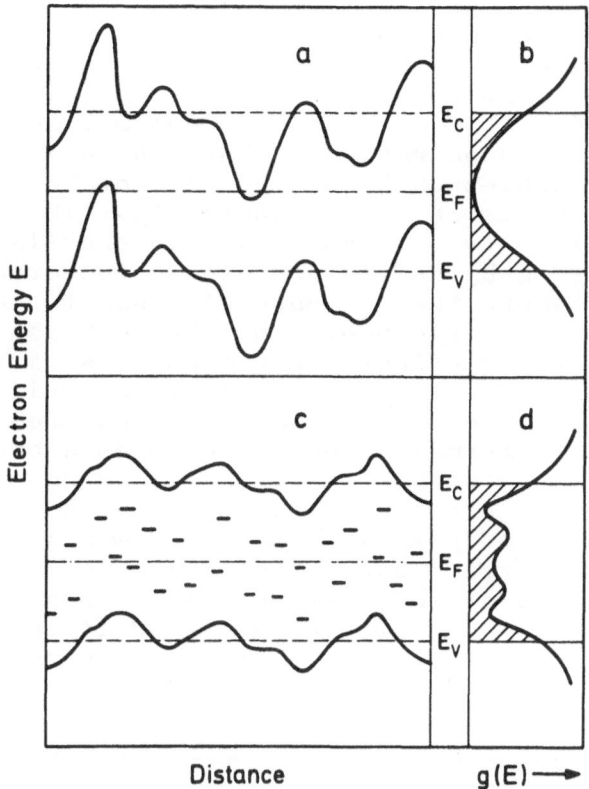

Fig. 2 Heterogeneous model (a) without gap states and
(c) including gap states. (b) and (d) density of states
distributions for models (a) and (b).

The most promising technique for the investigation
of localized gap states is the field effect. In such
an experiment an electric field is applied capacitively
to the surface of a semiconductor. Excess charge induced on the surface distributes among the available energy states. The resulting change in surface potential
manifests itself in a change of surface conductance,
which therefore contains information about the electronic structure of the surface and of the adjacent space
charge layer. For crystalline semiconductors field effect has been widely used for the investigation of surface states[10].

The change of surface conductance $\Delta\sigma = \sigma - \sigma_b$ as a

function of the bending of the energy bands at the surface ε is shown schematically in Fig. 3 for an n-type
sample. The band bending ε is taken to be positive for
an accumulation layer for which the bands are bent down.
It is obvious that both an accumulation ($\varepsilon > 0$) and an
inversion layer ($\varepsilon < 0$) have high conductances due to
their carrier concentrations. Between these two extreme
cases in the depletion layer the conductance is less and
exhibits a minimum $\Delta\sigma_m$ at an energy ε_m. The surface conductance is zero twice. For $\varepsilon = 0$ in the flat band position σ equals σ_b by definition and on the other side of
the minimum, where the decrease of σ due to the depletion
of electrons is just canceled by the contribution of the
holes. In a field effect experiment ε is varied externally by means of a transverse electric field and the
surface conductance of the sample is measured. In addition to these externally induced changes a band bending
can exist already without external field due to the presence of surface states. Since then one does not start
from the flat band position the problem arises how to
correlate the surface conductance with the position of
the energy bands at the surface. In crystalline semiconductors this can easily be achieved if the minimum
$\Delta\sigma_m$ is attained. The barrier height ε_m where this

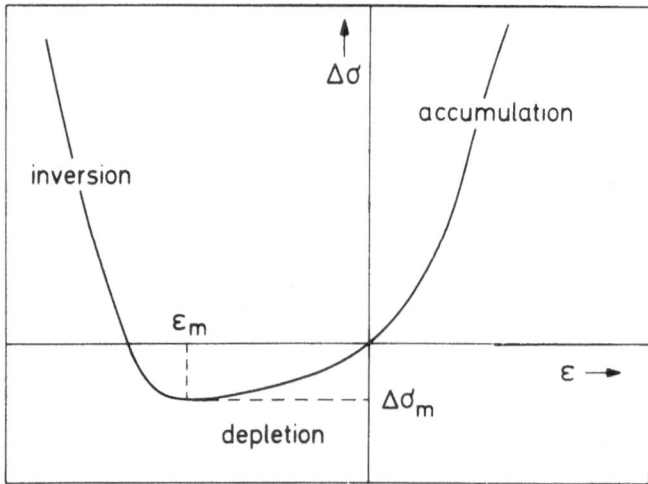

Fig 3 Dependence of surface conductance $\Delta\sigma = \sigma - \sigma_b$ of
an n-type crystal on band bending ε. (ε is positive for
an accumulation layer).

minimum occurs, namely is a known function of temperature and impurity concentration[10]. The band bending at the surface of a semiconductor can only be evaluated in a reliable manner from a measurement of conductivity, if this minimum is found.

In the case of amorphous semiconductors this method in principle enables a quantitative determination of the density of localized gap states.

In order to make the principle clear in Fig. 4 the energy diagram at the surface is shown without and with external voltage U_F. It is assumed that at zero bias no band bending exists. With the field electrode positively biased, the number N_i of electrons induced per cm^2 of the semiconductor surface is $N_i = \frac{1}{e} \frac{C}{A} U_F$, where U_F is the applied voltage and C the capacity of the arrangement. Let us assume that practically all of this charge is trapped in localized states. This necessitates that a corresponding number of states cross the Fermi level E_F, namely all states which lie between E_F and $E_F - \varepsilon$ in the shaded area in Fig. 4a,b. With the assumption that $g(E)$ is constant in this energy range, the induced charge density can be approximated by

$$N_i = \int_0^L g(E) \cdot \varepsilon(x) \, dx \tag{1}$$

It is obvious that thereby all energy states within the space charge region are shifted in energy by an amount $\varepsilon(x)$ which is the larger, the smaller the density of localized states near E_F is. Since by this shift the mobility edge of the conduction band approaches the Fermi energy, the conductivity $\sigma(x)$ at a distance x from the surface increases with respect to σ_b the conductivity outside the space charge region:

$$\sigma(x) = \sigma_b \exp \left(\frac{\varepsilon(x)}{kT} \right) \tag{2}$$

If the thickness of the sample D is small as compared to the extension of the space charge layer L, these relations can be simplified and allow a rough estimate of the density of states at the Fermi energy from the measured change in conductivity:

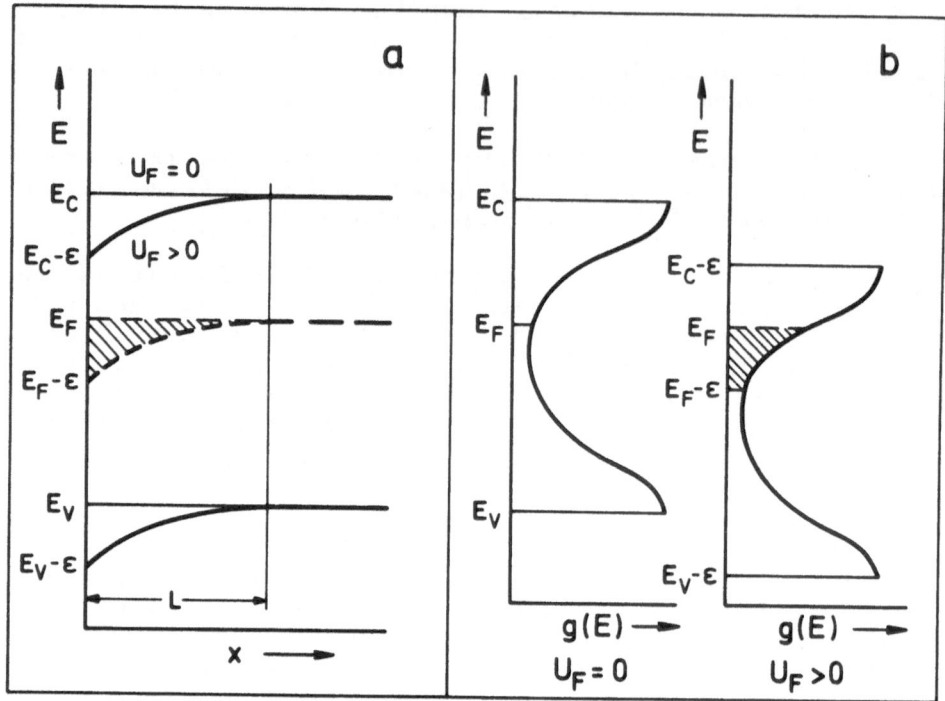

Fig. 4 Energy bands at the surface with and without external voltage U_F (a) and corresponding density of states distributions (b).

$$g(E_F) = \frac{N_i}{D \cdot \varepsilon} \qquad\qquad \frac{\sigma}{\sigma_b} = \exp\left(\frac{\varepsilon}{kT}\right) \qquad\qquad (3)$$

Egerton[11] has performed field effect measurements in some chalcogenide glasses and found only relatively small effects. The conductivity of these 500 Å thick films increased typically by a factor of 2 with a number of induced charges of 2.10^{12} cm^{-2}. That corresponds to a shift of the Fermi level by 0.02 eV and a density of states at E_F of 2.10^{19} cm^{-3} eV^{-1}. This estimate was obtained with rather crude simplifications. Generally one has to solve Poisson's equation and to integrate relation (1) and (2) over the whole space charge layer. Furthermore the influence of surface states was neglected. If present, such states could also trap some of the induced charge. The density of states determined by field effect measurements therefore can only be taken as an upper limit.

The only amorphous material which so far exhibits a large field effect is amorphous Si prepared by decomposition of silane in a radio frequency glow discharge This material has intensively been studied by Spear and LeComber[12] and Madan and Spear[13]. Fig. 5 shows typical room temperature measurements for differently prepared specimens. The current through the sample increases by several orders of magnitude when the field electrode is positively biased. With the field electrode negative, the changes are only small. The authors explain this by the assumption that predominantly electrons contribute to the transport. The steep increase of current with the applied voltage U_F indicates that the density of states $g(E)$ is relatively small. Since the slope of the curves at higher U_F decreases, $g(E)$ becomes larger when one approaches the conduction band. If one adopted this interpretation also for negative U_F one had to

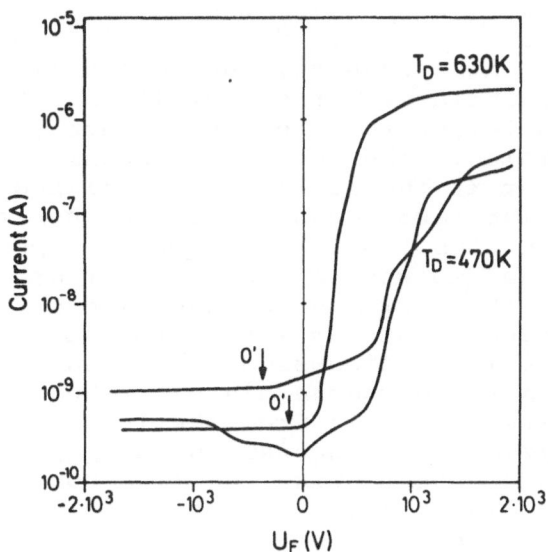

Fig. 5 Current through a layer of glow discharge Si as a function of voltage U_F applied between the field electrode and the specimen electrodes after Ref.·12. T_D denotes the substrate temperature and 0' the suggested flat band position.

assign the weak changes of the current to a particularly
high density of states. On the other hand, at high negative U_F also holes should contribute. In this case,
one expects a curve of the type shown in Fig. 3, which
is not observed. It is remarkable that practically all
these curves show the same behavior at negative U_F no
matter what the actual preparation conditions were. To
analyze these curves quantitatively it is essential to
decide at which value of U_F the energy bands are flat,
i.e., one has to find out the energy scale. As discussed above, this necessitates that the minimum of conductance is attained which is not the case here. The
authors suggest that at the transition from depletion layer to accumulation layer the current increases
rapidly. Consequently flat band position is assumed to
exist where this increase starts (points O'). A further
important assumption which has to be made is that the
density of states distribution is the same at the surface and within the specimen, i.e., that the field effect really probes bulk properties.

For an analysis Poisson's equation is solved for
a stepwise increasing potential of the field electrode
and the excess carrier distribution is averaged in a
suitable manner. The resultant density of states distribution is given in Fig. 6[4]. The full lines here represent experimental results, the dotted lines have been
drawn with a certain amount of intuition. Parameter is
the deposition temperature, curves E were obtained for
evaporated samples. The lower the deposition temperature, the higher is the density of localized states.
The highest values are found for the evaporated layers.
The preparation conditions are of major importance for
$g(E)$ deep in the gap whereas the density of states distribution near the former band edges seems to be only
weakly influenced. This is seen even better from drift
mobility measurements discussed below and indicates a
different character of these states. Those near E_c or
E_v are considered to arise from the loss of long range
order. Localized states deeper in the gap are attributed to certain defects like dangling bonds, vacancy
complexes etc. The density of states is not structureless but has a pronounced peak at about 0.4 eV below
the mobility edge of the conduction band E_c. Near the
former band edge at an energy E_A about 0.18 eV below E_c,
there is an abrupt increase in $g(E)$. The existence of
this edge in the density of states distribution throws
some doubt on the concept of mobility edges. These mobility edges, namely, have been introduced[5] to account

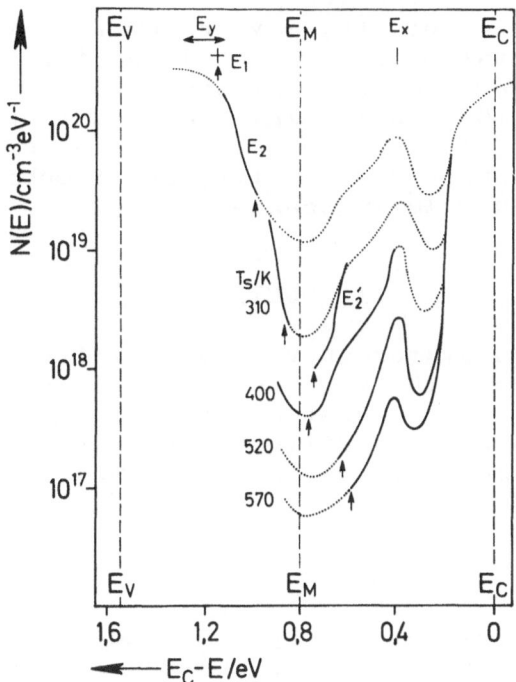

Fig. 6 Distribution of localized states of amorphous Si evaluated from field effect (full curves) Ref. 4. Curves E were obtained for evaporated layers, the other curves by decomposition of silane at the denoted substrate temperatures.

for the constant activation energy of conductivity in a structureless density of states model. If, however, there are edges in the density of states distribution this assumption may not be necessary. It is important that a minimum in the density of states is almost exactly in the center of the gap as can best be seen for the sample deposited at 400 K. The position of the Fermi energy as determined from the activation energy of conductivity is marked by arrows. Although there is some uncertainity in these values it seems that there is a transition from p-type to n-type conduction when the density of states decreases. This observation is in accordance with the results of thermoelectric power investigations which will be discussed later.

Owing to these field effect experiments amorphous Si is the only material where considerable details about

the distribution of the localized gap states are known.
The absolute value of g(E) may be somewhat too high
since surface states have been omitted for sake of simplicity. However, the general features of the density
of states distribution in Fig. 6 are quite well established and are also revealed in a number of other experiments like photoconductivity, drift mobility, conductivity and photoluminescence.

III. MOBILITY

In extended states above the mobility edges E_c and
E_v the mean free path of the carriers is supposed to be
of the order of the interatomic spacing a. Transport
then can be viewed as a diffusive process like Brownian
motion and the mobility can be expressed as[14]

$$\mu = \frac{1}{6} \frac{ea^2}{kT} \nu_{el} \qquad (4)$$

where a is the interatomic spacing and ν_{el} is an electronic frequency of the order 10^{15}Hz. From this formula one finds at 300 K a mobility of about 1 - 10 cm^2/
Vs.

Between E_c and E_v the electron states are localized.
Transport in these states can only occur by phonon assisted tunnelling usually called hopping. The hopping
mobility is given by an expression of the form[14]

$$\mu_H = \frac{1}{6} \frac{ea^2}{kT} \nu \, e^{-(2\alpha a + \frac{W}{kT})} \qquad (5)$$

Here the factor $e^{-2\alpha a}$ describes the overlap of the wave
functions, W is the energetic distance between the states
and ν is the so called phonon frequency (10^{12}Hz). If
one neglects the exponential, the upper limit to this
mobility is 10^{-2} to 10^{-3} cm^2/vS. Thus near E_c and E_v
the mobility is supposed to drop by about 3 orders of
magnitude.

A. Hall Mobility

Direct experimental information about the carrier
mobility would remove many difficulties in the understanding of the transport properties of amorphous solides. In case of crystalline semiconductors the combined measurement of conductivity and Hall effect has

been one of the most successful tools for the investigation of the transport mechanism. For amorphous semiconductors experimental difficulties arise mainly from the
high resistivity of the samples as well as from the low
values of the carrier mobility. However, besides these
experimental troubles, the major problem is the interpretation of the Hall effect of low mobility materials.
Typical data replotted from Ref. 15 are given in Fig. 7.
The observed Hall mobilities μ_H are between 10^{-1} and
10^{-2} cm^2/Vs and depend only slightly on temperature with
an activation energy of about 0.07 eV. An important
point is that the Hall effect is negative in contrast
to the thermoelectric power (n-p-anomaly). Thus if the
sign of the thermoelectric power is believed to give
the correct sign of the predominating carriers, the Hall
effect does not, it is negative here for transport by
holes.

Theoretical treatments of the Hall effect in disordered materials[16] indicate that it is the local geometry of the energy states which most influences the
magnitude, the temperature dependence and the sign of
the Hall mobility. In the random phase model used by
Friedman[16] it is assumed that the only effect of the

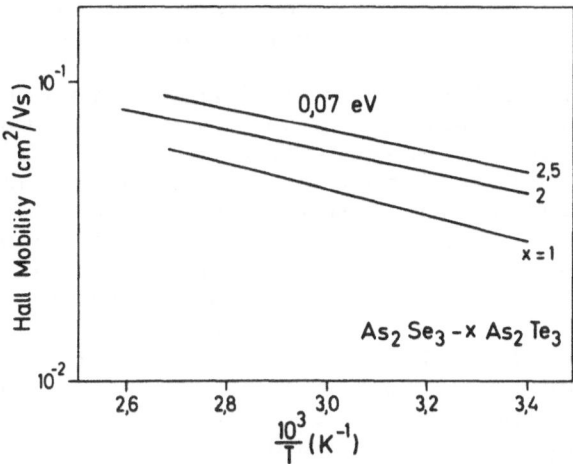

Fig. 7 Hall mobility of amorphous As_2Se_3-xAs_2Te_3 as a
function of temperature (Ref. 15).

magnetic field is to modify the phases of the charge
transfer integrals. For hopping transport as well as
for diffusive like motion in extended states the sign
of the Hall effect is that of an elementary jump process[17]. The sign for holes relative to that for electrons depends then on the site configuration. In case
of a three site configuration which is considered to be
characteristic of some chalcogenide glasses the Hall effect is negative for electrons and holes. If four sites
are involved, the sign for holes is opposite to that for
electrons. According to these theories the Hall mobility of disordered semiconductors is substantially different from the conductivity mobility. That is why
Hall effect measurements are much less informative for
amorphous than for crystalline semiconductors.

This problem is revealed in the few results published so far for amorphous Ge. In Fig. 8 the conductivity (dashed lines) and the Hall mobility (full lines)
obtained in different laboratories are plotted against
reciprocal temperature. Surprisingly large conductivity
values were found by Lomas et al[18]. Together with the
weak temperature dependence this raises some doubt on
whether the layers are really amorphous. The corresponding Hall effect is positive and the mobility increases considerably with temperature. If one combined σ
and μ_H, as is usually done for crystalline semiconductors,
one would obtain the peculiar result that the carrier
concentration decreases with rising temperature. An attempt to interpret these measurements in terms of the
recently developed theories was made. It is suggested
that the small values and the pronounced temperature dependence could arise from the competition of three and
four site contributions with the four site configurations
predominating. A negative Hall effect for amorphous Ge
has been reported by Seager, Knotek and Clark[19]. Since,
above about 350 K the thermoelectric power was positive,
these films exhibit the familiar n-p-anomaly. Although
below this temperature the thermoelectric power changed
sign, no corresponding change in the behavior of the
Hall mobility was observed. It is remarkable that the
Hall mobility of these authors (curve 1a) is not very
different from that of Lomas et al, although the corresponding conductivities (curves 1b and 2b) differ by
orders of magnitude. From these results on amorphous
Ge it is obvious that it is not possible to relate the
Hall mobility with the conductivity mobility. Hence
for amorphous materials Hall effect is not as useful
a method for the study of transport as it has been in

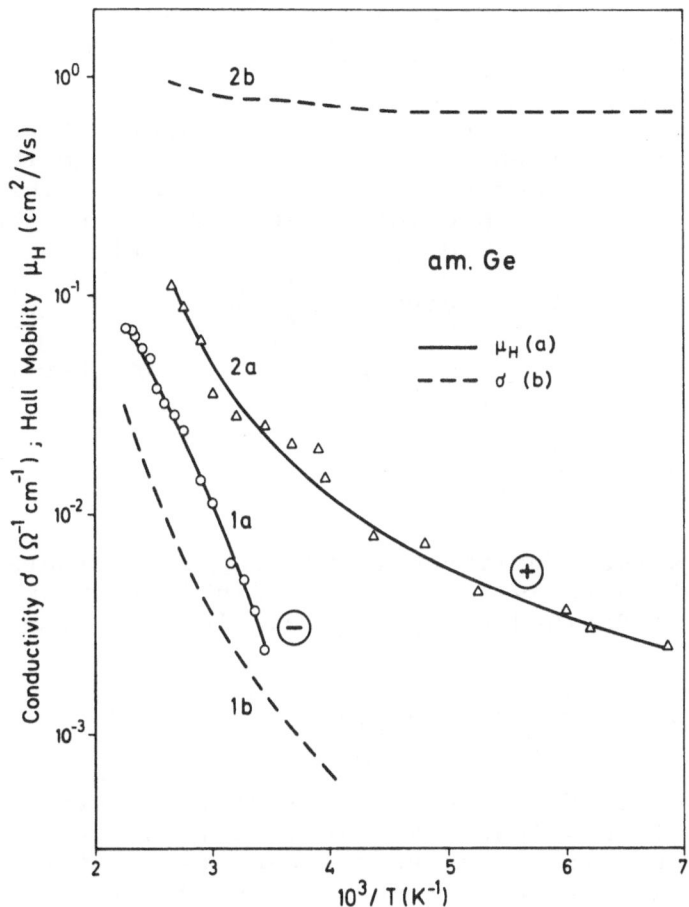

Fig. 8 Temperature dependence of Hall mobility and
conductivity of amorphous Ge-films (curves 1a and 1b
after Ref. 19, curves 2a and 2b after Ref. 18).

the past in case of crystalline semiconductors.

B. Drift Mobility

Drift mobility measurements yield direct information about the carrier transport particularly in materials which have high resistivities and low mobilities.
In these experiments carriers are generated near one
electrode by a short light or electron pulse. These
excess carriers are extracted by an electric field and

their transit time across the sample is measured. The
selection of the polarity of the field determines whether
electrons or holes are extracted. This technique, and
particularly its application to materials of high resistivity have been reviewed by Spear[20]. This type of
drift experiment is performed under conditions where the
dielectric relaxation time is much larger than the transit
sit time T. Trapping effects may be of major influence.
If shallow traps are present, which have short trapping
and release times t_t and t_r, the transit time will simply be enlarged by a factor $\dfrac{t_t + t_r}{t_t}$.

In this case of multiple trapping one finds for the
drift mobility

$$\mu_d = \mu_o \cdot \frac{t_t}{t_t + t_r} \tag{6}$$

where μ_o is the mobility without trapping effects. With
the assumption $t_t \ll t_r$ and for a single trap of density
N_t at a well defined energy E_t this leads to the expression

$$\mu_d = \mu_o \cdot \frac{N_c}{N_t} \cdot \exp\left(-\frac{E_t}{kT}\right) \tag{7}$$

In this formula N_c is the density of those states, where
conduction takes place. For instance, in crystalline
samples N_c equals the effective density of states. Deep
traps with long release times can produce transit pulses
which do not allow a meaningful interpretation. Therefore it is necessary that such experiments are performed under the condition $T < t_t$ when t_t is the trapping
time of the deep trapping levels.

It is not trivial to ask how this drift mobility
can be related to the conductivity mobility. A conductivity measurement, namely, is performed under equilibrium conditions. The excess carriers in a drift mobility experiment, on the other hand, introduce a strong
perturbation, since the local carrier density may be
larger by many orders of magnitude than the equilibrium
concentration. If the carrier mobility depends on the
energy of the states, such differences in the distribution of the carriers can lead to very different results.
In such a case there exists no correlation between the
conductivity mobility and the drift mobility and at the

worst both may refer to substantially different conduction mechanisms.

Drift mobility techniques have been applied to a number of amorphous substances. As an example in Fig. 9 the results are shown which have been obtained by LeComber, Madan and Spear[21], for amorphous Si prepared by glow discharge. The drift mobility (Fig. 9b) is found to be thermally activated with a clear kink near a temperature T_1. The two activation energies are 0.18 eV and 0.09 eV and are independent of the deposition temperature T_s. The striking feature of these results is that in the temperature dependence of the conductivity near the same temperature T_1 a kink is also observed (Fig. 9a, insert). Remarkably the difference of the activation energies in the two ranges is also about 0.1 eV, as in the temperature dependence of the drift mobility. This coincidence is taken as evidence for a change in conduction mechanism near T_1. It is suggested that above T_1 transport takes place by electrons in extended states above the mobility edge E_c. Below T_1 hopping in states close to E_A is assumed to predominate with a mobility of the type given by equ. (5), where the hopping energy is W = 0.1 eV.

The interesting question arises: how is it possible to have a single activation energy of the drift mobility in a density of states model like the one of Fig. 6? If one assumes that above T_1 conduction is by electrons above E_c, one has to attribute the activation energy of the drift mobility to a multiple trapping mechanism. The experimental result thus indicates that the electrons interact preferably with energy states 0.18 eV below E_c. In the density of states model at this energy E_A there is no sharp maximum but a strong gradient. However, if one calculates from this distribution the density of trapped electrons by multiplying g(E) with a Fermi distribution, one readily finds that excess electrons pile up near E_A (see for instance Fig. 20). With this result one obtains an estimate for the mobility near E_c from relation (7) by replacing $\frac{N_c}{N_t}$ by $\frac{g(E_c)}{g(E_A)}$. This procedure leads to a room temperature mobility at E_c of $\mu(E_c) \simeq 5$ cm^2/Vs which is in good accordance with theoretical predictions.

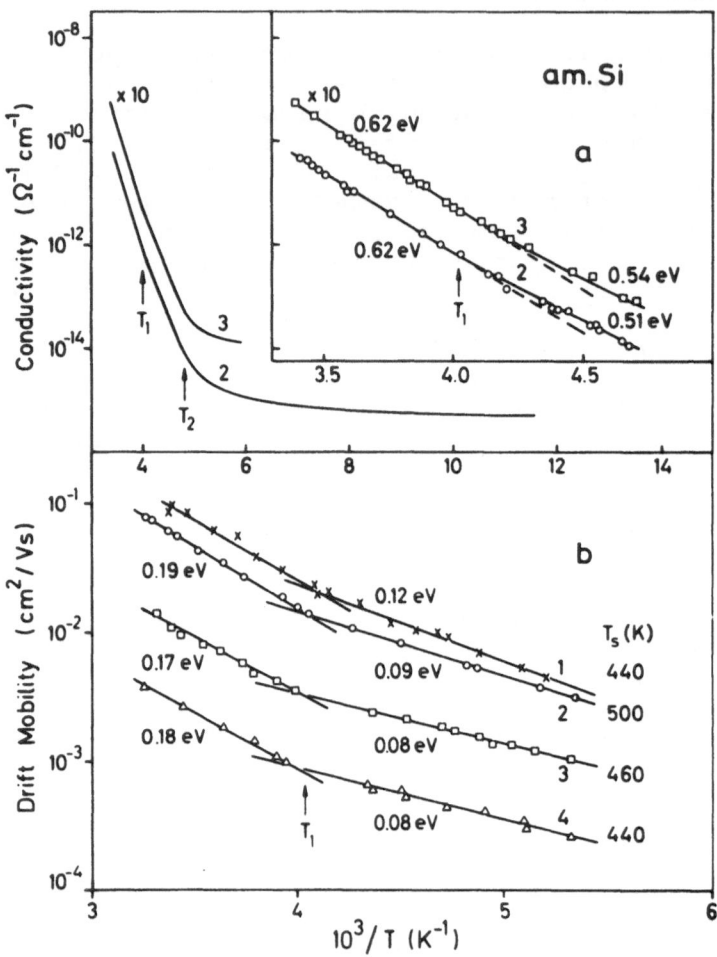

Fig. 9 Temperature dependence of conductivity (a) and drift mobility (b) of glow discharge Si (Ref. 21) (T_s denotes substrate temperature).

This interpretation has been criticized by Mell[3] who pointed out that it is difficult to understand why the values of μ_D for the four films in Fig. 9 differ appreciably although the activation energies are almost identical. Following formula (7), in the trap controlled region the decrease of μ_D from sample 1 to 4 has to be explained by an increase in the density of the trapping levels. In the low temperature range, transport is expected to occur in exactly these states by hopping.

The hopping mobility, therefore, should increase from sample 4 to 1 which is in contradiction to the experimental results.

Drift mobilities which are thermally activated with a well defined activation energy have been reported also for other amorphous semiconductors e.g. As_2Se_3[22]. It has been shown in the example of amorphous Si that this type of behavior does not necessarily need a defined trapping level or maximum in the density of localized states for its interpretation. This aspect may be of some importance also for other materials and will be discussed in connection with photoconductivity.

IV. CONDUCTIVITY AND THERMOELECTRIC POWER

A. Experimental Methods and Results

Combined investigations of conductivity and thermoelectric power have considerably contributed to the present knowledge about transport in amorphous semiconductors. In these measurements a small thermal gradient ΔT is established across the sample and the resulting voltage $U_{tn} = S \cdot \Delta T$ is measured as a function of temperature.

Results of Beyer and Stuke[23] obtained for two different evaporated amorphous Si-films are shown in Fig. 10. The only apparent difference between these samples is the evaporation rate which was 2 Å/s for sample 1 and 16 Å/s for sample 2. The curves 1a-1c and 2a-2d were obtained by annealing at successively increasing temperature. Thereby the conductivity decreases particularly in the low temperature range by several orders of magnitude. At high temperatures in all annealing states intrinsic-like behavior predominates; σ is thermally activated with an activation energy of 0.75 eV. This behavior is most pronounced for the highest annealed state (curve 2d). In the low temperature range no single activation energy can be defined. Here the conductivity follows Mott's relation $\ln\sigma \propto T^{-1/4}$. In Fig. 11, where σ of sample 1 is plotted logarithmically versus $T^{-1/4}$, straight lines emerge, which indicate that conduction occurs by variable range hopping. Obviously by annealing mainly the preexponential factor σ_0 decreases, whereas the slope of the straight lines remains approximately unchanged.

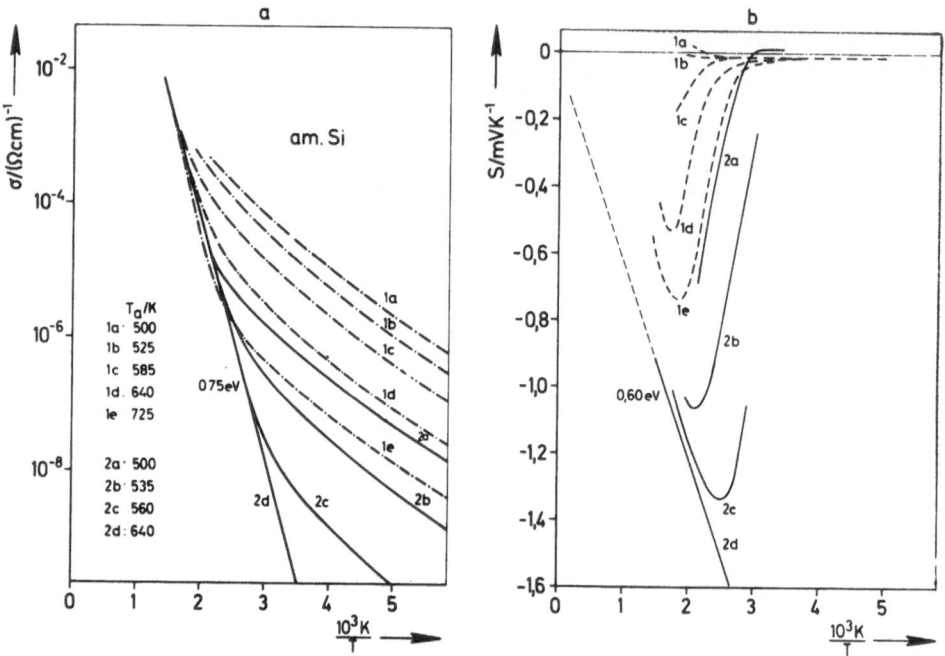

Fig. 10   Temperature dependence of conductivity and
thermoelectric power of two amorphous Si-layers[23].
Deposition rate: 2 Å/s for sample 1 and 16 Å/s for
sample 2. Substrate temperature 500 K, residual pressure, 2 · 10^{-6} Torr. T_a denotes annealing temperature.

         These different ranges in the temperature depen-
dence of σ are also revealed in the temperature dependence of the thermoelectric power S (Fig. 10). With
the exception of a small part of curve 2a, S is negative.
Hence in these films conduction by electrons predominates.
In the low temperature range where $\ln\sigma \propto T^{-1/4}$ is valid,
S has very small values and is approximately independent
of temperature. Moreover, at low temperatures S cannot
be changed markedly by annealing. With rising temperature, S increases strongly at the transition to the intrinsic-like region. In the high temperature range in
this plot a straight line behavior emerges, which is

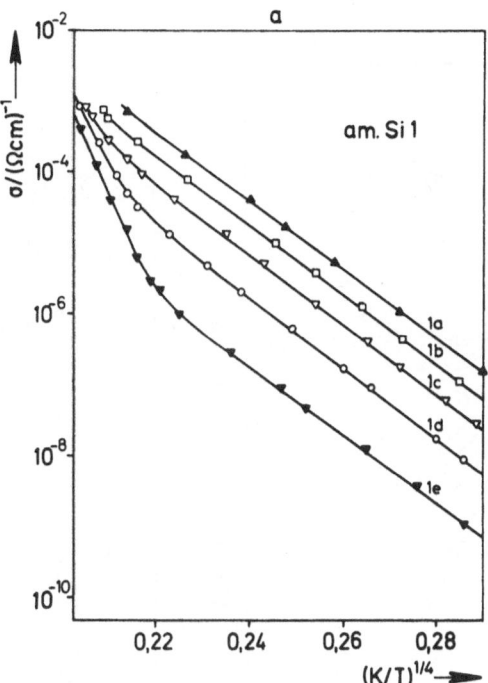

Fig. 11   Temperature dependence of conductivity of amorphous Si[23] (for preparation conditions see sample 1 in Fig. 10).

best developed for the highest annealed state 2d. From the slope of this straight line an activation energy of 0.60 eV is calculated. It is important to note that this energy is smaller than the activation energy of the conductivity by 0.15 eV.

A similar behavior is also found for other tetrahedrally bonded amorphous semiconductors[23,24]. In Fig. 12 measurements on two samples of amorphous Ge are shown. These specimens differ in the evaporation rate which was higher for sample 2. In the temperature dependence of σ again two ranges can be distinguished. At high temperatures there is intrinsic-like conduction with a single activation energy. At low temperatures a variation as $\ln\sigma \propto T^{-1/4}$ is found indicating the predominance of hopping transport near the Fermi level.

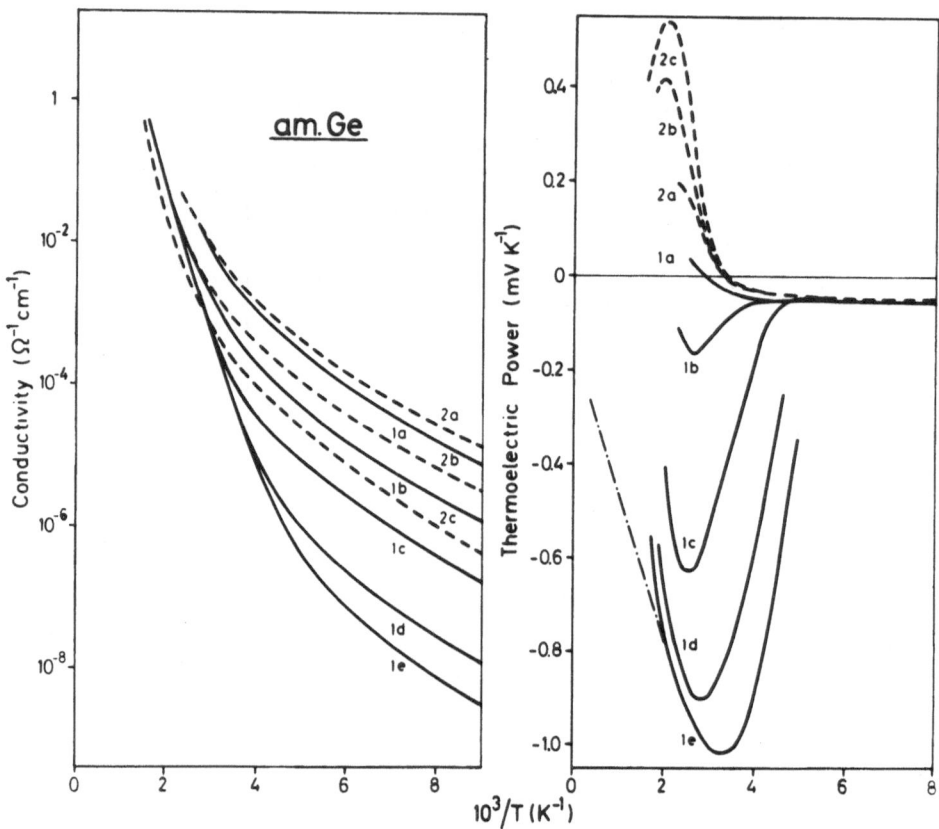

Fig. 12   Temperature dependence of conductivity and
thermoelectric power of two amorphous Ge-layers[23].

Sample 1: deposition rate 50 Å/s, T_s = 240 K,
residual pressure 2 · 10^{-6} Torr.

Sample 2: deposition rate 100 Å/s, T_s = 300 K,
residual pressure 6 · 10^{-8} Torr.

Annealing temperatures: (1a) 95°C; (1b) 170°C;
(1c) 220°C; (1d) 265°C; (1e) 310°C; (2a) 170°C;
(2b) 300°C; (2c) 385°C.

In this latter region the thermoelectric power has, as
in the case of amorphous Si, relatively small negative
values. At higher temperatures the slowly evaporated
film behaves like to amorphous Si. In the case of sample 2, however, S becomes positive indicating that here

hole conduction predominates. According to these results, it depends on the preparation conditions which carriers contribute most to the conduction. It is also important to realize that in the high temperature range of sample 1 no linear dependence of S on 1/T is observed, but with rising temperature the slope of these curves (1c-1e) increases considerably.

B. Interpretation

For an interpretation of these results it is helpful to discuss briefly the relevant formulas for the conductivity and the thermoelectric power. General expressions for these quantities are obtained as an integral over all electron states[14,25].

$$\sigma = e \cdot \int g(E)\mu(E) \cdot f(E) \{1-f(E)\} dE \qquad (8)$$

$$S = -\frac{k}{e} \int \frac{E-E_F}{kT} \cdot \frac{\sigma(E)}{\sigma} dE \qquad (9)$$

Here g(E) is the density of states, f(E) the Fermi distribution and $\mu(E)$ the energy dependent mobility. According to relation (9) the thermoelectric power has a simple physical meaning. S·T is namely the energy per unit charge of the carriers, measured relative to the Fermi energy E_F. Since carriers at an energy E contribute to the energy transport proportional to their contribution to the conductivity $\sigma(E)$, the thermoelectric power is a weighted average, the weighting factor being $\sigma(E)/\sigma$. If $E > E_F$, i.e., the transport takes place in states above E_F, S is negative. Thus the sign of S gives direct information about the sign of the carriers which contribute most to the transport.

It should be emphasized that this relation for the thermoelectric power is quite generally valid. It bases mainly on thermodynamic arguments and does not depend on special assumptions about the band structure and transport mechanism.

The above formula can easily be specified for some interesting cases:

a. If conduction occurs by one type of carrier only, for instance by electrons in states $E>E_c$ above a density of states edge or a mobility edge, one obtains with the assumption that Boltzmann statistics can be applied:

$$\sigma = \sigma_0 \exp(-\frac{E_c - E_F}{kT}) \tag{10}$$

$$S = -\frac{k}{e}(\frac{E_c - E_F}{kT} + A_c) \tag{11}$$

here A_c is a temperature independent constant of order unity. Following these relations σ as well as S include the same activation energy $E_c - E_F$. If, however, the mobility $\mu(E_c)$ is thermally activated the energies determined from σ and S will be different and this difference can be attributed to the activation energy of the mobility.

   b. Both electrons and holes at energies $E > E_c$ and $E < E_v$ respectively, contribute to conduction, and the Fermi level is fixed for some reason. In this case the above relations (8) and (9) yield

$$S = -\frac{k}{e}(\frac{\sigma(E_c)}{\sigma} \cdot \frac{E_c - E_F}{kT} - \frac{\sigma(E_v)}{\sigma}\frac{E_F - E_v}{kT} + A_{cv}) \tag{12}$$

   Although the first term as well as the second one vary linearly with $1/T$, S does not, unless the Fermi level is exactly fixed in the middle of the gap.

   c. Transport can also take place in states near the Fermi energy like in metallic conduction. The thermoelectric power then is given by[14]

$$S = -\frac{\pi^2}{3} \cdot \frac{k^2 T}{e} (\frac{1}{\sigma}\frac{d\sigma(E)}{dE})_{E_F} \tag{13}$$

From this formula S is expected to be relatively small and to be different from zero only if the energy derivative of the conductivity at the Fermi energy does not vanish. The sign of S may be positive or negative since it is determined by the slope of $\sigma(E)$ and therefore $g(E)$ at E_F.

   The main features of the experimental results (Fig. 10-12) can now be interpreted using the density of states models discussed above. Since in these models the Fermi level is effectively pinned, real intrinsic conduction, where there are equal numbers of electrons and holes above the edges in either band, can be excluded. In the high temperature range conduction takes place by carriers excited into states above the mobility edges E_c or E_v. In strongly annealed amorphous Si obviously only electrons participate in conduction (Fig. 10, curve 2d). Hence the expressions (10) and (11) should apply. With the help of these formulas one obtains from the conductivity curve an activation energy of 0.75 eV and from the temperature dependence of S an energy of 0.6 eV. This difference of 0.15 eV points to a thermally activated mobility. Consequently Beyer and Stuke[23] have suggested that high temperature conduction occurs by hopping processes among states near the energies E_A and/or E_B, i.e., near the former band edges. This interpretation is different from that given by LeComber et al.[21] for their drift mobility data where it is concluded that at high temperature conduction occurs in extended states above the mobility edge. Differences between the activation energies of conductivity and thermoelectric power have been reported also for a number of chalcogenide glasses[15], and have been explained by small polaron hopping[26]. In the low temperature range where $\ln \sigma \propto T^{-1/4}$ the thermoelectric power has small values and depends only a little on temperature. That is, what one expects if conduction is by hopping between localized states near the Fermi energy, a process which is likely to occur in all of the above density of states models. It should be stated that the small value of the thermoelectric power is a more convincing argument for the predominence of variable range hopping than the $T^{-1/4}$-plots over a rather limited temperature range.

As an alternative explanation to Mott's relation it was proposed that at low temperature ambipolar conduction in the tail states of the valence and conduction band dominates[27]. Since the integrant in relation (8) has a maximum value at an energy which shifts with rising temperature away from E_F, a temperature dependent activation energy can result. That may lead under certain conditions to a variation as $\ln \sigma \propto T^{-1/4}$. In order to account for the low values of the thermoelectric power in this model one has to assume conduction in both tails. Then in relation (12) the contributions of electrons and holes to S may cancel. This explanation requires almost perfect symmetry of the tails. Moreover this symmetry has to be preserved also in the various annealing stages, because the thermoelectric power thereby changes only little. Such high symmetry in the density of localized states is difficult to reconcile with the pronounced influence of annealing on the conductivity, which indicates strong changes in $g(E)$, as well as on the thermoelectric power in the high temperature range, which indicates a shift of the Fermi level.

The behavior of amorphous Ge obviously is more complex. Here also holes contribute to the conduction, and it depends on the preparation conditions which type of carrier dominates. This has led in the past to some confusion because the results of different laboratories seemed to be in contradiction. Fig. 13 gives a survey of the data published for amorphous Ge[3]. The great variety of results for the "same" material is at first sight confusing. However, it may be understood by the clear trend, that the higher the evaporation rate the larger is the contribution of holes to the conduction. This tendency is obvious from the results in Fig. 12. The contribution of holes has to be taken into account also in strongly annealed n-type samples. This is seen from curves 1c,d,e, in Fig. 12, where the increasing slope in the high temperature range indicates that with rising temperature the contribution of holes increases. Therefore relation (12) for ambipolar conduction applies. Indeed Beyer and Stuke[23] were able to fit their measurements by an expression which contained three types of contributions, namely conduction by electrons, holes and by hopping near the Fermi level.

$$S = \frac{\sigma_p}{\sigma} S_p + \frac{\sigma_n}{\sigma} S_n + \frac{\sigma_H}{\sigma} S_H$$

$$(14)$$

$$S_{n,p} = \mp \frac{k}{e} \left(\frac{E_{n,p}^s}{kT} + A \right)$$

In this formula σ_n, σ_p are the contributions of electrons and holes to the conductivity which are considered to be thermally activated and σ_H is that part of the conductivity which originates from hopping near E_F. It is further assumed that in accordance with the results on amorphous Si, the mobilities are activated with an activation energy of 0.12 eV. The only parameters in this fit then are the activation energies E_p^s, E_n^s of the thermoelectric power. Once these energy values are determined the size of what might be called conductivity gap $E_{ps} + E_{ns}$ as well as the location of the Fermi level in the gap are known. It is found that in rapidly evaporated samples, E_F is slightly below the middle of the gap in all annealing states, whereas in slowly evaporated layers of amorphous Ge, E_F is in the

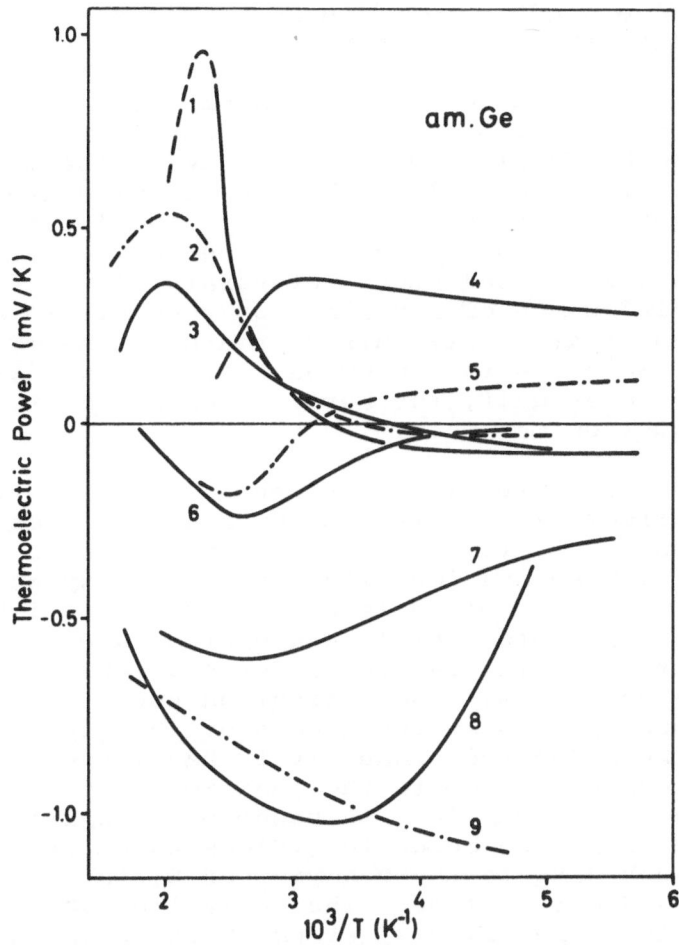

Fig. 13 Thermoelectric power of amorphous Ge prepared
by different groups. (1) Chopra et al (Ref. 28), (3)
Grigorovici (Ref. 29), (4) and (5) Buchy et al (Ref. 30),
(6) Vescan et al (Ref. 27), (9) Rockstadt et al (Ref.
31), (2), (7) and (8) Beyer (Ref. 24).

upper half of the gap, particularly after annealing.
In both cases the conductivity gap is found to increase
considerably on annealing. This behavior is consistent
with the observed shift of the absorption edge.

An unambiguous interpretation of this influence of
the evaporation rate on the transport properties in a

density of states model is extremely difficult. It is
clear that one needs a model which provides sufficient
structure like the one of amorphous Si (Fig. 6). Apparently the rapidly evaporated sample 2 is less contaminated. It is assumed that it contains a localized
states distribution, may be of the type of curve E in
Fig. 6 with the Fermi level slightly below the middle
of the gap. If by annealing the density of states $g(E)$
decreases approximately symmetrically, the location of
the Fermi level is not changed and the sample remains
p-type. In case of the slowly evaporated layer the more
pronounced influence of annealing on σ points to a stronger decrease of $g(E)$. If this decrease is unsymmetric,
i.e., more states in the upper half of the gap disappear
than in the lower half, the Fermi level is shifted and
n-type conduction may result.

A possible reason for this peculiar unsymmetrical
annealing effects in highly contaminated Ge-layers is
that some energy states disappear by reaction of defects
with chemically active impurity atoms like oxygen or
hydrogen. This assumption is confirmed by ion implantation of oxygen into rapidly evaporated layers. An example is given in Fig. 14. Here Curve 1 represents the
annealed p-type state. The activation energy is 0.55 eV
and therefore E_F is near the middle of the gap. After
oxygen implantation the conductivity has increased considerably (curve 2a) due to the additional structural
defects and the thermoelectric power S is negative. By
annealing σ decreases below the values which it had before and, according to the behavior of S, the sample becomes more and more n-type. The activation energy of
the conductivity now is 0.45 eV which is assigned to
E_c-E_F. This shift of the Fermi level by about 0.1 eV
is associated with the presence of oxygen, which may
saturate dangling bonds or form some kind of impurity-
vacancy complexes, and thus may reduce $g(E)$ unsymmetric-
ally. Similar results have been obtained by implantation of hydrogen. The structural defects which are created by ion bombardement cannot be responsible for this
shift. After implantation of He-ions, which do not react chemically, the sample remains p-type and approaches
the former state after subsequent annealing (Fig. 15)[24].

One can only speculate about the nature of the active structural defects. Spear[4] has tentatively assigned the two peaks in the density of localized states E_y
and E_x of amorphous Si to the donor and acceptor states
of divacancies or other vacancy complexes. In crystalline

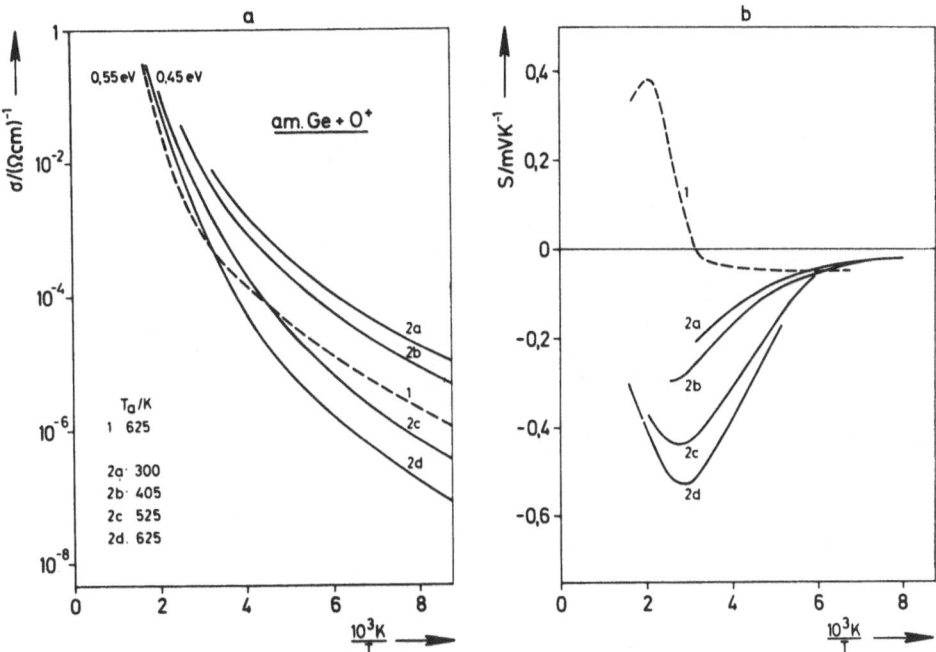

Fig. 14 Temperature dependence of conductivity and thermoelectric power of amorphous Ge before (curve 1) and after (curve 2) ion implantation of 20 keV oxygen-ions (Ref. 23).

Si the donor state of a divacancy is expected to be 0.32 eV above the valence band. There are two acceptor states associated with this defect, viz., 0.54 and 0.32 eV below the conduction band. Although it is tempting to assign the E_y and E_x - peaks to these divacancies, more complicated defects certainly have to be considered, particularly those containing impurities. In contaminated layers chemically active impurities may react with other defects forming new complexes. In crystalline Si for instance a divacancy-oxygen-complex is known to exist which has donor and acceptor states, respectively, about 0.35 eV above the valence band and 0.2 eV below the conduction band.

The results discussed in this section demonstrate that the investigation of the thermoelectric power gives very direct and meaningful information about the various

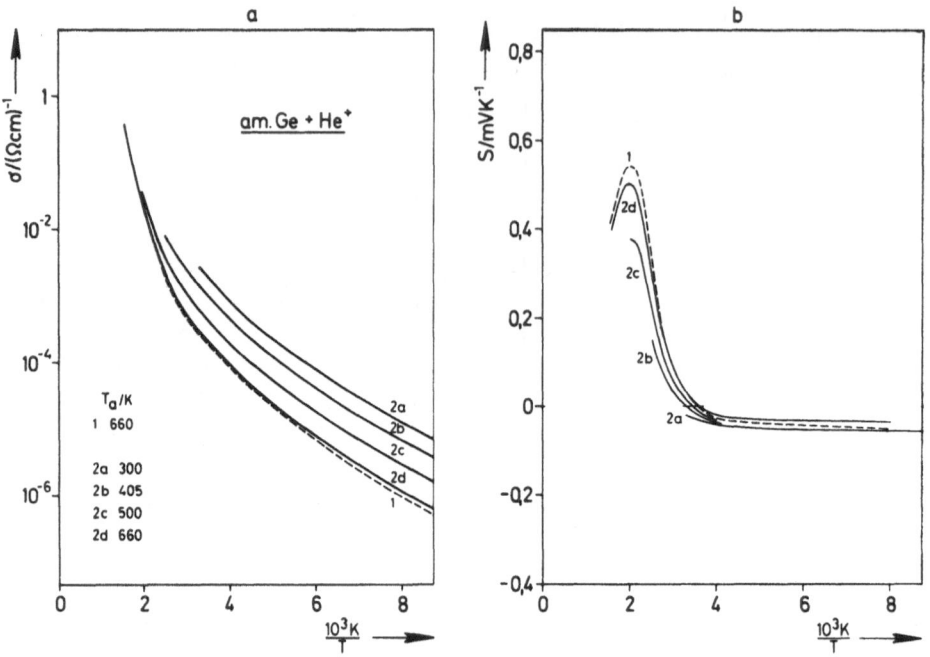

Fig. 15 Influence of Helium-implantation on the conductivity and thermoelectric power of amorphous Ge,
(Ref. 24). Curve 1 before, curves 2 after ion-bombarde-
ment.

types of conduction and their relative importance. For
the study of amorphous materials this quantity is particularly well suited, because relation (9) can be derived in a quite general way using thermodynamical arguments. The main features of the thermoelectric power
and conductivity of tetrahedrally bonded amorphous semiconductors can be explained in a density of states model
as proposed by Davis and Mott (Fig. 1b). The pronounced
influence of preparation conditions, annealing and ion
bombardement on the thermoelectric power of amorphous
Ge indicates, however, that a more complicated density
of states model may apply, which contains specific structure due to various kinds of defects. Hence for an interpretation of these measurements the type of model developed for amorphous Si mainly from field effect results
has some advantages. These homogeneous models are up
to now prefered by most authors in spite of the rather

convincing arguments in favor of the heterogeneous model (Fig. 2). The main features of conductivity and
thermoelectric power can also be qualitatively explained
in a heterogeneous model. For instance, the different
magnitude of the activation energy of conductivity and
thermoelectric power can be attributed to these fluctuations of the band edges. The disadvantage of this
model is that so far it is difficult to analyse the experimental data quantitatively. Nevertheless, it seems
that aspects of this heterogeneous model should be taken
more into account.

V. VARIABLE RANGE HOPPING

A considerable number of investigations deal with
the hopping regime in amorphous Ge and Si at low temperatures. In most papers it is examined whether the temperature dependence of conductivity can be fitted by the
theoretical expressions for variable range hopping using
reasonable parameters. These formulas for variable range
hopping are[32]

$$\sigma = \sigma_o \cdot \exp\left[-\left(\frac{T_o}{T}\right)^{1/4}\right]$$

$$\sigma_o = \sqrt{2\pi} \cdot \left(\frac{3}{2} e^2 \nu\right) \sqrt{\frac{g(E_F)}{\alpha kT}} \qquad (15)$$

$$T_o = C_3 \cdot \frac{\alpha^3}{3kg(E_F)}$$

Here ν is considered to be a typical phonon frequency
of the order 10^{12}-10^{13} Hz. Slightly different values
between 11 and 18 are given for the prefactor C_3 in T_o
by various authors[32,33,34]. In principle with a reasonable assumption for the attenuation distance of the wave
function, α, the density of states at E_F can be evaluated from the slope of the straight lines in the $\ln\sigma$ versus
$T^{-1/4}$ plot (Fig. 11). Knotek et al[35] were able to deduce α and $g(E_F)$ independently from an investigation of
variable range hopping in amorphous Ge as a function of
film thickness. For very thin films two-dimensional
percolation theory applies, where the conductivity is
given by

$$\sigma \propto \exp \left[-\left(\frac{T_0}{T}\right)^{1/3}\right]$$

$$T_0 = C_2 \frac{\alpha^2}{kdg(E_F)}$$

(16)

By observing with increasing film thickness d the transition from 2-dimensional to 3-dimensional hopping, the known slopes T_0 in both regions allow one to evaluate α and $g(E_F)$ without further assumptions. These authors find $\alpha^{-1}=10$ Å and $g(E_F)$ in the range $5 \cdot 10^{17}$ - $3 \cdot 10^{18}$ cm^{-3} eV^{-1}. Inserting $\alpha^{-1}=10$ Å into T_0 for 3-dimensional hopping, one finds for the evaporated amorphous Ge and Si layers of Figs. 10 - 12, a density of states of 1.5 - $4 \cdot 10^{18}$ cm^{-3} eV^{-1}. Much larger physically unreasonable values are determined very often from the preexponential factor σ_0[36]. In order to obtain consistent results for $g(E_F)$ from T_0 and σ_0 one has to insert extremely high values for the so called phonon frequency ν. This inconsistency also reveals in the influence of annealing on σ_0 and T_0. On annealing, for instance, the slope of the $\ln\sigma \propto T^{-1/4}$-curves remains practically constant, whereas σ_0 changes by orders of magnitude (see Fig. 11), although both σ_0 and T_0 depend explicitly on $g(E_F)$. It is also rather suspicious that the density of localized states near E_F determined from T_0 varies only little if one compares results from different laboratories, although the preparation conditions differ appreciably.

These results indicate that the above expressions do not give a quantitative description of the conductivity. In most theories it is assumed that the density of states $g(E)$ does not vary with energy. This is an unrealistic assumption in view of the above density of states models. Another argument against a constant density of states comes from the thermoelectric power, which is clearly different from zero. As is seen from relation (13) this is only possible if $\left(\frac{dg(E)}{dE}\right)_{E_F} \neq 0$.

Several authors [e.g. Ref. 36] have pointed out that the slope of $g(E)$ near E_F should be of major importance in the theory of variable range hopping. In fact recent theoretical investigations of this problem have shown that σ_0 as well as T_0 very sensitively depend on the gradient $\frac{dg(E)}{dE}$ at E_F[37]. This raises some doubt on the values of α and $g(E_F)$ obtained from the so far available formula for variable range hopping. In addition, the

recent work of Emin[38] has made the present interpretation of the $T^{-1/4}$-law somewhat uncertain. It has been
shown that such behavior may arise from the temperature
dependence of the jump rate itself rather than from the
percolation aspects of hopping conduction.

Also other methods like frequency dependence and
pressure dependence of conductivity were employed to
find direct evidence for this type of conduction. Strong
influence of pressure on σ, namely, can be expected to
originate from the tunnelling factor $\exp(-2\alpha a)$ if the
hopping mobility is of the type of relation (5). Instead of this, in the hopping range where $\ln\sigma \propto T^{-1/4}$
the pressure coefficients of conductivity were found to
be surprisingly small[39]. As an example a measurement
on amorphous Si using uniaxial stress is given in Fig.
16. In the upper part the conductivity and in the lower
part the elastoresistivity m are plotted versus $1/T$.
The latter quantity is defined by $m = \dfrac{1}{\varepsilon}\dfrac{\Delta\rho}{\rho}$ where ε is
the applied strain and $\dfrac{\Delta\rho}{\rho}$ the resulting relative change
of resistivity. The elastoresistivity does not depend
on the relative orientation of the uniaxial stress and
the direction of the current. Hence for the magnitude
of m the change in volume $\Delta V/V$ due to the uniaxial
stress is important. These measurements therefore provide the same type of information as measurements using
hydrostatic pressure. The region of interest here is
region III, where the conductivity follows Mott's relation. In this range the elastoresistivity is relatively small, the observed changes of resistivity are of the
order of the strain. From the tunnelling factor alone
a much larger value for the elastoresistivity of up to
10^2 were to be expected. This result, however, does not
contradict the assumption of variable range hopping. If
this process predominates, those jumps contribute most
to the conduction, for which the exponent $2\alpha a + W/kT$ is
a minimum. When external pressure is applied, the hopping distances decrease but this does not necessarily
lead to an equivalent decrease of the entire exponent,
since the hopping energy W of the most frequent jumps
thereby can increase. The competition of these two pressure dependences is suggested to be responsible for the
small pressure coefficients of conductivity in this range
of conduction. This interpretation is confirmed by the
temperature dependence of the elastoresistivity. If one
calculates this quantity from the expressions for variable range hopping, one finds $m \propto T^{-1/4}$. This type of
behavior is indeed found in experiment. This is seen

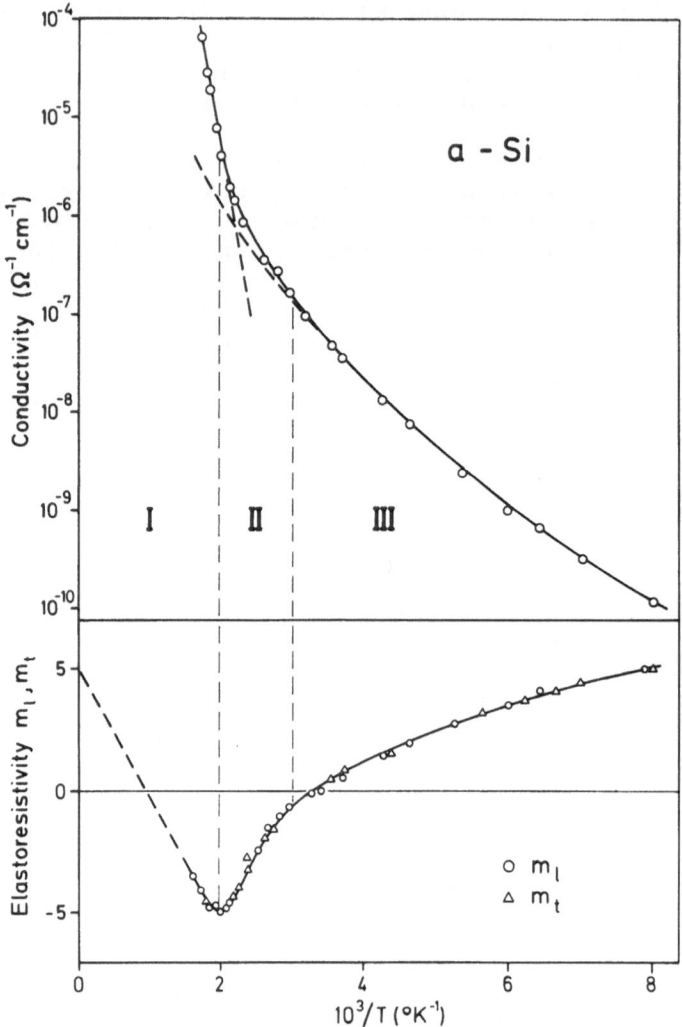

Fig. 16 Conductivity and elastoresistivity of amorphous Si as a function of temperature (Ref. 39).

from Fig. 17, where the data of Fig. 16 are replot-
ted in a $T^{-1/4}$-scale. In the temperature range where σ
increases with rising temperature following Mott's law,
the elastoresistivity decreases proportional to $T^{-1/4}$.
From these results it is concluded that the influence
of external pressure on the conductivity can be understood in the present model for the carrier transport in
amorphous semiconductors.

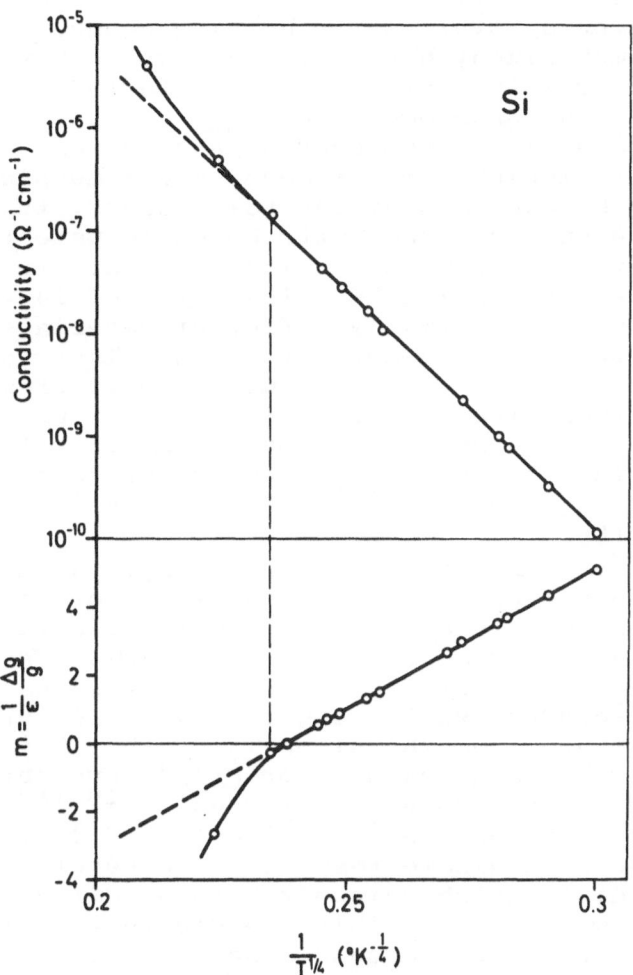

Fig. 17   Test of the $T^{-1/4}$ - dependence of conductivity and elastoresistivity (Ref. 39).

The frequency dependence of conductivity is considered to exhibit a characteristic form if conduction is by hopping in states near E_F. The ac-conductivity in the region of variable range hopping is given by[14,40]

$$\sigma(\omega) = \frac{\pi^3}{96} \, e^2 kT \, \alpha^{-5} \, g(E_F)^2 \, \omega \, (\ln\frac{\nu}{\omega})^4 \qquad (17)$$

For $\omega \ll \nu$ this leads to the well know variation as

$\sigma(\omega) \propto \omega^{0.8}$. This type of frequency dependence is often regarded as being typical for hopping conduction. It should be emphasized, however, that such behavior can also result from the influence of contacts[41] or inhomogeneities in the specimen[1]. Also in crystalline semiconductors a similar frequency dependence can be found. In trigonal selenium for instance, where hopping transport has to be excluded by several reasons, this frequency dependence has been attributed to potential fluctuations within the crystal which originate from lattice defects like dislocations[42]. This type of law thus is found in many substantially different materials and therefore seems to be more generally valid. The requirement for its existence is simply a sufficiently broad distribution of transition probabilities of the current carriers. $\sigma(\omega)$ in formula (17) depends more weakly on $g(E_F)$ than the dc-conductivity which varies exponentially with $\left(\frac{1}{g(E_F)}\right)^{1/4}$. $\sigma(\omega)$ therefore is less sensitive

to preparation conditions and annealing. Assumptions about the so called phonon frequency are of minor importance for the prefactor. For instance a factor of 10^3 in ν gives only a factor of 6 in $\sigma(\omega)$. Therefore one should expect that the values for $g(E_F)$ calculated from the ac-conductivity are not very different from those determined from the slope T_o of the $T^{-1/4}$-law. However, rather large values for $g(E_F)$ were obtained with this method for chalcogenide glasses ($10^{19}-10^{20}$ cm^{-3} eV^{-1})[14] and amorphous Ge (3.10^{21} cm^{-3} eV^{-1})[43]. These inconsistencies indicate that the ac-conductivity cannot be used to distinguish uniquely between different conduction-mechanisms. Possibly one has to take into account also a heterogeneous model for its interpretation.

VI. PHOTOCONDUCTION

Photoconductivity in amorphous semiconductors has been intensively studied by a number of groups particularly on simple or multicomponent chalcogenide glasses. It is not the aim of this lecture to give a review of all this interesting work. Rather it is attempted to discuss some aspects which seem to be typical for amorphous materials. Of particular interest in this context will be the models used for the density of gap states and the transitions involved. As an example in some more detail the results on glow-discharge silicon will be discussed because for this semiconductor there exists some

reliable information about the density of gap states
from other experiments.

Examples illustrating the main features of stationary photoconductivity of amorphous semiconductors are
given in Figs. 18 and 19. The results in Fig. 18 are
typical for chalcogenide glasses and may be summarized
as follows:

There is a maximum in the temperature dependence
of the photoconductivity σ_{ph} which generally occurs near
the temperature, where the dark conductivity σ_d exceeds
σ_{ph}. In the high-temperature range, where $\sigma_{ph} < \sigma_d$, σ_{ph}
varies linearly with light intensity and increases ex-
ponentially with $1/T$ with a well-defined activation en-
ergy. At lower temperatures, $\sigma_{ph} > \sigma_d$, σ_{ph} decreases
exponentially with $1/T$ and depends on the square root
of the intensity. At the lowest temperatures the curves
level off and it seems that they approach a constant
value. In Fig. 19 measurements on amorphous Si are presented. Clearly the general behavior resembles that of
amorphous chalcogenides.

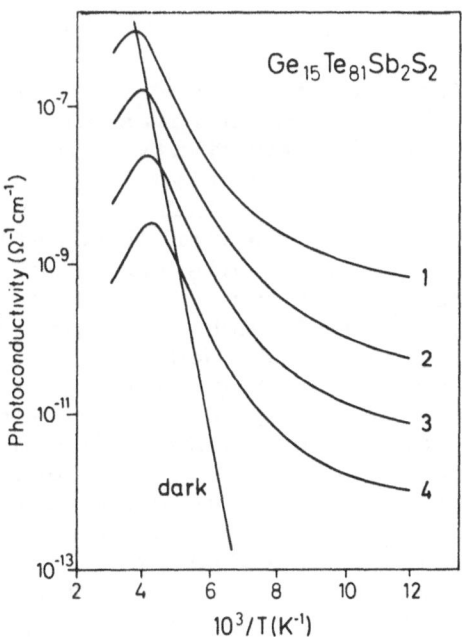

Fig. 18 Temperature dependence of photoconductivity of
amorphous $Ge_{15}Te_{81}Sb_2S_2$ after Ref. 44. Parameter illumination intensity: (1) 1; (2) 10^{-1}; (3) 10^{-2};
(4) 10^{-3}.

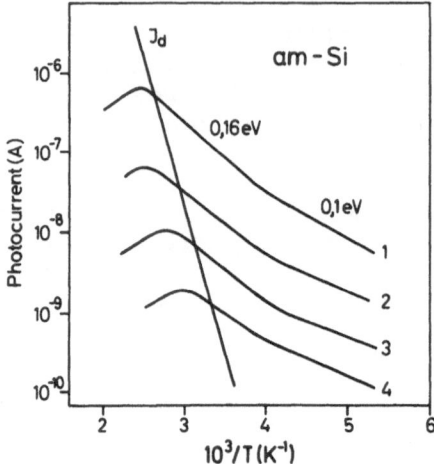

Fig. 19 Photoconductivity of amorphous Si as a function
of temperature after Ref. 45.   Parameter illumination
intensity: (1) 1; (2) 4.10^{-2}; (3) 10^{-2}; (4) 2.10^{-3}.

A striking feature of these results is that there
are well defined activation energies in the temperature
dependence of the photoconductivity. This behavior
strongly indicates that the active traps or recombination centers are located in a fairly narrow energy range.
One might ask, therefore, whether this reflects pronounced structure in the density distribution of localized
states. Indeed Marshall and Owen[22] interpreted their
photoconductivity and drift mobility results on amorphous
As$_S$Se$_3$ in a density-of-states model illustrated schematically in Fig. 20a. "Acceptor-like" localized states
are centered around an energy of E_p = 0.4 eV above the
mobility edge of the valence band. In order to explain
the intrinsic behavior of this semiconductor, a set of
compensating donors has to be assumed in the upper half
of the gap. These levels act as traps for electrons and
holes and confine the trapped carriers to a well defined
energy range. This distribution of trapped electrons
n_t and holes p_t is sketched in Fig. 20c. However, such
a distribution of trapped carriers can also arise in a
density of states model without distinct maxima. Such
a model has been used by Arnoldussen et al[44] for the interpretation of their data on multicomponent glasses
(Fig. 18). The important point is that in contrast to

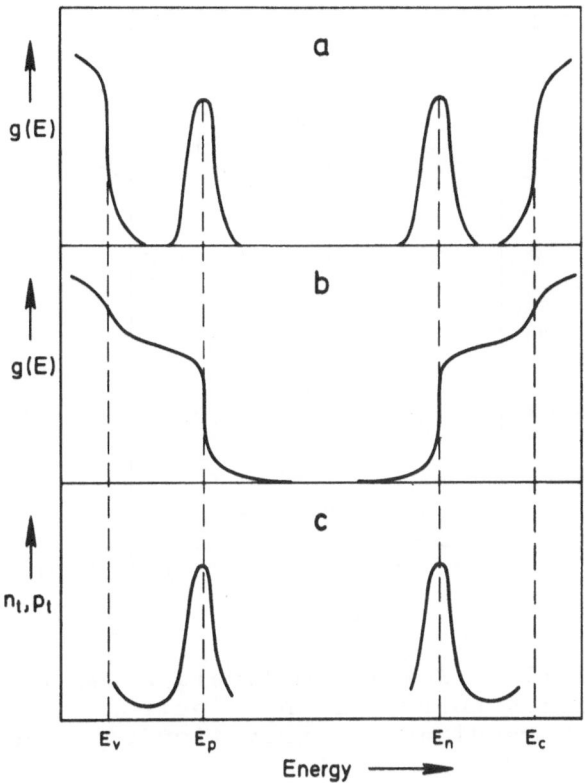

Fig. 20 Density of localized states (a) after Marshall and Owen[22], (b) after Arnoldussen et al (Ref. 44) and (c) distribution of trapped carriers n_t, p_t.

the Marshall and Owen model, near energies E_n and E_p the density of localized states has edges (Fig. 20b). Near these edges the distribution of trapped electrons n_t or holes p_t is expected to exhibit pronounced maxima.

An alternative model (Fig. 21) has been discussed by Fischer etal[46] who start from the density of states distribution of the CFO-model. Electrons excited above the mobility edge are supposed to have a much higher probability of loosing energy by thermalization p_{th} than by recombination p_R. That is as in crystalline materials, where thermalization of hot carriers occurs with time constants of about 10^{-13}s whereas the lifetime of the carriers typically is near 10^{-6}s. In a crystalline material thermalization ends near the band edge whereas in

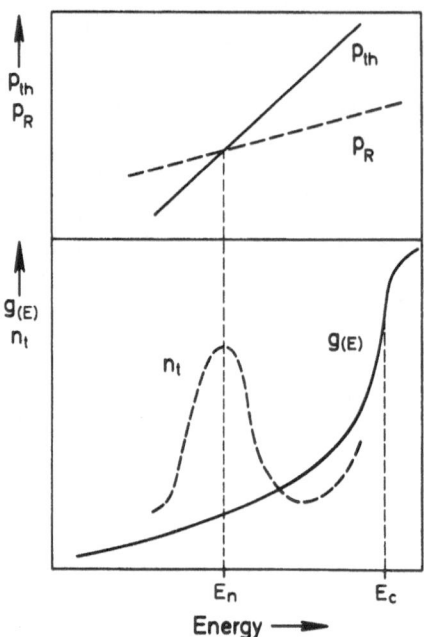

Fig. 21 Probability for thermalization p_{th} and recombination p_R versus energy of the electron states after
Ref. 46 (a) and resultant distribution of trapped carriers (b).

an amorphous photoconductor the excited carriers may
dribble down to lower energy states. Although the mobility edges play a dominant role in conduction, they
do not do so for recombination. At lower energies p_{th}
will decrease because the states will become more and
more localized. This localization does not so much influence the recombination probability p_R since the electrons may recombine with holes which are not so strongly
localized. Hence an energy should exist where $p_R = p_{th}$.
Near this recombination edge electrons pile up as indicated in Fig. 21b.

These three models are equivalent as far as the recombination processes are concerned and describe the main
features of photoconductivity of amorphous semiconductors.
What is required is, that the trapped carriers are accum-
ulated near certain energies. This can be achieved either
by a real variation of the density of localized states
or by an energy dependent recombination coefficient.

From this discussion it becomes clear that photoconductivity measurements alone cannot yield unambiguous
information about the gap states. In case of amorphous
Si detailed knowledge about the distribution of local-
ized-states comes from drift mobility studies and field
effect[12,21]. It is of interest, therefore, to look how
the photoconductivity results of this material fit into
the worked out model for the density of gap states, which
is presented in Fig. 22. This density of states distribution refers to a glow discharge specimen deposited at
a high substrate temperature. The states near E_A are
considered to be in thermal equilibrium with the states
above E_C at all tempertures such that trapped electrons
pile up near the edge at E_A. In the lower half of the
gap a maximum in the density of states distribution is
assumed to exist about 1.2 eV below E_C. For the following it is important that the temperature dependence of
the photoconductivity in Fig. 19 is apparently the same
as that of the drift mobility in Fig. 9. There it was
concluded that at high temperatures where the activation energy is 0.18 eV, conduction takes place by electrons in extended states above E_C whereas at low temperatures hopping in states near E_A predominates with an

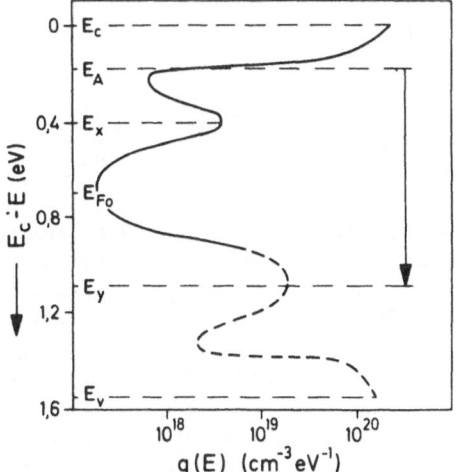

Fig. 22 Model for the density of states of amorphous
Si after Ref. 45. The arrow indicates the predominant
recombination path.

activation energy of about 0.1 eV. Spear et al[45] made
the reasonable assumption that the same interpretation
also applies for the photoconductivity. In the high
temperature range, therefore, on both sides of the maximum of σ_{ph} transport is by electrons in delocalized
states near E_c:

$$\sigma_{ph} = e \, \Delta n_c \cdot \mu_c \tag{18}$$

In case of thermal equilibrium between states at E_c and
E_A, the excess carrier density Δn_c is related to the concentration of electrons trapped near E_A by

$$\frac{\Delta n_c}{\Delta n_A} = \frac{N_c}{N_A} \exp\left(-\frac{E_A - E_c}{kT}\right) \tag{19}$$

When electrons trapped in states near E_A recombine with
holes trapped in states near E_y, as is indicated in Fig.
22, the steady state is determined by the equation

$$g = r\Delta n_A (p_y + \Delta p_y) \tag{20π}$$

where g is the generation rate, and p_y is the equilibrium concentration of holes at E_y. One has to consider
now the cases of small and large disturbance of the equilibrium. In the small-signal case $\Delta p_y \ll p_y$ and
$p_y = N_y \exp[-(E_y - E_F)/kT]$, since the Fermi level does
not move from its equilibrium position. With the above
relations:

$$\sigma_{ph} = e\mu_c \frac{N_c}{rN_A N_y} g \, \exp\left[\frac{E_y - E_F}{kT} - \frac{E_A - E_c}{kT}\right] \tag{21}$$

Inserting the known values of the energies one finds
that the photoconductivity increases with 1/T with an
activation energy of 0.2 eV and varies linearly with
light intensity g. This is in good accordance with the
behavior found in the high temperature range, where
$\sigma_{ph} < \sigma_d$. In the large signal case the Fermi level splits
into the quasi Fermi levels for electrons and holes. If
one assumes that most of the carriers are trapped near
E_A and E_y, one can use the neutrality condition in the
simple form $\Delta n_A = \Delta n_y$. Following the same procedure as
above, one obtains for the photoconductivity:

$$\sigma_{ph} = e\mu_c \cdot \sqrt{\frac{g}{r}} \cdot \frac{N_c}{N_A} \cdot \exp\left[-\frac{E_A - E_c}{kT}\right] \qquad (22)$$

In the large signal case, therefore, one expects a square-root dependence of σ_{ph} on light intensity and a thermally activated photoconductivity, what is indeed found in experiment. The activation energy is the same as in the temperature dependence of the drift mobility, namely 0.18 eV.

At lowest temperatures, in the third temperature range in Fig. 19, hopping transport in states near E_A is assumed to take over. Then conduction takes place in the same states which are the initial states in the recombination process. Since no thermal step is involved in the recombination process, the temperature dependence of σ_{ph} is attributed to the mobility which is of the type of relation (5) with an activation energy of W = 0.1 eV. It is remarkable that this type of hopping photoconductivity so far has not been observed in other amorphous photoconductors.

Thus the main features of the photoconductivity of amorphous Si can be described in the density of states model as developed from drift mobility and field effect measurements. Spear et al[45] have also discussed other recombination transitions. For instance they considered the case where the recombination occurs by transitions between states at E_c and E_y. However, the only transitions which lead to consistent results, are those between E_A and E_y.

The models used for the explanation of the $\sigma_{ph}(T)$ curves of chalcogenide glasses have some similarily with the one appropriate for amorphous Si. As was outlined by Arnoldussen et al[44] such behavior can only be understood in a model which contains at least one set of localized states near each band. Together with the extended states, therefore, at least 4 levels are involved in trapping and recombination. It is commonly agreed that the maximum of the $\sigma_{ph}(T)$ curves is not due to a change in the recombination mechanism but has to be explained by the turnover from mono - to bimolecular recombination which is associated with the transition from the low to the large signal case.

Although the kind of treatment given above might

explain most of the experiments, it gives only little
insight into the recombination process itself. It is
the recombination coefficient, which contains the unknown details of the mechanism. In the commonly used
models the transitions which lead to trapping and recombination may be of quite different character depending on the kind of states involved. The transitions
which are most frequently discussed in recombination models, particularly of crystalline photoconductors, are
those between extended and localized states. In the a-
bove outlined model for amorphous Si transitions are proposed to predominate where the initial and the final
states are localized. Also most other authors agree
that one needs localized-localized transitions for a
consistent explanation of the experimental results[44,46,47].
This type of recombination has been intensively discussed in case of the pair spectra in the luminescence of
GaP-crystals[48] where the recombination occurs between
isolated pairs of donors and acceptors. In the photoconduction of Se-crystals direct transitions between
traps for electrons and holes have been proposed to be
responsible for the slow nonexponential rise and decay
of the photoconductivity[49]. The main characteristic of
these transitions is, that the spatial distance between
the participating localized states determines the transition probability because the overlap of the wave functions strongly decreases with increasing distance. Since
the distances are distributed in a random manner, a distribution of the transition probabilities results. This
can lead to a recombination which cannot be described by
a single lifetime, but by a spectrum of lifetimes. The
consequence of this is that the decay of the photoconductivity is strongly nonexponential, and in special
cases, may follow a logarithmical time law[49]. It is important that this type of recombination does not directly involve thermal steps and therefore should not depend
very sensitively on temperature.

Transitions between localized states are more likely to occur in amorphous photoconductors than in crystalline ones because of the much higher density of localized gap states which can act as traps or recombination
centers. Direct evidence for the predominance of this
recombination channel comes from luminescence measurements. In amorphous As_2Se_3[50] and amorphous Si[51] strong
luminescence bands at energies below the mobility gap
have been assigned to transitions between localized states.
In view of the fact that recombination between localized
states is relevant, it is surprising that no striking

properties are observed due to this process. Of all predictions for this type of recombination obviously only
the independence of temperature is fulfilled. The temperature dependence of the photocurrent arises, because
of a multiple trapping mechanism, the recombination process itself thereby does not depend on temperature. In
the $\sigma_{ph}(T)$ curves of amorphous Si the weak influence of
temperature on the recombination can most clearly be
seen at the lowest temperatures. Here conduction takes
place among the localized states which also act as initial states in recombination. The observed temperature
dependence of photoconductivity then is only due to the
mobility which is thermally activated.

Besides the smallness of the recombination coefficient the spatial distances of the localized states
should show up in the decay curves as well as in the dependence of σ_{ph} on light intensity. However, in none
of the investigations known to the author, is there a
convincing indication for such effects. Several reasons
may explain why in amorphous materials this distance of
the recombining carriers is of minor importance: (i)
The recombination can take place by a transition
between two states of the same defect. (ii) Hopping
among the localized states can allow the carriers to
find states where they can easily recombine.

The latter argument is considered to be most important if one takes into account the conduction mechanism suggested in these materials. Since in crystalline semiconductors the traps are more strongly localized, the influence of the spatial distance is more pronounced there, if recombination occurs between such
states. However, recombination through states of the
same defect center has also to be considered. Such a
process has been discussed by Spear et al[45] in connection with the question, how the difference in energy can
be dissipated.

In amorphous and crystalline photoconductors the
transport of the excess carriers is generally assumed
to take place in extended states. The localized states
then act as traps or recombination centers. Only few
examples so far exist where photoexcited carriers are
considered to move by hopping among localized states,
namely, in partially compensated crystalline Ge and Si[52]
and in amorphous Si at low temperatures as discussed
above. In a recent investigation of the photoconductivity of amorphous Ge, Fischer and Vornholz[53] propose a

model where the photocurrent is produced by carriers near
the Fermi level by essentially the same mechanism as the
dark current. The main experimental results are shown
in Fig. 23. The dark conductivity σ_D exhibits at low
temperatures $\ln\sigma_D \propto T^{-1/4}$ behavior which indicates the
predominance of variable range hopping near the Fermi
level. An important point is that in this temperature
range the photoconductivity σ_{ph} depends only weakly on
temperature. At high temperatures σ_{ph} increases due to
heating of the sample by the laser beam. The influence
of annealing on σ_{ph} is surprisingly small as compared
to the behavior of σ_D which thereby decreases by many
orders of magnitude. For an explanation of these results the authors propose a model where the transport
in the dark as well as under illumination is by variable
range hopping. It is argued that due to the high density of localized states in evaporated amorphous Ge,
photoexcited carriers quickly thermalize by phonon emission down to states near the Fermi level. If the recombination is sufficiently slow, then excess carriers of
both signs will pile up there and split the Fermi level
into quasi-Fermi levels for electrons and holes. This
leads to an increase of the conductivity if the density
of localized states at the Fermi levels is higher than
that at the equilibrium Fermi level. Calculations based
on this model are in qualitative agreement with the experimental results. Particularly such a model can account for the weak influence of temperature and annealing on the photoconductivity. Although the agreement
of these calculations with the experimental $\sigma_{ph}(T)$-cur-
ves seems to be rather convincing, it is questionable
whether a considerable shift of the Fermi level can be
attained in view of the high density of localized states
in this material. An alternative explanation could be
that the conduction mechanisms are different in dark and
illuminated amorphous Ge. If then the states, where the
photoconduction takes place, are identical with the initial or final states of the recombination transition,
then the photoconductivity should not be very sensitive
to temperature, provided the mobility does not exhibit
pronounced temperature dependence. The difficulty with
such a model is, however, that the weak influence of annealing on the photoconductivity cannot be easily understood. In order to account for this feature too, one
has to assume additionally that the states which participate in the recombination process cannot be annealed.

The above discussion may be summarised as follows:
The models used to explain the photoelectrical properties

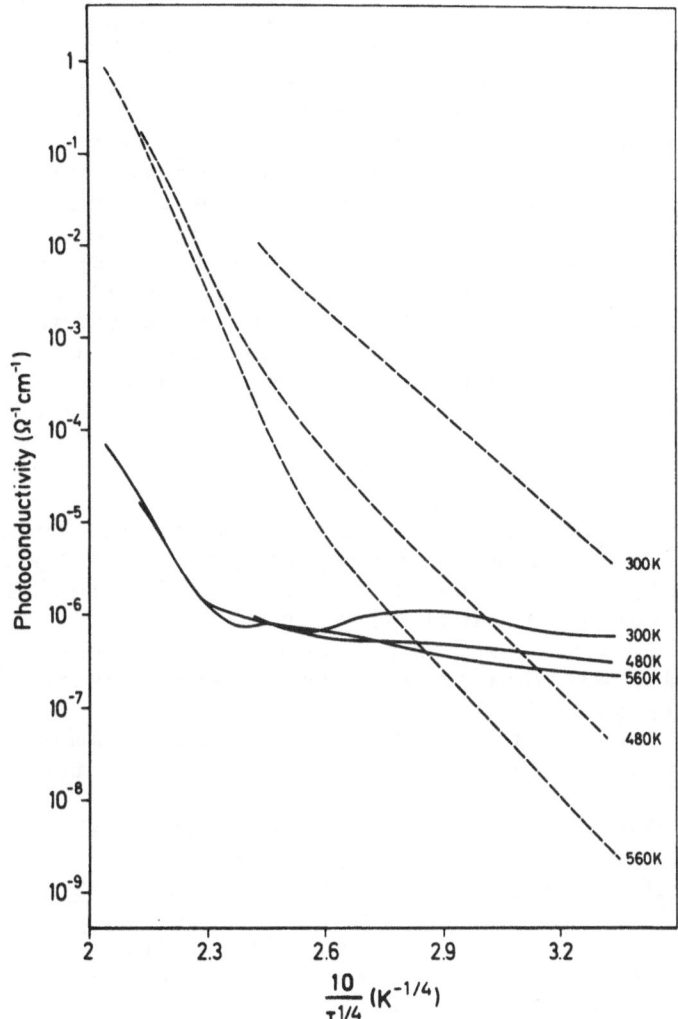

Fig. 23 Photoconductivity (solid lines) and conductivity
(dashed lines) of evaporated amorphous Ge in different
annealing states (Ref. 53).

of amorphous semiconductors do not differ appreciably
from the corresponding models for crystalline photoconductors. Four-level models (including the extended states)
are mostly applied. The resulting properties are rather
insensitive to details of the underlying model. It seems,
therefore, that the investigation of photoconductivity

alone does not give unambiguous information about the
distribution of the gap states.   In amorphous photoconductors recombination very often is supposed to occur
between localized states.   It is still not clear why no
effects are observed which originate from the influence
of the spatial distribution of the participating states.
With few exceptions, all authors assume that conduction
takes place in extended states and that the numerous localized states only act as traps or recombination centers.   Photoconduction by hopping among localized states
has been suggested to be relevant at low temperatures
for amorphous Si and Ge only.   Of particular interest is
a new type of photoconduction proposed to exist in amorphous Ge, namely hopping photoconduction of the excess
carriers in states around the Fermi level.

REFERENCES

1.   H. Fritzsche, Amorphous and Liquid Semiconductors,
      ed. J. Tauc, Plenum Press, Chapter 5 (1973).
2.   W. Paul, Proc. 11th Intl. Conf. on Physics of Semi-
      conductors, Warsaw (1972), p. 38.
3.   H. Mell, Proc. 5th Intl. Conf. on Amorphous and
      Liquid Semiconductors, Garmisch-Partenkirchen
      (1973), p. 203.
4.   W. E. Spear, Proc. 5th Intl. Conf. on Amorphous
      and Liquid Semiconductors, Garmisch-Partenkirchen
      (1973), p. 1.
5.   M. H. Cohen, H. Fritzsche and S. R. Ovshinsky,
      Phys. Rev. Letts 22, 1065 (1969).
6.   E. A. Davis and N. F. Mott, Phil. Mag. 22, 903
      (1970).
7.   N. F. Mott, Phil. Mag. 24, 911 (1971).
8.   W. Heywang and D. R. Haberland, Solid State Elec-
      tronics 13, 1077 (1970).
9.   H. Fritzsche, J. Non-Cryst. Solids 6, 49 (1971).
10.   A. Many, Y. Goldstein and N. B. Grover, Semicon-
      ductor Surfaces, North-Holland Publ. Comp. (1965).
11.   R. F. Egerton, Appl. Phys. Letts 19, 203 (1971).
12.   W. E. Spear and P. G. LeComber, J. Non-Cryst.
      Solids 8-10, 727 (1972).
13.   A. Madan and W. E. Spear, to be published.
14.   N. F. Mott and E. A. Davis, Electronic Processes
      in Non-Crystalline Materials, Clarendon Press,
      Oxford (1971).
15.   P. Nagels, R. Callaerts and M. Denayer, Proc. 11th
      Intl. Conf. on Physics of Semiconductors, Warsaw
      (1972), p. 549.

16. L. Friedman, J. Non-Cryst. Solids 6, 329 (1971).
17. T. Holstein, Phil. Mag. 27, 225 (1973).
18. R. A. Lomas, M. J. Hampshire, R. D. Tomlinson and K. F. Knott, Phys. Stat. Sol. (a) 16, 385 (1973).
19. C. H. Seager, M. L. Knotck and A. H. Clark, Proc. 5th Intl. Conf. on Amorphous and Liquid Semiconductors, Garmisch-Partenkirchen (1973), p. 1133.
20. W. E. Spear, J. Non-Cryst. Solids 1, 197 (1969).
21. P. G. LeComber, A. Madan and W. E. Spear, J. Non-Cryst. Solids 11, 219 (1972).
22. J. M. Marshall, A. E. Owen, Phil. Mag. 24, 1281 (1971).
23. W. Beyer and J. Stuke, Proc. 5th Intl. Conf. on Amorphous and Liquid Semiconductors, Garmisch-Partenkirchen (1973), p. 251.
24. W. Beyer, to be published, in Phys. Stat. Sol.
25. H. Fritzsche, Solid State Commun. 9, 1813 (1971).
26. D. Emin, C. H. Seager and R. K. Quinn, Phys. Rev. Letters 28, 813 (1972).
27. L. Vescan, M. Telnic and C. Popescu, Phys. Stat. Sol. (b) 54, 733 (1972).
28. K. L. Chopra and S. K. Bahl, Thin Solid Films 12, 211 (1972).
29. R. Grigorovici, Mat. Res. Bull. 3, 13 (1968).
30. F. Buchy, M. T. Clavaguera, Ph. Germain, Proc. Intl. Conf. on Low Mobility Materials, Eilat (1971) p. 235.
31. H. K. Rockstadt and J. P. deNeufville, Proc. 11th Intl. Conf. on Physics of Semiconductors, Warsaw (1972), p. 542.
32. N. F. Mott, J. Non-Cryst. Solids 1, 1 (1968).
33. V. Ambegaokar, B. I. Halperin and J. S. Langer, Phys. Rev. B4, 2612 (1971).
34. M. Pollak, J. Non-Cryst. Solids 11, 1 (1972).
35. M. L. Knotek, M. Pollak and T. M. Donovan, Phys. Rev. Letters 30, 853 (1973).
36. M. H. Brodsky and R. J. Gambino, J. Non-Cryst. Solids 8-10, 739 (1972).
37. K. Maschke, H. Overhof and P. Thomas, Phys. Stat. Sol. (b) 62, 113 (1974).
38. D. Emin, Phys. Rev. Letters 32, 303 (1974).
39. W. Fuhs, Phys. Stat. Sol. (a) 10, 201 (1972).
40. M. Pollak, Phil. Mag. 23, 519 (1971).
41. A. K. Jonscher, J. Phys. C 6, L235 (1973).
42. T. Salo, T. Stubb and E. Suosara, The Physics of Selenium and Tellurium, Pergamon Press (1969), p. 335.
43. K. L. Chopra and S. K. Bahl, Phys. Rev. B1, 2545 (1970).

44.   T. C. Arnoldussen, R. H. Bube, E. A. Fagen and S.
      Holmberg, J. Appl. Phys. 43, 1798 (1972).

45.   W. E. Spear, R. J. Loveland, A. Al-Sharbaty, to
      be published in J. Non-Cryst. Sol.

46.   R. Fischer, U. Heim, F. Stern and K. Weiser, Phys.
      Rev. Letters 26, 1182 (1971).

47.   W. Fuhs and D. Meyer, Phys. Stat. Sol. (a) 24
      (1974).

48.   D. G. Thomas, J. J. Hopfield and W. M. Augustyniak
      Phys. Rev. 140 A, 202 (1965).

49.   W. Fuhs and J. Stuke, Phys. Stat. Sol. 27, 171
      (1968).

50.   B. T. Kolomiets, T. N. Mamontova and A. A. Babaev
      J. Non-Cryst. Sol. 4, 289 (1970).

51.   D. Engemann and R. Fischer, Proc. 5th Intl. Conf.
      on Amorphous and Liquid Semiconductors, Garmisch-
      Partenkirchen (1973), p. 947.

52.   V. P. Dobrego and S. M. Ryvkin, Sov. Phys. Solid
      State 4, 402 (1962); Sov. Phys. Solid State 6,
      928 (1964).

53.   R. Fischer and D. Vornholz, Phys. Stat. Sol. (b)
      68, 561 (1975).

ELECTRICAL TRANSPORT IN SEMICONDUCTING NONCRYSTALLINE SOLIDS*

David Emin

Sandia Laboratories

Albuquerque, New Mexico 87110

I. INTRODUCTION

In the present lectures I shall try to outline the principal theoretical approaches and some specific calculations which appear to be relevant to an understanding of a number of electrical transport measurements in noncrystalline semiconductors. Since there is considerable evidence that the carriers in many of the frequently studied disordered semiconductors are characterized by very low mobilities $(\mu \lesssim 1 \text{ cm}^2/\text{V-sec})$, this review will primarily involve a discussion of the situations that are characteristic of low-mobility transport. In particular, much of this presentation will be concerned with the theory of phonon-assisted hopping motion. This topic has received increasing attention as a growing number of experiments have been interpreted as being indicative of hopping motion. However, the primary thrust of these lectures will be to address the generic features of hopping motion rather than to consider detailed attempts aimed at explaining specfic experiments. Hopefully, this will provide a vantage point which will be useful in viewing electrical transport experiments on low-mobility disordered semiconductors.

II. HIGH-MOBILITY TRANSPORT

To begin, consider the alternative situation of high-mobility charge carriers in crystalline semiconductors such as silicon and germanium. In these examples one

typically attempts to view charge transport from a scattering point of view. That is, one speaks of a particle
as moving freely through a material with its motion only
occasionally encumbered by a scattering event. These
scattering events are associated with the carrier's interaction with impurities (impurity scattering), defects
(defect scattering), and the lattice vibrations (phonon
scattering). Relating the drift mobility μ_D to the carrier's diffusion constant D via the Einstein relation,

$$\mu_D = \frac{e}{kT} D, \tag{1}$$

where e is the carrier's charge, k the Boltzmann constant,
and T the temperature, the diffusion constant can, in analogy to that of the kinetic theory of gasses, be expressed as a thermal average of the product of a particle's
speed v and its mean free path ℓ. Thereby one may write

$$\mu_D = \frac{e}{kT} < v \, \ell > ; \tag{2}$$

here the mean free path ℓ is defined as the product of
the particle's speed and a collision (relaxation) time
τ ; $\ell = v\tau$. Employing the oft-used approximation that
τ is independent of the particle's velocity and position,
one has the familiar expression

$$\mu_D = \frac{e}{kT} < v^2 > \tau = \frac{e\tau}{m*} , \tag{3}$$

where the particle's kinetic energy is simply $\frac{1}{2}m*v^2$.[1]
Typically, the dominant room-temperature scattering mechanism is taken to be phonon scattering. In this case,
with an increase in temperature, the number of phonons
present increases, the probability of electron-phonon
"collisions" therefore increases, and hence the collision
time τ decreases. Thus the mobility is typically (except
at very low temperatures) a decreasing function of increasing temperature.

When considering the scattering picture one is led
to contemplate situations in which this point of view will
break down. One approach to finding the minimum mobility
consistent with the scattering picture is to note that a
particle has a size associated with it, namely, the
deBroglie wavelength $\lambda \sim \hbar/p$, where p is the carrier's

momentum. By requiring that the mean free path be greater than the deBroglie wavelength, one finds that the minimum mobility consistent with the scattering picture is

$$\mu_{min}^{\lambda} \sim \frac{e}{kT} < v \; \lambda > \sim \frac{e}{kT} < v\hbar/p > \sim \frac{e}{kT} \left(\frac{Ja^2}{\hbar}\right), \qquad (4)$$

where J is an electronic transfer integral (the bandwidth
parameter in tight-binding theory) and a is the lattice
constant. Taking J $\sim$1 eV and a $\sim$5 A, one finds that μ_{min}
at room temperature is of the order of 10^2 cm^2/V-sec.

Another criterion which yields a lower limit of the
mobility consistent with the scattering picture is obtained by asserting that the scattering picture only loses
meaning when the mean free path is less than a lattice
constant. In the case of a band which is much wider than
kT this yields

$$\mu_{min}^{a} \sim \frac{e}{kT} \; a < v > \sim \frac{ea}{kT} \left(\frac{kT}{m^*}\right)^{\frac{1}{2}} \sim \frac{e}{kT} \left(\frac{Ja^2}{\hbar}\right)\left(\frac{kT}{J}\right)^{\frac{1}{2}}, \quad (5)$$

where it has been noted that the carrier's effective mass
at a band extremum is given by

$$m^* \sim \hbar^2/Ja^2 \; . \qquad (6)$$

Using the numbers of the previous example, one finds that
the room temperature value of μ_{min}^{a} is $\sim$20 cm^2/V-sec. Estimates of the room-temperature mobility in noncrystalline
insulators are frequently much smaller than this value.
Also, in the cases where Hall effect measurements have
been performed, the Hall mobility is typically much lower
than this minimum and is often even an increasing function
of temperature (as distinct from the prediction of the
scattering picture).[2-9] Thus one is led to investigate
alternatives to the simple scattering picture.

III. BROWNIAN MOTION: RANDOM-PHASE MODEL

An alternative approach to "extended-state motion"
is that proposed by Cohen[10]. In this approach one first
acknowledges that the scattering (from whatever source)
is sufficiently strong so that the mobility is less than
the lower limit arrived at by assuming that the mean free

path is comparable to a lattice constant. Writing the
diffusion constant as the square of the lattice constant,
a, divided by the time which characterizes the diffusion
of a particle from a site at which it is initially placed,
t_s, Cohen obtains the mobility expression

$$\mu_{BM} = \frac{e}{kT} \left(a^2/t_s \right) \quad , \tag{7}$$

where the Einstein relation has been utilized to relate
the mobility to the diffusion constant. Here the time of
stay, t_s, is taken to be $\sim \hbar/Jf$, where, as before, J is an
electronic transfer integral and f is a factor which em-
bodies the feature that with an increase in disorder the
diffusion of the carrier is impeded. In particular, while
f ~ 1 for an ordered material, with increasing disorder f
is taken to decrease; the time of stay increases with an
increase in disorder.[11,12] Concomitantly, with an increase in disorder the "Brownian-motion" mobility, μ_{BM},
is taken to decrease.[13] Finally, in this model if the
disorder is sufficiently great, being characterized by a
small value of f, Anderson localization is taken to have
occurred thereby rendering all the electronic states "localized". In this situation the formula is assumed to be
inapplicable. Thus, in the Mott-C.F.O. model the extended-state motion is assumed to be characterized by either
a scattering-type mobility, if the mobility is sufficiently high, or by a Brownian-motion type mobility if the mobility is lower than some value ~ 20 cm^2/V-sec.

Following an ad hoc procedure that is referred to as
the random-phase model, Hindley and Friedman have been
able to obtain this mobility formula.[14,15] However, this
formula, Eq. (7), has not (at least to my knowledge) been
obtained as a "strong-scattering" limit of a general mobility calculation in which the traditional weak-scattering picture also emerges. In the random-phase-model calculation the role of f in Eq. (7) is played by the factor
$[a^3(2zJ) \, g(\varepsilon_c)]$, where a is the interactomic spacing, 2zJ
the electronic bandwidth obtained from a tight-binding
calculation (z is the coordination number), and $g(\varepsilon_c)$ is
the density of states at an energy level near those which
are associated with the electrons that provide the dominant contribution to the conduction-band extended-state
conductivity.

While the d.c. drift mobility is a measure of a carrier's response to an applied d.c. electric field, the
Hall mobility is a measure of a carrier's response to the

additional application of a d.c. (Hall) magnetic field
perpendicular to the direction of the electric field.
Although the Hall mobility and the drift mobility are
equal in the free-electron illustrations of introductory
textbooks,[16] this is generally not the case. In particular, within the randon-phase model, Friedman finds that
the Hall mobility, μ_{Hall}, is related to the drift mobility essentially by a simple ratio:[15,17]

$$\frac{\mu_{Hall}}{\mu_D} \sim \frac{kT}{J} . \tag{8}$$

In other words, the Hall mobility within the random-phase-
model approach to "Brownian Motion" is temperature-independent. Furthermore, taking $J \gg kT$, Friedman asserts
that the Hall mobility will be considerably smaller than
the drift mobility.[18]

    Thus, although the random-phase model remains an ad
hoc model which is yet to be justified by a first-prin-
ciples calculation, its predictions are clearly stated.
The drift mobility should typically be smaller than
$\sim$20 cm^2/V-sec and fall with increasing temperature as T^{-1},
while the Hall mobility will be considerably smaller and
temperature-independent.[18] As pointed out in the previous
section these predictions are not consistent with a variety of experimental results.

IV. PHONON-ASSISTED TUNNELING

    Most other schemes for describing electrical trans-
port data involve the phenomenon of phonon-assisted tunneling of the charge carriers. In these instances one
speaks of "hopping in the localized (bandtail) states",
hopping in deep-defect states (such as "dangling-bond"
states), and/or "small-polaron hopping". While different words and different assumed values of the physical
parameters characterize the distinctions between these
situations, they all treat the carriers as moving via
phonon-assisted tunneling events. Therefore, it is useful to describe the fundamentals of the theory of phonon-
assisted tunneling. To accomplish this, attention will
first be directed toward understanding small-polaron hopping motion in a crystal. The extension of the theory to
include the general case of phonon-assisted tunneling between distinct spatial regions in a disordered solid will
then be treated. In particular, our concern will, for the
most part, be centered on elucidating the mechanism of an

individual phonon-assisted hop.

V. MOLECULAR CRYSTAL MODEL

An easily generalized model which provides a simple
way to visualize aspects of the small-polaron problem is
termed the Molecular Crystal Model.[12,19] In this model
one considers a single excess electron placed in a periodic array of deformable molecular units. Associated
with each molecule in the lattice is a configurational
coordinate which represents a distortion of the molecule
from its carrier-free equilibrium configuration. For instance, the configurational coordinate of a molecule at
a site g denoted by x_g, may be thought of as the molecu-
lar-distortion variable related to the breathing mode of
that molecule. In the absence of any coupling between
molecules, these local configurational coordinates are
taken to vibrate harmonically about their equilibrium positions. Furthermore, coupling between the vibrational
motion of one molecule and its neighbors is included in
this model, thereby providing the mechanism for the transport of vibrational energy through the lattice; it is this
intersite coupling which gives rise to dispersion of the
(optical-mode) lattice-vibrational frequencies. It will
be seen that the inclusion of this intersite vibrational
coupling, i.e., vibrational dispersion, is an essential
ingredient of a theory of hopping motion. Finally, the
energy of an excess carrier placed at a site in the lattice is taken to be linear function of the lattice displacements; this is a feature characteristic of most treatments of electron-lattice interaction. In particular, the
energy of a carrier placed at the origin is assumed to be
a linear function of the x_g's, x_g being the deviation from
equilibrium of the configurational coordinate associated
with the molecule at site g:

$$E(. . . x_g . . .) = E_o - \sum_g f(g) \, x_g \; ; \qquad (9)$$

$f(g)$ is a weighting function and g is the position vector
of a molecular site. For simplicity, most studies have
taken the electron-lattice interaction to be short-range
(deformation-potential-like); specifically, $f(g) = A\delta_{g,0}$.
In other words, in this approximation a carrier only
displaces atoms of the molecule on which it resides.

An excess electron placed at an isolated site (no

electronic overlap between molecules) in this molecular
crystal can lower its own energy and the energy of the
system as a whole by deforming the molecule on which it
resides. Concomitantly the atoms of the occupied molecule will assume new (displaced) equilibrium positions
about which they will oscillate. Specificially, if, for
simplicity, vibrational dispersion is ignored, the configurational coordinate associated with atomic displacements at the occupied site will be displaced by an amount
$A/M\omega_o^2$, where M and ω_o are respectively the reduced atomic
mass and the vibrational frequency associated with the
configurational coordinate under consideration. This will
lower the excess electron's energy from E_o to $E_o - A^2/M\omega_o^2$.
However, the distortion of the occupied molecule will be
associated with a strain which increases the energy of
the vibrational system by an amount $A^2/2M\omega_o^2$. Thus, the
net energy of the system comprising an excess electron
placed on an isolated molecule of the molecular crystal
is $E_o - A^2/2M\omega_o^2 + E_{vib}$, where $E_{vib} \equiv \hbar\omega_o(N + \frac{1}{2})$ is the
vibrational energy associated with vibrations of the con-
figurational coordinate about its "new" equilibrium value;
N is the vibrational quantum number related to the harmonic oscillations of the deformed (occupied) molecule.
The magnitude of the distortion-related lowering of the
energy of the coupled system is termed the small-polaron
binding energy, E_b:

$$E_b \equiv A^2/2M\omega_o^2. \tag{10}$$

This energy, one-half the ratio of the square of the electron-lattice force, A, to the stiffness of the individual
molecules, $M\omega_o^2$, is the energy which characterizes the
electron-lattice interactions. If one now takes cognizance of the presence of other identical molecules in the
molecular crystal one finds that the carrier can move between sites in the crystal. Associated with the ability
of the carrier to move, from a site on which it is arbitrarily placed to a neighboring site, is a transfer energy,
J. Specifically, in the tight-binding approach to the
formation of a small-polaron band, the bandwidth is proportional to this energy; in a simple cubic lattice structure the energy of a carrier is given by

$$E_{\underset{\sim}{k}} = E_o - E_b - 2Je^{-S}(\cos k_x a + \cos k_y a + \cos k_z a) + E_{vib}, \tag{11}$$

where $\underline{k}$ is the wave vector which characterizes the state
and $\exp(-S)$ is the previously discussed vibrational-over-
lap factor.[20]   In a disordered solid such parameters as
E_o, $J_{\underline{g}},\underline{g}$ ($\equiv$ J in a crystal), $h\omega_o$, A, and M will most
generally become functions of position in the material.
In most models of disorder only one or two of these parameters, typically E_o and/or J, are explicitly taken to be
altered by the presence of disorder. Finally, it should
be mentioned that since both the potential energy of the
system and the carrier's wavefunctions are altered as
the atoms are displaced, J is generally a function of the
lattice displacement. While one should be cognizant of
this fact, it is not necessary for an understanding of
the fundamental development of the theory of hopping motion.   Thus, it shall henceforth be assumed that the transfer integrals are constants.[21]

VI. SMALL-POLARON MOTION BETWEEN DEGENERATE SITES

A small polaron in an ideal crystal may move from
site to site via two distinct processes.[22,23] The first
involves the tunneling of a small polaron between neighboring sites with no change in the phonon population.
These so-called diagonal processes involve simply trans-
lating the carrier and its self-induced lattice distortion
between adjacent sites without any change in the atomic
vibratory motion (phonon population); this corresponds to
"polaron band motion".   The complementary processes are
those in which the phonon population changes with a site-
to-site transfer.   These processes, termed nondiagonal
processes, correspond to the phonon-assisted tunneling of
the carrier between adjacent sites. Thus, the small-
polaron mobility is a sum of two contributions: one associated with small-polaron band motion and the other with
small-polaron hopping motion. In an ideal crystal the
small-polaron band mobility will predominate at absolute
zero. With rising temperature the band contribution to
the mobility will diminish while the hopping component
will increase until, at sufficiently high temperatures,
small-polaron motion proceeds predominantly via phonon-
assisted hopping. The actual temperature of the change-
over depends on the details of the model [the details of
the coupling of the electron to both the optical and acoustic phonons of the crystal and the model for the "scattering" associated with small-polaron band motion]. However,
the extreme narrowness of the small-polaron band leads one
to suspect that the band motion typically will be washed
out by whatever disorder exists in a real crystal. Regardless, only small-polaron hopping motion will be

discussed here. More generally, in a material, a particle placed on a site may move to other sites either with
or without the assistance of phonons. These two types
of motion compete with one another for dominance of the
transport phenomena. In the case of Anderson localization, while a particle initially placed on an arbitrary
site may spread out somewhat, (make virtual transitions
to surrounding sites) it cannot actually escape from its
original location without the aid of phonons. With or
without Anderson localization, as illustrated by the
small-polaron situation, phonon-assisted transitions may
dominate the phonon-independent motion.

In the following discussion most of our concern will
be focused on calculations in which the electronic transfer integral associated with a small-polaron hop is sufficiently small so that it may be treated perturbatively;
the phonon-assisted site-to-site jump rate is then proportional to some power of the relevant electronic transfer integral. Physically, this regime corresponds to a
situation in which the electron cannot adiabatically follow an alteration of the atomic positions.[12,22] Thus this
is termed nonadiabatic small-polaron motion. The complementary regime, in which the electron can always adjust
to the atomic state, is termed the adiabatic regime and
will also be discussed here.

VII. PHONON-ASSISTED JUMP RATES BETWEEN DEGENERATE SITES

The fundamental quantity with which we shall now be
concerned is the rate that characterizes a phonon-assisted transition which transports a small polaron from one
site to a neighboring site. If one considers each hop of
a carrier to be uncorrelated with its prior hops(or hops
of other carriers), then the relevant jump rate is calculated by placing a small polaron on a particular site at
some initial time and computing the rate at which it moves
to a neighboring site. The lattice vibrations then act
as a thermal bath with which the electron can exchange
energy. It is clear that a necessary condition for the
lattice vibrations to fulfill this function is that vibrational energy be transferrable from lattice site to lattice site, i.e., there must be adequate dispersion of the
vibrational frequencies.[12,23]

The nonadiabatic jump rate may be calculated exactly.[23]
In Fig. 1 the phonon-assisted jump rate associated with
a small-polaron hop, in units of $2\pi J^2/\hbar^2\omega_o$ [J is the

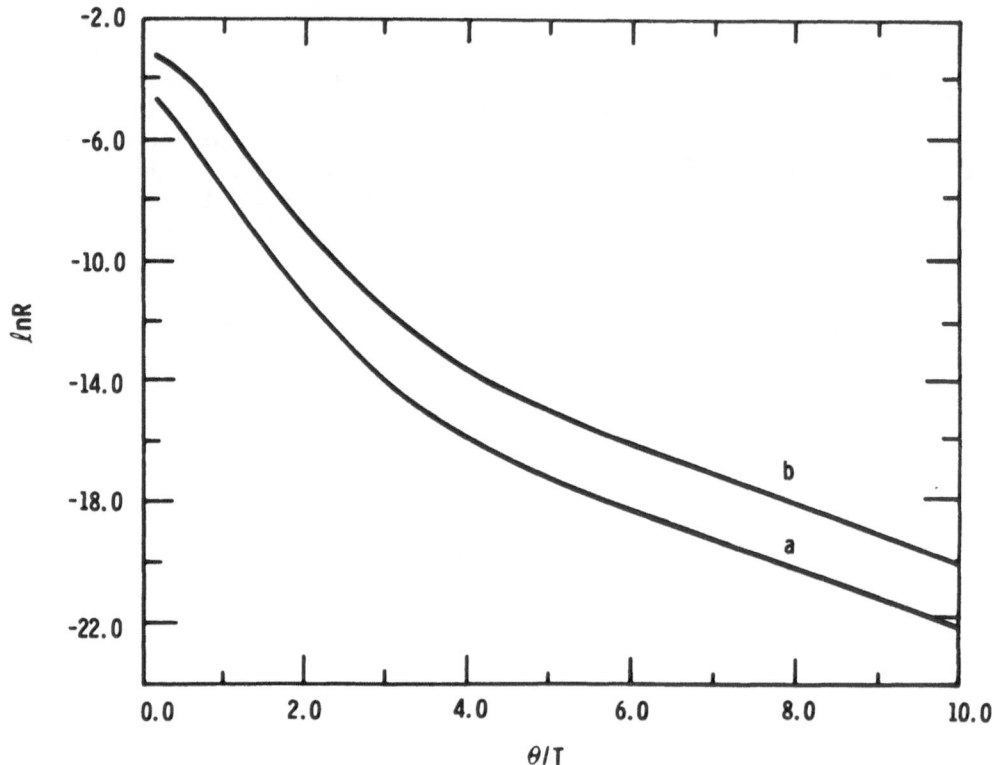

Fig. 1 The natural logarithm of the optical-phonon assisted jump rate in dimensionless units is plotted a-
gainst Θ/T $(\Theta = \hbar\omega_0/k)$ for $E_b/\hbar\omega_0 = 10$ and (a) $2\pi\omega_b/\omega_0 = 0.5$, (b) $2\pi\omega_b/\omega_0 = 0.05$.

electronic transfer integral and ω_0 is the mean optical-
mode frequency], is plotted versus reciprocal temperature,
in units of the optical mode temperature, $\Theta = \hbar\omega/K$, for
two values of the phonon-dispersion parameter, ω_b; the
width of the optical band is $6\omega_b$. In these curves the
electron-lattice coupling parameter, $E_b/\hbar\omega_0$, is taken to
be large compared with unity; this will be termed the
strong-coupling regime.

    Focusing attention on either one of the two curves
of Fig. 1, it is seen that two distinct temperature dependences are manifested. At sufficiently low tempera-
tures, $(2E_b/\hbar\omega_0)\mathrm{csch}(\beta\hbar\omega_0/2) \ll 1$ [in the case of small-

polaron hopping $(2E_b/\hbar\omega_0) \gg 1]$, multiphonon processes are
frozen out and the jump rate is dominated by the phonon-
assisted processes that involve the absorption of the minimum amount of vibrational energy which results in an energy-conserving phonon-assisted hop. In the case at hand,
a carrier's interaction is with only optical phonons; this
is a two-phonon process. Namely, a phonon of energy $\sim\hbar\omega_0$
is absorbed and another of energy $\sim h\omega_0$ is emitted. The
low-temperature jump rate is concomitantly activated with
the activation energy $\hbar\omega_0$, which is associated with the
probability of absorbing a phonon of energy $\hbar\omega_0$. In the
complementary high-temperature regime, multiphonon processes are no longer frozen out. Then the jump rate manifests a thermally-activated behavior with the activation
energy $\varepsilon_2 = E_b/2$. It should be noted that this high-
temperature activation energy is not associated with phonon energies but simply with the electron-lattice coupling
strength and the stiffness of the material, the parameters
involved in E_b. The fact that the activation energy depends on no quantum-mechanical quantities suggests that
a semiclassical interpretation of this high-temperature
activation energy is possible. Later, we shall see that
this is, in fact, the case. It is interesting to note
that, in the typical example illustrated in Fig. 1, the
transition between the low-temperature, and high-temperature regimes occurs over a relatively narrow range of temperature. This is a feature of both the values of the
parameters used in this plot and the fact that the electron-lattice interaction only involves optical phonons.
As the electron-lattice coupling strength is increased
the nonactivated region is extended to lower and lower
temperatures. Subsequently, it will be seen that systems
in which one includes the interaction between an electron
and acoustic phonons will be characterized by non-activated small-acoustic-polaron hopping at all temperatures below some fraction of the Debye temperature.

Some comment should also be made about the role of
vibrational dispersion in this calculation. The upper
curve of the two on Fig. 1 differs from the lower curve
solely in that the dispersion in this case is smaller.
The dispersion dependence becomes somewhat greater as the
temperature is lowered. If one proceeds to the limit of
zero vibrational dispersion the rate increases without
limit and becomes undefined. Alternatively, as the dispersion is increased, the dispersion dependence of the
jump rate becomes less and less. An interesting aspect
of the optical-phonon-assisted jump rate (which will not
be dwelt on here) is that it is undefined for a one-
dimensional system.[23,24] An attempt to calculate it

yields a divergent result for all values of the optical bandwidth; Fig. 1 pertains to a three-dimensional lattice.

Finally, a statement about the formal approach to this hopping calculation is in order. While the eigenstates of an electron in a periodic system are Bloch states, the hopping calculation typically views the basic states between which the electron hops (a phonon-induced transition occurs) to be the set of local (Wannier) states. Such a procedure is justified both on the grounds that it corresponds to the physical picture one typically adopts in thinking about hopping, and because it can be demonstrated that identical results are obtained by using the non-local (small-polaron band states) as the basis states when carrying out a perturbation calculation (to lowest order in J) of the phonon-assisted jump rate.[25]

If a carrier interacts with the atoms of a lattice via a short-range interaction, such as when a carrier on a site displaces adjacent atoms, the interaction between the carrier and the acoustic phonons may be, a priori, comparable to or greater than that between the carrier and optical phonons.[20] Thus one is led to consider the jump rate in the alternative situation where the carrier interacts solely with acoustic phonons. This situation differs from the optical-phonon problem in a fundamental manner. Namely, since the acoustic phonons with which a carrier can interact extend from the Debye energy down to zero energy, at no finite temperature are all acoustic phonons frozen out, i.e. $kT > \hbar\omega_q$ from some q. Because there is no special freezing-out energy for all the phonons, a thermally activated temperature dependence of the jump rate will not be observed at low temperatures. This is illustrated by the curve in Fig. 2, in which the logarithm of the phonon-assisted jump rate, in units of $J^2/\hbar^2\omega_m$, is plotted versus Θ_m/T, where Θ_m is the temperature corresponding to the energy of the maximum-energy phonon with which a carrier can interact, $\hbar\omega_m = k\Theta_m$. As in the optical-phonon case, the analogous electron-lattice coupling parameter is taken to be large compared with unity. At low temperatures the jump rate is nonactivated. As in the optical-phonon calculation, when the temperature is raised above the temperature corresponding to the maximum-energy phonon with which the carrier can interact, the jump rate becomes thermally activated. In fact, as in the optical-phonon problem, this high-temperature behavior can be understood via a semiclassical picture.

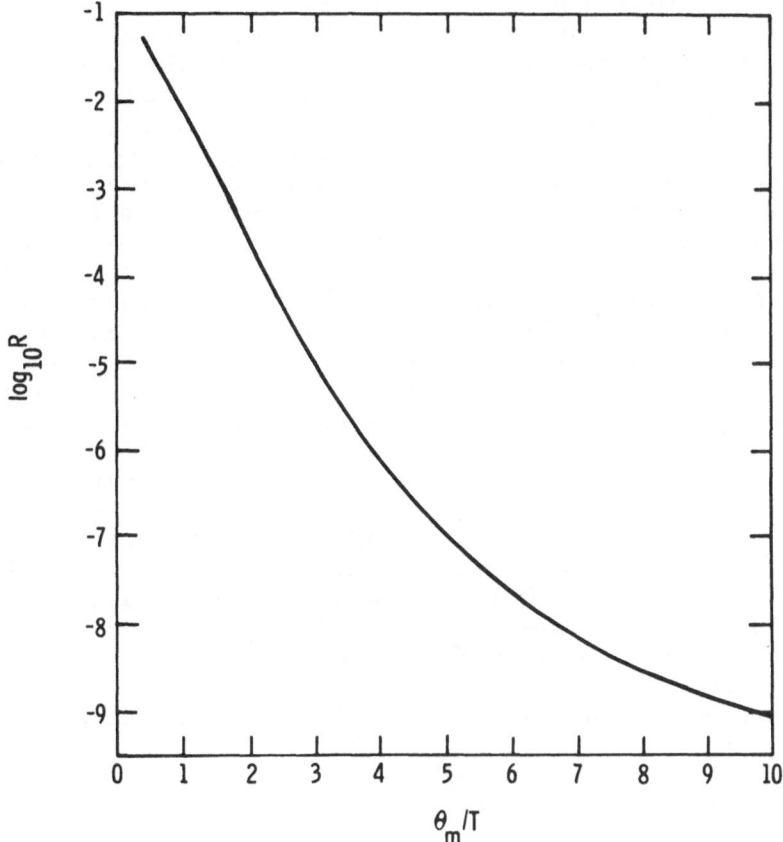

Fig. 2   The logarithm of the acoustic-phonon-assisted jump rate in dimensionless units is plotted against θ_{m}/T $(\theta \equiv \hbar\omega_m/k)$ for a typical value of the electron-lattice coupling strength.

VIII. SEMICLASSICAL APPROACH

A fundamental concept characteristic of this semi-classical regime is the notion of a <u>coincidence event</u>.[22] Specifically, taking the <u>electronic</u> energy level associated with a carrier occupying any site in a crystal to be a function of the instantaneous positions of the atoms of the crystal, it may be seen that, since the positions of the atoms are constantly changing (associated with the vibratory motion of the lattice), the electronic energy associated with a carrier occupying any given site is also

changing. Amidst the myriad of distortional configurations which are assumed by the vibratory atoms, occasionally a situation is encountered in which the electronic
energy of an electron at a given site "momentarily" equals
that which it would have if it occupied an adjacent site.
Such an occurrence is termed a <u>coincidence event</u>. While
an energy coincidence is viewed as instantaneous in terms
of classical physics, in quantum mechanics it has a finite
duration. If this duration time is long compared with the
time it takes an electron to transfer between coincident
sites, $\sim \hbar/J$, then the electron can always "follow the lattice motion" and avail itself of the opportunity to make
a hop.[12] This situation is characteristic of the so-called <u>adiabatic</u> regime. Alternatively, the time required
for an electron to hop may be long compared with the duration of a coincidence. Then an electron will not always
"follow the lattice motion" and hop when a coincidence
event presents itself; this is the <u>nonadiabatic</u> domain.
In this case, the jump rate, and hence the drift mobility,
is reduced from what it is in the adiabatic regime by a
factor P, P<1, where P is the probability that, given a
coincidence event, the carrier will hop. Finally, it is
noted that the minimum energy required to produce a lattice deformation associated with a coincidence event is
just the activation energy E_A. This ultimately yields
the formula,[12]

$$\mu_D = \left(\frac{ea^2}{kT}\right) \nu_{phonon} \, e^{-E_A/kT} \, P. \tag{12}$$

In the adiabatic regime (P = 1) the drift mobility is not
proportional to the square of the transfer integral, although there can be some nontrivial J-dependent contributions to the activation energy, E_A.[27] In the nonadiabatic regime the factor P and hence the drift mobility are
proportional to J^2. As J tends to zero E_A approaches the
activation energy found in the previously described jump-
rate studies, namely $\varepsilon_2 \equiv E_b/2$.

IX. CORRELATED HOPPING MOTION

In thinking about small-polaron hopping motion one
typically considers successive hops of a carrier to be
uncorrelated with one another. However, the preceding
discussion of a semiclassical view of a small-polaron hop
is suggestive of a mechanism via which small-polaron motion may be highly correlated. In particular, if a small
polaron's hop is to be considered independent of its previous hop the distortion associated with creating the
coincidence event of the first hop must relax, dissipating

an amount of energy comparable to the hopping activation
energy away from the involved sites, in a time which is
much shorter than the mean time between small-polaron
hops. If the carrier has a substantial probability of
hopping (either to a different neighbor or back to the
site it occupied previously) before the lattice relaxes,
then its motion will be highly correlated. In this case
the effective activation energy characterizing small-
polaron hopping motion will be substantially reduced from
that associated with uncorrelated hopping motion, since
much of the distortion needed to form a coincidence event
is present residually from the prior hop.[28-31] In fact,
in the highly correlated situation the carrier can be
viewed as frequently hopping back and forth between two
coincident sites. In this circumstance a contribution to
the net diffusion of the particle occurs when it alters
its back-and-forth jumping motion to hop to a third site.
Time is not adequate in the present review to develop a
detailed discussion of this phenomenon; for this the reader
is referred to the literature.[28-31] However, as illustrated in Fig. 3, in this type of correlated small-polaron
hopping situation the high-temperature small-polaron hopping mobility need not manifest a clear thermally activated behavior. In fact, the mobility in this domain may
even diminish slowly with increasing temperature.[32]

X. SMALL-POLARON JUMP RATE BETWEEN NONDEGENERATE SITES

Having very briefly reviewed the situation characterizing a hop of a small polaron between energetically
equivalent sites, it is now appropriate to comment on the
hop of a small polaron between energetically inequivalent
sites. Specifically, the concern here will, for the time
being, be restricted to the case in which the carrier only
interacts with optical phonons. Concomitantly, only the
situation in which the difference in the energy of the
(phononless) system between the carrier occupying one site
and another is a multiple of the phonon energy will be
considered. As in the case of nonadiabatic optical-phonon-
assisted small-polaron hopping in a crystal, this calculation may be performed exactly.[23] In Fig. 4 the logarithm
of the jump rate characterizing a jump between sites g and
g' [measured in units of $2\pi\, J_{g,g'}/\hbar^2\omega_o$] is plotted versus
reciprocal temperature [measured in units of the optical-
phonon temperature $\Theta \equiv \hbar\omega_o/k$] for hops downward in energy
by the amount $\Delta = m\,\hbar\omega_o$, where m is an integer. As in
the crystalline example two distinct regions are seen.

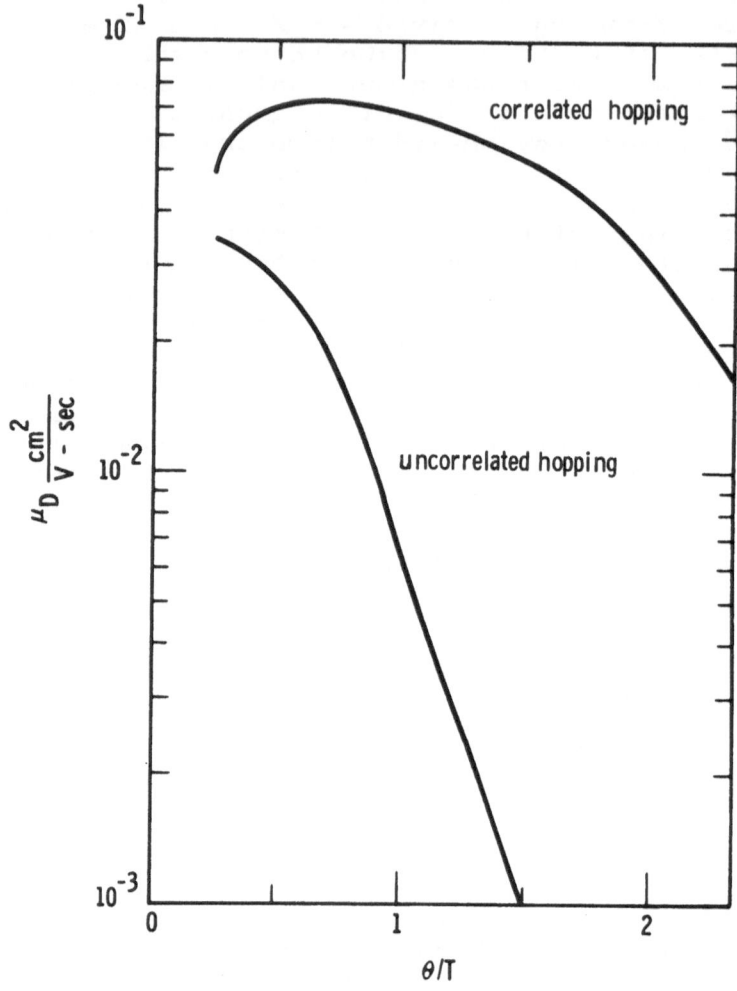

Fig. 3 The semiclassically calculated small-polaron drift mobility is plotted versus θ/T for $E_b/\hbar\omega_o = 10$ with correlation effects being included (upper curve) and with them being ignored (lower curve).

In the limit of sufficiently low temperatures the jump rate becomes temperature-independent as the hop proceeds via the spontaneous emission of phonons. As shown, with an increase in energy disparity the downward jump rate increases markedly. However, not shown in this figure is the feature that when $\Delta > 2E_b$ the probability of even the spontaneous emission of phonons <u>decreases</u> rapidly

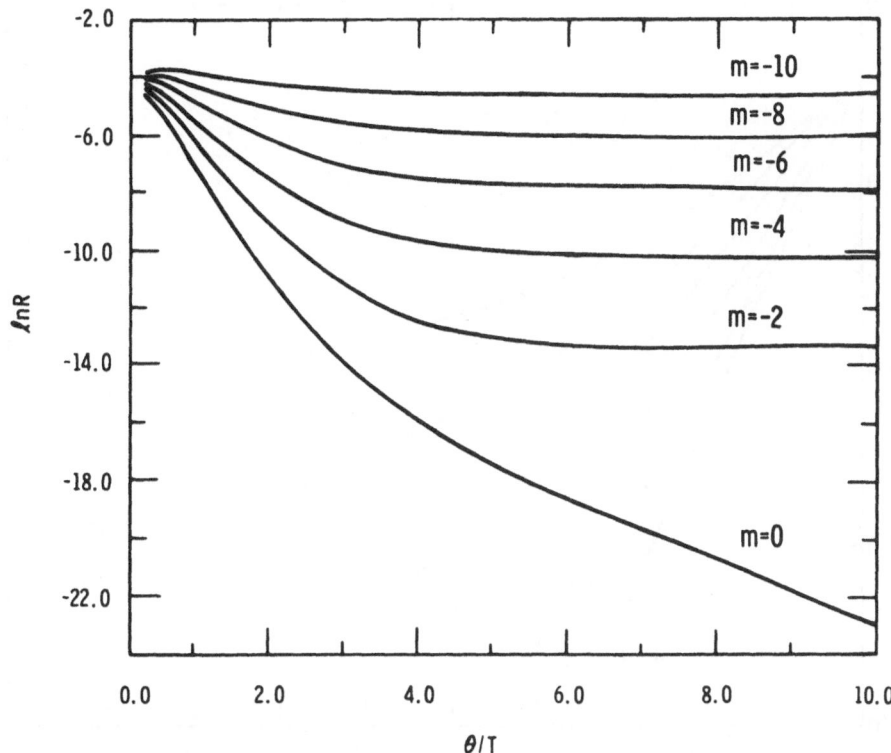

Fig. 4 The natural logarithm of the jump rate, in units of $2\pi J^2/\hbar^2\omega_0$, versus Θ/T is plotted for the situation in which the carrier hops downward in energy by the amount $\Delta = m\hbar\omega_0$. Here $E_b/\hbar\omega_0 = 10$ and $2\pi\omega_b/\omega_0 = 0.5$.

with increasing energy disparity. In the high-temperature regime the jump rate is qualitatively different; here the semiclassical approach attains validity. In Fig. 5, the analogous plots for a hop upward in energy are shown. Here the low-temperature behavior and high-temperature behavior are typically characterized by different activation energies. One can conclude that for both upward and downward hops the low-temperature behavior is characterized by the absorption of the minimum amount of vibrational energy consistent with the requirement of energy conservation. The low-temperature Δ-dependence is in part provided by the jump rate's proportionality to the factor $\exp[-\frac{1}{2} (\Delta + |\Delta|)/kT]$, which gives rise to temperature-

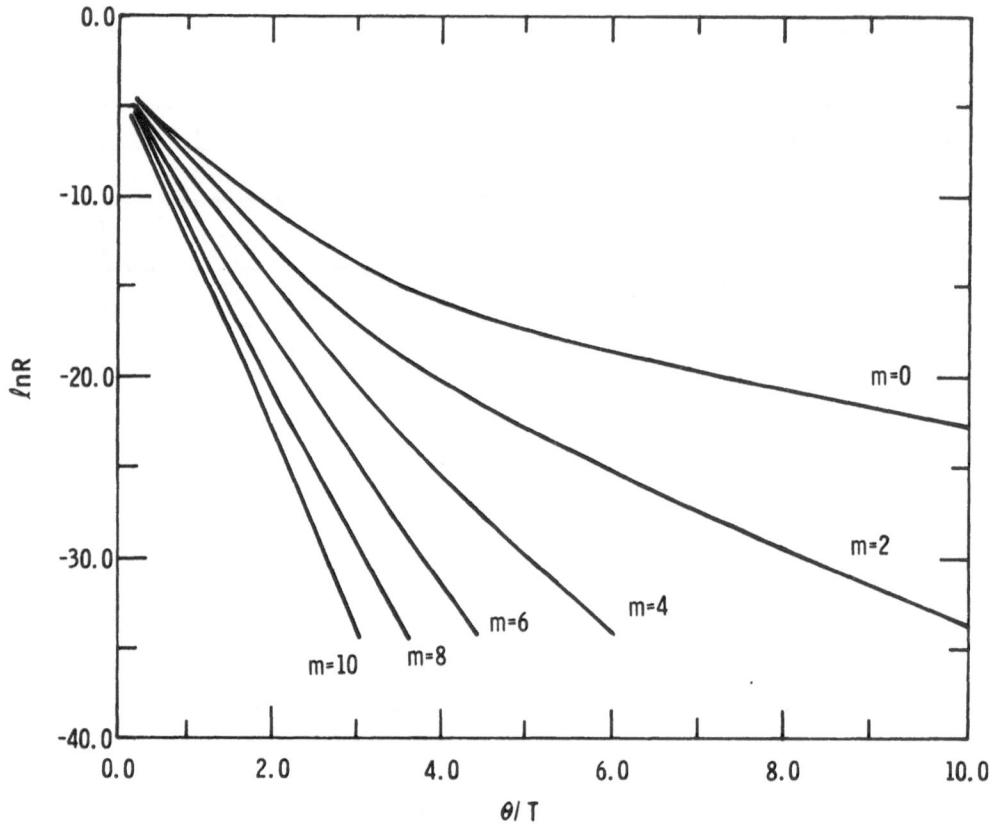

Fig. 5 A graph, analogous to Fig. 4, depicting the situation in which the carrier hops upward in energy. The
magnitude of the electronic energy change in units of
$\hbar\omega_o$ is m.

independent transition rates for hops downward in energy
$(\Delta < 0)$ and simply activated rates for hops upward $(\Delta > 0)$
in energy, proportional to $\exp(-\Delta/kT)$. A very important
additional factor which manifests itself in the Δ-depen-
dence of the magnitude of the jump rate is $[4E_b/\hbar\omega_o]^m/m!$,
where m is the number of phonons involved in the low-
temperature hop: $m \equiv |\Delta|/\hbar\omega_0$. For strong coupling,
$E_b/\hbar\omega_0 \gg 1$, this factor increases with increasing m, up
to a maximum at $m \sim 2E_b/\hbar\omega_0$, and then falls rapidly with
a further increase in m. Thus the magnitude of the jump
rate is, apart from the temperature-dependent factor, a

strong function of the magnitude of the disorder energy, Δ. In the high-temperature (semiclassical) limit the jump rate differs substantially from its low-temperature behavior; explicitly,

$$R(\Delta) \propto T^{-\frac{1}{2}} \exp[-(2E_b + \Delta)^2/8E_b kT] \ . \qquad (13)$$

Once again it is seen that for downward hops, $\Delta < 0$, the jump rate, at a particular temperature, is maximized when $|\Delta| = 2E_b$. In the regime where $\Delta \ll 2E_b$, the high-temperature (semiclassical) limit of the jump rate in the disordered situation differs from the $\Delta = 0$ rate by the added factor $\exp(-\Delta/2kT)$. In this case, the jump rate is characterized by an activation energy $\sim 1/2 \, (E_b + \Delta)$ with the major portion of this activation energy being independent of Δ.

In the optical-phonon strong-coupling examples which have been considered so far, those processes which involve the absorption of more than the minimum number of phonons necessitated by the requirement of energy conservation, dominate the jump rate except when they are frozen out at low temperatures. As mentioned previously, at no finite temperature can one completely freeze out acoustic phonons; some phonons are always associated with energies less than kT. Figure 6 illustrates the situation. Here some of the energy-conserving processes associated with acoustic-phonon-assisted transitions upward in energy by an amount Δ, taken to be less than the Debye energy, are depicted. In this case a single phonon can provide the required energy. However, as illustrated by the other diagrams of the first column, multiphonon processes can also contribute without increasing the net energy which is absorbed from the atomic vibrations. On the other hand, all of the processes other than those illustrared in the first column involve the absorption of additional vibrational energy. Thus, arbitrarily close to absolute zero the processes of the first column are the dominant ones. With an increase in temperature the processes other than those represented in the first column will make an increasing contribution to the total jump rate. In the regime in which the electron-acoustic-phonon coupling is large, multiphonon processes are preferred: hence one finds that as the temperature is raised the relative contribution to the jump rate of the non-first-column processes will increase and manifest themselves in jump rate. Thus, one has, somewhat in analogy with the

case of optical-phonon-assisted hops, several regimes.
At very low temperatures the first-column processes dominate and the upward jump rate is simply activated,
$\propto \exp(-\Delta/kT)$. As the temperature is raised the jump rate
manifests a nonactivated increase with temperature. When
the temperature is sufficiently high, comparable to some
(typically sizeable) fraction of the Debye temperature,
no phonon-assisted processes remain frozen out and the
jump rate regains the activated behavior characteristic
of the high-temperature strong-coupling (semiclassical)
regime. In the case of a downward hop in energy, the
arrows on Fig. 6 are to be reversed. This leads to a
situation where as the temperature is increased from absolute zero the jump rate rises in a nonactivated manner
until the semiclassical activated regime is reached. It
has been pointed out that for typical estimates of the
physical parameters one may expect the transition between
the low-temperature activated region and the nonactivated
regime to occur at very low temperatures (comparable to
10K).[26,33] Furthermore, for sufficiently (but not unreasonably) small values of Δ, the presence of a finite disorder energy only modifies the $\Delta = 0$ jump rate, above the
very-low-temperature-activated regime, by simply introducing the additional multiplicative factor: $\exp(-\Delta/2kT)$.
Finally, it is commented that, even in the low-temperature limit, the high-order (multiphonon) processes of the
first column can often dominate the one-phonon contribution. To illustrate this feature the logarithm of the
ratio of the $T = 0$ K n-phonon contribution to the jump
rate relative to the one-phonon contribution is plotted
versus n for several values of the physical parameters
in Fig. 7; for Δ equal to the Debye energy the parameter
B is typically of the order of several hundred.[33] In
this example the multiphonon contribution overwhelmingly
dominates the single-phonon contribution. However, for
very small values of Δ, B may be sufficiently small so
that the one-phonon contribution to the jump rate is dominant. This is the case which is encountered in the low-
temperature studies of impurity conduction with which
Miller and Abrahams were concerned;[34] despite the large
value of the electron-lattice coupling assumed in their
work, they only consider the one-phonon contributions to
the jump rate. This is justified in their case since they
restrict their attention to small values of $\Delta(\Delta \sim 10^{-4} eV)$
and very low temperatures (T ~2K). However, in noncrystalline solids one often considers hops which involve
large disorder energies ($\Delta \gg 10^{-2}$ eV) at very much higher temperatures. Thus the wholesale uncritical application of their low-temperature impurity-conduction study

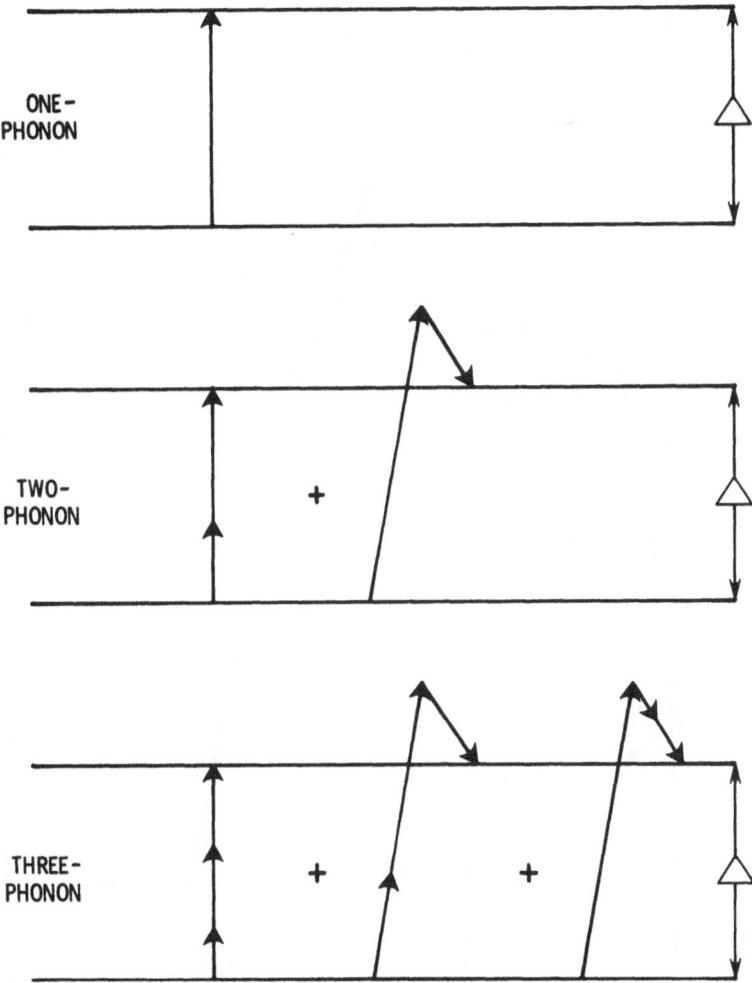

Fig. 6 A schematic representation of the processes which
correspond to the n=1, 2 and 3 terms in the series of
Eq. (1). It is assumed in this figure that the disorder
energy, Δ, is positive and less than that of the maximum-
energy phonon with which a carrier can interact, $\hbar\omega_m$.

to the situations envisioned in the case of phonon-assisted hopping in disordered solids is clearly inappropriate.

Considering the case of substantial coupling between

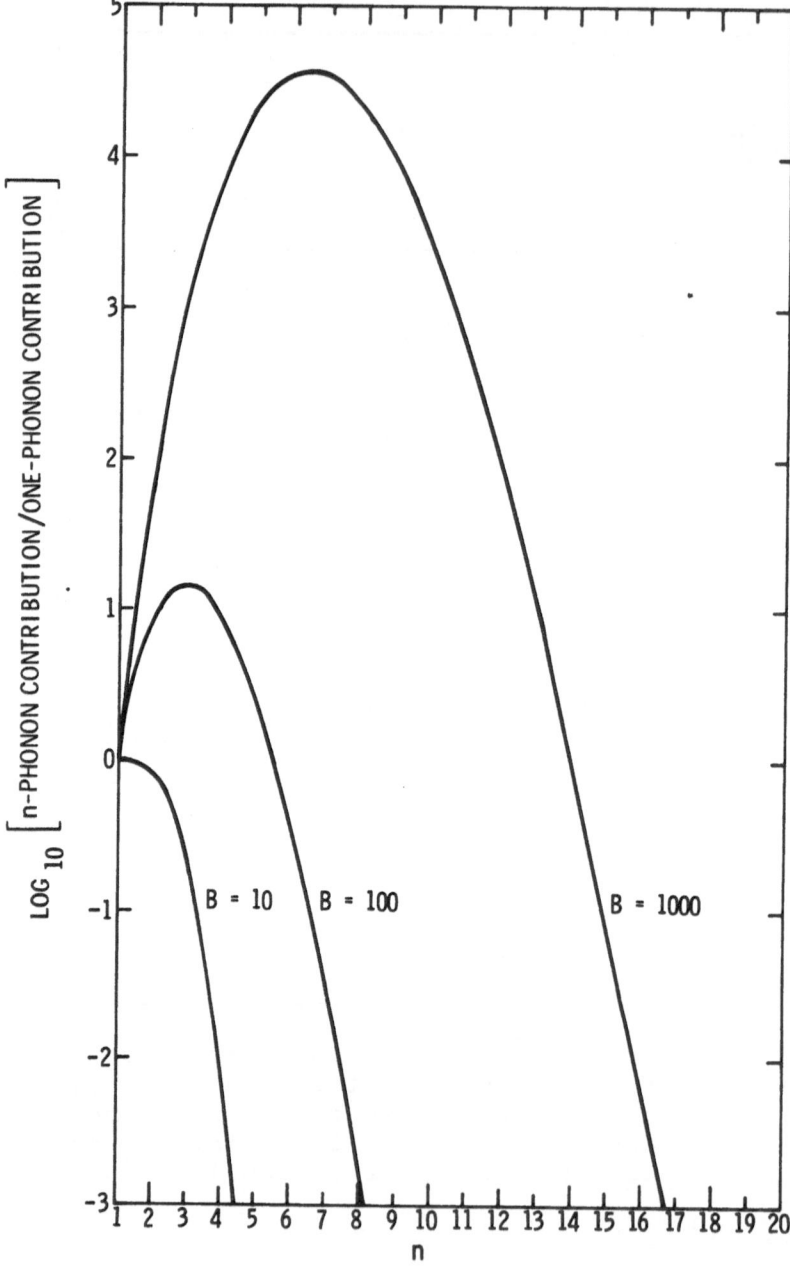

Fig. 7 The <u>logarithm</u> of the ratio of the contribution to
the jump rate of the n-phonon processes, to that of the
one-phonon process in the low-temperature limit, $\Delta/kT \to \infty$,
is plotted versus n for three values of $B(\Delta)$. The curves
only have meaning at integer values of n.

a carrier and both the optical and acoustic phonons, one expects that the dominant processes in the jump rate will in most instances involve both acoustic and optical phonons. In this situation the high-temperature semiclassical regime will not be qualitatively altered from what it is in either the purely optical-phonon-assisted or purely acoustic-phonon-assisted tunneling examples. In most of the low-temperature regime $kT \ll \hbar\omega_{Debye}/2$, $\hbar\omega_0/2$, the rate will not be strictly activated. However, as in the purely acoustic-phonon example, at sufficiently low temperatures, the rate characterizing an upward hop in energy $(\Delta > 0)$ will manifest a temperature dependence which is close to being activated with the activation energy Δ, i.e., the rate is nearly proportional to $\exp(-\Delta/kT)$. Between these temperature regimes it will be nonactivated. Again, as in the purely acoustic-phonon situation, for sufficiently small Δ the effect of Δ in this regime will be simply to multiply the $\Delta = 0$ by the factor $\exp(-\Delta/2kT)$.

In the preceding discussion attention has been focused on the elemental jump rate associated with a small polaron's motion in ordered and disordered materials. In these considerations the major role played by the spatial separation between initial and final sites is in determining the value of the transfer energy.[35] Thus the generalization of the preceding discussion so as to incorporate hopping between non-neighboring sites requires, in the main, simply that we alter the value of the transfer integral. Furthermore, the square of the transfer energy only enters as a proportionality constant into the calculations of the nonadiabatic jump rate. Thus the preceding discussion may be readily taken over to include the case of hopping between sites of disparate energy whose characteristic spatial separation is arbitrary, provided of course that the effective electron-lattice coupling strength is sufficiently large.

XI. ELECTRON-PHONON COUPLING STRENGTH

One is now led to consider how strong the electron-lattice coupling can reasonably be taken to be. The interpretation of experiments in terms of small-polaron motion typically yields estimates of $E_b/\hbar\omega_0$ that range from ~10 to ~100.[33] In addition, estimates of the corresponding coupling constant in the acoustic-phonon situation, obtained by utilizing textbook values of the deformation-potential constants, yield comparable figures.[26] More generally, if one notes that the coupling energy

[corresponding to the electron-lattice constant of the
molecular crystal model, A, multiplied by the lattice
constant, a] is essentially the change in the energy of
an excess electron on an atomic site per unit strain,
one expects that this energy be typically of the order
of an electronic energy (of the order of several electron volts); this again yields estimates for the electron-lattice coupling strength which put it into the
strong-coupling regime.

As implied in the discussion of "localized states,"
free carriers do not form small polarons in materials
such as Ge and Si not because this coupling parameter is
small (it presumably is not), but rather because the bandwidth parameter is sufficiently large. Similarly, the
effect of the electron-lattice interaction on large-radius
shallow donor states in these materials is small not because the fundamental coupling strength is small, but because the characteristic volume of these localized states
is very much larger than a unit cell size. Formally this
arises because a phonon of wavelength λ only interacts
appreciably with states of radii for which $1/2\lambda \gtrsim 2r$.[26],
[33,36,37] The fact that only a fraction of the totality
of phonons can effectively interact with an electron substantially reduces the _effective_ electron-lattice coupling strength which enters into the hopping calculations.
Specificially, in the optical-phonon-assisted hopping
problem the coupling constant $E_b/\hbar\omega_o$ is replaced by
$\sim(E_b/\hbar\omega_o)(a/r)^3$,[23] where $r > a$. A similar, but not i-
dentical, reduction occurs in the acoustic-phonon-assist-
ed hopping situation.[26,38] Therefore, whether one is
considering hopping between a pair of large-radius states,
and/or if one believes the fundamental electron-lattice
coupling strength to be small (the above estimates not-
withstanding), it becomes necessary to investigate the
phonon-assisted tunneling rate in the limit of a weak effective electron-lattice interaction.

XII. PHONON-ASSISTED JUMP RATE FOR WEAK
EFFECTIVE ELECTRON-LATTICE COUPLING

Returning to the problem of optical-phonon assisted
hops, the limit of weak effective electron-lattice interaction is one in which the jump process is always characterized by the involvement of the minimum number of phonons consistent with the requirement of energy conservation. Thus, in contrast to the strong-coupling regime, in
the weak-coupling limit the dominance of those processes
which involve a minimum number of phonons persists above

the optical phonon temperature. In the case of an optical-phonon-assisted transition which involves an energy
disparity Δ ($|\Delta| = m\hbar\omega_o$ with $m \neq 0$), one has, for the
m-phonon rate (with sufficiently large vibrational dis-
persion)[24]

$$R(\Delta) = \left(\frac{2\pi |J_{g',g}|^2}{\hbar^2 \omega_o}\right)\left(\frac{\gamma^m}{m!}\right)\left[N + \frac{1}{2} \mp \frac{1}{2}\right]^m \quad , \quad (14)$$

where γ is the effective electron-lattice coupling con-
stant, $\gamma \sim (2E_b/\hbar\omega_o) \times (a/r)^3 \ll 1$, N is the Bose factor,
$[\exp(\hbar\omega_o/kT)-1]^{-1}$, and the $\mp$ sign respectively refers to
a hop upward or downward in energy. The phonon-assisted
tunneling rate is proportional to the absolute square of
the electronic transfer energy, proportional to the effective coupling constant raised to the mth power divided by a factor which arises from the counting of phonon-
assisted processes, $m!$, and proportional to the mth power
of a factor related to the probability of absorbing or
emitting m optical phonons.

Addressing first the temperature-dependent (final)
factor, one notes that in the low-temperature limit it
approaches unity for a hop downward in energy, thereby
manifesting the fact that the jump proceeds via the spontaneous emission of phonons. In this low-temperature limit a hop upwards in energy ($\Delta > 0$) yields the characteristic activated factor $\exp(-\Delta/kT)$.[39] However, as the
temperature is increased ($kT \gtrsim \hbar\omega_o/2$) the temperature dependence of the weak-coupling rate approaches the form
$\exp(-\beta\Delta/2)(2kT/\hbar\omega_o)^m$ with $\beta \equiv (kT)^{-1}$. In this high-temperature regime the temperature dependence is dominated
by the power-law increase with temperature rather than
by the exponential (activated)factor. Thus in the weak-
coupling regime the jump rate will only be simply activated in the low-temperature limit.[23]

Since the value of γ characteristic of the weak-coupling situation is much less than unity, the factor $\gamma^m/m!$
is always much less than unity. Furthermore, it falls
very rapidly with a rise in the magnitude of the energy
disparity (with a rise in m). Thus a very major contribution to the Δ-dependence of the jump rate is contained
in this factor.

Turning now to the case of a carrier's weak interaction with acoustic phonons, one has one major new

ingredient to consider. Recall first that the effective
interaction between a carrier viewed as localized and
the atomic vibrations falls off rapidly for those normal
vibrational modes whose characteristic wavelengths in the
vicinity of the carrier are small compared with the characteristic spatial extent of the "localized" state. Thus,
a large-radius weakly-coupled state interacts most strongly with those acoustic phonons which are associated with
sufficiently large "wavelengths".[40] This produces a situation in which the dominant contribution to the weak-
coupling acoustic-phonon-assisted jump rate may involve
several low-energy acoustic phonons instead of a single
high-energy acoustic phonon. In general the effect of
this phenomenon is to limit the strict low-temperature
activated behavior to even lower temperatures than in the
(above-mentioned) optical-phonon case.[41] Even for very
localized deep states $(r \sim a)$ in a monatomic system ω_{max}
may be but a fraction of ω_{Debye}.[40]

It should be mentioned that the preceding discussion has implicitly assumed that initial and final sites
in a hop are both weakly or strongly coupled to the atomic
vibratory motion. However, one may investigate a transition between a strongly coupled state and a weakly coupled state. Such a situation may be characteristic of a
hop between a deep-lying small-radius state and a higher-
lying large-radius state. In this case the emission and
absorption of the phonons associated with a hop occurs
primarily at the strong-coupling site. The jump rate
will then be similar to that of the previously described
strong-coupling case but with a somewhat reduced jump
rate, since only one of the two sites provides a center
for the (relatively) easy emission and absorption of vibrational energy.

XIII. THE TRANSFER ENERGY

For hopping between distinct spatial regions in a
disordered solid, as in the case of small-polaron hopping in a crystal, the local states between which the
carrier is viewed as hopping are not eigenstates of the
system. For small-polaron hopping the local electronic
states can be taken as the Wannier states.[23] In the case
of hopping between widely separated impurity states the
zeroth-order states can be taken to be those of an isolated impurity. Similarly, in hopping between defect-
related states one can consider the basis states to be
those of a single (isolated) defect.[26] However, in the
case of a randomly fluctuating potential it is not always

clear how the zeroth-order states are to be chosen. Nonetheless, it has become a relatively common practice to
stipulate that the transfer energy associated with the
two sites involved in a hop varies with a spatial separation parameter, R, simply as $\exp(-\alpha R)$, where α is taken
to be a constant. The transfer integral associated with
an electron moving between the pair of one-dimensional
square wells of Fig. 8a (the potential between the two
wells is a constant) varies as $(1 + \alpha R) \exp(-\alpha R)$. In
the case of the two overlapping (three-dimensional) coulombic (r^{-1}) potentials of Fig. 8b, one finds that in
the asymptotic limit as R approaches infinity, the transfer energy varies as $R \exp(-\alpha R)$.[42-44]

However, if the potential barrier between the two
local states is peaked, as in the case of the two one-
dimensional harmonic wells of Fig. 8c, then the dependence of the transfer energy on R can be far from a simple exponential in R; in the case of the two harmonic
wells it varies with R principally as $\exp[-(R/r_0)^2]$ when
$R \to \infty$. Another feature about the transfer energy is its
general dependence on the energies of the two states that
it connects. This arises because deeper-lying "localized"
states typically are associated with more spatially compact wave functions. Thus the transfer integral connecting pairs of deep states separated by a distance R will
differ from that connecting a pair of relatively shallow
states which are also separated by the same distance R.
Similarly, the transfer integral connecting two nondegenerate sites will depend on the energies of both of
the individual zeroth-order local states. Thus one can
say little that is definitive about the transfer energies other than that they typically approach zero as the
spatial separation of the involved sites approaches infinity.

XIV. D.C. CONDUCTIVITY

Our discussion of the principal features of the phonon-assisted jump rates has not, as yet, yielded a quantity which is directly observed experimentally in a disordered solid. While the site-to-site small-polaron jump
rate provides sufficient information to determine the
carrier's (trap-free) d.c. mobility in a crystal, in a
disordered material a major statistical problem (at least,
in principle) remains. Namely, there are an overwhelmingly large number of inequivalent paths via which the
carrier can move through the sample. Most generally each
hop of each path will be characterized by a number of

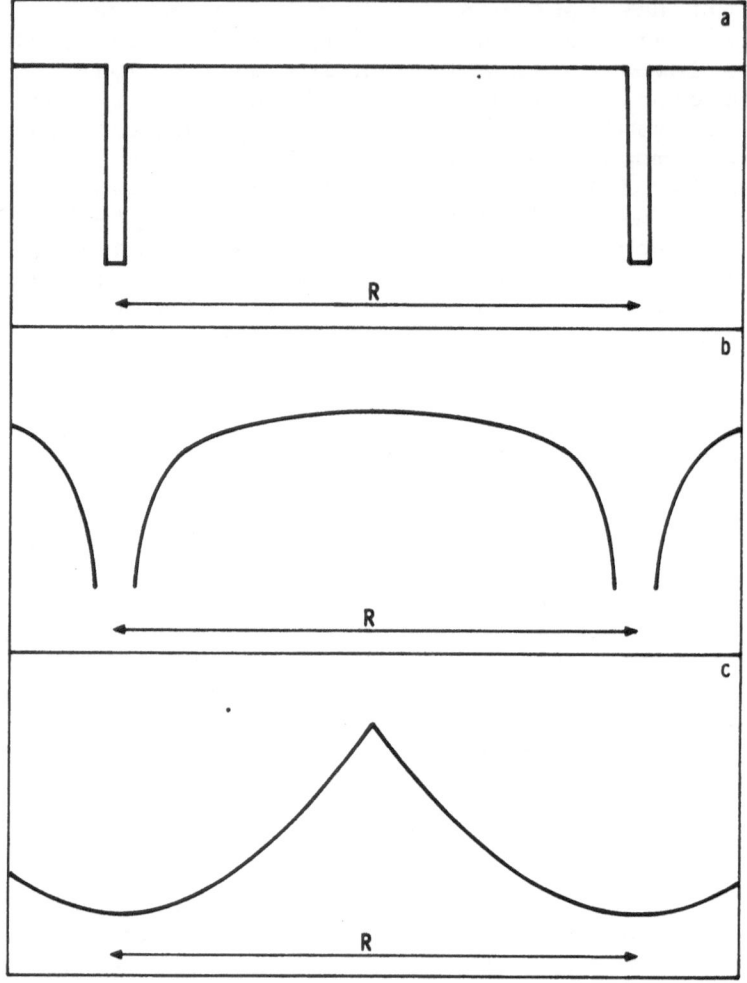

Fig. 8 The potential energy associated with two square
wells, two coulombic wells, and two harmonic wells is illustrated respectively in subfigures a, b, c.

unknown local microscopic physical parameters. To determine the d.c. conductivity in such a situation is one of
the tasks of so-called percolation theory.

     In approaching the statistical problem involved in
calculating the d.c. conductivity associated with hopping
motion one can, in the limit of a small applied electric

field, view the disordered solid as being a complicated conductance (resistance) network in which each conductance is related to a direct channel for hopping between a particular pair of sites.[34] Specifically, the conductance related to a hop from site $\underset{\sim}{g}$ to site $\underset{\sim}{g}'$, $G_{\underset{\sim}{g},\underset{\sim}{g}'}$ may be written as

$$G_{\underset{\sim}{g},\underset{\sim}{g}'} = \frac{e^2}{kT} f_{\underset{\sim}{g}}(1 - f_{\underset{\sim}{g}'})R_{\underset{\sim}{g},\underset{\sim}{g}'} \ , \tag{15}$$

where $f_{\underset{\sim}{g}}$ is the occupation probability (Fermi factor) associated with site $\underset{\sim}{g}$, $(1-f_{\underset{\sim}{g}'})$ is the probability that site $\underset{\sim}{g}'$ is unoccupied (for simplicity, one assumes that only one electron can occupy a site), and $R_{g,g'}$ is the elemental phonon-assisted jump rate associated with a hop from site $\underset{\sim}{g}$ to site $\underset{\sim}{g}'$. Typically the phonon-assisted transitions are assumed to be non-adiabatic, in which case, one generally has

$$R_{\underset{\sim}{g},\underset{\sim}{g}'} = |J_{\underset{\sim}{g},\underset{\sim}{g}'}/\hbar|^2 \exp(-2S_{\underset{\sim}{g},\underset{\sim}{g}'}) \exp[(E_{\underset{\sim}{g}} - E_{\underset{\sim}{g}'})/2kT]$$

$$\int_{-\infty}^{\infty} [e^{F_{\underset{\sim}{g},\underset{\sim}{g}'}(t)} - 1]\cos[(E_{\underset{\sim}{g}} - E_{\underset{\sim}{g}'})t/\hbar]dt, \tag{16}$$

where $J_{g,g'}$ is an electronic transfer integral associated with sites $\underset{\sim}{g}$ and $\underset{\sim}{g}'$. $S_{g,g'}$ is a vibrational overlap factor, $F_{g,g'}$ is related to the effective electron-lattice interaction of sites $\underset{\sim}{g}$ and g', and $E_{\underset{\sim}{g}}$ and $E_{g'}$ are the electronic energies of the two respective sites; explicit expressions for the functions in Eq. (16) are found in Eqs. (1)-(3) of Ref. (26). Combining Eqs. (15) and (16) and utilizing the explicit expressions for the Fermi factors, one readily obtains:

$$G_{\underset{\sim}{g},\underset{\sim}{g}'} = \frac{e^2}{4kT} \text{sech}[(E_{\underset{\sim}{g}} - E_F)/2kT]\text{sech}[(E_{\underset{\sim}{g}'} - E_F)/2kT]$$

$$\left| \frac{J_{\underset{\sim}{g},\underset{\sim}{g}'} \exp(-S_{\underset{\sim}{g},\underset{\sim}{g}'})}{\hbar} \right|^2 \ I, \tag{17}$$

where

$$I \equiv \int_{-\infty}^{\infty} [e^{F_{\underline{g},\underline{g}'}(t)} - 1] \cos [\frac{(E_{\underline{g}} - E_{\underline{g}'})t/h}{}]dt, \qquad (18)$$

and E_F is the Fermi energy. For a sufficiently weak electron-lattice coupling strength or low enough temperatures the first square-bracketed term of the integrand of Eq. (18) may be expanded in a power series in $F_{\underline{g},\underline{g}'}(t)$. If the lowest-order nonvanishing contribution to the integral I, Eq. (18), is its dominant term, then the hop proceeds with the involvement of the minimum number of phonons compatible with the requirement of energy conservation; this number is just the power to which the function $F_{\underline{g},\underline{g}'}$ is raised. More than the minimum number of phonons are involved when the electron-lattice coupling strength and/or temperature is sufficiently large so that higher order terms should not be neglected.

A comment about the effective resistance networks themselves is now perhaps in order. The role of the potential difference between a pair of sites (nodes) in this network is played by the electrochemical-potential difference between the pair. In an <u>ordered</u> system with a small steady-state current the electrochemical-potential difference is nothing more than the difference between the externally applied potential at the two sites. In a disordered system the application of an external electric field will, even with the drawing of a current, generally cause an altering of the occupation probabilities of the sites. It is for this reason that in a disordered system the electrochemical-potential difference between a pair of sites is generally <u>not</u> merely the potential difference associated with the externally applied field F, i.e., for the two sites $\underline{g}'$ and $\underline{g}$ it is not generally $eF \cdot (\underline{g}' - \underline{g})$. To be able to replace the hopping situation by an ohmic resistance network the applied field must be sufficiently small so that the field-induced changes in the occupation probabilities are very small compared to both the zero-field occupation and vacancy probabilities of the involved sites. Preliminary investigations indicate that this restriction of the strength of the applied field becomes more severe with an increase in disorder (as, say, measured by spreads in the local energies) and a decrease in temperature.[45]

In approaching the percolation problem, it has generally been <u>assumed</u>[46-55] that the elemental jump rate in

Eq. (15) is given by

$$R_{per}(R, \Delta) = \nu_{phonon} \, e^{-2\alpha R} \, e^{-(\Delta + |\Delta|)/2kT}, \qquad (19)$$

where ν_{phonon} is a constant, α is a constant independent of the energy of the states involved, R is their spatial separation, and Δ, the energy disparity between the two states, is independent of the other stochastic parameter, R. In this model an upward hop in energy is always activated and a downward hop in energy is always temperature-independent. This model may be appropriate for the study of shallow-state impurity conduction at helium temperatures.[26,34] However, as pointed out previously, the jump rate will in many circumstances deviate very substantially in R, Δ, and T dependence from this ad hoc ansatz. Nonetheless, one must appreciate the complexities of studying the statistical problem even for such an oversimplified model.

One important feature of this ad hoc model for the jump rate is that an individual jump can never by associated with a nonactivated behavior (a non-linear $\ln R_{per}$ versus T^{-1} curve). Thus in this model the observance of a non-activated d.c. conductivity is taken as being indicative of percolative effects. Namely, as the temperature is raised paths which involve hops of higher and higher activation energy become increasingly probable. This effect yields what may be thought of as an "effective activation energy" which increases with increasing temperature. In fact percolation calculations[46-55] based on the above model suggest that the conductivity should vary as $\exp[-(T_0/T)^{1/4}]$ at sufficiently low temperatures so that the percolation concepts are valid; T_0 is a constant. As shown in Fig. 9, the nonactivated portion of the acoustic-phonon-assisted jump rate in the absence of any disorder itself appears to fit a $T^{1/4}$-plot with a slope comparable to those observed experimentally.[56,57] Thus observance of the $T^{1/4}$ behavior below the Debye temperature by itself does not prove the existence of major percolative effects. In fact thermopower experiments on mixed-valency vanadate glasses indicate that there appears to be very little energetic disorder in these glasses,[58-60] yet a $T^{-1/4}$-like conductivity is observed below T_{Debye} but not above.[61] Furthermore, this behavior is consistent with the magnitude of the high-temperature, $T > T_{Debye}$, small-polaron activation energy

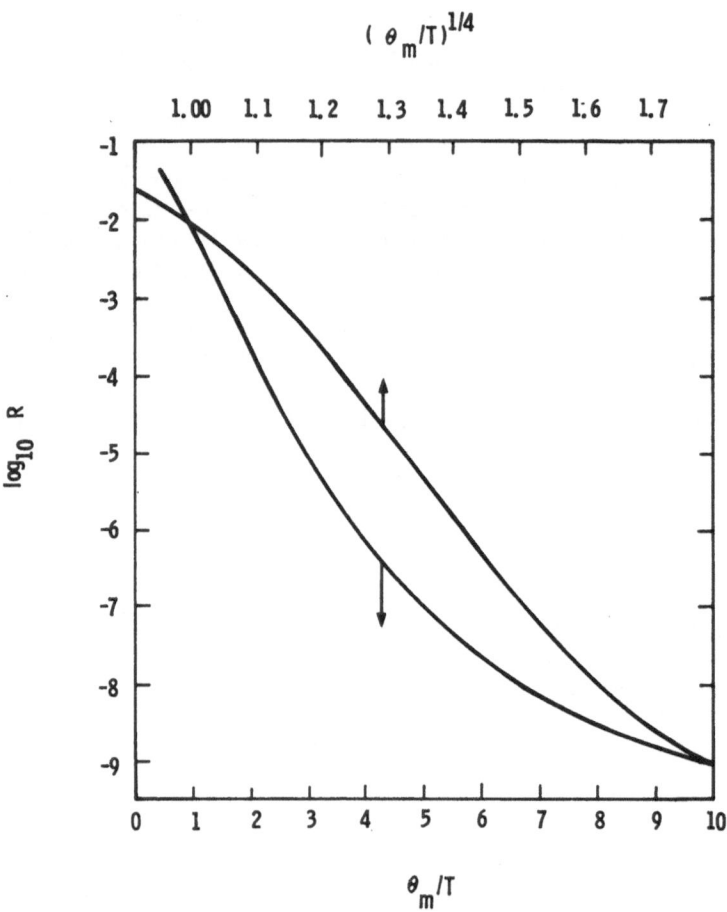

Fig. 9 The acoustic-phonon assisted jump rate of Fig. 2
is replotted here against Θ_m/T and $(\Theta_m/T)^{1/4}$. For
quantitative details see Refs. (26) and (33).

which is measured in these systems.[61]

An interesting prediction of the percolative studies
is that as a disordered film gets sufficiently thin so
that a significant fraction of the (contorted) relatively conducting paths, which in an infinitely thick sample
would be able to carry an electron through the material,
are terminated by reaching a "dead end" at the surface
of the film; the conductivity is reduced and its temperature

dependence increased. This is because the carrier is forced to follow "harder" paths. Some experiments on very thin films have been interpreted as manifesting this effect.[52]

It is useful to consider the circumstances in which utilization of the ad hoc jump-rate formula might be justified (setting aside the question of the R-dependence of J as well as its dependence on the energies of the initial and final sites). One such occurrence is at sufficiently low temperatures and sufficiently small values of the disorder energies so that the jump rate is activated in the manner of Eq. (19) with the additional disorder-energy dependence of the jump rate being neglectable. An example of such a case is that of shallow-impurity conduction at helium temperatures.[26,33] If the basic electron-lattice interaction parameter, estimated to be $\sim E_{electron}/E_{vibrational} \gg 1$, were for some (unknown) reason anomalously small (comparable or less than unity), the formula Eq. (19) might be approximately valid up to some fraction of the Debye temperature.[39-41] Another possibility occurs in the strong-coupling regime, at sufficiently high temperatures (above the very low-temperature activated regime) with Δ sufficiently small so that the principal Δ-dependence of the jump rate comes from the factor $\exp(-\Delta/2kT)$.[26,33] Then the rate characterizing the most difficult hops (upward in energy) possesses a Δ-dependence which, apart from a factor of one-half, is of the form of the ad hoc formula, Eq. (19). The total temperature dependence of the jump rate in this case however will not be that of Eq. (19). Specifically, an additional temperature-dependent factor emerges in this regime; it is simply that of the $\Delta = 0$ rate. The resulting conductivity can then be written as the product of that associated with small-polaron hopping motion for which $\Delta = 0$, multiplied by a percolation factor. Concomitantly, the logarithm of the jump rate will be a sum of two temperature-dependent factors, one of which (the jump-rate factor) will manifest an $\exp[-T_0/T)^{1/4}]$-type temperature dependence below some fraction ($\sim 1/3$ to $1/2$) of the Debye temperature (but above the low-temperature activated regime), while the other factor (the percolation factor) manifests such behavior up to some maximum temperature which is dependent on one's model of the disorder. Generally, one does not know which of the two factors provides the major part of the temperature dependence of the conductivity. Considering non-crystalline films of various thickness, it is even possible that the percolation factor dominates for sufficiently thin films (if the temperature dependence of the percolation factor is

greater than that of the $\Delta = 0$ jump-rate factor) but not
for thicker films, for which the temperature dependence
of the percolation factor is milder. Of course, the question as to the extent to which the simplified models of
the disorder on which the percolation calculations are
based can suitably represent real thin films remains as
a prevailing problem in attempting to obtain a detailed
comparison with experiment; elsewhere in this volume[62]
some of the conflicting experimental results which may
be relevant to this issue are briefly reviewed.

XV. THERMOELECTRIC POWER

In addition to a d.c. conductivity measurement, another basic type of transport measurement is that of the
thermoelectric power, or Seebeck coefficient. In this
experiment a small temperature gradient is applied a-
cross a material and the voltage induced in a circuit
which connects both ends of the sample is measured. The
thermoelectric power is the ratio of the induced voltage
to the temperature differential between both sides of the
sample; the thermoelectric power, usually denoted by α
or S, is typically studied as a function of the temperature and of the composition of the sample. In terms of
a kinetic-theory-type approach to the problem, one first
realizes that the imposition of a spatial temperature
gradient tends to induce a spatial gradient in both the
number of carriers and the diffusion constant. Carriers
are concomitantly shifted from one side of the material
to the other, thereby establishing an electric field and
a counter flow of charge, so that no net current flows
through the circuit. The voltage (associated with this
field) that is measured in the circuit generally arises
from the leads to the voltmeter as well as from the sample itself. Choosing leads whose thermoelectric power
is low relative to that of the sample, one typically speaks
of the measured thermoelectric power as being the thermoelectric power of the sample. Applying an analysis based
on the linearized Boltzmann equation and taking account
of the contact voltages at each of the junctions, one
finds that one may write the thermoelectric power as

$$S = \frac{kT}{e} \left[\frac{\partial}{\partial T} \ln\langle n \rangle + \frac{\partial}{\partial T} \ln D \right] \Bigg|_{E_F = \text{constant}}, \quad (20)$$

where $D \equiv \langle v \ell \rangle$, with v and ℓ respectively being the carrier's velocity and mean free path. In this form the relationship of the thermoelectric effect to the temperature-gradient-induced carrier-density gradient and diffusion-constant gradient is emphasized. Carrying out the indicated operations, one has

$$S = \frac{1}{eT} \left[\frac{\langle n \, v \, \ell \, (E - E_F) \rangle}{\langle n \, v \, \ell \rangle} \right] . \tag{21}$$

Here the square-bracketed term is interpretable as the average energy, relative to the Fermi energy, carried by a carrier; it is given a name: the Peltier heat. Evaluating this formula for motion in a wide-band semiconductor, one has for electrons in a conduction band whose edge is at E_c $[E_c - E_F \gg kT]$

$$S = \frac{k}{e} \left[\frac{E_c - E_F}{kT} + A \right] , \tag{22}$$

where A is a term which depends on the model for scattering; typically for bands very wide compared with kT, A varies slowly with temperature and is of the order of unity. In the opposite limit, of very narrow bands compared with kT, A tends to zero. For a system of several kinds of carriers which, as above, are presumed non-interacting, such as carriers in different bands, one may write

$$S = \frac{\sum_i \sigma_i S_i}{\sum_i \sigma_i} , \tag{23}$$

where σ_i and S_i are respectively the d.c. conductivity and thermoelectric power of the ith species of carriers.

Having briefly discussed the thermoelectric power characteristic of charge carriers in a high-mobility semiconductor, the thermoelectric power associated with hopping motion will now be succinctly addressed. In the simplest situation, in which the carrier hops between sites at which the strength of its interaction with the lattice is equal, no transport of vibrational energy is associated with a carrier when it makes a hop. In this situation the total contribution to the thermoelectric power arises from the temperature-induced carrier-density gradient. The resulting general formula for one species of carrier in a homogeneous material, written in terms of the conductances, G_j, and average site energies, $<E_j>$ associated with the current paths (labeled by j), is then[63]

$$S = \frac{1}{eT} \left[\sum_j <E_j> G_j / \sum_j G_j \right]. \tag{24}$$

Examples of situations in which initial and final sites may interact equally with lattice vibrations, so that Eq. (24) is valid, occur when the carriers are small polarons and when the hopping is between similar defects.[64] If the energy levels associated with the hopping are far from the Fermi level, Eq. (24) reduces to a formula of the form of Eq. (22) with E_c now representing the energy characteristic of the levels involved in the hopping. Equation (24) contains no contributions associated with the hopping <u>motion</u> of the carriers. If these levels are spread over a very narrow range in energy compared with kT then the term designated by A in Eq. (22) tends to zero. More generally, the magnitude of A depends on the density of states of the levels involved in the hopping; values envisioned for A may approach 10. For hopping between equivalent defects near the Fermi level the thermoelectric power will clearly be much smaller than in the above example. The density of states associated with the involved sites once again is an important parameter in determining the actual magnitude of the thermoelectric power; if the density of states is symmetric about the Fermi level the thermoelectric power vanishes.

If the hopping is between sites which interact unequally with the atomic vibrations, such as between a deep strongly coupled site and a shallow weakly coupled site, vibrational energy <u>is</u> transferred with a hop. In this case there is an additional (positive or negative) energy contribution within the square brackets of Eq. (24), the magnitude of which is some fraction of the energy associated (absorbed or emitted) with the hop. In general the magnitude of this energy will be an increasing function of temperature which reaches a maximum value within the high-temperature semiclassical regime. As a result of the prefactor T^{-1} outside the square brackets of Eq. (24) this contribution to the thermoelectric power will ultimately diminish as $T \to \infty$. This additional contribution to the thermoelectric power arises in several

distinct situations. If one proceeds beyond the linear
electron-lattice interaction of the molecular crystal model and incorporates a quadratic term (a carrier-induced
change of the stiffness of the material in the vicinity
of the carrier-occupied site) into the model then, due
to the difference in the local stiffness between the occupied and unoccupied sites, the effective electron-lattice interaction at an occupied site will differ from
that at an unoccupied site and some vibrational energy
will be transferred with a hop. If one considers hops
between deep strongly coupled states and shallow weakly
coupled states, then most of the energy transfer occurs
at the site of the deep state. For example, consider a
system characterized by only two types of state, a deep
highly localized state which interacts strongly with the
lattice and a shallow state which interacts negligibly
with the lattice. Then for hops between the deep states
and the relatively spread-out shallow states, the energy
transfer occurs principally at the deep state. If, as
is typically the case, the more localized states are the
energetically favored sites, then much of the characteristic disorder energy associated with the carrier's hopping motion is carried with it as it hops. In this example (perhaps applicable to "hopping within a bandtail")
the characteristic energy of the thermoelectric power
will relate to the upper-lying weakly-coupled level and
not to the deep-lying state. In the case of classical
over-the-barrier hopping the energy needed to hop is all
absorbed in the vicinity of the initial site and emitted
in the vicinity of the final site, resulting in the hopping energy being carried along eith the carrier. A
final example in which the transfer of vibrational energy
is associated with a hop is that of correlated small-
polaron hopping motion. Here, since the hopping energy
is primarily absorbed at the final site and emitted at
the initial site, the hopping energy is carried in a direction <u>opposite</u> to that of the carrier.

Finally it should be commented that in deriving the
results presented here and in standard treatments of the
thermoelectric power[65] the energies of the electronic levels are considered to be constants. However, it has been
suggested that in some situations it is important to consider the electronic levels themselves to be temperature-
dependent.[66] In order to include this effect one must
carry out the thermodynamic averages (involved in calculating such quantities as the average number of charge carriers at a given temperature) with respect to a Hamiltonian

which includes the coupling of the electron to some other
portion of the total system, the lattice vibrations, say,
from which this temperature dependence is presumed to a-
rise. Such a procedure is clearly model-dependent, is
often complicated, and typically does not yield a simple
result.

XVI. HALL MOBILITY

Another frequently employed measurement for studying the transport of charge in solids is that of determining the Hall mobility. As pointed out earlier, within the framework of the so-called random-phase model (RPM)
the Hall mobility has been calculated and is found to be
typically low and temperature-independent. The fact that
in a large number of materials the Hall mobility has been
measured to be an increasing function of temperature seems
to imply that the RPM (at least in its simplest form) is
not applicable in these situations. The general problem
of calculating the d.c. Hall mobility associated with hopping motion in a disordered solid has not been addressed.
However, the Hall mobility associated with small-polaron
hopping motion has been calculated for two different lattice structures within the nonadiabatic regime[67,68] and
for a triangular (hexagonal) lattice structure in the
adiabatic regime.[69] The small-polaron Hall mobility in
the semiclassical regime is typically low ($\mu_{Hall} \ll 1 cm^2/$
V-sec) and thermally activated with $E_A^{Hall} \approx E_A/3$. Furthermore, for hole-like small polarons moving within
"three-site" local structures the sign of the Hall mobility associated with their hopping is that which is typically ascribed to the motion of electrons.[70]   These
works are extensively reviewed in Ref.(12).

In addressing the calculation of the Hall mobility
associated with the hopping of charge carriers in a disordered solid one is confronted with an even more complicated statistical problem than that associated with the
conductivity. However, if it is assumed that the charge
carriers pass through the material via regions which are
characterized by relatively little energetic disorder
($1/2 \; \Delta \leq kT, \; \Delta \ll E_A$), then application of the small-
polaron calculations to such noncrystalline solids can
be justified. Based on these assumptions, one can consistently explain a substantial number of electrical
transport measurements on As-Te based chalcogenide glasses.[5,6]

XVII. HIGH-FREQUENCY ABSORPTION AND LUMINESCENCE

Associated with small-polaron hopping motion there is a rather distinctive a.c. conductivity. In particular, in the semiclassical regime, the hopping mobility is peaked about a frequency $\sim 2E_b/\hbar$ with a width $\sim (8kTE_b)^{1/2}/\hbar$.[12,14,71,72] This corresponds to the carrier hopping between sites with the absorption or emission of a photon. (While this phenomenon has been reported in crystals[73,77], it has not, to my knowledge, been studied in disordered solids). Since the a.c. conductivity is proportional to the number of charge carriers, in a regime in which the carrier number is activated the height of the associated absorption peak should reflect (in part) this number and any temperature dependence associated with it. In particular, this effect should presumably accompany hopping between deep well-localized states for which small-polaron formation may be expected.

Also associated with optical phenomena is the topic of luminescence. In particular, the Franck-Condon principle states that the radiation-induced generation of carriers will proceed without any substantial adjustment of the atoms surrounding the carriers. If small polarons are formed, at least a fraction of the radiative recombination will be expected to occur at a substantially lower energy than the excitation energy [presuming that the atoms have sufficient time to adjust to the presence of the carriers]. Thus an indication that the charge carriers form small polarons is the observation of a Stokes shift between excitation and luminescence.[6,12] Such effects have been reported in a number of chalcogenide glasses.[76-81] The interpretation of specific luminescence and absorption measurements in terms of major Franck-Condon relaxation effects (small-polaron formation) has recently been presented.[80,81]

XVIII. HIGH-FIELD D.C. CONDUCTIVITY

Finally, a comment about hopping conduction in high electric fields is in order. One aspect of the problem of high-field conduction in disordered insulators, as distinct from that in crystalline insulators, is that the effect of the field is not uniform throughout the material. In particular, even in the low-field (ohmic) limit the applied field (with a steady-state current flowing) generally produces a redistribution of the charge in the sample.[82] This effect is manifested by the fact that the

"driving force" in producing a current flow between a
pair of sites (in the low-field limit) is the electrochemical-potential difference between the two sites.[47]

     The representation of a disordered material as a re-
sistance network relies on the applied field being sufficiently small so that the effective voltage drop between a pair of sites is simply the difference in the
electrochemical potential divided by the electronic charge, e; at higher fields this approximation will break
down. Furthermore, the individual conductances of the
network, Eq. (15) will themselves generally manifest some
field dependence. Thus, the nonohmic passage of current
between a given pair of sites may be thought of as arising
either because the effective voltage drop between the
sites possesses a significant nonlinearity with respect
to the applied field and/or because the terms defined as
the conductance (at low fields), Eq. (15), manifest a
significant field dependence. Clearly, the field dependence of the d.c. conductivity will generally be associated more with some hops than others. While this is a
very complicated statistical problem, it has been argued,
based on a model of a chain of resistances in which the
dominant voltage drop in the chain is at the difficult
hops, that even a moderate degree of energetic disorder
($\sim$kT) can yield an enhanced field-dependent conductivity
which if interpreted without regard to this feature would
erroneously suggest rather long hopping distances.[60] In
particular, this calculation has been applied to iron and
vanadate-phosphate glasses in which the carriers are believed to form small polarons with the energetic disorder
being rather modest.[59,60] A similar enhanced field dependence of the d.c. conductivity has also been reported
in chalcogenide glasses.[8,83,84] Here again other transport data have been interpreted in terms of small-polaron
hopping motion with typical disorder energies less than
or comparable to .05 eV.[5,6]

     Thus while the fundamental physical processes which
govern hopping motion are far from simple, as seen by the
extensive discussion of the elemental jump rate, transport
in noncrystalline solids is further complicated by the
need to (at least in principle) stipulate a large number
of microscopic parameters. It is provocative to recognize that in a number of systems [chalcogenide and vanadate glasses] the simple assumption of a relatively small
amount of disorder of the parameters associated with hopping motion yields formulae which are consistent with a
large variety of transport experiments.[5,6,52,60] This
ansatz can be qualitatively justified on the basis of the

notion that "short-range order" exists in these noncrystalline solids. Nonetheless, the statistical problem
associated with the noncrystallinity of these insulators
introduces uncertainty into any attempt to explain transport experiments. In these talks I have attempted to
outline the less speculative aspects of the theory without dwelling on the task of relating the theory to specific experiments or to detailed models of the disorder.

REFERENCES

1. Here one has assumed that the carrier motion is as-
    sociated with the carrier's occupancy of electronic
    states which are sufficiently near the band extre-
    mum so that its kinetic energy can be written in
    terms of an effective mass, i.e., simply as $m*v^2/2$;
    this requires that kT be very much less than the re-
    levant carrier bandwidth.

2. M. Roilos, J. Non-Cryst. Solids $\underline{6}$, 5 (1971).

3. P. Nagels and M. Denayer, Mater. Res. Bull. $\underline{6}$, 1247
    (1971).

4. P. Nagels, R. Callaerts, M. Denayer, and R. DeConinde,
    J. Non-Cryst. Solids $\underline{4}$, 245 (1970).

5. D. Emin, C. H. Seager, and R. K. Quinn, Phys. Rev.
    Lett. $\underline{28}$, 813 (1972).

6. C. H. Seager, D. Emin, and R. K. Quinn, Phys. Rev.
    B $\underline{8}$, 4746 (1973).

7. M. Roilos and E. Mytilineou, _Amorphous and Liquid
    Semiconductors_, edited by J. Stuke and W. Brenig
    (Taylor and Francis, London 1974), p. 319.

8. A. J. Grant, T. D. Moustakas, T. Penny and K. Weiser,
    Amorphous and Liquid Semiconductors, edited by J.
    Stuke and W. Brenig (Taylor and Francis, London, 1974)
    p. 325.

9. C. H. Seager, M. L. Knotek and A. H. Clark, _Amorphous
    and Liquid Semiconductors_, edited by J. Stuke and W.
    Brenig (Taylor and Francis, London, 1974), p. 1133.

10. M. H. Cohen, _Amorphous and Liquid Semiconductors_,
    edited by N. F. Mott (North Holland, Amsterdam, 1970)
    p. 391.

11. Thus the presence of disorder is thought to effect-
    ively "slow down" a carrier. This has been suggest-
    ed also in Refs. (5) and (6), in the talk on local-
    ized states in this volume, and in Ref. (12) [see
    page 285 of Ref. (12)].

12. D. Emin, _Electronic and Structural Properties of
    Amorphous Semiconductors_, edited by P.G.LeComber
    and J.Mort (Academic, London, 1973), Chapt. 7.

13. The velocity which characterizes site-to-site motion in this wide-band strong-scattering model is Jfa/h. This velocity, however, may be greater than that associated with motion in a wide-band weak-scattering situation ($J \gg kT$) characteristic of a crystal for which $v \sim (Ja/\hbar)(kT/J)^{1/2}$.

14. N. K. Hindley, J. Non-Cryst. Solids $\underline{5}$, 17 (1970).

15. L. Friedman, J. Non-Cryst. Solids $\underline{7}$, 324 (1971).

16. C. Kittel, <u>Introduction to Solid State Physics</u>, 2nd Edition (Wiley, New York, 1962) p.296.

17. It should be commented that the above ratio, Eq. (8), is characteristic of tight-binding studies of the Hall mobility in which the primary contribution to the Hall mobility is associated with "three-site processes" (proportional to J^3) as distinct from higher-order n-site processes (proportional to J^n). In such calculations the smallness of J compared with other characteristic energies is assumed to justify the neglect of higher-order processes. However, in the random-phase model the only other characteristic energy, kT, is taken to be much smaller than J.

18. In the above numerical example μ_{Hall} is typically $\lesssim .5$ cm^2/V-sec.

19. T. Holstein, Ann. Phys. (N.Y.) $\underline{8}$, 325 (1959).

20. D. Emin, article in this volume.

21. However, in making numerical estimates it <u>is</u> misleading to assume that the transfer energies are those of a carrier-free (undeformed) material.

22. T. Holstein, Ann. Phys. (N.Y.) $\underline{8}$, 343 (1959).

23. D. Emin, Adv. Phys., to be published.

24. H. J. deWit, Philips Res. Repts. $\underline{23}$, 449 (1968).

25. L. Friedman, Phys. Rev. $\underline{135}$, A233 (1964).

26. D. Emin, Phys. Rev. Lett. $\underline{32}$, 303 (1974).

27. D. Emin and T. Holstein, Ann. Phys. (N.Y.) $\underline{53}$, 439 (1969).

28. D. Emin, Phys. Rev. Lett. $\underline{25}$, 1751 (1970).

29. D. Emin, Phys. Rev. B$\underline{3}$, 1321 (1971).

30. D. Emin, Phys. Rev. B$\underline{4}$, 3639 (1971).

31. D. Emin, J. Non-Cryst. Solids, $\underline{8\text{-}10}$, 511 (1972).

32. Even if the equilibrium electronic levels associated with the sites involved in a hop are nondegenerate the atoms surrounding both sites will still participate equivalently as regards their absorption and emission of vibrational energy. Thus, the preceeding theory is also applicable to hopping between nondegenerate sites: see Sec.(7.9) of Ref. (12). Disorder of the vibratory parameters will produce some modification of the theory: see Ref. (31).

33.   D. Emin, <u>Tetrahedrally Bonded Amorphous Semiconductors</u>, edited by M. H. Brodsky, S. Kirkpatrick, and D. Weaire, (American Institute of Physics, New York, 1974), p. 326.

34.   A. Miller and B. Abrahams, Phys. Rev. <u>120</u>, 745 (1960).

35.   In addition, the spatial separation enters into the theory through a correlation factor which is associated with the coherence of the phonon-induced alterations of the electronic levels of the sites involved in a hop. Specifically, this factor only arises when $q \cdot R \ll 1$, where q is a phonon wavevector and R the spatial separation between the sites. In the limit of large $|R|$ its effect disappears, and in the opposite limit of $|R|$ being an interatomic spacing, it produces no qualitative alteration of our results. Furthermore, in the absence of an explicit realistic microscopic model for the electron-lattice interaction detailed consideration of such correction factors is deemed inappropriate.

36.   For example, in the case of longitudinal acoustic phonons, the spatial extent of the region of a material associated with compression (or rarefaction), $\lambda/2$, must be greater than the spatial extent (diameter) of the electronic state, $2r$, or else the electron will experience regions of both compression and rarefaction, whose effect on the electron tend to cancel each other thereby yielding little net interaction between the "localized" electron and the lattice vibrations.

37.   P. W. Anderson, Nature <u>235</u>, 163 (1972).

38.   In the case of acoustic phonons a restriction of the values of q with which a carrier effectively interacts seriously restricts the vibrational frequencies with which the carrier interacts. However for optical phonons (in the weak-dispersion regime) a restriction of the q's poses little limitation on the vibrational frequencies associated with the carrier-lattice interaction.

39.   The fractional error in the jump rate involved in assuming that the upward jump rate is activated $[\exp(-\Delta/kT)]$ is $m \exp(-\hbar\omega_0/kT)$; with an increase in the phonon number the approximation is limited to lower and lower temperatures.

40.   In a monatomic lattice, characterized by a Debye phonon spectrum, the condition that an electron in a state of radius r only effectively interacts with phonons of wavelengths for which $\lambda/2 > 2r$, determines the maximum frequency phonon with which a carrier can interact to be $\omega_{max} = \omega_{Debye}(a/2r)$.

41. The fractional error arising in taking an m-phonon jump to be activated (for a hop upward in energy) or temperature-independent (for a hop downward in energy) is $\sum_{i=1}^{m} \exp(-\hbar\omega_i/kT)$, where the ω_i are the frequencies of the involved phonons and $\omega_i \leq \omega_{max}$.

42. T. Holstein, Westinghouse Research Report, 60-94698-3-RO (1955).

43. C. Herring, Rev. Mod. Phys. 34, 631 (1962).

44. The variation of the transfer energy with R as $R \rightarrow \infty$ is given, in the main, by the exponential term since the potential in the overlap region becomes increasing flat as $R \rightarrow \infty$.

45. D. Emin, unpublished.

46. N. F. Mott, Phil. Mag. 19, 835 (1969).

47. W. Ambegaokar, B. I. Halperin, and J. S. Langer, Phys. Rev. B4, 2612 (1971).

48. S. Kirkpatrick, Phys. Rev. Lett. 27, 1722 (1971).

49. W. Brenig, P. Wolfle and G. Döhler, Physik 246 (1971).

50. M. Pollak, J. Non-Cryst. Solids 11, 1 (1972).

51. R. Jones and W. Schaich, J. Phys. C5, 43 (1972).

52. M. L. Knotek, M. Pollak, T. M. Donovan and H. Kurtzman, Phys. Rev. Letters 30, 853 (1973).

53. M. Pollak and J. J. Hauser, Phys. Rev. Lett. 31, 1304 (1973).

54. K. Maschke, H. Overhof and P. Thomas, _Amorphous and Liquid Semiconductors_, edited by J. Stuke and W. Brenig (Taylor and Francis, London, 1974), p. 141.

55. G. Pike and C. Seager, _Amorphous and Liquid Semiconductors_, edited by J. Stuke and W. Brenig (Taylor and Francis, London, 1974), p. 147.

56. The curves of Fig. 9 are meant to illustrate both (1) that a nonactivated conductivity curve can resemble a linear $\ln\sigma$ versus $T^{-1/4}$ plot for $T < T_{Debye}$ if the $T^{-1/4}$ range is sufficiently small, and (2) that the slope of the jump rate, for hops between small-radius (strongly-coupled) centers, on a $T^{-1/4}$ plot is expected to be comparable with what is observed. The details of the curves are artifacts of the simple microscopic model used in the calculation and are therefore not meaningful.

57. To differentiate between a $\sigma \propto \exp[-(T_0/T)^{1/4}]$ temperature dependence and some other nonactivated temperature dependence, such as $\sigma \propto AT^n$ or $\sigma \propto \exp[-(T_0/T)^{1/n}]$, one is required to make measurements over a sufficiently large temperature interval. Specifically, in the first (power-law) example above, the requirement that the fractional deviation of the slope of $\log\sigma$ from being simply proportional to $T^{-1/4}$ over

the temperature interval, T_{min} to T_{max}, be smaller than some criterion f (typically $f > .3$) results in the condition $2(r - 1)/(r + 1) < f$, where $r \equiv (T_{max}/T_{min})^{1/4}$. When this criterion is met a power-law behavior will result in a curve which appears linear on a $T^{-1/4}$ plot. In the second example the criterion is slightly more complicated: $2|r/n - 1|(r - 1)/(r + 1) < f$. Similar arguments, of course, hold for $T^{-1/3}$ plots as well.

58. T. N. Kennedy and J. D. MacKenzie, Phys. Chem. Glasses 8, 169 (1967).

59. I. G. Austin and N. F. Mott, Adv. Phys. 18, 41 (1969).

60. I. G. Austin and M. Sayer, J. Phys. C7, 905 (1974).

61. G. N. Greaves, J. Non-Cryst. Solids 11, 427 (1973).

62. W. Fuhs, this volume, Sec. 5.

63. In the case of mixed-valency glasses, such as vanadate glasses, the carrier number is often only dependent on the concentration of ions of a given valency, c. In this case it is expedient to rewrite $(E_C - E_F)/kT$ as a function of c.

64. It is tacitly assumed here that the electron-lattice interaction is purely a linear one, as in the molecular crystal model.

65. R. R. Heikes and R. W. Ure, Jr., Thermoelectricity: Science and Engineering (Interscience, New York, 1961).

66. N. F. Mott and E. A. Davis, Electronic Processes in Noncrystalline Materials (Clarendon, Oxford, 1971), p. 220.

67. L. Friedman and T. Holstein, Ann. Phys. (N.Y.) 53, 439 (1969).

68. D. Emin, Ann. Phys. (N.Y.) 64, 336 (1971).

69. D. Emin and T. Holstein, Ann. Phys. (N.Y.) 53, 439 (1969).

70. T. Holstein, Phil. Mag. 27, 275 (1973).

71. M. I. Klinger, Phys. Lett. 7, 102 (1963).

72. H. G. Reik and D. Heese, J. Phys. Chem. Solids, 28 581 (1967).

73. V. N. Bogomolov, E. K. Kudinov, D. N. Mirlin, and Y. A. Firsov, Sov. Phys. Solid State 9, 1630 (1968).

74. V. N. Bogomolov and D. N. Mirlin, Phys. Status Solidi 27, 443 (1968).

75. E. K. Kudinov, D. N. Mirlin, and Yu. A. Firsov, Sov. Phys. Solid State 11, 2257 (1970).

76. R. Fischer, U. Heim, F. Stern, and K. Weiser, Phys. Rev. Lett. 26, 1182 (1971).

77. J. Cernogora, F. Mollot, and C. Benoit á la Guillaume, Phys. Status Solidi (ca.) 15, 401 (1973).

78. R. A. Street, T. M. Searle, and I. G. Austin, J. Phys. C6, 1830 (1973).

79.   S. G. Bishop and C. S. Guenzer, Phys. Rev. Lett.
       30, 1309 (1973).
80.   F. Mollot, J. Cernogora and C. Benoit à la
       Guillaume, Phy. Status Solidi (a) 21, 281 (1974).
81.   J. Cernogora, F. Mollot, and C. Benoit à la
       Guillaume, Proc. 12th Int. Conf. on the Physics of
       Semiconductors (Verlag, Stuttgart, 1974) p. 1027.
82.   This may be thought of as a field-induced change
       in the quasi-Fermi levels associated with each site.
83.   H. J. deWit and C. Crevecoeur, J. Non-Cryst. Solids,
       8-10, 787 (1972).
84.   J. M. Marshall and G. R. Miller, Phil. Mag. 27,
       1151 (1973).

INELASTIC COHERENT NEUTRON SCATTERING IN AMORPHOUS SOLIDS*

J. D. Axe

Brookhaven National Laboratory

Upton, New York 11973

I. INTRODUCTION

Inelastic neutron scattering is widely perceived to be more powerful than optical spectroscopy in investigating the dynamics of crystalline solids because light couples directly only to long wavelength fluctuations which comprise an insignificant fraction of the total, whereas thermal neutrons respond rather uniformly to the entire wavelength spectrum of the lattice excitations. The situation changes substantially if we consider amorphous solids. For simplicity, we shall consider explicitly only single component harmonic solids. There are of course well defined long lived normal modes of vibration in such a material which we will continue to call phonons, but since an amorphous solid lacks the translational periodicity of a crystal, these phonons no longer have a single characteristic propagation vector. Quite generally the frequency dependence of the one phonon cross section can be written in the form

$$S_1(\omega) = \frac{1 + n(\omega)}{\omega} g_\alpha(\omega),$$ (1.1)

where $n(\omega)$ is the Bose occupation number and

$$g_\alpha(\omega) = \frac{1}{3N} \sum_\lambda |\alpha_\lambda|^2 \delta(\omega - \omega_\lambda)$$ (1.2)

*Work performed under the auspices of the U. S. Atomic Energy Commission.

is an effective density of states of the material for the
particular spectroscopic technique in question. Each
mode, with eigenfrequency ω_λ, is weighted by the matrix
element $|\alpha_\lambda|^2$. For example, for Raman scattering α_λ
projects out the long wavelength optical polarizability
fluctuations associated with the λ'th normal mode.

Since in the amorphous case optical experiments
couple not just to long wavelength modes, but instead to
long wavelength Fourier components of all normal modes
we must reinvestigate the relative merits of optical and
neutron spectroscopy. We shall find that the principal
advantage of neutron spectroscopy in this case lies in
the fact that neutrons interact directly with the constituent nuclei rather than indirectly through phonon
induced modulation of the optical properties of the material. As a result calculations of the matrix elements
α_λ occurring in Eq. (1.2) are difficult and model dependent[1] for optical experiments. By contrast the neutron scattering cross section, which we derive in Section II, can also be cast in the form of Eqs. (1.1) and
(1.2) and has a rather simple structure. In particular
in the limit of large momentum transfer the appropriate
matrix element α_λ is independent of the mode index λ so
that the one phonon coherent scattering is proportional
to the true phonon density of states,

$$g(\omega) = \frac{1}{3N} \sum_\lambda \delta(\omega-\omega_\lambda).$$ (1.3)

In the opposite low momentum transfer region an amorphous solid behaves as an elastic continuum and the neutron cross section shows delta-function singularities
at frequencies which correspond to longitudinal sound
wave propagation.

At intermediate values of momentum transfer the coherent scattering cross section is too complex to characterize simply. In Section III we investigate a one-
dimensional model of an amorphous solid with very simple
dynamics for which an exact analytical expression for the
cross section at all momentum transfer can be written.
The solutions display broadening of acoustic phonon
groups and periodic modulation of the spectral response
which are purely structural rather than dynamic in origin,
and which are closely related to the structure of the
elastic scattering function. A discussion of the results is given in Section IV.

II.   COHERENT ONE PHONON SCATTERING

The derivation of the inelastic scattering cross section follows that for a harmonic crystal by making the familiar normal mode expansion[2] of displacements which are however no longer plane waves. The scattering cross section with an energy transfer $\hbar\omega$ and momentum transfer $\hbar Q$ is proportional to the time Fourier transform of the intermediate scattering function $I(\underline{Q},t)$

$$S(\underline{Q},\omega) = \frac{1}{2\pi} \int\limits_{-\infty}^{\infty} e^{-i\omega t} I(\underline{Q},t)\ dt, \qquad (2.1)$$

where

$$I(\underline{Q},t) = \sum_{\ell,\ell'} \ \langle e^{-i\underline{Q}\cdot\underline{r}_\ell(0)} \ e^{i\underline{Q}\cdot\underline{r}_{\ell'}(t)} \rangle. \qquad (2.2)$$

We can differentiate our ideal amorphous solid from a fluid by specifying each atomic position $\underline{r}_\ell(t)$ in terms of a small displacement away from a mean position $\underline{R}_\ell$

$$\underline{r}_\ell(t) = \underline{R}_\ell + \underline{u}_\ell(t). \qquad (2.3)$$

The elastic scattering is separated by examining the intermediate scattering function as $t \to \infty$. The displacements become uncorrelated in this limit so that

$$S_0(\underline{Q},\omega) = \delta(\omega)I(\underline{Q},\infty)$$

$$= \delta(\omega) \sum_{\ell,\ell'} e^{-i\underline{Q}\cdot(\underline{R}_\ell-\underline{R}_{\ell'})} \qquad (2.4)$$

$$\times \ \langle e^{i\underline{Q}\cdot\underline{u}_\ell(0)} \rangle \langle e^{i\underline{Q}\cdot\underline{u}_{\ell'}(0)} \rangle$$

$$= \delta(\omega) \ e^{-2W(\underline{Q})} \sum_{\ell,\ell'} e^{-i\underline{Q}\cdot(\underline{R}_\ell-\underline{R}_{\ell'})}.$$

An explicit expression for the Debye-Waller factor $e^{-W(Q)}$ is given together with certain other useful general results in Appendix I. The remaining inelastic scattering is given by

$$S_{inel}(\underline{Q},\omega) = \frac{1}{2\pi} \int\limits_{-\infty}^{\infty} dt e^{-i\omega t} [I(\underline{Q},t) - I(\underline{Q},\infty)]$$

$$= \frac{1}{2\pi} \int\limits_{-\infty}^{\infty} dt e^{-i\omega t} e^{-2W(Q)} \sum_{\ell,\ell'} e^{-i\underline{Q}\cdot(R_\ell-R_{\ell'})}$$

$$\times [e^{\underline{Q}\cdot<\underline{u}_\ell(0)\underline{u}_{\ell'}(t)>\cdot\underline{Q}} - 1]. \qquad (2.5)$$

To proceed further we use a normal mode expansion for the displacements,

$$m^{1/2}\underline{u}_\ell(t) = \sum_\lambda \underline{\xi}_\ell(\lambda)q_\lambda(t), \qquad (2.6)$$

where as usual $\underline{\xi}_\ell(\lambda)$ diagonalizes the harmonic dynamical matrix so that $q_\lambda(t)$ satisfies the harmonic equation of motion $\ddot{q}_\lambda(t) = \omega_\lambda^2 q_\lambda(t)$. See Appendix I. The one phonon cross section is obtained by expanding Eq. (2.5) to lowest order in $(\underline{Q}\cdot\underline{u})$ and using Eq. (2.6) and the well known[2] harmonic result for $<q_\lambda(0)q_\lambda(t)>$,

$$S_1(\underline{Q},\omega) = \frac{\hbar e^{-2W(\underline{Q})}}{2m} \sum_{\lambda,\ell,\ell'} \underline{Q}\cdot[\underline{\xi}_\ell(\lambda)\underline{\xi}_{\ell'}^*(\lambda)]\cdot\underline{Q}e^{-i\underline{Q}\cdot(\underline{R}_\ell-\underline{R}_{\ell'})}$$

$$\times [n_\lambda\delta(\omega+\omega_\lambda) + (n_\lambda+1)\delta(\omega-\omega_\lambda)] \qquad (2.7)$$

Finally we can cast Eq. (2.7) into the form of Eq. (1.1),

$$S_1(\underline{Q},\omega) = \frac{NhQ^2 e^{-2W(\underline{Q})}}{2m} (\frac{1+n(\omega)}{\omega})g_{eff}(\hat{\underline{Q}},\omega), \qquad (2.8)$$

where for $\omega > 0$

$$g_{eff}(\underline{Q},\omega) = \frac{1}{3N} \sum_\lambda |\sum_\ell (\hat{\underline{Q}}\cdot\underline{\xi}_\ell(\lambda))e^{-i\underline{Q}\cdot\underline{R}_\ell}|^2\delta(\omega-\omega_\lambda).$$

$$\qquad (2.9)$$

$\hat{Q} = \underline{Q}/|\underline{Q}|$ and $g_{eff}(-\omega) = g_{eff}(\omega)$.

Eq. (2.7) (or its equivalent Eqs. (2.8)-(2.9) is the desired general result for the one phonon cross section for arbitrary momentum transfer. However, the result can be considerably simplified in the large $\underline{Q}$ limit, by noting that the phase factor $e^{-i\underline{Q}\cdot(\underline{R}_{\ell}-\underline{R}_{\ell'})}$ will average to zero all contributions in $g_{eff}(\omega)$, except for self correlations, $\ell = \ell'$. Thus,

$$g_{eff}(\underline{Q} \to \infty, \omega) = \frac{1}{3N} \sum_{\lambda}\sum_{\ell} |\hat{\underline{Q}}\cdot\underline{\xi}_{\ell}(\lambda)|^2 \, \delta(\omega-\omega_{\lambda})$$

(2.10)

$$= \frac{1}{3N} \sum_{\lambda} \delta(\omega-\omega_{\lambda}) = g(\omega),$$

as a result of orthonormality (see Appendix I).

III. EXAMPLE: A ONE-DIMENSIONAL "RANDOM" LATTICE

Although the one phonon cross section is easily written down it is in general by no means simple to evaluate and there are both dynamical and structural difficulties mixed together in a rather complicated way. The dynamics makes itself felt through the appearance in Eq. (2.7) of not only the phonon frequencies but the phonon eigenvectors as well. It may be possible to make more or less plausible guesses from other experiments about the frequency distribution but it is much more difficult to "guess" meaningful eigenvectors. What is really required is a detailed lattice dynamical model including a specification of force constant variation (based upon uncertain interatomic potentials) and solved for a large representative number of particles. The structural aspects make themselves felt in a way quite independent of the dynamics through the appearance of the mean atomic positions in the phase factor $e^{i\underline{Q}\cdot\underline{R}_{\ell}}$. It is precisely the disappearance of this structural complication which accounts for the remarkable simplification of the cross section in the large Q limit, where each atom scatters independently.

It is however at least conceptually possible to dissociate dynamic and structural aspects of the problem by postulating a model amorphous solid in which the harmonic restoring forces are insensitive to equilibrium interatomic separation. The resulting effects on the phonon cross section are simple enough to discuss without recourse to extensive numerical study and, as we shall see, it is of interest to do so.

Let us consider a very simple but rather plausible structrual model of a one-dimensional amorphous solid which we define as a "random" lattice. In this model, the equilibrium distance between n'th nearest neighbors is the result of n individual steps which have themselves a Gaussian distribution.

$$p(x) = (2\pi\sigma)^{-1/2} \exp - [(x - a)^2/2\sigma^2]. \qquad (3.1)$$

The resulting distribution of n'th nearest neighbor distances is well known,

$$p_n(x) = (2\pi\sigma_n)^{-1/2} \exp - [(x-x_n)^2/2\sigma_n^2], \qquad (3.2)$$

where x_n = na and $\sigma_n^2 = n\sigma^2$. The model defines a lattice in the sense that the most probable interatomic spacings are a,2a,...,etc., but in contrast to a crystalline model each successive interatomic separation is more poorly defined, and thus the model structure lacks long range order. To see this, form the static density correlation function

$$G(x, t = \infty) = \frac{1}{N} \sum_{\ell,n} <\delta(x + x_\ell - x_{\ell+n})>$$

$$= 1 + \sum_{n=\pm 1}^{\pm\frac{1}{2}(N-1)} p_n(x). \qquad (3.3)$$

The elastic scattering is given by the Fourier transform

$$S_o(Q) = \int G(x,\infty) \, e^{iQx} dx$$

$$= 1 + \sum_{n=\pm 1}^{\pm \frac{1}{2}(N-1)} e^{iQx_n} e^{-\frac{1}{2}Q^2\sigma_n^2} \qquad (3.4)$$

$$= 1 + 2 \sum_{n=1}^{\frac{1}{2}(N-1)} \cos(Qan)e^{-\frac{1}{2}Q^2\sigma^2 n} .$$

The object behind introducing this particular model is that sums of the sort occurring in Eq. (3.4) can be performed exactly in the limit $N \to \infty$. The result for $S_o(Q)$ is[3]

$$S_o(Q) = \frac{\sinh(\frac{1}{2}\sigma^2 Q^2)}{\cosh(\frac{1}{2}\sigma^2 Q^2) - \cos(Qa)} . \qquad (3.5)$$

At large Q $S_o(Q) \to 1$, as expected for an amorphous material, but it shows more or less sharp diffraction maxima at $Qa = 2\pi$, 4π,..., so long as $Q^2\sigma^2 \ll 1$. A typical diffraction pattern for $(\sigma/a) = 1/5$ is shown in Fig. 1.

The dynamics of the model are particularly simple since the restoring forces depend only upon the displacements u_ℓ, independent of the local equilibrium configuration. The equations of motion, assuming only nearest neighbors interact, is

$$m\ddot{u}_\ell = \beta(u_{\ell+1} + u_{\ell-1} - 2u_\ell), \qquad (3.6)$$

and can be decoupled by a trial solution of the form of Eq. (2.6) with

$$\xi_\ell(\lambda) = N^{-1/2} e^{i\ell\phi_\lambda}, \qquad (3.7)$$

which reduces Eqs. (3.6) to

$$\ddot{q}_\lambda(t) = (\beta/m)(e^{i\phi_\lambda} + e^{-i\phi_\lambda} - 2)q_\lambda(t), \qquad (3.8)$$

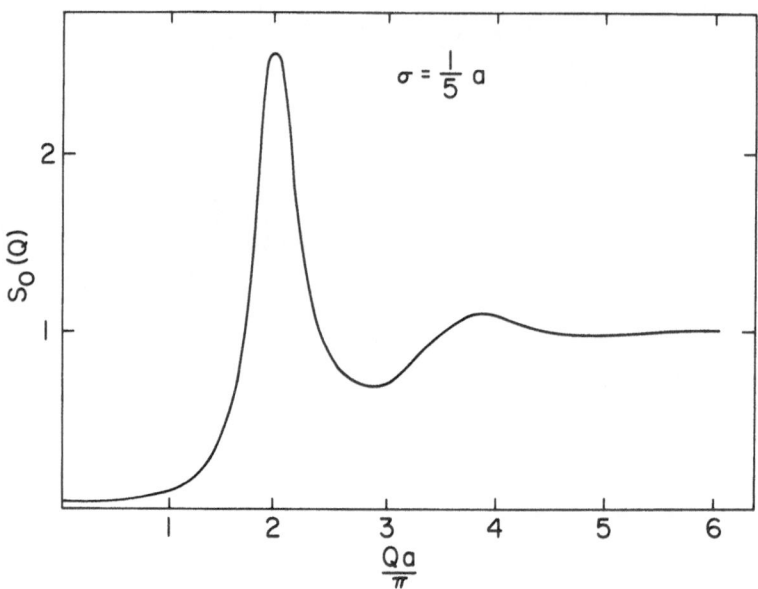

Fig. 1. The distribution of elastic scattering $S_0(Q)$ for a "random" one-dimensional lattice generated by iteration of a Gaussian single step distribution with $\langle x \rangle = a$ and $\langle x - \langle x \rangle \rangle^2 = \sigma^2$.

which has the solution

$$\omega_\lambda = \omega_M \sin \tfrac{1}{2}\phi_\lambda, \tag{3.9}$$

where $\omega_M = (4\beta/m)^{1/2}$. The allowed values of the phase shift ϕ_λ can be fixed most simply by applying periodic boundary conditions, $u_{\ell+N} = u_\ell$ so that $e^{iN\phi_\lambda} = 1$, or

$$\phi_\lambda = \left(\frac{2\pi\lambda}{N}\right), \quad \lambda = 0, \pm 1, \ldots, \pm\tfrac{1}{2}(N-1). \tag{3.10}$$

Note that the eigenfrequencies defined by Eqs. (3.9) and (3.10) are identical to that for the periodic linear chain as we would expect, but the eigenvectors are not plane waves because the atoms are not equally spaced.

Inserting Eq. (3.7) into Eq. (2.9) we find for the

effective density of states for one phonon scattering

$$g_{eff}(Q,\omega) = \frac{1}{N} \sum_{\lambda} [\sum_{\ell,\ell'} \frac{1}{N} e^{i\phi_\lambda(\ell-\ell')} e^{-iQ(x_\ell - x_{\ell'})}] \delta(\omega-\omega_\lambda)$$

$$= \frac{1}{N} \sum_{\lambda} [1 + \sum_{n \neq 0} e^{i\phi_\lambda n} \int dx\, e^{iQx} p_n(x)] \delta(\omega-\omega_\lambda)$$

$$= \frac{1}{N} \sum_{\lambda} [1 + \sum_{n \neq 0} e^{i\phi_\lambda n} e^{iQx_n} e^{-\frac{1}{2}Q^2 \sigma_n^2}] \delta(\omega-\omega_\lambda) \tag{3.11}$$

This last expression is easily recognized to be of the same form as (3.4) and can again be summed to yield

$$g_{eff}([,\omega) = \frac{1}{N} \sum_{\lambda} \gamma_\lambda(Q)\, \delta(\omega-\omega_\lambda), \tag{3.12}$$

with

$$\gamma_\lambda(Q) = \frac{\sinh(\frac{1}{2}Q^2\sigma^2)}{\cosh(\frac{1}{2}Q^2\sigma^2) - \cos(Qa+\phi_\lambda)}. \tag{3.13}$$

Eq. (3.13) is the central result of this section. Note that it is closely related in form to the expression for the elastic scattering $S_o(Q)$ given in Eq. (3.5). In particular, in the limit of large Q, $\gamma_\lambda(Q) \to 1$, so that

$$g_{eff}(Q \to \infty, \omega) = \frac{1}{N} \sum_{\lambda} \delta(\omega-\omega_\lambda) \equiv g(\omega), \tag{3.14}$$

which verifies the general result of Eq. (2.10) (the loss of the factor 1/3 occurs in passing to one dimension).

For purposes of numerical calculation it is convenient to reexpress Eq. (3.13) explicitly in terms of the phonon density of states,

$$g_{eff}(Q,\omega) = Y(Q,\omega)g(\omega), \tag{3.15}$$

where $y(Q,\omega)$ is given by (A2.1). This form is also useful in investigating the small Q limit of g_{eff}. In Appendix II we show that for $Q^2\sigma^2 \ll 1$

$$g_{eff}(Q,\omega) = \frac{1}{\pi} \{\frac{\Gamma(Q,\omega)}{(\Delta\omega)^2 + \Gamma(Q,\omega)^2}\}, \tag{3.16}$$

where

$$\Delta\omega = (\omega - \omega_m \sin\tfrac{1}{2}Qa)$$

and   (3.17)

$$\Gamma(Q,\omega) = \sigma^2 Q^2/4\pi g(\omega).$$

Thus at small Q, g_{eff} no longer bears any resemblance to
$g(\omega)$, but is a single Lorentzian peak centered at the
frequency appropriate to dispersion of the ordered linear
chain but with a width $\sim Q^2/g(\omega)$. We stress again that
this width is not due to any damping of the individual
phonons, but rather due to the fact that phonons with a
range of energies have a momentum component Q which can
be exchanged with the neutron beam. As it must, Eq.
(3.16) recovers the exact result for an ordered array in
the limit $\sigma \to 0$.

In order to observe how $Y(Q,\omega)$ evolves from a sharply peaked function of ω at small Q to frequency independent behavior at large Q, we have plotted $Y(Q,\omega)$ vs Q for the
arbitrary ratio $\sigma/a = 1/5$. Fig. 2 is evaluated at Q-
values corresponding to successive Brillouin zone boundaries of the ordered lattice. The broadening and disappearance of structure in $Y(Q,\omega)$ is closely related to the
approach of $S_0(Q)$ to its asymptotic limit, $S_0(\infty) = 1$.
Fig. 3, showing the behavior of $Y(Q,\omega)$ at successive
"pseudo-Brillouin zone" centers, is even more interesting
since it shows at the first diffraction maximum a single
peak (rather broad for this choice of parameters) centered at $\omega \approx 0$. This is rather similar to the familiar "roton"

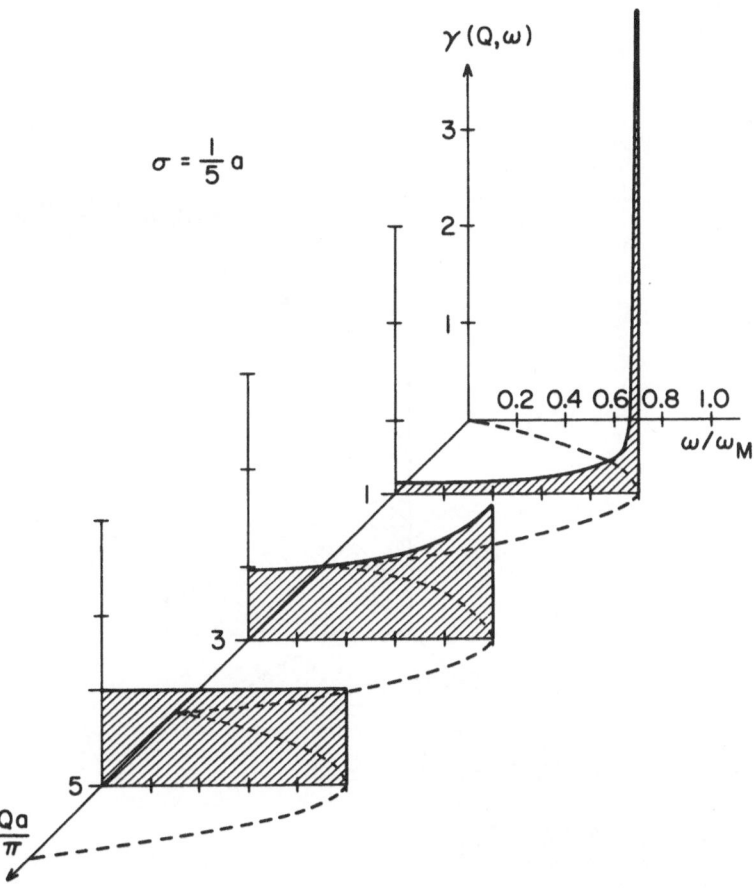

Fig. 2. The scattering function $\gamma(Q,\omega)$ (see text) for the random one-dimensional lattice with $\sigma/a = 1/5$. The dotted line shows the dispersion relation for the ordered linear lattice. $\gamma(Q,\omega)$ is evaluated at successive Brillouin-zone boundary points for the ordered lattice.

minimum which occurs in the inelastic scattering of liquid He at the value of Q corresponding to the diffraction maximum. The effect here is in a sense trivial, since it is structural not dynamical in origin.

Although the inelastic scattering width at small Q (Eq. (3.16)) was derived for a lattice built by a Gaussian random walk, the result is more generally valid.

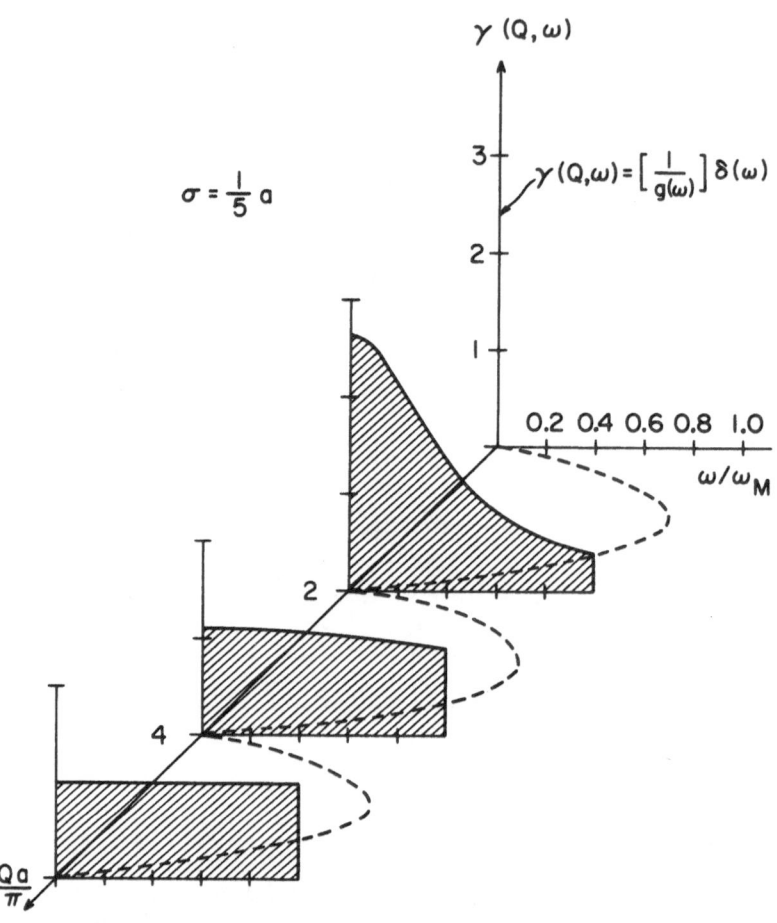

Fig. 3. $\gamma(Q,\omega)$ as in Fig. 2, evaluated at successive
Brillouin-zone centers of the ordered lattice.

This is because the central limit theorem guarantees that
the iteration of any single step characterized by $\langle x \rangle = a$
and $\langle x - \langle x \rangle \rangle^2 = \sigma^2$ will in the limit of large step, n,
be given by Eq. (3.2). These long range correlations
dominate $\gamma(Q,\omega)$ at small Q.

IV. DISCUSSION

There is not a large body of experimental inelastic neutron scattering results on amorphous systems, and most of what does exist is incomplete time-of-flight data which, although it can be usefully compared with similar data taken in the liquid or polycrystalline state, is not satisfactory for a quantitative discussion. This is perhaps just as well, because by and large the systems chosen for study have been structurally complex multi-component glasses which only intensifies an already difficult job of interpretation. Much of this data (prior to 1970) has been discussed in a review by Leadbetter, et.al.[4]. More recent work, of a more complete nature, or on structurally simpler systems include attempts to determine the phonon density of states in the amorphous metal Co_4P by Moss et.al.[5] and in amorphous Ge by Axe et.al.[6], and in vitreous silica by Leadbetter and String-fellow[7].

The development in Section II may be insufficient to interpret inelastic neutron scattering data in real materials in at least three respects:
1) Most glasses are multicomponent. This introduces additional complexity into the expression which can be handled without difficulty in a formal way, but the practical complications are severe.
2) We have assumed that the constituent nuclei are purely coherent scatterers. In general the scattering will contain an incoherent contribution as well, which is difficult to separate, but such a separation is not necessary at high Q where both coherent and incoherent scattering are proportional to $g(\omega)$.
3) We have neglected scattering processes involving more than a single vibrational quantum, and this is increasingly difficult to justify at large $\underline{Q}$. The fraction of the total inelastic scattering that is one phonon in character, $f(Q)$, depends upon the Debye-Waller factor[2] $W(\underline{Q})$.

$$f(\underline{Q}) = \frac{2W(Q)e^{-2W(Q)}}{1 - e^{-2W(Q)}} \cdot \qquad (4.1)$$

Fig. 4 shows the behavior of the multiphonon fraction $1-f(Q)$ vs. Q for a value of $W(Q)$ thought to be appropriate to amorphous Ge at room temperature. It can be seen that there is a fundamental competition between the requirement for large Q in order that $S_1(Q,\omega)$ be

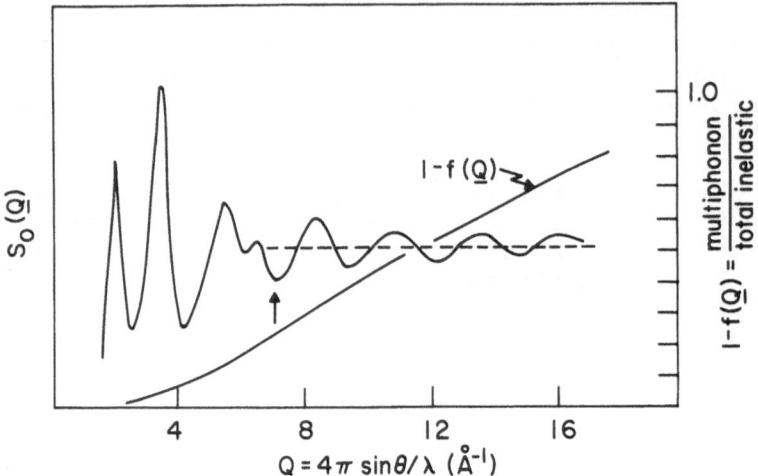

Fig. 4. $S_0(\underline{Q})$ and the multiphonon inelastic fraction, $(1-f(\underline{Q}))$, for amorphous Ge at room temperature. The arrow at $Q \sim 7\text{Å}^{-1}$ represents a reasonable compromise between approach to the incoherent limit $S_0(Q \to \infty) = 1$ and small multiphonon corrections to the one phonon scattering.

proportional to $g(\omega)$, and the necessity to make extensive corrections for multiphonon effects. This can be partially alleviated by measuring at low temperatures, which is in any case desirable to avoid large anharmonic effects.

Finally, it can be noted that although we have discussed exclusively vibrational excitations in amorphous materials, ferromagnetically ordered glasses do exist, and it is obvious that much of the present discussion is also applicable to collective spin excitations in such materials. In particular continuum mechanics predicts the existence of well defined long wavelength spin waves in the same way that macroscopic elastic theory predicts sound waves in an amorphous elastic medium. Inelastic neutron scattering experiments have recently detected sharp spin wave excitations with a normal quadratic dispersion $(\omega(Q) = DQ^2)$ in a metallic glass of composition $Fe_{.75}P_{.15}C_{.10}$.[8]

ACKNOWLEDGMENT

I would like to thank D. T. Keating for discussions
of the approximations employed in Section III. After
this manuscript was completed I learned that the model
of Section III had been previously discussed by Kim and
Nelkin[9] from a somewhat different point of view.

APPENDIX I

Some Exact Results for an Harmonic Amorphous Solid

Here we collect together mostly without proof some
useful results of normal mode theory as applied to an
amorphous solid, which for present purposes can be thought of as a large polyatomic molecule. The general equations of motion are

$$m_\ell \ddot{u}_{\ell\alpha}(t) = - C_{\ell\alpha,\ell'\beta} u_{\ell',\beta}(t), \qquad (A1.1)$$

where α, β indicate cartesian components of vector quantities and summation over repeated indices is implied.
The normal mode transformation is given by Eq. (2.6)
where the eigenvectors $\underline{\xi}_\ell(\lambda)$ diagonalize the (3Nx3N) dynamical matrix $D_{\ell\alpha,\ell'\beta} = C_{\ell\alpha,\ell'\beta}(m_\ell m_{\ell'})^{-1/2}$, i.e.

$$\xi^*_{\ell\alpha}(\lambda) D_{\ell\alpha,\ell'\beta} \xi_{\ell'\beta}(\lambda') = \omega_\lambda^2 \delta_{\lambda,\lambda'} . \qquad (A1.2)$$

$\underline{\xi}_\ell(\lambda)$ satisfy the orthogonality and completeness relations

$$\xi_{\ell\alpha}(\lambda) \xi^*_{\ell\alpha}(\lambda') = \delta_{\lambda,\lambda'} \qquad (A1.3)$$

$$\xi_{L\alpha}(\lambda) \xi^*_{\ell'\beta}(\lambda) = \delta_{\ell,\ell'} \delta_{\alpha,\beta}$$

The Debye-Waller factor for the ℓ'th atom is $e^{-2W_\ell(\underline{Q})}$
where[2]

$$2W_\ell(Q) = \langle[\underline{Q}\cdot\underline{u}_\ell]^2\rangle$$

$$= \sum_{\lambda\lambda'} (\underline{Q}\cdot\underline{\xi}_\ell(\lambda))(\underline{Q}\cdot\underline{\xi}_\ell^*(\lambda')) \langle q_\lambda q_{\lambda'}\rangle \qquad (A1.4)$$

$$= \frac{\hbar Q^2}{2m} \sum_\lambda (\frac{2n_\lambda+1}{\omega_\lambda}) |\hat{\underline{Q}}\cdot\underline{\xi}_\ell(\lambda)|^2$$

$$= \frac{\hbar Q^2}{2m} \cdot 3N \int_0^{\omega_m} d\omega \, (\frac{n(\omega)+1}{\omega})\langle|\hat{\underline{Q}}\cdot\underline{\xi}_\ell(\omega)|^2\rangle_{AV} g(\omega),$$

where we have used the result $\langle q_\lambda q_{\lambda'}\rangle = (\hbar/\omega_\lambda)(n_\lambda+\frac{1}{2})\delta_{\lambda,\lambda'}$ and $\langle...\rangle_{AV}$ indicates an average over modes with eigenfrequencies lying between ω and $\omega + d\omega$. If we assume that the local symmetry about any atom can be made equivalent to that of any other by spatial rotations, the $\langle...\rangle_{AV}$ operation can be replaced with an average over ℓ, in which case

$$2W_\ell(\underline{Q}) = N \frac{\hbar Q^2}{2m} \int_0^{\omega_m} d\omega \, (\frac{n(\omega)+1}{\omega})g(\omega), \qquad (A1.5)$$

and W_ℓ becomes independent of ℓ. We have made this assumption throughout this work, but it must be regarded with some skepticism.

APPENDIX II

$Y(Q,\omega)$ in the Small Q Limit

In rewriting Eq. (3.13) in the form

$$g_{eff}(Q,\omega) = Y(Q,\omega)g(\omega) \qquad (3.15)$$

$Y(Q,\omega)$ takes the form

$$Y(Q,\omega) = \frac{1}{2}[Y_+(Q,\omega) + Y_-(Q,\omega)], \qquad (A2.1)$$

where

$$Y_{\pm}(Q,\omega) = \frac{\sinh(\frac{1}{2}Q^2\sigma^2)}{\cosh(\frac{1}{2}Q^2\sigma^2) - \cos(\Delta\phi_{\pm}(\omega))} \qquad (A2.2)$$

and

$$\Delta\phi_{\pm}(\omega) = [2\sin^{-1}(\frac{\omega}{\omega_m}) \pm Qa]. \qquad (A2.3)$$

The two terms are needed to properly account for the degeneracy of modes with eigenphases $\pm\phi_\lambda$. In the small Q limit it is permissible to expand (A2.2) for small values of the arguments $\varepsilon \equiv \frac{1}{2}Q^2\sigma^2$ as well as $\Delta\phi_{\pm}(\omega)$.

$$Y_{\pm}(Q,\omega) \simeq \frac{\varepsilon}{\frac{1}{2}\varepsilon^2 + \frac{1}{2}(\delta\phi_{\pm})^2} = \frac{2\varepsilon}{\varepsilon^2 + |\frac{d\Delta\phi_{\pm}}{d\omega}|(\Delta\omega)^2}$$

$$(A2.4)$$

where $\Delta\omega = (\omega - \omega_M \sin\frac{1}{2}Qa)$.

Using the relation $Ng(\omega)d\omega = g(\phi)d\phi$, where $g(\phi)$ is the density of eigenphases, we find that $|d\phi/d\omega| = 2\pi g(\omega)$ which upon substitution into (A2.4) gives

$$\gamma_+(\underline{Q},\omega) = \gamma_-(\underline{Q},\omega) = \frac{1}{g(\omega)} \cdot \frac{1}{\pi}\{\frac{\Gamma(Q,\omega)}{(\Delta\omega)^2 + \Gamma(Q,\omega)^2}\},$$

$$(A2.5)$$

with $\Gamma(Q,\omega)$ given by Eq. (3.17).

REFERENCES

1. R. Alben, J. E. Smith, Jr., M. H. Brodsky, and D. Weaire, Phys. Rev. Letters 30, 1141 (1973).
2. See, for example, W. Marshall and S. W. Lovesey, "Theory of Thermal Neutron Scattering," (Oxford, 1971), Chapter 4.
3. I. S. Gradshteyn and I. M. Ryzhik, "Tables of Integrals, Series, and Products," (Academic, New York, 1965), p. 42.
4. A. J. Leadbetter, A. C. Wright and A. J. Apling, in "Amorphous Materials," R. W. Douglas and B. Ellis, ed. (Wiley, London, 1972) p. 423.
5. S. C. Moss, D. L. Price, J. M. Carpenter, D. Pan and D. Turnbull, Bull. Am. Phys. Soc. 19, 321 (1974).
6. J. D. Axe, D. T. Keating, G. S. Cargill III, and R. Alben, Proceedings of International Topical Conference on Tetrahedrally Bonded Amorphous Semiconductors, to be published.
7. A. J. Leadbetter and M. W. Stringfellow, "Neutron Inelastic Scattering (IAEA, Vienna, 1972) p. 501.
8. L. Passell, C. C. Tsuei, and J. D. Axe, to be published.
9. K. Kim and M. Nelkin, Phys. Rev. B7, 2762 (1973)

INFRARED AND RAMAN SPECTROSCOPY OF AMORPHOUS SEMICONDUCTORS

J. Tauc

Division of Engineering and Department of Physics

Brown University, Providence, R. I. 02912

1. INTRODUCTION

Infra-red and Raman spectra are an important tool for studies of vibrations, chemical bonding and atomic arrangement. In crystals, the long-range order allows one-phonon absorption or scattering processes only for k-vectors close to $\vec{k} = 0$. Disorder makes it possible to couple to k-vectors far from the origin, and the spectra probe in an indirect way (through the density distribution function) the phonon dispersion relations. We will discuss this topic using amorphous Ge and As_2S_3 and related materials as examples. In addition, we will consider the problem of "wrong" bonds in amorphous compounds, and the relationships of the molecular and crystalline models for the vibrations of As_2S_3 and As_2Se_3.

2. FUNDAMENTAL CONSIDERATIONS

Infra-red and Raman spectroscopy are based on the interaction of electromagnetic radiation with vibrations, and both give some information about the phonon spectra.

Infrared spectroscopy is associated with an absorption process. The absorption depends on the formation of an electric dipole by the vibrations. When the center of the positive charge is displaced by vector $\vec{u}$ from the center of the negative charge of a molecule an electric dipole of moment $\vec{M} = e^*\vec{u}$ is formed (e^* is an effective charge). The probability that electromagnetic radiation is absorbed is proportional to $|\vec{M}.\vec{E}|^2$. In a crystalline solid, the resultant $|\vec{M}.\vec{E}|$ can be non-zero only if the k-vector of the electromagnetic radiation $\vec{k}_r$ is equal to the k-vector of the vibrations $\vec{k}$:

$$\vec{k}_r = \vec{k} \tag{1}$$

As $|\vec{k}_r| = 2\pi/\lambda_{rad}$ is very small compared to the Brillouin zone boundary k-vector $|\vec{k}| = \pi/a$ (a = lattice constant), the radiation practically couples to $\vec{k} = 0$. From energy conservation it follows that the frequency of the absorbed radiation is equal to the frequency of vibration. It is easy to see why it is so because for $\vec{k} \neq 0$ the average dipole moment $\vec{M}$ over a distance $\approx \lambda_{rad}$ averages to zero. In addition to this rule which follows from the long-range order, the lattice vibrations must produce a non-zero dipole moment in each unit cell. In monoatomic crystals which have 1 or 2 atoms in unit cell it can be shown by symmetry that this dipole moment is zero, and therefore there is no infrared one-phonon absorption in these materials (e.g. in crystalline Ge and Si).

Raman spectra are associated with a scattering process and are produced by the modulation of the electrical polarizability by vibrations. As for all scattering processes we have conservation of the k-vector and energy:

$$\vec{k}_s = \vec{k}_i \pm \vec{k} \tag{2}$$

$$\omega_s = \omega_i \pm \omega \tag{3}$$

where the + sign refers to anti-Stokes, - sign to Stokes lines; $\vec{k}_s$, ω_s refers to scattered, $\vec{k}_i, \omega_i$ to incident radiation. From eq. (2) it follows again that in a crystal we couple practically to phonons with $\vec{k} = 0$.

The polarizability of a medium is described by the polarizability tensor α_{mn}

$$P_m = \Sigma \; \alpha_{mn} E_n \tag{4}$$

where P_m, E_n are components of the polarizability vector $\vec{P}$ and electric field $\vec{E}$ (m,n = x, y, z). Raman scattering produced by a normal vibrational mode Q_j is described by a tensor R_{mn}^j with components $\partial\alpha_{mn}/\partial Q_j$. The terms R_{mn}^j relate the scattered radiation polarized parallel to n to the intensity of the incident radiation polarized parallel to m. This ratio of intensities is proportional to $|R_{mn}^j|^2$. The Raman tensor R_{mn}^j defined in this way is symmetrical.

The basic differences between the infra-red and Raman spectra are associated with the difference of the symmetry which is described by a vector (dipole moment) for the absorption process and by a symmetrical tensor (Raman tensor) for the scattering. The selection rules are different. The difference is most pronounced if the crystal (or the molecule) has a center of symmetry. Then the vibrations can be classified as either even or odd. The dipole transitions are forbidden for even symmetry, the scattering for odd symmetry. Therefore the infra-red spectra show the odd vibrations, the Raman spectra the even vibrations, and both spectra are complementary. If there is no center of symmetry, the selection rules are more complicated. But still the most intensive lines of the infra-red spectrum are often weak in the Raman spectrum, and vice-versa.

The other consequence of the different symmetry of both effects is the different averaging in disordered systems. Infra-red absorption in a non-cubic crystal depends on light polarization relative to the crystal axis. But if we grind an anisotropic crystal and measure the absorption in the powdered sample (or do this on a microscopic scale, e.g. molecules in a gas), the absorption will not depend on light polarization.

The Raman spectra are sensitive to the local assymmetry, and by scattering polarized light on a disordered sample (e.g. powder) we can obtain some information on the local anisotropy in the scattering material.

Let us assume that the incident beam is polarized vertically (along x-direction). In the path of the scattered beam we put a polarizer and measure the intensity for polarization parallel to $x(I_{\parallel}$ = VV configuration) and perpendicular to $x(I_{\perp}$ = VH configuration). The depolarization ratio is defined

$$\rho = I_{\perp}/I_{\parallel} \qquad\qquad (5)$$

It is determined[1] by the components of the polarizability tensor α_{mn}:

$$\rho = \frac{3\beta^2}{45\alpha^2 + 4\beta^2} \qquad (6)$$

where the isotropic part α and the anisotropic β are rotational invariants of the polarizability tensor of the molecule (or of the small crystals in the powder):

$$\alpha = \frac{1}{3}(\alpha_{xx} + \alpha_{yy} + \alpha_{zz}) \qquad (7)$$

$$\beta^2 = \frac{1}{2}[(\alpha_{xx} - \alpha_{yy})^2 + (\alpha_{yy} - \alpha_{zz})^2 + (\alpha_{zz} - \alpha_{xx})^2$$
$$+ 6(\alpha_{xy}^2 + \alpha_{yz}^2 + \alpha_{zx}^2)] \qquad (8)$$

For molecules[1], α is different from zero only if the vibration is totally symmetric (it transforms with the full molecular symmetry). For all other modes $\alpha = 0$, and $\rho = 3/4$. These modes are called depolarized. The symmetric modes have $\rho < 3/4$ and are polarized (the scattered radiation tends to keep the same polarization as the incident wave).

3. AMORPHOUS Si AND Ge

Crystalline Ge and Si have no infra-red one-phonon absorption band because the transitions are forbidden by symmetry. The one-phonon Raman spectrum is allowed and consists of a line at the optical phonon frequency at $\vec{k} = 0$.

In the amorphous solid, the crystalline selection rules are relaxed. A simple way of looking at this relaxation of the selection rules is to say that the k-vector is not conserved during the transitions, since it is no longer a good quantum number in an amorphous solid. If the short-range order is the same in the amorphous and crystalline case, one can try to use the phonon dispersion curves of the crystal, and assume that phonons with the k-vector in the whole first BZ contribute. If we further assume that the $\vec{k}$-dependent weighting factors and the matrix elements are approximately constant then the spectrum should be in the first approximation similar to the density of phonon states $g(\omega)$ which may be a broadened version of $g_c(\omega)$ in the crystal. In addition, there should be a similarity between the infra-red and Raman spectra because the disorder tends to make the selection rules ineffective.

This crude reasoning explains the main qualitative features of the infra-red and Raman spectra of amorphous Si and Ge[2-5,15] (Fig. 1). For a meaningful comparison of the Raman spectra with the infra-red spectra and with $g(\omega)$ one must reduce the Raman spectra by a factor which is related to the population of the vibrational states by phonons. In the case of Stokes lines

$$I_{reduced} = I_{measured} \left[\omega(n+1)^{-1} \omega_s^{-4}\right],$$

where $n = \left[\exp h\omega/k_B T - 1\right]^{-1}$. We see that the infra-red and Raman spectra have a similar structure. In detail, they differ considerably. The factors by which one must multiply $g(\omega)$ to obtain the infra-red and Raman spectra depend on ω, and are different for both cases. A more sophisticated theory is obviously needed for the explanation of the actual shape of the spectra and is discussed in the papers by Mitra[31] and Thorpe[32].

Nevertheless, the simple explanation is very useful since the maxima in the spectra are usually close to the maxima of the density of states curves of the crystal. Therefore the spectra of amorphous semiconductors give some information about the frequencies of phonons in the crystal at $\vec{k} \neq 0$, in particular those frequencies for which the densities are high. This often occurs at the BZ boundaries.

Brodsky and Lurio[5] note that the integrated absorption strength of the a-Ge and a-Si IR spectra are only a little weaker than that of related polar materials. They also comment that the Raman spectra of a-Ge and a-Si are more weighted towards the frequency which is close to the frequency of the Raman active modes in the crystal. The infra-red spectra in the amorphous states are weighted more equally through the spectrum because they are directly due to disorder.

Prettl et.al.[7] have suggested that the square of the matrix element in amorphous semiconductors is proportional

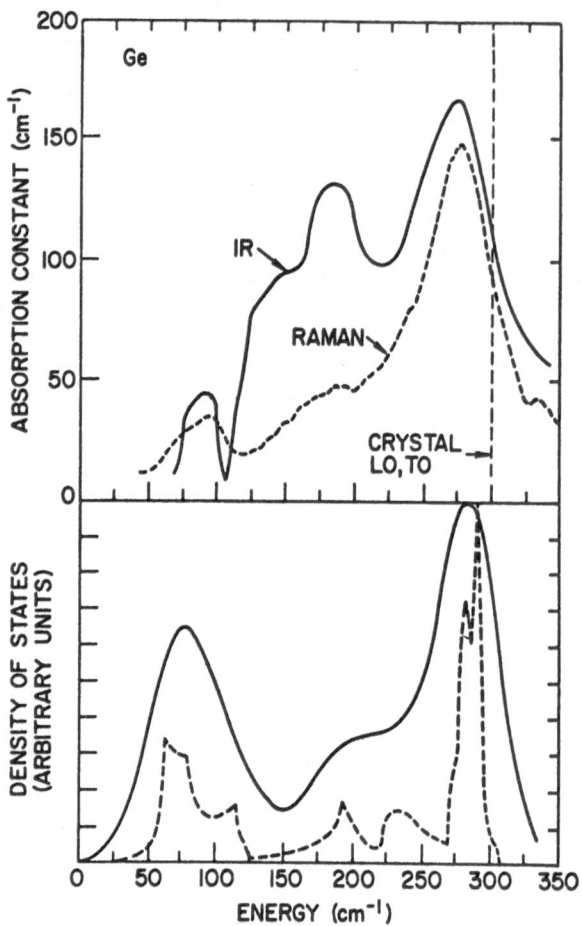

Fig. 1 (a) Infrared absorption constant (solid line) and reduced Raman spectrum (dashed line) of a-Ge. (b) Density of states (dashed line) of c-Ge from neutron scattering data. The solid line is broadened density of states (after Ref. 5).

to ω^2 for the infra-red absorption but to ω^4 for the Raman scattering. Their argument holds for very low frequency modes for which the material can be treated as a continuum. The model appears to reproduce correctly[13] the IR[15], Raman and neutron scattering[12] data up to 100cm^{-1}. In particular, it is in agreement with the observation that the lowest peak of $g(\omega)$ in sputtered a-Ge occurs at

77 cm^{-1} which is a lower frequency than that of the TA peak observed at 88 cm^{-1} in $g_c(\omega)$. The density of states $g(\omega)$ in a-Ge was found proportional to ω^2 (Debye model) up to 35 cm^{-1}. $g(\omega)$ determined from these data is in good agreement with the specific heat data of a-Ge[14].

The depolarization ratio in the Raman spectra of a-Si was reported[3] to be constant, and equal within the experimental error to 0.75. Solin[6] suggested that a constant depolarization ratio is an indication of a "non-molecular" amorphous solid; in a molecular amorphous solid (e.g. As$_2$S$_3$) ρ depends on ω (see section 4).

4. AMORPHOUS TETRAHEDRALLY BONDED ALLOYS AND COMPOUNDS

The Raman spectra of the solid solutions Ge$_x$Si$_{1-x}$ are consistent with the random mixing of Ge and Si atoms in a continuous random network structure[8,16]. In these spectra one finds features which one can associate with Ge-Ge, Si-Si and Si-Ge bonds: the latter occur approximately at the same frequency as in Ge-Si crystalline alloys. The same structural model applies to Ge$_{1-x}$Sn$_x$[9]; however, no Sn-Sn modes were observed, and it appears that Sn atoms shift and broaden the a-Ge spectrum.

Crystalline III-V compounds have an infra-red absorption band corresponding to the TO phonon at $\vec{k} = 0$, and two Raman lines corresponding to TO and LO modes at $\vec{k} = 0$. However, in the amorphous form the spectra are similar to the spectra of a-Ge and a-Si, in agreement with the similarity of the phonon densities of states[4,7,9,10,15].

Fig. 2 shows an example of a good agreement between the maxima in the reduced Raman spectra of a-InP and the crystalline phonon energies. In this figure we see another peak denoted P which it is impossible to ascribe to InP, and which was ascribed[4-10] to P-P bonds in this material. Wrong bonds in III-V compounds were first observed by Raman spectroscopy by Wihl et.al.[4], and have been recently discussed by Lannin[10]. There are several difficulties associated with the studies of wrong bonds by Raman scattering. The bands of wrong bonds are sometimes overlapped by the spectrum of the compound (e.g. GaAs). When this does not happen it is still difficult to estimate the concentration of wrong bonds because it is difficult to determine the ratio of the scattering cross-sections. For example, Sb is a much more efficient Raman scatterer than GaSb. Therefore the presence of a

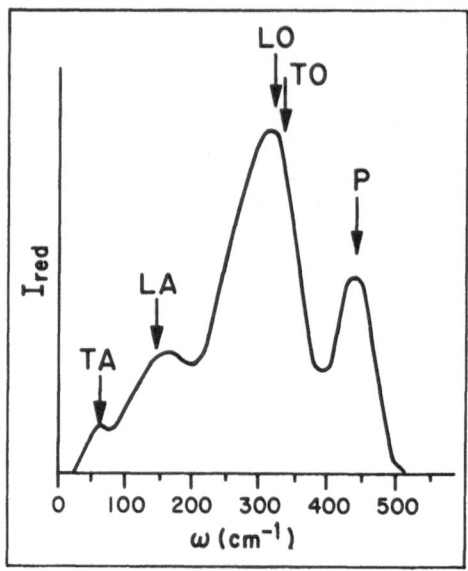

Fig. 2   Reduced Raman intensity I_{red} of a-InP. The arrows indicate energies[27] in c-InP (after Ref. 10).

few percent of Sb-Sb bonds produces bands of comparable
strength to the main bands. Te is an extremely strong
scatterer and Raman spectra of tellurium compounds can
be dominated by Te-lines (crystalline or amorphous, depending on preparation conditions).

     Lannin[10] concludes that the data on III-V compounds
points to a few percent of wrong bonds. The question is,
of course, whether this is due to microscopic phase separation or a random mixing of bonds on atomic scale
(this must happen in the Polk model with odd numbered
rings). It seems that Raman scattering is not a method
sensitive to the presence of odd-numbered rings, as photoemission appears to be.

     Gorman and Solin[11] studied Raman spectra of amor-
phous carbon-deficient sputtered SiC films. They found
a strong band at 1430 cm^{-1} which they attributed to the
presence of C-C bonds, and another band at about 525 cm^{-1}
attributed to Si-Si bonds. The authors argue that the
C-C bonds must be partly due to graphite-like bonding,

and that therefore the material is not completely tetrahedrally coordinated (i.e. the structure is not described
by the Polk model).

5. CHALCOGENIDE GLASSES As_2S_3 and As_2Se_3

Chalcogenide glasses (compounds containing S, Se
and Te) are an important group of amorphous semiconductors. They differ from the tetrahedrally bonded semiconductors by being more "molecular" (in particular the
S and Se compounds). The amorphous Ge_xTe_{1-x}[28] and amorphous Ge_xS_{1-x}[29] systems are examples of these as discussed by Fisher[30]. We shall discuss here the simple compounds As_2S_3 and As_2Se_3 which can be prepared in the
bulk form by quenching the melt.

The vibrations of the crystalline form of As_2S_3 (orpiment) and As_2Se_3 were studied by Zallen et al.[17,18].
The crystals have a layer structure. The layers are
built of strongly bonded $As-X_3$ pyramids (Fig. 3) and the
interaction between the layers is weak. Therefore, the
vibrations can be described as vibrations in the layer
(slightly changed by the weak inter-layer coupling), and
vibrations between the layers. The amorphous form is
built practically of the same $As-X_3$ pyramids, but there
is a controversy about the ordering of these units. One
point of view stresses the existence of the remnants of
the layer structure in the amorphous form; there are several experimental facts supporting this model. Perhaps
the most important is the preservation of two inequivalent sites of As (Fig. 3) in the glass as found in spin-
spin relaxation measurements[19]. This indicates the existence of the inter-layer coupling in the glass. However,
the periodic atomic arrangement in the layer appears to
be much less perfect than in the crystal[20].

From the opposite point of view, the glass is considered to be a 3-dimensional interconnected network composed of pyramidal $As-X_3$ units[21,22].

The infra-red spectra of the crystal and the glass
are shown in Fig. 4 Vibrations with E||a and E||c are in
the layer, vibrations with E||b are perpendicular to the
layer. We see that the infra-red band in the glass is
similarly situated as the strongest bands in the layer.
The Raman spectrum of crystalline As_2S_3 is shown in Fig.
5. The Raman spectrum of the glass with two polarization
directions is shown in Fig. 6 where also the depolarization ratio ρ is plotted. The dips in ρ at 162, 218 and

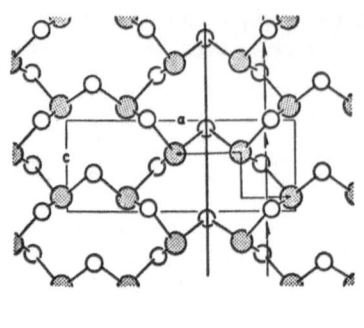

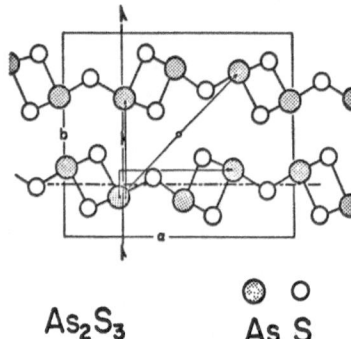

Fig. 3 Crystal structure of As₂S₃ and As₂Se₃ (after
Ref. 17). a,b,c are crystal axis.

344 cm⁻¹ show the predominantly symmetrical modes. The
band dominating the $I_{\parallel}$ spectrum has a maximum at 344 cm⁻¹.
In the $I_{\perp}$ spectrum the maximum is at 310 cm⁻¹; it is at
this frequency where the infra-red band has its maximum.

Lucovsky and coworkers[21,22,26] have suggested that
this property of the main band, and other features of the
spectra of the glass can be explained by considering the
vibrations of As-X₃ (a pyramidal molecule) and As-X-As
(a water-like molecule). They estimated the frequencies
of vibrations by scaling the corresponding frequencies
observed in compounds with the same structural elements,
and obtained good agreement with the observed data. For
example, the band at 344 cm⁻¹ is a symmetrical mode of
the pyramid (a strong Raman mode, polarized), the band at
310 cm⁻¹ an antisymmetrical mode (a strong infra-red band,
depolarized).

Let us now discuss the comparison with the crystalline

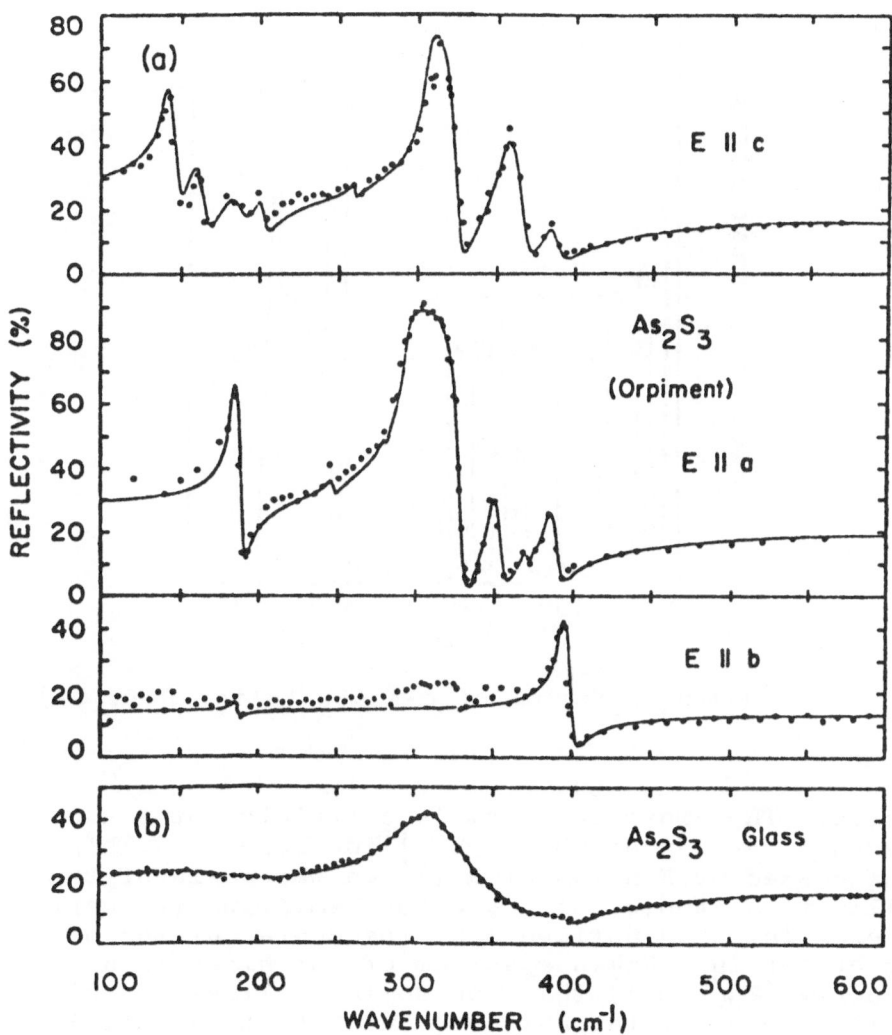

Fig. 4 Infrared spectra of c-As₂S₃ for 3 different polarizations, and of a-As₂S₃ (after Ref. 19).

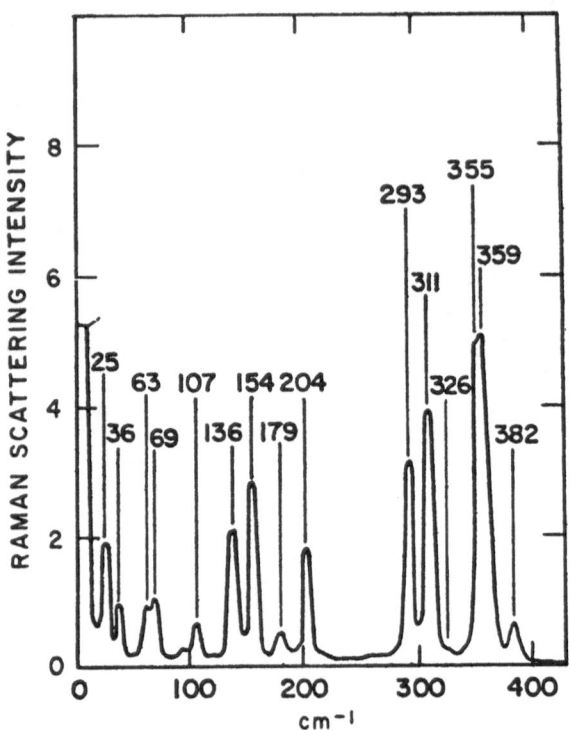

Fig. 5   Raman spectrum of c-As$_2$S$_3$ (after Ref. 17).

spectra. The amorphous form is certainly not a simple
broadened version of the crystalline spectra. This point
is discussed by Finkman et al.[23] on the basis of the crystallization studies of As$_2$Se$_3$ by Raman spectroscopy.
For example, they noticed that the center of the dominant
band of the I$_{\parallel}$ Raman spectrum of the glass is at a higher energy (232 cm^{-1}) than the dominant lines of the crystalline spectrum (at 202 cm^{-1} and 215 cm^{-1}) (cf. Figure
8 in Ref. 24). These two lines appear to contribute little to the spectrum of amorphous As$_2$Se$_3$ which appears to
be a broadened version of the crystal lines at 230 cm^{-1}
and 247 cm^{-1} (the peak at 272 cm^{-1} is weak). This is easy
to understand on the basis of the crystalline model if
we assume that the frequencies of the crystal modes at
230 cm^{-1} and 247 cm^{-1} are only slightly k-dependent, and
that of the modes at 202 cm^{-1} and 215 cm^{-1} are strongly
k-dependent as is shown symbolically in Fig. 7. In fact,

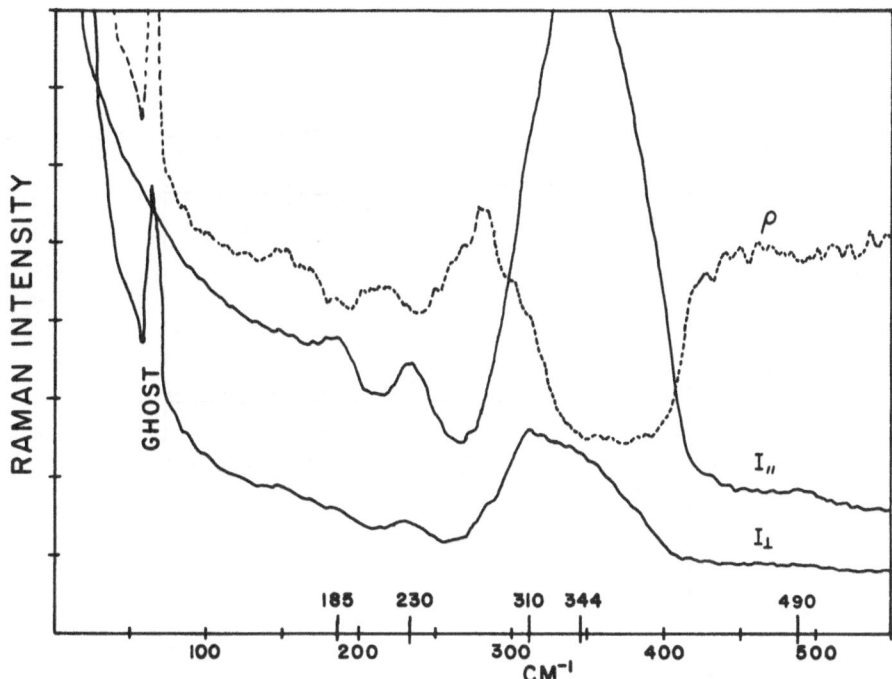

Fig. 6 Raman spectrum[25] of a-As$_2$S$_3$ for two polarization configurations I$_{||}$ and I$_\perp$. ρ is the depolarization ratio I$_\perp$ /I$_{||}$.

the calculations of DeFonzo[25] on crystalline As$_2$S$_3$ and As$_2$Se$_3$ support this model.

Let us note that in the molecular model the modes at 230 cm^{-1} and 247 cm^{-1} correspond to the vibrations ν_3 and ν_1 of the pyramid. We note a connection between the "molecular" and "crystalline" models: The molecular modes are such that the interaction between vibrating units is small. They can therefore be approximately treated as isolated units, as "molecules". In the crystalline model, because of the smallness of the interaction, the frequency of these modes is little $\vec{k}$-dependent, and they contribute most of the strength in the glass spectra. From the calculations starting with the crystalline vibrations one obtains a justification why certain modes can be treated as molecular[25]. It follows from

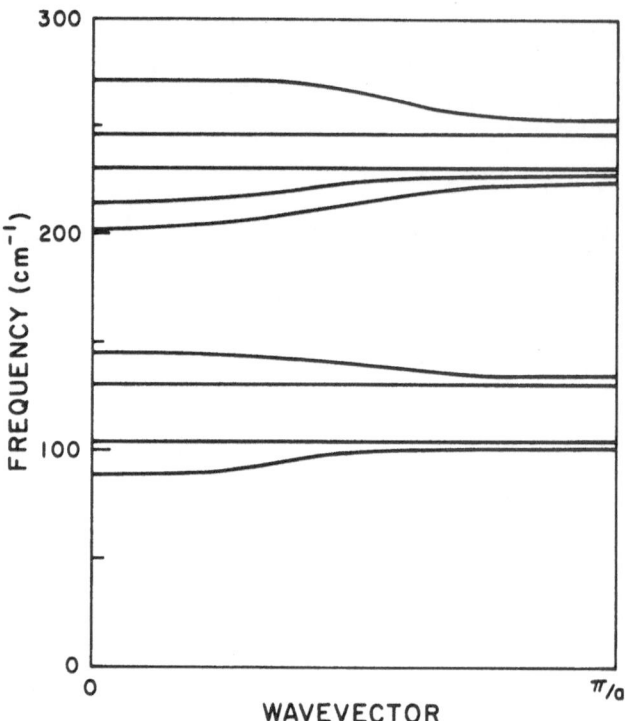

Fig. 7   Schematic dispersion relations[23] for As_2Se_3.

these considerations that the success of the molecular model of Lucovsky and Martin[21,22] for the interpretation of the infra-red and Raman spectra of As_2S_3 and As_2Se_3 cannot be considered as an argument against the model of Taylor et al.[19] based on the remnants of layers in the glass. The main qualitative features of these spectra can be deduced from the crystalline spectra as it was the case for tetrahedrally bonded semiconductors.

ACKNOWLEDGMENT

This paper is partly based on work supported by a grant from the National Science Foundation in which E. Finkman and A. P. DeFonzo participated. I thank also G. B. Fisher for his comments on the manuscript.

REFERENCES

1. E. B. Wilson, J. C. Decius and P. C. Cross, Molecular Vibrations, McGraw-Hill, New York, 1955; A. Anderson (editor), The Raman Effect, Dekker, New York 1971.

2. J. Tauc, A. Abraham, R. Zallen and M. Slade, J. Non-Crystalline Solids $\underline{4}$, 279 (1970).

3. J. E. Smith, Jr., M. H. Brodsky, B. L. Crowder, M. I. Nathan and A. Pinczuk, Phys. Rev. Lett. $\underline{26}$, 642 (1971).

4. M. Wihl, M. Cardona and J. Tauc, J. Non-Crystalline Solids $\underline{8\text{-}10}$, 172 (1972).

5. M. H. Brodsky and A. Lurio, Phys. Rev. B$\underline{9}$, 1646 (1974).

6. R. J. Kobliska and S. A. Solin, Phys. Rev. B$\underline{8}$, 756 (1973).

7. W. Prettl, N. J. Shevchik and M. Cardona, Phys. Status Solidi B$\underline{59}$, 241 (1973).

8. N. J. Shevchik, J. S. Lannin and J. Tejeda, Phys. Rev. B$\underline{7}$, 3987 (1973).

9. J. S. Lannin, Sol. State Communications $\underline{11}$, 1523 (1972).

10. J. S. Lannin, in Tetrahedrally Bonded Amorphous Semiconductors, (M. H. Brodsky and S. Kirkpatrick, editors), AIP 1974, p. 260.

11. M. Gorman and S. A. Solin, Solid State Communications, $\underline{15}$, 761 (1974).

12. J. D. Axe, D. T. Keating, G. S. Cargill and R. Alben in Tetrahedrally Bonded Amorphous Semiconductors, (M.H. Brodsky and S. Kirkpatrick, editors) AIP 1974, p. 279.

13. G. A. Connell, Proc. 12th Int. Conference on Physics of Semiconductors, Teubner, 1974, p. 1003.

14. C. N. King, W. A. Phillips and J. P. de Neufville, Phys. Rev. Lett. $\underline{32}$, 538 (1974).

15. R. W. Stimets, J. Waldman, J. Lin, T. S. Chang, R. J. Temkin and G. A. N. Connell, Amorphous and Liquid Semiconductors (ed. J. Stuke and W. Brenig), Taylor and Francis, London 1974, p. 1239.

16. J. S. Lannin, Amorphous and Liquid Semiconductors (ed. J. Stuke and W. Brenig), Taylor and Francis, London 1974, p. 1245.

17. R. Zallen, M. L. Slade and A. T. Ward, Phys. Rev. B$\underline{3}$, 4257 (1971).

18. R. Zallen and M. L. Slade, Phys.Rev.B$\underline{9}$, 1627 (1974).

19. The experimental evidence supporting the remants of the layer structure was collected and analyzed in the paper by P. C. Taylor, S. G. Bishop and D. L. Mitchell, Amorphous and Liquid Semiconductors (ed.

J. Stuke and W. Brenig), Taylor and Francis, London 1974, p. 1267.

20. S. G. Bishop and N. J. Shevchik, Sol. State Communications, $\underline{15}$, 629 (1974).

21. G. Lucovsky, R. M. Martin, J. Non-Crystalline Solids $\underline{8-10}$, 185 (1972).

22. G. Lucovsky, Phys. Rev. B$\underline{6}$, 1480 (1972).

23. E. Finkman, A. P. DeFonzo and J. Tauc, in Proc. of 12th Intl. Conference on Physics of Semiconductors, Teubner, 1974, p. 1022.

24. J. Tauc, lecture on Light Scattering in Liquid Semiconductors, these Proceedings.

25. A. P. DeFonzo, Thesis (Brown University, 1975).

26. G. Lucovsky, Amorphous and Liquid Semiconductors (ed. J. Stuke and W. Brenig), Taylor and Francis, London 1974, p. 1099

27. S. S. Mitra, Phys. Rev. $\underline{132}$, 936 (1963).

28. G. B. Fisher, J. Tauc and Y. Verhelle, Amorphous and Liquid Semiconductors (ed. J. Stuke and W. Brenig), Taylor and Francis, London 1974, p. 1259.

29. G. Lucovsky, F. L. Galeener, R. C. Keezer and R. H. Geils, Bull. APS 19, 315 (1974).

30. G. B. Fisher, these Proceedings.

31. S. S. Mitra, these Proceedings.

32. M. F. Thorpe, these Proceedings.

DYNAMICS OF STRUCTURALLY DISORDERED SOLIDS

J. F. Vetelino
Department of Electrical Engineering
University of Maine, Orono, Maine 04473

and

S. S. Mitra[*]
Department of Electrical Engineering
University of Rhode Island, Kingston, R.I. 02881

I. INTRODUCTION

In this chapter the dynamics of disordered lattices is discussed. The extent of disorder is assumed small such that the properties of the disordered system can, in principle, be obtainable as an appropriate extention of the corresponding properties of the ideal ordered system. Such an approach, albeit limiting, affords a convenient starting point for disordered systems, since the information regarding the corresponding ordered systems is assumed known.

The presence of even a small concentration of defects may significantly affect properties of a solid through the destruction of the translational symmetry of the unperturbed perfect lattice. Such changes are often sufficient to allow interaction with light in cases where it would not otherwise occur. The defects may be of a chemical nature such as an impurity atom, or they may be of a mechanical nature such as vacancies at particular lattice sites, and extended defects such as dislocations. If the impurity content of a solid is increased to such an extent that interactions between impurity atoms become significant the system is then termed a disordered solid rather than an impure crystal. Disordered lattices are of two main types: the disordered alloy (isotopic mixture

or mixed crystals) and the glass-like substance in which
the disorder is spatial rather than configurational.

It is the purpose of these lectures to discuss the
atomic vibrations of the following three types of disordered solids: (i) Crystals with substitutional impurities, (ii) mixed crystals or alloy systems, and (iii)
amorphous materials. In particular, the lattice dynamical theory will be presented and the theoretical results
will be compared with experimental data whenever available.

II. CRYSTALS WITH SUBSTITUTIONAL IMPURITIES

It is well known that the introduction of impurity
atoms modifies the potential energy function in a lattice. Hence, the phonon spectrum of the crystal is modified and consequently any physical property of the
crystal in which the lattice vibrations play a central
role. In the case of a substitutional impurity whose
mass is different from that of the atom it replaces, the
modification of the phonon spectrum is usually manifested
in the appearance of localized frequencies. These frequencies may appear within the allowable frequency range
of the perfect crystal in which case they are known as
resonant modes or outside the frequency spectrum in which
case they are known as gap or local modes.

The first theoretical discovery of localized vibrational modes in crystals was made by Lifshitz[1] in 1943.
This work was followed by the work of Montroll and Potts[2]
and Mazur[3]. However, this early work was restricted to
one and two dimensional lattices and therefore was only
of qualitative value.

The first experimental discovery of the existence
of localized vibrational modes was made by Schaefer[4] in
1960. He observed a strong peak in the infrared lattice
vibrational absorption spectra of alkali halide crystals
containing U-centers which are H^- ions substituted for
the halogen ions. It was after this experimental discovery that much of the theoretical work on the lattice
dynamical behavior of point defects in three dimensional
cystals was undertaken. This initial experimental discovery was also followed by the appearance of a large in-
flux of experimental data[5] in the literature.

In addition to infrared techniques to study localized modes, various other experimental techniques were

developed and utilized. Among the most prominent of these techniques are: Specific heat measurements, tunneling between superconductors, the Mössbauer effect, neutron scattering and Raman spectroscopy. However, the method which has yielded the most detailed and accurate information about the frequency and amplitude of the impurity modes is the infrared technique, and to some extent Raman scattering.

The theoretical understanding of the impurity mode absorption is based on the lattice dynamical behavior of crystals with defects. In 1962, Dawber and Elliott[6] presented the formalism of a Green's function method, introduced by Lifshitz[7], to calculate the frequency and amplitude of vibration of a substitutional impurity in a cubic crystal. This method requires a complete knowledge of the frequency and amplitude of vibration of all the phonon states in the perfect crystal. This technique was then applied by Dawber and Elliott[8] to calculate the impurity mode frequency in silicon. In 1965, a molecular model was proposed by Jaswal[9] to describe the dynamical behavior of the impurity atom. This model was localized in that it considered the motion of the impurity ion and its nearest neighbors while the remainder of the lattice was assumed to be at rest. The results obtained from this technique were not as good as the results obtained from the Green's function technique. Since 1965, the Green's function technique or modifications thereof, have been used to calculate the frequency and amplitude of the phonon states of substitutional impurities in various alkali halide[10-13] and zinc-blende crystals[14-16].

A. Linear Diatomic Chain with a Substitutional Impurity

The perfect linear diatomic chain modified by the replacement of one of the constituent atoms by an impurity particle is shown in Figure 1. The index "0" designates the impurity particle of mass m_I. It is assumed in this analysis that only nearest neighbor forces are operative. The perfect lattice near neighbor force constant is defined as "f" while the nearest neighbor force constant due to the impurity is denoted by f_I. Writing the appropriate equations of motion for this lattice and assuming a standard traveling wave solution, the following system of equations result for the center of the Brillouin zone ($\underline{k} = 0$),

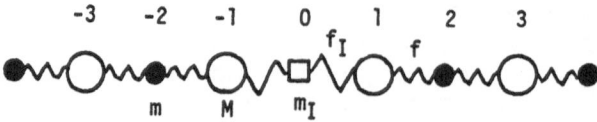

Fig. 1 Linear diatomic lattice containing a substitutional impurity particle with mass defect and force constant defect.

$$
\begin{bmatrix}
(-f) & (2f-m\omega^2) & (-f) & & & & 0 \\
 & (-f) & (f_I+f-M\omega^2) & (-f_I) & & & \\
 & & (-f_I) & (2f_I-m_I\omega^2) & (-f_I) & & \\
 & & & (-f_I) & (f_I+f-M\omega^2) & (-f) & \\
0 & & & & (-f) & (2f-m\omega^2) & (-f)
\end{bmatrix}
\begin{bmatrix}
\sigma'_3 \\
\sigma'_2 \\
\sigma'_1 \\
\sigma'_0 \\
\sigma'_{-1} \\
\sigma'_{-2} \\
\sigma'_{-3}
\end{bmatrix}
= 0
$$

$$(1)$$

The σ'_i's are the appropriate eigen-amplitudes of the imperfect lattice.

It will be shown that the solution to the diatomic chain with an impurity can be expressed in terms of the solution to the perfect diatomic chain. The equations for the perfect diatomic chain can be obtained by setting $m_I = m$ and $f_I = f$ in equation (1). This yields the following matrix equation

$$[[\underline{F}] - \omega^2[I]][\sigma] = 0 \qquad\qquad (2)$$

$[F]$ is the coupling coefficient matrix of the perfect lattice $[\sigma]$, the associated column matrix representing the eigen-amplitudes and $[I]$ the identity matrix. The eigenvalues, ω^2, are determined by the solution to the secular equation,

$$|[F]-\omega^2[I]| = 0, \qquad\qquad (3)$$

In matrix notation the equations of motion for the linear diatomic chain with a substitutional impurity can be written as,

$$[F][\sigma'] - \omega'^2[I][\sigma'] = [S][\sigma'] \tag{4}$$

$[\sigma']$ and ω'^2 are the eigen-vectors and eigen-values respectively of the imperfect diatomic chain, and $[S]$ is a perturbation matrix of low rank incorporating the change of mass and force constants due to the impurity. The rank of the perturbation matrix, $[S]$, is determined by the spatial extension of the defect. For example, if one considers the second neighbor interactions from the impurity particle also, the rank of the perturbation matrix would naturally increase. An examination of eqs. (1) and (2) reveals that the following expression may be written for the perturbation matrix,

$$[S] = \begin{bmatrix} \ddots & & & 0 & & & \ddots \\ & (\Delta f) & (-\Delta f) & & 0 & & \\ 0 & (-\Delta f) & (2\Delta f - \Delta m\omega^2) & (-\Delta f) & & 0 \\ & 0 & & (-\Delta f) & (\Delta f) & & \\ \ddots & & & 0 & & & \ddots \end{bmatrix} \tag{5}$$

where $\Delta f = f - f_I$ and $\Delta m = m - m_I$. Δf and Δm denote the force constant defect and the mass defect parameters respectively. In the isotropic case, where $\Delta f = 0$, or what is commonly called the mass defect approximation, the perturbation matrix $[S]$ has only the element $(-\Delta m\omega^2)$ at the center.

Due to the limited spatial extension of the perturbation matrix $[S]$, a Green's function method[7] can be used to calculate the frequency and the amplitude of vibration of the impurity particle. Since this technique will be used to calculate the impurity modes in the three dimensional crystal, it is convenient to use matrix notation. Furthermore, since the basic equations of motion in the three dimensional lattice are of the same form as the linear diatomic chain, the analysis will be formulated so as to apply to the three dimensional case as well.

The equation of motion for the linear diatomic chain

with a substitutional impurity may be written as follows,

$$[L][\sigma'] = [S][\sigma'], \tag{6}$$

where

$$[L] = [F] - \omega'^2[I].$$

A Green's function matrix $[G]$ is then defined in the following fashion,

$$[L][G] = [I]. \tag{7}$$

With the use of the defining equation for the Green's function matrix, eq. (6) may be written as

$$[\sigma'] = [G][S][\sigma'], \tag{8}$$

where

$$[G] = [L]^{-1}.$$

In terms of the matrix elements, eq. (8) can be written as,

$$\sigma'_x = \sum_{x'} g_{xx'} \sum_{x''} S_{x'x''}\, \sigma'_{x''} \;. \tag{9}$$

The eigen values ω'^2 can be obtained by solving the secular equation,

$$\left| [L]^{-1}[S] - [I] \right| = 0 \tag{10}$$

or

$$\left| [G][S] - [I] \right| = 0.$$

In order to solve eqs. (9) and (10) for the amplitude and frequency of vibration of the impurity particle, one needs an explicit representation of the elements $g_{xx'}$, of the Green's function matrix. Premultiplying eq. (6) by the eigen-vector $[\sigma] = [\sigma(\underline{k},j)]$ of the perfect lattice where j denotes the particular branch and $\underline{k}$, the corresponding wave vector, one obtains

$$<[\sigma],[F][\sigma']> - \omega'^2<[\sigma],[\sigma']> = <[\sigma],[S][\sigma']>.$$

$$(11)$$

$[F][\sigma']$ and $[S][\sigma']$ are column matrices and the product $<,>$ is defined such that for two column matrices $[A]$ and $[B]$,

$$<[A],[B]> = (a_1^* a_2^* \cdots) \begin{bmatrix} b_1 \\ b_2 \\ \vdots \\ \vdots \\ \vdots \end{bmatrix} = \sum_i a_i^* b_i,$$

$$(12)$$

where a_i and b_i are the elements of $[A]$ and $[B]$ respectively and a_i^* is the complex conjugate of a_i.

In the case of the linear diatomic chain, the matrix $[F]$ is symmetric, therefore the first term in eq. (11) can be transformed as,

$$<[\sigma],[F][\sigma']> = <[F][\sigma],[\sigma']>.$$

$$(13)$$

From eqs. (3) and (13) one obtains

$$<[\sigma],[F][\sigma']> = <\omega^2[\sigma],[\sigma']>.$$

$$(14)$$

Substitution of equation (14) into equation (11) yields,

$$<[\sigma][\sigma']> = \frac{<[\sigma],[S][\sigma']>}{\omega^2 - \omega'^2}.$$

$$(15)$$

Since the eigen vectors $[\sigma(\underline{k},j)]$ of the perfect lattice can be represented as a set of orthonormal vectors[17], an arbitrary vector $[\sigma']$ can be expressed uniquely as a linear combination of this orthonormal set, $[\sigma(\underline{k},j)]$, as

$$[\sigma'] = \sum_{\underline{k},j} \alpha(\underline{k},j)[\sigma(\underline{k},j)],$$

$$(16)$$

where $\alpha(k,j)$ are the expansion coefficients.

Premultiplying eq. (16) by the eigen-vectors $[\sigma(\underline{k},j)]$ of the perfect lattice and using the orthonormality condition of $[\sigma(\underline{k},j)]$, the constants $\alpha(\underline{k},j)$ can be expressed as,

$$\alpha(\underline{k},j) = <[\sigma],[\sigma']>. \qquad (17)$$

From eqs. (15) and (17) one obtains

$$\alpha(\underline{k},j) = \frac{<[\sigma[,[s][\sigma']>}{\omega^2-\omega'^2} . \qquad (18)$$

Therefore eq. (16) can be written as

$$[\sigma'] = \sum_{\underline{k},j} \frac{<[\sigma],[s][\sigma']>}{\omega^2-\omega'^2} [\sigma] \qquad (19)$$

or in terms of the matrix elements,

$$\sigma_x' = \sum_{\underline{k},j} \sum_{x'} \frac{\sigma_{x'}^*([s][\sigma'])_{x'}}{\omega^2-\omega'^2} \sigma_x \qquad (20a)$$

$$= \sum_{x'} \left(\sum_{\underline{k},j} \frac{\sigma_{x'}^*\sigma_x}{\omega^2-\omega'^2} \right)([s][\sigma'])x' \qquad (20b)$$

$$= \sum_{x'} \left(\sum_{\underline{k},j} \frac{\sigma_{x'}^*\sigma_x}{\omega^2-\omega'^2} \sum_{x''} S_{x'x''}\sigma_{x'} \right. . \qquad (20c)$$

A comparison of eqs. (20c) and (9) results in the following representation of the Green's function matrix elements

$$g_{xx'} = \sum_{\underline{k},j} \frac{\sigma_{x'}^*(\underline{k},j)\sigma_x(\underline{k},j)}{\omega^2-\omega'^2} . \qquad (21)$$

Having obtained the Green's function matrix $[G]$, eqs. (9) and (10) can now be used to calculate the frequency and the amplitude of vibration of the impurity atom. The essential point of this technique is the limited extension of the perturbation matrix $[S]$. This results in a solution of equations (9) and (10) in the perturbed region only. For a linear diatomic lattice of Fig. 1, the matrix [S], viz., eq. (5) has the elements,

$$S_{0,0} = 2\Delta f - \Delta m\omega^2$$

$$S_{1,1} = S_{-1,-1} = \Delta f \qquad\qquad (22)$$

$$S_{-1,0} = S_{1,0} = S_{0,-1} = S_{0,1} = -\Delta f$$

while the other elements are zero. Substituting the matrix elements of [S] into eq. (9) one obtains

$$\sigma'_x = \sigma'_{-1} \Delta f(g_{x,-1} - g_{x,0}) + \sigma'_1 \Delta f(g_{x,1} - g_{x,0})$$

$$+ \sigma'_0[(2\Delta f - \Delta m\omega^2)g_{x,0} - \Delta f(g_{x,1} + g_{x,-1})]. \qquad (23)$$

By setting x equal to -1, 0, 1 three equations for σ'_{-1}, σ'_0 and σ'_1 evolve. Since the impurity is a center of inversion for the linear diatomic chain shown in Fig. 1,

$$\sigma_{-1} = \pm \sigma_1. \qquad\qquad (24)$$

Also since the Green's function matrix elements g_{xx}, defined by equation (21) have the same symmetry property, two equations in σ'_0 and $\sigma'_{\pm 1}$, will result. After obtaining these, all other σ'_x's are obtained by recursion. The eigenfrequencies ω'^2 of the perturbed lattice which are the solution of the secular equation $|[G][S] - [I]| = 0$, are now determined by the determinant of the coefficients of two homogeneous equations for σ'_0 and $\sigma'_{\pm 1}$.

The frequency of vibration of the substitutional atom may occur either within or outside the vibrational spectrum of the perfect crystal. If the impurity atom is lighter in mass than the atom it replaces in the crystal, it is possible for certain exceptional vibrational modes called localized modes to appear. These modes are characterized by the fact that their frequencies lie above the maximum frequency of the unperturbed crystal. The displacement amplitudes of the atoms of a crystal with a localized mode decay very quickly with increasing distance from the impurity site.

If the frequency distribution function of the perfect crystal has a gap between the acoustic and optic branches, it is possible for an allowed localized vibrational mode to occur in this gap. This localized vibrational mode is usually referred to as a gap mode. Gap modes occur when the mass of the impurity atom may be lighter or heavier than one or both of the host lattice masses. Since a gap mode has a frequency which lies in the forbidden frequency band of the perfect crystal, the displacement of the atoms vibrating in the neighborhood of the impurity decay rapidly with increasing distance from the impurity site.

Due to the fact that the frequency of a local or a gap mode lies outside the allowed frequency range of the perfect crystal, there is no mixing between the impurity mode and the unperturbed modes of the crystal. Therefore the localized and gap mode may occur at any wave vector point in any direction and there is no dispersion for these modes.

A third kind of vibrational mode, introduced into the crystal by the presence of impurities, is the resonance mode. Unlike localized and gap modes, resonance modes are not true normal modes of a perturbed crystal. These modes are localized in frequency but not in space. Since the frequency of a resonance mode falls in the range of the allowed frequency band of the host crystal, it can decay into the continuum of unperturbed band modes. A crude physical explanation of the nature of these modes can be given as follows. In the low frequency acoustic branches all the atoms are moving in phase with each other independent of their masses. If the impurity atom is very heavy or if it is coupled very weakly to the surrounding host crystal, it will be vibrating in phase with its neighbors at very low frequencies. However, as the frequency of the lattice vibrations increases the impurity atom begins to lag behind its neighbors because of its heavy mass or weak binding to the crystal. This continues until a frequency is reached at which the impurity atom vibrates 180° out of phase with its neighbors. This frequency is defined as the frequency of a resonance mode. As might be expected, the mean square vibrational amplitude of the impurity atom as a function of frequency is sharply peaked at the resonance mode frequency, in comparison to the amplitudes of its neighbors. Resonant modes may occur in the acoustic or the optic band of the frequency distribution function of the perfect crystal. The impurity masses associated with resonant modes may

be heavier or lighter than the host crystal atom masses.
The force constants with which the impurity particle
couples to its nearest neighbors play an important role
in determining the frequency of the resonant mode.

Mazur and co-workers[3] have solved the problem of
impurity modes in a linear diatomic chain with an impurity atom of mass m_I which replaces either the heavy mass,
M, or the light mass, m, of the chain as shown in Fig. 1.
Since in this analysis it is assumed that m<M, there is
a gap present in the frequency distribution function.
The solutions of this impurity problem are shown in Fig.
2 for m_I replacing M and m respectively. For the case
where m_I<M, a local mode above the top of the optic spectrum and a gap mode between the acoustic and optic branches is predicted. For the case where m_I>M neither local

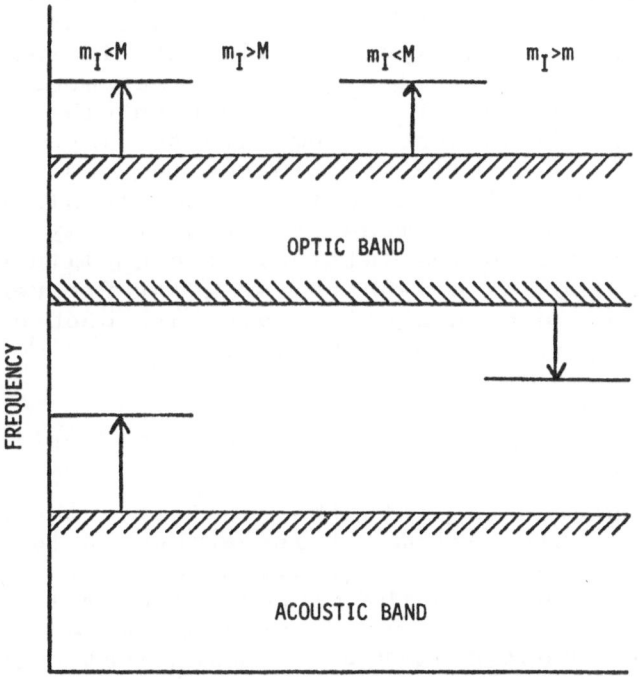

Fig. 2 Impurity modes in a linear diatomic chain (m<M).
Four cases shown are:
(1) a light impurity atom replaces the heavy atom of the
    host lattice (m_I<M),
(2) a heavy impurity atom replaces the heavy atom (m_I>M)
(3) a light impurity atom replaces the light atom (m_I<m)
and (4) a heavy impurity atom replaces the light atom
(m_I>m).

nor gap mode is predicted. When the impurity atom replaces the lighter host atom such that $m_I < m$, a local
mode above the optic spectrum is predicted. Finally,
when $m_I > m$, a gap mode between the optic and acoustic
branches is predicted.

It will be shown that three dimensional crystals
having a gap in the frequency distribution function con-
form with the predictions for the case 1, 3, and 4 for
Figure 2. For the second case where $m_I > M$, a gap mode is
predicted which constitutes a new condition hitherto un-
noticed.

The amplitude of an impurity ion and its neighbors
in a linear diatomic chain have been calculated by Renk[18]
for the local and gap modes. These are shown in Figure
3. For local modes, the negative ions move in one direction while the positive ions move in the opposite direction. The maximum amplitude is associated with the impurity ion and decays exponentially with increasing distance from the impurity ion. For the gap mode similar
ions vibrate 180° out of phase with each other. The
maximum amplitude is associated with the impurity ion
and decays exponentially with increasing distance from
the impurity ion. In contrast with the localized and
gap modes, the resonant mode is not spatially localized
but extends far into the lattice. The amplitude of the
impurity ion and its neighbors for a low frequency resonant mode in a linear diatomic chain have been calculated
by Weber[19] and are shown in Fig. 3.

B. Three Dimensional Lattice with a Substitutional
   Impurity.

Before digressing into the explicit mathematical
formalism for this system, it is worthwhile to spend sometime reviewing the lattice dynamics for perfect crystals
in three dimensions and the concept of normal coordinates.
The total potential energy of the atoms in a crystalline
solid will be denoted as Φ . Assuming that each atom un-
dergoes a small displacement, $\underline{u}\binom{\ell}{K} = [u_x\binom{\ell}{K}, \; u_y\binom{\ell}{K}, \; u_z\binom{\ell}{K}]$

from its equilibrium position, the potential energy function Φ can be expanded in a Taylor series about the equilibrium position in terms of the powers of the displacements:

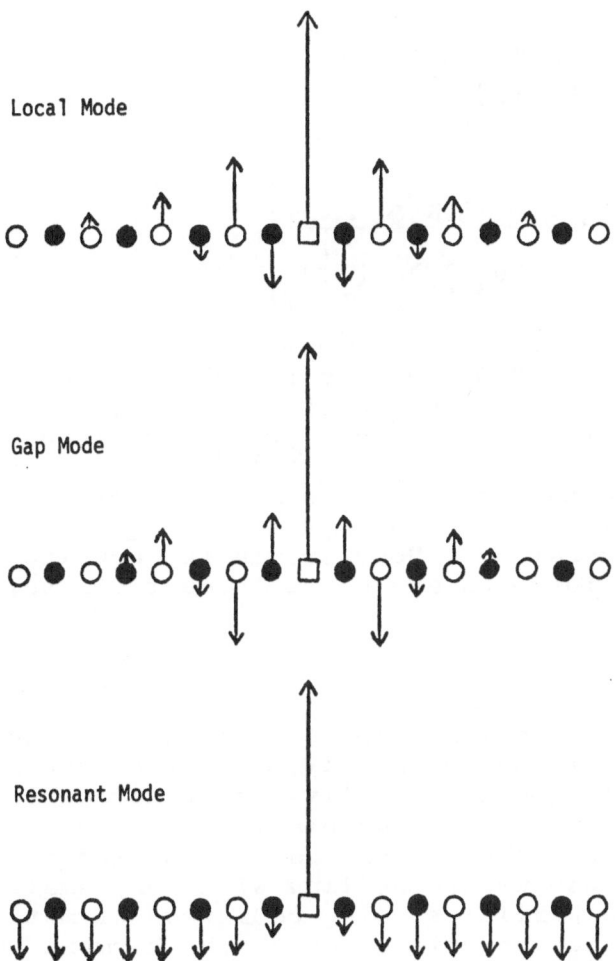

Fig. 3 Eigenvectors for local, gap and resonant modes in a linear diatomic chain.

$$\Phi = \Phi_o + \sum_{\ell, K, \alpha} \Phi_\alpha\binom{\ell}{K}\, u_\alpha\binom{\ell}{K}$$

$$+ 1/2 \sum_{\ell, K, \alpha} \sum_{\ell', K', \beta} \Phi_\alpha\binom{\ell \ell'}{KK'} u_\alpha\binom{\ell}{K} u_\beta\binom{\ell'}{K'} + \cdots , \tag{25}$$

where the subscripts α and β refer to the x, y and z component of the displacement and

$$\Phi_\alpha\binom{\ell}{K} = \frac{\partial \Phi}{\partial u_\alpha\binom{\ell}{K}}\Bigg|_0 \tag{26}$$

and

$$\Phi_{\alpha\beta}\binom{\ell\ell'}{KK'} = \frac{\partial^2 \Phi}{\partial u_\alpha\binom{\ell}{K}\partial u_\beta\binom{\ell'}{K'}}\Bigg|_0. \tag{27}$$

The subscript "0" means that the derivatives are evaluated at the equilibrium configuration of the lattice. Physically, the coefficient, $\Phi_\alpha\binom{\ell}{K}$, defined in eq. (26) can be explained as being the negative of the force acting in the α-direction on the atom at $\underline{r}\binom{\ell}{K}$ in the equilibrium configuration. However, in the equilibrium configuration the force on any particle must vanish, therefore,

$$\Phi_\alpha\binom{\ell}{K} = 0. \tag{28}$$

Under the harmonic approximation the potential energy expansion contains terms only up to the second power in displacement. Mathematically the resulting equation of motion could not be solved exactly if higher ordered terms were retained. The harmonic approximation is quite reasonable because we are dealing with very small displacements of the particles. The equations of motion of the lattice can now be written under the harmonic approximation as,

$$m_K \ddot{u}_\alpha\binom{\ell}{K} = -\frac{\partial \Phi}{\partial u_\alpha\binom{\ell}{K}} = -\sum_{\ell',K',\beta} \Phi_{\alpha\beta}\binom{\ell\ell'}{KK'} u_\beta\binom{\ell'}{K'}. \tag{29}$$

It is obvious from the form of eq. (29) that the coefficient, $\Phi_{\alpha\beta}\binom{\ell\ell'}{KK'}$, is a force constant. Physically it describes the force exerted in the α-direction on the atom at $\underline{r}\binom{\ell}{K}$ when the atom at $\underline{r}\binom{\ell'}{K'}$ is displaced in the β-direction. It is also interesting to point out that from eq. (27) the coefficient $\Phi_{\alpha\beta}\binom{\ell\ell'}{KK'}$ satisfies the following symmetry condition,

$$\Phi_{\alpha\beta}\binom{\ell\ell'}{KK'} = \Phi_{\beta\alpha}\binom{\ell'\,\ell}{K'K}. \tag{30}$$

The periodicity of the lattice requires that if the lattice as a whole is translated relative to itself by a lattice vector $\underline{r}\binom{\ell}{K}$, it coincides with itself. Due to this fact, the same triplet of integers (ℓ_1, ℓ_2, ℓ_3) can be added to the cell index ℓ in the coefficient $\Phi_\alpha\binom{\ell}{K}$ and to both of the cell indices in the coefficient $\Phi_{\alpha\beta}\binom{\ell\ell'}{KK'}$ without changing their value. Thus $\Phi_\alpha\binom{\ell}{K}$ must be independent of ℓ, while $\Phi_{\alpha\beta}\binom{\ell\ell'}{KK'}$ can only depend on the relative cell index $\ell-\ell'$ and not on ℓ and ℓ' separately. These results can be expressed as,

$$\Phi_\alpha\binom{\ell}{K} = \Phi_\alpha\binom{0}{K}$$

and

$$\Phi_{\alpha\beta}\binom{\ell\ell'}{KK'} = \Phi_{\alpha\beta}\binom{\ell-\ell'}{KK'}. \tag{31}$$

Due to the translational symmetry within the lattice, the equation of motion should be invariant for rigid translations of the lattice as a whole, namely,

$$\sum_{\ell'K'} \Phi_{\alpha\beta}\binom{\ell\ell'}{KK'} = \sum_{\ell K} \Phi_{\alpha\beta}\binom{\ell\ell'}{KK'} = 0, \tag{32}$$

and also invariant for a rigid rotation of the lattice[20].

The equations of motion form an infinite set of simultaneous linear differential equations. The solution of eq. (29) can be expressed in terms of a traveling wave given by,

$$u_\alpha\binom{\ell}{K} = \frac{1}{\sqrt{m_K}} \sigma_\alpha(K) \exp[-i\omega t + 2\pi\underline{k}\cdot\underline{r}(\ell)]. \tag{33}$$

ω is the angular frequency of the wave, $\dfrac{\sigma_\alpha(K)}{\sqrt{m_K}}$, the wave amplitude which is independent of ℓ and $\underline{k}$, the wave vector. Substitution of eq. (33) into eq. (29) yields,

$$\omega^2 (m_K)^{1/2} \sigma_\alpha(K) = \sum_{\ell' k' \beta} \Phi_{\alpha\beta}\binom{\ell\ell'}{KK'} \frac{\sigma_\beta(K')}{(m_{K'})^{1/2}}$$

$$\exp[i2\pi\underline{k}\cdot r\binom{\ell'}{K'}) - r\binom{\ell}{K})]. \qquad (34)$$

Using the periodic symmetry condition of the lattice, given in eq. (31), eq. (34) may be shown to reduce to the following,

$$\omega^2 \sigma_\alpha(K) = \sum_{k'\beta} F_{\alpha\beta}(\frac{k}{KK'})\sigma_\beta(K'), \qquad (35)$$

where

$$F_{\alpha\beta}(\frac{k}{KK'}) = \frac{1}{(m_K m_{K'})}^{1/2} \sum_\ell \Phi_{\alpha\beta}\binom{\ell}{KK'}$$

$$\times \exp[-2\pi i\underline{k}\cdot r\binom{0\ell}{KK'})] \qquad (36)$$

and

$$r\binom{0\ell}{KK'}) = r\binom{\ell}{K'}) - r\binom{0}{K}). \qquad (37)$$

$F_{\alpha\beta}(\frac{k}{KK'})$ is defined as the coupling coefficient.

Due to the fact that $\Phi_{\alpha\beta}\binom{\ell\ell'}{KK'}$ is a function of $\ell - \ell'$ only and does not depend on ℓ and ℓ' separately, the coefficients $F_{\alpha\beta}(\frac{k}{KK'})$ are independent of ℓ. Therefore, the problem of solving the infinite set of equations of motion (29) is now reduced to the problem of solving a set of 3s linear homogeneous equations in 3s unknowns, $\sigma_\alpha(K)$, given in eq. (35), where s is the number of particles per unit cell.

The equations of motion can be written more concisely in matrix notation as,

$$[F][\sigma] - \omega^2[\sigma] = 0. \qquad (38)$$

$[\sigma]$ is a 1 x 3s matrix written as,

$$[\sigma] = \begin{bmatrix} \sigma_x(1) \\ \vdots \\ \sigma_x^{\cdot}(S) \\ \sigma_y(1) \\ \vdots \\ \sigma_y^{\cdot}(S) \\ \sigma_z(1) \\ \vdots \\ \sigma_z^{\cdot}(S) \end{bmatrix} \tag{39}$$

and $[F]$ is a 3s x 3s matrix defined as

$$[F] = \begin{bmatrix} F_{xx}(\tfrac{k}{11}) \cdots F_{xx}(\tfrac{k}{1S})F_{xy}(\tfrac{k}{11}) \cdots F_{xy}(\tfrac{k}{1S})F_{xz}(\tfrac{k}{11}) \cdots F_{xz}(\tfrac{k}{1S}) \\ F_{xx}^{\cdot}(\tfrac{k}{S1}) \cdots F_{xx}^{\cdot}(\tfrac{k}{SS})F_{xy}(\tfrac{k}{S1}) \cdots \cdots \cdots \\ F_{yx}(\tfrac{k}{11}) \cdots F_{yx}(\tfrac{k}{1S}) \cdots \cdots \cdots \\ F_{yx}^{\cdot}(\tfrac{k}{S1}) \cdots \cdots \cdots \\ F_{zx}(\tfrac{k}{11}) \cdots \cdots \cdots \\ F_{zx}^{\cdot}(\tfrac{k}{S1}) \cdots \cdots \cdots F_{zz}(\tfrac{k}{S1}) \quad F_{zz}^{\cdot}(\tfrac{k}{SS}) \end{bmatrix} \tag{40}$$

The condition for these 3s homogeneous simultaneous equations to have a non-trival solution is

$$|[F] - \omega^2[I]| = 0. \tag{41}$$

Eq. (41) is an equation of degree 3s in ω^2. The 3s solutions for each value of wave vector $\underline{k}$ are denoted by $\omega_j(\underline{k})$ where $j = 1, 2,\ldots, 3s$. Examination of eq. (36) reveals that the matrix $[F]$ is hermitian, and hence the eigenvalues, $\omega_j^2(\underline{k})$, are real[17]. The physics of the problem also requires $\omega_j(\underline{k})$ to be positive.

For each of the 3s values of $\omega_j(\underline{k})$ corresponding to a given value of $\underline{k}$ there exists a vector $[e(K|\tfrac{k}{j})]$. The vector $[e(K|\tfrac{k}{j})]$ is not to be confused with $[\sigma(K)]$. The

former refers to the eigenvector associated with a par-
ticular eigenfrequency, $\omega_j(\underline{k})$, whereas $[\sigma(K)]$ refers to
the general set of eigenvalues, ω^2. The components of
the vector $[e(K|\frac{k}{j})]$ are the solutions to the set of equa-
tions (35), which can be written as

$$\omega_j(\underline{k}) = e_\alpha(K|\tfrac{k}{j}) = \sum_{k'\beta} F_{\alpha\beta}(\tfrac{k}{KK'}) \, e_\beta(K'|\tfrac{k}{j}). \qquad (42)$$

Eq. (42) defines $[e(K|\frac{k}{j})$ to within a constant factor and
this factor can be chosen in such a way that $[e(K|\frac{k}{j})]$
satisfies the orthonormality conditions,

$$\sum_{k,\alpha} e_\alpha^*(K|\tfrac{k}{j}) \, e_\alpha(K|\tfrac{k}{j}) = \sigma_{jj'} \ , \qquad (43)$$

and

$$\sum_j e_\beta^*(K'|\tfrac{k}{j}) \, e_\alpha(K|\tfrac{k}{j}) = \sigma_{\alpha\beta}\sigma_{KK'} \ , \qquad (44)$$

where $\sigma_{\alpha\beta}$ and $\sigma_{jj'}$ are Kronecker deltas.

The values which the wave vector $\underline{k}$, introduced in
eq. (33), one can assume are determined by the boundary
conditions imposed on the components of the displacement
vectors $\underline{u}(^\ell_K)$. The equations of motion are derived for an
infinitely extending lattice. This lattice can then be
normalized to a finite volume by partitioning the lattice into cubes of dimension L^3 each. It is assumed that
the cube contains N lattice cells with L lattice cells
along each edge. These partitions form a macro-lattice
with the cube of N cells being the macro-cell with $L\underline{a}_1$,
$L\underline{a}_2$, $L\underline{a}_3$ as basis vectors. The periodic boundary condition postulates that the atomic displacement be periodic
with the periodicity of the macro-cells, that is,

$$\underline{u}(^{\ell+L}_K) = \underline{u}(^\ell_K). \qquad (45)$$

This cyclic boundary condition simplifies the lattice
dynamical behavior which does not depend explicitly on
the crystal's surface. When applied to the components
of the displacement vector given by eq. (33) the cyclic
boundary condition requires that

$$e^{2\pi i\underline{k}\cdot L\underline{a}_1} = e^{2\pi i\underline{k}\cdot L\underline{a}_2} = e^{2\pi i\underline{k}\cdot L\underline{a}_3} = 1. \qquad (46)$$

The reciprocal lattice vector can be represented as

$$\underline{g}(h) = h_1\underline{b_1} + h_2\underline{b_2} + h_3\underline{b_3} , \qquad (47)$$

where h_1, h_2, h_3 are arbitrary integers which can be positive, negative or zero. The scalar product between a direct lattice vector and a reciprocal lattice vector is just an integer:

$$\underline{r}(\ell)\cdot\underline{g}(h) = \ell_1 h_1 + \ell_2 h_2 + \ell_3 h_3. \qquad (48)$$

Therefore an expression for $\underline{k}$ which satisfies equation (46) is given by

$$\underline{k} = \frac{1}{L}\,\underline{g}(h) = \frac{h_1}{L}\,\underline{b_1} + \frac{h_2}{L}\,\underline{b_2} + \frac{h_3}{L}\,\underline{b_3} . \qquad (49)$$

The values of the integers h_1, h_2, h_3 are now restricted by eqs. (45) and (48) such that if any reciprocal lattice vector $\underline{y}(h)$ is added to $\underline{k}$, the value of $u_\alpha(^\ell_K)$ remains the same. Hence, it is possible to obtain all distinct solutions if the value of $\underline{k}$ is restricted to lie in one unit cell of the reciprocal lattice:

$$\underline{k} = \frac{h_1}{L}\,\underline{b_1} + \frac{h_2}{L}\,\underline{b_3} + \frac{h_3}{L}\,\underline{b_3} ,$$

where $\qquad\qquad\qquad\qquad\qquad\qquad\qquad\qquad (50)$

$$h_1,\ h_2,\ h_3 = 1,\ 2,\ \ldots$$

Thus there are $L^3 = N$ allowed values of wave vector $\underline{k}$. For most calculations it is important that these values are uniformly distributed. From eq. (50) the volume in reciprocal space which contains all the allowed values of the wave vector $\underline{k}$ can be seen to be that generated by a reciprocal lattice vector $\underline{b}$. An equivalent volume in reciprocal space which displays a higher degree of symmetry than the reciprocal lattice can be obtained by drawing vectors from the origin of the reciprocal lattice to all lattice points and then constructing planes which are the perpendicular bisectors of these vectors. The smallest volume in the reciprocal space which is enclosed by these planes can be shown to be completely equivalent to the unit cell in that every allowed value of $\underline{k}$ in the unit cell differs from a corresponding point in the symmetric polyhedron only by a translation vector of the

reciprocal lattice, and hence the two are equivalent.
The symmetric polyhedron constructed in this fashion and
containing all allowed values of $\underline{k}$ is known as the first
Brillouin zone of the reciprocal lattice.

Due to the symmetry of the first Brillouin zone,
one need only consider $\underline{k}$ values in a small region (the
irreducible element) of this zone to evaluate any vibrational property of the lattice. For cubic crystals, the
volume of the irreducible element is $1/48^{\text{th}}$ of the volume
of the Brillouin zone. Thereafter, the totality of frequencies, $\omega_j(\underline{k})$ can be obtained by solving equation (35)
for values of $\underline{k}$ in $1/48^{\text{th}}$ of the irreducible volume of
the total Brillouin zone.

Since the displacements $\underline{u}(\begin{smallmatrix}\ell\\K\end{smallmatrix})$ are small and the potential energy expansion (25) contains quadratic terms
in displacement, the general motion of the lattice can
be given by the principle of superposition. Each of the
waves given by equation (33) travels through the lattice
independent of the others. The whole configuration of
the lattice may then be expressed in terms of the displacements due to this set of independent waves, just as
well as in terms of the coordinates of the individual
lattice points. Therefore, the waves can be considered
as themselves constituting an independent set of coordinates known as normal coordinates. Each normal coordinate describes an independent mode of vibration of the
crystal with only one frequency and phase and is referred
to as a normal mode. In general there are 3Ns normal
modes of vibration. In terms of these normal coordinates,
$Q(\begin{smallmatrix}k\\j\end{smallmatrix})$, the general motion of the lattice can be defined
in the following fashion,

$$u_\alpha(\ell) = \frac{1}{(Nm)}1/2 \sum_{\underline{k},j} e_\alpha(\tfrac{k}{j})Q(\tfrac{k}{j})\exp[i2\pi\underline{k}\cdot\underline{r}(\ell)] \; , \quad (51)$$

where the subscript k of equation (33) has been suppressed and the time dependence is understood in the $Q(\tfrac{k}{j})$
term. m and $e_\alpha(\tfrac{k}{j})$ in equation (51) are the mass and the
eigenamplitude corresponding to the frequency $\omega_j(\underline{k})$, of
both the positive and the negative ions, respectively.
By defining the function, $\chi_\alpha(\ell|\tfrac{k}{j})$ as

$$\chi_\alpha(\ell|\tfrac{k}{j}) = (Nm)^{1/2}e_\alpha(\tfrac{k}{j})\exp[i2\pi\underline{k}\cdot\underline{r}(\ell)] \qquad (52)$$

equation (51) can be written in the following fashion,

$$u_\alpha(\ell) = \sum_{\underline{k},j} \chi_\alpha(\ell|{\textstyle\frac{k}{j}})\underline{Q}({\textstyle\frac{k}{j}}) \tag{53}$$

Substitution of the displacements $u_\alpha(\ell)$, defined in equation (51), into equation (29) yields,

$$\sum_{\beta\ell'} \Phi_{\alpha\beta}(\ell\ell')\chi_\beta(\ell'|{\textstyle\frac{k}{j}}) = m\omega_j^2(\underline{k})\chi_\alpha(\ell|{\textstyle\frac{k}{j}}) \tag{54}$$

The terms $\chi_\alpha(\ell|{\textstyle\frac{k}{j}})$ are eigenvector terms with respect to the normal coordinates and are normalized as follows:

$$\sum_{\alpha\ell} m\left|\chi_\alpha(\ell|{\textstyle\frac{k}{j}})\right|^2 = 1. \tag{55}$$

Since the characteristic frequency of the $({\textstyle\frac{k}{j}})$ normal mode is $\omega_j(\underline{k})$, which is the same as the eigenfrequency of the matrix $[F]$, and since there is a one to one correspondence between these eigenvalues and the normal coordinates, the normal mode frequencies of the lattice can also be identified as $[\omega_j(\underline{k})]$ where the brackets indicate the complete set of lattice frequencies.

It can be seen from equations (42), (32) and (36), that out of the 3s solutions for each $\underline{k}$, three tend toward zero as $\underline{k}$ goes to zero. Such modes are called acoustic modes and are characterized by the condition.

$$\frac{[e(K|{\textstyle\frac{0}{j}})]}{\sqrt{m_K}} = \frac{[e(K'|{\textstyle\frac{0}{j}})]}{\sqrt{m_{K'}}} = \underline{u}({\textstyle\frac{\ell}{K}}|{\textstyle\frac{0}{j}}) = \underline{u}({\textstyle\frac{\ell}{K'}}|{\textstyle\frac{0}{j}}), \tag{56}$$

where the expression $\underline{u}({\textstyle\frac{\ell}{K}}|{\textstyle\frac{k}{j}})$ refers to the displacement from equilibrium of the Kth atom in the ℓth unit cell when it is vibrating with frequency $\omega_j(\underline{k})$. Therefore, all s particles in each unit cell move in parallel and with equal amplitudes which is a characteristic of the displacement associated with an acoustic wave in an elastic continuum.

The remaining 3s - 3 modes whose frequencies do not vanish at $\underline{k}$ = 0 are called optic modes. For diatomic crystals ($\bar{s}$ = 2), i.e., equation (43) for $\underline{k}$ = 0 can be written in vector form as

$$[e(A|_j^0)] \cdot [e(A|_{j'}^0)] + [e(B|_j^0)] \cdot [e(B|_{j'}^0)] = 0 \qquad (57)$$

where A and B refer to the positive and negative ions, respectively, j refers to any one of the acoustic branches and j' refers to any one of the optic branches. Equations (56) and (57) can be solved to obtain,

$$[e(A|_j^0)] \cdot \{[e(A|_{j'}^0)] + \frac{m_B}{m_A}[e(B|_j^0)]\} = 0 \qquad (58)$$

Excluding the trival solutions, $[e(K|_j^k)] \equiv 0$, one gets

$$\sqrt{m_A}\,[e(A|_j^0)] = -\sqrt{m_B}\,[e(B|_j^0)]. \qquad (59)$$

Therefore for the optic mode the two ions in each unit cell vibrate 180° out of phase, while the center of mass of the cell remains stationary.

For increasing values of k, the ratio of the amplitudes of the lighter to heavier atom in the optic branch in general increases. The corresponding ratio in the branch eq. (56) decreases from its value to unity for k = 0.

The dispersion of the lattice is expressible by the equation,

$$\omega = \omega_j(\underline{k}), \quad j=1, 2 \ldots 3s. \qquad (60)$$

In diatomic crystals the number of atoms in each unit cell is two, therefore in general there are three acoustic and three optic branches of phonon dispersion.

The introduction of a substitutional impurity for one of the atoms of the host lattice modifies the dynamical behavior of the lattice in the region of the substitution. It is assumed in this analysis that the force constants remain unchanged from the perfect lattice in the vicinity of the defect and that only the mass parameter is changed. Due to the presence of a substitutional impurity, the mass m in eq. (29) will depend on ℓ. Writing this in terms of its deviation from the perfect lattice value, the general equation of motion which gives the normal modes and frequencies of the perturbed lattice can be written as

$$m(\ell)\ddot{u}_\alpha(\ell) + \sum_{\beta\ell'} \Phi_{\alpha\beta}(\ell\ell')u_\beta(\ell') = \Delta m(\ell)\ddot{u}_\alpha(\ell) \qquad (61)$$

The new normal modes for the perturbed lattice can be defined by a transform similar to eq. (53),

$$u_\alpha(\ell) = \sum_f \chi'_\alpha(\ell|f)df. \qquad (62)$$

These new normal modes are labeled by a quantum number, f, which takes 3Ns values. The new normal coordinates of the perturbed lattice, df, are similar to the normal coordinates, $Q(\frac{k}{j})$, of the perfect lattice and also include the time dependence. The $\chi'_\alpha(\ell|f)$ terms are the eigenvector terms corresponding to the normal coordinates of the perturbed lattice. Substitution of eq. (62), into eq. (63) yields,

$$-\omega'^2 m \chi'_\alpha(\ell|f) + \sum_{\beta\ell'} \Phi_{\alpha\beta}(\ell\ell') \chi'_\beta(\ell'|f)$$

$$= \sum_{\beta\ell'} S_{\alpha\beta}(\ell\ell') \chi'_\beta(\ell'|f), \qquad (63)$$

where $S_{\alpha\beta}$ now incorporates the change of mass and is given as,

$$S_{\alpha\beta}(\ell\ell') = -\Delta m(\ell)\omega'^2 \delta_{\alpha\beta}\delta_{\ell\ell'} \qquad (64)$$

The eigenvectors $\chi_\alpha(\ell|f)$ of the perturbed lattice, which correspond to the normal coordinates df, are normalized by a relation similar to that given in eq. (55), namely,

$$\sum_{\alpha\ell} (m(\ell) + \Delta m(\ell))|(\chi'_\alpha(\ell|f)|^2 = 1. \qquad (65)$$

In matrix notation the equations of the perfect lattice [eq. (54)], can be written as,

$$[L][\chi] = [[D] - \omega^2[I]][\chi] = 0, \qquad (66)$$

where [D] incorporates the particle masses, m, and the force constants $\Phi_{\alpha\beta}(\ell\ell')$. For the lattice with a substitutional impurity, eq. (66) becomes

$$[L][\chi'] = [[D] - \omega'^2[I]][\chi'] = [S][\chi'] \qquad (67)$$

where ω'^2 and $[\chi']$ are the eigenvalues and the eigenvectors of the imperfect lattice and $[S]$ is the perturbation matrix incorporating the change of mass.

A Green's function technique, introduced earlier
can be used to calculate the new normal mode frequencies
and the amplitude of the lattice with substitutional impurities. Eq. (66) is the perfect crystal equation while
(67) is identical to equation (6) with the symbol D replaced by F and χ by σ. Therefore, the Green's function
technique described before can be applied for the three
dimensional case. The Green's function matrix $[G]$ defined as in eq. (7) may be written as,

$$-\omega^2 m g_{\alpha\gamma}^{\omega}(\ell, \ell'') + \sum_{\beta\ell'} \Phi_{\alpha\beta}(\ell, \ell') \, g_{\beta\gamma}^{\omega}(\ell', \ell'') = \delta_{\alpha\gamma}\delta_{\ell\ell''}$$

$$(68)$$

where $g_{\alpha\gamma}^{\omega}(\ell, \ell'')$ are the elements of $[G]$. The explicit
representation of the elements, $g_{\alpha\gamma}^{\omega}$, of the Green's function matrix, as given by eq. (21) can be written as

$$g_{\alpha\gamma}^{\omega}(\ell, \ell'') = \sum_{\underline{k}, j} \frac{\chi_{\alpha}^{*}(\ell|\frac{\underline{k}}{j}) \, \chi_{\beta}(\ell''|\frac{\underline{k}}{j})}{\omega_{j}^{2}(\underline{k}) - \omega'^2} \qquad (69)$$

Eq. (9) describing the eigenvectors of the perturbed lattice can be written as

$$\chi_{\alpha}'(\ell|f) = \sum_{\beta\ell'\gamma\ell''} g_{\alpha\beta}^{\omega}(\ell, \ell'')S_{\beta\gamma}(\ell'', \ell')\chi_{\gamma}'(\ell'|f). \quad (70)$$

The eigenfrequencies of the perturbed lattice are determined by the secular equation,

$$\left| \sum_{\beta\ell''} g_{\alpha\beta}^{\omega}(\ell\ell'')S_{\beta\gamma}(\ell''\ell') - \delta_{\alpha\gamma}\delta_{\ell\ell'} \right| = 0. \qquad (71)$$

If the cell containing the single substitutional impurity of the perturbed lattice is chosen to be at the
origin, eq. (64) can be written as,

$$S_{\alpha\beta}(\ell\ell') = -m\varepsilon\omega'^2 \, \delta_{\alpha\beta}\delta_{\ell 0}\delta_{\ell'0}, \tag{72}$$

where ε is the dimensionless mass defect parameter defined as

$$\varepsilon = \frac{m-m'}{m} . \tag{73}$$

m is the mass of the particle to be substituted for and m' is the mass of the substitutional impurity.

Using eq. (52) the element $g_{\alpha\beta}^{\omega}(0,0)$ of the Green's function matrix (69) can be written as

$$g_{\alpha\beta}^{\omega}(0,0) = \frac{1}{Nm} \sum_{\underline{k},j} \frac{e_{\alpha}^{*}(\frac{k}{j})e_{\beta}(\frac{k}{j})}{\omega_{j}^{2}(\underline{k})-\omega'^2} \tag{74}$$

Substitution of eq. (72) into eq. (71), reduces it to a 3 x 3 matrix

$$\left| m\varepsilon\omega'^2 \, \delta_{\alpha\beta} g_{\alpha\beta}^{\omega}(0,0) + \delta_{\alpha\beta} \right| = 0. \tag{75}$$

Due to the $\delta_{\alpha\beta}$ term, this equation contains only diagonal elements and the solution is triply degenerate. Substituting eq. (74) into eq. (75) one obtains

$$1 + \frac{\varepsilon(K)\omega'^2}{3N} \sum_{\underline{k},j} \frac{\left| \underline{e}(K|\frac{k}{j}) \right|^2}{\omega_{j}^{2}(\underline{k})-\omega'^2} = 0 \tag{76}$$

The above equation is used to calculate the impurity mode frequency due to a substitutional impurity replacing either the cation or the anion of the host lattice, where K refers to the ion which has been replaced by the impurity and $\varepsilon(K)$ is defined as

$$\varepsilon(K) = \frac{m(K) - m'(K)}{m(K)} .$$

It is also possible in the framework of this analysis to obtain an expression for the amplitude of vibration of the impurity. Substitution of eqs. (72) and (74) into eq. (70) yields the following expression for the amplitude of vibration in the imperfect lattice,

$$\chi_{\beta'}(\ell|f) = \frac{1}{N} \sum_{\alpha, \underline{k}, j} \frac{e_{\beta}^{*}(\frac{k}{j}) \; e_{\alpha}(\frac{k}{j})}{\omega_j^2(\underline{k})-\omega'^2} \chi_{\alpha}'(0|f) \varepsilon \omega'^2(f) \quad x$$

$$\exp[-i\pi\underline{k}\cdot\underline{r}(\ell)] \; . \tag{77}$$

The term $\chi_{\beta}'(\ell|f)$ represents the modification of the amplitude of vibration of the particles in the lattice due to the defect mass. For the case $\ell = 0$, one gets $\chi_{\beta}'(0|f)$ which corresponds to the amplitude of the defect atom itself.

The normalization condition given in eq. (65), can be written as

$$\sum_{\beta\ell} m |\chi_{\beta}'(\ell|f)|^2 + \sum_{\alpha} \varepsilon m |\chi_{\alpha}'(0|f)|^2 = 1 \tag{78}$$

Substitution of eq. (78) along with the use of the orthonormality conditions given in eqs. (43) and (44) into eq. (77) yields the following equation for the amplitude of vibration for the impurity atom, $|\chi'(0|f|^2$,

$$\frac{1}{m(k) |\chi'(0|f)|^2} = \frac{\varepsilon^2(K)\omega'^4(f)}{3N} \sum_{\underline{k}j} \frac{|\underline{e}(K|\frac{k}{j})|^2}{[\omega'^2(f)-\omega_j^2(\underline{k})]^2} \varepsilon(K). \tag{79}$$

Eqs. (76) and (79) therefore provide the basic description of the single mass defect problem in cubic crystals.

The eigenfrequencies and eigenamplitudes of the perfect lattice can be substituted into eq. (76) in order to determine the triply degenerate localized and gap modes. The local modes correspond to the condition that the substitutional impurity mass, $m'(k)$ is smaller than the corresponding host lattice atomic masses. Gap modes occur when $m'(k)$ may be heavier or lighter, than one or both of the host lattice masses.

Utilizing the modified rigid ion model[39] for the perfect lattice, calculations were performed for the impurity modes of zinc-blende type crystals. The variation of the local mode frequencies as functions of the mass defect parameter is presented in Figs. 4 and 5 for GaP and ZnSe, respectively.

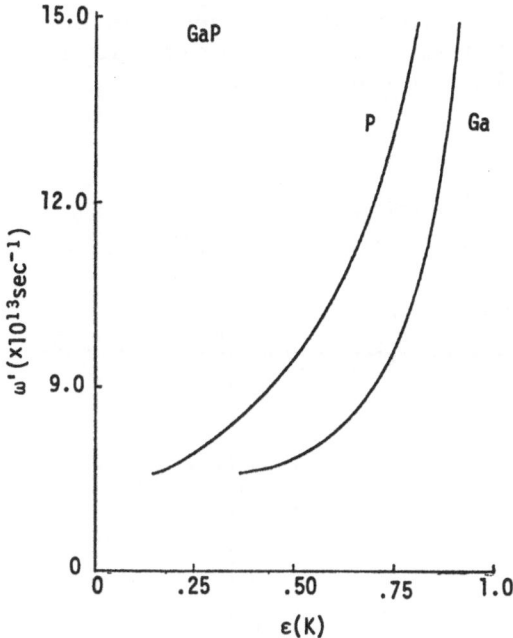

Fig. 4 Localized mode frequency, ω', as a function of the mass defect parameter, ε(K), for GaP.

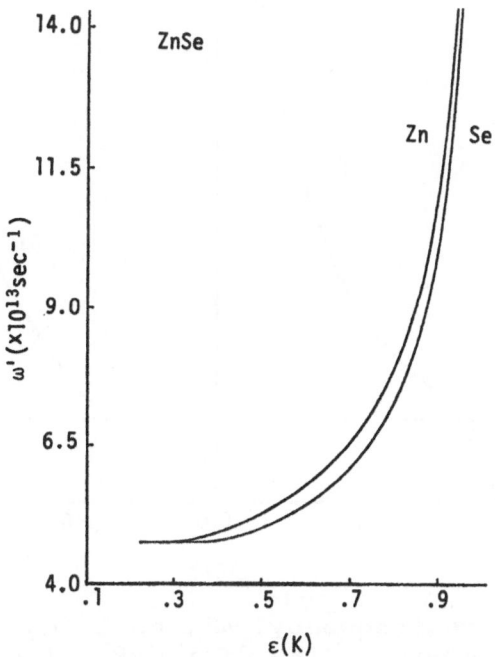

Fig. 5 Localized mode frequency, ω', as a function of the mass defect parameter, ε(K), for ZnSe.

The variation of gap mode frequencies as a function of
mass defect parameter are presented in Figs. 6 and 7,
for GaP and ZnTe, respectively.

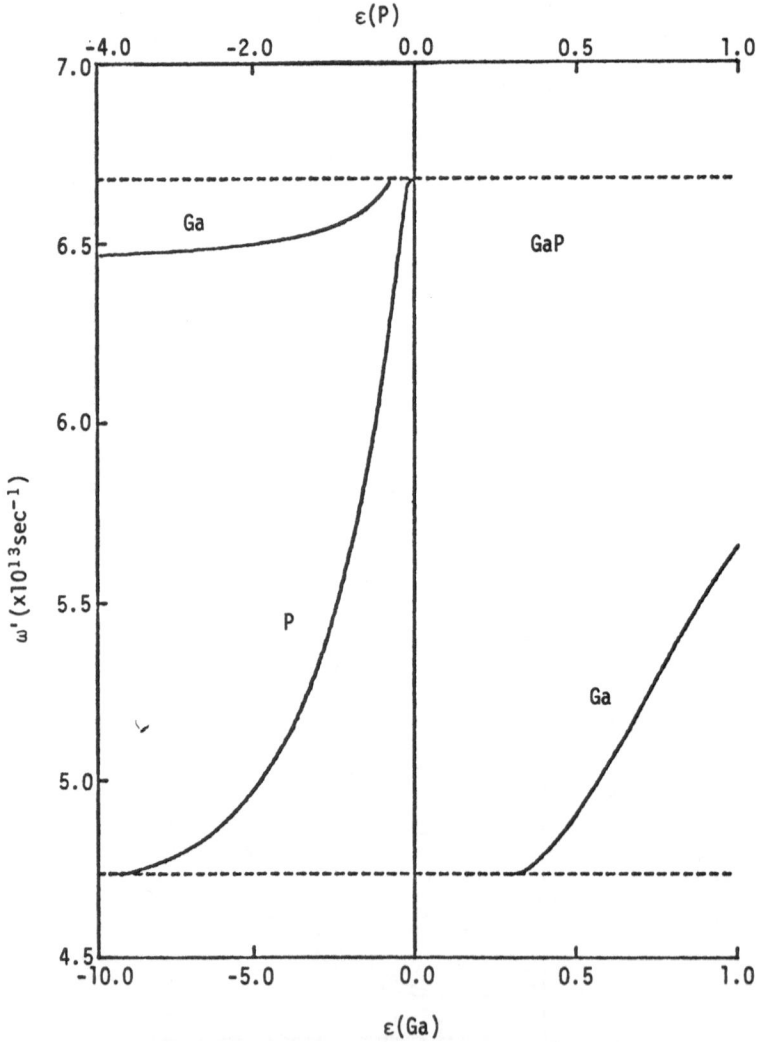

Fig. 6 Gap mode frequency, ω', as a function of the
mass defect parameter, $\varepsilon(K)$, for GaP. The gap in the
frequency spectra is shown by the dotted lines.

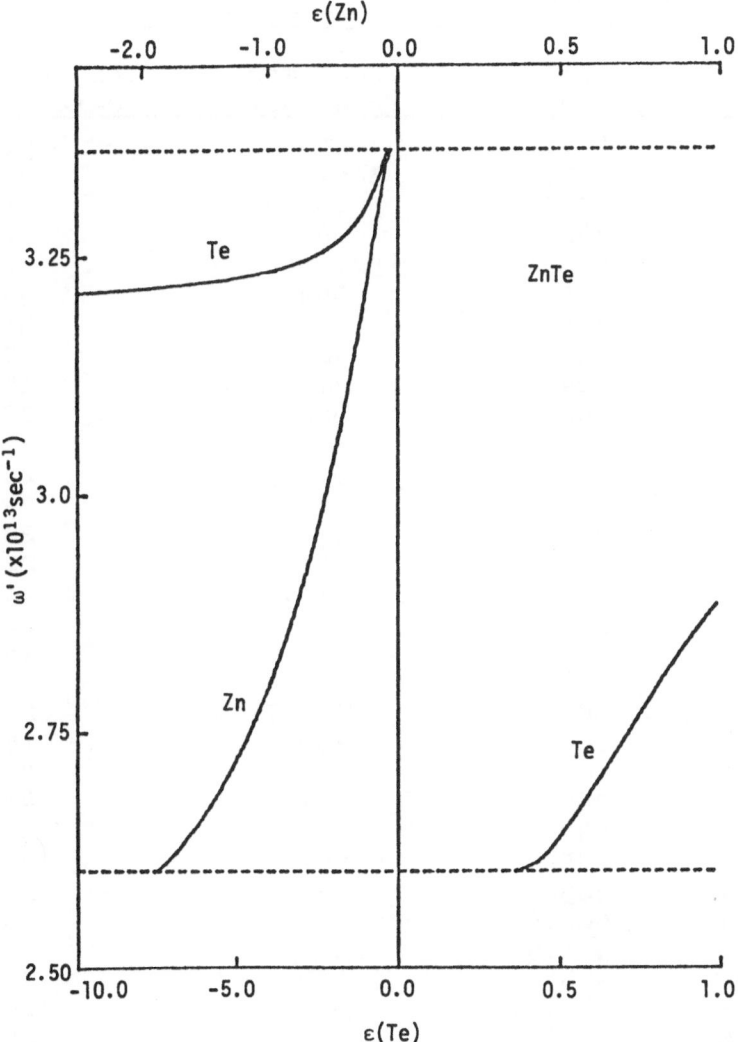

Fig. 7 Gap mode frequency, ω', as a function of the mass defect parameter, $\varepsilon(K)$, for ZnTe. The gap in the frequency spectra is shown by the dotted lines.

The available experimental data are compared to some calculated values in Tables 1 and 2.

Table I: Comparison of calculated and experimental values of local mode frequencies (in cm^{-1}) for various zine-blende compounds.

Compound[a]	Calculated local mode frequency	Experimental local mode frequency	Reference for experimental data
$\underline{Ga}P:Si$	437.4	453	31
$\underline{Ga}P:^{11}B$	611.6	569,571	31,15
$\underline{Ga}P:^{10}B$	636.6	592,594	31,15
$\underline{Ga}P:Al$	442.2	443	31
$\underline{Ga}P:^{12}C$	563.1	527,606.2	31,33
$Ga\underline{P}:^{14}C$	525.2	564	33
$Ga\underline{P}:O$	494.4	464	31
$Ga\underline{P}:^{14}N$	525.2	488	15
$Ga\underline{P}:^{15}N$	509	472	15
$\underline{GaAs}:Si$	362.9	384	27
$\underline{GaAs}:Si$	365	399	27
$Ga\underline{As}:P$	351	355.4,351	25,29
$\underline{GaAs}:Al$	368.9	362	24
$\underline{GaAs}:^{7}Li$	685.9	422.2	22
$\underline{GaAs}:^{6}Li$	739	451.4	22
$Ga\underline{Sb}:Al$	333	316.7	21
$Ga\underline{Sb}:P$	327.3	324	21
$Ga\underline{Sb}:As$	248	240	16
$\underline{In}As:Ga$	244	240	30
$\underline{In}Sb:Al$	307	295.7	23,25
$In\underline{Sb}:As$	206.6	200	16
$\underline{In}Sb:Ga$	209.8	196	36
$\underline{Zn}S:Al$	362.6	437.9	41
$\underline{Zn}S:Be$	543	486	37
$Zn\underline{Se}:S$	285.9	297	28
$\underline{Zn}Se:^{7}Li$	563	383	26
$\underline{Zn}Se:^{6}Li$	606.6	412	26
$\underline{Zn}Se:Al$	301.9	359	42
$\underline{Zn}Se:Be$	498	450	26
$\underline{Zn}Se:^{24}Mg$	317.2	352	26
$\underline{Zn}Se:^{25}Mg$	312	345	26
$\underline{Zn}Se:^{26}Mg$	306.6	334	26
$\underline{Zn}Te:Al$	274.5	312.6	41
$Zn\underline{Te}:S$	266.5	269 to 272	35
$\underline{Zn}Te:Be$	461.5	411	37
$\underline{Cd}Te:Be$	426	391	32
$Cd\underline{Te}:Se$	171.6	170	35
$Cd\underline{Te}:Mg$	266.3	248	34

a   Underlined atom has been replaced.

Table 2: Comparison of calculated and experimental values of gap mode frequencies (in cm^{-1}) for impurities in GaP and SiC.

Compound[a]	Calculated gap mode frequency	Experimental gap mode frequency	Reference for experimental data
$\underline{GaP}:^{10}B$	289.2	293.8	15
$\underline{GaP}:^{11}B$	287.9	284.2	15
$\underline{GaP}:As$	275	270	15
$\underline{SiC}:He(?)$	676	670	38

a   Underlined atom has been replaced.

III. MIXED CRYSTAL OR ALLOY SYSTEM

If the concentration of the substitutional impurity in a crystal increases significantly, the resulting lattice is defined as a mixed crystal system. Symbolically, a mixed crystal may be represented as $AB_{1-x}C_x$ where $0 \leq x \leq 1$. A successful lattice dynamical model for a mixed crystal system should predict the appropriate impurity modes for a substitutional impurity as x approaches either zero or unity.

The study of mixed crystals can be dated back to as early as 1928[43]. Since then several phenomenological models[44-49] have been proposed to study the vibrations of mixed crystal systems. It has been observed that the models[46,49] using the concept of a unit cell do provide some basic understanding of mixed crystals. Verleur and Barker[46] considered a cluster model to account for the two mode behavior of the mixed crystals, $GaP_{1-x}As_x$ and $CdS_{1-x}Se_x$. This model assumed that like negative ions clustered around positive ions or vice versa depending on whether the impurity was an anion or a cation. In the random element isodisplacement (REI) model, Chen, Shockley and Pearson[47] assumed that in a mixed crystal, $AB_{1-x}C_x$, the B and C atoms are distributed on the anion sublattice and the anions of like species vibrate in phase with identical amplitudes against the cations which also vibrate as a rigid unit. Later Chang and Mitra[48] modified the REI model to include the polarization field. Subsequently, Chang and Mitra[49] proposed the pseudo unit cell model and predicted the zone boundary phonons of systems exhibiting two mode behavior at the zone center.

Recently, Kutty[50] applied the Green's function technique
to derive the phonon dispersion relations in mixed crystals as a function of wave vector. Sen and Hartman[51]
explained the switching from one mode to two mode type
behavior observed in some III-V mixed crystals by using
the coherent potential approximation technique in one
dimension. At present no adequate theory exists to describe the complete lattice dynamics of mixed crystal systems.

A natural "brute force" technique to calculate the
lattice dynamics of a mixed crystal is to start with a
finite chain and then permute the various number of ways
to configurationally orient the crystal depending upon the concentration, x. For each configuration, the
lattice dynamics would be performed and the resulting
ensemble would then be averaged to obtain the lattice
dynamics of the mixed crystal system. This technique
would however be very tedious, long and perhaps impossible due to the number of configurational orientations,
and the violation of lattice periodicity in many of the
configurations. For values of x approaching either zero
or unity, the Green's function technique described before
could be applied. However, as the impurity concentration
increases, the impurity-impurity interactions become significant. This causes the Green's function approach to
become very complicated.

In this presentation a pseudo unit cell model developed by Chang and Mitra[48,49] will be used to describe
the lattice configuration. The pseudo unit cell is formed
by the ions A, $(1-x)$ B and xC. Although this unit cell
is simple it is obviously unphysical as far as the exact
representation of the lattice. The fact that it obeys
translational symmetry is a physical drawback. In spite
of these deficiencies, however, this model has proven[46-49]
to be a simple approach by which one can estimate general
features of the optical properties of mixed diatomic crystals. In particular this model has theoretically verified[49] the experimental behavior of certain zone center
and zone boundary phonons in mixed diatomic crystals.
In view of the simplicity and previous success of this
model, it was decided to extend the calculations to all
wave vectors to see if the model has merit in mixed crystal lattice dynamical calculations. In the linear chain
lattice and the mixed zinc-blende lattice, the modified
rigid ion (MRI) model[39] is used to predict the lattice
dynamics. The MRI model consists of short range central
and noncentral repulsive interactions and long range

Coulomb interactions among ions of appropriate effective
ionic charge. The model parameters of the mixed crystal of the form $AB_{1-x}C_x$ are deduced from the elastic constants, optical mode frequencies and impurity mode frequencies of the host crystals. The model parameters
along with the lattice constant are assumed to vary linearly as a function of concentration. Explicit calculations have been performed for a hypothetical mixed linear chain and for $ZnS_{1-x}Se_x$ and $GaP_{1-x}As_x$. The phonon
dispersion in various symmetry directions and the frequency distribution function are obtained as a function
of concentration.

A. Pseudo Unit Cell

The unit cell as defined in perfect crystals cannot
be uniquely defined for the mixed crystals. However, the
mixed crystal problem may be treated in a manner similar
to the pure crystal case, if certain assumptions are made
on the distribution of the ions in the lattice. In a
mixed crystal, $AB_{1-x}C_x$ where $0 \leq x \leq 1$, the B and C ions
are assumed to be distributed randomly in their corresponding sublattice and to obey the law of statistics. A
corresponding pseudo unit cell is then formed by ions A,
$(1-x)$ B and xC. The resulting mixed crystal system can
be thought of as a repetition of such cells. Probabilis-
tically the pseudo unit cell may be thought of as a configurational average unit cell. A schematic representation of the pseudo unit cell for a mixed crystal in one
dimension is given in Figure 8. The fractional amount
of the B and C ions located at the same lattice site is
proportional to the mixing ratio in the crystal. This
means that the corresponding forces involving these ions
are weighted by these factors.

B. Mixed Linear Diatomic Chain

The generalized equations of motion in a mixed diatomic linear chain may be written as,

$$m_A \ddot{u}_{2n}^A = (1-x) \sum_m \left(F_{AB}^m + \Phi_{AB}^m \right) \left(u_{2n+m}^B - u_{2n}^A \right)$$

$$+ x \sum_m \left(F_{AC}^m + \Phi_{AC}^m \right) \left(u_{2n+m}^C - u_{2n}^A \right)$$

$$+ \sum_\ell \left(F_{AA}^\ell + \Phi_{AA}^\ell \right) \left(u_{2n+\ell}^A - u_{2n}^\ell \right),$$

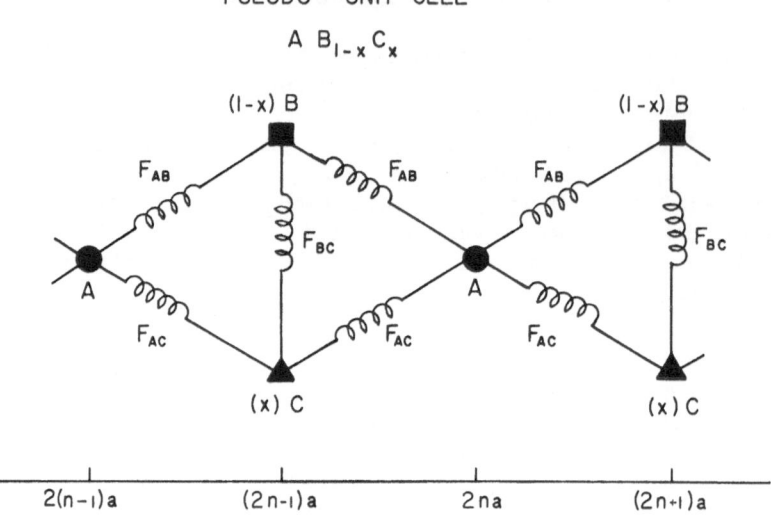

Fig. 8 Pseudo unit cell for a mixed linear diatomic
chain, $AB_{1-x}C_x (0 \leq x \leq 1)$.

$$(1-x)m_B\ddot{u}^B_{2n+1} = (1-x) \sum_m (F^m_{AB} + \Phi^m_{AB}) (u^A_{2n+1+m} - u^B_{2n+1})$$

$$+ (1-x)^2 \sum_\ell (F^\ell_{BB} + \Phi^\ell_{BB}) (u^B_{2n+1+\ell} - u^B_{2n+1})$$

$$+ x(1-x) F_{BC} (u^C_{2n+1} - u^B_{2n+1})$$

$$+ x(1-x) \sum_\ell F^\ell_{BC} + \Phi^\ell_{BC}) (u^C_{2n+1+\ell} - u^B_{2n+1})$$

$$(80)$$

and

$$(x)m_C\ddot{u}^C_{2n+1} = x \sum_m (F^m_{AC} + \Phi^m_{AC}) (u^A_{2n+1+m} - u^C_{2n+1})$$

$$+ x^2 \sum_\ell (F^\ell_{CC} + \Phi^\ell_{CC}) (u^C_{2n+1+\ell} - u^C_{2n+1})$$

$$+ x(1-x) F_{BC}(u^B_{2n+1} - u^C_{2n+1})$$

$$+ x(1-x) \sum (F^\ell_{BC} + \Phi^\ell_{BC}) (u^B_{2n+1+\ell} - u^C_{2n+1}),$$

where $\ell = \pm 2, \pm 4, \pm 6 \ldots$ and $m = \pm 1, \pm 3, \pm 5 \ldots m_{A,B,C}$ are the atomic masses, and u_{2n}^{A} and $u_{2n+1}^{B,C}$ are the atomic displacements. The subscripts represent the ion's position. $F_{KK'}^{j}$ and $\Phi_{KK'}^{j}$ are the force constants representing the jth neighbor repulsive and the electrostatic (Coulomb) interactions respectively between the Kth and K'th ions.

The electrostatic force constants are defined as,

$$\Phi_{KK'}^{j} = - \frac{d^2}{dx^2} \left(\frac{e_K e_{K'}}{x}\right) \quad , \qquad x = |r_j| \quad \tag{81}$$

where

$$e_K = - e_{K'} = ze,$$

z is the effective ionic charge and r_j is the distance between the ions K and K'. The B and C ions are assumed to have the same effective ionic charge, therefore,

$$\Phi_{AB}^{m} = \Phi_{AC}^{m} = \Phi_{12}^{m} \tag{82}$$

and

$$\Phi_{AA}^{\ell} = \Phi_{BB}^{\ell} = \Phi_{CC}^{\ell} = \Phi_{BC}^{\ell} = \Phi_{11}^{\ell} .$$

Assuming a traveling wave solution of the form

$$u_{2n}^{A} = u_A e^{i(\underline{k}2na - \omega t)} \tag{83}$$

and

$$u_{2n+1}^{B,C} = u_{B,C} e^{i(\underline{k}(2n+1)a - \omega t)},$$

where ω is the angular frequency, k, the wave vector and a, the lattice spacing, eq. (80) reduces to the following matrix equation

$$[\underline{D} - \omega^2 \underline{I}][\underline{u}] = \underline{0}. \tag{84}$$

$\underline{I}$ is the identity matrix and $\underline{u}$ is the displacement matrix, the transpose of which is,

$$\underline{u}^{T} = [u_A \quad u_B \quad u_C]. \tag{85}$$

The dynamical matrix $\underline{D}$ is a 3 x 3 symmetrical matrix defined as,

$$\underline{D} =$$

$$
\begin{bmatrix}
m_{AA}[(1-x)T_{mAB} + xT_{mAC} & -\sqrt{1-x}\, m_{AB}[T'_{mAB} + F_{12}] & -\sqrt{x}\, m_{AC}[T'_{mAC} + F_{12}] \\
\quad + T_A + F_{11}] & & \\[2ex]
-\sqrt{1-x}\, m_{AB}[T'_{mAB} + F_{12}] & m_{BB}[T_{mAB} + (1-x)(T_B+F_{11}) & -\sqrt{x(1-x)}m_{BC}[F_{BC} + T'_{\ell BC} \\
& \quad + x(F_{BC} + T_{\ell BC} - \Phi^{\circ})] & \quad -F_{11} - \Phi^{\circ}] \\[2ex]
-\sqrt{x}\, m_{AC}[T'_{mAC} + F_{12}] & -\sqrt{x(1-x)}m_{BC}[F_{BC} + T'_{\ell BC} & m_{CC}[T_{mAC} + x(T_C + F_{11}) \\
& \quad - F_{11} - \Phi^{\circ}] & \quad +(1-x)(F_{BC}+T_{\ell BC} - \Phi^{\circ})]
\end{bmatrix}
$$

where

$$\Phi^{\circ} = -\frac{4z^2e^2}{a^3}\left[\sum_m m^{-3} - \sum_{\ell} \ell^{-3}\right] = -3.6062\,\frac{z^2e^2}{a^3}\,,$$

$$F_{12} = \frac{4z^2e^2}{a^3}\sum_m m^{-3}\cos(kma),$$

$$F_{11} = -\Phi^{\circ} + \frac{4z^2e^2}{a^3}\sum_{\ell} \ell^{-3}\cos(k\ell a),$$

$$m_{st} = (m_s m_t)^{-1/2}\,,$$

$$T_{rst} = 2\sum_r F_{st}^{\,r}\,,$$

$$T_s = 2\sum_{\ell} F_{ss}^{\,\ell}(1-\cos(k\ell a)),$$

$$(86)$$

and

$$s,t = A,B,C$$

$$r = \ell,m$$

The indices ℓ and m are restricted to have only positive values. The normal modes of vibration are obtained by solving the secular equation

$$\left| \underline{D} - \omega^2 \ \underline{I} \right| = 0. \tag{87}$$

The three roots of eq. (87) give two optic modes and one acoustic mode. In the limit of infinite dilution (as $x \to 0$ and $x \to 1$), two of the above solutions become the acoustic and optic modes of the host lattice while the third mode becomes the impurity mode.

Explicit lattice dynamical calculations were performed considering nearest neighbor repulsive interactions and long range Coulomb interactions. It was assumed that the force constants vary linearly with concentration from one end member to the other as,

$$\frac{F_{AB}}{F_{ABO}} = \frac{F_{AC}}{F_{ACO}} = \frac{F_{BC}}{F_{BCO}} = 1 - \Theta x. \tag{88}$$

The solution of eq. (87) in the long wavelength limit gives three vibrational frequencies. One of these frequencies is zero and refers to the acoustic mode of the system. The remaining two modes pertain to the optic mode of the host lattice and the impurity mode. These solutions are given as,

at $x=0$

$$\omega^2_{LO,AB} = \frac{1}{\mu_{AB}} \left(2F_{ABO} + 4.2072 \ \frac{z_1^2 e^2}{a_1^3} \right)$$

and

$$\omega^2_{I,C} = \frac{1}{m_C} \left(2F_{ACO} + F_{BCO} + 3.6062 \ \frac{z_1^2 e^2}{a_1^3} \right.$$

$$= \text{Impurity mode of C in AB}, \tag{89}$$

at x = 1,

$$\omega^2_{LO,AC} = \frac{1}{\mu_{AC}} \left(2F_{ACO}(1-\theta) + 4.2072 \frac{z_2^2 e^2}{a_2^3} \right)$$

and

$$\omega^2_{I,B} = \frac{1}{m_B} \left((2F_{ABO}+F_{BCO})(1-\theta)+3.6062 \frac{z_2^2 e^2}{a_2^3} \right)$$

$$= \text{Impurity mode of B in AC.} \qquad (90)$$

z_1 and a_1 refer to the AB system, z_2 and a_2 refer to the AC system and μ_{AB} and μ_{AC} are the reduced masses. The term $\frac{z^2 e^2}{a^3}$ is a force constant type term and is assumed to vary linearly with x, as,

$$\frac{z^2 e^2}{a^3} = F_Z = F_{ZO}(1-\theta x). \qquad (91)$$

It was assumed that $z_1 = 1.0$. This enables the complete determination of the force constant parameters, F_{ABO}, F_{ACO}, F_{BCO}, F_{ZO} and θ.

The phonon dispersion curves can now be obtained by solving the secular eq. (87) at different points in the k space. Moreover, the phonon behavior can also be obtained for the entire composition range $0 \leq x \leq 1$.

The frequency distribution function as a function of the squared frequencies is defined as,

$$D(\omega^2) = \frac{dM(\omega^2)}{d(\omega^2)}, \qquad (92)$$

where the integrated frequency spectrum, $M(\omega^2)$, is the fraction of phonon states the squares of whose frequencies are less than or equal to ω^2. This can be expressed for the mixed crystal case as,

$$M(\omega^2) = \frac{1}{N} \int_o^{\omega^2} \sum_{k,j} |\underline{e}(k|j)|^2 \delta(y-\omega_j^2(k))dy, \qquad (93)$$

where

$$|\underline{e}(k|j)|^2 = |u_A(k|j)|^2 + (1-x)|u_B(k|j)|^2$$
$$+ x|u_C(k|j)|^2.$$

$u_A(k|j)$, $u_B(k|j)$ and $u_C(k|j)$ are the eigenvectors for the frequency $\omega_j(k)$ and N is the total number of phonon states.

The model parameters were determined using the values of physical observables appropriate to $ZnS_{1-x}Se_x$. This was done so as to get physically realizable values for the results. The corresponding lattice dynamical results in no way are related to the actual results of the three dimensional mixed crystals but may provide insight into the actual three dimensional behavior. The physical observables used as input parameters and the resulting force constant parameters are summarized in Table 3. The phonon dispersion curves for the $ZnS_{1-x}Se_x$ case are presented in Figure 9. In general one expects three branches of

Table 3. Physical properties used and deduced force constant parameters for the $An_{1-x}Se_x$ case (ω in cm^{-1}, F in 10^5 dynes/cm.).

Physical Properties used	$\omega_{LO,AB}$	350^{52}
	$\omega_{LO,AC}$	252^{52}
	$\omega_{I,C}$	220^{52a}
	$\omega_{I,B}$	297^{52b}
Model Parameters	F_{ABO}	0.402950
	F_{ACO}	0.473864
	F_{BCO}	0.665076
	F_{ZO}	0.177979
	Θ	0.210241
	Z_2	0.945792

a: Extrapolated gap mode

b: Local mode

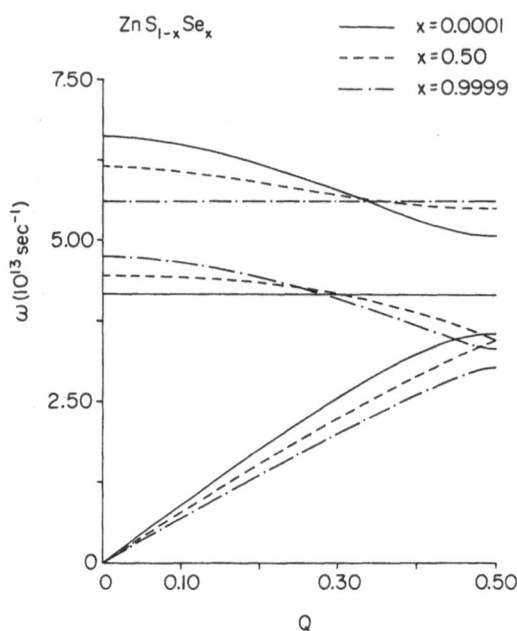

Fig. 9 Phonon dispersion for the $ZnS_{1-x}Se_x$ case at concentrations of x = 0.0001, x = 0.5 and x = 0.9999 as a function of reduced wave vector $(Q = ka/\pi)$.

phonon dispersion in the k-space for the mixed linear diatomic chain. One of these branches refers to the acoustic mode of the system. The two remaining branches are the optic mode of the host lattice and the impurity mode of the system. The mixed $ZnS_{1-x}Se_x$ type systems exhibits two mode behavior and indeed one observes the gap mode at x = 0.0001 and the local mode at x = 0.9999. At these values of concentration the nondispersive behavior of the impurity modes is quite evident. As the concentration departs from these limits of infinite dilution, the impurity mode starts to show dispersion while the optic mode becomes less and less dispersive.

The frequency distribution function for the $ZnS_{1-x}Se_x$ case are presented in Figs. 10 and 11. At x = 0.0001, which is a case of isolated impurities of Se in ZnS, the frequency distribution function is essentially that of the perfect ZnS system. But as x is increased, the impurities begin to play an increasing role in the

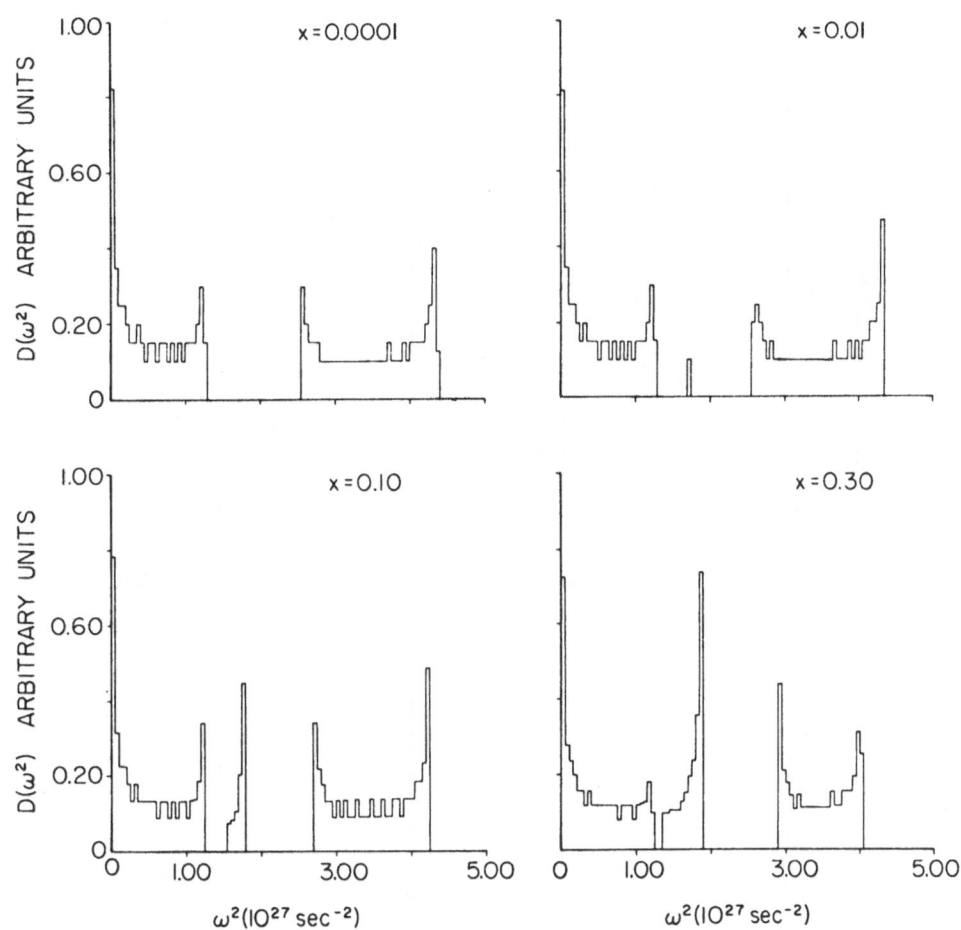

Fig. 10 Frequency distribution function for the $ZnS_{1-x}Se_x$ case at concentrations of x = 0.0001, 0.01, 0.10, and 0.30.

spectrum and consequently one notices a gradual emergence of the impurity frequency band in the gap of the optic and acoustic band. With increasing concentration, the frequency distribution in the impurity band increases while it goes on decreasing in the optic band. As x approaches unity one also notices the turning of the optic mode into the impurity mode and vice versa. And finally, at x = 0.9999, the frequency distribution function is essentially that of the pure ZnSe system.

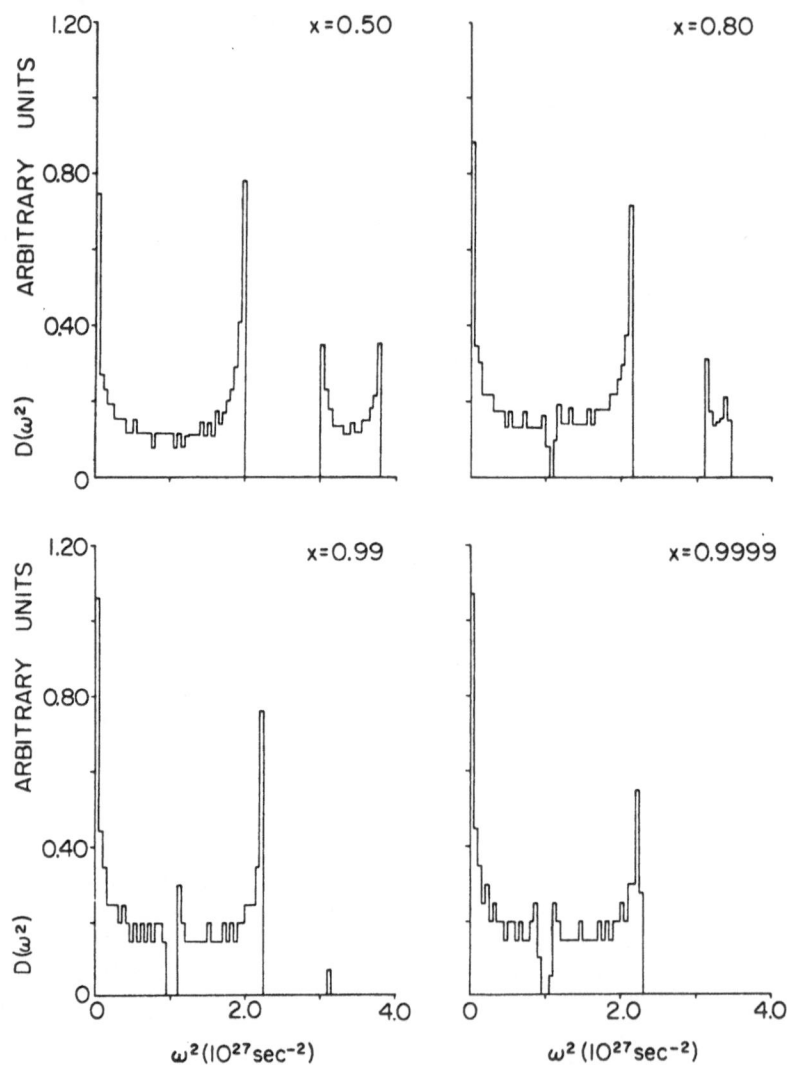

Fig. 11 Frequency distribution function for the
$ZnS_{1-x}Se_x$ case at concentration of x = 0.50, 0.80, 0.99
and 0.9999.

C. Three Dimensional Mixed Crystal System

In order to fully appreciate the mathematical complexity in describing a three dimensional mixed crystal system in the framework of the pseudo unit cell model, it is necessary to present the details of the three dimensional lattice dynamical model used. In these lectures the details of the lattice dynamical model will not be given, but rather the general approach will be outlined. Those who are interested in the details of the lattice dynamical model for the perfect crystal may refer to the appropriate literature[39].

The explicit calculations described herein utilized a modified rigid ion model[39] for the perfect lattice. Corresponding calculated results are presented for various zinc-blende crystals. The modified rigid ion model consists of short-range central and non-central interactions and long-range Coulomb interactions among ions of appropriate effective ionic charge. The complete details of the three dimensional calculation may be found elsewhere[53].

The equation of motion for a three dimensional mixed crystal may be written as

$$[\underline{D} - \omega^2 \underline{I}][\underline{u}] = 0 \qquad\qquad (94)$$

$[I]$ is the familiar identity matrix and ω, the eigenfrequencies. $[u]$ is a nine by one column eigenvector matrix defined as

$$[u] = \begin{bmatrix} u_x(1) \\ u_x(2) \\ u_x(3) \\ u_y(1) \\ u_y(2) \\ u_y(3) \\ u_z(1) \\ u_z(2) \\ u_z(3) \end{bmatrix} \qquad\qquad (95)$$

where particles A, B and C are designated by the numbers 1, 2 and 3 respectively. The matrix D is the nine by

nine dynamical matrix of the form given in eq. (40) relating to the substitutional impurity systems. In this
analysis[53] "s" may take the values 1, 2 or 3. The mixed
crystal coupling coefficient matrices are similar to
those of the perfect three dimensional crystal with the
appropriate concentration weighting factors. A similar
sort of reasoning applies for the Coulomb coupling coefficients. The force constant parameters, the ionic charge
and the lattice constant are assumed to vary in a linear
fashion from one end member to the other.

     The model parameters are determined from the two
optic mode frequencies, the local, gap or resonant mode
frequency and the three elastic constants evaluated at
$x = 0$ and $x = 1$. This gives a total of twelve equations.

     Examples of the three dimensional phonon dispersion
for the system $ZnS_{1-x}Se_x$ are presented in Figs. 12 and
14, for various values of concentration and wave vector

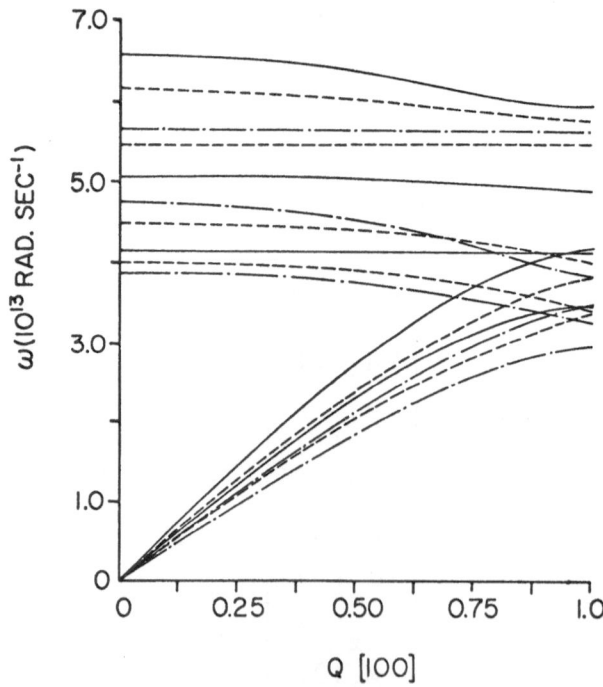

Fig. 12   Phonon dispersion for $ZnS_{1-x}Se_x$ in the $[100]$ direction.   Solid line, $x = 0.0001$; dotted line, $x = 0.50$;
dot-dash line, $x = 0.9999$.

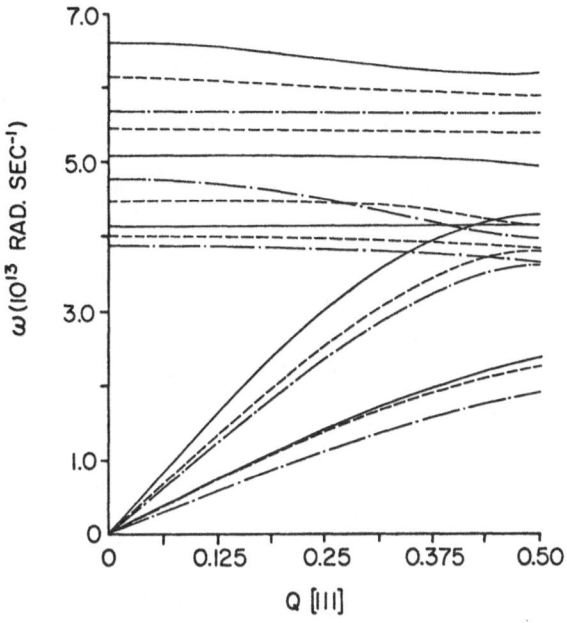

Fig. 13 Phonon dispersion for $ZnS_{1-x}Se_x$ in the $[111]$ direction. Solid line, $x = 0.0001$; dotted line, $x = 0.50$; dot-dash line, $x = 0.9999$.

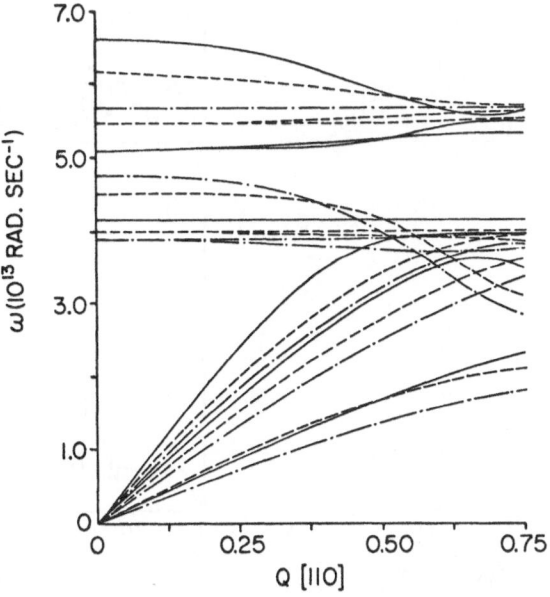

Fig. 14 Phonon dispersion for $ZnS_{1-x}Se_x$ in the $[110]$ direction. Solid line, $x = 0.0001$; dotted line, $x = 0.50$; dot-dash line, $x = 0.9999$.

direction. Appropriate distribution functions are presented for $ZnS_{1-x}Se_x$ and $GaP_{1-x}As_x$ in Figs. 15 - 18. It
is observed that $GaP_{1-x}As_x$ contains local and gap modes
at $x = 1$ and $x = 0$ respectively, while $ZnS_{1-x}Se_x$ has a
local mode at $x = 1$ but no gap mode at $x = 0$.

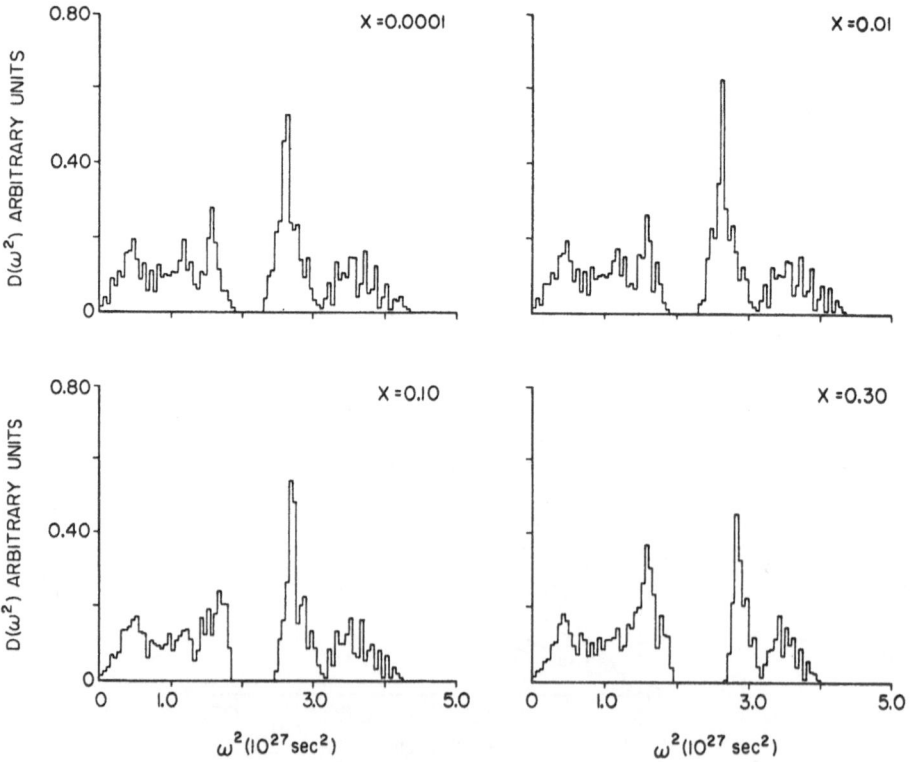

Fig. 15 Frequency distribution function for $ZnS_{1-x}Se_x$
for various values of concentration, x_1 = 0.0001, 0.010,
0.10, 0.30.

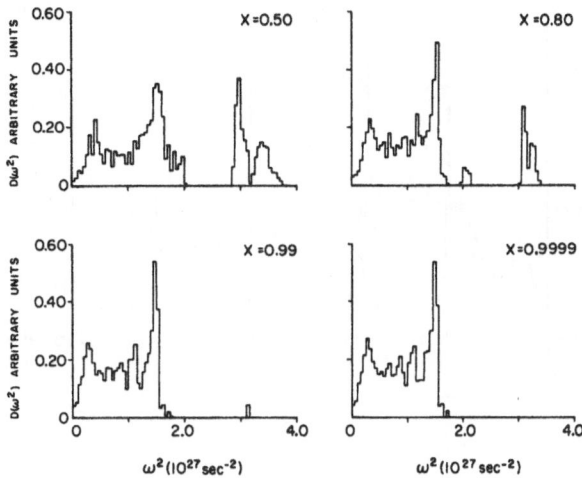

Fig. 16 Frequency distribution function for $ZnS_{1-x}Se_x$ for various values of concentration , x = 0.50, 0.80, 0.99, 0.9999.

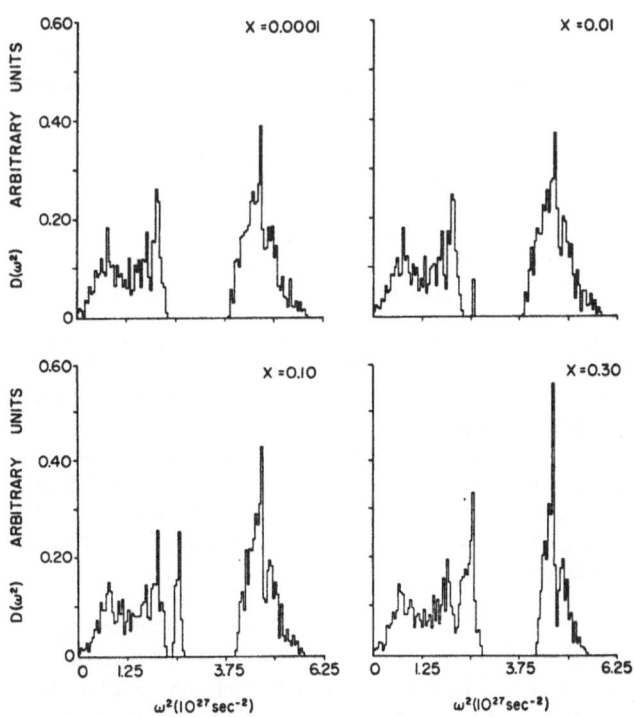

Fig. 17 Frequency distribution function for $GaP_{1-x}As_x$ for various values of concentration, x = 0.0001, 0.010, 0.10, 0.30.

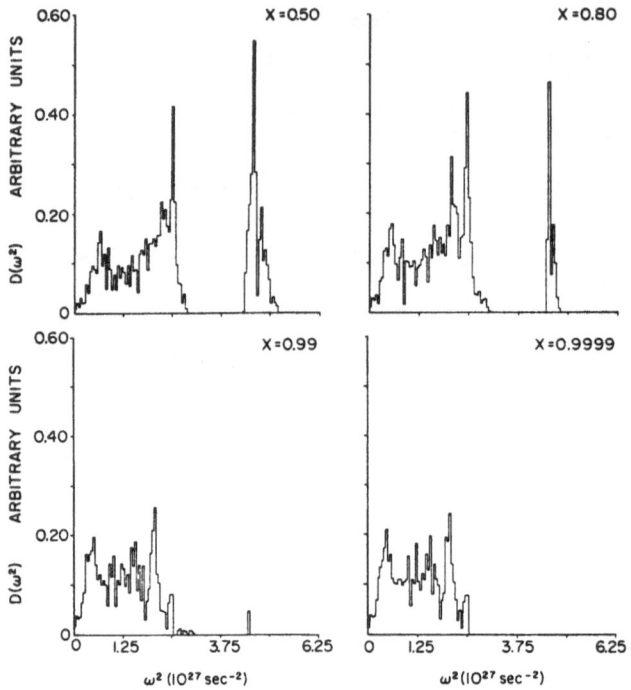

Fig. 18 Frequency distribution function for $GaP_{1-x}As_x$ for various values of concentration, x = 0.50, 0.80, 0.99, 0.9999.

IV. AMORPHOUS SYSTEMS

A. Brief Background

The lack of lattice periodicity greatly complicates the mathematics involved in calculating the dynamical properties of amorphous materials. The complexity increases with randomness in atomic type and geometry. Theoretically, if one knows the exact structure of an amorphous material, one can in principle study the dynamical properties in direct space. However, this technique becomes mathematically cumbersome as the size of the lattice increases.

The majority of the work done on amorphous materials has been in tetrahedrally bonded semiconductors[54] such as germanium and silicon. These materials have been studied by both experimental measurements and by theoretical model simulations. Shevchik and Paul[55,56] have examined the structure of amorphous Ge using small angle X-ray scattering along with the analysis of the radial

distribution function (RDF) which is defined as the number of atoms at different radii in three dimensional
space. The RDF is obtained by a Fourier transform of the
coherently scattered radiation measured by diffraction
experiments. Moss and Graczyk[57] have investigated the
structure of Si by the scanning electron diffraction
technique. The proposed theoretical models for amorphous
Ge and Si[58] might be classified as being microcrystallite models[59], dense random packed models[60], random network models[61] and amorphous cluster models[62].

Comparing the theoretical model RDF with the corresponding experimental data, it is found that the structure of amorphous Ge and Si is a random network lattice[58]
with short-range order (SRO) and long-range disorder.
The tetrahedral bond, with a coordination number of four,
characteristic of the perfect crystal, is retained in the
amorphous phase. The bond lengths and bond angles vary
slightly from the corresponding perfect crystalline values.
The atomic density of amorphous Ge and Si[58] is a few percent lower than the corresponding crystalline value.

Much of the early work[63-68] in the lattice dynamics
of structurally disordered solids was done in one and
two dimensional glasses which are essentially amorphous
compounds. In 1964, Dean[63] was the first to compute the
vibrational density of states for a glass-like disordered
linear chain. The disorder is introduced by a continuous
probability distribution function for interatomic distances, hence for interatomic force constants. The resulting phonon spectra showed a broad feature which not
only smeared out the spectral singularities but extended
the frequency range of the related ordered chain. Bell[65]
has calculated the vibrational spectrum in a topologically disordered chain and examined the mode localization
associated with it. All the atomic masses and force constants were taken to be equal, but the neighbors interacting with each atom were randomly selected to produce
topological disorder. The calculated spectrum is essentially a smoothed-over version of the corresponding crystal. Bell et al[69] have investigated the vibrational
spectra in a two dimensional glass-forming model. The
influence of topological disorder and geometrical disorder was discussed separately.

Lattice dynamical calculations in three dimensions
becomes much more difficult due to the complexity of the
structure. Bell et al[70] have calculated the lattice dynamics of glasses from the three dimensional random net-

work models built by Bell and Dean[71]. Weaire and
Alben[72-74] have calculated the phonon spectra of amorphous Ge and Si. In their work, the calculation was done
for a 61-atom random network constructed by Henderson[75].
The model was obtained by perturbating the atoms in a
distorted diamond lattice. Thorpe[76] has calculated the
phonon density of states for a 5-atom single tetrahedron
to show the effect of local order on the frequency spectra. Bell[77] also indicated the effect of topological
disorder alone on the vibrations of three dimensional
systems. It was pointed out[69] that, in amorphous SiO_2,
topological disorder mainly alters the detailed spectral
profile whereas geometrical disorder extends the frequency
range. Very recently, Alben et al[78] have extensively reviewed the vibrational properties of amorphous Ge and Si.
In their work, the vibrational densities of states of
tetrahedrally bonded amorphous semiconductors have been
calculated for various physical models. These models however have a finite size of about 90 atoms only.

Experimental data[79-84] on the vibrational density
of states of amorphous Ge and Si have been obtained by
optical spectroscopy such as Raman scattering and infrared absorption measurements. The phonon spectra of crystalline and amorphous forms are found to be similar relative to their qualitative feature. Inelastic neutron
scattering experiments[85] have also been used to measure
the vibrational density of states. Additionally, tunneling spectroscopy[86] and electron-energy-loss spectroscopy[87] have recently been employed to obtain the phonon
spectra of amorphous Ge and Si. At the present time,
only the low-energy spectra can be obtained due to the
intensity limitation[78,85].

To date, it is impossible to obtain all the phonon
frequencies present in a three dimensional amorphous
solid. Theoretically, the reason is the inadequate model
representation of the amorphous structure. Experimentally, the phonon spectra cannot be obtained simply by Raman
or infrared measurement because of the uncertainties in
the matrix elements. Although, the neutron scattering
method avoids the dependence of matrix elements, it can
only supply the low-frequency part of the phonon spectra
in amorphous materials.

In order to calculate the physical properties associated with an amorphous material, one must have a
physical model for the structure. In the last few years,
the random network model has been generally accepted[58,88]

for representing tetrahedrally bonded amorphous semiconductors. This approach has more physical appeal than the
approach based on a perturbation of the perfect crystal.
In the simulation of the real amorphous structure, a model
with numerous atoms is usually desirable[78]. However, a
large amount of work is required to generate a truly random network of any significant size. Most of the existing networks are not truly random in nature, but rather
a perturbation of the perfect crystalline structure.
Even though the RDF of the resulting network agrees fairly well with the experimental RDF, the resulting structural representation, itself, might not be truly indicative of the real amorphous structure[88,89]. Also, the use
of a finite size random network along with the invocation
of the periodic boundary condition is at best an approximation to the actual amorphous system.

   In this work, the lattice dynamics of amorphous materials is calculated using a statistical method in which
the lattice dynamics is performed at various lattice spacings and the ensemble is statistically averaged. Statistical models have often been used[90] in describing disordered systems. A phenomenological model was utilized by
Tsay et al[91] to statistically calculate the electronic
spectra for tetrahedrally bonded amorphous semiconductors.
Mitra et al[92] have also developed a statistical approach
to calculate the Raman scattering intensity for amorphous
Ge and Si. The calculated results agree quite well with
experimental values and indicate that the dynamical disorder, as manifested by the change in lattice spacing is
almost entirely responsible for the Raman spectrum[92].
The experimental results[79,80,82,84] on the phonon spectrum show that the general features of the crystalline
and amorphous materials are not too different.

   To illustrate the application of the statistical
averaging technique, the lattice dynamics of an amorphous
chain is calculated. This is followed by the calculation
of the lattice dynamics of amorphous germanium and silicon. The density of phonon states of the corresponding
crystal with different neighbor distances was statistically averaged. The weighting function in the lattice spacings is determined by the experimental RDF[55-57], which
consists of two Gaussian distributions corresponding to
the nearest and next-nearest neighbor distances centered
around the crystalline values. The calculations are compared to experimental data obtained by Raman and infrared
measurements and other theoretical calculations.

B. Linear Monatomic Chain Model

In a one dimensional linear monatomic chain with periodicity, the dispersion relation is given by

$$\omega = 2 \sqrt{\frac{\beta}{m}} \sin \frac{ka}{2} , \qquad (96)$$

where
    m = particle mass
    β = interatomic first neighbor force constant
    a = interparticle spacing
and
    k = wave vector.

In order to determine a value for the force constant β, eq. (96) is evaluated in the long wavelength limit, therefor ka<<1. The dispersion relation reduces to the following,

$$\omega = \sqrt{\frac{\beta}{m}} ak \qquad (97)$$

In the long wavelength case or for a continuum the disperison relation is given as

$$\omega = \sqrt{\frac{c}{\rho}} k \qquad (98)$$

where
    c = elastic constant
and
    ρ = mass density.

Comparison of eqs. (97) and (98) yields the following expression for the force constant,

$$\beta = ca \qquad (99)$$

For a one dimensional chain of N atoms with the end points fixed, the phonon frequencies, ω_i may be written as

$$\omega_i = \sqrt{\frac{2}{m}} \sqrt{ca} \sin\frac{i\pi}{2N} , \qquad (100)$$

where
    i = 1--------N - 1.

The elastic constant, c, is a function of the lattice spacing in the crystal. Intuitively one would expect the chain to become stiffer when the atoms are closer together.

In order to get an order of magnitude result, the physical constants associated with germanium are used.

The dependence of the elastic constants on lattice spacing is deduced from the pressure dependence of the elastic constants, which is known experimentally. The relationship between the lattice spacing and the elastic constants can be expressed in terms of Murngahan's equation given as

$$a = a_0 [1 + \frac{(dB/dp)_0}{B}]^{-\frac{1}{3(dB/dp)_0}} \tag{101}$$

where a_0 = lattice constant at atmospheric pressure
and

$$B = \text{bulk modulous} = \frac{1}{3}(C_{11} + 2C_{12})$$

The lattice dynamics may now be determined for a particular value of lattice spacing. In particular for each different lattice constant the Murngahan's equation is used to get the pressure and hence the appropriate elastic constant value.

The lattice constant values are weighted according to the radial distribution function. This is given in terms of a Gaussian distribution function as

$$p(a) = \frac{1}{2\pi\sigma} e^{-[\frac{(a-a_0)^2}{2\sigma^2}]}, \tag{102}$$

where σ^2 = mean square deviation.

The calculation was performed for 50 different lattice spacings. The phonon density of states of the amorphous material was obtained by associating the appropriate weighting factor with the density of states at each lattice spacing.

Figure 19, depicts the phonon density of states in a perfect and amorphous material. Qualitatively the phonon density of states in the amorphous chains is just a smoothed over version of the phonon density of states in the perfect crystal.

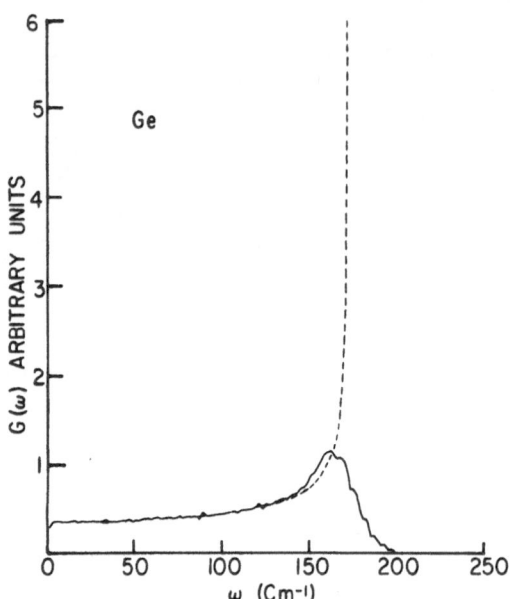

Fig. 19 Density of phonon states of Ge. Solid line -
amorphous; dashed line - crystal.

C. Lattice Dynamics of Amorphous Germanium and Silicon

Experimental studies on the radial distribution function have shown that some sort of short-range order
exists in amorphous germanium and silicon. Owing to the
short range order in these materials plus the fact that
they do not have appreciable Coulomb interactions, a
short range model is not a bad approximation. A modified
rigid ion model[39] is used as the lattice dynamical model.
The lattice dynamics of a perfect crystal is performed
for various values of radial distance and then statistically weighted according to the radial distribution functions in order to obtain the frequency distribution function.

The phonon frequencies, $(\omega_1 \ldots \omega_n)_i$, corresponding
to the i-th lattice spacing are given by the solution of
the following determinantal equation,

$$|D - \omega^2 I|_i = 0, \tag{103}$$

where D is the dynamical matrix and I, the identity matrix. The model parameters needed in the D matrix can be expressed in terms of the elastic constants and the Raman frequency as follows,

$$C_{11} = \frac{\alpha}{a} + \frac{8\mu}{a'} \tag{104}$$

$$C_{12} = \frac{1}{a} (-\alpha + 2\beta) + \frac{4\mu}{a'} , \tag{105}$$

$$C_{44} = \frac{1}{a} (\alpha - \beta^2/\alpha) + \frac{4\mu}{a'} , \tag{106}$$

and

$$f_R = \frac{1}{2\pi c} \sqrt{\frac{8\alpha}{m}} . \tag{107}$$

α and β are near neighbor force constants and μ is a central second neighbor force constant. a is the lattice constant appropriate to near neighbor interaction while a' is appropriate to second neighbor interactions. Obviously a = a' in a perfect crystal.

In order to obtain the model parameters at each lattice spacing, it is necessary to know the values of the physical observables at the spacing. The input physical observables, which include the elastic constants and the Raman frequency, are known[93-95] experimentally for both Ge and Si. They are simply linear functions of the pressure for reasonable pressure ranges. Figure 20 shows the pressure dependence of the elastic constants and the Raman frequency for Ge. Similar results are known[94] for Si. The pressure dependence of the physical observables can be written as,

$$c_{ij}(p) = c_{ij}(p=0) + (dc_{ij}/dp) p, \tag{108}$$

$$B(p) = B(p = 0) + (dB/dp) p, \tag{109}$$

and

$$f_R(p) = f_R(p = 0) + (df_R/dp) p, \tag{110}$$

where the permutation of the indices i and j is understood to indicate C_{11}, C_{12} and C_{44}. The elastic constants and Raman frequency at each lattice constant can be evaluated if the corresponding pressure at this lattice constant is known. This is done by using Murnaghan's

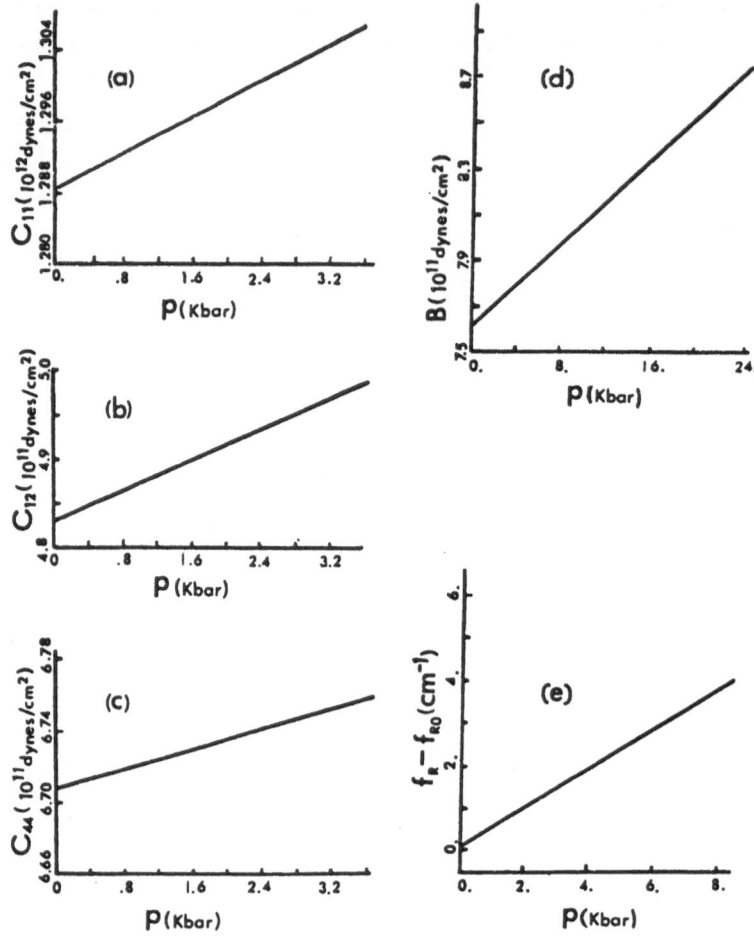

Fig. 20 Pressure dependence of the elastic constants and the Raman frequency for Ge. (a) C_{11} (ref. 93); (b) C_{12} (ref. 93); (c) C_{44} (ref. 93); (d) B (ref. 93) and (e) f_R (ref. 95).

equation,

$$p = \left[(a_0/a)^{3(dB/dp)} - 1 \right] \frac{B_0}{(dB/dp)} . \tag{111}$$

so as to give a pressure corresponding to each lattice constant.

The first and second near neighbor distances, x and y, are known experimentally from the RDF data,

Table 4: Summary of Structural Data of Germanium and Silicon

	Ge[*]	Si[**]
r_1^c (Å)	2.45	2.35
r_1^a (Å)	2.46 ± 0.02[+] 2.47 ± 0.01	2.35
σ_1^c (Å)	0.080 ± 0.008	0.141
σ_1^a (Å)	0.088 ± 0.008	0.167
σ_1 (Å)	0.037 + 0.026 − 0.037	0.09
r_2^c (Å)	4.00[++]	3.84[++]
r_2^a (Å)	4.00 ± 0.04	3.86
σ_2^c (Å)	0.12 ± 0.02	0.200
σ_2^a (Å)	0.28 ± 0.01	0.319
σ_2 (Å)	0.25 ± 0.02	0.25

[*] Experimental data due to Temkin et al. (Ref. 96)

[**] Moss and Graczyk (Ref. 57).

[+] Shevchik and Paul (Ref. 55)

[++] Calculated from nearest neighbor distance

$$(r_1^c), \quad r_2^c = \sqrt{8/3} \ r_1^c$$

$$P(x,y) = \exp \left\{ - 0.5 \left[\left(\frac{x-x_0}{\sigma_1} \right)^2 + \left(\frac{y-y_0}{\sigma_2} \right)^2 \right] \right\} \qquad (112)$$

where

$$x = \sqrt{3} \, a/4$$

and

$$y = a / \sqrt{2} \ .$$

x_0, y_0 and σ_1, σ_2 are the mean and standard deviation respectively of the first and second neighbor distance and K is a proportional constant. The means are chosen to be the corresponding crystalline values. The standard deviation used in this context is the static deviation, σ_1 derived[96] from the experimental crystalline and amorphous deviations, $\sigma_i{}^C$ and $\sigma_i{}^a$ respectively, according to

$$\sigma_i = \left[(\sigma_i{}^a)^2 - (\sigma_i{}^C)^2 \right]^{1/2} \ . \qquad (113)$$

It is assumed that thermal broadening is the same for both the crystalline and amorphous forms. The values of the means and standard deviations of Ge and Si are presented in Table 4. Since the lattice constant is known from x and y, the corresponding value for the pressure is obtained from eq. (111). The value of the elastic constants and Raman frequency for the appropriate lattice spacing is then deduced from the pressure dependence of these quantities which is presented in Table 5. Since there is an unbalance in the number of physical observables and model parameters, the model parameters are deduced by solving equations (104) - (107) in a least square sense[39].

To avoid the complexity of performing lattice dynamical calculations at many lattice spacings, a linear approximation which calculates the lattice dynamics only three times is used. The linear Taylor expansion used to obtain the phonon frequencies at each i-th spacing is

$$\omega_i(x,y) = \omega(x_0, \ y_0) + \frac{\partial \omega(x,y)}{\partial x} \Big|_{x_0, y_0} (x-x_0)$$

$$+ \frac{\partial \omega(x,y)}{\partial y} \Big|_{x_0, y_0} (y-y_0) \ ,$$

Table 5: Summary of Physical Observables * of Germanium and Silicon

Material	C_{110}**	C_{120}**	C_{440}**	B_0**	ν_0***	$\dfrac{dC_{11}}{dp}$	$\dfrac{dC_{12}}{dp}$	$\dfrac{dC_{44}}{dp}$	$\dfrac{dB}{dp}$	γ	$\dfrac{d\omega}{dp}$*****
Ge	1.2886	0.4829	0.6710	0.7515	301	4.99	4.32	1.25	4.51	0.89[69]	0.356
Si	1.65779	0.639365	0.796246	0.9788	520	4.33	4.19	0.80	4.24	0.90[70]	0.478

* Experimental data due to Mcskimin (ref. 93,94) except otherwise indicated

** Elastic constants and bulk moduli are in units of 10^{12} dyne/cm^2.

*** Raman frequencies (ref. 97) are in units of cm^{-1}.

**** Raman shifts, as calculated from Grüneisen constant in this paper, $\gamma = B/\nu \, (d\nu/dp)$, are in units of cm^{-1}/kbar.

where

$$\frac{\partial \omega(x,y)}{\partial x}\Big|_{x_o,y_o} = \frac{\omega(x_1,y_o) - \omega(x_o,y_o)}{x_1-x_o} \ ,$$

$$\frac{\partial \omega(x,y)}{\partial y}\Big|_{x_o,y_o} = \frac{\omega(x_o,y_1) - \omega(x_o,y_o)}{y_1-y_o}$$

(114)

and

$$x_1 = x_o + 0.01$$

$$y_1 = y_o + 0.02$$

In the present calculation, the variation of x around x_o is chosen to be about $\pm 5^o/o$, whereas the y variation around y_o is about $\pm 12.5^o/o$. These spreads are about twice the Gaussian spread for both amorphous Ge and Si. The phonon frequencies are calculated at the perfect lattice spacing (x_o,y_o) and at the slightly different spacings (x_1,y_o), $x_o,y_1)$. This was done by calculating the force constant set $[\alpha,\beta,\mu]$ from the physical observables determined at the average value of two pressures corresponding to the first and second neighbor distances. Using eq. (114), the phonon frequencies at various lattice spacings for first and second neighbors are then obtained.

The density of phonon states, $G(\omega)$, is defined as the following,

$G(\omega)d\omega$ = number of phonon states whose frequencies
                are located between $\omega = d\omega/2$ and
                $\omega + d\omega/2$.   (115)

The $G(\omega)$ of a crystal is obtained by categorizing all the allowed phonon frequencies in the first Brillouin zone. According to the symmetry of the Brillouin zone associated with the diamond structure, the k points need only be confined to a region in which

$$16 > k_x > k_y > k_z > 0.$$ (116)

This region, named the irredicuble Brillouin zone, is only 1/48th of the first Brillouin zone. Since the phonon states are uniformly distributed in k space, it is reasonable to select an equally distributed mesh of points

in the irreducible Brillouin zone so that the phonon density of states will be obtained. In this work, 149 independent $\underline{k}$ points along with their degeneracy factors[98]
were used for the frequency calculations. A total of
24,576 frequencies were used to calculate phonon density
of states for the crystalline structure.

The phonon density of states for the amorphous material was calculated by statistically averaging each
$g_i(\omega)$ for the i-th lattice spacing according to the weighting function appropriate to the RDF. Mathematically,
this can be represented as

$$G(\omega) = \sum_{i=1}^{n} P_i(x,y) \, g_i(\omega) \tag{117}$$

$P_i(x,y)$ is the associated RDF weighting function defined
by eq. (112) and n is the number of lattice spacings used.
In the present calculation, n was equal to 1250. The
resulting densities of phonon states for both crystalline
and amorphous Ge and Si are presented in Figs. 21 and 22
respectively.

Experimental data for the density of phonon states
of various amorphous materials can be obtained by Raman
and infrared measurements. Due to the lack of long-range
order, the $\underline{k}$-selection rules break down in amorphous
solids. Therefore, all the vibrational modes can take
part in the Raman or infrared process. Under the assumption that all vibrational modes couple equally to the
light, the Raman or infrared spectrum can then be used
as a measure of the density of phonon states in an amorphous material[79,80]. Fig. 23 and Fig. 24 present the
optical and phonon spectra for Ge and Si respectively[84].
Fig. 23a is the reduced Raman and infrared spectra obtained by experimental measurements for amorphous Ge.
Fig. 23b shows the density of phonon states. The dashed
line is the crystalline spectrum obtained by Dolling and
Cowley[99] using an eleven-parameter model to fit the neutron inelastic scattering data. The solid line for amorphous Ge[84] is a Gaussian-broadened version of the crystalline density of states. This is done by convoluting the
crystalline density of states with a Gaussian distribution. The Gaussian spread was chosen[79] by comparing the
reduced Raman spectrum in the amorphous case with the
corresponding crystalline Raman frequency. Fig. 24 shows
the equivalent results for Si. However, to be more physically plausible, one should really determine the

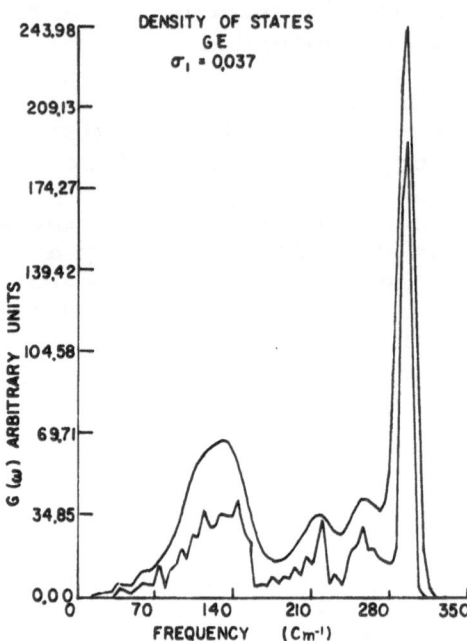

Fig. 21 Density of phonon states of Ge. Top curve -
amorphous; bottom curve - crystal.

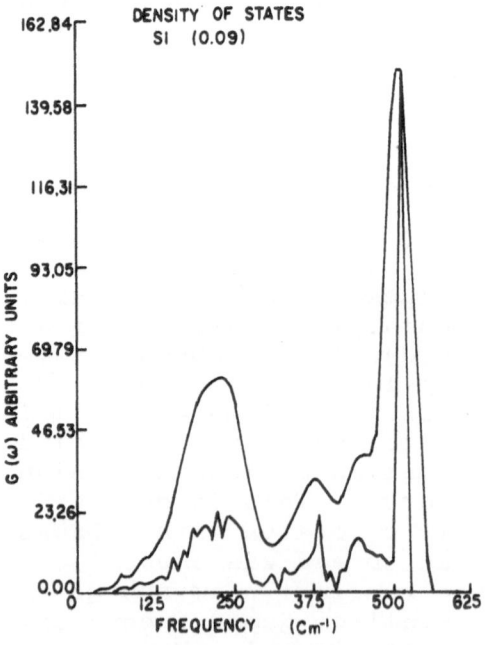

Fig. 22 Density of phonon states of Si. Top curve -
amorphous; bottom curve - crystal.

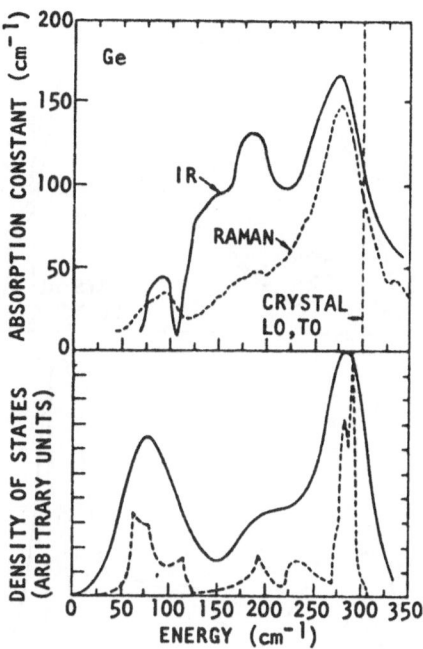

Fig. 23 (a) The infrared and Raman spectra for amorphous Ge. Solid line - infrared absorption constant (ref. 84). Dashed line - Raman spectrum (ref. 82). (b) The density of phonon states for Ge. Solid line - amorphous (ref. 84); Dashed line - crystal (ref. 99).

amorphous density of states from a lattice dynamical model rather than just use a broadening of the crystalline density of states[84].

Figs. 25 and 26 compares the phonon density of states of Ge and Si calculated using the statistical method, to the results of other work. The present results, as well as the previous results, indicate that the amorphous phonon density of states is a smoothed over version of the crystalline density of states. The singularities in the crystalline density of states are smeared out by the disorder in the amorphous state. The similarity between the crystalline and amorphous spectra is due to the fact that the short-range model used for the lattice dynamical calculations is essentially the same in both the amorphous and crystalline phases. Comparing this work to

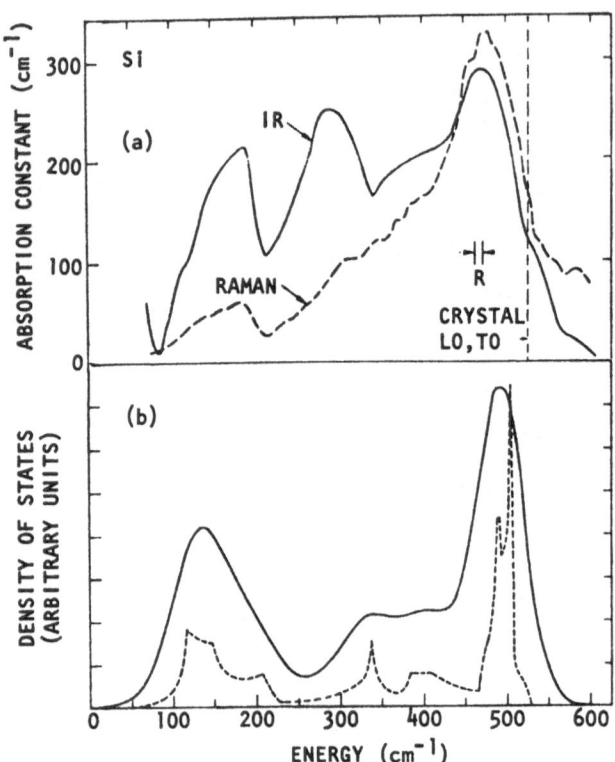

Fig. 24 (a) The infrared and Raman spectra for amorphous
Si. Solid line - infrared absorption constant (ref. 84);
Dashed line - Raman spectrum (ref. 79). (b) The density
of phonon states for Si. Solid line - amorphous (ref.
84); Dashed line - crystal (ref. 99).

Brodsky's results[84], the agreement is fairly good. However, the frequency associated with our first peak is
slightly higher. This is because the rigid ion model was
used in our calculation whereas Dolling's model[99] was
used in the other work. Our work also compares favorably
to Alben's theoretical calculation[73] using a 61-atom random network[75]. The finite size of the model along with
the periodic boundary conditions might be the reason that
the phonon density of states in Alben's work is not as
smooth as the others.

In Figure 27, the calculated phonon density of
states of Ge is compared to various experimentally obtained Raman and infrared data[83,84]. It may be pointed

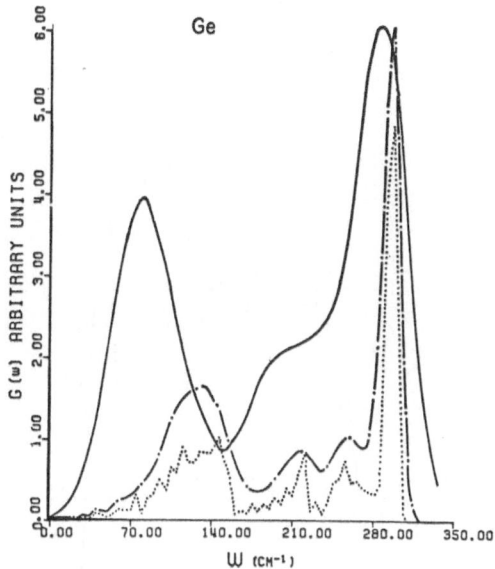

Fig. 25 Density of phonon states for Ge. Dash-dotted
line - amorphous form as calculated in this work; Dotted
line - crystalline form as calculated in this work;
Solid line - Brodsky's result (ref. 84).

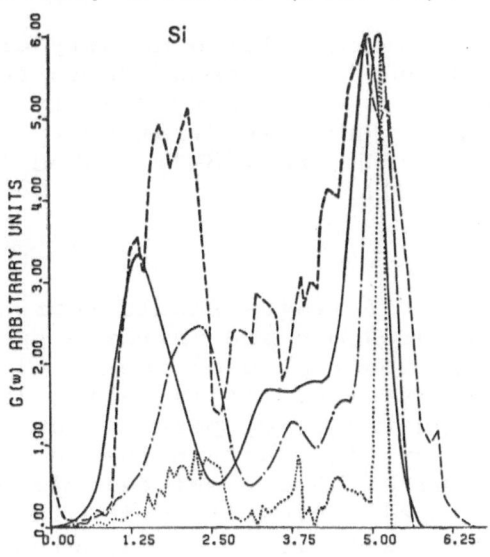

Fig. 26 Density of phonon states for Si. Dash-dotted
line - amorphous form as calculated in this work; Dotted
line - crystalline form as calculated in this work; Dashed
line - Alben's calculation (ref. 73); Solid line -
Brodsky's result (ref. 84).

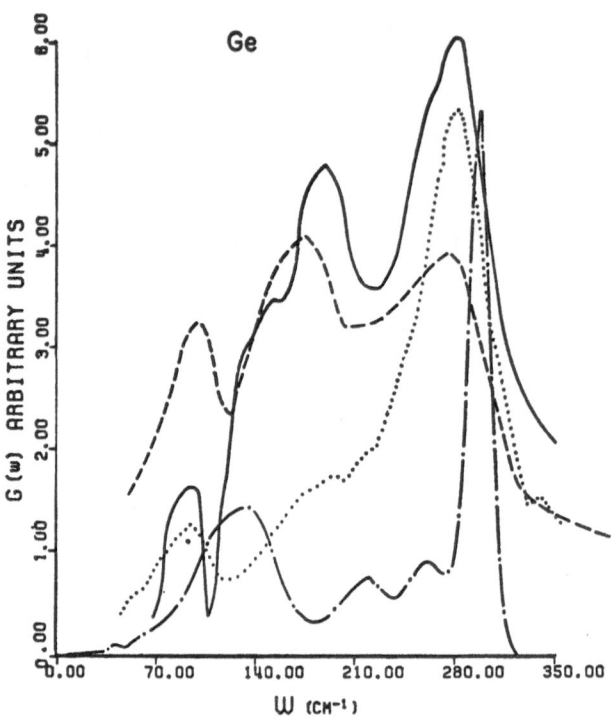

Fig. 27 Comparison of calculated density of phonon
states to infrared and Raman measurements for amorphous
Ge.   Dash-dotted line - calculated density of phonon
states; Dotted line - Raman spectrum (ref. 84); Solid
line - infrared spectrum (ref. 84); Dashed line - infrared spectrum (ref. 83).

out that the density of states should represent the features observed in the Raman and infrared spectra.   However, they are not expected to be identical because of
the matrix element effects.   It is particularly worth
noting that the high frequency peak in the Raman spectrum is much stronger than that in the infrared spectrum,
and compares well with the present calculated result.
This becomes even more obvious from the infrared results
of Stimets et al[83].   This is to be expected since in crystalline Ge, the first order Raman spectrum for the long-
wavelength optical phonon is allowed whereas it is forbidden for the infrared spectrum.   It has been pointed
out before[92] that the allowed Raman spectrum in the crystalline case, although somewhat broadened, persists in
the amorphous phase.

V. ONE-PHONON INFRARED AND RAMAN SPECTRA
OF TETRAHEDRALLY BONDED AMORPHOUS SEMICONDUCTORS

A. Introduction

Most existing works[73,82,100] on the optical properties of amorphous solids emphasize the role of $\underline{k}$-selec-
tion rule breakdown, and what is consequently expected
to be a strong similarity between densities of states and
optical spectra of such solids. Recently, the role of
crystalline-allowed analogue (CAA) processes in tetrahedrally bonded amorphous semiconductors (TBAS) was pointed out by Tsay et al[91] for electronic spectra, and by
Mitra et al[92] for lattice spectra. In this section, we
develop a statistical approach to the interpretation of
lattice spectra in amorphous solids, which accounts in
a natural way for the role of CAA processes. The formalism is then applied to interpret available experimental
first-order infrared and Raman spectra of TBAS.

We view[101] the properties of an amorphous solid as
derived from an ensemble of atomic (structural) configurations (labeled by "n") with probabilities p_n, rather
than from a single unique structural entity. Then for
a particular property Θ we require the configurational
average

$$<\Theta> = \sum_n p_n \, \Theta(n) \tag{118}$$

where $\Theta(n)$ is the expectation value of Θ in configuration
n. Clearly, for any relistic system, it is difficult if
not impossible to evaluate $<\Theta>$ from first principles.
Rather, for the statistical approach to be useful one
must be able to identify a restricted class of configurations which elicit the principal features of the properties concerned; moreover, of course, one must be able to
calculate the corresponding $\Theta(n)$'s for these configurations. We here propose to parameterize TBAS in terms of
two classes of principal configurations, namely those
possessing (a) both short and long range order {denoted
by $(n)_{SL}$} and (b) just short range order {denoted by
$(n)_S$}. The motivation for this parameterization for TBAS
is that the experimental evidence[83,102,103] indicates
that to a large extent they are "nearly crystalline";
i.e., spectral features that are prominent in the crystal
remain prominent in the corresponding amorphous solid,
and are expected to arise from the $(n)_{SL}$. Features which
are weak or absent for the crystal are generally also
weak in the amorphous counterpart, and are expected to

arise principally through the non-crystalline configurations $(n)_S$. Each of the configurations in the set $(n)_{SL}$
will lead individually to features obeying crystalline
k-selection rules, i.e., $\theta^{SL}(n) = 0$ for forbidden processes. The final spectra will in general however, be broadened out because of the averaging process. If, on the
other hand, the p_n possess a considerable spread, then
the crystalline features may be largely washed out. The
configurations $(n)_S$, on the other hand, are not subject
to crystalline selection rules, and always contribute to
the spectrum. In addition to broad, featureless contributions arising from the breakdown of the k-selection
rule, various peaks evident in the crystalline density
of states, but suppressed by selection rules in optical
spectra, will reappear for the amorphous solid.

In quantum mechanical terms, this is equivalent to
the introduction of a perturbation Hamiltonian in an
otherwise ordered solid such that,

$$H_{disord} = H_{ord} + (H_{disord} - H_{ord}) = H_{ord} + H' \tag{119}$$

When the solid is disordered, such a perturbation will
introduce changes in the reststrahlen frequencies, ω_{TO},
or ω_{LO} and also in the frequency-dependent matrix elements for dipole moment, $M(\omega)$ and induced polarizability,
$\alpha(\omega)$. For ordered solids, $M(\omega)$ and $\alpha(\omega)$ depend purely
on the structural symmetry. For example, for the rock-
salt and zinc-blende structures $M(\omega) = 0$ for all ω except $\omega = \omega_{TO}$, where it is large. On the other hand, for
NaCℓ-type crystals $\alpha(\omega) = 0$ for all ω, and for the diamond structure $\alpha(\omega) \neq 0$ for $\omega = \omega_{LO}$. In the structurally disordered solids $M(\omega)$ is now finite for all ω because of the removal of the translational symmetry, and
probably small, except for $M_a(\omega_{TO})$, corresponding to the
allowed transition in the case of an ordered solid, which
is still large. Similar remarks hold for $\alpha(\omega)$. If the
perturbation limit is valid, we shall expect differences
between the perturbed and unperturbed cases to be finite
and small, which is indeed observed in many cases, for
example, mixed crystals systems[49] where translational
symmetry is substantially disturbed. We shall demonstrate
later that the viewpoint described here does indeed provide a basis for a consistent interpretation of observed
spectra of TBAS.

With the above prescriptions, one may write,

$$<\Theta> = <\Theta>^{SL} + <\Theta>^{S}$$

$$= \sum_{n} [\rho^{SL} p_n^{SL} \Theta^{SL}(n) + \rho^{S} p_n^{S} \Theta^{S}(n)] \qquad (120)$$

$$\sum_{n} p_n^{SL} = \sum_{n} p_n^{S} = 1, \quad \rho^{SL} + \rho^{S} = 1$$

where we have chosen a convenient normalization for the p_n's. The parameters $\eta \equiv \rho^{S}/\rho^{SL}$, which we here deduce from experiments, is the fraction of disordered relative to ordered configurations, and thus provides a measure of the amorphicity of the solid ($\eta = 0$ for a crystal; $\eta \gg 1$ for a highly disordered system). In order to proceed, one must specify the $(n)_{SL}$ and $(n)_{S}$ to be employed above. Among the configurational parameters n suggestive for TBAS, are the interatomic spacings, local densities and bond angles, all known to deviate from those of their crystalline analogues and to display characteristic spreads as well. As a first step in the calculation of $<\Theta>^{SL}$ we here restrict n to correspond to variations in local density ρ alone. While the calculation of $<\Theta>^{SL}$ for a given value of ρ will be relatively straight forward, this is not the case for $<\Theta>^{S}$. A reasonable, but not unique, possibility is to employ just a nearest neighbor (nn) unit immersed in a continuum to calculate the $(n)_{S}$. This would account for the principal features representative of short-range order, yet suppress those features characteristic of the periodic lattice structure beyond the nn cell. In this paper, we calculate only $<\Theta>^{SL}$ for IR and Raman spectra; analogous calculations for $<\Theta>^{S}$ will be reserved for future work.

B. Local Density Distribution in TBAS

One of the most important quantitative data on the structure of amorphous semiconductors is the RDF which is defined as

$$F(r) = 4\pi r^2 \rho(r) \qquad (121)$$

where $\rho(r)$ is the density of atoms at a distance r from an arbitrarily chosen central atom $(r = 0)$. The experimental data on the RDF of amorphous tetrahedrally-bonded semiconductors indicate the following:

(i) The RDF for both the crystalline and amorphous forms of the same material are very similar, at least up to the second neighbor distance, implying that the short-range order is preserved for the amorphous material.

(ii) The positions of the first and second peaks of the amorphous RDF are slightly shifted (a few percent) toward higher values as compared with the crystalline counterparts. This is consistent with the fact that the amorphous form of a material usually has lower density than that of the crystalline form.

(iii) Aside from thermal broadening, there are intrinsic widths, σ_i^s, of the peaks in the amorphous RDF. This is due to the introduction of static distortions in bond lengths and/or angles resulting in the amorphous structure. The first and second peaks of the RDF of an amorphous material are fairly well-defined, and may normally be expressed as a Gaussian of the form

$$\exp\left[-\frac{(r-r_i^a)^2}{2\sigma_i^{a\,2}}\right], \quad i = 1, \text{ or } 2 \tag{122}$$

where r_i^a and σ_i^a are the position and width, respectively, of the ith peak. The spreads σ_i^a, however, include thermal as well as disorder-induced effects. In order to consider disorder-induced effects only, the static spread, σ_i^s, should be obtained from σ_i^a after correcting for the thermal spread.

If one regards the spreads in the first and second peaks in the amorphous RDF as distributions of the first and second nn distances which exist in an amorphous material, the "local" density at a point is expected to depend on the extent of the distortions of the local environment from the crystalline case. We define local density as the number of atoms per unit volume, for a volume whose linear dimension is of the order of a few multiples of the nn distances. Given the statistical distributions in the local distortions, (i.e., the spreads in first and second nn distances), the distribution in local density can be approximately calculated.

In crystalline semiconductors where the distance between any pair of atoms is fixed, the local density should be uniform throughout the crystal. For diamond and zinc-blende crystals, there are two particles per unit cell. The cell volume is $a^3/4$, where a is the lattice constant. Defining the nearest neighbor distance as

$$r_1 = \frac{\sqrt{3}}{4}\, a \tag{123}$$

the density in terms of this distance is

$$\rho = \frac{3\sqrt{3}}{8} \left(\frac{1}{r_1^3}\right) \qquad (124)$$

In amorphous solids, both r_1 and r_2 have spreads about their respective mean values, r_1^a, r_2^a. In order to include the second neighbor atoms in defining the local density, one may write

$$\rho(r_1, r_2) = \frac{3\sqrt{3}}{8} \left[\frac{1}{r_1^3} + C \frac{k^3-1}{r_2^3-r_1^3}\right]/(1+C) \qquad (125)$$

where $k = r_2^a/r_1^a$ and C is a weighting factor determining the relative importance of the contribution of the second nn to the local density. The reasons for defining the local density in the form of eq. (125) are: (a) The terms

$\frac{1}{r_1^3}$ and $\frac{k^3-1}{r_1^3-r_1^3}$ define a correct density in the crystal-

line case; and (b) If an amorphous semiconductor is such that its first nn distance, r_1^a, and the next neighbor distance, r_2^a do not have any spread, eq. (125) reduces to eq. (124).

    In order to calculate the distribution in local density, a range of r_1 (or r_2) about $3\sigma_1$ (or $3\sigma_2$) from $r_1^a(r_2^a)$ is considered. The probability distributions of r_1 and r_2 in these ranges are assumed to be independent. Therefore,

$$p(r_1, r_2) \propto \exp\left\{-0.5\left[\left(\frac{r_1-r_1^a}{\sigma_1}\right)^2 + \left(\frac{r_2-r_2^a}{\sigma_2}\right)^2\right]\right\} \qquad (126)$$

From eq. (124) and (125), the distribution of local density, $p(\rho)$ can be determined, provided the value of C is known. A reasonable value of C may be $\frac{r_1^a{}^2}{r_2^a{}^2}$.

In the crystalline case, this value of C is equal to 0.375. This definition for C implies that the relative importance of atoms in contributing to the local density is inversely proportional to the distance from the origin. Fig. 28, shows the local density distribution for various values of C. It is seen that for $C \leq 0.5$, the curve does

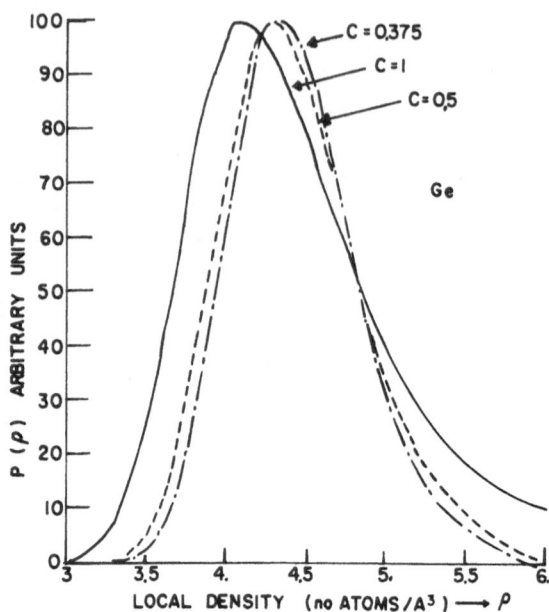

Fig. 28 Changes in the distribution of local density, $p(\rho)$, due to variations in C (see text).

not depend strongly on C, and has a Gaussian shape except for a slight asymmetry on the high density side.
The local density distribution for various values of σ_2,
as shown in Fig. 29 are not affected appreciably, indicating that the local density is most likely determined
by σ_1, the spread in the nearest neighbor distance. The
most reasonable value for local density in this analysis
seems to be for C = 0.375.

   C. Method of Calculation

   To calculate the contribution of $\langle\Theta\rangle^{SL}$ to infrared
absorption we take for the imaginary part of the lattice
susceptibility of a crystal with a single reststrahlen
frequency ω_{TO}

$$\varepsilon_2^c \propto \frac{\Gamma\omega}{(\omega_{TO}^2 - \omega^2)^2 + \Gamma^2\omega^2} \tag{127}$$

with $\Gamma \sim$ constant in the vicinity of the reststrahlen spectrum. Then the SL contribution to the amorphous solid is

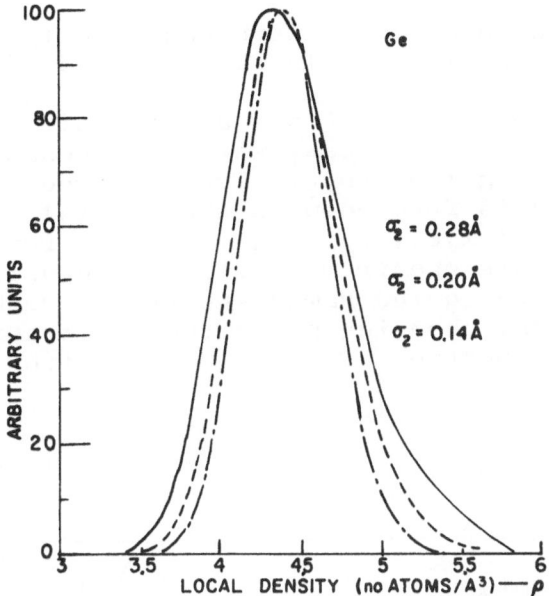

Fig. 29 Changes in the distribution of local density, $p(\rho)$, due to variations in σ_2 (see text).

$$\varepsilon_2^{SL} = \rho^{SL} \int_0^\infty d\rho \ p(\rho) \ \varepsilon_2^c(\rho) \qquad (128)$$

In the simplified calculations carried out here we assume that ε_2^c depends on ρ only through ω_{TO}, and that the distribution $p(\rho)$ is a Gaussian. This spread will be relatively narrow, and is taken to be peaked around the equilibrium nn distance r_0 of the amorphous solid. Such assumptions seem reasonable in view of the local density distribution as discussed in the previous section. Values of $\omega_i(\rho)$ may be inferred directly from the Grüneisen approximation, which should be adequate for small departures in ρ_0 from crystalline values. Thus[104]

$$\omega_i(\rho) = \omega_i(\rho_0)(\rho/\rho_0)^{\gamma_i} \qquad (129)$$

where γ_i is the mode Grüneisen parameter[105,95]. An obvious analogous calculation yields the SL contribution to the Raman intensity for an amorphous solid.

D. Results and Discussions

The spread in $p(\rho)$ has been adjusted to provide a best fit with experiment. In Fig. 30a and 30b we display the first order Raman and infrared spectra of amorphous Ge. If one chooses $\eta = 0$ a good fit is obtained for the principal Raman peak, with the remaining features presumably arising from $(n)_S$. The infrared one-phonon spectrum, which is forbidden in the crystal, arises entirely from $(n)_S$; i.e., $\varepsilon_2 = \varepsilon_2^S$. Apart from matrix element effects, this spectrum should be proportional to the $(n)_S$ contribution to the Raman spectrum, which may be determined by first choosing η and then subtracting the calculated (SL) portion of the Raman spectrum from the

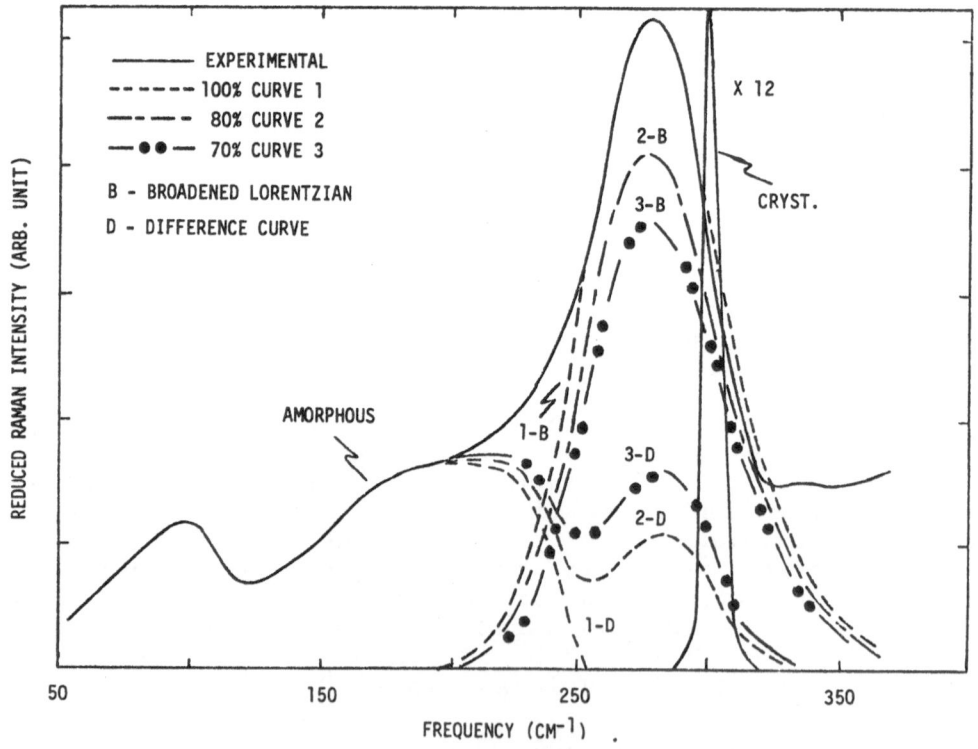

Fig. 30a Comparison of calculated crystalline like contribution and experimental Raman spectrum of amorphous Ge. Experimental data (ref. 82). The difference curves represent the $(n)_S$ contributions similar to the vibrational density of states.

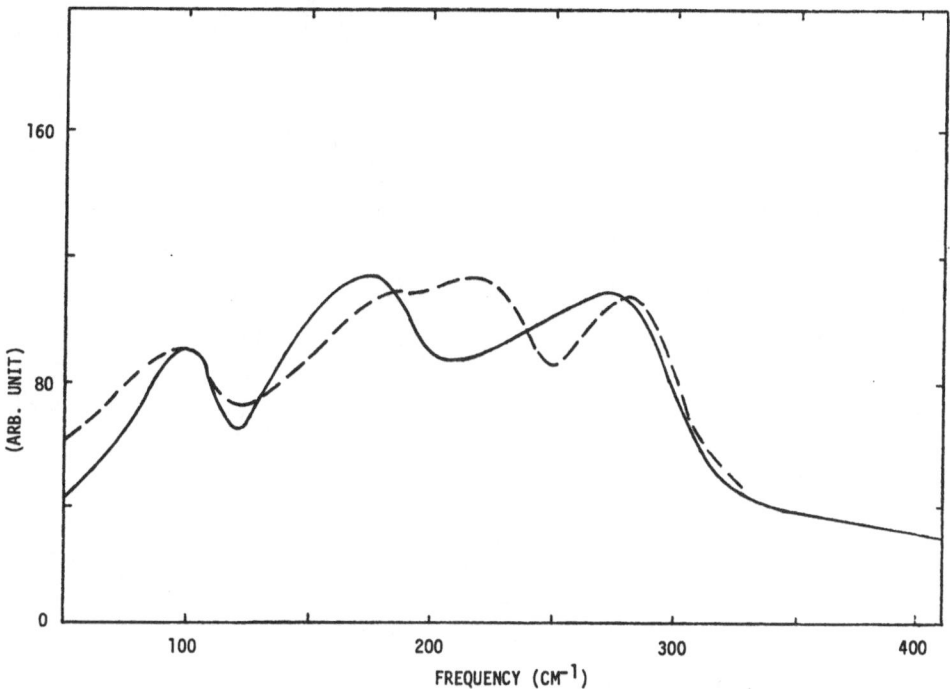

Fig. 30b Infrared data (solid curve) of ref. 83 is compared to the $(n)_S$ contribution (dashed curve) corresponding to $\eta = 0.7$ [curve 3-D of (a) above].

experimental one. Comparison of the infrared data with ε_2^S corresponding to $\eta = .7$ is indicated in Fig. 30b. Similar results may be shown to follow for the case of Si as well. It appears that the peak positions and shapes of the Raman spectra of these solids are well represented by a simple sum of $(n)_{SL}$ and $(n)_S$ contributions, and infrared spectra by $(n)_S$ contributions. For binary TBAS the crystalline counterparts display allowed spectra both for infrared and first order Raman. Fig. 31 displays the $\eta = 0$ fit to the IR spectrum for amorphous GaAs; Fig. 32 shows the same for SiC, but for a geometry where both TO and LO were excited (Berreman effect[106]). Very similar results were also obtained for the infrared spectra of amorphous GaP, GaSb and InAs. The conclusion reached is that the $(n)_S$ contribute very little to the spectrum of typical binary TBAS; the principal features arise from just $(n)_{SL}$ alone.

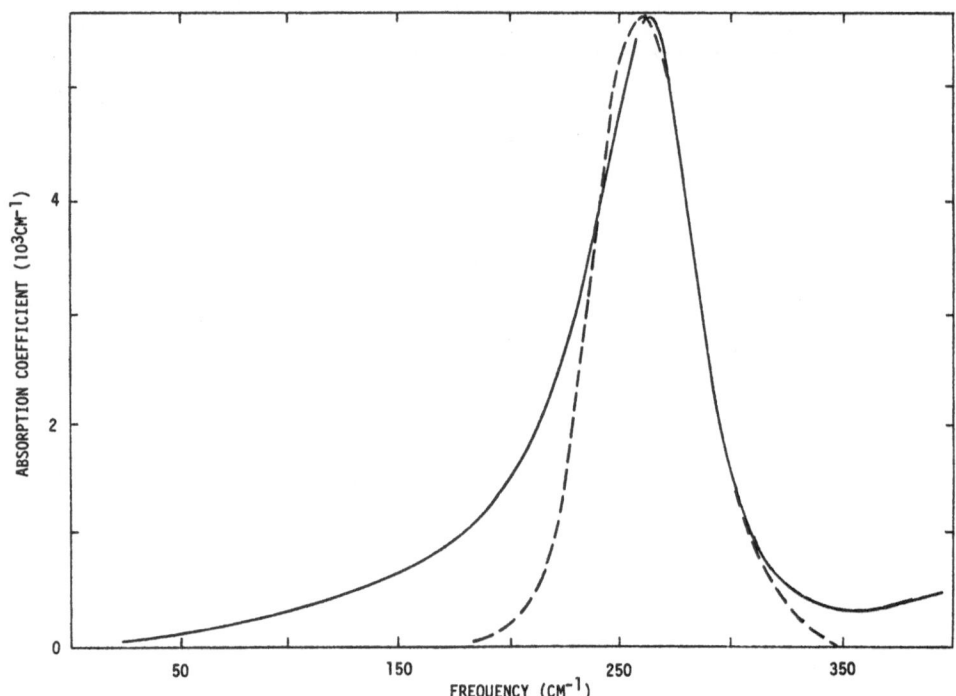

Fig. 31 Infrared absorption in amorphous GaAs. Solid
curve: experimental (from ref. 102); dashed curve:
broadened Lorentizian calculated in this work with
$\sigma_\rho^a = 0.076\ \rho_{ao}$, and $\rho_{ao}/\rho_c = 0.96$, where σ_ρ^a is the Gaussian spread in local density, ρ_{ao}/ρ_c is the ratio of
mean density of the amorphous phase to the crystalline
density.

The present results thus suggest that CAA processes
play an important role in the spectra of TBAS, and that
a statistical approach incorporating such processes can
account successfully for the principal features of measured spectra.

ACKNOWLEDGMENTS

The authors gratefully acknowledge the close collaboration of and helpful discussions with Drs. B.
Bendow, D. K. Paul and Y. F. Tsay, and Messers J. Chen,
S. P. Gaur and S. Varshney at various stages of this
work.

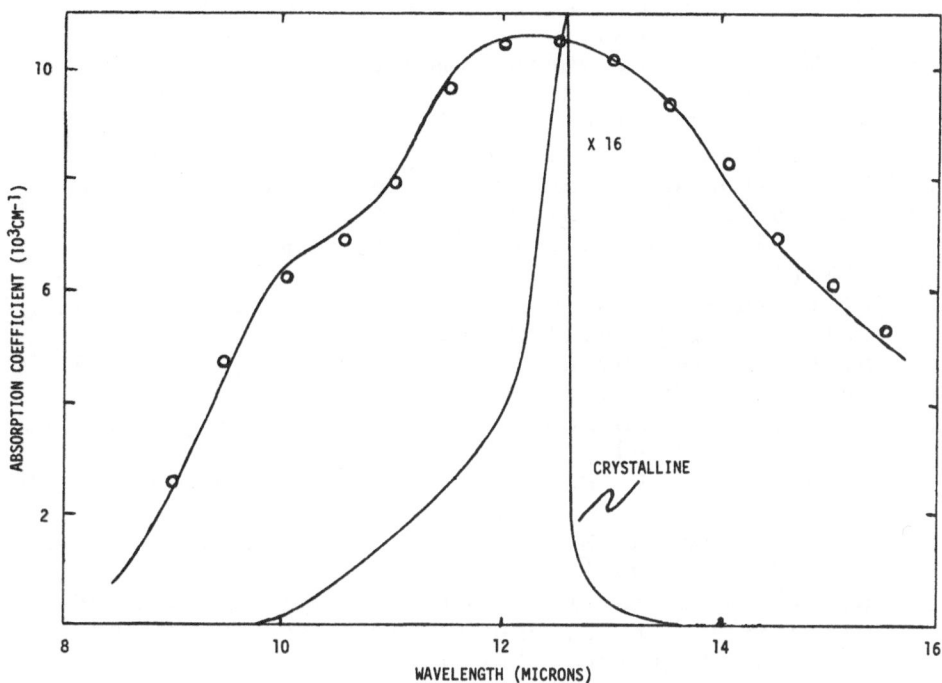

Fig. 32 Infrared absorption in amorphous SiC. Solid curve: calculated in this work. Relevant crystalline Lorentz oscillator parameters were taken from D. W. Feldman, et al, Phys. Rev. 170, 698 (1968); 173, 787 (1968); and mode Grüneisen parameters from ref. 105; Circles: experimental data from ref. 103.

REFERENCES

*Work supported in part, by Air Force Cambridge Research Laboratories (AFSC), Contract No. F19628-72-C-0286.

1. I. M. Lifshitz, J. Phys. USSR, 7, 215 (1943); 7, 249 (1943); 8, 89 (1944).
2. E. W. Montroll and R. B. Potts, Phys. Rev. 100, 525 (1955); ibid, 102, 72 (1965).
3. P. Mazur, E. W. Montroll and R. B. Potts, J. Wash. Acad. Sci. 46, 2 (1956).
4. G. Schaefer, Phys. Chem. Solids 12, 233 (1960).
5. See for example, L. Genzel, "Optical Properties of Solids", ed. S. Nudelman and S. S. Mitra (Plenum, N. Y., 1969), p. 453.

6.   P. G. Dawber and R. J. Elliott, Proc. Roy. Soc. A273, 222 (1963).

7.   I. M. Lifshitz, Nuovo Cimento Suppl. 3, 716 (1956).

8.   P. G. Dawber and R. J. Elliott, Proc. Phys. Soc. 81, 521, 453 (1963).

9.   S. S. Jaswal, Phys. Rev. 140, 2A, 687 (1965).

10.   N. Krishnamurthy and T. M. Haridasan, Indian J. Pure Appl. Phys. 4, 255 (1966).

11.   J. B. Page and D. Strauch, Phys. Stat. Sol. 24, 469 (1967).

12.   A. P. Zhernov and G. R. August, Soviet Physics, Solid State 9, 8, 1724 (1968).

13.   N. Krishnamurthy and T. M. Haridasan, Indian J. Pure Appl. Phys. 7, 89 (1969).

14.   J. Govindarajan and T. M. Haridasan, Phys. Letters 29A, 387 (1969).

15.   W. Hayes, H. F. MacDonald and C. T. Sennet, J. Phys. C (Solid State Phys.), 2 2402 (1969).

16.   G. Lucovsky, M. H. Brodsky and E. Burstein, Phys. Rev. B, 2, 3295 (1970).

17.   See for example, F. B. Hilderbrand, Methods of Applied Mathematics, 2nd edition, Sec. 1.12 and 1.17, Prentice Hall Inc., New Jersey.

18.   K. F. Renk, Z. Physik 201, 445 (1967).

19.   R. Weber, Ph. D. Thesis, Physikalisches Institut der Universität, Freiburg im Breisgau (1967).

20.   J. F. Vetelino, Ph. D. Dissertation, University of Rhode Island (1969), (unpublished).

21.   W. Hayes, Phys. Rev. Letters 13, 275 (1964).

22.   W. Hayes, Phys. Rev. 138, 1227 (1965).

23.   A. R. Goodwin and S. D. Smith, Phys. Letters 17, 203 (1965).

24.   O. G. Lorimor, W. G. Spitzer and M. Waldner, J. Appl. Phys. 37, 2509 (1966).

25.   S. D. Smith, R. E. V. Chaddock and A. R. Goodwin, Proc. Intern. Conf. Phys. Semicond., Kyoto, 67 (1966).

26.   A. Mitsuishi, A. Manabe, H. Yoshinaga, S. Ibuki and H. Komiya, Progress of Theoretical Phys., Supp. 45, 21 (1970).

27.   O. G. Lorimor and W. G. Spitzer, J. Appl. Phys. 37, 3687 (1966).

28.   O. Brafman, I. F. Chang, G. Lengyel and S.S. Mitra, Phys. Rev. Letters 19, 1120 (1967).

29.   W. G. Spitzer, J. Phys. Chem. Solids 28, 33 (1967).

30.   M. H. Brodsky and G. Lucovsky, Phys. Rev. Letters, 21, 990 (1968).

31   W. G. Spitzer, W. Allred, S. E. Blum and R. J. Chicotka, J. Appl. Phys. 40, 2589 (1969).

32. C. T. Sennet, D. R. Bosomworth, W. Hayes and A. R. L. Sray, J. Phys. C. (Solid State Phys.), $\underline{2}$, 1137 (1969).

33. W. Hayes, M. D. K. Wiltshire, P. J. Dean, J. Phys. C. (Solid State Phys.), $\underline{3}$, 1762 (1970).

34. D. T. F. Marple, unpublished.

35. M. Balkanski, R. Beserman and L. K. Vodopianov, Localized Excitation in Solids, edited by R. F. Wallis, (Plenum, N. Y., 1968). p. 154.

36. M. H. Brodsky, G. Lucovsky, M. F. Chen and T. S. Plaskett, Phys. Rev. B $\underline{2}$, 3303 (1970).

37. A. Mitsuishi, unpublished.

38. W. J. Choyke and L. Patrick, Bull. Am. Phys. Soc. $\underline{16}$, 397 (1971).

39. J. F. Vetelino and S. S. Mitra, Phys. Rev. $\underline{178}$, 1349 (1969).

40. S. P. Gaur, J. F. Vetelino and S. S. Mitra, J. Phys. Chem. Solids $\underline{32}$, 2737 (1971).

41. S. Ibuki, H. Komiya, A. Mitsuishi, A. Manabe and H. Yoshinaga, International Conference on Semiconductors, Moscow, (1968), p. 1047.

42. S. Ibuki, H. Komiya, A. Mitsuishi, A. Manabe and H. Yoshinaga, II-VI Semiconductor Compounds (Edit. by D. G. Thomas) (Benjamin, New York, 1967), p. 1140.

43. F. Krueger, O. Reinkober and E. Koch-holm, Annln. Phys. $\underline{85}$, 110 (1928).

44. F. Matossi, J. Chem. Phys. $\underline{19}$, 161 (1951).

45. P. Dean, Proc. Phys. Soc. $\underline{73}$, 413 (1959); ibid Proc. Roy. Soc. $\underline{254A}$, 507 (1960); ibid Proc. Roy. Soc. $\underline{260A}$, 263 (1961).

46. H. W. Verleur and A. S. Barker, Phys. Rev. $\underline{149}$, 715 (1966); ibid Phys. Rev. $\underline{155}$, 750 (1967).

47. Y. S. Chen, W. Shockley and G. L. Pearson, Phys. Rev. $\underline{151}$, 648 (1966).

48. I. F. Chang and S. S. Mitra, Phys. Rev. $\underline{172}$, 924 (1968).

49. I. F. Chang and S. S. Mitra, Adv. in Phys. $\underline{20}$, 359 (1971).

50. A. P. G. Kutty, Solid St. Commun., $\underline{14}$, 213 (1974).

51. P. N. Sen and W. M. Hartmann, Phys. Rev. B $\underline{9}$, 367 (1974).

52. O. Brafman, I. F. Chang, G. Lengyel, S. S. Mitra and E. Carnall, Phys. Rev. Letters $\underline{19}$, 1120 (1967): Proc. Intl. Conf. on Localized Excitations in Solids (New York, Plenum Press, 1968), p. 602.

53. S. C. Varshney, J. F. Vetelino and S. S. Mitra, Phys. Rev. (to be published).

54. See for example, "AIP Proc. No. 20 on, Tetrahedrally Bonded Amorphous Semiconductors", ed. M. H. Brodsky,

S. Kirkpatrick and D. Weaire (1974).

55. N. J. Shevchik and W. Paul, J. Non-Cryst. Sol. 8-10, 381 (1972).

56. N. J. Shevchik and W. Paul, J. Non-Cryst. Sol. 13, 1 (1973/74).

57. S. Moss and J. Graczyk, "Proc. Tenth Intern. Conf. on the Physics of Semiconductors, Cambridge, Mass., 1970, ed. S. P. Keller et al, Conf.-700801 (USAEC Division of Technical Information, Springfield, Va., 1970), p. 658.

58. D. Turnbull and D. E. Polk, J. Non-Cryst. Sol. 8-10, 19 (1972).

59. M. L. Rudee, Thin Solid Films 12, 207 (1972).

60. J. D. Bernal, Proc. Roy. Soc. 280A, 299 (1964).

61. W. H. Zachariasen, J. Am. Chem. Soc. 54, 384 (1932).

62. R. Grigorovici and R. Manaila, J. Non-Cryst. Sol. 1, 371 (1969).

63. P. Dean, Proc. Phys. Soc. 84, 727 (1964).

64. P. Dean, Rev. Mod. Phys. 44, 127 (1972).

65. R. J. Bell, J. Phys. C5, L315 (1972).

66. R. J. Bell, P. Dean and D. C. Hibbins-Butler, J. Phys. C3, 2111 (1970).

67. J. Hori, "Spectral Properties of Disordered Chains and Lattices", Intern. Series of Monographs in Natural Philosophy, Vol. 16, Pergamon Press, (1968).

68. M. Goda, J. Phys. C6, 3047 (1973).

69. R. J. Bell, N. F. Bird and P. Dean, J. Phys. C7, 2457 (1974).

70. R. J. Bell, N. F. Bird and P. Dean, J. Phys. C1, 299 (1968).

71. R. J. Bell and P. Dean, Nature 212, 1354 (1966).

72. D. Weaire and R. Alben, Phys. Rev. Letters 27, 1505 (1972).

73. R. Alben, J. E. Smith, Jr., M. H. Brodsky and D. Weaire, Phys. Rev. Letters 30, 1141 (1973).

74. R. Alben, "Tetrahedrally Bonded Amorphous Semiconductors", AIP Proc., No. 20 (1974).

75. D. Henderson and F. Herman, Jr. J. Non-Cryst. Sol. 8-10, 359 (1972); D. Henderson, "Computational Solid State Physics", ed. F. Herman (Plenum, N.Y. 1972), p. 175.

76. M. F. Thorpe, Phys. Rev. B8, 5352 (1973).

77. R. J. Bell, J. Phys. C7, L265 (1974).

78. R. Alben and D. Weaire, J. E. Smith and M. H. Brodsky, Phys. Rev. B11, 2271 (1975).

79. J. E. Smith, Jr., M. H. Brodsky, B. L. Crowder, M. I. Nathan and A. Dinczyk, Phys. Rev. Letters, 26, 642 (1971).

80. J. E. Smith, Jr., M. H. Brodsky, B. L. Crowder, M. I. Nathan and A. Dinczyk, J. Non-Cryst. Sol. 8-10, 179 (1972).
81. M. Wihl, M. Cardona and J. Tauc, J. Non-Cryst. Sol. 8-10, 172 (1972).
82. J. E. Smith, Jr., M. H. Brodsky, B. L. Crowder and M. I. Nathan, "Proc. of the 2nd Intern. Conf. on Light Scattering in Solids", ed. M. Balkanski (Flammarion, Paris, 1971) p. 330.
83. R. W. Stimets, J. Waldman, J. Lin, T. S. Chang, R. J. Temkin and G. A. N. Connell, Solid St. Commun. 13, 1485 (1973).
84. M. H. Brodsky and A. Lurio, Phys. Rev., B9, 1646 (1974).
85. J. D. Axe, D. T. Keating, G. S. Cargill, III and R. Alben, AIP Proc., No. 20 (1974), p. 279.
86. F. R. Ladan and A. Zylberstein, Phys. Rev. Letters 29, 1198 (1973).
87. B. Schröder and J. Geiger, Phys. Rev. Letters 28, 301 (1972).
88. D. E. Polk, J. Non-Cryst. Sol. 5, 365 (1971).
89. R. J. Temkin, Solid St. Commun. 15, 1325 (1974).
90. R. Grigorovici, J. Non-Cryst. Sol. 1, 303 (1969).
91. Y. F. Tsay, D. K. Paul and S. S. Mitra, Phys. Rev. 138, 2827 (1973).
92. S. S. Mitra, D. K. Paul, Y. F. Tsay and B. Bendow, "Tetrahedrally Bonded Amorphous Semiconductors," AIP Proc. No. 20 (1974), p. 284.
93. H. J. McSkimin, J. Acous. Soc. of Am. 30, 314 (1958).
94. H. J. McSkimin and P. Andreatch, Jr., J. Appl. Phys. 35, 2161 (1964).
95. C. J. Buchenauer, F. Cerdeira and M. Cardona, "Proc. of the 2nd Intern. Conf. on Light Scattering in Solids", ed. M. Balkanski (Flammarion, Paris, 1971) p. 280.
96. R. J. Temkin, W. Paul and G. A. N. Connell, Adv. in Phys. 22, 581 (1973).
97. S. S. Mitra, "Optical Properties of Solids", ed. S. Nudelman and S. S. Mitra (Plenum, N.Y. 1969), p.333.
98. S. P. Gaur, M.S. Thesis, University of Maine (1971).
99. G. Dolling and R. A. Cowley, Proc. Phys. Soc., London, 88, 463 (1966).
100. H. Shuker and R. W. Gammon, Phys. Rev. Letters 25, 222 (1970).
101. S. S. Mitra, Y. F. Tsay, D. K. Paul, B. Bendow, Phys. Stat. Solidi, to be published.
102. W. Prettl, N. J. Shevchik and M. Cardona, Phys. Stat. Sol. (b) 59, 241 (1973).

103. E. A. Fagan, in Proc. of the Fifth Intl. Conf. on
      Amorphous and Liquid Semiconductor (Garmisch-
      Partenkirchen, West Germany, September 1973),
      pp. 601-7.
104. See for example, S. S. Mitra in "Optical Properties
      of Solids", ed. S. Nudelman and S. S. Mitra, Plenum
      Press, New York (1969).
105. S. S. Mitra, O. Brafman, W. B. Daniels and R. K.
      Crawford, Phsy. Rev. 186, 942 (1969).
106. D. W. Berreman, Phys. Rev. 130, 2193 (1963).

PHONONS IN AMORPHOUS SOLIDS

M. F. Thorpe

Department of Engineering and Applied Science
Becton Center, Yale University
New Haven, Connecticut, 06520

1. INTRODUCTION

Lattice vibrations, or phonons, in crystalline solids have been studied since the beginning of the century[1]. Until about 15 years ago they were seen experimentally rather indirectly but since then, inelastic neutron scattering has made the direct measurement of phonon dispersion curves almost routine.[2] The phonons in a solid are a basic ingredient in understanding such things as specific heat, melting, ferroelectricity and superconductivity. Phonons also occur in amorphous solids or glasses and play similar roles in determining their physical properties. It is therefore important to understand them from a fundamental, i.e. microscopic, viewpoint. It is only recently with the advent of good experimental data on simple systems that this has been undertaken and the subject is still in its infancy. In these lectures we will attempt to outline the progress that has been made and point out some of the problems that still remain.

It is important to bear in mind from the outset, that this is a difficult theoretical problem and a direct frontal attack is rarely possible. Although this is somewhat unsatisfactory intellectually, the ingenuity of the theoretician is challenged and progress, although slow is rewarding. It is our purpose in these lectures to establish a sound theoretical framework and others (J. D. Axe, and J. Tauc) will discuss the experimental situation. However, the implications of some of the more important experiments will be discussed.

Since the general acceptance in the last few years of the continuous random network model of amorphous semiconductors[3], it has been realized that amorphous materials are not "messed up crystals" but intrinsic structures that are very appealing intellectually. This has the important consequence that it is not possible to derive amorphous properties by perturbing the corresponding crystal. It is necessary to have a good understanding of the structure (as discussed by J. F. Graczyk, W. Paul and D. Turnbull) before the phonons, which are small deviations from the equilibrium structure, can be understood. In the next section we discuss the general principles involved in constructing a potential energy to use for calculating phonon frequencies. These principles are the same for crystalline and amorphous solids. In Sec. 3 we review the calculation of phonon frequencies in crystals using both the rigid ion and shell models. Just as it is impossible to paint a Picasso without being able to draw a horse, so it is impossible to appreciate phonons in amorphous solids without a thorough understanding of the situation in crystalline solids. Indeed we have found that it is imperative to understand, say crystalline Si, preferably in a number of different structures, before attempting to understand amorphous Si. In Sec. 4 we discuss densities of states and thermal properties which again is quite general. In Sec. 5 we show that simple potentials are required if progress is to be made in amorphous materials and then discuss two techniques that have led to useful insights, numerical diagonalization and the structural potential approximation. In Sec. 6 we review the experimental and theoretical situation regarding the anomalously large low temperature specific heat and possible explanations for it. Finally in the conclusion, we make some conjectures about future progress. We have tried to reference mainly review articles and books although this has not been possible for the more recent work.

2. HARMONIC APPROXIMATION

The dynamics of a solid is a complicated problem and it must be simplified if progress is to be made. The first of these simplifications is the adiabatic or Born-Oppenheimer approximation which allows us to write down a potential energy containing only the nuclear coordinates, the second is the harmonic approximation which truncates this potential after the second order terms.

The Hamiltonian, or energy, of any solid is a func-
tion of the nuclear coordinates $\underline{u}_i$ where i labels the
ion, and the electron coordinates $\underline{r}_e$ where e labels the
electron.   A curly bracket denotes a set

$$H = H(\{\underline{u}_i\}, \{\underline{r}_e\}) \tag{1}$$

So for example in GaAs, there are 2 nuclear and 64
electronic labels per molecular unit and each one of
these is a vector that contributes three degrees of freedom.

The adiabatic approximation is possible because the
electron mass m is much smaller than the nuclear mass M,

$$m/M \simeq 10^{-3}$$

The electronic motion is therefore very rapid compared
with the nuclear motion and so when a nucleus has completed only a small fraction of a cycle, the electrons
have been through many cycles. Because the nuclear motion is so slow, the electrons come into equilibirum at
each nuclear configuration. Therefore the electron coordinates may effectively be eliminated from (1)

$$H = H(\{\underline{u}_i\}) \tag{2}$$

Another way to state the approximation is that the electronic excitations ($\sim$1 eV) are much higher in energy than
the phonon energies ($\sim 10^{-3}$ eV) it is desired to calculate.
One must, however, be careful with this approximation in
metals where there is no energy gap between occupied and
unoccupied electron states so that electronic excitations
with arbitrarily small energy are possible. This is also
true in amorphous semiconductors where there is finite density of states at the Fermi level. However, in this case
the number of states in the "gap region" is so small
($< 1^o/o$ of total states) that one can make the adiabatic
approximation with some confidence. In the rigid ion model, Eq. 2 forms the starting point but in the shell model the electron coordinates are included at the beginning of the calculation and eliminated later.

The simplest example of the Born-Oppenheimer approximation is a diatomic molecule. The two nuclei are held
at a fixed separation $\underline{R}$ and the electronic problem is
solved (The Schrödinger equation can be solved exactly

for the H_2^+ ion). This can be done at each $\underline{R}$ and the
ground state energy of the molecule $W(\underline{R})$ obtained. Thus,
the electronic degrees of freedom are eliminated and only
the nuclear separation $\underline{R}$ occurs in $W(\underline{R})$ which may now be
regarded as a potential energy. The minimum in $W(\underline{R})$ determines the equilibrium separation and the vibrational
excited states may be found directly from $W(\underline{R})$ without
further reference to the electrons.

In solids, the equilibrium positions of the atoms
are usually well defined, and so it is reasonable to ex-
pand (2) as a Taylor series in the $\{\underline{u}_i\}$ about the equilibrium positions. The linear terms vanish as the net force
on each atom must be zero in equilibrium. In the harmonic approximation the series is truncated at the second
order terms.

$$H = \sum_i \frac{p_i^2}{2m_i} + V \qquad (3)$$

where

$$V = \frac{1}{2} \sum_{ij} \underline{u}_i \, \underline{\underline{V}}_{ij} \underline{u}_j \qquad (4)$$

This can be shown, a posteriori, to be a good approximation if the mean square displacement is small compared
to the nearest atomic separation a,

$$\langle u^2 \rangle \ll a^2$$

This is usually so, but breaks down as the melting temperature of the solid is approached, and also in the quantum
crystals H and He where there is a large zero point motion due to the light nuclear masses. It also breaks down
if the potential energy V has two neighboring minima between which tunelling can take place. The best example
of this is a ferroelectric material like KH_2PO_4 where the
proton can tunnel between the two minima associated with
the hydrogen bond. A similar phenomena may occur in glasses, that manifests itself at low temperatures. This will
be discussed in Sec. 6.

The word _phonon_ as used in these lectures means simply a quantum of energy in a vibrational mode. The fact
that the phonon is an eigenstate presupposes that we are
working with a harmonic Hamiltonian. It is of course possible to write down Hamiltonians in both crystalline and
amorphous solids if the structure is locally stable Thus

phonons are well defined in amorphous solids - the main
qualitative difference from a crystal is that amorphous
phonons do not have a well defined $\underline{k}$ vector.

3. PHONONS IN CRYSTALLINE SOLIDS

Having arrived at the harmonic form of the potential given in Eq. 4, the problem is to evaluate the elements of the 3 x 3 tensor $\underline{V}_{ij}$ that couples ions at sites
i and j. In principle, these may be calculated from a
knowledge of the electronic band structure, however, this
band structure must be known for the ions in their equilibrium positions <u>and</u> for arbitrary small displacements a-
way from them. It is therefore not surprising that a
first principles calculation of the $\underline{V}_{ij}$ is rarely achieved.
Most progress has been made in metals using a weak pseudopotential so that perturbation theory may be used.[4]

In semiconductors and insulators it is usual to treat
the parameters in $\underline{V}_{ij}$ as unknown and to be determined from
the experimental dispersion curves. It is necessary to
limit the number of parameters and the interactions are
usually assumed to be zero for ions separated by more
than a few nearest neighbor distances. Of course, full
use is always made of symmetry to reduce further the number of parameters. There is no problem to include the
long range Coulomb forces as their form is known.

Sometimes, it is possible to adopt an approach intermediate between the two above like the valence force
method, which has been quite successful in systems where
the bonding is primarily covalent. Atoms form bonds with
characteristic lengths, e.g. the length of the Si = Si
covalent bond is always around 2.35 Å. This bond length
is not affected by more than a few percent by environment.
Of course, the potential, $\underline{V}_{ij}$ is rather sensitive to even
small changes in the bond length but nevertheless forces
can often be used interchangeably between different structures as a reasonable starting point, small adjustments
can always be made later. Chemists have been particularly successful in exploiting this technique. The C=C
bond has a characteristic stretching frequency ~ 1650 cm^{-1};
small changes $\sim 2^o/o$ in this frequency can even be used to
identify the environment of the bond. Whilst the C=C
bond is particularly insensitive to changes in the local
environment because of its high frequency, the idea holds
for other bonds also. The simplest application to phonon
frequencies is to take valence forces from molecules

(obtained by fitting observed frequencies) and use them in solids.[5] Similar methods may also be used to relate force constants in polymorphs like diamond cubic Si and Si III.[6]

a) Rigid ion model

As an instructive and simple example, we use the harmonic potential to calculate the longitudinal modes of a linear chain. We assume that the forces act between nearest neighbors only and use periodic boundary conditions so that atom i and atom i + N are identical in a ring with N atoms. In the thermodynamic limit N → ∞ it is well known, and reasonable, that the answer is independent of the particular boundary conditions, e.g. free ends, cyclic, etc.[7] Cyclic or Born von Karman boundary conditions are the simplest mathematically. The potential V for the chain is given by

$$V = \frac{1}{2} \sum_{j=1}^{N} f(u_j - u_{j+1})^2 \tag{5}$$

where u_j is the longitudinal displacement of the j atom.

All the symmetry is built into this potential; the translational symmetry tells us that all bonds are equivalent so that the force constant f is independent of i. Another "symmetry" that is present is that a rigid translation of the whole chain costs no energy. This is clearly so in (5) as only the differences of coordinates appear and this has the direct consequence that the acoustic mode comes in at zero frequency. The simplest procedure now is to calculate the modes classically and quantize later. The frequencies are the same using either classical or quantum mechanics for harmonic potentials. The classical equation of motion

$$M\ddot{u}_j = - \partial V / \partial u_j \tag{6}$$

leads to

$$M\ddot{u}_j = -f(u_j - u_{j+1}) - f(u_j - u_{j-1}) \tag{7}$$

From Bloch's theorem we know that the solutions to this set of N linear equations are running waves

$$u_j = u \exp i(\omega t - kja) \tag{8}$$

where a is the spacing between adjacent atoms. Eq. 7 yields the eigenfrequencies.

$$\omega^2 = 2f/M(1 - \cos ka) \tag{9}$$

which are shown in Fig. 1 as curve R. At small ω, k the dispersion is linear and these are the longitudinal sound waves.

Before proceeding to look at the phonon frequencies in silicon it will be useful to set up the dynamical matrix for a general case. The phonon frequencies are obtained directly from the dynamical matrix. Notation is somewhat of a problem and we will try and keep it as compact as possible. The starting point is the potential energy

$$V = \frac{1}{2} \sum_{\substack{\ell\ \alpha \\ \underline{\ell}'\ \beta}} V(\underline{\ell},\underline{\ell}',\alpha,\beta)u_\alpha(\underline{\ell})u_\beta(\underline{\ell}') \tag{10}$$

where $\underline{\ell}, \underline{\ell}'$ label the N unit cells and α,β label the <u>degree of freedom</u> within the unit cell, i.e. both the atom and its component of displacement. Thus, if we have s atoms in the unit cell, we have 3s possible values for α,β.

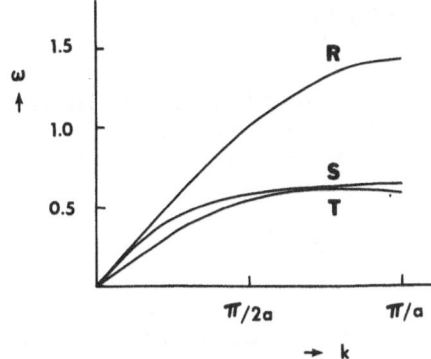

Fig. 1 Phonon dispersion relations for a linear chain with nearest neighbor distance a and 2f/M = 1 for the rigid ion model (R), Eq. 9; the shell model (S), Eq. 24 with f = g; and a truncated shell model (T) Eq. 25.

We do the summation so that there is no double counting.
A mass M_α is associated with each degree of freedom although it only depends on the atom and not its component
of displacement. Using a classical equation of motion
like (6) we find

$$M_\alpha \ddot{u}_\alpha(\underline{\ell}) = - \sum_{\underline{\ell}'\beta} V(\underline{\ell},\underline{\ell}',\alpha,\beta)u_\beta(\underline{\ell}') \qquad (11)$$

Bloch's theorem demands solutions of the form

$$u_\alpha(\underline{\ell}) = (M_\alpha)^{-1/2}u_\alpha \exp i(\omega t - \underline{k}\cdot\underline{\ell}) \qquad (12)$$

which leads to

$$\omega^2 u_\alpha = \sum_{\underline{\ell},\beta} (M_\alpha M_\beta)^{-1/2}V(\underline{\ell},\underline{\ell}', \alpha,\beta)e^{i\underline{k}(\underline{\ell}-\underline{\ell}')}u_\beta \qquad (13)$$

Bloch's theorem therefore allows us to go from the set of
3Ns linear equations in (11) to the 3s linear equations
in (13). Eq. 13 can be tidied up if we define the dynamical matrix $D_{\alpha\beta}(\underline{k})$

$$D_{\alpha\beta}(\underline{k}) = \sum_{\underline{\ell}-\underline{\ell}'} (M_\alpha M_\beta)^{-1/2}V(\underline{\ell}-\underline{\ell}',\alpha,\beta)\exp i\underline{k}(\underline{\ell}-\underline{\ell}') \qquad (14)$$

so that

$$\omega^2 u_\alpha = \sum_\beta D_{\alpha\beta}(\underline{k})u_\beta \qquad (15)$$

and we have used the translational invariance $V(\underline{\ell},\underline{\ell}',\alpha,\beta)$
$V(\underline{\ell}-\underline{\ell}',\alpha,\beta)$. The condition for the 3s homogeneous linear
Eq. 15 to have a solution is that

$$\mathrm{Det}[D_{\alpha\beta}(\underline{k}) - \omega^2\delta_{\alpha\beta}] = 0 \qquad (16)$$

which determines the 3s eigenfrequencies $\omega^2(\underline{k})$ at each $\underline{k}$.
There are exactly N distinct $\underline{k}$ vectors which are usually
chosen to be in the first Brillouin zone so that we have
found all the 3Ns solutions to the problem; this is the
total number of degrees of freedom of the whole solid.
The equilibrium condition that the starting structure be
stable leads to positive definite matrices in (11) and

(15) and this leads to positive eigenvalues $\omega^2(\underline{k}) > 0$.
Strictly speaking, the matrices are only positive semi-
definite as the three rigid translations lead to three
zero frequency modes.   The condition for stability requires only a <u>local</u> <u>minimum</u> of the potential energy as
shown in Fig. 2.   Thus, for example, diamond cubic Si is
almost certainly the most stable form of Si, but Si III
which has 8 atoms/unit cell, and various samples of amorphous Si are <u>metastable structures</u> which, nevertheless,
have well defined vibrational excitations which may be
quantized to give phonons.   As the amplitude of vibration
gets larger, the solid may tunnel through a potential barrier to a more stable structure.   This possibility of tunnelling to neighboring structures is an anharmonic effect
and so outside our present theory.   It is taken up in
Sec. 6.

We now consider the original Born model applied to
the diamond structure.   Extensive notes may be found in
Cochran[8] and so we pick out only the essential points.
The diamond structure consists of two interpenetrating
f.c.c lattices so that there are two atoms/unit cell.
Each atom has four neighbors positioned at the corners of

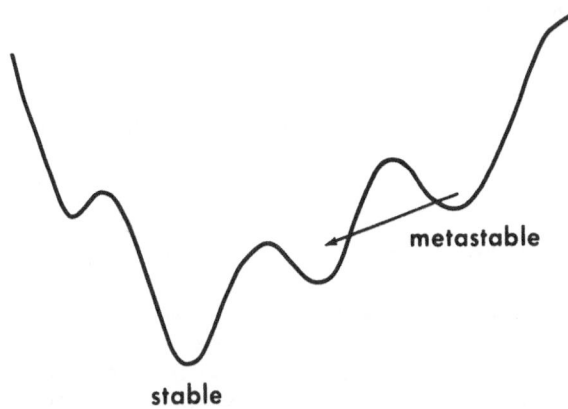

Fig. 2   A sketch of the potential energy of a solid a-
gainst some generalized coordinate.   The metastable states
may be crystalline or amorphous and are stable against
small displacements.   However large displacements or quantum mechanical tunnelling can take the solid to a neighboring minima as indicated by the arrow.

a tetrahedron. We assume that the forces act only between near neighbors and the symmetry of the lattice demands that these forces have axial symmetry about the
bond joining the two neighbors so that the potential V_{ij}
between a pair of atoms may be written

$$V_{ij} = \frac{3}{2} \beta [(\underline{u}_i - \underline{u}_j) \cdot \underline{r}_{ij}]^2 + \frac{1}{2}(\alpha-\beta)(\underline{u}_i - \underline{u}_j)^2 \tag{17}$$

where $\underline{r}_{ij}$ is a unit vector joining sites i and j and
$\alpha + 2\beta$, $\alpha - \beta$ are the central, non-central force constants. By summing over all pairs in the crystal, the
complete potential may be formed from which the 6 x 6
dynamical matrix is constructed. In general this matrix
must be diagonalized numerically but along the (100) and
(111) directions in reciprocal space the solutions are
easy to obtain.

$k = (k,0,0)$

L: $M\omega^2(\underline{k}) = 4\alpha[1 \pm \cos(ka/4)]$

T: $= 4\alpha \pm \sqrt{[4\alpha \cos(ka/4)]^2 + [4\beta \sin(ka/4)]^2}$

$k = (k,k,k)$

L: $M\omega^2(\underline{k}) = 4\alpha \pm \sqrt{[4\alpha\cos(ka/2)]^2+[2(\alpha-2\beta)\sin(ka/2)]^2}$

T: $= 4\alpha \pm \sqrt{[4\alpha\cos(ka/2)]^2+[2(\alpha+\beta)\sin(ka/2)]^2}$

$$\tag{18}$$

The distance a is the side of a cubic unit of the diamond
structure so that the nearest neighbor separation is a
$\sqrt{3}a/4$. A plot of these modes for $\beta/\alpha = 0.6$ is shown in
Fig. 3. Longitudinal L and transverse T refer to the direction of the eigenvector, i.e. displacement of the atoms,
with respect to $\underline{k}$. In a general $\underline{k}$ direction, modes are
mixed L and T but along high symmetry directions such as
those shown in Fig. 3, they are entirely one or the other.
By convention the three branches that go to zero frequency
are called acoustic and the others optic. By examining
the linear behavior of ω against $\underline{k}$ in the two directions
given, it is easy to get the three elastic constants[1]

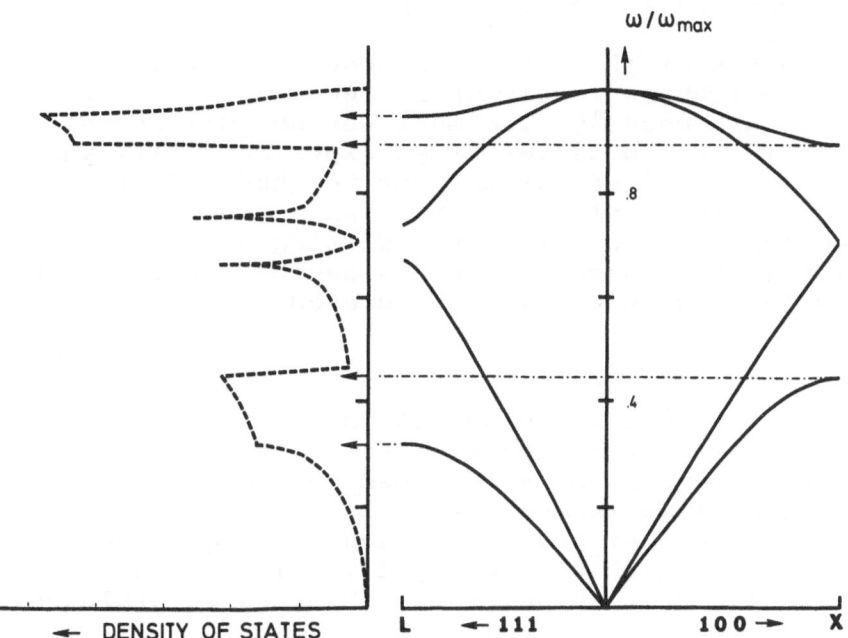

Fig. 3 The phonon dispersion and density of states for diamond cubic Si/Ge using the Born model potential, Eq. 17, with $\beta/\alpha = 0.6$. The dotted lines to some of the Van Hove singularities in the density of states indicate their origin in $\underline{k}$ space. The maximum frequency is given by $M\omega^2_{max} = 8\alpha$.

$$c_{11} = \alpha/a$$

$$c_{12} = (2\beta - \alpha)/a \qquad\qquad (19)$$

$$c_{44} = (\alpha^2 - \beta^2)/\alpha a$$

Soundwaves also exist in an amorphous solid of course as the solid acts like an elastic continuum for lone wavelengths (see also Sec. 6) and it does not matter much how the atoms are arranged. Because continuous random networks are presumably isotropic if they are large enough. there are only two elastic constants corresponding to longitudinal and transverse waves.

Although qualitatively the dispersion curves of Fig. 3 look like those measured experimentally in Ge[9] by elastic neutron scattering (see Fig. 4) there are important discrepancies that cannot be entirely corrected within the framework of the rigid ion model even when forces out to fifth neighbors involving fifteen adjustable force constants are used.[10] The most serious discrepancy is that the TA branch is very flat experimentally, which demands that forces go out to many neighbors in the rigid ion model. This introduces more parameters and the whole problem degenerates into curve fitting and little insight is gained. Fortunately there is another approach that leads naturally to the flat TA branch.

b) Shell model

The shell model, which was introduced about 15 years ago, does not discard the electronic degrees of freedom from the beginning but keeps then initially.[8] The ideas can again be illustrated with a linear chain as shown in Fig. 5. The nuclei couple to each other via the electronic shells which behave rigidly. This is a classical model and the precise quantum mechanical equivalence is

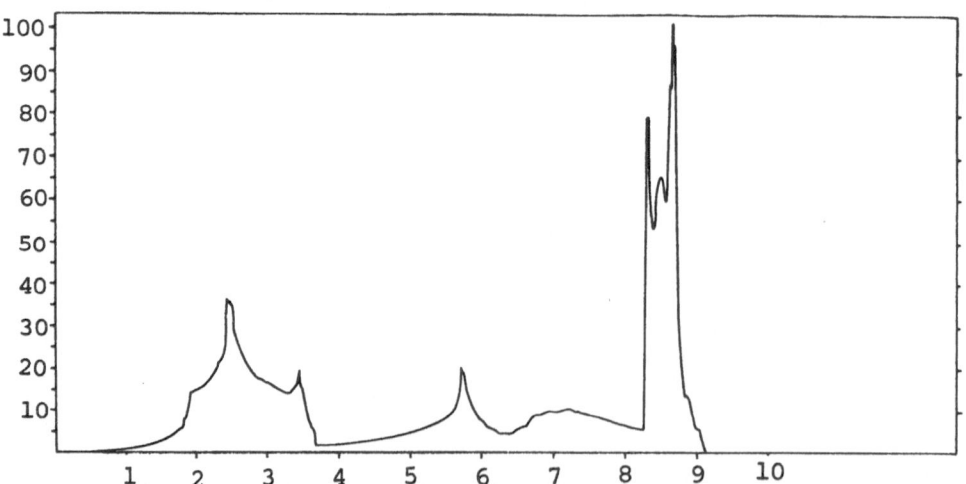

Fig. 4 The density of states for diamond cubic Ge[9]. These results are obtained by measuring ω against $\underline{k}$, using inelastic neutron scattering over the whole Brillouin zone and not just along high symmetry directions. The density of states is therefore obtained directly from experiment by doing a Brillouin zone integration (see Eqs. 26 and 27). The horizontal scale gives frequency in terahertz.

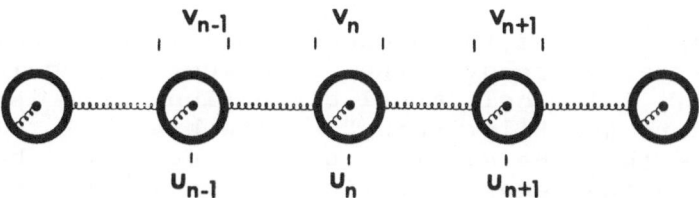

Fig. 5 Sketch of forces in a linear chain shell model.
The atoms are represented by dots with coordinates u_n,
etc., and the shells by circles whose centers have coordinates v_n, etc.

not clear except insomuch as the attempt is made to simulate the electronic degrees of freedom. If u,v denote
the nuclear, shell displacements, the potential is

$$V = \frac{1}{2} \sum_j \{ f(v_j - v_{j+1})^2 + g(v_j - u_j)^2\} \qquad (20)$$

where f is the shell-shell force constant and g is the
nucleus-shell force constant. If the mass of the electron is m we find

$$M\ddot{u}_j = g(v_j - u_j) \qquad (21)$$

$$m\ddot{v}_j = -g(v_j - u_j) - f(v_j - v_{j+1}) - f(v_j - v_{j-1})$$
$$\qquad (22)$$

we put

$$u_j = u \exp i(\omega t - kja)$$
$$\qquad (23)$$
$$v_j = v \exp i(\omega t - kja)$$

and also <u>at this point</u> we note that m << M, and make the
adiabatic approximation m = 0. Eq. 21 may be solved for
v which when inserted into (22) leads to the eigenfrequencies

$$\omega^2 = \frac{2f/M(1 - \cos ka)}{1 + 2f/g(1 - \cos ka)} \tag{24}$$

It can be seen that as the nucleus-shell force constant g becomes large, we recover the rigid ion frequencies given by (9). This shell model dispersion curve is shown in Fig. 1 where the "flattening" of the branch is clearly seen (note that the elastic constant does not change). This is due to the renormalization factor in the denominator of (24). In order to get such a dispersion curve in a rigid ion model, forces f_n would have to be evaluated up to very large n

$$\omega^2 = \frac{2f/M(1 - \cos ka)}{1 + 2f/g(1 - \cos ka)} = \sum_{n=1}^{\infty} f_n(1 - \cos kna)$$

where $\tag{25}$

$$f_n = \frac{4f/M}{s(2s+1)^{1/2}} \{\frac{(s + 1) - (2s + 1)^{1/2}}{s}\}^n$$

and s = 2f/g. In Fig. 1 we have also plotted the dispersion given in (25) but truncated after f_1 and f_2. The difficulties that are encountered in trying to fit the flat TA branches in Ge with a rigid ion model are nicely illustrated by this example. Although the flat part of the dispersion is fairly well reproduced in T, the sound velocity is too small by ~35°/o. The rigid ion model with short range forces has difficulty in giving both the flat dispersion and the high elastic constants.

The interested reader may be wondering why we do not get twice as many modes with the shell model as by including the electron coordinates we have doubled the number of degrees of freedom. If we had not made the adiabatic approximation we would have found these other (predominantly electronic) modes at $\omega^2 \sim \max(f,g)/m$. However, as m → 0, these modes go off to infinity.

4. DENSITY OF STATES AND THERMAL PROPERTIES

Many physical properties involving phonons may be obtained from the density of states $\rho(\omega)$ defined so that $\rho(\omega)d\omega$ is the number of states with frequencies between ω and $\omega + d\omega$. In the thermodynamic limit, $\rho(\omega)$ is a continuous function. Not all physical properties can be expressed in terms of $\rho(\omega)$ alone; sometimes a knowledge of the phonon eigenvectors is necessary. An example of this would be a scattering experiment in which phonons with a particular $\underline{k}$ are selectively excited. This happens in a neutron scattering experiment in which a $\underline{k}$ vector is "imposed" on the system by the momentum transfer of the neutron. In a crystal this illicits a sharp response at certain discrete frequencies. In an amorphous solid, there is a broad response over a range of frequencies as the excitations are not eigenstates of $\underline{k}$ and it is unnatural to describe them as such (except perhaps at very long wavelengths).

In a crystal the density of states is often expressed as an integral in $\underline{k}$ space

$$\rho(\omega) = \frac{V}{(2\pi)^3} \int_{S_\omega} \frac{dS_\omega}{|\nabla_{\underline{k}}\omega|} \tag{26}$$

where S_ω is a surface of constant ω. This density of states exhibits Van Hove singularities, which are discontinuities in the slope of $\rho(\omega)$ in 3 dimensions, whenever $\nabla_{\underline{k}}\omega = 0$. This happens most frequently at high symmetry points as illustrated in Fig. 3. An alternative way of writing $\rho(\omega)$ is

$$\rho(\omega) = \sum_{\underline{k}} \delta(\omega - \omega_{\underline{k}}) = \frac{-1}{\pi} \operatorname{Im} \sum_{\underline{k}} \frac{1}{\omega - \omega_{\underline{k}}} \tag{27}$$

where ω is assumed to have a small positive imaginary part and the sum over $\underline{k}$ includes all branches and goes over the first Brillouin zone. This form, (27), also applies in amorphous solids, however $\underline{k}$ becomes merely a label (no longer a vector) that enumerates all the phonon eigenstates, and has no other significance or physical interpretation. Thus we can define the density of states directly from the force constant matrix

$$\rho(\omega) = \frac{2\omega}{\pi} \operatorname{Im} \operatorname{Tr} (\tilde{V} - \omega^2)^{-1} \tag{28}$$

where the matrix elements of $\tilde{V}$ between vector displacements at sites i and j are just given by the elements of the 3 x 3 tensor $(M_i M_j)^{-1/2} \underline{\underline{V}}_{ij}$ defined in (4). Thus the density of states can be obtained directly from the

potential via (28), without any reference to unit cells
or Brillouin zones. This form is appropriate therefore
in all harmonic solids. Quantities, like the heat capacity C, can be expressed in terms of $\rho(\omega)$ in the usual
way using Bose statistics

$$C/k_B = \int (\frac{\hbar\omega}{k_BT})^2 \ \frac{\exp(\hbar\omega/k_BT) \ \rho(\omega)d\omega}{[\exp(\hbar\omega/k_BT) - 1]^2} \tag{29}$$

or changing variables to $x = \hbar\omega/k_BT$

$$C/k_B = (k_BT/\hbar) \int \frac{x^2 \exp x}{[\exp x - 1]^2} \ \rho(k_BTx/\hbar)dx \tag{30}$$

if $\rho(\omega) \sim \omega^P$ at low frequencies, then at low temperatures

$$C \sim T^{1+p} \tag{31}$$

In a crystal at low frequencies $\omega \sim |k|$ and so using
$\rho(\omega)d\omega = V/(2\pi)^3 \ 4\pi k^2 dk$ we find $\rho(\omega) \sim \omega^2$; giving p=2
and the usual Debye T^3 heat capacity. Low temperature
experiments in amorphous SiO_2, GeO_2 and Se have shown an
anomalously large heat capacity between 70 mK and 1 K,
that seems to go as T and so dominates the T^3 contribution at these low temperatures. If these extra excitations were harmonic and so obeyed Bose statistics, then
the above arguments show us that p = 0 and the density of
states approaches a constant at low frequencies. The situation is actually rather more complex and will be taken
up in Sec. 6.

5. PHONONS IN AMORPHOUS SOLIDS

In the previous sections we have discussed the way
in which the potential is set up and diagonalized in a
crystalline solid where we eventually have to solve a matrix whose size is equal to the number of degrees of freedom in the unit cell. As the unit cell gets larger the
number of branches increases and the volume of the Brillouin zone gets smaller. An amorphous solid can be thought of as the limit of this process as the unit cell becomes infinite in size and the Brillouin zone becomes
just the point k = 0 so that k is no longer useful in clas-
sifying the modes. The high pressure phases of Si III
and Ge III provide useful examples of these "intermediate

structures". Si III has eight atoms/unit cell and hence
24 phonon branches. There are nine distinct finite frequencies at $\underline{k}$ = 0 as the high symmetry (cubic) makes
many modes degenerate. Of these nine frequencies, five
are Raman active.[6] This is to be contrasted with diamond
cubic Si where there is a single Raman active mode at
520 cm^{-1}. The frequencies in Si III are all below this -
roughly speaking we are seeing more of the density of
states at $\underline{k}$ = 0 - eventually when the unit cell becomes
infinite in size, we see the complete density of states.
It would be wrong, however, to think that we merely relax the $\underline{k}$ selection rules as the unit cell becomes larger; the density of states itself changes slowly. Some
features in the density of states remain and some are
washed out. This fact, that the density of states looks
rather similar in crystalline and amorphous solids, has
led a number of authors to attempt to try and describe
amorphous solids by applying perturbations to crystals.
This approach is fundamentally wrong, as there is no continuous structural path between crystalline and amorphous
solids. The proper and most satisfying way to approach
the problem is to make calculations for amorphous solids
from scratch - without invoking any concepts peculiar to
crystals like unit cells and Brillouin zones. This of
course is difficult in practice although simple in principle as it involves the diagonalization of the matrix $\tilde{V}$
defined for Eq. (28). Only two techniques, one numerical
and one more analytical have so far evolved and we shall
discuss them in detail in this section.

It is useful to recognise two kinds of disorder that
must occur in an amorphous material. The first is quantitative disorder, that is the variation of force constants,
like α, β in (17), with site. The environment of every
bond in an amorphous solid is slightly different, and so
α, β will vary from site to site. This kind of disorder
has not been discussed much in amorphous solids, except
for localized states in the vicinity of the band gap.
Even these states are usually discussed in terms of the
alloy problem. Localized states may occur near band edges
where the density of states is very small and arises from
unlikely, i.e. improbable, atomic configurations. This
subject is still largely unresolved and we shall not
pursue it further here. The second kind of disorder is
unique to amorphous solids. Topological disorder neglects
the variation of force constants from site to site and
concentrates on the random network which manifests itself
through the summation sign in (10). We shall concentrate
on this as it is most important except for these few
(<1%)states around the band edges. For the rest of this

section we shall keep α, β fixed in Eq. 17 and not let them vary from site to site.

It is important to remember that we can only gain a knowledge from the crystal of the forces between certain configurations of atoms. Forces between nearest and next-nearest neighbors in various forms of Si will be about the same if one adopts a local bond picture. However, this will not be so for 3rd nearest-neighbors as this force will surely depend on the dihedral angle. From analyzing experimental data on diamond cubic Si we can at best only determine the forces between 3rd nearest-neighbors in the staggered configuration. We therefore see that in the absence of a knowledge of the general potential (e.g. the screened Coulomb potential), it is necessary to only have very short range forces in our model if progress is to be made.

There are a number of rather general theorems that are useful in finding allowed frequency regions. Gerschgorin's theorem[12] is perhaps the best known of these however it is hard to apply directly in lattice dynamics to obtain tight bounds. A number of more useful theorems are stated and proved in the appendix. Theorem 1 may be used to find the maximum frequency of the potential (17).

The mean square frequency is given by

$$M<\omega^2> = \beta \sum_{j\alpha} (r_{ij}^\alpha)^2 + \sum_j (\alpha - \beta) \tag{32}$$

where j goes over the four neighbors of i. This reduces to

$$M<\omega^2> = 4\alpha \tag{33}$$

We may apply theorem 1 if the bonding is _perfectly_ tetrahedral to obtain an upper bound for the maximum frequency

$$M\omega_{max}^2 \le 8\alpha \tag{34}$$

which is what we found by explicit calculation in crystalline Si (see Eq. 18). The above derivation shows it to be true in diamond, cubic, wurtzite and all other possible polytypes corresponding to different stacking along the c axis. It is not quite true in amorphous Si because the bonding is not perfect tetrahedrally. However, bond angle distortions $\sim 10^o/o$ will only increase the upper limit 8α by a few percent. We can say this with some confidence

as long range periodicity was not used in establishing
the above inequality. The lower limit is fixed at zero
because the potential is positive definite. Similar arguments can also be applied to multicomponent glasses
like SiO_2 to give a number of spectral regions as discussed briefly in Sec 5b. Similar electronic problems
have been tackled using Gerschgorin's theorem and equations of motion.

Knowing that the regions where the densities of states
are finite are similar in crystalline and amorphous solids when the local bonding is essentially the same, the
next question is obviously what happens within these spectral regions.

a) Numerical techniques

These techniques have been pioneered by Dean and coworkers and are reviewed in Ref. 13. Most of this work
is on SiO_2 and related glasses. Weaire and Alben used
the Keating potential (similar to the Born potential (17)
but with the non-central force replaced by an angle bending force involving the coordinates of three atoms) with
a ratio of angle bending to central force constants of
·2, similar to the ratio 2/11 in the Born potential previously discussed. It is necessary to have this small
non-central piece in the potential to achieve stability.
If there are N atoms in the system, there are 3N degrees
of freedom. Central forces may be thought of as a constraint on the system involving keeping all the 4 · N/2
bonds fixed in length. It can be shown that these constraints are linearly independent and so we are left with
N degrees of freedom. This means that there must be N
eigenfrequencies at zero frequency for all random networks and crystals with nearest neighbor central forces
only. Detailed calculations confirm this (see Fig. 6).

Similar calculations for other random networks with
free surfaces confirm the general picture shown in Fig. 6.
Present day computers can handle matrices up to about
300 x 300 without using special techniques so that the
phonon frequencies in random networks with about 100 atoms
may be found numberically. The problem is that the number of surface atoms is high in three dimensional networks. Even in a network with 500 atoms, about 35°/o are
on the surface. Dean and coworkers[13] have used different boundary conditions to see which parts of the frequency spectrum are surface sensitive. The time involved in
diagonalizing a matrix goes roughly as the third power of

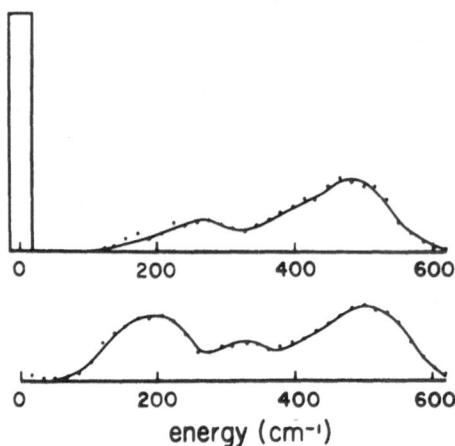

Fig. 6 Calculations of phonon frequencies for Si using Henderson's 61 atom random network with periodic boundary conditions. The upper curve is for nearest neighbor central forces where 1/3 of the weight comes in the delta function at zero frequency and the lower curve is with an additional small angle bending force. From Weaire and Alben[14].

its size and so doubling the size of the matrix means of factor eight in time. There are special techniques for sparse matrices and also for banded matrices that have been exploited by Dean.[13]

It can be seen by comparing Figs. 3 and 6 that the TA and TO peaks in the crystal are rounded but retained, although the TO peak is much reduced in height, especially in the Henderson model which is a more distorted structure than the 85 atom cluster which we discuss later in Fig.13. The two sharp longitudinal peaks in the crystal coalesce into a rather flat region. These general conclusions are also reached using the S.P.A. to be described later. Comparison with experiment is difficult because the density of states is not measured directly but modulated by matrix element effects that describe the interaction between the probe and the system. The one phonon infrared and Raman scattering from amorphous Ge are shown in Fig. 7.

A comparison with Fig. 6 (with the frequency scale

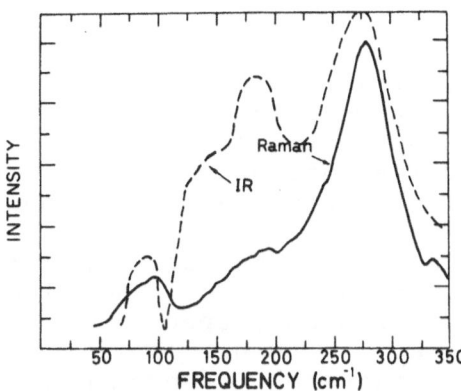

Fig. 7 The infrared absorption and Raman scattering from amorphous Ge.[15] The Raman scattering has the Bose factor $[n(\omega) + 1]/\omega$ removed.

suitably adjusted to go from Si to Ge; this can be done by noting that the optic modes at $\underline{k} = 0$ in diamond cubic Ge and Si are at 301 cm^{-1} and 520 cm^{-1}, respectively) shows that matrix element effects are indeed important particularly in the infrared experiment where the central portion is apparently amplified by matrix element effects.[14] Attempts to include matrix element effects[6,13,14] using a local bond interaction have not been terribly success- ful and often give a weighted density of states that os- cillates up and down wildly. This may or may not be a real effect and can be smoothed out for comparison with experiment. However, because of this uncertainty we will stay with the density of states as being a fundamental quantity. The matrix elements for inelastic neutron scat- tering are known and so this could be used to yield a density of states directly. However, so far only the low frequency part of the spectrum for Ge has been measured (see lecture by J.D.Axe) but the whole spectrum should be available soon.

Numerical calculations have been made by Dean[13] for SiO$_2$ glasses. The structure is more complicated than Si. Each Si atom is bonded to four O atoms and each O atom

to two Si atoms. The structure can be brought into a
one to one correspondence with Si if all the O atoms are
removed. Unfortunately the disorder is not entirely
topological in this case as the Si-O-Si bond angle
is thought to vary between about 120° and 180° and
this may be thought of as adding quantitative disorder
to the network. The Born potential between nearest neighbor Si-O pairs is used (Eq. 17 with $\beta/\alpha = 14/23 = \cdot61$).
Because of the Si,O mass difference, the computed spec-
tra fall in several distinct regions as shown in Fig. 8.

The spectrum in Fig. 8 was calculated for a 330 atom
cluster which involves matrices of dimension almost 1000.
Bell and Dean have developed a special technique based
on the negative eigenvalue theorem for finding frequency
histograms of banded matrices. Banded matrices have all
their finite elements close to the diagonal and occur naturally in one dimensional systems. By clever labelling
of atoms they also can be constructed for finite three
dimensional networks although with rather wide bands,
which naturally increases computer time. The method
gives all the eigenfrequencies less than a given one and
so by varying the given frequency a histogram is obtained as in Fig. 8a. The method does not give the eigenvectors but "typical" eigenvectors may be found in a separate calculation. From these the character of the vibrations may be found and this is shown in Fig. 8b where
C, R, B, S refer to the Cation (Si), and the Rocking,
Bending, and Stretching of the anion (O). The participation ratio is defined as

$$p = \left[3N \sum_{i=1}^{3N} u_i^4 \right]^{-1} \tag{35}$$

at each eigenfrequency where u_i is the eigenvector (normalized to 1) and i goes over all 3N degrees of freedom
of the system. For illustration, suppose the eigenvector
has 3Nr equal values $(3Nr)^{-1/2}$ and the rest zero; then
p = r. Thus p is a reasonable definition of the fraction
of atoms participating in a particular mode. It is shown
by Fig. 8c for SiO_2 where it can be clearly seen that p
is small near the band edges - except the origin. Where
it approaches 1/3 as one would expect. It would be interesting to calculate p for a Si random network where there
is much less quantitative disorder and so localization
at the band edges will be much harder to achieve.

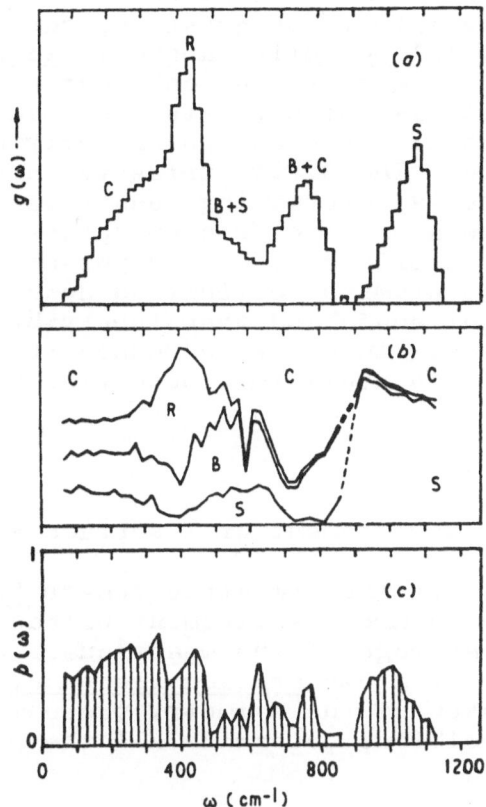

Fig. 8 (a) Frequency spectrum, (b) assignment diagram
and (c) participation ratio for an SiO_2 model with fixed
end boundary condition. The designations C, R, etc. refer to types of atomic motion described in the text.
Note that it is the distance between curves in (b) rather
than the heights of the curves themselves, which indicates
the proportion of energy arising from different types of
motion. From Bell and Dean.[16]

b) Structural potential approximation

Although the numerical work described in the previous section provides us with a lot of information, there
are two objections that can be raised. The first is that
the influence of the surface is rather unclear, and second, little understanding is achieved as to why certain

spectral features in the crystal remain and some disappear. Some answers to these questions can be achieved
using the <u>structural potential approximation</u> (S.P.A).
This is rather analagous to the coherent potential approximation in alloys which treats the quantitative disorder when two atomic species are randomly arranged on
a regular lattice. The S.P.A. tries to understand the
role of structure in determining densities of states.
In Fig. 9 we show a piece of random network. The forces
along bonds are described by a Born potential (17). However, the surface atoms by definition do not have their
full complement of bonds and therefore vibrate with too
large an amplitude. In order to reduce this amplitude,
we place the surface atoms in a potential well

$$V_s = \frac{1}{2} \underline{u}_s \, \underline{\underline{K}}_s \, \underline{u}_s \tag{36}$$

where $\underline{u}_s$ is the displacement of a surface atom.

The elements of the symmetric tensor $\underline{\underline{K}}_s$ are chosen
so that the mean-square displacement of each surface atom
is equal to the average of the mean-square displacements
of all the interior atoms <u>at each frequency</u> - appropriate
account being taken of the symmetry present in the pro-
blem. These conditions determine all the elements of $\underline{\underline{K}}_s$

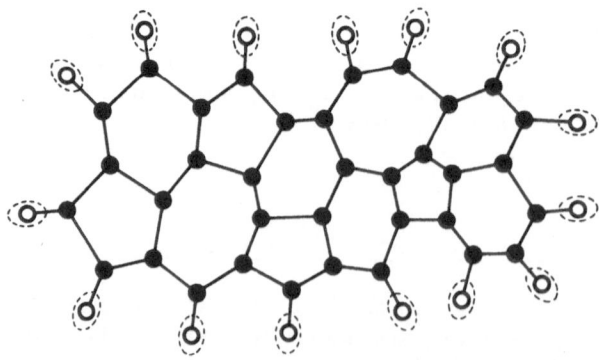

Fig. 9 A piece of random network with interior atoms
(solid circles) and surface atoms (open circles). The
dashed ellipse on the surface atoms represent the self-
consistent potentials.

and the density of states is defined as the average of the local density of states over the whole cluster. This procedure is straightforward in principle but leads to a set of self-consistent equations for the elements of $\underline{K}_s$ which must be solved at each frequency ω. However, by adopting this procedure, the abrupt distinction between the surface atoms and the interior atoms is lost as they are all moving with the same mean-square amplitudes. This simulates the medium outside the cluster which can be regarded as having been replaced by an effective medium.

The mean square amplitude of vibration is most easily calculated from the Green function (see also Eq. 27).

$$G = (M\omega^2 - V)^{-1} \tag{37}$$

If we write the displacement corresponding to the u_i degree of freedom as $|i>$, then the local density of states $\rho_i(\omega)$ and the contribution to the mean square amplitude $<u_i^2>_\omega$ at a frequency ω can be easily derived from G.

$$\rho_i(\omega) = -(2M\omega/\pi)\,\text{Im}\,<i|G|i> $$
$$= -(2M\omega/h)\,<u_i^2>_\omega \tag{38}$$

The relations (38) are true at zero temperature. However, it is trivial to include the temperature dependence as the excitations are Bosons. Thus by achieving self-consistency at zero temperature, below, we automatically achieve it at all temperatures.[17] The self-consistency condition is

$$\frac{1}{N_i} \sum_i <u_i^2>_\omega = \frac{1}{N_s} \sum_s <u_s^2>_\omega \tag{39}$$

where N_i, N_s are the number of interior surface atoms. The elements of the tensor $\underline{K}_s$ are <u>complex</u> and <u>frequency dependent</u>. Thus the cluster potential is non-Hermitian - this can lead to difficulties as the Green function may not have the right analytic properties. This "problem" is also encounted in the C.P.A. In the case of a single tetrahedron, see Fig. 10, the S.P.A. is equivalent to constructing a Bethe lattice, that is a lattice with no closed loops. One can show in this case that the Green function does have the right analytic properties.

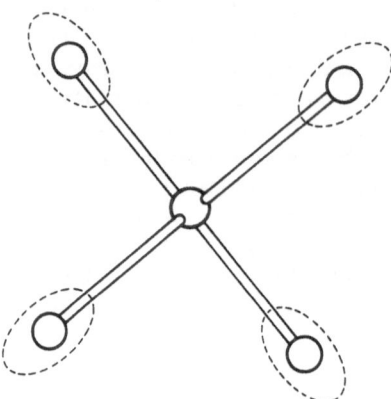

Fig. 10   A single tetrahedron with one interior and four
surface atoms.   The dashed ellipsoids represent the surface potentials.

        The simplest application of the S.P.A. is to the
five Si atoms shown in Fig. 10.   By symmetry the surface
potentials are all equivalent and are ellipsoids with one
parameter describing motion along the bond and another
motion perpendicular to it.   The mean square vibrations
of the interior atom are spherically symmetric and by de-
manding (Eq. 39) that the surface vibrations are also
spherically symmetric and equal to those of the interior
atom - we obtain two conditions.   Thus both parameters
in the surface potential may be eliminated and an equation formed for the Green function G.

$$\frac{4}{3}\{1 + [2(\alpha - \beta)G]^2\}^{1/2} + \frac{2}{3}\{1 + [2(\alpha + 2\beta)G]^2\}^{1/2}$$

$$= 1 + G(M\omega^2 - 4\alpha). \tag{40}$$

This equation may be multiplied out to give a quartic in
G.   The roots are real except for one complex pair in a
certain frequency range.   The imaginary part gives the
density of states (Eq. 38) and is shown in Fig. 11.

        It can be seen that the broad TA and the sharp TO
peaks are retained in the cluster   (although the designa-

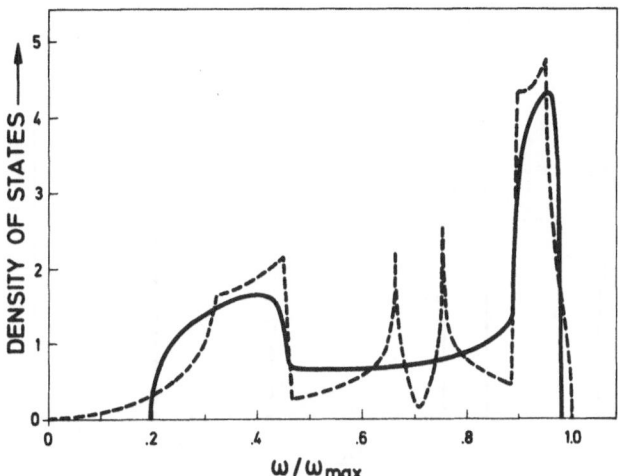

Fig. 11 Comparison of the density of states for Si/Ge
in the diamond cubic structure (dashed line) and for a
tetrahedral cluster (solid line). The same Born potential is used in both cases.

tions TA, TO lose their meanings). The two sharp longitudinal peaks in the center of the crystal spectra are
totally absent in the cluster. If we examine the origin
of the crystalline peaks in k space we can see qualitatively what is happening. The TA, TO peaks have contributions from all over the Brillouin zone; it can be seen
in Fig. 3 that the X and L points make contributions.
The longitudinal peaks, on the other hand, arise from
states extremely close to the hexagonal faces on the
Brillouin zone boundary. As the zone boundary is produced by long range structural order, the states on and
very close to the zone boundary should be very dependent
on long range order. The cluster calculation shows very
nicely that the TA, TO peaks are produced by the four-
fold local coordination and should therefore be present
in the amorphous solid. This conclusion is supported by
calculations using the S.P.A. on clusters with up to 18
atoms[17] and is also confirmed by experiment(see Fig. 7).

To demonstrate the large effect of the surface potential we show in Fig. 12 the eigenfrequencies of the
five atom tetrahedron without the surface potential.

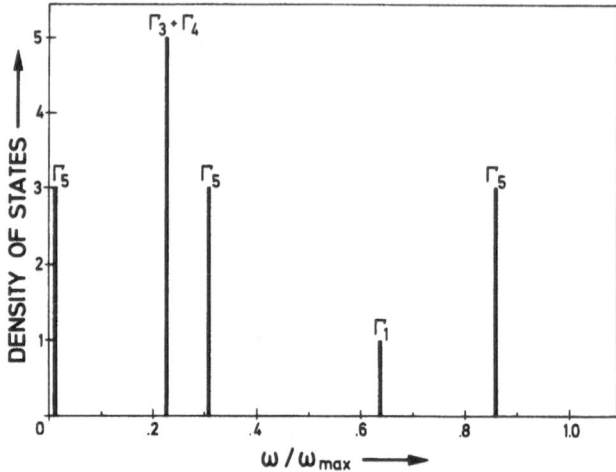

Fig. 12   The eigenfrequencies of the five atom tetrahedron of Fig. 10 with <u>no</u> surface potential.

It can be seen that the surface potential broadens the delta functions of the "molecule" into a continuous spectrum at higher frequencies.

Both calculational techniques described in this section fail to give the very long wavelength modes. It is clear that these can not be defined in any natural way for a finite system. A composite picture of various calculations for Si type solids is shown in Fig. 13. In comparing the two cluster calculations it can be seen that the spectral shape is essentially completely determined by the tetrahedral unit. The 85 atom cluster has a broadened peak at high frequencies and more weight in the central region - this is probably due to the angular distortions. Calculations show that the peak at high frequencies is sharper in models with smaller angular distortions.

In treating the vibrational modes of multicomponent glasses, Lucovsky[18] has tried to isolate typical atomic configurations containing a few atoms. The vibrational spectra of these "molecules" is then calculated or obtained from experiments on similar molecules in the gaseous phase. This technique has been rather successful

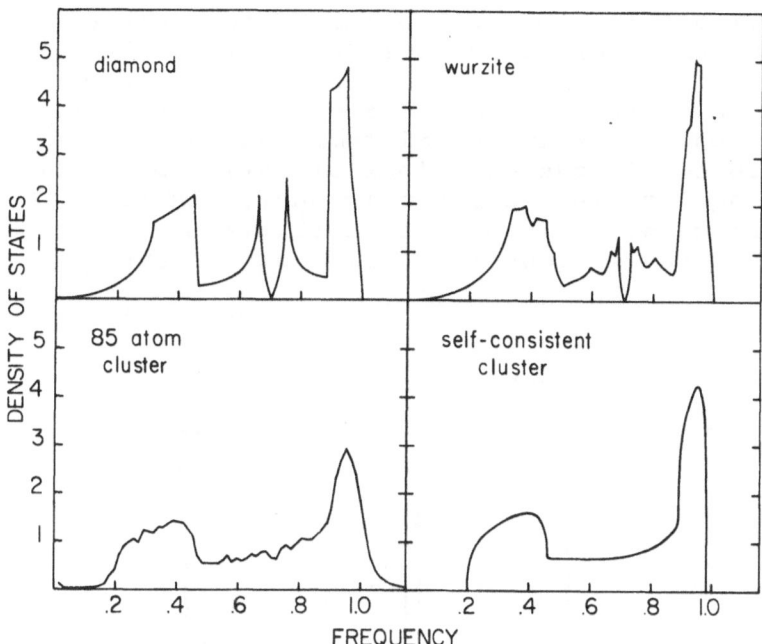

Fig. 13 The density of states for Si with the maximum
frequency in diamond cubic normalized to 1. All calculations are for the Born model with β/α ≈ .6 which is equivalent to a ratio of 2/11 between the non-central and central forces, except for the 85 atom cluster which uses
the Keating force constants with a ratio of 2/10 between
the angle bending and bond stretching force constants
(R. Alben, private communication). This cluster uses
"quasi-periodic" boundary conditions in which surface
atoms are connected to other surface atoms so that every
atom is four fold coordinated. The diamond and self-consistent cluster are the same as in Fig. 11. Wurtzite is
a hexagonal crystal with 4 atoms in each unit cell.

particularly in the chalcogenide glasses like As_2Se_3.
The main objection is that the bonds between the "molecule" and the rest of the matrix are neglected. In many
cases, these bonds are as strong as the bonds within the
"molecule" and so one has no feeling for the width (i.e.
localization in a loose sense) of the "molecule bands".
This difficulty may be largely overcome by using the S.P.A.

We have done this for an SiO$_2$ tetrahedron like Fig. 14,
but with four extra oxygen atoms on the Si-Si bonds. In
order to illustrate the broadening of the molecular modes,
the Si-O-Si units are assumed to be linear. The forces
used are the same as in Fig. 8. Ellipsoidal potentials
are placed upon the surface Si atoms and the same conditions are applied as in the Si tetrahedron - the O atoms
being included via the equations of motion but otherwise
not entering into the self-consistency. In Fig. 14 we
show the results. It can be seen that some "molecular
modes" are strongly modified whilst others are not. The

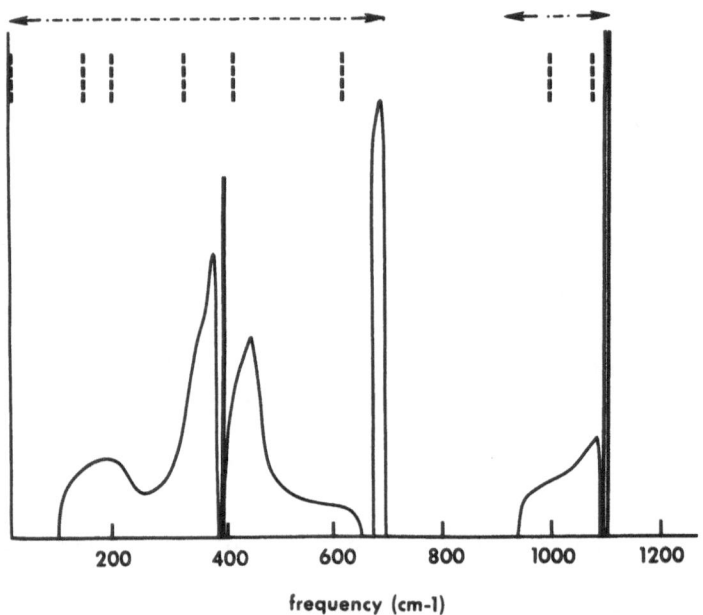

frequency (cm-1)

Fig. 14a The frequency spectrum of SiO$_2$. The molecular
modes of a single tetrahedron with five Si and four O a-
toms are shown as vertical dashed lines near the top of
the diagram. A similar S.P.A. to that discussed for Si
yields the spectrum shown. It can be seen that there are
some very narrow bands and some broader bands. The two
horizontal arrows at the top of the diagram indicate the
allowed spectral regions for all SiO$_2$ solids with <u>perfect</u>
tetrahedral bonding. These were found using the theorems
in the appendix. All such solids described by a Born model have a delta function near 400cm^{-1} which corresponds
to the oxygen atoms moving perpendicular to the bonds and
the silicon atoms stationary.

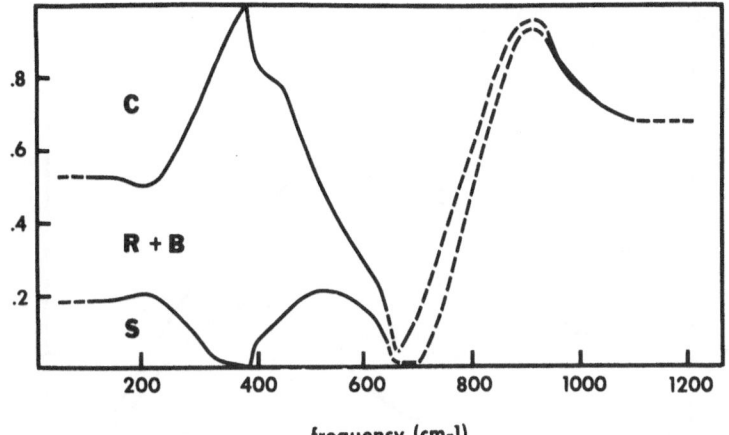

frequency (cm-1)

Fig. 14b   The assignment diagram where C, R, B, S are
explained in the text.   There is no distinction between
R and B when the Si-O-Si bond is straight.   The dotted
lines are for the guidance of the eye and are not calculated or even defined.

assignment diagram, which characterizes the modes, is analogous to that of Fig. 8b, except that there is no distinction between R and B for linear Si-O-Si units.   It
can be seen that most of the character of the modes is
determined by entirely local considerations.   Because of
the linear Si-O-Si unit it is unfair to make a direct
comparison between Figs. 8 and 14 although similarities
are striking.

6. ANOMOLOUS LOW TEMPERATURE BEHAVIOR

It has been known for some time that the low temperature specific heat and thermal conductivity of most
amorphous materials is anomolous.   We shall concentrate
on the specific heat; the thermal conductivity also requires a knowledge of a phonon mean free path.   Fig. 15
shows some measurements of the specific heat in SiO_2

Careful work[11] down to .01°K shows that the specific
heat becomes linear and may be fit with a form
$C \sim aT + bT^3$, this is shown in Fig. 16.

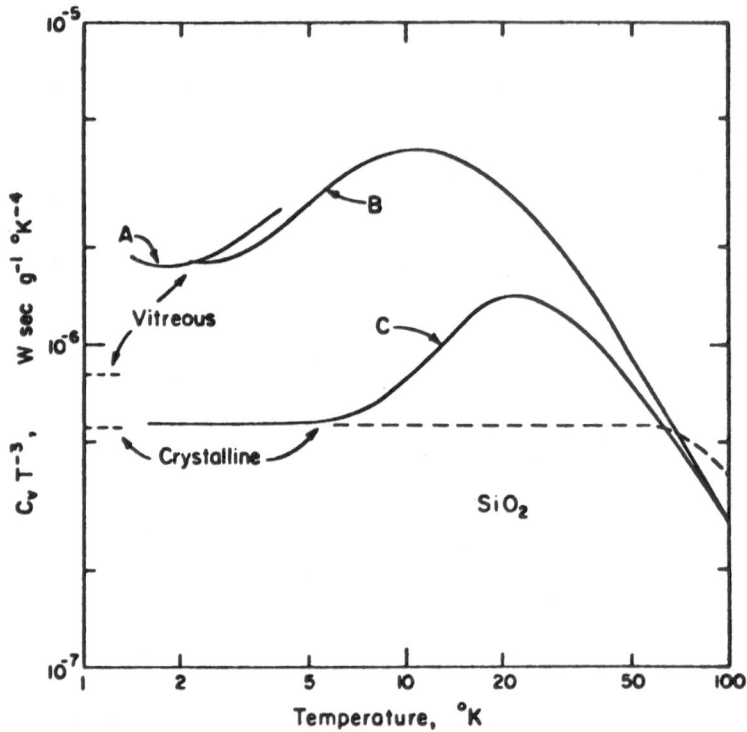

Fig. 15 The specific heat of two amorphous forms (A,B)
and crystalline (C) SiO_2.[11] At low temperatures the crystalline specific heat obeys the usual Debye T^3 law where
the coefficient is derivable from the elastic constants
(dashed line). In vitreous SiO_2 the specific heat is
much larger than would be predicted by the elastic constants.

Recent experiments in amorphous Ge down to $2°K$ also
give an enhanced heat capacity but do not go low enough
in temperature to shown whether the heat capacity becomes linear in temperature.[19]

There have been a number of attempts to try and explain this rather general phenomona. Fulde and Wagner[20]
suggest that the sound waves are damped and depending on
the temperature dependence of the damping, they can
achieve many forms of low temperature specific heat -
including linear. A more promising explanation has been

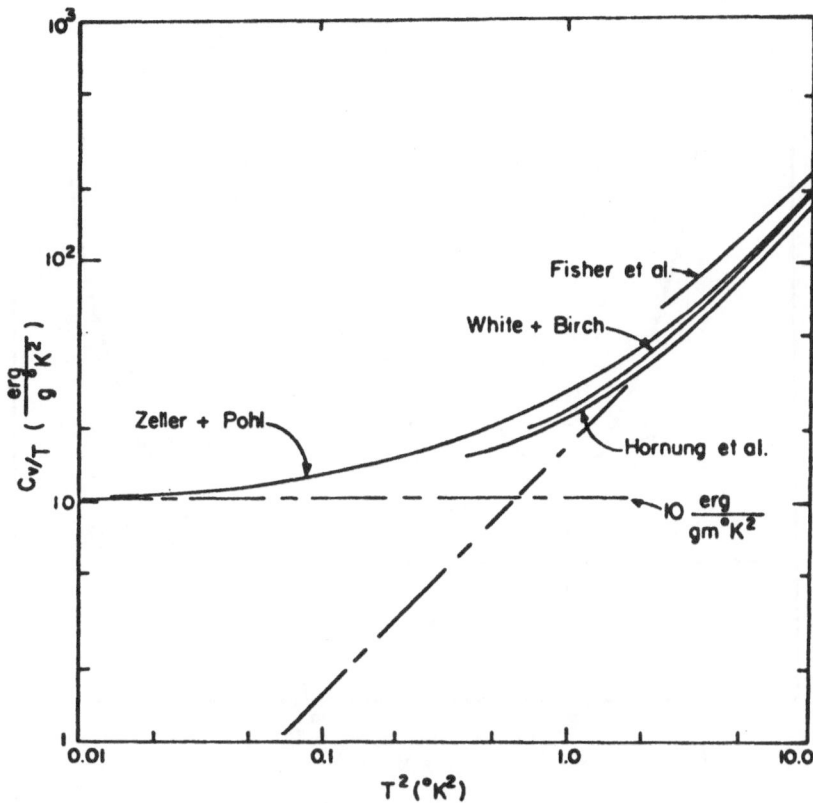

Fig. 16 Measurements[11] to lower temperatures in vitreous SiO_2 show that a linear specific heat is approached.

put forward independently by Anderson, Halperin and
Varma[21] and by Phillips.[22] They suggest that when the
energy of a glass is plotted against some generalized
coordinate, there may be neighboring minima as shown in
Fig. 17.

The precise origin of these neighboring potential
minimas is not known. In multicomponent glasses like
SiO_2 it may be the flipping of an O atom between two
roughly equivalent positions. This sort of behavior is
known to be possible because of the high-low transition
in quartz. Conceivably, many more atoms may be involved
in the transition. Although it is quite easy to visualize
this behavior in rather open structures with two fold

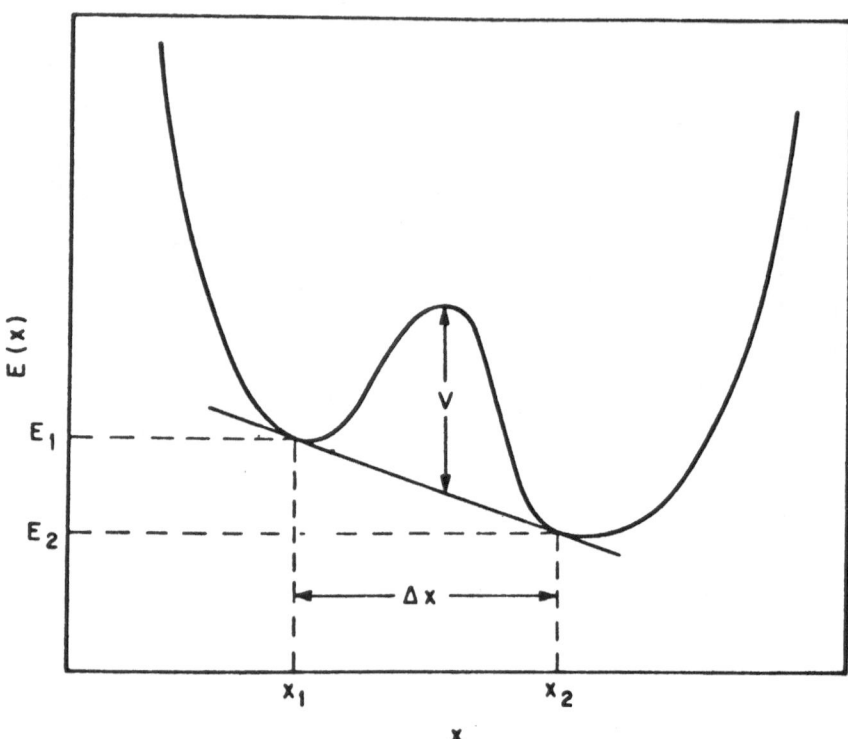

Fig. 17 The energy E plotted against some generalized
coordinate x.[21] This is similar to Fig. 2 except that
now we are considering only local structural rearrangements instead of macroscopic ones.

coordinated atoms like SiO_2, it is much harder in Si/Ge
where there is not much local structural flexibility.
For this reason heat capacity measurements in Ge at lower
temperatures would be very interesting. However, for the
moment we will accept Fig. 17. Each well will have a
lowest energy state and these states will be separated
by some energy ΔE which depends primarily on the difference in depth $E_1 - E_2$. Only these two levels are involved at the lowest temperatures and so the free energy
$F(\Delta E)$ is simply

$$F(\Delta E) = -k_B T \, \ln[1 + \exp(-\Delta E/k_B T)] \tag{41}$$

and from this the heat capacity $C = -T(\partial^2 F/\partial T^2)$ is given by

$$C/k_B = (\Delta E/2k_B T)^2 \, \text{sech}^2(\Delta E/2k_B T) \qquad (42)$$

There will be a distribution $n(\Delta E)$ in the amorphous solid and so the total heat capacity will be

$$C/k_B = \int_0^\infty (\Delta E/2k_B T)^2 \, \text{sech}^2(\Delta E/2k_B T)n(\Delta E)d\Delta E \qquad (43)$$

and at low temperatures only states with $\Delta E \sim k_B T$ contribute and so

$$C/k_B = \pi^2/6(k_B T) \, n(\Delta E = 0) \qquad (44)$$

The observed linear heat capacity in SiO_2 may be used in (44) to calculate $n(\Delta E = 0) \approx 1/25$ states per SiO_2 group per eV which seems like a reasonable number.

An alternative explanation for the linear specific heat is obtained by noticing that (44) is similar to the form of the <u>electronic heat capacity</u> in metals except that $n(\Delta E = 0)$ is replaced by the density of states at the Fermi level. We expect a small but finite number of states at the Fermi level due to band tailing but the number 1/25 per SiO_2 group per eV seems unreasonable in a large band gap (~ 9 eV) material like SiO_2. Other arguments against this explanation are a) the density of states in the "gap" region is sensitive to impurities whereas the linear specific heat is not, b) the anomolous thermal conductivity, presumably due to phonons, cannot also be explained.

Linear heat capacities have also been found for polymers at low temperatures,[11] and so it may be a rather general phenomenon in non-crystalline solids that results from anharmonicity. If this were the case, it would not be necessary to postulate the extreme form of anharmonicity of the double potential well. More work needs to be done in this area.

7. CONCLUSIONS

We have tried to cover, somewhat subjectively, some of the interesting ideas that are currently being dis-

cussed concerning phonons in amorphous materials using
Si/Ge and SiO$_2$ as illustrations. We have pointed out
that within the harmonic approximation the problem is
well defined and in a sense its complete solution awaits
only the advent of larger computers. However, this statement is reminiscent of the more extreme viewpoint that
all of solid state physics merely involves solving the
Schrödinger equation. Numerical solutions do not always
lead to insight and other techniques like the S.P.A. are
useful in this regard. The treatment of longer range
forces in amorphous materials remains a problem, it may
be possible to include them in some mean field way, or
perhaps more simply by using a shell model. One of these
alternatives is necessary if a detailed comparison with
the TA phonons in Si/Ge is to be made. The anomolous
low temperature specific heat and thermal conductivity
remain a puzzle, especially as there seems to be some
rather general principle involved.

    Some of the concepts in this paper may also be re-
levant in the social and political sciences where coupled networks are often encountered as shown in Fig. 18.

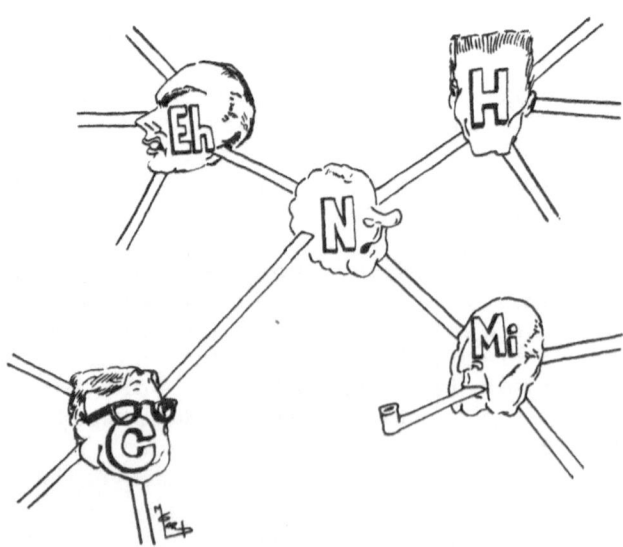

Fig. 18 A coupled political network with bonds of varying
strength, dangling bonds, and other analogies with solid
state networks.

ACKNOWLEDGMENTS

The author acknowledges the support of the N.S.F. in preparing these notes and also many helpful discussions with colleagues at Yale.

APPENDIX

In this appendix, we show how certain theorems about spectral limits that are useful in lattice dynamics are established. These theorems are similar to well known ones like Gerschgorin's theorem[12] - however, we give them here in forms that are more easily applied in lattice dynamics.

Theorem 1

The maximum frequency ω_{max} and the mean square frequency $<\omega^2>$ are related by the following inequality

$$\omega_{max}^2 \leq 2<\omega^2> \qquad\qquad A1$$

provided that

a) The harmonic potential from which the frequencies are derived may be written

$$V = \frac{1}{2} \sum_{\substack{i,j \neq i \\ \alpha\beta}} V(i,j,\alpha,\beta)(u_i^\alpha - u_j^\alpha)(u_i^\beta - u_j^\beta) \qquad A2$$

b) For every bond i,j; the matrix $V(i,j,\alpha,\beta)$ must be positive semidefinite (i.e. if this matrix is diagonalized, the eigenvalues are non-negative.)

c) The solid must contain a single atomic species with mass M and the matrix $\sum_{j \neq i} V(i,j,\alpha,\beta)$ must be isotropic at each site i - that is

$$\sum_{j \neq i} V(i,j,\alpha,\beta) = a\delta_{\alpha\beta}. \qquad\qquad A3$$

where a is independent of i.

This theorem is extremely useful in giving an upper bound for ω_{max} which is often quite difficult to obtain without solving the whole problem. The mean square frequency is very easy to find and indeed using A3 is just

$$<\omega^2> = a/M \qquad\qquad\qquad A4$$

Condition a) is somewhat restrictive as this is <u>not</u> the most general form of the potential for the rigid ion model as written in Eq. 10 even if the translational invariance is incorporated. However, the Born potential (17) is in this form and indeed any force constants derived from a central potential $V(r)$ would be in this form. The second condition b) states that each pair of atoms <u>considered separately</u> must be in stable equilibrium and condition c) restricts us to cubic and tetrahedral symmetries with a single atomic species.

It is quite simple to prove this theorem using the equations of motion (11)

$$\left[M\omega^2 - \sum_{j\neq i} V(i,j,\alpha,\alpha)\right]u_i^\alpha = - \sum_{j\neq i} V(i,j,\alpha,\beta)u_j^\beta$$

$$A5$$

Because the potential **V** in Eq. A2 is positive semi-definite, the eigenvalues of the matrix $V(i,j,\alpha,\beta)$ on the right hand side of A5 must be bounded from above by

$$\sum_{j\neq i} V(i,j,\alpha,\alpha) = \frac{1}{3} \sum_{\substack{j\neq i \\ \alpha}} V(i,j,\alpha,\alpha) = <\omega^2>/M \qquad A6$$

in order to ensure that all the $\omega^2 \leq 0$. We now define a <u>new system</u> which is described by a potential similar to A2 but with the two minus signs replaced by plus signs. The potential for this new system is not translationally invariant but is positive semi-definite because of condition b. The change of signs in A2 means that the sign of the right hand side of A5 is changed but the <u>rest of the equation is unaltered</u>. Therefore the eigenvalues of the matrix $V(i,j,\alpha,\beta)$ must be bounded from <u>below</u> by $<\omega^2>/M$ and we have the theorem that the eigenvalues ω^2 of the original equation A5 are bounded from above by $2<\omega^2>$.

<u>Theorem 2</u>

If the eigenvalues λ of a Hermitian matrix M_1 are bounded by A_1 and B_1 so that

$$A_1 \leq \lambda_1 \leq B_1 \qquad\qquad\qquad\qquad A7$$

and similarly if the eigenvalues λ_2 of another Hermitian matrix M_2 are bounded by A_2 and B_2

$$A_2 \leq \lambda_2 \leq B_2 \qquad\qquad\qquad\qquad A8$$

then eigenvalues λ of the matrix $M_1 + M_2$ are bounded by $A_1 + A_2$ and $B_1 + B_2$

$$A_1 + A_2 \leq \lambda \leq B_1 + B_2 \qquad\qquad\qquad\qquad A9$$

Note that the matrices M_1, M_2 and M must be defined in the same space.

This theorem is useful when a matrix can be decomposed into two pieces whose bounds are known. The proof goes as follows. <u>Any</u> eigenstate $|i>$ of M may be expanded either in eigenstates $|j1>$ of M_1 with eigenvalues λ_1^j <u>or</u> $|j2>$ of M_2 with eigenvalue λ_2^j so that

$$|i> = \sum_j a_{j1}^i \; |j1> = \sum_j a_{j2}^i \; |j2>$$

so that the corresponding eigenvalue λ^i is given by

$$\lambda^i = \sum_j \left| a_{j1}^i \right|^2 \lambda_1^j + \sum_j \left| a_{j2}^i \right|^2 \lambda_2^j$$

It is clear that by using the inequalities in A7 and A8 and by using closure

$$\lambda^i \leq \sum_j \left| a_{j1}^i \right|^2 B_1 + \sum_j \left| a_{j2}^i \right|^2 B_2$$

i.e. $\lambda^i \leq B_1 + B_2$

Noting that λ^i is any eigenvalue and using a similar argument for the lower bound the result A9 follows.

<u>Theorem 3</u>

If the eigenvalues λ_1 of a matrix M_1 lie within the disc

$$|\lambda_1 - c_1| \leq |D_1| \qquad\qquad\qquad A10$$

and similarly if the eigenvalues λ_2 of a matrix M_2 are within the disc

$$|\lambda_2 - c_2| \leq |D_2| \qquad\qquad\qquad A11$$

then the eigenvalues of λ of $M_1 + M_2$ lie within the disc

$$|\lambda - c_1 - c_2| \leq |D_1| + |D_2| \qquad\qquad\qquad A12$$

This can be proved from Theorem 2. The conditions A10 and A11 may be rewritten as

$$c_1 - |D_1| \leq \lambda_1 \leq c_1 + |D_1|$$

$$\qquad\qquad\qquad A13$$

$$c_2 - |D_2| \leq \lambda_2 \leq c_2 + |D_2|$$

and so by applying Theorem 2 we easily prove this theorem.

REFERENCES

1.   An elementary introduction may be found in C. Kittel, Introduction to Solid State Physics, J. Wiley and Sons, Inc., New York, 1971, 4th Ed. More exhaustive treatments may be found in M. Born and K. Huang, Dynamical Theory of Crystal Lattices, Oxford University Press, London and New York, 1954; and in Solid State Physics, Supplement 3, 1971, 2nd ed., A. A. Maradudin, E. W. Montroll, G. H. Weiss and I. P. Ipatova, Theory of Lattice Dynamics in the Harmonic Approximation.
2.   B. N. Brockhouse in Phonons in Perfect Lattices and in Lattices with Point Imperfections, ed. by R. W. H. Stevenson, Oliver and Boyd, Edinburgh and London, 1966, p. 110.
3.   W. H. Zachariasen, J. Am. Chem. Soc. 54, 3841 (1932), and D. E. Polk, J. Non-Cryst. Solids 5, 365 (1971).
4.   W. A. Harrison in Phonons in Perfect Lattices and in Lattices with Point Imperfections, ed. by R. W. H. Stevenson, Oliver and Boyd, Edinburgh and London, 1966, p. 73

5.   See for example: A. W. Solbrig, Jr., J. Phys. Chem.
     Sol. $\underline{32}$, 1761 (1971).
6.   R. J. Kobliska, S. A. Solin, M. Selders, R. K. Chang,
     R. Alben, M. F. Thorpe, and D. Weaire, Phys. Rev.
     Letts. $\underline{29}$, 725 (1972).
7.   W. Ledermann, Proc. Roy. Soc. A$\underline{182}$, 362 (1944).
8.   W. Cochran in <u>Phonons in Perfect Lattices and in
     Lattices with Point Imperfections</u>, ed. by R. W. H.
     Stevenson, Oliver and Boyd, Edinburgh and London,
     1966, p. 53.
9.   For some very careful recent work see G. Nilsson
     and G. Nelin, Phys. Rev. B $\underline{5}$, 3151 (1972).
10. F. Herman, J. Phys. Chem. Sol. $\underline{8}$, 405 (1959).
11. R. C. Zeller and R. O. Pohl, Phys. Rev. B$\underline{4}$, 2029
     (1971).
12. For a general discussion of Gerschgorin's theorem
     see R. S. Varga, <u>Matrix Iterative Analysis</u> (Engle-
     wood Cliffs, NJ, Prentice Hall, 1962), p. 16.
13. P. Dean, Rev. Mod. Phys. $\underline{44}$, 127 (1972).
14. D. Weaire and R. Alben, Phys. Rev. Lett. $\underline{29}$, 1505
     (1972); and R. Alben in Proc. of Conf. on Tetra-
     hedrally Bonded Amorphous Semiconductors, Yorktown
     Heights, 1974, AIP Conf. Proc. 20, p. 249.
15. A. Lurio and M. H. Brodsky, Bull. Am. Phys. Soc.,
     $\underline{17}$, 322 (1972)
16. R. J. Bell and P. Dean, Disc. Faraday Soc. $\underline{50}$, 55
     (1970).
17. M. F. Thorpe, Phys. Rev. B$\underline{8}$, 5352 (1973) and M. F.
     Thorpe in, Proc. of Conf. on Tetrahedrally Bonded
     Amorphous Semiconductors, Yorktown Heights, 1974,
     AIP Conf. Proc. 20, p. 267.
18. G. Lucovsky in the Proc. of the 5th Intl. Conf. on
     Amorphous and Liquid Semiconductors, Garmish-
     Partenkirchen, (Taylor and Francis, London, 1974),
     Vol. 2, p. 1099.
19. C. N. King, W. A. Phillips and T. P. deNeufville,
     Phys. Rev. Lett. $\underline{32}$, 538 (1974).
20. P. Fulde and H. Wagner, Phys. Rev. Lett. $\underline{27}$, 1280
     (1971).
21. P. W. Anderson, B. I. Halperin and C. M. Varma,
     Phil. Mag. $\underline{25}$, 1 (1972).
22. W. A. Phillips, J. Low Temp. Phys. $\underline{7}$, 351 (1972).

LOCAL ORDER AND LOW FREQUENCY MODES IN AMORPHOUS SOLIDS:

MAGNETIC RESONANCE TECHNIQUES

P. C. Taylor, E. J. Friebele[+] and
   Mark Rubinstein

Naval Research Laboratory
Washington, D. C. 20375

1. INTRODUCTION

The absence in amorphous materials of long range periodic order, and hence of many of the selection rules present in crystalline materials, makes the interpretation of x-ray, neutron or electron scattering experiments much more difficult. Information concerning the static and dynamic properties of these materials must therefore be extracted using many different experimental techniques. One technique which has proven quite useful is magnetic resonance spectroscopy.

About 16 years ago the first nuclear magnetic resonance (NMR) studies of amorphous materials were carried out using continuous wave (cw), broad-line techniques.[1] These measurements involved the identification of planar BO_3 and tetrahedral BO_4 units in alkali borate glasses by ^{11}B NMR. The first nuclear quadrupole resonance (NQR) measurements of amorphous materials were performed using pulse techniques about two years ago.[2]

Many nuclei possess magnetic moments $\vec{\mu} = \gamma \hbar \vec{I}$ where $\vec{I}$ is the nuclear spin and γ the nuclear gyromagnetic ratio. If I is greater than 1/2, then these nuclei also possess electric quadrupole moments Q which result from a nonspherically symmetric nuclear charge distribution.

+ NAS-NRC Postdoctoral Associate.

Either the interaction of the nuclear moment with a magnetic field H (Zeeman interaction) or the interaction of
a nucleus which possesses a quadrupole moment with the
gradient of the electric field at the nuclear site can
lead to sets of nuclear energy levels whose separations
fall in the radio frequency (rf) range. Resonant absorption can then occur when incident rf radiation is tuned
to these energy differences.

The magnetic resonance technique allows one to select out of the total magnetic susceptibility a particular contribution of interest, for example, a particular
nucleus. This technique provides information on a microscopic or atomic basis since the resonant nuclei serve
as sensitive probes of their local environments. The importance of nuclear magnetic moments in studying amorphous
materials is that rf spectroscopy allows the use of some
nuclei as probes of the internal magnetic fields. The
importance of nuclear electric quadrupole moments is that
they allow certain nuclei to be used as probes of the internal electric fields. When the magnetic moment dominates the nuclear energy levels through the Zeeman interaction, then the resonant absorption of rf energy is called nuclear magnetic resonance (NMR). On the other hand,
when the quadrupole interaction makes the dominant contribution to the nuclear energy levels, then the resonance phenomenon is called pure or nuclear quadrupole resonance (NQR).

NMR and NQR spectroscopy yield information about
both the static and dynamic properties of amorphous materials, especially when comparisons can be made with measurements in corresponding crystalline materials. Static
properties which can be investigated using magnetic resonance techniques include local structural order and details of local bonding configurations. As a probe of local order magnetic resonance measurements are usually
only sensitive to first nearest neighbor interactions although in some cases details out to third or fourth nearest neighbors can be inferred. Dynamic properties of amorphous solids which can be investigated by magnetic resonance include dipole-dipole or spin-spin interactions, the
motion or diffusion of atoms or ions, and interactions
of nuclear spins with vibrational modes which can probe
details of the density of vibrational states.

The purpose of these lecture notes is to summarize
the types of information obtainable from NMR and NQR studies of amorphous materials. The usefulness of the magnetic resonance technique is illustrated in some detail

with a specific example: ^{75}As pulsed NQR in vitreous and crystalline As_2S_3. No attempt will be made to describe any experimental equipment or to discuss the details of any experimental technique. The experimental aspects of NMR and NQR are discussed in several excellent sources.[3,4] The theoretical foundations of the interactions important in magnetic resonance are summarized in Section II. Sections III and IV describe NMR and NQR in amorphous materials, respectively. Future prospects for the use of magnetic resonance techniques in amorphous solids are briefly discussed in Section V.

II. MAGNETIC RESONANCE INTERACTIONS

The theory of magnetic resonance is well established and comprehensive, and many detailed references on NMR and NQR theory are readily available.[3,5,6] The emphasis of this section will be to discuss briefly those interactions which will be necessary to understand the examples which follow.

A. Zeeman Interaction

The energy of interaction of a nuclear moment in a magnetic field H is termed the Zeeman energy and is governed by the following Hamiltonian

$$H_Z = - \vec{\mu} \cdot \vec{H} \tag{1}$$

Since the magnetic moment is related to the nuclear spin I by

$$\vec{\mu} = \gamma \hbar \vec{I} \tag{2}$$

one can write the Zeeman term as

$$H_Z = - \gamma \hbar \vec{I} \cdot \vec{H} \tag{3}$$

If H is taken in the Z direction, then the eigenvalues of Eq. (3) are given by $E_m = -\gamma \hbar H m$ where m is the Z-component of the nuclear spin and ranges from -I to I in integral steps. The nuclear energy levels for this interaction are thus equally spaced with a separation

$$\Delta E = \gamma \hbar H \tag{4}$$

When an rf electromagnetic wave of frequency ν_0 is applied to the nuclear system whose Hamiltonian is given by Eq. (1) then energy is absorbed when $h\nu_0 = \Delta E$ or

$$\nu_0 = \frac{\gamma H}{2\pi} \tag{5}$$

Equation 5 defines the nuclear magnetic resonance condition for the Zeeman interaction.

B. Quadrupolar Interaction

Nuclei with $I > 1/2$ can possess in addition to a magnetic dipole moment, an electric quadrupole moment which reflects the departure of the nucleus from spherical symmetry. For a nucleus of charge density $\rho(\vec{r})$, the energy of interaction between the nucleus and the electrostatic potential $V(\vec{r})$ due to all sources external to the nucleus is

$$\mathcal{K} = \int \rho(\vec{r}) \ V(\vec{r}) \ d\tau \tag{6}$$

where the integral is taken over the nuclear volume. The usual procedure is to expand $V(\vec{r})$ in a power series about the nuclear site

$$V(\vec{r}) = V_0 + \sum_{i=1}^{3} \left(\frac{\partial V}{\partial X_i}\right)_0 X_i + \frac{1}{2} \sum_{i,j=1}^{3} \left(\frac{\partial V}{\partial X_i \partial X_j}\right)_0 X_i X_j + \cdots \tag{7}$$

where the subscript o indicates that the quantity is evaluated at the nuclear site and may thus be taken outside the integral of Eq. (6).

Substitution of Eq. (7) into Eq. (6) yields an interaction energy

$$H = V_0 \int \rho(\vec{r}) \ d\tau \ + \sum_{i} \left(\frac{\partial V}{\partial X_i}\right)_0 \int \rho(\vec{r}) X_i \ d\tau$$

$$+ 1/2 \sum_{i,j=1}^{3} \left(\frac{\partial^2 V}{\partial X_i \partial X_j}\right)_0 \int \rho(\vec{r}) \ X_i X_j \ d\tau + \cdots \tag{8}$$

where $\int \rho(\vec{r}) d\tau$ is just the nuclear charge Ze. The first term of Eq. (8) is the electrostatic energy of a point nucleus and is independent of nuclear size and shape. We thus can ignore this term. The second term is proportional to the nuclear electric dipole moment which can be shown to be identically zero because the nucleus is in a state of definite parity. The third term is the interaction between the nuclear quadrupole moment tensor given by

$$Q_{ij} = \int \rho(\vec{r}) \, X_i X_j \, d\tau \qquad (9)$$

and the electric field gradient tensor $V_{ij} = (\partial^2 V/\partial X_i \partial X_j)_0$.

If all higher moments are neglected the electrostatic interaction reduces to the nuclear quadrupole interaction

$$H_Q = 1/2 \sum_{i,j=1}^{3} Q_{ij} V_{ij} \qquad (10)$$

Equation (10) can be simplified by symmetry considerations and can be written in the form

$$H_Q = \frac{eQ}{6I(2I-1)} \sum_{i,j=1}^{3} V_{ij} [3/2(I_i I_j + I_j I_i - \delta_{ij} \vec{I}^2] \qquad (11)$$

where the nuclear quadrupole moment Q is defined as

$$Q = 1/e \int (3z^2 - r^2) \, \rho(\vec{r}) \, d\tau \ . \qquad (12)$$

A final simplification of Eq. (11) is possible by transforming to a coordinate system in which V_{ij} is diagonal. One obtains

$$H_Q = \frac{e^2 q Q}{4I(2I-1)} [3I_z^2 - I(I+1) + \eta/2(I_+^2 + I_-^2)] \qquad (13)$$

where x, y, z now represent the principal coordinate system of V_{ij} and where

$$q = (1/e) V_{ZZ} \ ; \quad \eta = (V_{xx} - V_{yy})/V_{ZZ} \ ;$$
$$I_\pm = (I_x \pm i \, I_y) \qquad (14)$$

When the Zeeman interaction of Eq. (3) is much greater than the quadrupolar interaction of Eq. (13), then
the experiment is in the province of NMR. When Eq. (13),
dominates, the experiment is in the province of NQR. To
the experimentalist this is more than an academic distinction, since it determines whether a fixed frequency
spectrometer may be used in which the external magnetic
field is swept to search for the resonance or whether a
variable frequency spectrometer is required.

C. Magnetic Shift Interactions

There are several magnetic shift interactions which
are important in certain NMR applications in disordered
solids. These include the chemical shift, the Knight
shift and the paramagnetic shift interactions. The chemical shift arises when the external magnetic field induces a current in the electronic orbitals in a solid.
This current in turn produces a magentic field at the
nucleus, so the nuclear magnetic field differs from the
applied field by a certain amount. The Knight shift is
a magnetic shift arising from the spin polarization of
the conduction electrons in a metal, while the paramagnetic shift interaction results from the existence of
paramagnetic centers in a solid which are polarized by
a magnetic field. All three of these interactions can
be expressed in the same functional form, and all three
have the same angular dependence in a magnetic field.
Specifically, the chemical shift Hamiltonian can be written in the form

$$H_{cs} = \gamma \hbar \sum_{i,j=1}^{3} I_i \sigma_{ij}(H)_j \tag{15}$$

where H is the applied field and σ_{ij} a component of the
chemical shift tensor. When only the Zeeman and chemical
shift interactions are present (Eqs. (3) and (15)), the
effect of the chemical shift is to alter the energy separations of Eq. (4) and to make them a function of the
orientation of the magnetic field with respect to the
three principal axes of the chemical shift tensor.

D. Dipole - Dipole Interactions

Resonance conditions of the form of Eq. (5) predict
that rf power is absorbed precisely at a frequency ν_0 and
at no other. In realistic physical systems there is a
finite linewidth to any resonance absorption. Several

possible mechanisms include instrumental broadening, interactions of nuclear spins with each other and interactions of nuclear spins with thermal lattice vibrations.
In this section we consider the effect of magnetic dipole-
dipole interactions.

The Hamiltonian which describes the interaction of
a nulcear spin with the magnetic fields arising from all
other nuclei in the sample is given by

$$H_d = 1/2 \sum_{i,j}^{N}{}' \frac{\vec{\mu}_i \cdot \vec{\mu}_j}{r_{ij}^3} - \frac{3(\vec{\mu}_i \cdot \vec{r}_{ij})(\vec{\mu}_j \cdot \vec{r}_{ij})}{r_{ij}^5} \qquad (16)$$

where the sums over i and j are taken over all N nuclei
in the system and the prime indicates that the term
i = j is excluded. The factor 1/2 assures that each pair
of nuclei is counted only once.

Although it is impossible in general to calculate
the Hamiltonian of Eq. 16 for practical systems, one can
approximate its effects by computing the second moment
of the contribution of Eq. (16) to the magnetic resonance
linewidth. Details of the second moment calculations are
available elsewhere[6,7]. We list for reference the result
for dipole-dipole interactions between like nuclei for
the specific case of I = 3/2 which will be of interest
in the following sections.

$$<\Delta\omega^2> = \frac{\gamma^4 h^2}{96} \sum \frac{1}{r_{ij}^6} [207(1-3\gamma_{ij}^2)^2 + 1512\gamma_{ij}^2(1-\gamma_{ij}^2)$$

$$+ 459(1-\gamma_{ij}^2)^2 - 108(1-3\gamma_{ij}^2)(\alpha_{ij}^2-\beta_{ij}^2)]$$

$$(17)$$

The direction cosines of the radius vector r_{ij} between
nuclei i and j in the principal axis system are given by
α_{ij}, β_{ij}, γ_{ij}.

E. Spin - Lattice Interactions

For spin - 1/2 nuclei, nuclear spin-lattice relaxation can occur via relaxation due to the dipolar inter-

action of the magnetic moments of the nuclei or relaxation due to paramagnetic impurities. For nuclei with
$I > 1/2$, however, the dominant relaxation mechanism involves the coupling of the nuclear electric quadrupole
moment to the electric field generated by the charges
outside the nucleus (quadrupolar relaxation). Since the
examples to be presented in Sections III and IV involve
nuclei with $I > 1/2$, we indicate the essential features
of quadrupolar relaxation in solids. The relaxation can
be expressed in terms of transition probabilities which
indicate the rate at which thermal fluctuations in the
crystal field induce transitions between the various spin
states of a given nucleus.

The Hamiltonian can be written as a sum of a Zeeman
term, H_Z, a lattice term H_L, and a term due to the presence of the nuclear electric quadrupole moment H' which
is a function of $\vec{I}$ and of the coordinates of the centers
of gravity of <u>all</u> nuclei in the solid.[8] One then expands
H' in a power series in displacements of the nuclei relative to the displacement of the central nucleus. In
this formalism the first term in the expansion is a constant term which is just the quadrupolar Hamiltonian H_Q
of Eq. (10). The higher order terms give rise to the
relaxation processes. The linear term gives rise to direct processes where a nuclear spin makes an upward (or
downward) transition and one of the lattice oscillators
is de-excited(or excited). Since the probability for
the process is proportional to the density of phonon states,
which is negligible at energies corresponding to rf waves,
this term is unimportant. The third term (quadratic term)
gives rise to Raman processes which are in fact the effective relaxation processes.

The probability that the nuclear spin makes a transition from the state m to the state $m + \mu$ as a result
of a Raman process can be written as[8]

$$P(m, m+\mu) = \frac{2\pi |Q_{\mu m}|^2}{M^2} \int_0^{\omega_m} \frac{\rho(\omega)^2}{\omega^2} \frac{e^x}{(e^x-1)^2} M_\mu(ka) d\omega \tag{18}$$

where $x = \hbar\omega/kT$, $Q_{m\mu}$ is the quadrupolar matrix element,
M is the sample mass and $M_\mu(ka)$ is a complicated sum of
products of displacement coefficients constituting each
two phonon process. The quantities ω_m, a and k are the
maximum phonon frequency, the interatomic spacing and a
reciprocal lattice vector respectively. The function $\rho(\omega)$

represents the phonon density of states.

At high temperatures (measurement temperatures much greater than the temperature of the modes effective in relaxing the nuclear spins or $\hbar\omega \ll kT$) Eq. (18) predicts that the spin lattice relaxation rate T_1, which is proportional to P, is proportional to T^2 regardless of the functional form of $\rho(\omega)$. At low temperatures, T_1^{-1} (or P) is proportional to T^7 or $e^{-\Theta E/T}$ depending on whether the relaxation is via acoustic or optic phonons, respectively.[8,9]

III. NMR IN AMORPHOUS MATERIALS

In NMR the energy levels of the spin system are determined by performing a diagonalization of Eqs. 1, 13 and 15. Once the energy levels are determined, the energy level separations that correspond to allowed transitions can be calculated. These energy level separations represent the quantum of energy (designated by $h\nu$) absorbed by the spin system. Such an expression is referred to as a _resonance condition_ and is of the form $h\nu = f(H)$. Under most experimental conditions, the frequency ν is kept constant, and the magnetic field varied, so the resonance condition is solved for H, the magnetic field at which resonance will occur.

In a single crystal, the resonance field or frequency depends on the orientation of the single crystal with respect to the applied magnetic field. The resonance condition is calculated from the energy eigenvalues obtained from the Hamiltonian. In a polycrystalline, vitreous, or amorphous sample, the nuclear sites are randomly oriented with respect to the applied field. Then the magnetic resonance spectrum, referred to as a "powder pattern", is an average over the resonance conditions for all possible orientations of the nuclear site. A review of powder patterns in NMR is available elsewhere,[10] and we merely summarize the essential features in this section.

It is an extremely lengthy process to obtain the necessary eigenvalues for all members of an ensemble of randomly-oriented crystallites by exact diagonalization of the Hamiltonian. Fortunately, in many instances the Zeeman term in the Hamiltonian predominates, and perturbation theory may be employed to derive approximate solutions. If the magnetic shift term is neglected and the axially symmetric quadrupolar term is treated to second

order in perturbation theory, then the following NMR resonance condition is obtained for the m = 1/2 → m = -1/2
transition:

$$\nu = \frac{\gamma H}{2\pi} + \frac{\nu_Q^2 [I(I+1) - 3/4]}{16\left(\frac{\gamma H}{2\pi}\right)} (9\cos^2\theta - 1)(1 - \cos^2\theta)$$

(19)

where

$$\nu_Q = \frac{3e^2 q_{mol.} Q/h}{2I(2I-1)}$$

In Eq. (19), θ is the angle between the principal axis
of the electric field gradient (EFG) tensor and the applied magnetic field, Q is the quadrupole moment of the
nucleus and eq_{mol} is the z-component of the electric
field gradient for the nucleus in a given molecular site.
The quantity $(e^2 q_{mol} Q/h)$ is the quadrupole coupling constant (in frequency units) for the given molecular configuration.

The NMR resonance condition of Eq. (19) can be in-
verted and written as $H_m = H_m(\mu)$ where $\mu = \cos\theta$. The
powder pattern is then the ensemble average of this resonance condition over all equally probable elements of
solid angle $d\Omega = d\mu$. The absorption at field H in the
interval dH may be written as[10]

$$S(H)dH = \frac{1}{4\pi} \int_H^{H+dH} I_m(\Omega)d\Omega(H_m)$$

(20)

The quantity S(H) is termed a shape function or
powder pattern and represents the normalized amplitude
of the magnetic resonance signal at field H. Equation
10 is integrated over those elements of solid angle
$d\Omega(H_m)$ such that $H < H_m < H + dH$ where H_m is the resonance condition. In general $d\Omega$ is a multi-valued function
of H_m, there being more than one value of μ for some resonance fields H_m. The quantity $I_m(\Omega)$ is the transition
probability, and may be taken outside the integral sign
in most cases.[10]

In a disordered or glassy material an additional complication arises due to the basic randomness of the structure itself. In these materials, there often exists a
continuously random variation in the local environments
surrounding any particular nuclear site. This variation

of local environments can result in a continuous variation of the Hamiltonian parameters describing the magnetic resonance spectra of these sites. While the absorption in a powdered crystalline material can be accurately represented by a randomly-oriented ensemble of otherwise identical sites, the absorption in a glassy or disordered solid must often be characterized by an additional ensemble of sites with differing local environments.
Computer simulation techniques, which numerically evaluate integrals of the form of Eq. 20, are essential to
the evaluation of magnetic resonance spectra observed in
glasses and disordered materials because closed algebraic
expressions are not generally possible to describe the
more complicated averages over both orientation and local
environment. One uses a finite number of sites possessing discrete Hamiltonian parameters to approximate the
continuously random variation of local environments and
a finite number of elements of solid angle to approximate the ensemble of random orientations.

In favorable situations, only some of the Hamiltonian parameters (and thus only some of the singular points
of the powder spectrum) are sensitive to the existing
variations in local environments. In the most favorable
of situations the variations in Hamiltonian parameters
can be calculated from an assumed model of the nuclear
site. A general review of all the NMR investigations in
glasses and amorphous materials is beyond the scope of
the present article; however, excellent summaries are a-
vailable in the articles of Wong and Angell[11] and Bray.[12]
We confine our discussion here to one specific example
which illustrates the use of magnetic resonance in non-
crystalline materials.

One basic principle has proved to be extremely useful in evaluating the magnetic resonance spectra of glasses, namely that the best starting point for fitting a
glass spectrum is the spectrum observed in the corresponding crystalline compound, when one exists. The success of this procedure is well illustrated by the ^{11}B NMR
spectra observed in the alkali borate glass system
(XMe$_2$O (1-X)B$_2$O$_3$, where Me is an alkali metal) in which
three distinct boron sites are observed in three crystalline materials of the system. The relative prominence
of each site in any given glass composition can be determined using computer simulation techniques.[13]

In the crystalline alkali borate system, for example,
four compounds exist up to 33 1/3 mole ^{o}/o alkali oxide

in which four distinct structural groupings are present
(boroxol, pentaborate, triborate and diborate groups).
It was first proposed by Krough-Moe[14] that the alkali
borate glasses with less than 33 1/3 mole o/o alkali
oxide contain mixtures of these four crystalline structural groups and that the relative percentage of each
type of group depends on the exact composition of the
glass with respect to the crystalline compounds.

NMR measurements of the relative fraction of boron
atoms which are four-coordinated (tetrahedrally coordinated) in alkali borate glass provide strong support for
the Krough-Moe, or "discrete site", hypothesis. The [11]B
NMR spectra of alkali borate glass of low alkali oxide
content ($\leq$ 30 mole o/o Me_2O) have been interpreted as
the sum of a sharp line due to tetrahedral BO_4 units and
a broad line due to planar BO_3 units. Spectra attributable to BO_3 units with one non-bridging oxygen atom
(metaborate groups) have been found to occur in alkali
borate glasses with greater than about 30 mole o/o alkali
oxide. In the glasses of low alkali-oxide content, the
fraction of boron atoms which are tetrahedrally-coordin-
ated, N_4, has been found to follow the relation $N_4 \approx \dfrac{x}{1-x}$
where x ($\lesssim$ 1/3) is the mole fraction of alkali oxide in
the glass. This behavior is exactly what one would expect if each oxygen added as Me_2O converts two planar
BO_3 units into two tetrahedral BO_4 units which are incorporated into pentaborate, triborate or diborate groups.

The general features of the NMR spectra of alkali
borate glasses have been fit using simple chemical bonding arguments and assuming the spectra are sums of the
responses due to discrete sites which also occur in crystalline compounds of the system. More accurate computer
simulations of these spectra reveal that the discrete crystalline-like sites can be represented by ensembles of
slightly distorted sites in the glass due to random fluctuations in the local glass environments. The "half -
widths" of the random fluctuations in local environments
in an ensemble of sites of a given genre are found to be
small compared to the differences in local environments
between generically different crystalline-like sites
BO_3 and BO_4 sites). A simple model for the electric field
gradient in BO_3 structural units is used to illustrate
that the NMR spectra are well explained by the introduction of discrete sites whose average distortions are
small.[15] Using this model, the deviations in average O-B-O
bond angles in both BO_3 and BO_4 groups are determined

by computer lineshape simulations to be less than $\pm \sim 2^{\circ}$ in alkali borate glasses.

In B_2O_3 glass the boron atoms are bonded to three essentially equivalent oxygen atoms forming a triangular molecule. For all practical purposes, the two inequivalent BO_3 sites in crystalline B_2O_3 are planar and can be assumed for the present purposes to be identical and to possess axial symmetry about an axis through the boron atom perpendicular to the molecular plane.[16] We consider the effect on the electric field gradient at the boron nucleus as this molecule is progressively distorted. For simplicity, we constrain the distortions to retain axial symmetry. That is, the O-B-O molecular bond angles of Fig. 1 remain identical but are allowed to decrease toward the tetrahedral value for a BO_4 molecule. The departure from planarity is conveniently represented by the angle α of Fig. 1.

The electric field gradient at the boron nucleus can be calculated fairly accurately using the Townes and Dailey theory[17] in terms of the distortion angle α. We then consider the effect on the NMR spectrum of the Gaussian ensemble of distorted BO_3 units as a function of the

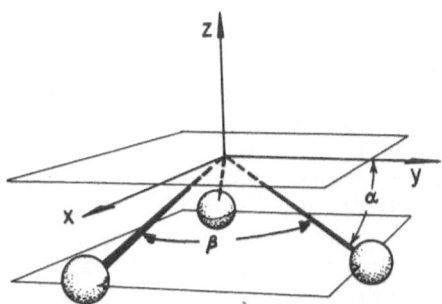

Fig. 1   Schematic diagram of a distorted BO_3 unit constrained to maintain axial symmetry. The boron atom is at the origin of the coordinate system. The three oxygen atoms are represented by open circles. There are three sp^3-type boron molecular orbitals along the three B-O bonds and a fourth (normally empty) lone-pair orbital along the z-axis. The angle α represents the departure of the B-O bonds from the x-y plane. The O-B-O bond angle is represented by β.

half-width σ_α of the distribution in α. As will be shown below, the Townes and Dailey result is essentially independent of whether the boron is coordinated to three or four oxygen atoms and depends only on the angle α.

The NMR Hamiltonian for a ^{11}B nucleus in a magnetic field, which contains both nuclear Zeeman and nuclear quadrupolar terms, is given by the sum of Eqs. (3) and (13) of Section II. For the case of axial symmetry, the NMR resonance condition for the $m = 1/2 \to m = -1/2$ transition (where m is the z-component of the nuclear spin) can be obtained from the Hamiltonian. The result, which contains quadrupole terms correct to second order in perturbation theory, is given by Eq. (19).

In the Townes and Dailey theory the molecular coupling constant $e^2q_{mol}Q/h$ is related to an atomic coupling constant $e^2q_{at}Q/h$ which is a unique (although empirically determined) property of a given nucleus[17]. The atomic coupling constant represents the coupling between an atomic nucleus and a single p electron. The molecular coupling constant is evaluated in the Townes and Dailey approximation by tensor addition of the contribution of each hybrid p electron to the field gradient.

For a pyramidal molecule with three s-p hybrid bonds (B-O bonds) and no "lone-pair" electrons in the fourth orthogonal orbital (along the symmetry axis in Fig. 1) one obtains[3]

$$e^2q_{mol}Q/h = \frac{(1-\mathcal{I})(3\sin^2\alpha-1)}{\cos^2\alpha}\,(e^2q_{at}Q/h) \qquad (21)$$

where $\mathcal{I}$ is the ionic character of the bond ($\mathcal{I}$ = half the electro-negativity difference between boron and oxygen ≈ 0.7 for a B-O bond) and α is the angle of deviation of the B-O bonds from the planar configuration (Fig. 1). The value of the atomic coupling constant for boron is $e^2q_{at}Q/h \approx -5.39$ MHz.[3] The values of the constants $(1-\mathcal{I})$ and $e^2q_{at}Q/h$ of Eq. (21) are not accurately known and we thus normalize to the case of planar BO_3 units ($\alpha=0$) in crystalline B_2O_3 where experimentally $(e^2q_{mol}Q/h) = 2.53$ MHz assuming axial symmetry. This compares with the approximate Townes and Dailey value of $(1-\mathcal{I})(e^2q_{at}Q/h) \approx 1.6$ MHz.

As the BO_3 "molecule" of Fig. 1 becomes distorted

enough to approach the tetrahedral configuration one would expect by simple chemical arguments to form a BO_4 unit, when there is extra oxygen present, by adding one extra oxygen atom to the distorted BO_3 unit which contributes two electrons to the fourth orthogonal orbital of Fig. 1. In this case the coupling constant is still given by Eq. (21) with the exception that the factor $(1-\mathcal{I})$ is replaced by $(\mathcal{I}'-\mathcal{I})$ where $\mathcal{I}'$ is the ionic character of the fourth B-O bond. The factors $(\mathcal{I}'-\mathcal{I})$ and $(1-\mathcal{I})$ are of the same order of magnitude everywhere except near the tetrahedral bond angle $(\alpha_T = 19.5^\circ)$ where $\mathcal{I}' \approx \mathcal{I}$ by symmetry and chemical considerations. In crystalline materials where the boron is known to be four coordinated, the coupling constant falls in the range $e^2 q_{mol}Q/h \sim 100$ to 900 kHz.[18]

It is apparent from Eq. (21) that once the molecular coupling constant has been normalized to fit planar BO_3 units, the values to be used in the evaluation of the NMR resonance condition (Eq. 19) are determined solely by the geometry of the molecule in the Townes and Dailey approximation. The normalized quadrupole coupling constant distribution function $F(e^2 q_{mol}Q/h)$ for a Gaussian ensemble of sites can be obtained from

$$F(\alpha) = \frac{e^{-\alpha^2/2\sigma_\alpha^2}}{\sqrt{2\pi}\ \sigma_\alpha} \tag{22}$$

where α is obtained by inverting Eq. (21).

The ^{11}B NMR spectrum for an ensemble of randomly distorted sites is calculated using the resonance condition of Eq. (19) and the distribution function of Eq. (22) with $d\alpha$ expressed in terms of $d(e^2 q_{mol}Q/h)$ using Eq. (21). The NMR absorption spectrum or powder pattern can be calculated numerically. One may randomly choose $e^2 q_{mol}Q/h$ according to the probability expressed in Eq. (22), randomly choose an orientation θ in Eq. (19) and calculate the frequency ν of the NMR absorption. The absorption spectrum is then obtained by repeating the above process many times and plotting a histogram of the absorption versus frequency. In practice several mathematical short cuts, which are described elsewhere, can be used.[10,13]

The fit to the experimental derivative spectrum for ^{11}B NMR in glassy B_2O_3 for $\sigma_\alpha = 0.1$ rad. is very good as can be seen from Fig. 2. This value of σ_α corresponds to an average deviation in the apex bond angle β (see

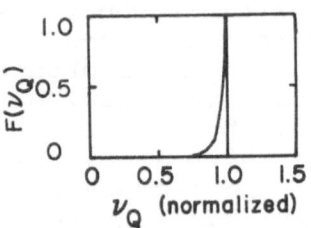

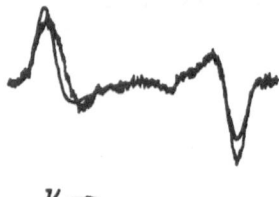

Fig. 2   Distribution function F (ν_Q) as a function of ν_Q
(top) for the case of $\sigma_\alpha = 0.1$ and the resulting best-
fit to the B_2O_3 glass spectrum at $\nu_0 = 16$ MHz (bottom).
The solid line represents the theoretical fit.

Fig. 1) of a BO_3 group of $\sim \pm 2°$. Thus an ensemble of
nearly planar, relatively undistorted BO_3 groups provides
an entirely adequate explanation for the observed ^{11}B NMR
spectrum in B_2O_3 glass.

Glassy $(Ag_2O)_5 (B_2O_3)_{95}$ yields an experimental example with which to compare the contributions of BO_4 and
BO_3 units to the NMR spectrum. This derivative spectrum
has a sharp central line whose derivative intensity is
roughly equal to that of the two outlying features of the
broad line. In Fig. 3 the fit to the experimental spectrum is shown assuming two discrete axial sites. The
distribution function for the quadrupole coupling constants is also indicated in the inset. The small distribution widths for the broad and narrow lines are centered about the best fits to the ^{11}B NMR responses observed
in crystalline B_2O_3 and $Ag_2O.2B_2O_3$ or $Ag_2O.4B_2O_3$ assuming
axial symmetry.[19] In crystalline B_2O_3 the boron atoms

are coordinated to three oxygens in a planar triangular arrangement, and in crystalline $Ag_2O \cdot 2B_2O_3$ or $Ag_2O \cdot 4B_2O_3$ some of the borons are coordinated to four oxygens in a nearly tetrahedral arrangement. Thus the narrow line corresponds to four coordinated borons and the broad line to three coordinated borons. The shapes of the two narrow distributions in the inset to Fig. 3 were determined using the model described above where the narrow line distribution was centered about the value of α inferred for crystalline $Ag_2O \cdot 4B_2O_3$ ($\alpha_0 \approx 33^o$). The values of σ_α for both the BO_3 and BO_4 sites correspond to deviations in the average O-B-O bond angles β of $\sim \pm 2^o$. The agreement between the model and experiment in Fig. 3 is everywhere quite adequate.

One can refine the agreement between experimental and computer simulated traces by allowing for non-zero asymmetry ($\eta \neq 0$) in the quadrupolar tensor. In Fig. 4, a compendium of traces[13,19,20] is presented which shows the agreement typically obtained between experimental and computer simulated ^{11}B NMR derivative spectra for representative borate glasses whose fractions of tetrahedrally coordinated boron atoms N_4 range from zero to 0.45. The relative intensities of the sites are given by the relation $N_4 = x/(1-x)$ where x is the mole fraction of metal oxide provided $x \lesssim 0.33$. For the trace of Fig. 4d where $x = 0.4$, an additional site corresponding to a three-coordinated boron with one oxygen not bonded to another boron also contributes to the intensity. In this case, the relative intensities of the three sites are given by additional, but simple, chemical bonding considerations.[13] The Hamiltonian parameters used to simulate the spectra of Fig. 4 are available in the literature[15]. One will note the improvement of the fit to the silver borate spectrum of Fig. 3 when the $\eta = 0$ restriction is removed (compare to Fig. 4b).

IV. NQR IN AMORPHOUS MATERIALS

NQR is difficult to detect in amorphous materials. In fact, the first NQR experiments in these materials were performed about two years ago.[2] These experiments involved pulsed measurements of [75]As in amorphous and crystalline As_2S_3. The magnitude of the electric field gradient at a nuclear site is an extremely sensitive function of the nuclear near-neighbor environment. Since the configuration and position of the nearby atoms vary from site to site in a disordered solid, the overall linewidth of

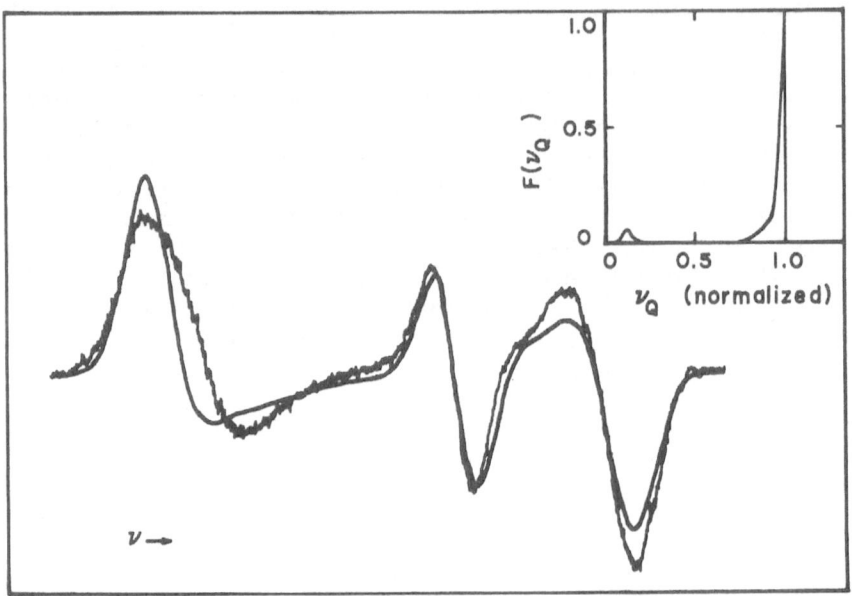

Fig. 3 Experimental and computer-simulated ^{11}B NMR derivative spectra for glassy $(Ag_2O)_{0.5}$ $(B_2O_3)_{0.95}$ at $\nu_0 = 16$ MHz. The inset represents the distribution function used in the simulation.

the resonance is considerably broadened. As will be discussed below, the half-width of the ^{75}As NQR observed in crystalline As_2S_3 is ~50 kHz, while the half-width in glassy As_2S_3 is ~3.5 MHz. In As_2Se_3 glass the half-width is ~6 MHz.

The ^{75}As nucleus is 100°/o abundant, and has a spin of I = 3/2. For this case, Eq. (13) predicts the existence of a single resonance line when no external magnetic field is applied to the sample at a resonance frequency[3],

$$\nu = \frac{1}{2} \; \frac{e^2 q_{mol} Q}{h} \; (1 + \frac{\eta^2}{3})^{1/2} \tag{23}$$

Since there is only one resonance frequency for I = 3/2, the two parameters e^2qQ/h and η cannot be determined uniquely. In crystalline materials η can be determined

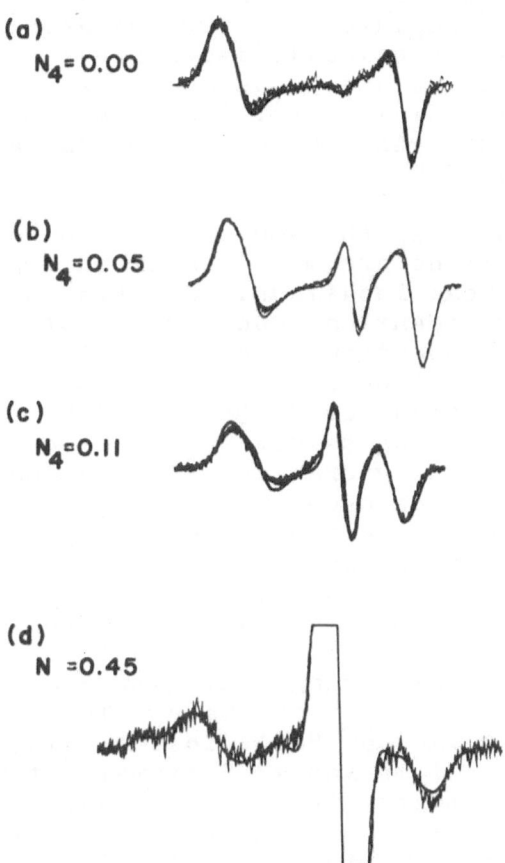

Fig. 4 Experimental and computer-simulated ^{11}B NMR derivative spectra at ν_o = 16 MHz for various borate glasses whose fractions of four-coordinated boron atoms, N4, vary from 0 to 0.45. (a) B_2O_3 glass; (b) $(Ag_2O)_{0.05}$ $(B_2O_3)_{0.95}$ glass; (c) $(Cs_2O)_{0.1}$ $(B_2O_3)_{0.9}$ glass; (d)$(Cs_2O)_{0.4}$ $(B_2O_3)_{0.6}$ glass.

uniquely by applying a perturbing magnetic field. Note that Eq. 23 does not contain any angular dependence such as exists in the presence of the nuclear Zeeman term. Thus the additional complication of calculating a powder pattern, which was necessary in the preceding NMR section, is not necessary here.

Crystalline As_2S_3 is a yellow, layered crystal which is strongly anisotropic and can be easily cleaved parallel to the plane of the layers. Vitreous As_2S_3 is deep red in color and macroscopically isotropic. The availability of As_2S_3 in both its amorphous and crystalline forms makes this material especially useful because the information obtained from the [75]As NQR in the crystal and the glass can be compared.

We first discuss the NQR spectra in natural As_2S_3 samples (crystalline orpiment) obtained from several different geographical locations. Spectra for all crystalline samples are identical and consist of a pair of [75]As resonance lines, one from each of the two inequivalent As sites in the unit cell. At 4.2K, the nuclear resonance frequencies occur at 70.38 and 72.86 MHz, and decrease with increasing temperature due to the thermal effects of the lattice vibrations.[21] The measured linewidths are all of the order of 50 kHz and are due to broadening caused by sample inhomogeneities or strains and by instrumental effects.

Monoclinic orpiment has a layer structure in which each As atom is covalently bonded to three S atoms in a triangular pyramidal arrangement with an As atom at the apex and three S atoms at the base. The layers are held together by weak van der Waals forces. The unit cell consists of 8 As atoms and 12 S atoms, but many of these are related by symmetry transformations, and hence have identical quadrupole coupling constants. There are two nonequivalent As atoms in the unit cell, which are labeled I and II in Fig. 5a. To a first approximation, the quadrupole resonance frequencies may be interpreted in terms of isolated AsS_3 pyramids which exist in both the crystal and the glass and are shown in Fig. 5b.

The AsS_3 "molecule" contains sp^3 hybridized electron orbitals, of which three form covalent As–S bonding orbitals and the fourth forms an unshared or lone pair orbital. In the Townes and Dailey approximation discussed in the preceding section, the NQR frequency can be related to the [75]As bonding configuration of a <u>regular</u> pyramidal molecule as expressed in Eq. 21 with the factor $(1 - \vartheta)$ replaced by $(1 + \vartheta)$. Equation 21 can be extended to cover the case of an asymmetric pyramidal molecule, which is applicable to As_2S_3.[2] In this case one obtains the following parameters for the two non-equivalent sites in orpiment: $\nu_1 = 117$ MHz, $\eta = 0.15$; $\nu_{II} = 114$ MHz, $\eta_{II} = 0.42$. Better agreement with the

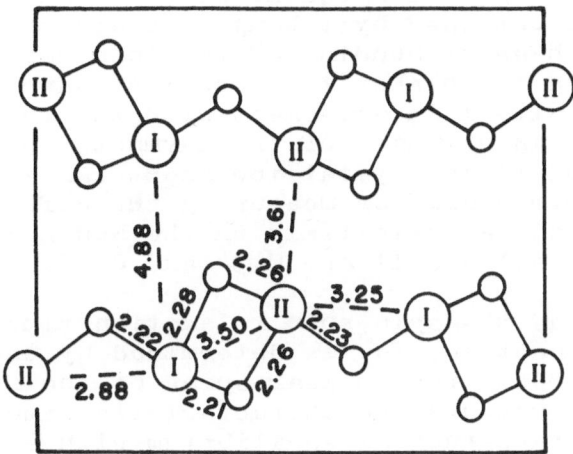

Fig. 5a   The unit cell of orpiment (As_2S_3) projected on
a (001) plane (ab face).   The indicated distances between
atoms are in Angstroms.   Small circles denote S atoms.

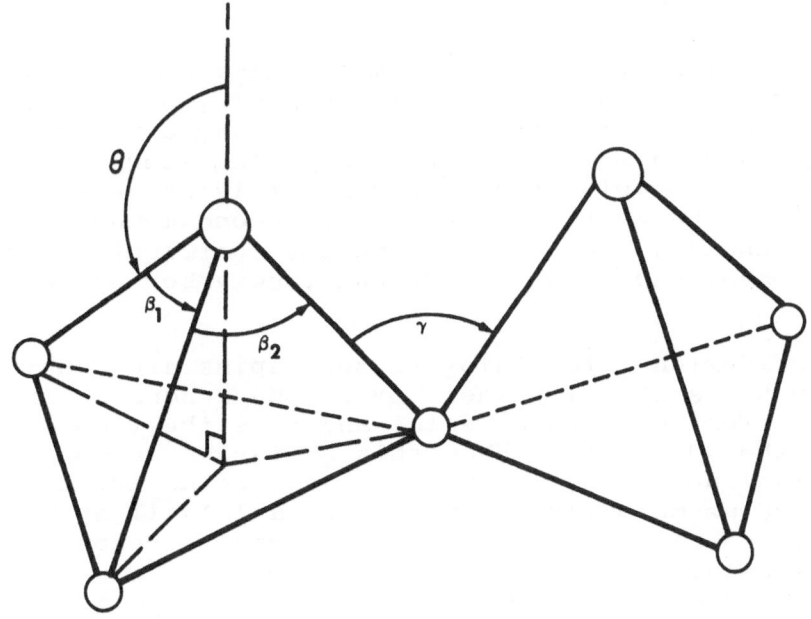

Fig. 5b   The linked AsS_3 pyramidal units of orpiment.
Small circles denote S atoms; large circles denote As atoms.

experimentally observed frequencies (72.9 and 70.4 MHz) can perhaps be obtained by taking into account other contributions such as pi bonding of the lone pair orbital. (We have chosen the higher frequency resonance to be associated with site I in agreement with the theoretical calculation.) An estimate of the asymmetry parameters for sites I and II in crystalline As_2S_3 can also be determined experimentally by measuring the effect of an applied external magnetic field on the NQR spectrum. The values for sites I and II are 0.34 and 0.37, respectively.

In a solid, the spin-spin relaxation time, or transverse relaxation time, T_2, is determined by the dipolar interaction between the nuclear spins of the form of Eq. 16, and represents the characteristic time with which the nuclei achieve thermal equilibrium with each other. A graph of the spin-echo amplitude as a function of pulse separation, τ, is a measure of T_2. Such a plot is shown in Fig. 6, for site II of the orpiment crystals at 300 K. The initial decrease of the echo amplitude with   is Gaussian $[\exp(-2\tau/T_2)^2)]$ for small τ. Evidence of an oscillation is seen at $2\tau \approx 1$ millisecond. At larger values of the pulse separation, a slower, exponential decay is indicated.

A complete theoretical description of the time evolution of the transverse nuclear magnetization following a spin-echo pulse sequence is not available, but the initial Gaussian decay of the spin-echo can be explained by the method of second moments.[7] When applying second moment calculations to spin echo phenomena it is necessary to retain only the contribution to the second moment due to resonant nuclei. One must ignore the contribution to the second moment caused by non-resonant neighboring nuclear spins.

A criterion determining whether spins are considered as resonant is $\Delta\nu < 1/t$ where $\Delta\nu$ is the resonance frequency difference between spins and t is the time duration of the 90° pulse. When this criterion is applied to crystalline As_2S_3, we find all As I spins are to be considered as mutually resonant, and all As II spins are mutually resonant, but that no As I spin is resonant with an As II spin.

The second moment calculations predict linewidths which are orders of magnitude less than the experimentally determined linewidth, since random strains and crystal defects are much more effective causes of line broadening

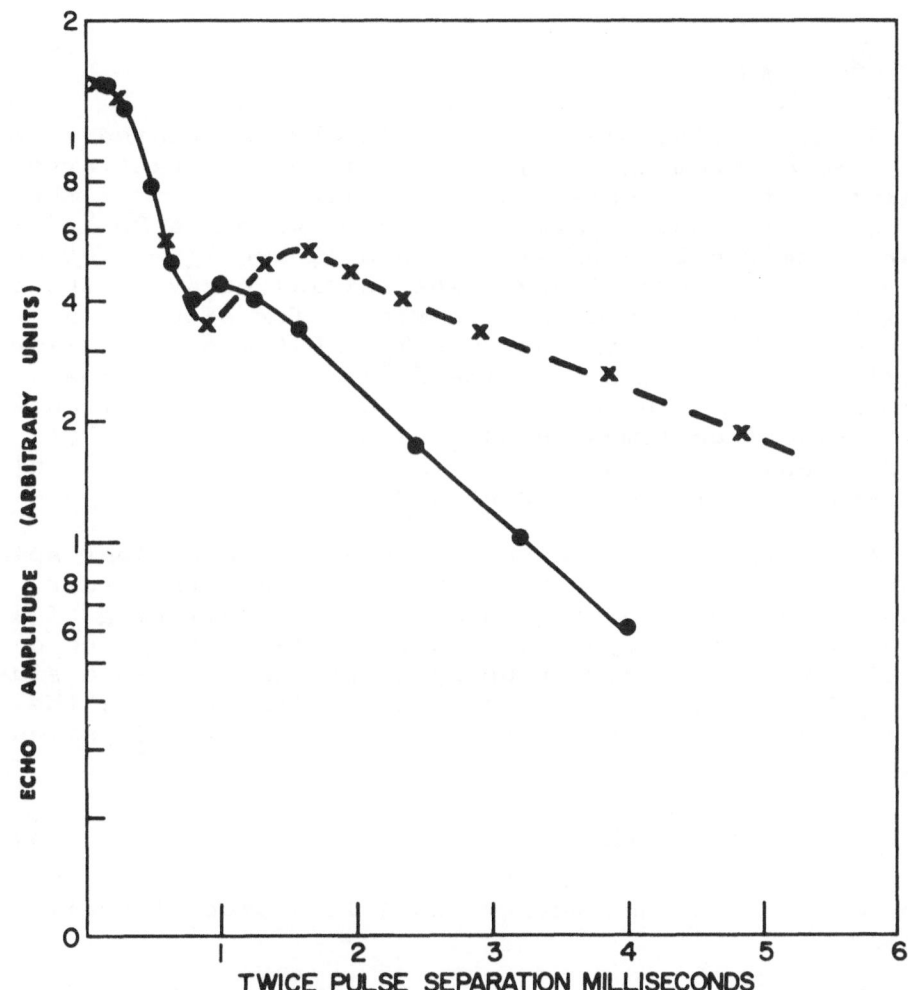

Fig. 6 Echo amplitude of ^{75}As NQR in site II of crystalline As_2S_3 (solid line) and glassy As_2S_3 (dashed line),
following a 90° - 180° pulse sequence, as a function of
twice the pulse separation. Glass data are at 4.2K,
crystal data are at 300K.

than dipolar interactions. However, the second moment $<\Delta\omega^2>$ of Eq. 17 does apply to the transverse relaxation of the spin-echo signal in the limit when the pulse separation τ is small. It can be shown that[2]

$$T_2 = \frac{2}{<\Delta\omega^2>} \tag{24}$$

Equation (24) may be used with the calculated values of $<\Delta\omega^2>$ from Eq. (17) to determine the transverse relaxation time expected in crystalline As_2S_3. If the sum is carried out to all like nuclei within a 20 Å sphere, the results for sites I and II are $T_2 = 0.95$ msec and 0.7 msec, respectively. The initial decay of the experimental spin-echo signal in Fig. 6 for site II is Gaussian with a relaxation time $T_2 = 0.6$ msec. Similar results were obtained for site I $\left(T_2 \approx 1 \text{ msec}\right)$. The agreement between the calculated and experimental transverse relaxation times is sufficient to conclude that the dominant contributions to the relaxation in crystalline As_2S_3 are the dipolar effects.

As mentioned in Section II, quadrupole nuclear spin-lattice relaxation results primarily from first-order Raman processes involving the inelastic scattering of a phonon by the spin system. The relaxation can proceed via either acoustical[8] or optical[9] phonons. When a single optical mode dominates the relaxation process, the temperature dependence of the relaxation rate T_1^{-1} given by Eq. (18) reduces to

$$T_1 = A \sinh(\epsilon/2kT) \tag{25}$$

where ϵ is the phonon energy and A is a proportionality constant.

The spin-lattice relaxation times in orpiment at 4.2, 77 and 297K are displayed in Fig. 7. The decay curves at all three temperatures are exponential. The data for orpiment are well fit by Eq. (25) with $A = 30.64$ sec and $\epsilon = 29$K, which corresponds to an optical lattice vibration of frequency 20 cm^{-1} and is consistent with the lowest frequency mode observed in Raman spectroscopy.[22] The deviation from the theoretical curve in Fig. 4 near 300K probably results from the influence of higher lying optical phonons.

We now discuss the NQR results in vitreous As_2S_3.

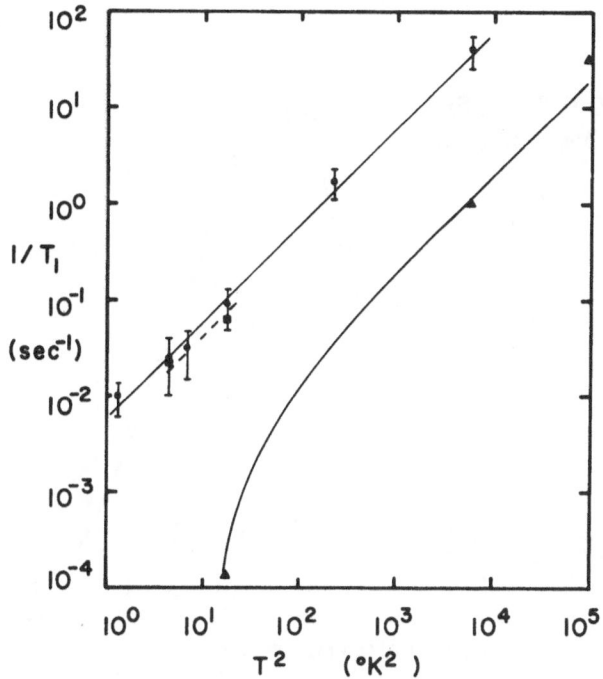

Fig. 7 Relaxation rate $(1/T_1)$ vs temperature squared for vitreous As_2S_3 (circles), crystalline As_2S_3 (triangles), vitreous As_2Se_3 (squares).

The proposed structural models for As_2S_3 and As_2Se_3 glasses are the subject of considerable debate and fall into two categories: those which are based on a continuous three dimensional random network of well defined AsS_3 or $AsSe_3$ pyramidal units[23] and those which invoke two-dimensional layer-like correlations between pyramidal units such as exist in the crystal modifications.[24] The NQR results in vitreous As_2S_3 and As_2Se_3 are best interpreted by assuming that layers are still retained in the amorphous phases of these materials.

The [75]As NQR spin-echo spectrum of amorphous As_2S_3 is shown in Fig. 8 with the spectrum of crystalline As_2S_3. In contrast to crystalline As_2S_3, the spectrum in glassy As_2S_3 consists of a single extremely broad line of approximately 3.5 MHz half-width at 1/e. The width of the line

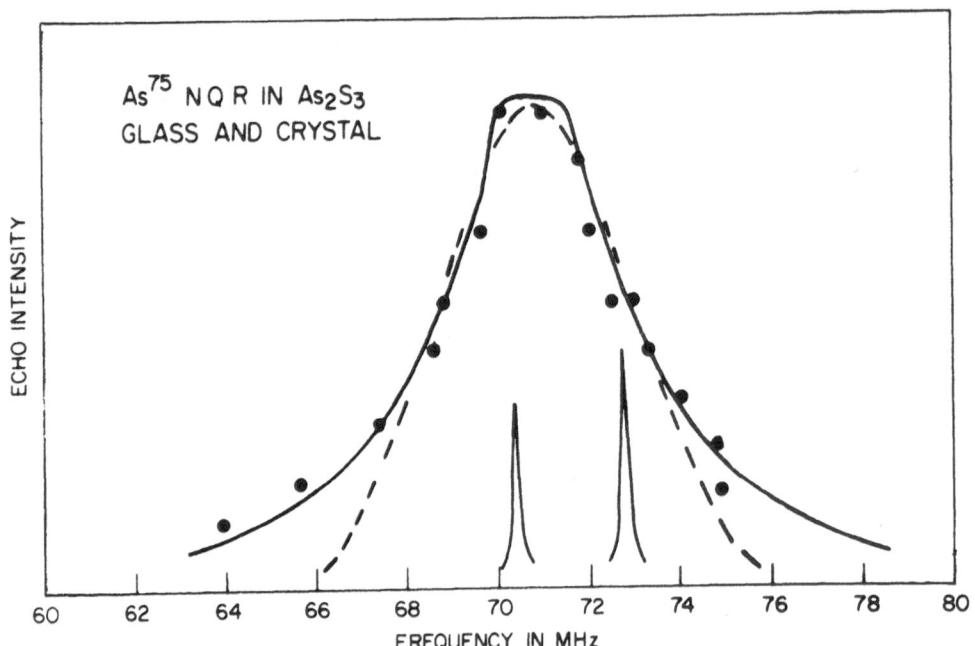

Fig. 8 Nuclear Quadrupole Resonance spectrum of vitreous
As_2S_3 at 4.2K. The experimental data are shown by the
points. The solid and dashed lines are fits to the data
by Lorentzian and Gaussian distributions, respectively,
of resonance frequencies. The spectrum of crystalline
As_2S_3 is also displayed for purposes of comparison.

is caused by a distribution in the EFG tensors due to
bond angle distortions present in the glass.

    The NQR linewidth in the glass is primarily due to
deviations in the apex bonding angles (β_1 of Fig. 5b) of
the AsS_3 pyramidal units. Since the transverse relaxation time measurements to be discussed below indicate
that the two inequivalent As sites of the crystal remain
in the glass, we assume that two distributions of equal
width are necessary to fit the data of Fig. 8. In As_2S_3
we estimate the magnitudes of these deviations in β by
comparing the width of the glass line to the frequencies
of the two As sites in the crystal, for which the apex
angles are known from x-ray data. The average apex

bonding angle of the first As site in crystalline orpiment exceeds that of the second by 1.3°. A difference in NQR frequencies of 2.48 MHz is observed between these two sites. If the spread in frequencies is assumed to vary linearly with β over the region of interest, then one calculates a distribution width of average apex bonding angles for the two sites in the glass of between ± 0.8 and ± 1.1°, depending on whether the distributions are Lorentzian (solid line of Fig. 5) or Gaussian (dashed line).

A second estimate of the average apex bond angle deviations in the glass can be obtained from Eq. (21) (which assumes that η = 0) by calculating the best fit to the spectrum assuming two Gaussian distributions of average apex bonding angles β. This method yields a distribution width for the average β of each glass site of ± 0.6°. Although the exact width of the distribution of average apex bond angles in glassy As_2S_3 varies somewhat (0.6 to 1.1°) depending on the method of estimation, the pyramidal units are very well defined in the glass.

The spin-spin relaxation measurements in vitreous As_2S_3, which were obtained at 4.2K, are independent of frequency within experimental error. As in the crystalline case, the initial decay is Gaussian with a transverse relaxation time T_2 = 0.6 msec as shown in Fig. 6.

In the second moment calculation for crystalline As_2S_3, approximately half the contribution to $<\Delta\omega^2>$ is due to the nearest neighbor resonant As Nuclei, (second nearest neighbor As nuclei of Fig. 5a), and the dipolar fields from second and third nearest resonant neighbors are responsible for most of the reminder. From the near equality of the crystalline and amorphous initial relaxation behavior, we conclude that the near-neighbor environments of the crystal and the glass are quite similar. Specifically, these results demonstrate the existence in the glass to two inequivalent As sites analogous to those that exist in the crystalline unit cell.

Only the interactions between mutually resonant nuclei are effective in contributing to transverse relaxation as measured by the spin-echo technique; i.e., only those spins that are simultaneously flipped by the applied r.f. pulses. For the pulse width used, (~ 7 micro seconds) near-neighbor nuclei in the glass must have quadrupole resonance frequencies within $\Delta\nu \sim 1/t \sim 150$ kHZ in order to be mutually resonant. Thus even though the two As sites in the glass have resonance frequencies which vary by $\Delta\nu \sim 1.5$ MHz throughout the bulk sample, the

variance locally is nearly always 10 times less than this
spread. In terms of the variation of average apex bond
angles, this restriction implies that on the average
near neighbor like As nuclei have average apex bond angles differing by no more than 0.1° even though the total deviation in the bulk sample is on the order of 1°.

A second criterion which must be satisfied in order
for the transverse relaxation times of the crystal and
the glass to be nearly equal is that near neighbor resonant nuclei must be separated by approximately the same
distances in both materials. This fact, coupled with the
preservation of the two inequivalent As sites in the glass,
strongly suggests that the two dimensional correlations
between AsS_3 pyramidal units which exist in the crystal
are by and large present in the glass. These results are
at variance with the predictions of the random network
model, which as applied to As_2S_3 glass in its simplest
form[23], postulates the existence of a random distribution
of As-S-As bond angles (γ). A completely random and uncorrelated distribution of γ-type bonding angles would
result in an inital spin-spin relaxation time in the glassy state ~10 times longer than the initial spin-spin relaxation time in the crystalline state. (The factor of
10 is obtained from the ratio of the NQR linewidth in the
glass to the amplitude of the applied r.f. field).

Measurements of the specific heat[25], thermal conductivity[25] acoustic attenuation[26], far infrared and mi-
crowave absorption[27] and Raman scattering[28] of glasses
at low temperature have indicated the existence of an enhanced density of low frequency vibrational modes which
appear to be characteristic of the amorphous state. Since
nuclear spin-lattice relaxation proceeds via vibrational
modes, one might expect greater low temperature spin lattice relaxation rates in glassy As_2S_3 over those in the
crystalline material.

Increased nuclear spin-lattice relaxation rates may
in fact be a general feature of disordered materials.
Greatly increased spin-lattice relaxation rates have been
observed for protons in organic glasses as compared to
the rates in crystalline counterparts[29]. A similar increase in the relaxation rate of ^{11}B in glassy B_2O_3 over
the crystalline material has also been observed[30]. The
present measurements of the temperature dependence of
spin-lattice relaxation in glassy and crystalline As_2S_3
also show greater relaxation rates in the glass, which
indicate an enhanced density of low frequency modes in

the vitreous material. The normalized relaxation curves
for glassy As_2S_3 are shown in Fig. 9 at three temperatures. One difference between spin-lattice relaxation
in the glass and the crystals is that the decay cannot
be represented by a single exponential characterized by
one relaxation time. A second difference is the rate of
decay, which is ~15 seconds to reach 1/e in the glass
compared to ~2 hours in the crystal at 4.2K.

Even though the decay curve in the glass is not exponential, one may still define a characteristic relaxation time as the time for the nuclear spin system to at-
tain (1-1/e) of its equilibrium magnetization after the
application of a saturating comb of r.f. pulses (i.e.,
the time at which the decay curve of Fig. 9 falls to 1/e).
This definition is particularly meaningful because the
shape of the decay curves is independent of temperature,
as shown in Fig. 9.

In crystalline materials the measured relaxation is
characterized by two competing processes -- direct interaction of nuclei with phonons (T_1 or spin lattice processes) and interactions between nuclear spins (T_2 or
spectral or spin diffusion processes). Spin diffusion
processes are effectively removed in crystals by completely saturating the nuclear spin system with the r.f. pulses. However, the glass linewidth is 3.5 MHz, and it is
not possible to saturate the entire nuclear quadrupole
resonance line by applying a train of r.f. pulses. With
the available r.f. power, we only "burn a hole" in the
broad resonance line, and spectral or spin diffusion to
nearby non-resonant spins, must be considered as a possible relaxation mechanism. However, the weight of evidence indicates that spectral diffusion is not a significant factor in the glass and that direct spin-lattice relaxation is the dominant process. Spectral diffusion occurs when a fraction of the spins is selectively excited
by an r.f. pulse, and the excitation spreads out to un-
excited spins with different resonant frequencies through
spin-spin coupling. This mechanism leads to an apparent
loss of excitation at the frequency of measurement, but
does not involve the lattice. Hence it is expected to
be relatively temperature independent. However, a strong
dependence of relaxation rate on temperature is observed
in As_2S_3 which indicates that spectral diffusion is unimportant. Additional arguments also support this conclusion.[2]

The temperature dependence of an effective T_1, as

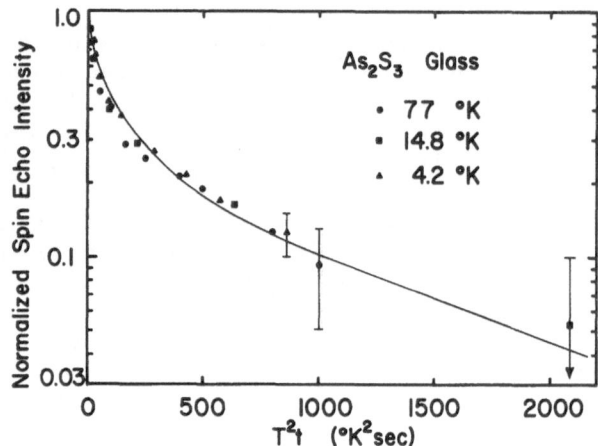

Fig. 9 Normalized spin-lattice decay for vitreous As_2S_3
taken at 4.2K, 14.8K and 77K vs T^2t, where T is temperature and t is time. The solid line is a theoretical fit
to the data as described in the text.

defined above, in glassy As_2S_3 is presented in Fig. 7.
The effective relaxation rate, T_1^{-1}, is proportional to
T^2 from 77 to 1.1K and can be described fairly accurately by the empirical relation $T_1^{-1} = 8 \times 10^{-3} T^2$. The relaxation rate measured in As_2S_3 glass is much greater
than the rate in the crystal at any given temperature,
and this difference is enhanced at lower temperatures
where the crystal relaxation rate decreases exponentially with temperature, while the glass rate remains proportional to T^2.

In a crystalline material, a T^2 dependence of the
relaxation rate occurs whenever the dominant modes relaxing the nuclear spin are at energies which correspond to
temperatures lower than the measurement temperatures.
(see the discussion of Eq. 18, in Section II). The T^2
temperature dependence in glassy As_2S_3 is strong evidence
for the existence of very low frequency vibrational modes
in the glass which are thermally excited even at ~1K, and
are strongly coupled to the nuclear spins.

It is informative to divide out the T^2 temperature

dependence of the relaxation curves in glassy As_2S_3.
Figure 9 plots the decay curves at 4.2, 14.8 and 77K as
a function of T^2t. The data taken at all three temperatures fall on a universal curve, whose shape is determined by a temperature independent distribution of relaxation centers. The maximum and minimum relaxation
rates, as given by the maximum and minimum slopes of the
curve, differ by a factor of ~100. Since the distribution of relaxation rates is governed by the slope of the
decay curve, all relaxation rates in the distribution
have the same T^2 temperature dependence.

The models proposed to explain the anomalous low
temperature thermal properties of amorphous materials
fall into two general categories -- those which invoke
atomic tunneling between two near-equilibrium positions[31]
and those which assume low frequency vibrational modes
which obey Bose statistics.[25,32] With either type of model one can obtain a distribution of relaxation times
which could yield the observed non-linear decay curves.
In the tunneling models, the magnitude of the spin-lattice relaxation depends on the relaxation time of a two-
level state to the phonons τ_C.[2] A distribution of nuclear spin lattice relaxation times results from a distribution in τ_C's. In the models which assume low frequency vibrational modes which obey Bose statistics, one
obtains a distribution of relaxation times if the vibrational states are to some extent localized and vary from
region to region in the glass. In either model, the observed T^2 temperature dependence of the relaxation rates
can be obtained provided that the interaction of the nuclear spins with either the phonons or the tunneling modes
proceeds via a first order Raman process.

In the case of nuclear relaxation via the two level
states there are several competing rate processes to consider. We assume as in the crystalline case that the extended phonons which relax the two level states are everywhere in thermal equilibrium with the bath. According to
the tunneling models,[31] states are dilute and relaxation
processes between them can be neglected (the two level
states are non-interacting). Relaxation is either limited by the direct interaction of the nuclear spins with
the two level states, which is a local interaction or it
is limited by nuclear spin diffusion processes (nuclear
spin-spin interactions) outside of a small region around
the two level states.[6] In either situation the magnitude
of the relaxation for those nuclear spins surrounding a
given two level state depends on the relaxation rate of
that two level state to the phonons, τ_C^{-1}.

To test the applicability of the tunneling models to the explanation of the [75]As nuclear spin lattice relaxation in glassy As_2S_3, we consider relaxation via these two level localized states, which should follow in form relaxation via spin 1/2 paramagnetic impurities[26] except that the interaction is via a modulation of the electric field gradient instead of the magnetic field. In both cases, the 'impurity' relaxation rates are much greater ($10^5 - 10^6$ sec^{-1}) than typical nuclear relaxation rates so that only their average properties are important. Ultrasonic attenuation measurements in As_2S_3 glass from 1 to 140K,[33] when combined with the relaxation calculations of Jackle,[34] yield average relaxation rates for the two level systems of $\tau_c^{-1} \sim 10^{+5}$ sec^{-1} at 1K.

As in the case of relaxation via paramagnetic impurities, reasonable nuclear spin lattice relaxation rates are obtained only if the relaxation is assumed to proceed via nuclear spin diffusion processes (outside of a small region of radius r_0 surrounding the two level state), and not by the direct interaction of the localized two level state with the nuclear spins. In order for spin diffusion processes to occur, the radial fall off of the electric field gradient generated by the two level state must be sufficient so that at r_0 this effect is at most comparable to the effects of the average random dipolar field at the As sites in the glass. For paramagnetic impurities the radial dependence of the dipolar field is known and r_0 can be estimated easily. In the case of the two level states, this restriction implies that these states must primarily involve motions of neutral groups of atoms (no net motion of charge), such as AsS_3 pyramids and is consistent with arguments based on dielectric relaxation data in SiO_2.[31]

A simple calculation demonstrates that the number of two level states necessary to relax the nuclear spins is of the same order of magnitude as the number predicted by the linear term in the specific heat of glass As_2S_3. (Data on the magnitude of the linear term in the low temperature specific heat imply that the number of tunneling states per °K per AsS_3 group is $n \sim 2 \times 10^{-5}$).[25] Relaxation at sites nearest the two level states proceeds via a direct interaction between the [75]As nuclei and the two level states and can be accounted for following an approach originally taken by Bloembergen.[35] In the limit where $\tau_c \nu_0 \gg 1$, the average relaxation rate at sites nearest the two level state is well approxiamted by $T_{1\,max}^{-1} \sim (V_{ij})^2/\nu_o^2 \tau_c$ where V_{ij} is the fluctuating

field gradient (in sec^{-1}) at nuclear sites nearest the
tunneling mode and includes antishielding and covalent
effects, and ν_0 ($\sim$70 MHz) represents the static field
gradient primarily due to the three As-S pyramidal bonds.
A conservative upper bound on $|\beta V_{ij}|$ at an ^{75}As site due
to motion of an adjacent pyramidal unit is $|\beta V_{ij}| < 0.1 \nu_0$
(the order of the observed NQR linewidth). This esti-
mate places an upper bound on the relaxation of sites adjacent to a two level state of $T_{1\,max}^{-1} \lesssim 10^3$. At sites
far away from the localized two level state the magnitude
of the relaxation via the direct process depends on the
radial dependence of the electric field gradient of the
state.

When nuclear spin diffusion is dominant at sites
outside r_0, the average relaxation rate is given by[36]
$\bar{T}_1^{-1} \sim n z T_{1\,max}^{-1}$ where z is the number of nearest neighbor nuclei at identical sites (z $\sim$1). The upper bound
on $T_{1\,max}$ and the experimentally observed average relaxation rate at 1K ($\bar{T}_1^{-1} \sim 10^{-2}$ sec^{-1}) provide a lower

bound on the concentration of n $\geq 10^{-5}$ states per AsS$_3$
group per K which is consistent with the number obtained
from specific heat data[37] (2 x 10^{-5}). One difficulty
with this estimate, which must be mentioned, is that the
very effect which creates the large fluctuating field gradient may make it energetically impossible for the surrounding nuclear spins to interact via spin diffusion processes.

We now examine the possibility that the relaxation
rates observed in the glass could also result from interactions with extended vibrational modes instead of the
two level tunneling states. In crystalline materials the
dominant relaxation process involves a Raman scattering
of phonons and is expressed in terms of transition probabilities as given by Eq. 18. At high temperatures (measurement temperatures much greater than the temperature
of the modes effective in relaxing the nuclear spins),
Eq. 18 predicts that the spin lattice relaxation rate
T_1^{-1} is proportional to T^2 regardless of the functional
form of the phonon density of states $\rho(\omega)$. The temperature dependence of the relaxation rates in glassy As$_2$S$_3$
is well explained by relaxation via low frequency vibrational modes since the T^2 behavior of the Raman process,
as expressed in Eq. 18, depends solely on the high temperature assumption and is independent of the details of
the density of states. The detailed shape of the As$_2$S$_3$

decay curve (Fig. 9) can also be explained by this mechanism if one assumes that inhomogeneities in the glass
structure produce localized vibrational modes.

If one assumes a constant density of low frequency
modes (a necessary condition to obtain a linear term in
the specific heat), Eq. 18 in the high temperature regime
implies that the transition probability for a Raman process involving phonons of frequency $\omega' \approx \omega$ is
$P(m, m + \mu, E) = A N_o^2 k^2 T^2 / E^4$ where $E = \hbar\omega$, N_o is the
constant density of phonons per unit E, and A is a con-
stant easily determined from Eq. 18.

In general, the intensity decay curve for a distribution of relaxation rates can be expressed as a sum over
volume elements within which there exist given densities
of (localized)modes. If all low frequency modes are localized to roughly the same extent, and the relaxation
in any given region of glass is dominated by modes within
$d\omega$ of some frequency ω, then the expression for the decay in intensity becomes

$$I(t,T) \propto \int_0^{E_{max}} e^{-\gamma P(m,m+\mu,E)t} \, dE \qquad (26)$$

where γ is a factor to take account of anti-shielding
and covalent effects in the spin-lattice coupling. Using
P as approximated above, one obtains a result in terms of
an exponential integral function of fractional order.
The solid line in Fig. 9 is the theoretical fit to the
experimental data where the asymptotic behavior at long
times has been fit by a suitable value of the one adjustable parameter $\gamma A N_o^2 / E^4_{max}$. The agreement with experiment is remarkable considering the simplifying assumptions
where were made.

A crude estimate of N_o can be obtained from the value
of $\gamma A N_o^2 / E^4_{max}$ necessary to fit the data. If one assumes (1) that the value of γA necessary to fit the relaxation rate in crystalline As_2S_3 in the high temperature approximation is appropriate to the low frequency
modes in the glass, (2) that there is one As atom moving
per unit cell in the crystalline mode and (3) that
$E_{max} \sim 1K$ at 1K, then the number of modes per AsS_3 group
with energies less than 1K is $N_o \sim 10^{-4}$. Although this
number is an order of magnitude greater than the density
predicted from specific heat measurements ($n \sim 10^{-5}$), it
represents an upper bound imposed by the experimental

limitations. If the T^2 temperature dependence in Fig. 2 exists down to 0.3K, then the estimated value of N becomes $\sim 10^{-5}$.

The NMR and NQR measurements on amorphous B_2O_3 and As_2S_3 discussed in the last two sections point to the same general conclusion - the fundamental structural units in these two materials are very well defined. In amorphous As_2S_3, the NQR results indicate in addition that even between AsS_3 pyramidal units there is substantial crystalline-like order preserved. These results may be contrasted with materials like amorphous Ge and SiO_2 where the random network model appears to provide a quite adequate description of the local structural order.

V. FUTURE PROSPECTS OF NMR AND NQR IN DISORDERED SOLIDS

There has not been space in this set of lecture notes to describe all of the NMR measurements in amorphous materials over the last 15 years, and certainly no attempt will be made to detail all of the measurements which might be attempted in the future. It is reasonable, however, to mention the general directions which future research might profitably take.

Most NMR measurements in disordered solids to date have been performed using cw techniques. The use of pulsed NMR techniques to study disordered solids, similar to the NQR results presented in Section IV, would allow information on spin-spin and spin-lattice relaxation processes to be extracted from a wider variety of amorphous materials. Measurements into the liquid state, where possible, would provide useful probes of atomic motion in vitreous liquids

In NQR, measurements could profitably be extended to amorphous materials containing other favorable quadrupolar nuclei, for example ^{121}Sb, ^{123}Cl, or even ^{11}B. Here again measurements at lower ($<1K$) or higher ($>77K$) temperatures, if possible, should provide new and interesting information concerning low frequency modes frozen in distortions, and molecular motions in amorphous materials.

REFERENCES

1.   A. H. Silver and P. J. Bray, J. Chem. Phys. $\underline{29}$, 984 (1958).

2.   M. Rubinstein and P. C. Taylor, Phys. Rev. Letters $\underline{29}$, 119 (1972); M. Rubinstein and P. C. Taylor, Phys. Rev. $\underline{89}$, 4258 (1974).

3.   T. P. Das and E. L. Hahn, Solid State Physics Suppl. 1, Edited by F. Seitz and D. Turnbull, Academic Press, N. Y. (1958).

4.   E. R. Andrew, Nuclear Magnetic Resonance, Cambridge Univ. Press, London (1958); G. E. Pake, Solid State Phys. $\underline{2}$, 1 (1956).

5.   M. H. Cohen and F. Reif in Solid State Physics $\underline{5}$, Edited by F. Seitz and D. Turnbull, Academic Press, N. Y. (1957).

6.   A. Abragam, The Principles of Nuclear Magnetism, Clarendon Press, Oxford (1961).

7.   J. H. Van Vleck, Phys. Rev. $\underline{74}$, 1168 (1948); A. Abragam and K. Kambe, Phys. Rev. $\underline{91}$, 894 (1953).

8.   J. Van Kranendonk, Physica $\underline{20}$, 781 (1954).

9.   K. R. Jeffrey and R. L. Armstrong, Phys. Rev. $\underline{174}$, 359 (1968).

10.   P. C. Taylor, J. F. Baugher and H. M. Kriz, Chem. Reviews $\underline{75}$, 203 (1975).

11.   J. Wong and C. A. Angell, Applied Spectroscopy Reviews, Vol. 4, E. G. Brame, Jr., Ed., Marcel Dekker, New York (1971), pp. 200-232.

12.   See for example, P. J. Bray in Magnetic Resonance (Plenum, N. Y. (1970)), p. 11.

13.   P. C. Taylor and P. J. Bray, J. Mag. Res. $\underline{2}$, 305 (1970).

14.   J. Krogh-Moe, Phys. Chem. Glasses $\underline{3}$, 101 (1962). $\underline{6}$, 46 (1965).

15.   P. C. Taylor and E. J. Friebele, J. Non-Cryst. Solids, $\underline{16}$, 375 (1974).

16.   S. L. Strong and R. Kaplow, Acta Cryst. $\underline{B24}$, 1032 (1968); G. E. Gurr, P. W. Montgomery, C. K. Knutson and B. T. Gorres, Acta Cryst. $\underline{B26}$, 906 (1970).

17.   C. H. Townes and B. P. Dailey, J. Chem. Phys. $\underline{17}$, 782 (1949).

18.   H. M. Kriz and P. J. Bray, J. Mag. Res. $\underline{4}$, 76 (1971).

19.   H. M. Kriz and P. J. Bray, J. Non-Cryst. Solids $\underline{6}$, 27 (1971); K.S. Kim and P.J. Bray, J. Non-Metals (1974), in press.

20.   C. Ree and P. J. Bray, Phys. Chem. Glasses $\underline{12}$, 165 (1971).

21.   H. Bayer, Z. Phys. $\underline{130}$, 227 (1951).

22.   R. Zallen, M. L. Slade and A. T. Ward, Phys. Rev. $\underline{B3}$, 4257 (1971).

23.   G. Lucovsky and R. M. Martin, J. Non-Cryst. Solids
      8-10, 185 (1972).

24.   P. C. Taylor, S. G. Bishop, D. L. Mitchell and D.
      Treacy, Proc. 5th Int. Conf. on Amorph. and Liquid
      Semicon. (Taylor and Francis, London, 1974), p.1267.

25.   R. C. Zeller and R. O. Pohl, Phys. Rev. B4, 2029
      (1971); R. B. Stephens, Phys. Rev. B8, 2896 (1973).

26.   B. Golding, J. E. Graebner, B. I. Halperin and R.
      J. Schultz, Phys. Rev. Letters 30, 223 (1973); S.
      Hunklinger, W. Arnold, S. Stein, R. Nava and K.
      Dransfeld, Phys. Letters A 42, 253 (1972).

27.   U. Strom and P. C. Taylor, Proc. 5th Int. Conf. on
      Amorph. and Liquid Semicon. (Taylor and Francis,
      London, 1974), p.375.

28.   R. Shuker and R. W. Gammon, Phys. Rev. Letters, 25
      222 (1970).

29.   J. Haupt and W. Müller-Warmuth, Z. Naturforshg.,
      239, 208 (1968); J. Haupt, Proc. XVI Coll. Ampere.
      Bucharest, 630 (1970).

30.   M. Rubinstein, H. Reising and J. R. Hendrickson,
      Bull. Am. Phys. Soc., 19, 202 (1974).

31.   P. W. Anderson, B. I. Halperin and C. M. Varma,
      Phil. Mag. 25, 1 (1972).; W. A. Phillips, J. Low
      Temp. Phys. 7, 351 (1972).

32.   H. B. Rosenstock, J. Non. Cryst. Solids 7, 123 (1972).

33.   D. Ng and R. J. Sladek, Fifth International Conf.
      on Amorphous and Liquid Semiconductors, (Taylor and
      Francis, London, 1974), p.1173

34.   J. Jackle, Z. Physik 257, 212 (1972).

35.   N. Bloembergen, Physica 15, 386 (1949).

36.   J. Hatton and B. V. Rollin, Proc. Roy. Soc. A 199,
      222 (1949).

37.   Paramagnetic impurities are known to exist in vitre-
      ous As_2S_3 and could relax the nuclear spins. Mag-
      netic susceptability measurements (J. DiSalvo, A.
      Menth, J. V. Waszczak and J. Tauc, Phys. Rev. B6,
      4574 (1972))indicate that the paramagnetic impuri-
      ties in As_2S_3 effective at low temperatures are Fe^{3+}
      ions in concentrations of $n \sim 10^{-6}$. Since relaxation
      times for Fe^{3+} ions at low temperature are typically
      $\tau_c \sim 10^{-5}$ sec, the concentration of Fe^{3+} ions is a-
      bout four orders of magnitude too small to account
      for the observed average relaxation rates.

BONDING IN NON-TETRAHEDRALLY COORDINATED AMORPHOUS SOLIDS[*]

Galen B. Fisher[+]

Department of Physics and Division of
Engineering
Brown University, Providence, R. I. 02912

1. INTRODUCTION

Most of the amorphous materials studied in the past decade have contained elements from Groups IVA, VA, and VIA of the periodic table. Typically they are semiconductors which have been of interest from both a basic and a practical standpoint. The most commonly studied elemental amorphous semiconductors have been placed in boxes in the portion of the periodic table shown in Table I. The tetrahedrally bonded amorphous solids which are isomorphic with the Group IV materials (e.g., Ge, Si, GaAs) have been discussed by others. My purpose is to review some of the properties of several of the amorphous materials from Group V and VI, shown within the solid line in Table I.

While the tetrahedrally bonded amorphous solids can usually be made only by quenching from the vapor phase by evaporation or RF-sputtering, many of the chalcogenide materials are excellent glass formers, quenchable from the melt. An extreme case of a stable glass is As_2S_3 which can be found as a crystal (orpiment) only in nature.

---

*Work funded in part by AROD and from the general support of Materials Science at Brown University by NSF.

+Present Address: Surface Processes and Catalysis Section, National Bureau of Standards, Washington, D. C. 20234

TABLE I.   Three Groups in the Periodic Table

	IVA	VA	VIA	
...	C	N	O	...
...	Si	P	S	...
...	Ge	As	Se	...
...	Sn	Sb	Te	...
n =	4	3	2	

The table shows a portion of the periodic table where the outer atomic electron configuration is given by $...s^2p^{6-n}$, where n is both the typical valence <u>and</u> the usual coordination of the element, particularly in the amorphous state. The elements placed in boxes are discussed in the text.

To a great degree specific facets of the non-tetrahedrally bonded materials have been introduced in previous discussions of structure,[1,2,3] transport properties,[4,5,6] optical properties,[7,8] photoemission[8,9] and lattice dynamics.[7,10,11] This presentation will attempt to bring many of the experimental facts together to discuss the bonding in non-tetrahedrally coordinated materials. In this discussion we emphasize the role of the local coordination in relation to the electronic structure and vibrational states.

We can group the non-tetrahedrally bonded materials we will discuss as follows:
   a. Group VI elements (S, Se, Te) are nearly always two-fold coordinated and in the free atom have a valence electron structure of $...s^2p^4$ and are treated in Section 2.
   b. Group V elements (As, Sb), are nearly always three-fold coordinated and have a free atom electron configuration of $...s^2p^3$ (Section 3).
   c. Chalcogenide compounds (Section 4).

   i. V-VI compounds, which include As_2S_3, As_2Se_3, Sb_2S_3, and Sb_2Se_3, maintain their atomic coordination in the compound.

ii. The IV-VI compounds, whose coordination has been
the topic of some discussion[3], include the germanium-
chalcogenide materials. In particular, we will examine
the Ge-Te alloy system in Section 5.
    iii. The III-VI compounds (e.g., In_2Te_3 and Ga_2Se_3)
can maintain the structural units of the amorphous form
in the liquid phase. They are mentioned in discussions
of liquid semiconductors[12] and will not be considered
further here. Section 6 contains concluding remarks.

In amorphous materials structural studies with X-
ray, neutron, or electron diffraction can only tell us
the average atomic separations and the average number of
nearest neighbors. More local structural probes, such
as magnetic resonance studies[13] or, more recently, extended x-ray absorption fine structure (EXAFS) measurements[14] are available, although the former can require
significant amounts of sample material and the latter
has not yet come into general use. Often much structural
information comes from the methods which probe the vibrational modes of amorphous systems and the optical and
electron spectroscopies, which examine their electron
states. Certainly the relation between vibrational modes
and the coordination and local symmetry is a long established tool of structural analysis which has been applied successfully to amorphous materials.[7,10] From the
arrangement of Table I, it is clear we will also emphasize the relationship between valence electronic structure and the coordination. In Table I, the valence, n,
of the elements we consider is also the usual coordination, following the 8-n rule. What this implies then
is that, as the spatial arrangement of the bonds becomes
less uniform in moving from Group IV to Group VI, the local order and vibrational modes can to a better approximation be considered molecular. In the same progression
from Group IV to Group VI as the number of valence electrons increases and the coordination decreases, fewer
electrons are involved in bonding and the valence band
density of electron states is affected in a significant
way. We will now examine in particular the vibrational
and electronic structure of the elements and compounds involving Group V and Group VI materials. The clearer relationship between the vibrational modes, electronic
states, and the local order that emerges may allow us to
study new disordered systems with which typical structural methods have difficulty.

2. GROUP VI MATERIALS

The Group VI amorphous solids (S, Se, Te) may be the most studied of all amorphous materials. Their structure[2] and transport properties[5] are discussed elsewhere. Unlike some amorphous solids, they typically have very reproducible properties (e.g. the a-Se absorption edge).

The vibrational modes of amorphous Se as seen in infrared transmission measurements[15] are compared in Figure 1 with the spectrum of crystalline α-monoclinic Se which consists of puckered rings of Se_8 molecules. This was used along with a comparison with trigonal Se to show the existence of both rings and chains in vitreous Se.[15] A molecular model incorporating the two-fold coordination of the chalcogen in the symmetry of the polymer chain and the S_8 or Se_8 molecule has been successful.[10,16] This approach contrasts to the effective-media method[17] that works well for Ge, where the bonding is more homogeneous in space.

X-ray (XPS) and ultraviolet (UPS) photoemission studies of chalcogens have been described.[7-9] In Figure 2 are XPS valence band spectra of disordered S, Se, and Te. The spectrum of S, the first reported, is treated elsewhere.[18] These spectra can be understood in terms of two-fold coordination of the chalcogen. The most notable aspect of the spectra is the double peaking in the upper 7 eV. This is understandable in terms of the model of Mooser and Pearson[19] if we assume that the upper 7 eV of each spectrum consists largely of the four p electrons in the chalcogen. The lower energy structure is associated with the two valence s electrons of each material. Two-fold corrdinated chalcogens then have two p electrons which form a filled orbital (largely non-bonding or lone pair orbital). The remaining two p electrons form bonds and have their energy lowered relative to the other p electrons, forming a double peak. The peak at the top of the density of states largely due to the lone pair electrons is narrower, as expected, reflecting their smaller overlap with neighboring atoms relative to the bonding electrons. The increasing energy separation between the lone pair and the bonding orbitals shown by the dashed lines in Figure 2 as one goes from Te to S scales with the increase in bond strength[20] from Te to S.[18] This supports further the division of the valence band into mostly non-bonding p, bonding p, and s electrons as one goes deeper into the valence band. The general view then is that the difference in the spatial distribution of p electrons leads to their separation in energy. These

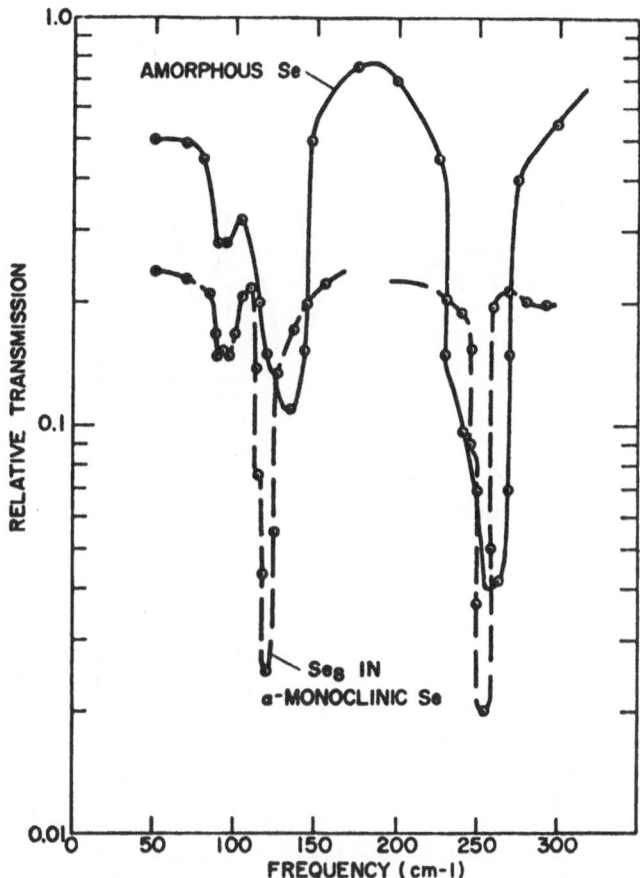

Fig. 1   A comparison of the room-temperature infrared transmission spectra of Se_8 in α-monoclinic selenium and amorphous selenium [from Ref. 15].

basic principles also carry over to crystals.[21]   This is a simplistic view since there is clearly hybridization within the p orbitals and between the s and p orbitals. But at a first glance it gives us a point from which to start.   A further correlation of interest is that the spread in the s electron level in Figure 2 increases as the interatomic spacing decreases from 2.86 Å (Te) to 2.06 Å (S).   This indicates that S may have relatively more overlap with its neighboring valence s level than occurs in Te.[18]

3.   GROUP V MATERIALS

Relative to the Group VI materials, the Group V elements (As,Sb) have been very little studied in the

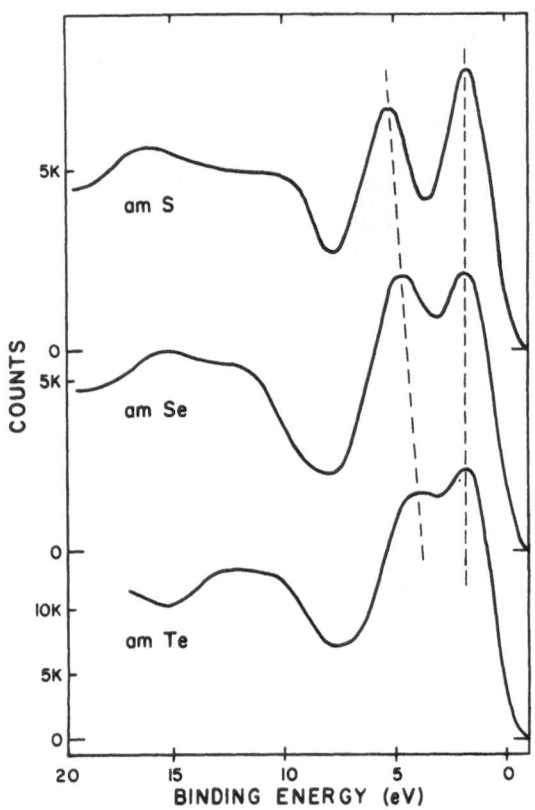

Fig. 2   X-ray photoelectron spectra of the valence band
region in amorphous sulfur, amorphous selenium, and amorphous tellurium [Ref. 18].   For S MgKα X-rays (hν =
1253.6 eV) were used, while AlKα X-rays (hν = 1486.6 eV)
were used for Se and Te.   Note the persistent double-
peaked upper valence band.

amorphous phase.   We know the interatomic spacing changes
less that 1°/o from that of the three-fold coordinated
crystals,[2] although they do change from semimetals to
semiconductors.

Raman scattering measurements have been made on amorphous Sb.[22]   Electronic properties of amorphous As have

been measured,[23] and recently infrared absorption measurements of a-As have been reported.[24] a-As has two
main regions of spectral strength as do the amorphous
chalcogens. The experimentally determined half-widths
($\Delta\nu/\nu$) of the dominant spectral features increase uniformly in going from a-Se (0.15) to a-As (0.23) to a-Ge
(~0.3),[24] which is related to the greater number of nearest neighbor interactions as the number of bonds increase.
Unlike a-As or a-Se, a-Ge has three main spectral features in the infrared, but this is simply due to the fact
that a tetrahedron has one mode in which the central atom
does not move, while the molecular units of As and Se
have no such mode.[24] The symmetry properties of molecules will be very important in establishing structures
as we shall see.

     The X-ray photoemission spectra of amorphous As and
Sb have been measured[25] and are shown in Figure 3. UPS
spectra of a-Sb have also been measured.[26] If we again
assume a separation of the two s electrons into the lower
peak and the three p electrons into the upper peak, then
the result that the three p electrons would be in similar bonds is consistent with the appearance of a single
peak of p electrons. This is also in agreement with the
XPS data of amorphous Ge and Si (Fig. 4),[27] where those
valence band electrons in the four bonds are expected to
be in the same electronic state and thus one peak is seen
in the upper 5 eV. Obviously hybridization is important
here but one can distinguish the more p-like (upper 5 eV)
and the s-like (lower 8 eV) regions of the valence band.
The difference at X-ray wavelengths between the 3p/3s
and 4p/4s matrix element ratios explains the difference
in strength of the two parts of the valence band between
Si and Ge. The single-peaked upper valence band seen in
Group V and IV is in contrast to the double peak seen in
the chalcogenides where the four p electrons are in two
different spatial configurations. The equivalence of
the bonding seems to explain the single-peaked upper density of states in the Group V and Group IV materials.

4. CHALCOGENIDE COMPOUNDS

     The main groups of chalcogenide compounds are the
V-VI and IV-VI compounds. In the first group amorphous
As_2S_3 has importance as a window material in the infrared. It has been extensively studied and its properties
have been recently reviewed.[2,5,7,28] A molecular model
seems to explain its vibrational modes, as it does for
amorphous As_2Se_3,[29] although a layer model works well

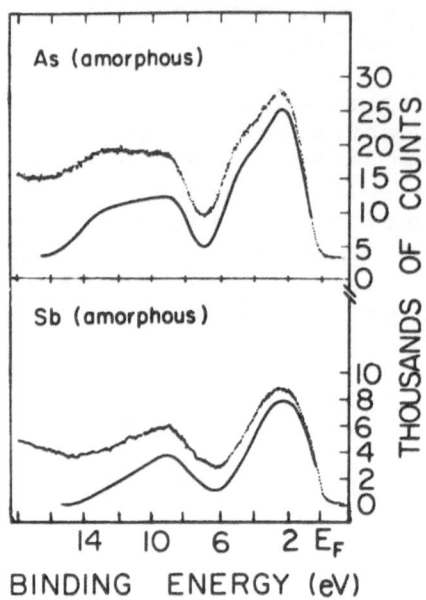

Fig. 3 Uncorrected (dots) and corrected (line) photoelectron spectra of the valence region of amorphous As
and Sb [Ref. 25]. The correction is for inelastically
scattered electrons. Monochromatized AlKα X-rays were
used. Note the generally single-peaked upper valence
bands.

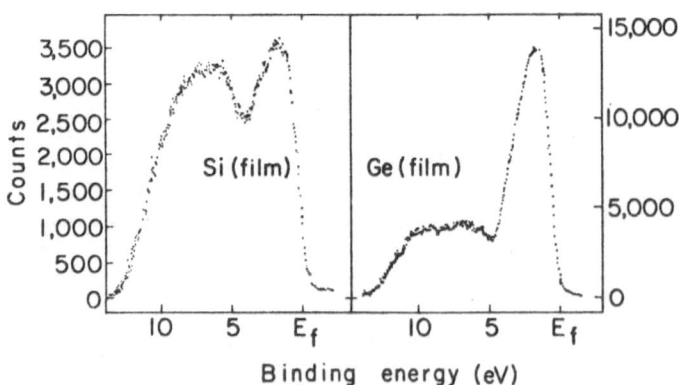

Fig. 4 Photoelectron spectra of the valence region of
films of amorphous Si and Ge already corrected for the
inelastically scattered electrons [from Ref. 27]. Monochromatized AlKα X-rays were used. Note the generally
single-peaked upper valence band in each case.

also[30]. The short-range structure in each material is very close to that of the crystal, with each As atom three-fold coordinated and each chalcogen two-fold coordinated. Thus, it is not surprising that the X-ray photoemission results from the valence band for amorphous and crystalline As_2Se_3 shown in Figure 5 are quite similar.[31] What is interesting is that with 1 eV resolution the valence band spectra of As_2S_3[31] in its crystalline and amorphous forms are as different as they are in the upper 6 eV, as seen in Figure 6. As in the case of the

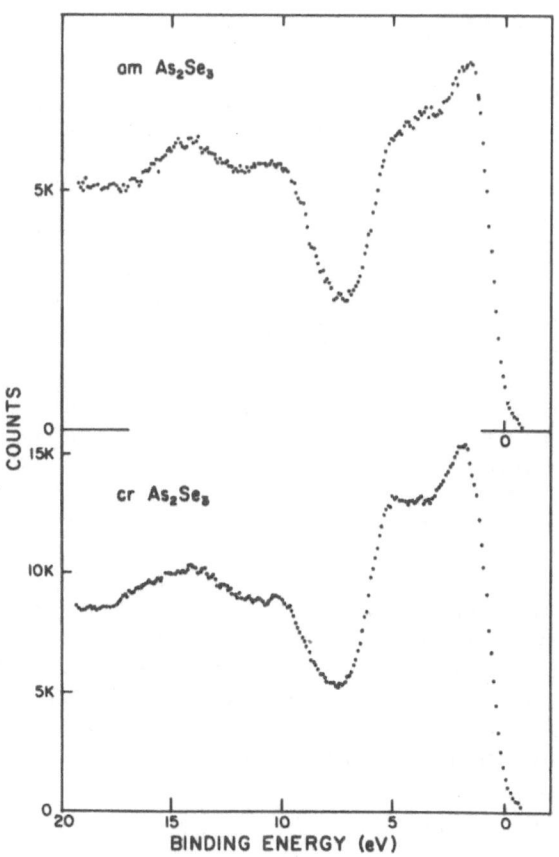

Fig. 5 X-ray photoelectron spectra of the valence band region of crystalline and amorphous As_2Se_3. MgKα X-rays were used.

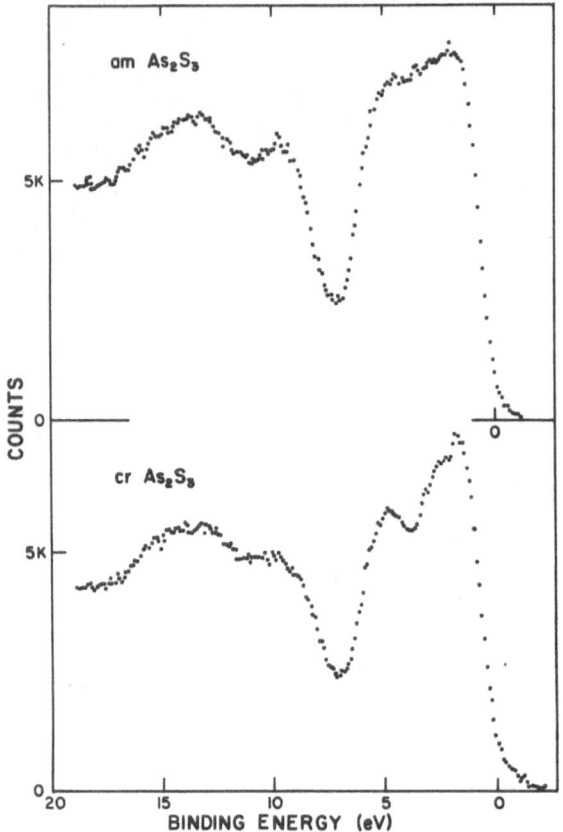

Fig. 6 X-ray photoelectron spectra of the valence band region in amorphous and crystalline As_2S_3. $MgK\alpha$ X-rays were used.

chalcogenides the valence band seems separable into an upper region of p-like states from both As and S(Se) and a lower region of s-like electrons from both As and S(Se). The general double peaking in the upper valence band in several of the spectra is related to the two-fold coordination of the chalcogenide atoms in these materials. The interpretation is generally similar for photoemission work on Sb_2Se_3.[32]

One undeniable conclusion from Figures 5 and 6 is that the valence band has a total width of about 17 eV. This is much wider than that predicted in Hückel molecular orbital calculations[33] which postulate considerable s-p hybridization. The apparent separability of the valence band into two largely atomically based regions and the difference in electron emission from the lower valence band when using uv[34] or X-ray radiation indicates that the s-p hybridization in the valence band may not be as large as the molecular orbital calculation suggests.[31] More recent work[35] corroborates this view, and more realistic calculations should bear it out.

The IV-VI amorphous compounds are interesting since they make up a large class of high temperature infrared window materials and in the Ge-Te system have shown interesting switching properties when mixed with a few percent of some other elements. However, from a more fundamental point of view they are interesting because at the $50^{o}/o$ composition, GeTe, GeSe, and GeS each have a different coordination in the amorphous form than is found in the associated crystal. They constitute one of the few cases where the short-range order is different in the amorphous and crystalline forms of the same material. The difference in RDF's between crystalline and amorphous GeTe is dramatic (Fig. 7).[36] The vibrational modes of the amorphous Ge_xS_{1-x} and Ge_xSe_{1-x} systems have been explored,[10,37,38,39] but only in the amorphous Ge_xTe_{1-x} alloy system are both vibrational studies[40] and high resolution photoemission studies with ultraviolet and X-ray radiation[41-43] available for comparison.

Thus far, we have seen some of the information obtainable from vibrational studies using both infrared and Raman spectroscopies about the structural units in a system and hence the bonding in that system. Now we use this information obtained mostly for elements combined with that available from the photoemission experiment about the electronic density of states to resolve the picture of the bonding in a previously poorly understood system, the amorphous Ge-Te alloy system.

5. THE AMORPHOUS Ge-Te ALLOY SYSTEM: A CASE STUDY

In the case of amorphous GeTe, interest arose when structural studies[44,45] showed that the coordination is lower than that of the crystal, an average of three in amorphous GeTe versus six in the slightly distorted rock salt structure of the crystal (Fig. 7). However, structural studies have been unable to resolve the nature of the short-range order in the amorphous Ge_xTe_{1-x} system.[46]

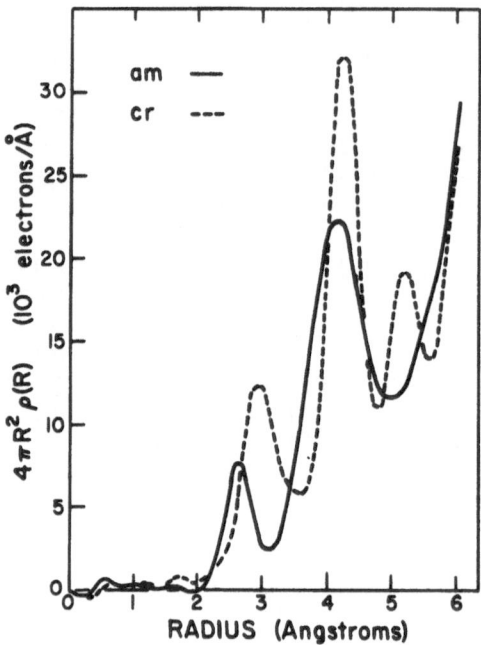

Fig. 7   Radial distribution functions (RDF's) of amorphous and crystalline GeTe [from Ref. 36].   The dashed
curve is the RDF of polycrystalline GeTe, while the solid curve is one of several similar RDF's of amorphous
Ge_xTe_{1-x}, where in this figure x = 0.57.   The area under
the nearest-neighbor peak of the amorphous form indicates
an average of three nearest neighbors centered about
2.7 Å from each other.   The area under the first peak in
the crystalline RDF indicates six nearest-neighbors with
a separation of about 3 Å, nearly the distance of the
first minimum in the amorphous GeTe RDF.

Two models have been proposed which are consistent with
the structural data.   One model[47] proposes a random covalent network of Ge and Te atoms, always 4- to 2-fold
coordinated, respectively, throughout the composition
range.[48]   The other view[49] suggests that while compositions from Te to $Ge_{0.33}Te_{0.67}$ ($GeTe_2$) may be 4- and 2-
fold coordinated, at the $Ge_{0.50}Te_{0.50}$ composition Ge and
Te atoms are three-fold coordinated, possibly in the same
short-range order as black phosphorous.[49]   As we shall
see, ultraviolet and X-ray photoemission measurements on
GeTe[41-43,50,51] tend to support the presence of two-fold

coordinated Te and photoemission studies of the amorphous Ge_xTe_{1-x} system[42,43] have given support to a model that does not involve changes in bonding throughout the composition range. However, the ability to obtain detailed information about bonding from the valence band density of states is just beginning to develop.

A. Vibrational Studies

Infrared and Raman scattering measurements can often provide a sensitive means of gaining detailed information about the structural units in a system from the vibrational modes. We can see this with the far-infrared absorption and Raman studies of the amorphous Ge_xTe_{1-x} system. In addition to resolving the conflict among models for the bonding through the system, there is strong evidence for the existence of particular structural species in the system. We will now consider this evidence in some detail as an example of this analysis.

Experimental Observations

Figure 8 shows the infrared transmission spectra of two films of amorphous $Ge_{0.50}Te_{0.50}(GeTe)$ noted as 1 and 2, and one film of amorphous $Ge_{0.33}Te_{0.67}(GeTe_2)$. Because of the difficulty of obtaining very thick films, maxima and minima due to interference were seen in the spectra measured from 3800 cm^{-1} to 30 cm^{-1}. The minima were found at a uniform spacing over this range in several samples, indicating no major changes in the index of refraction as the frequency approaches the far-infrared. The far-infrared absorption peaks of each spectrum in Figure 8 are added to the broad first interference minimum for each film.

The most prominent feature of Figure 8 is the strong absorption band at about 210 cm^{-1} which appears in every sample of a-GeTe and the a-GeTe$_2$ film. The full width at half maximum (FWHM) of this band is about 28 cm^{-1}, slightly larger in the a-GeTe samples. One can draw a baseline corresponding to the interference minima itself in each sample, which indicates that the absorption at 210 cm^{-1} is about twice as strong in GeTe$_2$ as in GeTe. The absorption coefficient for this mode in GeTe$_2$ is on the order of 10^3 cm^{-1}. The second strongest absorption in the infrared is in the region of 75 cm^{-1} and is about 20 cm^{-1} wide in each sample. Again, the FWHM is apparently somewhat wider in the GeTe sample and about half as strong as in GeTe$_2$ using a crude baseline. There is

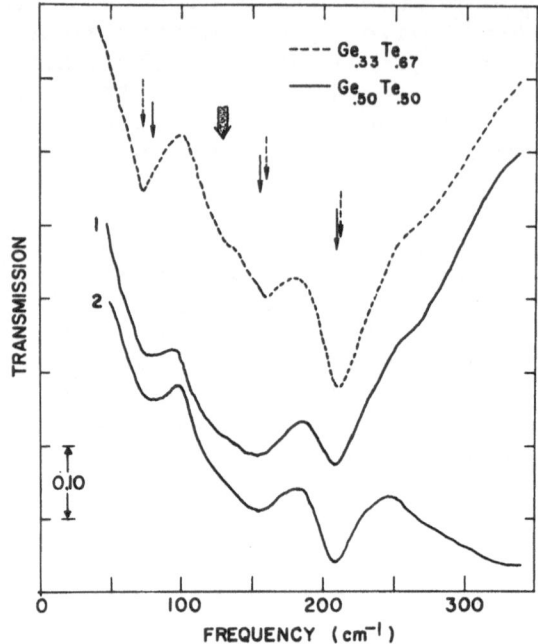

Fig. 8   Infrared transmission spectra of amorphous Ge$_{.33}$
Te$_{.67}$(GeTe$_2$) and amorphous GeTe (films 1 and 2) [from
Ref. 40].   The absorption peaks of each spectra are marked by arrows (GeTe$_2$ - dashed, GeTe - solid) and are superimposed in each film on a broad interference minimum.
The vertical scale is absolute.   The origin of each curve
is shifted such that the lower abscissa is 0.0, 0.2, and
0.4 for GeTe$_2$, GeTe (1), and GeTe (2), respectively.
Note the similarity in peak positions despite the change
in composition.

also absorption in the region between 110 cm^{-1} and
170 cm^{-1} in each sample.   In the GeTe$_2$ sample this may
arise from two absorption bands, a broader one near
130 cm^{-1} and a better defined one at about 160 cm^{-1}.   In
a-GeTe any separation is difficult to make.   Film 2 of
a-GeTe shows that the absorption at about 160 cm^{-1} is
not solely due to the minimum in the interference fringe
in the other films.

Raman scattering measurements were made on films of
amorphous Ge$_x$Te$_{1-x}$, where x = 0.20, 0.33, 0.42, 0.50, and

0.67. The reduced Raman intensity of the extreme compositions and amorphous GeTe are shown in Figure 9. The
33°/o and 42°/o Ge samples (not shown) look much like
the 50°/o composition, but detailed comparisons are difficult. The measured intensities are reduced by multi-
plying by the factor $\omega(n+1)^{-1}(\omega_1 - \omega)^{-4}$, where ω is the

frequency of vibrations, ω_1 the laser frequency and n
the Bose-Einstein occupation number. All of the infrared
and Raman data are listed in Table II.

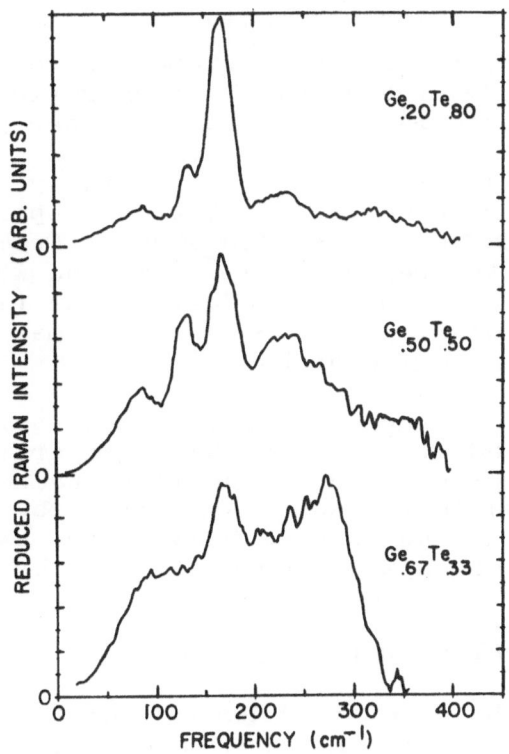

Fig. 9 Unpolarized room temperature reduced Raman spectra of amorphous Ge_xTe_{1-x}, where x = 0.20, 0.50, and
0.67 [from Ref. 40].

Starting at the lowest vibrational frequencies in Figure 9, there is a small peak [band (1) in Table II] centered at about 85 cm^{-1}. It appears strongest in the GeTe composition of those shown and is weakened and shifted to higher frequencies in the Ge-rich sample. The second peak (2) is centered at about 131 cm^{-1} and is about 15 cm^{-1} wide. It also appears strongest at the 50°/o composition and is not apparent in the Ge-rich composition. The third band at 165 cm^{-1} is present in all compositions but is strongest in the Te-rich sample. It is about 25 cm^{-1} in width. Band (4) is centered at about 227 cm^{-1} and is about 30 cm^{-1} wide. As with peaks (1) and (2) it appears strongest in the 50°/o composition and is not apparent in the Ge-rich sample. In the Ge-rich sample there is a new and strong peak centered at about 275 cm^{-1} above which the spectrum drops off sharply to low values, an effect seen in several measurements.

The depolarization ratio, $\rho(\omega) = I_{HV}/I_{VV}$, can give information about the symmetry properties of the vibrational modes.[52,53] The depolarization ratio of the $Ge_{0.20}Te_{0.80}$ sample is plotted in the upper part of Figure 10 for spectra taken with (dashed) and without (solid) an iodine gas cell which preferentially absorbs the exciting laser frequency. The component spectra taken without the iodine cell are also shown. From the drop of the ratio in the region of 130 cm^{-1} and 165 cm^{-1}, these appear to be symmetric modes. Bands (1) and (4) appear to be depolarized, although there may be some rise in $\rho(\omega)$ after the region of band (4).

In Table II it seems plausible to associate the lowest frequency band, (1), in the Raman experiment with the lowest absorption band in the infrared experiment, and also band (4) in the Raman work with the infrared mode at 210 cm^{-1}. They are each about the same width, but the Raman frequencies seem shifted up by about 10 - 15 cm^{-1}.

Discussion of Vibrational Studies

Reviewing the data of Figures 8 and 9, there are clearly similarities in the structure over the whole range of compositions studied. The three main modes in Figure 8 coincide almost exactly. In the Raman data, the GeTe sample has peaks at the same places that the Te-rich sample has. These data along with recent photoemission work[41] suggest the bonding makes gradual changes as the

TABLE II. Observed Raman and Infrared Vibrational Frequencies of Amorphous Ge_xTe_{1-x}

Composition	Experiment	Frequency, cm^{-1}				
		Band 1	Band 2	Band 3	Band 4	Band 5
$Ge_{.20}Te_{.80}$	R	87(D)	133(P)	166(P,s)	228(D)	na
$Ge_{.33}Te_{.67}$	R	~80	131(s)	167(s)	~230	na
	IR	73(S)	127(w)	160	211(s)	na
$Ge_{.42}Te_{.58}$	R	78	128(s)	158(s)	225	na
$Ge_{.50}Te_{.50}$	R	87	131(s)	169(s)	228	~275(w)
	IR	78	~125(w)	157	209(s)	na
$Ge_{.67}Te_{.33}$	R	91	na	169(s)	na	275(s)

P – polarized   s – strong   na – not apparent
D – depolarized   w – weak

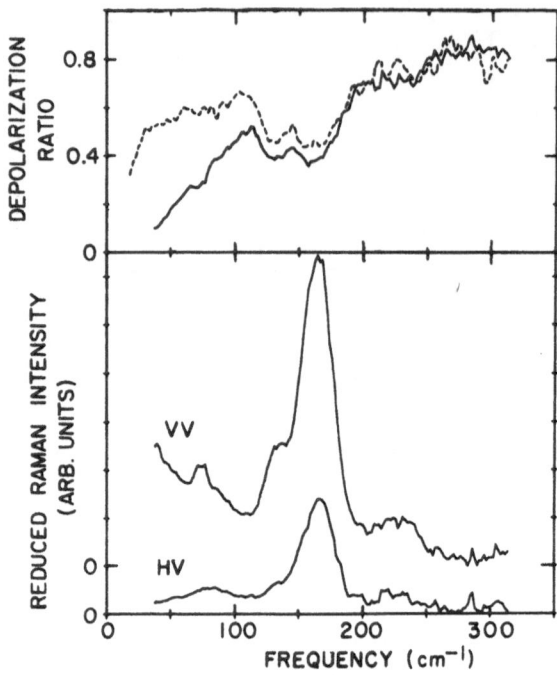

Fig. 10 The depolarization spectrum of amorphous
Ge $_{.20}$Te $_{.80}$ [from Ref. 40]. The polarized reduced Raman
spectra, I_{HV} and I_{VV}, taken at room temperature are shown
in solid curves in the lower portion of the figure. The
solid line in the upper part of the figure is their depolarization ratio. The dashed line is the depolarization ratio of spectra taken with an iodine cell. Note
that the modes at 131 cm^{-1} and 165 cm^{-1} are polarized.

composition changes. These results then favor a model
that does not require changes in bonding. The random
network model of two-fold coordinated Te and four-fold
coordinated Ge is the only proposed model that meets
these criteria. Within this view, one expects Ge-Te,
Ge-Ge, and Te-Te bonds at all compositions. But whatever
the bonding, the clear indication is that it has a similar form over a large part of the composition range.

    Before proceeding further it may be helpful to re-
view the available Raman data on elements related to our
study. The main Raman frequency of amorphous Ge is about
270 cm^{-1}.[54,55] The second reference shows a low energy
peak at about 95 cm^{-1}. The Raman spectrum of amorphous

Te[56] has a strong broad peak at 157 cm^{-1} and a weaker structure at about 85 cm^{-1}. The main peak is interpreted as a shift of the main symmetric mode (A_1) in the crystal (122 cm^{-1})[57] due to disorder in the amorphous phase, which reduces the relatively strong interchain bonding in trigonal Te.[10]

In the Raman data of Figure 9 the Ge-rich sample has a peak band (5) at about 275 cm^{-1} coinciding with that found in amorphous Ge. Also the strength at 275 cm^{-1} grows gradually from a minimum in the Te-rich sample, to a shoulder in the $Ge_{0.5}Te_{0.5}$ sample, to a peak in the Ge-rich sample. It is plausible to associate this peak with the same interactions in Ge-Ge bonding that make it the dominant Raman mode in amorphous Ge. In addition, Figure 9 shows that band (3) at 165 cm^{-1} and very near the frequency of the main Raman mode in amorphous Te at 157 cm^{-1} is strongest in the $Ge_{0.20}Te_{0.80}$ sample and weakest in the Ge-rich sample, although still clearly evident. It is also polarizable as expected for the main mode of amorphous Te. However, to associate this frequency only with the Te-Te bond in a sample that is two thirds Ge relies on the high Raman cross-section for this Te vibration.[58]

If we look for bands (5) and (3) in the infrared data of Figure 8, they are not as apparent. Band (5) is not really expected because of the weak infrared activity of Ge. There is a band at 160 cm^{-1}, but the two dominant modes are at 210 cm^{-1} and about 80 cm^{-1}.

The main stretching mode of the Ge-Te bond in amorphous GeTe compounds has been associated with an infrared reflectivity peak at about 196 cm^{-1}.[59] This corresponds to the strong absorption peak in the infrared at 210 cm^{-1}. More recently, Lucovsky and White[60] have noted from Hilton's data that the main frequencies of amorphous GeS, GeSe, and GeTe are in the same ratio to each other as the main infrared active frequency of the germanium tetrahalides, $GeCl_4$, $GeBr_4$, and GeI_4, suggesting that Ge and its chalcogen form tetrahedra in the amorphous material.

In the tetrahedral molecule, there are four modes. All are Raman active; two are infrared active. For masses like those of Ge and Te, the strongest infrared line (ν_3) is the highest frequency, while the strongest Raman active frequency (ν_1) is the next highest. The ν_1 mode in which the central atom does not move, is symmetric

(A_1) and strongly polarized in the isolated molecule
while the other vibrations are depolarized. The low frequency bending modes are ν_2 (Raman active) and ν_4 (Raman
and infrared active). The experimentally determined
modes[61] of $GeCl_4$, $GeBr_4$ and GeI_4 are listed in Table III.
Comparing with a tetrahedral model, one polarized Raman
active mode (ν_1) and a higher frequency mode (ν_3) which
is dominant in the infrared and depolarized in the Raman
spectrum are consistent with the infrared mode in Ge_xTe_{1-x}
at 210 cm^{-1} and the polarized Raman modes at either
131 cm^{-1} or 165 cm^{-1}. The data can be understood with
either Raman mode being assigned to the ν_1 mode, but in
either case tetrahedral bonding with Ge atoms four-fold
coordinated and Te atoms two-fold coordinated is supported.
Other molecular analogs (e.g. MX_3 and MX_2) don't have the
correct number and combination of Raman and infrared active modes to explain the data.

Let us first pursue a direct analogy with GeI_4, the
closest tetrahedra in mass ratio to one consisting of Ge
and Te. From Table III we see that the ν_1/ν_3 ratio is
0.60. If we associate the 227 cm^{-1} Raman mode with ν_3
of a $GeTe_4$ tetrahedra, we would expect a mode at 136 cm^{-1},
quite close to band (2) at 131 cm^{-1} which is strongly polarized. Further, it appears that bands (1), (2), and
(4) all grow and disappear together. Without postulating
further modes due to the coupling of molecules, this assignment leaves band (3) due to Te-Te interactions, a
high Raman cross-section causing it in the Ge-rich material. Both the ν_2 and ν_4 modes may be reasonably included
in band (1).

In the infrared the tetrahedral model is consistent
with the two strong peaks in the data being ν_3 and ν_4
modes. The weaker bands near 160 cm^{-1} and 125 cm^{-1} may
be due to Te-Te bonding and the perturbation of the symmetry of the ν_1 mode of the $GeTe_4$ tetrahedra in the solid
due to second nearest neighbor interactions. However,
the 10 to 15 cm^{-1} difference in frequency between the
centers of bands (1) and (4) in the Raman and infrared
experiment is still unexplained.

Recently, studies of the Ge_xS_{1-x} system[37,38] have
also shown a difference in the Raman and infrared frequencies in the region of the ν_3 mode. These may either
be due to intermolecular coupling or the introduction of
entirely new modes due to the three dimensional network
of the amorphous solid. They should depend sensitively
on the relative masses of the constituents.

TABLE III. Vibrational Frequencies of the Germanium Tetrahalides*

Molecule	Activity	$\nu_1(A_1)$	$\nu_2(E)$	$\nu_3(F_2)$	$\nu_4(F_2)$	ν_1/ν_3	ν_2/ν_3	ν_4/ν_3
		R	R	R, IR	R, IR			
$GeCl_4$	R	397	132	451	171	.88	.29	.38
$GeBr_4$	R	234	78	328	111	.71	.24	.34
GeI_4	R	159	60	264	80	.60	.23	.30

Frequency, cm^{-1}

*from Reference 61.

Conclusions from Vibrational Studies

From the infrared and Raman data the similarity in mode frequencies over wide ranges of composition strongly suggests that whatever form of bonding exists in the amorphous Ge_xTe_{1-x} system, it is maintained throughout the phase diagram. This result is independent of any model.

Of the models proposed in the literature the random covalent network model of Betts, Bienenstock, and Ovshinsky[47] is most favored. We see evidence of Ge-Ge, Te-Te, and Ge-Te bonding as it predicts. The evidence for $GeTe_4$ tetrahedra is one of the first clear indications of the structural units in this system and supports the view that Ge and Te are four- and two-fold coordinated, respectively. In this analysis it is clear that we have exploited the different symmetries which relate back to the coordination preferences of atomic species to arrive at a result difficult to obtain otherwise.

The data[38] in Figure 11 further demonstrate the structural information present in vibrational states. The Raman spectra from the GeS_2 composition has been shown to agree with the random covalent network model in which Ge is in tetrahedral coordination as in $GeTe_2$. Here the main peak at 342 cm^{-1} and the shoulder at 375 cm^{-1} correspond to the ν_1 and ν_3 modes, respectively. However, as the concentration of sulfur goes beyond $80°/o$, the rapid increase in strength of lines due to Raman modes from S_8 rings indicates the growth of these strictly two-fold coordinated structures at the expense of GeS_4 tetrahedra and the breakdown of the form of random covalent network that apparently exists for the GeS_2 compound and those compositions containing less sulfur that GeS_2.[38] The mode at 435 cm^{-1} may be related to intermolecular bonding within the network.

We can extend our results to draw conclusions in the field of dark-field transmission electron microscopy where recent studies of several amorphous materials (Ge, Si, SiO_2, and Ge-Te alloys) have shown they contain coherent scattering regions or ordered domains.[62,63,64] Some workers[64] have interpreted these data as evidence of microcrystallites of about 12 Å in diameter as the structural basis for amorphous Ge and Si. Others have interpreted the results in terms of a random network model[65] and have shown such a structure can give coherent scattering.[66] Amorphous Ge_xTe_{1-x} and amorphous GeTe have coherent

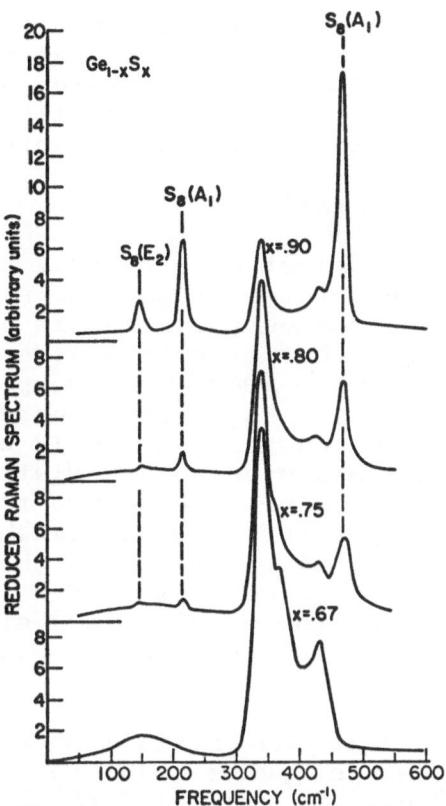

Fig. 11 Comparison of the reduced Raman spectra of the
S-rich amorphous $Ge_{1-x}S_x$ alloys (x = 0.75, 0.80, and
0.90) with the spectrum of the compound composition, GeS_2
(x = 0.67, where the subscript indices of composition are
complimentary to those already used in the Ge-Te alloy
system) [from Ref. 38]. The sharp line features in the
three S-rich compositions are assigned to vibrational
modes (symmetries indicated) of S_8 molecules.

scattering regions.[63] With the strong case for a random
network in amorphous GeTe, coordinated so differently from
the crystal, this is the first system in which one can
convincingly say that microcrystallites cannot cause the
coherent scattering. This result may be generalizable
to other related amorphous solids.

    B. Photoemission Studies

    Photoemission is especially suited to this investi-
gation of the amorphous Ge-Te alloy system because of its

ability to probe the upper valence bands of a solid to
obtain the density of states. For a system in which
bonding changes, the upper valence states should be most
affected. If changes occur in the density of states, they
should be observed by the photoemission experiment.

Experimental Observations

Photoemission was first used on this system to examine the densities of states of amorphous and crystalline GeTe[41,42], since the coordination in amorphous GeTe
is markedly lower than that of crystalline GeTe as we saw
in Figure 7. Thin films (~2000 Å) of amorphous and crystalline GeTe evaporated in situ were measured in 10^{-11}
Torr vacuum with photon energies up to 12 eV. Similarly
prepared samples were measured in a Hewlett-Packard 5950A
ESCA Spectrometer after a brief exposure to air during
transfer from the evaporator. Curves taken at $h\nu$ = 9.2 eV
and $h\nu$ = 11.7 eV in the ultraviolet and the valence band
density of states (VBDOS) seen with XPS using monochromatized AlKα radiation ($h\nu$ = 1486.6 eV) are shown in
Figure 12. The UPS resolution is about 0.1 eV and the
XPS about 0.55 eV.

If we first look at the crystalline data, the 9.2 eV
curve has three prominent peaks below E_F at -0.7 eV,
-1.75 eV, and -2.4 eV. At $h\nu$ = 11.7 eV the two lower
peaks have merged into one broad peak, because of final
state effects[67]. The X-ray photoemission data show one
pronounced peak at -2.5 eV, one distinct shoulder at
-0.7 eV and one weaker shoulder at -1.6 eV. The overall
agreement between the UPS ($h\nu$ = 9.2, 11.7 eV) and XPS
($h\nu$ = 1486.6 eV) data in the top 5 eV is remarkable considering the difference in photon energy.

An OPW band structure calculation of room-temperature
GeTe, which has a slightly distorted NaCl structure, gives
a density of states which gives encouraging agreement with
the first three peaks (dashed arrows) in the experimental
crystalline VBDOS and with the position of the two lower
levels of largely s-like character.[67]

For amorphous GeTe a double peak is observed (Figure 12, solid arrows) both in the UPS and XPS data. For
a range of photon energies in UPS no structure modulation
is observed; this indicates a lack of final state effects
as is usually found for amorphous materials. The main
features are the narrow leading peak 1.7 eV below the
Fermi level, a minimum at -2.25 eV, and a broader peak

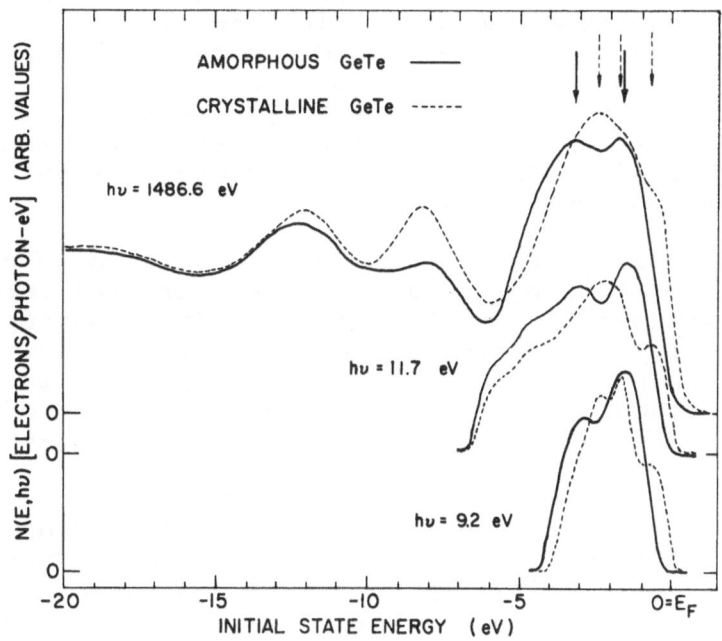

Fig. 12 Valence band density of states of amorphous and crystalline GeTe, as seen by XPS (hν = 1486.6 eV) and UPS with hν = 11.7 eV and hν = 9.2 eV [from Ref. 43]. The curves are plotted with respect to the Fermi level which is referenced to the Fermi level in Au. The double peak at the top of the valence band in amorphous GeTe (solid arrows) contrasts with the three structures for crystalline GeTe (dashed arrows).

at about -3.25 eV. The zero is defined by E_F in Au.

Other XPS work on GeTe using non-monochromatized X-rays and a resolution of 1.5 eV shows almost identical valence band spectra for the amorphous and crystalline phases[50] and had been interpreted as evidence for little difference existing between the density of states (DOS) of amorphous and crystalline GeTe, despite the difference in coordination. However, Figure 12 clearly shows that high resolution X-ray photoemission (0.5 - 0.6 eV) can easily distinguish between amorphous and crystalline GeTe and that there are important differences in the density of states. The curves in Figure 12 are referenced to the Fermi level and also show that amorphous GeTe is a semi-

conductor with the top of the valence band about 0.4 eV
below the Fermi level whereas crystalline GeTe is a p-
type extrinsic semiconductor. Instead of being a smeared
form of the crystalline spectrum, in the DOS of amorphous
GeTe there is an additional minimum at -2.25 eV for amorphous GeTe which cannot be attributed to loss of long-
range order per se. Instead the observed differences between crystalline and amorphous GeTe are attributed to
differences in their bonding.[41,43,51]

Since differences in the DOS are observable due to
the change in bonding in amorphous and crystalline GeTe,
it is reasonable to use photoemission for studying the
amorphous Ge_xTe_{1-x} system, especially at the amorphous
$GeTe_2$ and GeTe compositions. For this purpose thin films
($\sim$3000 Å) of amorphous Ge_xTe_{1-x}, where x = 0.15, 0.33,
0.42, 0.50, and 0.67, were examined by the UPS and XPS
techniques.

In Figure 13 we see the valence bands and upper core
levels of amorphous Ge_xTe_{1-x} films of four different compositions. Film b is within 3°/o of $Ge_{0.50}Te_{0.50}$. Films
a, c, and d were argon sputter cleaned to remove oxygen.
In so doing the films lost Te. We estimate that the relative composition may have changed on the order of 10°/o.
The trends within the Ge_xTe_{1-x} system are clear. There
do not appear to be any abrupt changes in the density of
states. First, most samples exhibit a double-peaked upper valence band consisting of a leading peak always a-
bout 1.7 - 1.9 eV below the Fermi level and a lower,
broader peak which approaches the valence band maximum
with decreasing Te content. For the extreme Ge-rich
sample, a single-peaked valence band is seen in XPS.
As Te content decreases, the width of the upper valence
band ($\sim$6 eV) also decreases by about 1 eV over the range
studied. Also, the lower peak of the upper valence band
density of states remains well defined for all compositions and is not smeared, as a sum of Te and Ge levels
in this region would be if Ge and Te were phase separated. We note also that the strength of the leading peak
($\sim$-1.7 eV) and the s-like band near -12.5 eV correlate
well with the strength of the Te 4d levels, while the
peak at about -8 to -9 eV grows with the Ge 3d level.

The upper core levels are the $4d_{3/2}$ and $4d_{5/2}$ levels
of Te and the 3d levels of Ge. The spin-orbit splitting
of the Te levels is 1.4 ± 0.1 eV, while the splitting in
Ge is not resolved. The relative intensity of these peaks
can be used as a measure of the composition. For many

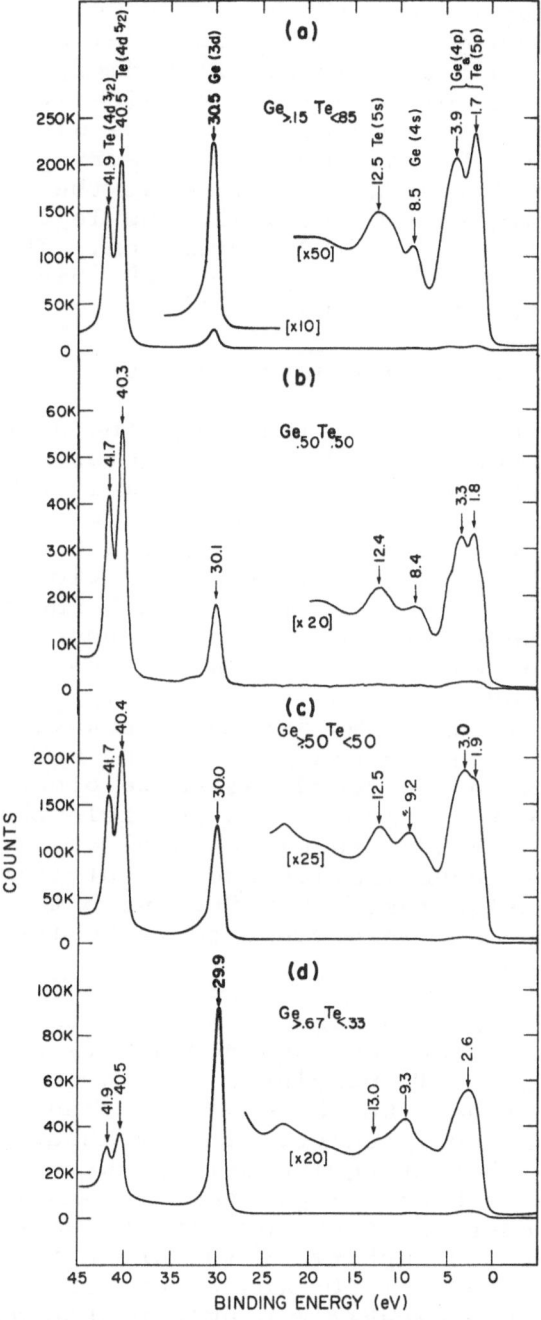

Fig. 13 XPS valence band and upper core level spectra of four compositions of amorphous Ge_xTe_{1-x} [Ref. 43]. Note the relative compositional trends of the core levels and the various valence band peaks. Monochromatized AlKα X-rays were used.

samples of amorphous GeTe the Te $4d_{5/2}$/Ge 3d ratio was
3:1. We should point out there was no significant relative shift of these core levels from amorphous to crystalline GeTe. The plasmon loss energies for amorphous
and crystalline GeTe were measured to be 17.0 and 17.6
eV, respectively. Furthermore, our observations have
shown a preferential oxidation of Ge at the surface, especially in crystalline GeTe, while the Te levels exhibit
no apparent shift due to oxide formation. These data appear to explain the previously reported shift in Ge levels
from amorphous to crystalline GeTe[68].

Discussion of Photoemission Studies

Reviewing the data of Figure 13, the general result
is that the valence band density of states in the amorphous Ge_xTe_{1-x} system is similar over most of the phase
diagram and changes in it with composition are gradual.
Hence, whatever the character of the electron states and
the bonding, they appear similar over a wide range of
composition. This result is model independent, but of
the proposed models for bonding in amorphous Ge_xTe_{1-x} it
favors the random covalent network of two- and four-fold
coordinated Te and Ge atoms, respectively, because this
model can apply to all compositions. However, the model
involving three-fold coordination at the GeTe composition
predicts changes in the coordination to occur going away
from this composition, implying a possible change in the
density of states. The gradual changes are more consistent with the 2-4 coordination model, but this does not
provide a conclusive test. However, with the infrared
and Raman data we are more prepared to believe the 2-4
coordination model.

We may speculate in some detail about the applicability of the 2-4 model by considering its prediction of
Ge-Ge, Te-Te, and Ge-Te bonding with 2-4 coordination and
how this might effect the shape of the density of states.
As we have already pointed out, in tetrahedrally bonded
germanium with its four equivalent sp^3 hybridized bonds,
one might expect a broad single-peaked density of states
because the valence electrons are in the same sort of
bond. This view has support in the XPS data of amorphous
Ge and Si in Figure 4 [27]. On the other hand, in chalcogenides like Te one expects a double-peaked density of
states, since only two of the four p electrons are included in bonding, while the other two form a filled p
orbital (non-bonding or lone pair electrons). The peak
in the density of states due largely to the lone pair

electrons should be narrower, reflecting the smaller overlap with neighboring atoms, and be at a smaller binding
energy than the broader peak largely due to electrons in
bonds.[19] Other photoemission data from amorphous and
crystalline S,[69,70] Se,[70,72] and Te[70,72-75] show a double-
peaked structure supporting this view. We have already
seen this clearly in Figure 2. Therefore, it can tentatively be stated that the relatively narrow leading peak
may be a signature of two-fold coordination in these and
related materials. In the case of bonding between Ge and
Te with Te two-fold coordinated, the 2-4 model would predict that the lone pairs would still not be involved in
bonding and would form a narrow leading peak as in Te.
The main difference should be in the lower energy bonding peak which might shift its position, reflecting the
presence of covalent bonds between Ge and Te atoms.

If we now look at the data on amorphous Ge_xTe_{1-x} the
existence of a double-peaked upper valence band is clear,
from Te-rich samples to a few that are somewhat Ge-rich.
The leading peak is narrower than the lower energy one
and remains fixed about 1.8 eV below the Fermi level in
most samples. The leading peak's strength scales with
the Te 4d peaks in Figure 13, giving strong support to
its assignment as being due largely to Te lone pairs.
The double peak seen over such a wide composition range
is consistent with the presence of both Te-Te and Ge-Te
bonding. The gradual shift in binding energy of the lower
peak from 3.9 eV to 2.6 eV in Figure 13 suggests a progression in the relative concentration of bonds from Te-
Te, to Ge-Te, and finally more Ge-Ge bonding. The single-
peaked upper valence band of curve d in Figure 13 is much
like that of amorphous Ge.[27] Attempts to relate the accompanying shift in the minimum near 6 eV to a change in
the overlap of the bonding orbitals are complicated by
the presence of the lone pair peak and the possibility of
changes in s-p hybridization. The trends in the data appear consistent with the shape of the upper valence band
expected within the 2-4 coordination model. In fact,
they show just the progression in electronic structure
that is seen for the elements in going from Group VI to
Group IV that is related to a greater coordination.

If we speculate again about the implication of three-
fold coordination, some ideas may be obtained from the
density of states of Group V elements which have three
p electrons on each atom and are generally three-fold coordinated. It is reasonable to assume that the three similar bonds will give rise to a large hump at the top of

the valence band, a fact which has been observed in recent XPS measurements,[25] but this is not necessarily true
in general. However, high resolution UPS and XPS data
in Figure 12 do indicate that the upper valence band of
amorphous GeTe is not single-peaked.

Before summarizing our present conclusions about
amorphous GeTe and the non-tetrahedrally bonded glasses,
it should be clear that in our rudimentary discussions
above we have barely alluded to the effects of hybridization between s and p states in the solid. Differences
in the amorphous and crystalline GeTe valence bands may
to some extent be related to changes in the s-p hybridization with the bonding change. However, in amorphous
Ge_xTe_{1-x} we see that to a first approximation the levels
with more s-like character at 12-13 eV and 8-9 eV scale
well with Te and Ge content, respectively. This implies
that s-p hybridization as seen by XPS may not be as large
as suggested in some other amorphous solids.[33] In Figures 12 and 13 primary valence band emission appears to
cease approximately 15 eV below the Fermi level; the rise
near 19 eV in all samples is attributed to a plasmon loss
by electrons emitted from the upper valence band.

6. CONCLUSIONS

With high resolution XPS and UPS measurements we observe a clear difference in the densities of states of
amorphous and crystalline GeTe. It is interesting that
while amorphous crystalline Se and Te and amorphous GeTe
all have a double-peaked valence band with a relatively
narrow leading lone pair peak, crystalline GeTe does not.
This is consistent with its crystal structure in which
the Te is not two-fold coordinated, but in which both Te
and Ge are nearly six-fold coordinated. We relate the
basic differences in the densities of states, not to the
loss of long-range order per se, but to the differences
in coordination and bonding between the two materials.

In the amorphous Ge_xTe_{1-x} system the main results
are the similarity in the valence band spectra over a wide
composition range and the generally double-peaked shape
of those spectra, suggesting the presence of two-fold coordinated Te. Together these determinations strongly support the description of coordination and bonding by the
random covalent network model[47] of two- and four-fold coordinated Te and Ge, respectively. This interpretation
is also consistent with the study of the vibrational modes

in this system indicating the presence of $GeTe_4$ tetrahedra.[21] Support is lacking for the existence of threefold coordination in amorphous GeTe. In fact, in amorphous Ge_xTe_{1-x} it appears that Ge and Te satisfy their local valence requirements throughout the phase diagram. From the data on the Ge-S and Ge-Se systems this result can probably be generalized to all of the amorphous IV-VI alloy systems.

Thus we can see the power of using both vibrational and photoemission studies on the same system. In the case of all of the chalcogenide-containing materials the appropriateness of a molecular model in describing the vibrational structure and the lone pair concept in describing the electronic structure have put us closer to understanding their bonding. We have also seen that some properties of the chalcogenides and nearby members of the periodic table scale with coordination as it decreases in going from Group IV to Group VI. Thus, at the same time the linewidths of the vibrational modes decrease, the lone pair electrons replace the single peak of bonding p electrons at the top of the valence band density of states (e.g. Figure 13). In general, as the spatial distribution of bonds becomes less homogeneous, molecular models generally increase in their applicability to the understanding of the vibrational and electronic structures of amorphous systems.

ACKNOWLEDGMENTS

For useful discussions over the past years which have contributed to this paper, I would like to thank A. Bienenstock, J. P. deNeufville, M. Kastner, L. Laude, G. Lucovsky, P. Nielsen, W. E. Spicer, J. Tauc and R. Zallen. I thank F. Betts for his permission to use the unpublished data in Figure 7. I very much appreciate careful readings of the manuscript by S. R. Nagel and M. Zanini.

REFERENCES

1.   W. Paul, in these proceedings.
2.   R. Grigorivici, in <u>Amorphous and Liquid Semiconductors</u>, ed. by J. Tauc (Plenum Press, London, 1974), p. 45; and in <u>Electronic and Structural Properties of Amorphous Semiconductors</u>, ed. by P.G. Lecomber

and J. Mort (Academic Press, London, 1973), p. 191.

3.   A. Bienenstock, in <u>Amorphous and Liquid Semiconduc-</u>
     <u>tors</u>, Proceedings of the 5th International Confer-
     ence on Amorphous and Liquid Semiconductors, ed. by
     J. Stuke and W. Brenig (Taylor and Francis, London
     1974), p. 49.
4.   W. Fuhs, in these proceedings.
5.   H. Fritzche, in <u>Amorphous and Liquid Semiconductors</u>,
     ed. by J. Tauc (Plenum Press, London, 1974),p. 221.
6.   D. Emin, in these proceedings.
7.   J. Tauc, in <u>Amorphous and Liquid Semiconductors</u>, ed.
     by J. Tauc (Plenum Press, London, 1974), p. 159.
8.   E. A. Davis, in <u>Amorphous and Liquid Semiconductors</u>,
     Proceedings of the 5th International Conference on
     Amorphous and Liquid Semiconductors, ed. by J.
     Stuke and W. Brenig (Taylor and Francis, London,
     1974), p. 519.
9.   L. Laude, in these proceedings.
10. G. Lucovsky, in <u>Amorphous and Liquid Semiconductors</u>,
     Proceedings of the 5th International Conference on
     Amorphous and Liquid Semiconductors, ed. by J.
     Stuke and W. Brenig (Taylor and Francis, London,
     1974), p. 1099.
11. M. F. Thorpe, in these proceedings.
12. J. Tauc, in these proceedings.
13. P. C. Taylor, in these proceedings.
14. D. E. Sayers, E. A. Stern, and F. W. Lytle, Phys.
     Rev. Letters 27, 1204 (1971); D. E. Sayers, F. W.
     Lytle and E. A. Stern, in <u>Amorphous and Liquid Semi-</u>
     <u>conductors</u>, Proceedings of the 5th International
     Conference on Amorphous and Liquid Semiconductors,
     ed. by J. Stuke and W. Brenig (Taylor and Francis,
     London, 1974), p. 403.
15. G. Lucovsky, in <u>Physics of Selenium and Tellurium</u>,
     ed. by W. C. Cooper (Pergamon Press, Oxford, 1969),
     p. 255.
16. G. Lucovsky and R. M. Martin, J. Non-Cryst. Solids,
     8-10, 185 (1972).
17. M. F. Thorpe, Phys. Rev. B 8, 5352 (1973).
18. G. B. Fisher and R. B. Shalvoy, to be published.
19. E. Mooser and W. B. Pearson, in <u>Progress in Semi-</u>
     <u>conductors</u>, (Heywood and Company, Ltd., London,
     1960), vol. 5, p. 104; M. Kastner, Phys. Rev. Let-
     ters 28, 355 (1972).
20. L. Pauling, <u>The Nature of the Chemical Bond</u>, 3rd Ed.
     (Cornell University Press, Ithaca, 1960), p. 85.
21. M. Schlüter, J. D. Joannopoulos, and Marvin L. Cohen,
     Phys. Rev. Letters 33, 89 (1974).

22.   M. Wihl, P. J. Stiles, and J. Tauc, in <u>Proceedings</u>
       <u>of the 11th International Conference on the Physics</u>
       <u>of Semiconductors</u>, Warsaw, Poland (Polish Scientific
       Publishers, Warsaw, 1972), p. 484.

23.   G. N. Greaves, J. C. Knights and E. A. Davis, in
       <u>Amorphous and Liquid Semiconductors</u>, Proceedings of
       the 5th International Conference on Amorphous and
       Liquid Semiconductors, ed. by J. Stuke and W. Brenig
       (Taylor and Francis, London, 1974), p. 369.

24.   G. Lucovsky and J. C. Knights, Phys. Rev. B $\underline{10}$,
       4324 (1974).

25.   L. Ley, R. A. Pollak, S. P. Kowalczyk, R. McFeely,
       and D. A. Shirley, Phys. Rev. B $\underline{8}$, 641 (1973).

26.   I. Abbati, L. Braicovich, B. DeMichelis, and A.
       Fasana, Phys. Letters $\underline{48A}$, 33 (1974).

27.   L. Ley, S. Kowalczyk, R. A. Pollak, and D. A. Shirley
       Phys. Rev. Letters $\underline{29}$, 1088 (1972).

28.   See several articles in <u>Electronic and Structural</u>
       <u>Properties of Amorphous Semiconductors</u>, ed. by P. G.
       LeComber and J. Mort (Academic Press, London, 1973).

29.   G. Lucovsky, Phys. Rev. B $\underline{6}$, 1480 (1972).

30.   P. C. Taylor, S. G. Bishop, D. L. Mitchell and D.
       Treacy, in <u>Amorphous and Liquid Semiconductors</u>,
       Proceedings of the 5th International Conference on
       Amorphous and Liquid Semiconductors, ed. by J. Stuke
       and W. Brenig (Taylor and Francis, London, 1974),
       p. 1267.

31.   G. B. Fisher, Bull. Am. Phys. Soc. $\underline{18}$, 1589 (1973).

32.   Z. Hurych, D. Davis, D. Buczek, C. Wood, G. J.
       Lapeyre, and A. D. Baer, Phys. Rev. B $\underline{9}$, 4392 (1974).

33.   I. Chen, Phys. Rev. B $\underline{8}$, 1440 (1973).

34.   P. Nielsen, Bull. Am. Phys. Soc. $\underline{17}$, 113 (1972).

35.   S. G. Bishop and N. J. Shevchik, in <u>Proceedings of</u>
       <u>the 12th Interntl. Conf. on the Physics of Semicon-</u>
       <u>ductors</u>, Stuttgart, Germany, ed. by M.H. Pilkuhn (B.
       G. Teubner, Stuttgart, 1974), p. 1017; K.S. Liang, A.
       Bienenstock and C.W. Bates, Phys.Rev. B $\underline{10}$, 1528 (1974).

36.   F. Betts, Ph.D. Thesis, Stanford University, 1972
       (unpublished). The data in Figure 7 are from
       Figures 3.2 and 3.3 of this work.

37.   G. Lucovsky, J. P. deNeufville, and F. L. Galeener,
       Phys. Rev. B $\underline{9}$, 1591 (1974).

38.   G. Lucovsky, F. L. Galeener, R. C. Keezer, R. H.
       Geils, and H. A. Six, Phys. Rev. B $\underline{10}$, 5134 (1974).

39.   P. Tronc, M. Bensoussan, A. Brenac, and C. Sebenne,
       Phys. Rev. B $\underline{8}$, 5947 (1973).

40.   G. B. Fisher, J. Tauc, and Y. Verhelle, in <u>Amorphous</u>
       <u>and Liquid Semiconductors</u>, Proceedings of the 5th
       International Conference on Amorphous and Liquid
       Semiconductors, ed. by J. Stuke and W. Brenig,

(Taylor and Francis, London, 1974), p. 1259.

41. G. B. Fisher and W. E. Spicer, J. Non-Cryst. Solids 8-10, 978 (1972); and to be published.

42. G. B. Fisher, B. A. Orlowski, I. Lindau, and W. E. Spicer, Bull. Am. Phys. Soc. 18, 390 (1973).

43. G. B. Fisher, I. Lindau, B. A. Orlowski, W. E. Spicer, Y. Verhelle, and H. E. Weaver, in Amorphous and Liquid Semiconductors, Proceedings of the 5th International Conference on Amorphous and Liquid Semiconductors, ed. by J. Stuke and W. Brenig (Taylor and Francis, London, 1974), p. 621.

44. A. Bienenstock, F. Betts, and S. R. Ovshinsky, Non-Cryst. Solids 2, 347 (1970).

45. D. B. Dove, M. B. Heritage, K. L. Chopra and S. K. Bahl, Appl. Phys. Letters 16, 138 (1970).

46. F. Betts, A. Bienenstock, D. T. Keating, and J. P. deNeufville, J. Non-Cryst. Solids 7, 417 (1972).

47. F. Betts, A. Bienenstock, and S. R. Ovshinsky, J. Non-Cryst. Solids 4, 554 (1970).

48. This model has been modified [see the second citations in Refs. 14 and 35] and extended [see Ref. 38].

49. A. Bienenstock, J. Non-Cryst. Solids 11, 447 (1973).

50. N. J. Shevchik, J. Tejeda, D. W. Langer, M. Cardona Phys. Rev. Letters 30, 659 (1973).

51. N. J. Shevchik, J. Tejeda, D. W. Langer, M. Cardona Phys. Stat. Solidi B 57, 245 (1973).

52. G. Herzberg, Infrared and Raman Spectra of Polyatomic Molecules (D. Van Nostrand Co., Inc, New York, 1945), p. 270

53. R. J. Kobliska and S. A. Solin, Phys. Rev. B 8, 756 (1973).

54. J. E. Smith, Jr., M. H. Brodsky, B. L. Crowder, M. I. Nathan and A. Pinczuk, Phys. Rev. Letters 26, 642 (1971).

55. M. Wihl, M. Cardona, and J. Tauc, J. Non-Cryst. Solids 8-10, 172 (1972).

56. M. H. Brodsky, R. J. Gambino, J. E. Smith, Jr. and Y. Yacoby, Phys. Stat. Solidi B 52, 609 (1972).

57. A. S. Pine and G. Dresselhaus, Phys. Rev. B 4, 356 (1971).

58. J. S. Lannin, in Proceedings of Int. Conf. on Tetrahedrally Bonded Amorphous Semiconductors, Yorktown Heights, 1974, ed. by M. H. Brodsky, et al. (American Institute of Physics, New York, 1974), p. 260.

59. A. R. Hilton, C. E. Jones, R. D. Dobrott, H. M. Klein A. M. Bryant and T. D. George, Phys. Chem. Glasses 7, 116 (1966).

60. G. Lucovsky and R. M. White, Phys. Rev. B 8, 660 (1973).

61. J. R. Ferraro, <u>Low Frequency Vibrations of Inorganic and Coordination Compounds</u> (Plenum Press, London, 1971), pp. 128-129.

62. M. L. Rudee, Phys. Stat. Solidi B <u>46</u>, K1 (1971).

63. P. Chaudhari, J. F. Graczyk, and S. R. Herd, Phys. Stat. Solidi B <u>52</u>, 801 (1972).

64. A. Howie, O. Krivanek, and M. L. Rudee, Phil. Mag. <u>27</u>, 235 (1973).

65. D. E. Polk, J. Non-Cryst. Solids <u>5</u>, 365 (1971).

66. P. Chaudhari, J. F. Graczyk, H. P. Charbnau, Phys. Rev. Letters <u>29</u>, 425 (1972).

67. G. B. Fisher, I. B. Ortenburger, and W. E. Spicer, to be published.

68. F. Betts, A. Bienenstock, C. W. Bates, J. Non-Cryst. Solids <u>8-10</u>, 364 (1972).

69. P. Nielsen, Phys. Rev. B <u>10</u>, 1673 (1974).

70. P. Nielsen, in <u>Amorphous and Liquid Semiconductors</u>, Proceedings of the 5th International Conference on Amorphous and Liquid Semiconductors, ed. by J. Stuke and W. Brenig (Taylor and Francis, London, 1974), p. 639.

71. P. Nielsen, Phys. Rev. B <u>6</u>, 3739 (1972).

72. N.J. Shevchik, M. Cardona, and J. Tejeda, Phys. Rev. B <u>8</u>, 2833 (1973).

73. R. A. Powell and W. E. Spicer, Phys. Rev. B <u>10</u>, 1603 (1974).

74. R. A. Pollak, S. P. Kowalczyk, L. Ley and D. A. Shirley, Phys. Rev. Letters <u>29</u>, 274 (1972).

75. M. Schlüter, J. D. Joannopoulos, M. L. Cohen, L. Ley, S. P. Kowalczyk, R. A. Pollak and D. A. Shirley, Solid State Commn. <u>15</u>, 1007 (1974).

LIGHT SCATTERING IN LIQUID SEMICONDUCTORS

J. Tauc

Division of Engineering and Department of Physics

Brown University, Providence, R. I. 02912

1. INTRODUCTION

Structure and electrical properties of liquid semiconductors have been reviewed in Refs.1 and 2. In these lectures, I will discuss the light scattering in liquid semiconductors with a particular emphasis on the temperature range close to the glass transition rather than the critical point to which most studies of this kind have been devoted. I will first present a brief description of the relevant properties of the melts. An attempt is then made to present a discussion of the scattering processes from a unified viewpoint based on the correlation functions. The aim of the approach is to show the physical principles and the potential usefulness and limitations of the methods rather than a development of a precise theory and a detailed analysis of actual experiments. The limited length of this paper makes it impractical to quote all original papers in this broad field, and I often refer to review papers if available.

Glass formation is not restricted to certain kinds of materials, and representatives of all classes of materials have been prepared in the amorphous form. However, the differences in the glass- forming tendency are enormous. In these lectures, we shall deal with materials which can be prepared in the amorphous form by quenching the liquid. These materials form an important class of amorphous materials because they can be prepared in the bulk form (as distinguished from thin films), and are sometimes referred to as "real" glasses. A basic question

739

is what properties of the liquid are essential for the
glass formation. From the macroscopic point of view,
the most important property is the viscosity of the super
cooled liquid which prevents the growth of crystals as
discussed in lectures by Turnbull[40]. Much less is known
however, about the microscopic conditions, in particular
the atomic structure of the supercooled liquid, which is
the underlying reason for these properties[3].

2. STRUCTURE OF LIQUIDS

The classification of liquids has been done from
various points of view[1,3]; for our purposes it is sufficient to mention the following points:

(a) <u>Simple liquids</u>, either monoatomic (e.g. Na, Ar),
or molecular, but the molecules are simple (in the simplest case spherical and interacting with van der Waals
forces).

<u>Complex liquids</u> (composed of large structural units,
such as chains in Se, or large anisotropic molecules).

The theory of liquids[4] has been able to deal with
the first case. The second case is much more difficult
and also from the fundamental point of view, complicated
by effects not necessarily typical for the liquid state.
Unfortunately, good glass-formers fall into this category, and it is these liquids which we shall be considering.

(b) <u>Metallic liquids</u> (with electrical conductivities
above $300\,\Omega^{-1}cm^{-1}$). All metals, and also typical semiconductors (Si, Ge, $A^{III}B^{V}$ compounds and similar materials) fall into this category.

<u>Insulating and semiconducting liquids</u>. This cate-
gory contains simple liquids (e.g. Ar) but also complex
liquids.

Electrical conductivity σ of metals changes little
when one passes through the melting point T_m. The temperature dependence of σ for Ge and As_2Se_3 is shown in
Fig. 1. It jumps at T_m for the crystalline → liquid transition in Ge but is almost constant at the glass transition in As_2Se_3. Also shown in Fig. 1 is Ga_2Te_3 which is
an intermediate case: close to T_m the liquid is semiconducting ($\sigma(T)$ increases with temperature[5]), but at higher

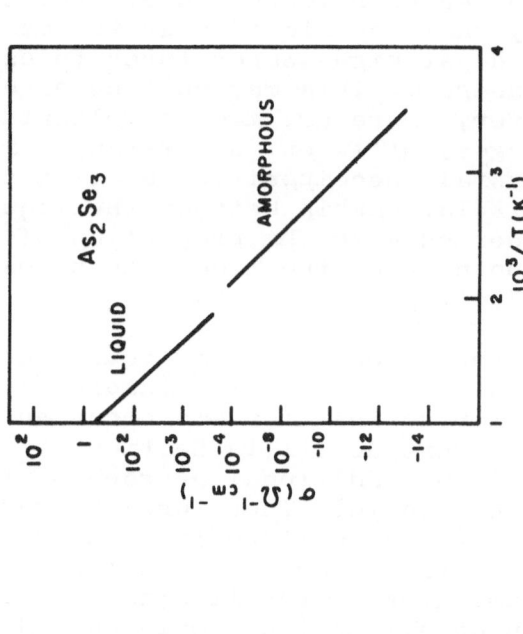

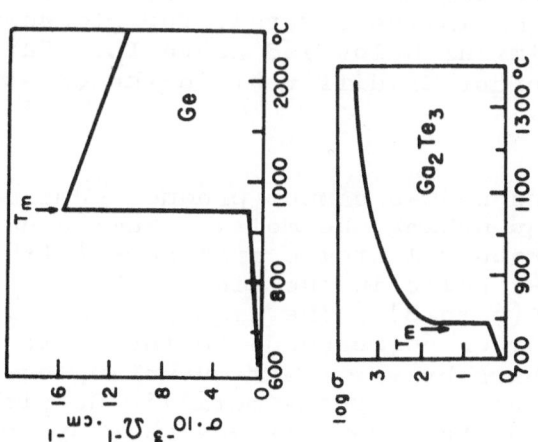

Fig. 1 Temperature dependence of electrical conductivity close
to the melting point T_m in Ge, Ga_2Te_3 (after Ref. 2) and through
the glass transition in As_2Se_3 (after Ref. 29).

temperatures it becomes metallic (in the sense that $\sigma(T)$ decreases with temperature).

(c) The example of Ga_2Te_3 shows that the character of a liquid may change with temperature. For simple liquids, one may say that the liquid near the melting point resembles the solid, at high temperatures (close to the critical point) the gas. This may be true of complex liquids too, however, there are many complications, such as structural changes, different short-range order in the liquid and the crystal, decomposition etc. Since we are interested in the relationship between the liquid and the glass, we are concerned with the properties of a liquid near the melting point when the liquid is normally referred to as the "melt".

For studying the structure of a liquid one can use X-ray and neutron diffraction[1]. The information one obtains is very similar to that for amorphous solids. As we shall discuss in more detail in Section 3, a liquid differs from a solid by additional degrees of freedom, which are frozen in a solid. Therefore the radial distribution function $g(r)$ determined from the diffraction measurements is a spatial average for amorphous solids but a spatial - time average for liquids. From $g(r)$ of the liquid one can determine (in simple cases) the short-range order in the liquid. For metals, it is usually the same in the melt as in the crystal; for glasses, one expects it to be similar below and above T_g. But for Ge, the short-range order is different in the crystal and and in the melt.

This explains why we cannot produce semiconducting amorphous Ge by quenching the melt. Semiconducting a-Ge has the tetrahedral atomic arrangement (the coordination number 4) while in the liquid the coordination number is higher (6 to 8). The liquid is metallic, and the short-range order corresponds to the white tin structure[6]. Perhaps by very fast quenching one might be able to produce solid amorphous metallic Ge with white tin structure which has been already made by applying high pressures to a-Ge[7].

When we have a 2 component system, then we observe the tendency of atoms to form chemical species corresponding to their chemical nature; e.g. in a system of A and B atoms one finds in the radial distribution function evidence for A-A, B-B and A-B bonds[1]. However, the radial distribution function is difficult to interpret in these systems, and it is easier to study chemical bond formation by measuring the electrical conductivity σ, viscosity η and other effects. In Fig. 2, σ and η are shown as functions of composition of Aℓ-Sb melts. We see that these functions have maxima and minima respectively when the valency of Aℓ and Sb is satisfied. These extrema are more pronounced at lower temperatures which indicates that the chemical bonding is more pronounced the closer we are to T_m. Warren[8] has studied these effects in In$_2$Te$_3$ and similar liquids by MNR.

When we have a liquid A_xB_{1-x} $(x \neq 1/2)$ we may ask whether the molecules AB corresponding to the chemical bonding are isolated in the liquid or whether they form clusters (regions composed of several AB molecules holding together). These regions are rigid in an amorphous solid but in the liquid they would appear as fluctuations and their dimensions and location will change with time.

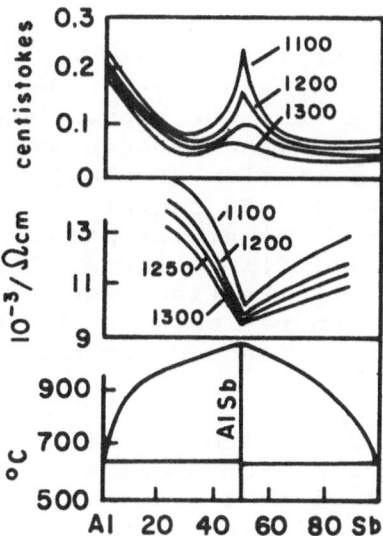

Fig. 2   Kinematic shear viscosity $(=\eta/\rho_o)^{34}$, electrical conductivity σ and the phase-diagram of the system Aℓ-Sb (after Ref. 2). The numbers in the two upper curves denote temperatures in ^oC. η is dynamic shear viscosity[34].

There are indications that in the glass forming liquids the clustering is essential. Unfortunately, the
diffraction methods are not sufficiently sensitive to
these atomic correlations extending over distances of
the order 10A, and they have been studied by less direct
methods. The classical approach is to measure the viscosity which is a fundamental parameter in the phenomenological theories of glass-formation and crystallization
(cf. Turnbull[40],Bagley[42]). If the viscosity close to T_m
is small the crystallization is difficult to overcome,
and this is the case of metallic liquids. If the viscosity is high then it is easier for the liquid to be
supercooled without crystallization, and eventually pass
through T_g and form a glass. From this point of view the
question to ask is what atomic arrangements make the viscosity high.

I cannot go in any detail into this difficult problem with often conflicting experimental data. Let me
only mention an effect of interest for the glass formation: At temperatures much higher than T_m, log η is found
to be a linear function of T^{-1}. This agrees with a simple theory of viscosity[9] (cf Section 3). In most liquids
(except SiO_2 and GeO_2) one observes a deviation from this
behavior at T close to T_m (and below T_m is supercooled
liquids): η is here higher as shown in Fig. 3.

One explanation of this excess of viscosity (others
are discussed in Ref. 40 and 41) has been ascribed to the
formation of clusters when one approaches T_m[10]. Complex
liquids tend to have a higher viscosity, and to form clusters. However, the formation of clusters does not necessarily imply that the liquid cannot crystallize. On the
contrary, the theory of crystallization is based on the
idea that the crystallization starts if the energy gained
by the correlations effects (long-range order) is larger
than the interfacial energy between the cluster and the
melt; this situation is reached for a certain minimum volume of the cluster.

To overcome this difficulty, it has been suggested[30]
that the clusters are non-crystallizable. This means that
the basic structural element cannot be repeated to form
a periodic lattice.[30,31] There are several examples of
such cases. E.g. Grigorovici[11] showed that a very thin
film of Ge (a few atomic layers) has a lower energy if
the atomic arrangements contain 5 fold rings than if it
has the regular crystallizable structure with 6 fold rings.
Of course if the film becomes thicker, the regular arrangement is energetically more favorable. But if the

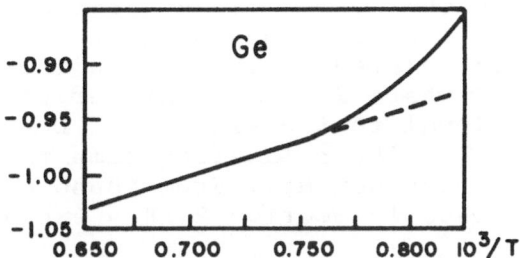

Fig. 3 Temperature dependence of the logarithm of the
kinematic shear viscosity of molten Ge (after Ref. 2).

viscosity is high, the material cannot rearrange itself
into the state of minimum energy. So by depositing Ge
a few atomic layers at a time we obtain amorphous Ge films
containing five fold rings. A similar process may be effective in quenching a liquid with non-crystallizable
clusters which are formed when we approach T_m. Simultaneously, their formation increases η so that the rebuild-
ing of the clusters from the "non-crystallizable" form
into the energetically more favorable state of "crystallizable" form is no longer possible.

From these considerations one expects that glass-
forming liquids are not "quasi-crystalline" in the same
sense that the structure of amorphous solids is in general not quasi-crystalline. For this reason, for glass-
forming liquids it is of primary importance to study the
clustering (correlation effects).

3. ATOMIC MOTIONS IN LIQUIDS

Atoms in a solid vibrate about their equilibrium
positions which are fixed in space. In a liquid, the
atoms (or molecules) undergo also a translational motion.
The feasibility of this motion is the basic difference
between the liquid and the solid.

In simple liquids, it is perhaps meaningful to ima-
gine that a spherical molecule weakly interacting with
other molecules via van der Waals forces moves freely for
a certain period of time between two collisions. However,
for the kind of liquids we have in mind, it is much more

appropriate to assume that the atoms (or molecules) have certain possible positions, and they jump from one position to another by overcoming a potential barrier. The time they spend on the average in a position is the relaxation time τ_o; the frequency of jumps is $1/\tau_o$. This relaxation time is related to the self-diffusion coefficient $D \approx \ell^2/6 \tau_o$, where ℓ is the barrier width. D is inversely proportional to the viscosity $D \approx kT/\ell\eta$, and consequently $\tau_o \sim \eta$. The relaxation time $\tau_o \sim \exp W/k_BT$ where W is the barrier height. From these considerations the relationship used in Section 2, $\eta \sim \exp W/k_BT$ follows.

For times shorter than τ_o the liquid is unable to adjust itself to external stresses and behaves mechanically as a solid. The same conclusion holds for the high-frequency effects, in particular the optical modes of vibration as long as their frequency $\nu_{opt} \gg \tau_o^{-1}$.

Another simple case is effects at low frequencies ($\nu \ll \tau_o^{-1}$) since in this case we may assume that the liquid is everywhere in equilibrium, and the fluctuations obey thermodynamic and hydrodynamic equations.

In complex liquid there are usually several relaxation mechanisms with different relation times τ_o. Before we start discussing the light scattering in liquids we will introduce some concepts[4] which can be used to describe the atomic motion in liquids, and which will be useful for the further discussion.

From the scattering experiments on amorphous solids one determines the radial distribution function $g(r)$ which is the probability of finding two atoms (or molecules) at a distance $r = |\vec{r}|$:

$$\rho_o g(r) = \frac{1}{N} < \sum_{i \neq j=1}^{N} \delta(\vec{r} + \vec{r}_i - \vec{r}_j)> \tag{1}$$

where N is the number of particles in volume V, $\rho_o = N/V$ is the average particle density, $\vec{r}_j$, $\vec{r}_i$ are the positions of the particles and the brackets denote the average over the whole system. The correctness of the formula can be seen intuitively because the δ-function makes a unit contribution for whenever $\vec{r} = \vec{r}_j - \vec{r}_i$. The normalization factor is chosen so that for $|\vec{r}| \to \infty$ $g(r) \to 1$.

If the positions of particles change with time we must consider a generalized function $G(r,t)$ (Van Hove correlation function) which is proportional to the

probability of finding the particle at $\vec{r}_o + \vec{r}$ at time t
if it was at $\vec{r}_o$ at t = 0:

$$G(\vec{r},t) = \frac{1}{N} < \sum_{i,j=1}^{N} \delta[\vec{r} + \vec{r}_i(o) - \vec{r}_j(t)]> \qquad (2)$$

where the averaging is over the system in thermal equilibrium. The formula holds for a classical system for
which we have no problems with the commutability of operators.

For t = 0

$$G(\vec{r},t) = \delta(r) + \rho_o g(r) \qquad (3)$$

The delta function corresponds to the terms i=j (self-
correlation of a particle). For $|\vec{r}| \to \infty$ and/or
$|t| \to \infty$ $G(\vec{r},t) \to \rho_o$. If we use the particle density
function

$$\rho(\vec{r},t) = \sum_{i=1}^{N} \delta[\vec{r} - \vec{r}_i(t)] \qquad (4)$$

we can write:

$$\rho_o G(\vec{r},\tau) = < \rho(0,0) \rho(\vec{r},t)> . \qquad (5)$$

This shows that $G(\vec{r},t)$ is related to the density fluctuations, both in space and time. The correlation function (5) is written in a short-hand form for the equiva-
lent form $< \rho(\vec{r},t') \rho(\vec{r}' + \vec{r}, t' + t)>$ since it can be
shown[23] that the correlation function depends only on the
differences $\vec{r}' - \vec{r}$, $\vec{t}' - \vec{t}$.

4. INTENSITY OF SCATTERED LIGHT

It is known that a medium in which the dielectric
constant ε is the same everywhere does not scatter electromagnetic radiation if $\lambda \gg a$ (= lattice constant).
Light is scattered by fluctuations of ε of the medium.

Van Hove[12] developed a general theory of scattering
processes based on the correlation function $G(\vec{r},t)$. We
will use it here only in a simple formulation in which
quantum effects are neglected.

For simplicity, let us first assume that ε is a function of density ρ only. We can then calculate the scattered radiation from Maxwell's equations. Electric field of the incident electromagnetic wave is

$$\vec{E}_i = \vec{E}_{io} \ \exp[i\vec{k}_i\vec{r} - i\omega_i t] \ . \tag{6}$$

The spectral density I_s of the scattered wave at a distance $R = |\vec{R}|$ large in comparison with the dimensions of V is given

$$I_s(\vec{k}_s, \ \omega_s) = \frac{I_i}{R^2} \ (\frac{d\varepsilon}{d\rho})^2_{\rho = \rho_o} \ \frac{\sin^2\psi}{16\pi^2 c^4} \ \omega_s^4 \ N \ S(\vec{k}_s, \ \omega_s)$$

$$S(\vec{k}_s, \ \omega_s) = \frac{1}{2\pi} \ \int \ (G(\vec{r},t) - \rho_o) \ \exp(-i\vec{K}\cdot\vec{r} - i\omega t)d^3 r dt \tag{7}$$

where ρ_o is average particle density, $\vec{k}_s$ is the propagation vector of the scattered wave and ω_s its frequency, ψ is the angle between $\vec{E}_{io}$ and $\vec{k}_s$, $\vec{K} = \vec{k}_s - \vec{k}_i$ is the scattering vector and $\omega = \omega_s - \omega_i \ (<<\omega_i)$. I_i is the intensity of incident radiation in volume V containing N atoms (molecules).

Eq. (7) is a simple example of the general result of Van Hove that the scattered radiation is determined by the space-time Fourier transform of the Van Hove correlation function. Since

$$< \rho(0,0) \ \rho(\vec{r},t) \ > \ = \ <\Delta\rho(0,0) \ \Delta\rho(\vec{r},t)> + \ \rho_o^2$$

equation (7) for the scattering function $S(\vec{k}_s,\omega)$ can be written in the following form:

$$NS(\vec{k}_s, \ \omega_s) = \frac{V}{2\pi} \int <\Delta\rho(0,0)\Delta\rho(\vec{r},t)> \ \exp(-i\vec{K}\cdot\vec{r} - i\omega t)d^3 r dt \tag{8}$$

We can rewrite Eq. (7) and (8) in another form with $\Delta\varepsilon = \varepsilon(\rho) - \varepsilon(\rho_o)$:

$$I_s(\vec{k}_s,\omega_s) = \frac{I_i}{R^2} \ \frac{\sin^2\psi}{32\pi^3 c^4} \ \omega_s^4 \ V\int < \Delta\varepsilon(0,0) \ \Delta\varepsilon(\vec{r},t)>$$

$$\exp(-i\vec{K}\cdot\vec{r} - i\omega t) \ d^3 r dt \tag{9}$$

This formula should be valid in the classical approximation even if $\Delta\varepsilon$ is not a function of $\Delta\rho$ only.

For unpolarized light we must average over all angles ψ keeping the angle Θ between the incident and scattered beams fixed.

This results in replacing $\sin^2\psi$ in Eq. (7) and (9) by its average[13]

$$<\sin^2\psi> = \frac{3}{4}(1 + \cos^2\Theta) \tag{10}$$

We shall apply Eq. (7) and (9) for discussing the Rayleigh, Brillouin and Raman scattering in liquids. The first two cases correspond to the situation when the frequency of fluctuations in ε is smaller than τ_0^{-1}. For the Rayleigh scattering the fluctuations do not propagate, for the Brillouin scattering the modes propagate as sound waves (collective motion in the liquid). The frequency of the Raman scattering is usually considerably above τ_0^{-1} and there is not much difference between the amorphous solid and the liquid, except for the interaction with the low frequency modes characteristic for the liquid.

5. RAYLEIGH SCATTERING

If we have "frozen-in" fluctuations of ε in the solid then ε is a function or $\vec{r}$ only and in the correlation function in Eq. (9) we have to consider the coordinates only. The frequency of the radiation scattered by frozen-in inhomogeneities is equal to the frequency of incident radiation $\omega_s = \omega_i$, $\omega = 0$.

The correlation of the fluctuations of ε extends over a certain distance called the correlation length Λ. Fluctuations at two sites $\vec{r}_1$ and $\vec{r}_2$ for which $|\vec{r}_1 - \vec{r}_2| \gg \Lambda$ are uncorrelated, and the correlation function is zero. From this consideration it is easy to see then if $\Lambda \ll \lambda_{light}$ the phase factor $e^{i\vec{K}\cdot\vec{r}}$ in Eq. (9) is equal to one. This is a good approximation if $\Lambda < \lambda_{light}/20$; this means for visible light $\Lambda < 200\text{Å}$. The correlation function then gives a number

$$(\Delta\varepsilon)^2 = V^{-1} \int (\Delta\varepsilon)^2 d^3\vec{r}$$

and the angular dependence is given by Eq. (10). The

total scattered intensity for unpolarized light is obtained from Eq. (9) and (10) by integration over all angles Θ:

$$I_{s,tot} = I_i \frac{\omega^4}{6\pi c^4} \; V <(\Delta\varepsilon)^2> \qquad (11)$$

These results[14] are often expressed in a different way. The scattering produces an attenuation of the I_i beam when it passes through the medium, in addition to the absorption process. The attenuation constant α_{sc} can be defined from the relationship for a beam with cross section A passing in x-direction

$$I_{s,tot} = \frac{\alpha_{sc} I_i dV}{A} = \alpha_{sc} I_i dx \qquad (12)$$

where $dV = Adx$ is the scattering volume. For α_{sc} we obtain from Eq. (11)

$$\alpha_{sc} = \frac{\omega^4}{6\pi c^4} \; V <(\Delta\varepsilon)^2>. \qquad (13)$$

It seems strange that α_{sc} should depend on the illuminated volume V. In thermodynamical equilibrium $<(\Delta\varepsilon)^2> \sim V^{-1}$ and the volume cancels out. However, we may have a medium with non-homogeneities which do not correspond to thermal equilibirum, e.g. we can have a material with the dielectric constant $\varepsilon + \Delta\varepsilon$ embedded in a material with the dielectric constant ε. Then the scattered radiation will depend on the average volume of the clusters of the first material.

Light scattering in amorphous materials is very important in optical applications of glasses. In the last century, the discovery of methods for mass production of glasses with small scattering was essential for the development of the optical industry. In recent years, much lower levels of attenuation (absorption and scattering) have been required for the application of glasses for optical fiber communications. The studies[35-38] of attenuation have shown that losses by scattering can be very low in carefully prepared samples of As_2S_3 and some oxide glasses, and extremely low in SiO_2 glass (3×10^{-6} cm^{-1} at 1.06 μm[36,43]).

If the correlation length $\Lambda > \lambda_{light}/20$ the factor

$e^{i\vec{K}\cdot\vec{r}}$ cannot be neglected. Its presence produces an angular dependence of I_S superimposed on factor (10). In principle, by measuring this angular dependence we can learn about the spatial correlation function. As it is apparent from Eq. (9) the correlation function is the inverse Fourier transform of $I_S(\vec{K})$. This approach is of course limited to correlations extending over distances larger than $\lambda/20$ ($\approx 200\text{\AA}$). This method is used in macromolecular chemistry to determine the shape and other parameters of very large molecules[39]. For these investigations, it is necessary to relate the scattering to polarizabilities on an atomic scale, and, if necessary, make corrections for the interactions between atoms or molecules. In our equations, these interactions are included in the dielectric constant ε figuring in Eq. (9) which holds generally for classical systems.

In the liquid, the atoms (or molecules) are in motion, and the frequency of the scattered light is no longer exactly equal to ω_i. The Rayleigh line is broadened by the motion. If we measure the shape of the Rayleigh light we can learn something about this motion. Knowing $I_S(\vec{k}_s, \omega_s)$ as a function of $\omega (= \omega_s - \omega_i)$ we can determine the time-dependent part of the Van Hove correlation function which is the Fourier transform of $I_s(\omega)$.

The determination of $G(\vec{r}, t)$ as the Fourier transform of I_S is of fundamental importance for the understanding of the liquids, and may be useful for the understanding of the glass-transition in some systems. The attempts for the direct determination of $G(\vec{r}, t)$ are relatively recent.

The phenomenological way to describe Rayleigh scattering is based on thermodynamic equations. They are applicable if the thermodynamic qualities can be defined at each point in the liquid. This is the case if the frequency of the motion is small compared to τ_0^{-1} and if the wavelength of the motion is large compared to the interatomic distances.

The finite width of the Rayleigh line is associated with certain modes of motion which do not propagate in the liquid. As Frenkel[15] discusses in detail, the density fluctuations can be decomposed into two components (1) pressure (P) fluctuations at constant entropy and (2) entropy (S) fluctuations at constant pressure.

One obtains from thermodynamics[16]:

$$< (\Delta \varepsilon)^2 > = \left(\frac{\partial \varepsilon}{\partial P} \right)_S^2 < (\Delta P)^2 > + \left(\frac{\partial \varepsilon}{\partial S} \right)_P^2 < (\Delta S)^2 > \qquad (14)$$

$$\left(\frac{\partial \varepsilon}{\partial P} \right)_S^2 < (\Delta P)^2 > = \left(\rho \frac{\partial \varepsilon}{\partial \rho} \right)_S^2 \frac{k_B T \beta_S}{V} \qquad (15)$$

$$\left(\frac{\partial \varepsilon}{\partial S} \right)_P^2 < (\Delta S)^2 > = \left(T \frac{\partial \varepsilon}{\partial T} \right)_P^2 \frac{k_B}{C_{p^0} {}_0 V} \qquad (16)$$

where C_p is the specific heat at constant pressure, $\beta_S = \left(\frac{1}{0} \frac{\partial \rho}{\partial P} \right)_S$ the adiabatic compressibility. The first term corresponds to the two Brillouin components, the second term to the Rayleigh line (Fig. 4). For this component we may write

$$\left(\frac{\partial \varepsilon}{\partial T} \right)_P = \left(\rho \frac{\partial \varepsilon}{\partial \rho} \right)_T \alpha + \left(\frac{\partial \varepsilon}{\partial T} \right)_P \qquad (17)$$

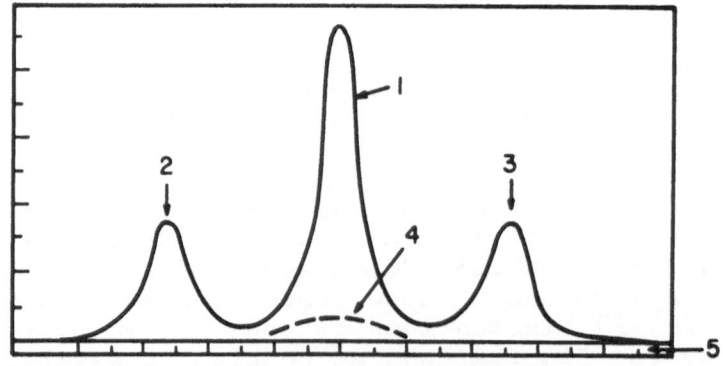

Fig. 4 Rayleigh and Brillouin Scattering:
1--Rayleigh line 2,3 -- Brillouin doublet
4--Mountain component 5--Background

where α is the coefficient of cubic expansion. The second term is usually much smaller than the first one, and we will neglect it.

When $<(\Delta\epsilon)^2>$ from Eq. (14) is put into Eq. (11) we obtain the total scattered intensity. An important relation can be derived from the above thermodynamic equations:

$$\frac{I_{Rayleigh}}{2I_{Brillouin}} = \frac{C_P}{C_V} - 1 \tag{18}$$

where C_V is the specific heat at constant volume.

The pressure fluctuations (first term in Eq. (14)) propagate in the liquid and give rise to the Brillouin scattering which is discussed in the next section. The entropy fluctuations (second term in Eq. (14)) do not propagate in normal liquids and are dispersed by thermal conductivity. We shall discuss them now.

The incident wave and the scattered wave have different wave vectors. The fluctuation which they produce will have zero real part of frequency but a non-zero wave vector $\vec{K} = \vec{k}_s - \vec{k}_i$. The spatial dependence of their amplitude will be $\exp i\vec{K}.\vec{r}$. As $|\vec{k}_s| = |\vec{k}_i|$ one obtains for $K = |\vec{K}|$:

$$K = 2k_i \sin\theta/2 = 2 \frac{\omega_i c}{n} \sin\theta/2 \tag{19}$$

where θ is the angle between the directions of the incident and scattered beams (between $\vec{k}_s$ and $\vec{k}_i$). The index of refraction n figures in the expression since $\vec{k}_s$ and $\vec{k}_i$ are electromagnetic waves in a medium with $n \neq 1$. The return to equilibirum of these fluctuations is governed by the thermal diffusion law

$$(\frac{\partial\rho}{\partial t})_P = D_{th} \nabla^2 \rho \tag{20}$$

where D_{th} is thermal diffusivity $= K_{th}/\rho_o C_p$ (K_{th} is thermal conductivity). Let's assume that

$$\Delta\rho(\vec{r},t) = (\Delta\rho)_o \exp(i\vec{K}.\vec{r}-\gamma t) \tag{21}$$

These fluctuations will decay as $e^{-\gamma t}$. If we put Eq. (21) into Eq. (20) we determine $\gamma = D_{th}K^2$. K is given by

Eq. (19).

The Fourier transform of the time dependent part

$$I_s \sim \frac{1}{\pi} \frac{\gamma}{(\omega_s-\omega_i)^2 + \gamma^2} \tag{22}$$

This means that the Rayleigh line is centered about ω_i and has a finite width

$$\gamma = D_{th} \, 4\frac{\omega^2 n^2}{c^2} \, \sin\theta/2.$$

By measuring it we can determine the thermal diffusivity of the liquid. The integral over $\omega_s-\omega_i$ from 0 to ∞ is equal to one regardless of the value of γ.

We shall now consider the spatial part of the correlation function in the thermo-hydrodynamic theory. We shall essentially follow Van Kempen's analysis[17]. In a liquid far from the critical point the spatial correlations extend only over distances small compared to λ_{light}. In Eq. (8) it is sufficient to consider only the $K = 0$ components of the Fourier transform of $<\Delta\rho(0)\Delta\rho(\vec{r})>$ for which it holds

$$\int <\Delta\rho(0)\Delta\rho(\vec{r})>d^3\vec{r} = \rho_o \, k_B T(\frac{d\rho}{dP})_T V^{-1} \tag{23}$$

where k_B is the Boltzman constant. For the total intensity of the Rayleigh line we obtain from Eq. (11)

$$I_{s,tot} = I_i \frac{\omega^4}{6\pi c^4} \cdot (\frac{d\varepsilon}{d\rho})^2 \cdot \rho_o k_B T(\frac{d\rho}{dP})_T \tag{24}$$

However, when we approach the critical temperature, the correlations in the densities extend over large distances, and it becomes necessary to consider also $K > 0$, as the following analysis shows.

The long-range correlations in the classical Van der Waals theory of condensation are described by

$$<\Delta\rho(0)\Delta\rho(r)> = \frac{k_B T}{2\pi b} \frac{e^{-r/\Lambda}}{r} \tag{25}$$

where b is a constant. The correlation length Λ obeys the relation

$$\Lambda^2 = \frac{1}{2} \rho_o \, b \left(\frac{\partial \rho}{\partial P}\right)_T , \qquad (26)$$

and hence tends to infinity at the critical point. The Fourier transform of this correlation function is

$$I_s \sim \frac{k_B T}{2\pi b} \cdot \frac{4\pi}{K^2 + \Lambda^{-2}} \cdot \qquad (27)$$

When $\Lambda \ll 1/K$ (a short correlation length) we can neglect K^2 in the denominator and we obtain again Eq. (24). Using Eq. (26) we can write Eq. (24) in the form

$$I_{s,tot} = I_i \, \frac{\omega^4}{6\pi c^2} \left(\frac{d\varepsilon}{d\rho}\right)^2 \cdot \frac{2k_B T}{b} \cdot \Lambda^2 \qquad (28)$$

which shows that when Λ is much smaller than the wavelength of light the intensity increases proportional to Λ^2.

For larger values of Λ Eq. (27) must be used which saturates for $\Lambda > K^{-1}$ (that is when the correlation lengths becomes comparable with the wavelength of light). Since K is small we obtain a very large intensity for $\Lambda \to \infty$.

When Λ becomes large the conditions for the applicability of hydrodynamics cease to be valid, but the theory gives a reasonable description of the observed effects when the temperature approaches the critical point, T_c. In particular, it shows that the intensity dramatically increases (critical opalescence), and it correctly describes the physical reason for it: increase of the average scattering volumes. The Rayleigh line also sharpens significantly when $T \to T_c$ since its width $\gamma = K_{th}/\rho_o C_p$, and C_p diverges at T_c (usually more strongly than K_{th}).

Another non-propagating mode in a liquid was considered by Mountain[18]. It is related to the relaxation processes associated with the internal degrees of freedom of molecules. It has been observed as a line centered at $\omega = 0$, usually much broader than the Rayleigh line (Fig. 4).

Experimental studies of the Rayleigh line are difficult because its width is very small (usually much less than 100 MHz). It has been made possible by a special technique, optical mixing spectroscopy[19,21]. The incident and scattered radiation with slightly different frequencies fall on a nonlinear detector, and the beat frequency is amplified and recorded (heterodyne principle). A modification of this method is self-beating spectroscopy (homodyne principle). In this case only scattered light is allowed to fall on the detector which mixes the spectral components of the signal itself. A resolution of 10^{14} (relative to the incident beam frequency) has been attained in this method.

These techniques have been worked out primarily for studying the Rayleigh line in the vicinity of the critical point where the total intensity increases and the line sharpens significantly. For studies close to the glass-transition, the line is expected to be rather broad. Some correlation processes such as formation of clusters close to T_g as discussed in section 2 might produce fluctuation in $\Delta\varepsilon$ and would enhance the intensities. However, if the structural differences between the clusters and the medium are such that ε is the same in both, no scattering will be produced. This seems to be the case for many glasses.

6. BRILLOUIN SCATTERING

Brillouin scattering is associated with collective motions in the liquid. These are waves propagating through the liquid with a certain wave vector $\vec{k}$ and frequency Ω. Assuming that $\Omega \ll \tau_0^{-1}$ these motions can be described by hydrodynamic equations as fluctuations of density of constant entropy (thermal sound waves). The correlation function of undamped propagating waves is proportional to

$$e^{i\vec{k} \cdot \vec{r} - i\Omega t}$$

and its Fourier transform is proportional to

$$\delta(\vec{k}_s - \vec{k}_i \pm \vec{k})\delta(\omega_s - \omega_i \pm \Omega) \tag{29}$$

which gives conservation of the $\vec{k}$-vector and conservation of energy

$$\vec{k}_s = \vec{k}_i \mp \vec{k} \tag{30}$$

$$\omega_s = \omega_i \mp \Omega \tag{31}$$

We can interpret these conservation laws as due to the interferences between the fluctuations $\vec{K}$, ω produced by the scattered light wave, and the sound waves $\vec{k}$,Ω which is constructive only when $\vec{K} = \pm\vec{k}$, $\omega = \pm\Omega$. We therefore observe a doublet symmetrically situated about ω_i (Fig. 5) The dispersion relation for sound waves with phase velocity v_s is

$$\Omega = v_s k \tag{32}$$

As $\Omega \ll \omega_i$ Eq. (30) again reduces Eq. (19) and we obtain for the frequency Ω observed in an experiment with fixed Θ

$$\Omega = \frac{2v_s \omega_i n}{c} \sin \Theta/2 \tag{33}$$

In reality, the waves are damped. Similarly as in Rayleigh scattering, the correlation function contains the factor $\exp(-\gamma t)$. The shape of the line is again described by Eq. (22), however, the lines are centered about $\omega_i = \Omega$ and $\omega_i + \Omega$ respectively. Their width is determined by $\gamma = \frac{1}{2} D_s K^2$ where K is given by Eq. (19) and

$$D_s = D_{th}(C_p/C_v - 1) + \rho_o^{-1}(\frac{4}{3}\eta + \xi) \tag{34}$$

where η and ξ are shear and bulk viscosity[22,34]. With increasing viscosity the Brillouin lines broaden and it is difficult to observe Brillouin scattering in high viscosity liquids.

We shall discuss the spatial damping by considering a wave propagating in x-direction with the correlation function proportional to $\exp(ikx - x/\Lambda_B)$. Its Fourier transform is

$$\frac{1}{\pi} \frac{\Lambda_B^{-1}}{(K_x - k)^2 + \Lambda_B^{-2}} \tag{35}$$

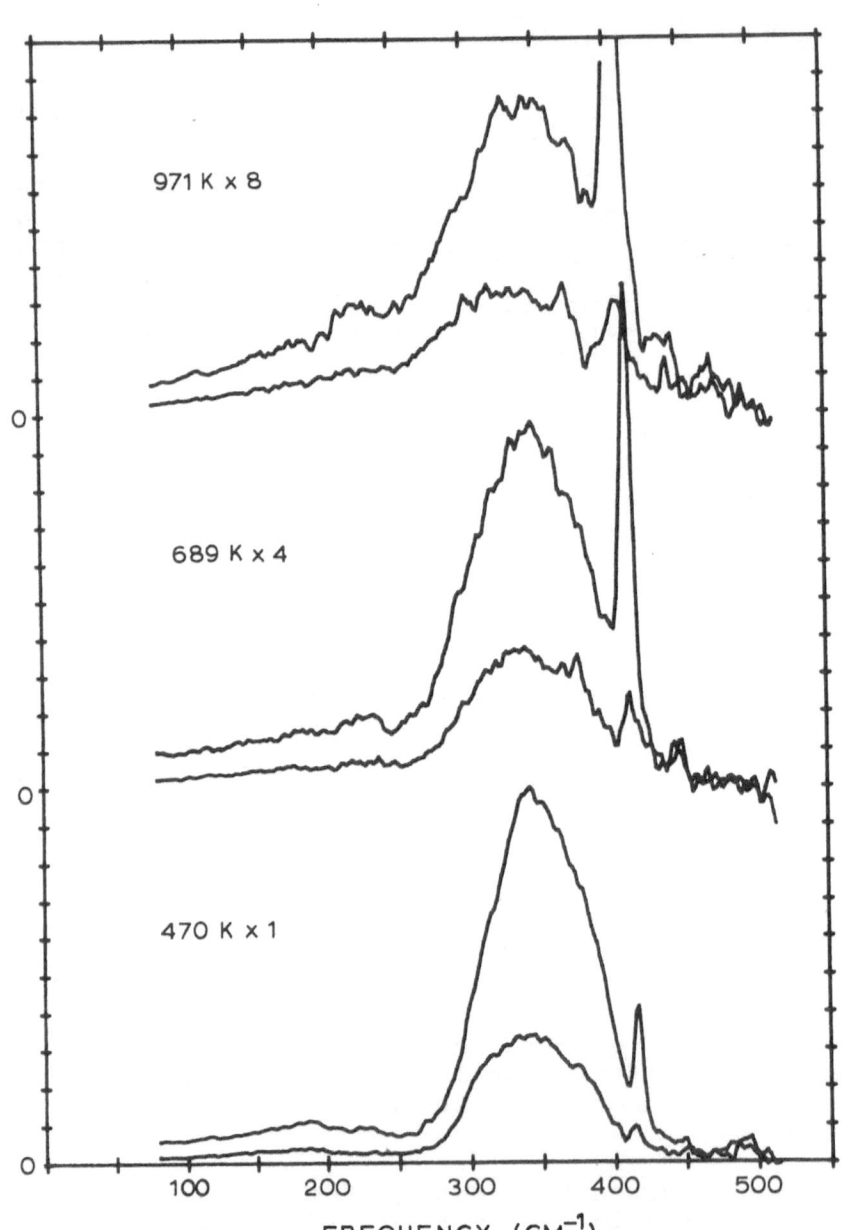

Fig. 5 Polarized Raman spectra of liquid As_2S_3. The intensities were reduced as explained in the text. The upper curves correspond to the VV, the lower curves to the
HV polarization configurations. The intensities are in
arbitrary units, however, to compare different temperatures
one must multiply them by the factors shown in the figure.
The sharp lines are the sapphire lines which originate
from the window (after Ref. 25).

K_x is the x-component of the scattering vector $\vec{K} = \vec{k}_s - \vec{k}_i$; the y and z components are unchanged in this scattering process. For $\Lambda_B \rightarrow \infty$ Eq. (35) gives the δ-function in Eq. (29).

If Λ_B is finite, the δ-function broadens, the radiation couples to a band of k-values about K_x. However, the total intensity of the line is an integral $\int\limits_0^\infty dk$

over the factor (35), and this integral gives one regardless of the values Λ_B. Therefore, the intensity of the Brillouin line does not depend on the correlation length, in sharp contrast to the Rayleigh line. This conclusion is in agreement with experimental results. The spectacular increase in intensity and the sharpening of the Rayleigh line when one approaches the critical point is absent in Brillouin scattering (although one may observe some sharpening of the Brillouin line with increasing temperature which is associated with the decrease of viscosity).

These conclusions follow also from Eq. (18) deduced for the hydrodynamic theory. The ratio C_P/C_V diverges at $T \rightarrow T_c$, and the Rayleigh line becomes more and more prominent.

In the Rayleigh scattering the $\vec{k}$-vector of the fluctuations in the medium is fixed ($=\vec{K}$). If we could measure the Brillouin scattering with a fixed $\vec{k}$, we may observe the dependence of the intensity on the correlation length Λ_B.

However, even in this case we would not see anything similar to the critical opalescence. The reason is that Λ_B is different from Λ. The correlation length Λ is associated with the dimensions of density fluctuations, and the critical opalescence is due to the fact that the electrical polarizability of larger domains is larger, and therefore, the light scattering is larger. However, when light is scattered on acoustic waves, the correlation length Λ_B characterizes the coherence of the waves (the average distance over which the phase relationship of vibrations is maintained). This length is obviously not expected to increase when $T \rightarrow T_c$.

Our discussion of the scattering processes has been simplified to make the basic processes clear. For more complicated liquids, even the starting equation (7) must be written in a more general form taking into account the

local structural asymmetries of the liquid. In such liquids, there may be several relaxation processes associated with orientational and structural changes, and these
would be reflected in the generalized correlation function $G(\vec{r},t)$. Let us also note that in crystals the Brillouin scattering gives more lines than two because in
the solid also transverse waves can be propagated. These
waves were observed also in amorphous solids[16] (e.g.
fused quartz). The velocity and damping of sound in both
crystalline and amorphous solids have been recently discussed on the basis of a new theory of hydrodynamics of
solids.[32,33] This theory also suggests an additional
contribution to Rayleigh scattering due to vacancy diffusion.

7. RAMAN SCATTERING

In a crystal, Raman scattering is associated with
optical phonons with $\vec{k} \approx 0$. In an amorphous solid, one
can couple also to phonons with $k \neq 0$; therefore, acoustic phonons can contribute also. In a semiconducting
melt Raman scattering is produced by motions whose frequencies are much larger τ_0^{-1}. At these frequencies the
melt resembles an amorphous solid, and the problems of
interest are the changes in the Raman spectra produced
by interactions of the low frequency modes[e] with the high
frequency vibrational modes. If these interactions are
understood, Raman spectroscopy may provide a probe for
studying these interactions. Raman spectra are also sensitive to structural changes in the liquid.

The classical thermodynamics and hydrodynamics are
not applicable and quantum mechanical approach must be
used. Nevertheless, we shall assume that a modified
Eq. (9) is still applicable, although there are cases when
the proper quantum mechanical approach is necessary.

In a crystal the correlation function in Eq. (9) is
replaced by

$$G_{\alpha\beta,\gamma\delta}(\vec{r},t) = \langle \Delta\varepsilon_{\alpha\beta}(\vec{r}',t')\Delta\varepsilon_{\gamma\delta}(\vec{r}' + \vec{r}, t' + t)\rangle$$

(36)

where $\Delta\varepsilon_{\alpha\beta}$ is the fluctuation of the local dielectric tensor of the medium produced by vibrations (phonons). Ob-
viously, in this case $\Delta\varepsilon$ cannot be simply calculated from
the density fluctuations. It is proportional to the

displacement of atoms from the equilibrium positions, and the proportionality factor is determined by the change of the electrical polarizability produced by the displacement. The description of this process involves photon-electron, electron-phonon and electron-photon interactions.

The fluctuations in a crystal can be described as an assembly of plane waves $e^{i\vec{k}_j\vec{r} - i\omega_j t}$ with the quantum numbers $\vec{k}_j$, ω_j and s (= band index) which are completely independent of each other in the harmonic approximation. As previously, the time and space correlation can be calculated independently.

The time correlation function is determined from the quantum mechanical formula for an operator $A(t)$:

$$\langle A(0)A(t)\rangle = \frac{1}{Z}\ \mathrm{Tr}\ e^{-\beta H}Ae^{itH/\hbar}Ae^{-itH/\hbar} \tag{37}$$

where H is the Hamiltonian of the system, $Z = \mathrm{Tr}\ e^{-\beta H}$, $\beta = 1/k_B T$. For phonons, the operator A is phonon creation or annihilation operator, and the eigen energies of the Hamiltonian are $\hbar\omega$. The calculation is straightforward[23], and gives

$$\langle e^{i\omega_j t'} e^{i\omega_j(t'+t)}\rangle =$$

$$\frac{\hbar}{2\omega_j}[n(\omega_j)e^{i\omega_j t} + [1 + n(\omega_j)]e^{-i\omega_j t}] \tag{38}$$

where

$$n(\omega_j) = [\exp(\hbar\ \omega_j/k_B T - 1)]^{-1} \tag{39}$$

(Planck distribution function). The Fourier transform of this gives $\delta(\omega-\omega_j)$ or $\delta(\omega+\omega_j)$ weighted with the factors $\omega_j^{-1}n(\omega_j)$ (anti Stokes, $\omega_s > \omega_i$) or $\omega_j^{-1}(n(\omega_j) + 1)$ (Stokes, $\omega_s < \omega_i$). These factors are temperature dependent and are related to the population of the phonon states. We may divide the measured Raman intensities of the Stokes or anti-Stokes spectrum by them. The "reduced" Raman spectrum will then correspond to the spatial part of the correlation function only.

For the spatial part it is easy to see that it is proportional to $\delta(\vec{K} - \vec{k}_j)$; the scattering process involves only those phonons whose $\vec{k}$ vector is equal to the scattering vector $\vec{K}$.

Let us consider a branch of vibrations in a certain direction in the k-space. The k-vectors span the values $k = 0$, π/L, $2\pi/L$ up to the maximum value at the first BZ boundary π/a. L is the dimension of the sample in the corresponding direction in real space, a is the lattice constant. We do not use negative values for k (we consider rigid boundary conditions instead of the periodic boundary conditions). There are L/a vibrations in the branch, however, when L/a >> 1 light couples only to vibrations with $k = K \approx 0$.

If we have a crystalline sample, the waves can be described by wavepackets. In the k-space we will need an interval $\Delta k \approx L^{-1}$ to describe a wave localized over a distance L. In this case, all waves with $k_j < \Delta k$ will significantly contribute to the scattering since each wave packet centered about k_j has a component at K. In the extreme case $L \approx a$ all k-vectors in the BZ contribute to the scattering, with different weighting factors, of course.

A mathematically simple description of waves localized over a distance L is a damped wave with the correlation function $\exp(ik_j x - x/\Lambda)$ where $\Lambda \approx L$. Its Fourier transform is given by Eq. (35).

$$J_j = \frac{1}{\pi} \frac{\Lambda^{-1}}{(k_j - K)^2 + \Lambda^{-2}} \qquad (40)$$

Shuker and Gammon[24] suggested that such waves be a rough description of the Raman spectra in amorphous solids. If $\Lambda \approx a$ then each mode k_j contributes with almost the same weight and the spectrum approximately reflects the density of states of the corresponding crystal as discussed in paper.[41]

We shall now consider the difference between the amorphous solid and the liquid. We did some experiments[25] on liquid As_2S_3 and As_2Se_3 whose vibrational spectra are discussed in paper[41]. The experimental facts which we observed can be summarized as follows:

In both materials, the width of the main Raman band

changes very little when one passes through T_g; at higher
temperatures it increases somewhat (Fig. 5). The total
Raman intensity in the band decreases with temperature
as $\exp E/k_B T$:

In As_2S_3[35], $E \approx 0.18eV$ (above T_g),   in As_2Se_3[26]

$E \approx 0.14eV$ (above T_m).

These results show that the high frequency vibrations
in these materials are similar in the melt as in the solid. This justifies our expectation that the liquid behaves as a solid for frequencies $\gg \tau_0^{-1}$. The decrease
of the total band intensity may be due either to structural changes in the liquid, and/or interactions of the
high frequency modes with the low frequency modes[27]. The
constancy of the band width (Fig. 6) appears to rule out
a strong influence of anharmonicity.

Another effect which we have been studying is the
laser catalyzed crystallization of As_2Se_3 which can be
continuously followed by Raman spectroscopy[28]. Figure
7 shows the development of the crystalline spectrum from
the amorphous spectrum of the melt kept at T between T_g
and T_m. The changes in the spectra suggested that they
might be due to the changes of the correlation length according to Eq. (40); for small $\Lambda (\approx a)$ the whole BZ contributes, with increasing Λ the contributions closest to
$k = 0$ become more important. However, a closer inspection
shows that the spectrum is a superposition of the crystalline I_c and amorphous bands I_a: $I_{total} = xI_c + (1-x)I_a$,
$(0 \le x \le 1)$. This means that Λ changes rapidly during the
initial stages of crystallization.

Let us conclude these lectures with a note on the
structure of liquid As_2S_3 and As_2Se_3. The analysis of
our results appears to support the similarity of the melt
of these materials to the amorphous solid. This means
that it is not an assembly of moving molecules but an interconnected network with atomic distances as required by
chemical bonding, probably with some remnants of the layer
structure in the low temperature range. The low viscosity of the liquid is due to the free volume in the melt
which makes it possible for the atoms to jump from one
potential minimum to another. However, in the liquid some
structural configurations may develop which have a higher
entropy and are less Raman active than the "crystal-like"
configurations, and these configurations may prevail at
high temperatures[25,27].

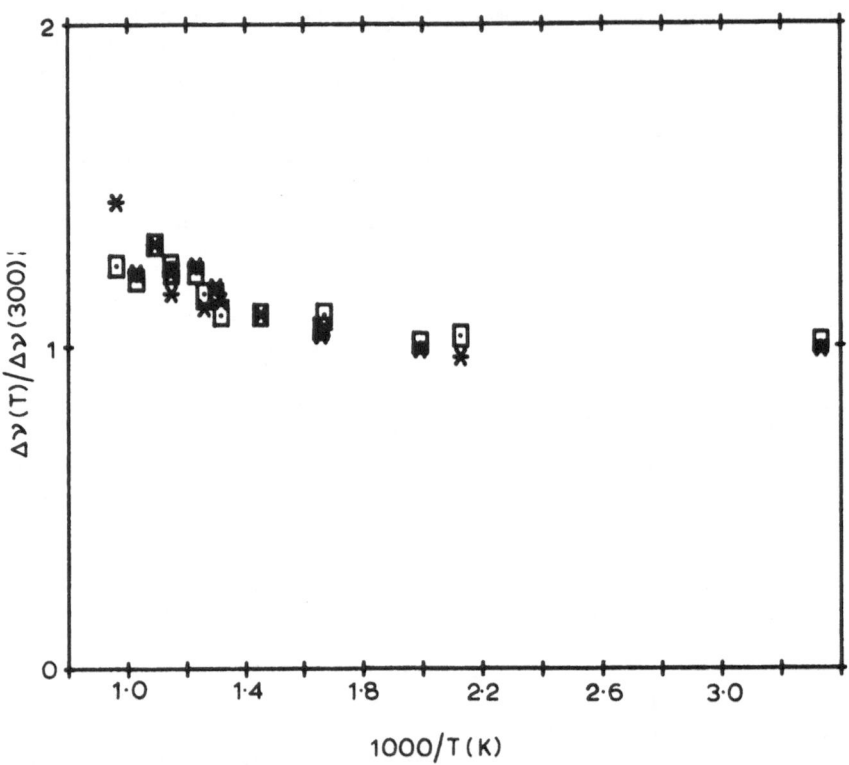

Fig. 6   Temperature dependence of the width of the dominant bands of a-As$_2$S$_3$.   * VV, □ HV polarization configurations.   The widths are normalized to their value at
300 K ($\Delta\nu(330)$ = 90 cm^{-1} at the half of the maximum reduced intensity).   (After Ref. 25).

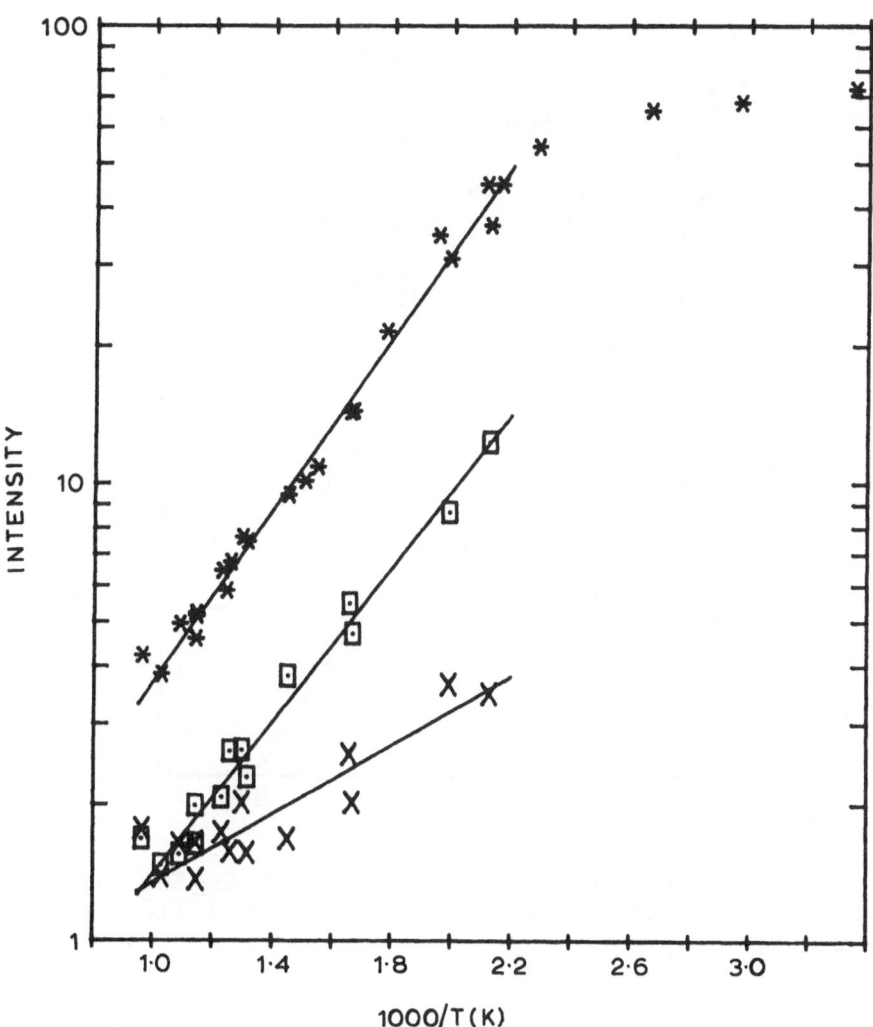

Fig. 7 Temperature dependence of the reduced Raman intensity (in arbitrary units) of the dominant peak in the
VV (*) and HV (□) polarization configurations. The lines
are least square fits to the experimental points. The
crosses correspond to the 230 cm^{-1} band (VV polarization
configuration). (After Ref. 25).

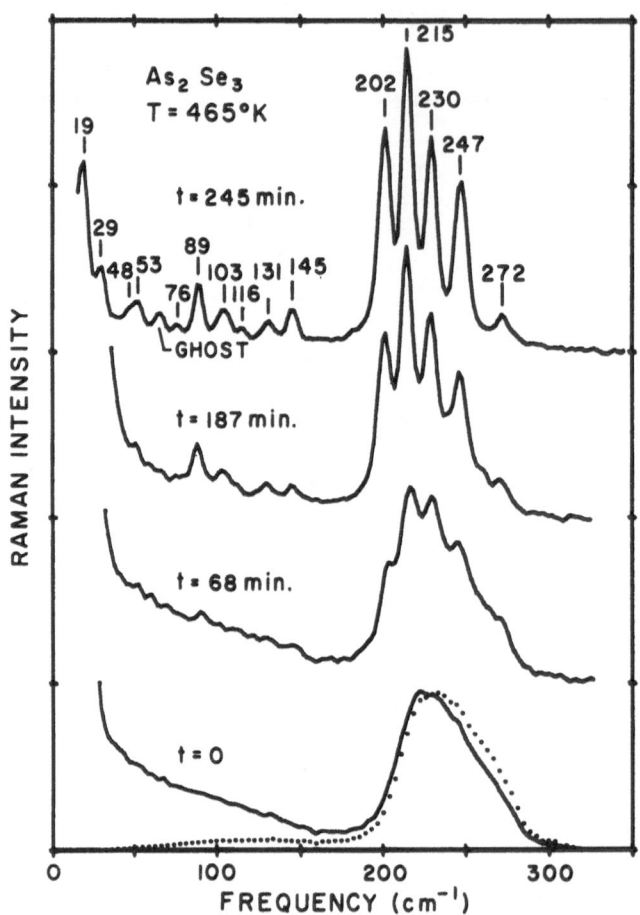

Fig. 8 Changes in the Raman spectrum of As2Se3 during
crystallization (after Ref. 28) (the dotted line is the
reduced spectrum).

ACKNOWLEDGMENT

     I acknowledge very helpful discussions with B. G.
Bagley and P. D. Fleming and their comments on the manuscript. I benefited also from discussions in the Inorganic
glasses group at Brown University, organized under the
Materials Research Laboratory program of the NSF.

     This paper is partly based on work supported by a

grant from the National Science Foundation in which E. Finkman and A. P. DeFonzo participated.

REFERENCES

1. J. E. Enderby, in Amorphous and Liquid Semi-Conductors (ed. J. Tauc), Plenum Press, London, 1974, p. 361.

2. V. M. Glazov, S. N. Chishevskaya and N. N. Glagoleva, Liquid Semiconductors, Plenum Press, New York 1969.

3. H. Rawson, Inorganic Glass Forming Systems. Acad. Press, New York 1967.

4. P. A. Egelstaff, An Introduction to the Liquid State, Acad. Press, New York 1967.

5. As J. E. Enderby[1] has pointed out, this criterion may be misleading[1]. In some binary metallic alloys $d\sigma/dT > 0$ while the Hall constant and thermoelectric power are constant. This indicates that these liquids are metallic (concentration of carriers is temperature independent) but the mobilities increase with T.

6. S. P. Isherwood, B. R. Orton and R. Manaila, J. Non-Crystalline Solids 8-10, 691 (1972).

7. S. Minomura et al., in Tetrahedrally Bonded Amorphous Semiconductors, (M.H. Brodsky and S. Kirkpatrick, editors), AIP 1974, p. 234.

8. W. W. Warren, Jr., J. Non-Crystalline Solids, 8-9, 241 (1972).

9. S. Glasstone, K.J. Laidler and H. Eyring, Theory of Rate Processes, McGraw Hill, New York 1941.

10. A. R. Ubbelohde, Melting and Crystal Structure, Clarendon Press, London 1965.

11. R. Grigorovici, in Amorphous and Liquid Semiconductors (ed. J. Tauc), Plenum Press, London 1974 p.45.

12. L. Van Hove, Phys. Rev. 95, 249 (1954); L. I. Komarov and I. Z. Fisher, Soviet Physics JETP 16, 1358 (1963); R. Pectora, J. Chem. Phys. 40, 1604 (1964).

13. L. D. Landau and E. M. Lifshitz, Electrodynamics of Continuous Media, Pergamon Press, Oxford 1960.

14. I. L. Fabelinskii, Molecular Scattering of Light, Plenum Press, New York 1968.

15. J. Frenkel, Kinetic Theory of Liquids, Oxford 1960.

16. R. S. Krishnan in the Raman Effect, Vol. 1 (ed. A. Anderson), Dekker, New York 1971.

17. N. G. Van Kampen, in Quantum Optics, (ed. R.J. Glauber), Acad. Press, New York 196 , p. 235.

18. R. D. Mountain, J. Nat. Bureau of Standards, 70A, 207 (1966).

19.   G. B. Benedek, in Statistical Physics, Phase Transi-
       tions, and Superfluidity, Vol. 2 (eds. M. Chretien,
       E. P. Gross and S. Deser), Gordon and Breach, New
       York, 1968, p. 1.
20.   G. B. Benedek, in Polarization, Matter and Radiation,
       Presses Universitaire de France, Paris 1969, p. 49.
21.   M. J. French, J. C. Angus and A. G. Walton, Science
       163, 345 (1969).
22.   H. E. Stanley, Introduction to Phase Transitions and
       Critical Phenomena, Oxford University Press, New
       York 1971.
23.   A. A. Maradudin, E. W. Montroll, G. H. Weiss and I.
       P. Ipatova, Theory of Lattice Dynamics in the Har-
       monic Approximation, Acad. Press, New York 1971
24.   R. Shuker and R. W. Gammon, Phys. Rev. Lett. 25,
       222 (1970).
25.   E. Finkman, A. P. DeFonzo and J. Tauc, in Amorphous
       and Liquid Semiconductors, (ed. J. Stuke and W.
       Brenig), Taylor and Francis, London 1974, p. 1275.
26.   E. Finkman, A. P. DeFonzo and J. Tauc, Bull APS 19,
       212 (1974).
27.   A. P. DeFonzo, Thesis (Brown University 1975).
28.   E. Finkman, A. P. DeFonzo and J. Tauc, Proc. 12th
       Intl. Conference on the Physics of Semiconductors,
       Teubner, 1974, p. 1022.
29.   J. Stuke, J. Non-Crystalline Solids 4, 1 (1970).
30.   F. C. Frank, Proc. Roy. Soc. (London) A215, 43
       (1952).
31.   B. G. Bagley, J. of Crystal Growth 6, 323 (1970).
32.   P. D. Fleming and C. Cohen, (to be published).
33.   C. Cohen, P. D. Fleming and J. H. Gibbs, (to be
       published).
34.   L. D. Landau and E. M. Lifshitz, Fluid Dynamics,
       Pergamon Press, London, 1959.
35.   D. L. Wood and J. Tauc, Phys. Rev. 5, 3144 (1972).
36.   T. C. Tich and D. A. Pinnow, Appl. Phys. Lett. 20,
       264 (1972).
37.   D. A. Pinnow, T. C. Rich, F. W. Ostermayer, Jr.,
       and M. DiDomenico, Jr., Appl. Phys. Lett. 22, 527
       (1973).
38.   D. B. Keck, R. D. Maurer and P. C. Schultz, Appl.
       Phys. Lett. 22, 307 (1973).
39.   M. G. Huglin, Light Scattering from Polymer Solu-
       tions, Academic Press, New York 1972.
40.   D. Turnbull, these Proceedings.
41   J. Tauc. these Proceedings.
42.   B. G. Bagley, in Amorphous and Liquid Semiconductors
       (ed. J. Tauc), Plenum Press, London 1974, p. 1.
43.   J. Tauc, in Optical Properties of Highly Transparent
       Solids (ed. S.S.Mitra and B.Bendow), Plenum Press,1975.

AUTHOR INDEX

Abaav, D.A., 94,100
Abbate, I., 709,735
Abeles, F., 123,129,165,384
Abragam, A., 667,671,695,
  700
Abraham, A. 529,539
Abrahams, B., 480,489,491,
  503
Adler, D., 90
Agarwal, S.C., 162,169
Alben, R., 32,33,43,58,60,
  62,64,66,75,79-81,90,97,
  98,100,105,107,108,151,
  167,359,363,365,367,508,
  510,519,524,530,539,590,
  591,604,605,607,620,621,
  628,639,641,643,651,663
Allen, F.G., 245,246,289
Allred, W., 570,618
Al-Sharbaty, A., 157,177,
  179,196,448,451,453,460
Ambegaokar, W., 441,459,
  490,491,500,504
Anderegg, M., 258,289
Anderson, A., 295,332,348,
  349,386,407,409,410,484,
  503,528,529,695,696,701,
  767
Anderson, P.W., 655,656,663
Andreatch, P.,Jr., 595,596,
  621
Andrew, E.R., 667,700
Andreyevski, A.I., 50,89
Angell, C.A., 675,700
Angew, Z., 128,165
Angus, J.C., 756,768

Apling, A.J., 519
Armstrong, R.L., 673,688,
  700
Arnold, W., 692,696,701
Arnoldussen, T.C., 447-449,
  453,454,460
Ashcroft, N.W., 216,289
Ashley, E.J., 133,146,147,
  155,157,166,168
Ast, D.G., 38,42
August, G.R., 618
Augustyniak, W.M., 454,460,
Austin, I.G., 181-183,185-
  188,190-193,196,197,407,
  410,491,499,500,505
Averbach, B.L., 22,30,47,88
Axe, J.D., 507,519,520,524,
  530,539,590,621,623,655

Babaev, A.A., 180,186,196,
  454,460
Badzian, A.R., 49,89
Baer, A.D., 711,735
Bagley, B.G., 1-5,744,7.66,
  768
Bahl, S.K., 157,168,437,446,
  459,713,736
Balkanski, M., 570,619,621
Bansil, A., 307,317,318,348
Barker, A.S., 571,572,619
Barnes, M.R., 250,251,289
Bashara, N.M., 165
Bassoni, F., 117,164
Bates, C.H., 70,90,103,108
Bates, C.W., 730,735,737
Bauer, R.S., 127,146,150,

SUBJECT INDEX